CWN

Certified Wireless
Network Administrator

Study Guide
Fifth Edition

CWNA®
Certified Wireless Network Administrator

Study Guide
Exam CWNA-107

Fifth Edition

David D. Coleman, CWNE #4

David A. Westcott, CWNE #7

SYBEX®
A Wiley Brand

Senior Acquisitions Editor: Kenyon Brown
Development Editor: Kim Wimpsett
Technical Editor: Ben Wilson
Production Manager: Kathleen Wisor
Copy Editor: John Sleeva
Content Enablement and Operations Manager: Pete Gaughan
Associate Publisher: Jim Minatel
Book Designer: Judy Fung
Proofreader: Nancy Carrasco
Indexer: Johnna VanHoose Dinse
Project Coordinator, Cover: Brent Savage
Cover Designer: Wiley
Cover Image: ©Getty Images Inc./Jeremy Woodhouse

Copyright © 2018 by John Wiley & Sons, Inc., Indianapolis, Indiana

Published simultaneously in Canada

ISBN: 978-1-119-42578-6
ISBN: 978-1-119-47743-3 (ebk.)
ISBN: 978-1-119-47750-1 (ebk.)

Manufactured in the United States of America

Library of Congress Control Number: 2018953990

V10003724_082718

"An investment in knowledge always pays the best interest."

Benjamin Franklin, one of the Founding Fathers of the United States.

Almost fifteen years ago, we were approached by our publisher to co-author a study guide for the Certified Wireless Network Administrator (CWNA) certification exam. At the time, the phrase "Wi-Fi" was only a few years old and had yet to become ingrained into our culture. 802.11g technology had begun to emerge and we were so excited by the blazing speeds of 54 Mbps available on the 2.4 GHz frequency band. We agreed to author the book. Five editions later, 802.11 WLAN technology has drastically evolved and Wi-Fi is now an integral part of our day-to-day lives.

The CWNA certification has long been recognized as the foundation-level certification for network professionals looking to validate their knowledge of 802.11 WLAN technology. As authors, we have been humbled by the tens of thousands of individuals who have purchased the CWNA Study Guide to assist in their pursuit of the CWNA certification. We are also humbled that many universities and colleges have selected the book as part of their curriculum for wireless technology classes. In our travels, we have met and become friends with many of our readers of the past four editions of the book. We discovered that a large number of people who purchase the book use it as a reference guide in the workplace and not just as a study guide. We have also had many people tell us that the book has helped them advance in their Wi-Fi careers. Once again, we are very humbled, and we would like to dedicate the 5th edition of the CWNA Study Guide to our readers.

Our goal has always been to educate as many people as possible about WLAN technology. If you are a newcomer to 802.11 wireless networking, we hope this book will be your first investment in Wi-Fi knowledge. If you are a veteran WLAN professional, we hope that when you are done reading this book, you will pass it along to a friend or a colleague. Sharing the Wi-Fi knowledge will be a sound investment.

Sincerely,
David Coleman and David Westcott

Acknowledgments

When we wrote the first edition of the CWNA Study Guide, David Coleman's children, Brantley and Carolina, were young teenagers. Carolina now holds a master's degree in public policy from the University of Southern California (USC). Brantley graduated from Boston University and recently earned his Ph.D. in biochemistry at the University of Washington. David would like to thank his now adult children for years of support and for making their dad very proud. David would also like to thank his mother, Marjorie Barnes, stepfather, William Barnes, and brother, Rob Coleman, for many years of support and encouragement. David also thanks Valla Ann for being so special and making him laugh sideways.

David Coleman would also like to thank his friends and family at Aerohive Networks (www.aerohive.com). There are many past and present Aerohive employees he would like to thank, but there simply is not enough room. So thank you to all of his Aerohive co-workers. It has been one wild ride the past nine years!

David Westcott would like to thank Janie and Savannah for the smiles and hugs he receives when arriving home after being away delivering a training class. He would also like to thank Janie for her patience and understanding of his travel and writing demands.

David Westcott also would like to thank the training department at Aruba Networks. In 2004 Aruba Networks hired him as their first contract trainer. Much has changed over the years, but it is still a fun and exciting journey.

Writing *CWNA: Certified Wireless Network Administrator Study Guide* has once again been an adventure. We would like to thank the following individuals for their support and contributions during the entire process.

We must first thank Sybex acquisitions editor Jim Minatel for reaching out to us and encouraging us to write this fifth edition of our book. We would also like to thank our development editor, Kim Wimpsett, who has been a pleasure to work with on multiple book projects. We also need to send special thanks to our editorial manager, Pete Gaughan; our production editor, Katie Wisor; and John Sleeva, our copyeditor.

We also need to give a big shout-out to our technical editor, Ben Wilson of Fortinet (www.fortinet.com). Ben has accumulated years of Wi-Fi experience working for three major WLAN vendors. The feedback and input provided by Ben was invaluable.

Special thanks must also go out to both Andrew vonNagy, CWNE #84, and Marcus Burton, CWNE #78, for their expertise as technical editors in earlier editions of the book.

Andrew Crocker has again provided us with wonderful photographs and some amazing editing of some not so wonderful photographs that we provide him. You can see much more of his work and talent at www.andrew-crocker.com.

Thanks to Proxim and to Ken Ruppel (kenruppel@gmail.com) for allowing us to include the video *Beam Patterns and Polarization of Directional Antennas* with the book's online resources, which can be accessed at www.wiley.com/go/cwnasg.

Special thanks goes to Andras Szilagyi, not only for creating the EMANIM software program but for all the extra assistance he provided over the past thirteen years by creating customized versions of the program for the different editions of the book. You can reach Andras at www.szialab.org.

Thanks to Chris DePuy of the technology research firm, 650 Group (www.650.group.com), for the WLAN industry trend analysis.

Very special thanks to Marco Tisler, CWNE #126, for his content contribution about APIs. Thank you to Chris Harkins for his content about cloud networking. Thanks to Gregor Vucajnk, CWNE #96, for his copy regarding LTE. Thanks to Karl Benedict for his input and content about directional antennas. Thanks to Perry Correll for his input regarding 802.11ax.

Most gracious appreciation to Rick Murphy, CWNE #10, for his content regarding future 5 GHz U-NII bands. Rick offers some outstanding WLAN training resources at howwirelessworks.com.

We need to send thanks to Joel Crane, CWNE #233, of both MetaGeek and Ekahau fame for his contributions and the spectrum analyzer screenshots.

Very special thanks to Adrian Granados for all of his contributions to the wireless community. Be sure and check out his cool Wi-Fi applications at www.adriangranados.com.

Several other WLAN rock stars that are mentioned in the copy of this book: Mike Albano, CWNE #150; Eddie Forero, CWNE #160; James Garringer, CWNE #179; Jerome Henry, CWNE #45; and François Vergès, CWNE #180.

We would also like to thank the following individuals and companies for their support and contributions to the book:

Devin Akin, CWNE #1, of Divergent Dynamics (www.divergentdynamics.com).

Dennis Burrell, Product Innovation Technologist, and Tauni Odia, Marketing Manager of Ventev (www.ventev.com).

Kelly Burroughs, Product Marketing Manager of iBwave (www.ibwave.com).

Mike Cirello, Co-Founder of HiveRadar (www.hiveradar.com).

Jaime Fábregas Fernández, R&D Manager of Tarlogic Research S.L. (www.acrylicwifi.com).

Tina Hanzlik, Director, Marketing Communications at the Wi-Fi Alliance (www.wi-fi.org).

James Kahkoska, CTO - Handheld Network Test, and Julio Petrovitch, Principal Technical Marketing Engineer, Netscout (www.netscout.com).

Brian Long, CWNE #159, Senior Director, Global Professional Services at Masimo (www.masimo.com).

Bruce Miller, VP, Product Marketing, Riverbed Technology (www.riverbed.com).

Jerry Olla, CWNE #238, Technical Solutions Architect, and Jussi Kiviniemi, Senior VP of Ekahau (www.ekahau.com).

Scott Thompson, President of Oberon, Inc. (www.oberoninc.com).

Ryan Woodings, Founder, and Peter Vomocil, CMO of MetaGeek (www.metageek.com).

We also need to thank Keith Parsons, CWNE #3, and his team at wirelessLAN Professionals. Keith has built a worldwide community of WLAN experts that share knowledge. You can learn more about the wirelessLAN Professionals conferences at www.wlanpros.

We would also like to thank, Tom Carpenter, CWNE #104, of the CWNP program (www.cwnp.com). All CWNP employees, past and present, should be proud of the internationally renowned wireless certification program that sets the education standard within the enterprise Wi-Fi industry. It has been a pleasure working with all of you for the past two decades.

Finally, we would like to thank Chuck Lukaszewski, CWNE #112, for his gracious foreword that he wrote for this book.

About the Authors

David D. Coleman   is the Senior Product Evangelist for Aerohive Networks (www.aerohive
.com). David collaborates with the Aerohive Marketing team and travels the world for WLAN
training sessions and speaking events. He has instructed IT professionals from around the
globe in WLAN design, security, administration, and troubleshooting. David has written
multiple books, blogs, and white papers about wireless networking, and he is considered an
authority on 802.11 technology. Prior to working at Aerohive, he specialized in corporate
and government Wi-Fi training and consulting. In the past, he has provided WLAN training
for numerous private corporations, the US military, and other federal and state government
agencies. When he is not traveling, David resides in Atlanta, Georgia. David is CWNE #4,
and he can be reached via email at mistermultipath@gmail.com. Please follow David on
Twitter: @mistermultipath.

David Westcott   is an independent consultant and technical trainer with more than 32 years
of experience. David has been a certified trainer for more than 25 years, and specializes in
wireless networking, wireless management and monitoring, and network access control.
He has provided training to thousands of students at government agencies, corporations,
and universities in more than 30 countries around the world. David was an adjunct faculty
member for Boston University's Corporate Education Center for more than 10 years. David
has written multiple books as well as numerous white papers, and has developed many
courses on wired and wireless networking technologies and networking security.

David was a member of the original CWNE Roundtable. David is CWNE #007 and has
earned certifications from many companies, including Aruba Networks, Cisco, Microsoft,
Ekahau, EC-Council, CompTIA, and Novell. David lives in Concord, Massachusetts with
his wife Janie. David can be reached via email at david@westcott-consulting.com. Please
follow David on Twitter: @davidwestcott.

Contents at a Glance

Contents

Table of Exercises

Foreword

Congratulations! Your purchase of this book means that you have decided to take the first step to truly formalize the expertise you have already developed working with 802.11 wireless local area networks (WLANs). You have chosen to dig deeper to understand the mechanics of the technology and to improve your troubleshooting skills. As the world transitions to an exclusively wireless access layer, it needs more well trained engineers to build and operate wireless networks.

When I passed my Certified Wireless Network Administrator (CWNA) exam over a decade ago in 2007, I could not have imagined that it would lead me to a senior strategy role at a major wireless equipment manufacturer, much less the honor of writing this foreword. At the time, I was leading a team of wireless engineers deploying WLANs for enterprise customers, and few of us had any formal training. The CWNA curriculum made an immediate difference in the quality of our networks, and I was hooked.

Over time, I achieved all the professional level CWNP certifications, and then went on to earn and become CWNE #112. These certifications enabled me to continue to not only grow professionally and deliver better performing systems to my customers, but also to begin to give back to the community by participating in the IEEE 802.11 Working Group and Wi-Fi Alliance. These standards development organizations (SDOs) are responsible for evolving the technology behind Wi-Fi. CWNP certification also led to publishing my own books on topics like high-density and outdoor Wi-Fi networks, as well as technical research to help explain the behavior of various aspects of the 802.11 protocols.

You will find that radio technology is an infinitely deep rabbit hole. There is always another layer to the onion. For example—just as atoms are made up of protons and electrons, which, in turn, are made up of quarks and leptons—learning about 802.11 data rates at the Physical layer will eventually lead you to symbols and subcarriers and then to modulation and coding techniques. Studying the Medium Access Control (MAC) layer will inevitably lead you to the statistical elegance that underpins the basic rules that govern how Wi-Fi devices manage airtime and determine who has the right to transmit. And on and on.

There is no limit to how far your interest can carry you. David Coleman and David Westcott—whom I have had the pleasure of knowing for many years now—are excellent guides. They have been educating wireless engineers for many years, and have structured this book to provide an in-depth overview of all the key areas that must be mastered to be truly effective as a wireless administrator.

Wi-Fi is here to stay, and the industry needs many more certified engineers. The Wi-Fi industry shipped over 3 billion chipsets in 2017, and the installed base is estimated to be over 10 billion worldwide. There are now more Wi-Fi devices than humans on Earth. A study issued earlier this year found that unlicensed spectrum contributes over $830 billion to the United States economy alone, to say nothing of other countries. Other studies have found that Wi-Fi networks are carrying between 50 and 80 percent of all IP traffic originating on mobile devices.

At the same time, the technology is continuously advancing. 802.11ax equipment will begin shipping in late 2018, bringing true multi-gigabit data rates, scheduled access, and new multi-user techniques, enabled by a major evolution in the PHY and MAC layers. Wi-Fi Protected Access 3 (WPA3) has just been announced, bringing a major revision to the encryption and authentication used to protect WLANs. So, we can never stop learning. After you have earned your CWNA, I encourage you to push further to become the best WLAN engineer you can possibly be, and in time to make your own contributions to the field.

Chuck Lukaszewski
CWNE #112, CWSP, CWAP, CWDP, CWNA
Vice President, Wireless Strategy & Standards
Aruba, a Hewlett Packard Enterprise Company
June 2018

Introduction

If you have purchased this book or if you are thinking about purchasing this book, you probably have some interest in taking the CWNA® (Certified Wireless Network Administrator) certification exam or in learning more about what the CWNA certification exam encompasses. We would like to congratulate you on this first step, and we hope that our book can help you on your journey. Wireless networking is one of the hottest technologies on the market. As with many fast-growing technologies, the demand for knowledgeable people is often greater than the supply. The CWNA certification is one way to prove that you have the knowledge and skills to support this growing industry. This Study Guide was written with that goal in mind. This book was written to help teach you about wireless networking so that you have the knowledge needed not only to pass the CWNA certification test but also to be able to design, install, and support wireless networks. The CWNA certification is a required prerequisite for the training classes offered by many of the major WLAN vendors. We have included review questions at the end of each chapter to help you test your knowledge and prepare for the test. We have also included labs and an online learning environment to further facilitate your learning.

Before we tell you about the certification process and requirements, we must mention that this information may have changed by the time you take your test. We recommend that you visit www.cwnp.com as you prepare to study for your test to determine what the current objectives and requirements are.

 WARNING
Do not just study the questions and answers! The practice questions in this book are designed to test your knowledge of a concept or objective that is likely to be on the CWNA exam. The practice questions will be different from the actual certification exam questions. If you learn and understand the topics and objectives, you will be better prepared for the test.

About CWNA® and CWNP®

If you have ever prepared to take a certification test for a technology that you are unfamiliar with, you know that you are not only studying to learn a different technology but probably also learning about an industry that you are unfamiliar with. Read on and we will tell you about CWNP.

CWNP is an abbreviation for *Certified Wireless Network Professional*. There is no CWNP test. The CWNP program develops courseware and certification exams for wireless LAN technologies in the computer networking industry. The CWNP certification program is a vendor-neutral program.

The objective of CWNP is to certify people on wireless networking, not on a specific vendor's product. Yes, at times the authors of this book and the creators of the certification will talk about, demonstrate, or even teach how to use a specific product; however, the goal is the overall understanding of wireless, not the product itself. If you learned to drive a car, you had to physically sit and practice in one. When you think back and reminisce, you probably do not tell someone you learned to drive a Ford; you probably say you learned to drive using a Ford.

The CWNP program offers the following nine wireless certifications:

CWS: Certified Wireless Specialist   CWS is an entry-level WLAN certification exam (CWS-100) for those in sales, marketing, and entry-level positions related to Wi-Fi. CWS teaches the language of Wi-Fi and is an excellent introduction to enterprise Wi-Fi.

CWT: Certified Wireless Technician   CWT is an entry-level WLAN certification exam (CWT-100) for teaching technicians to install and configure Wi-Fi at the basic level. CWT provides the skills needed to install and configure an AP to specifications and configure a client device to connect to and use the WLAN.

CWTS: Certified Wireless Technology Specialist   CWTS is an entry-level WLAN certification exam (PW0-071) for sales professionals, project managers, and networkers who are new to enterprise Wi-Fi. Learn what Wi-Fi is before you learn how it works.

CWNA: Certified Wireless Network Administrator   The CWNA certification is an administration-level Wi-Fi certification exam (CWNA-107) for networkers who are in the field and need to thoroughly understand RF behavior, site surveying, installation, and basic enterprise Wi-Fi security. CWNA is where you learn how RF and IP come together as a Wi-Fi network. The CWNA certification was the original certification of the CWNP program and is considered to be the foundation-level certification in the Wi-Fi industry. CWNA is the base certification for enterprise Wi-Fi within the CWNP family of certifications and a springboard toward earning CWSP, CWDP, CWAP, and CWNE certifications.

CWSP: Certified Wireless Security Professional   The CWSP certification exam (CWSP-205) is a professional-level Wi-Fi certification for network engineers who seek to establish their expertise in enterprise Wi-Fi security. Contrary to popular belief, enterprise Wi-Fi can be secure, if the IT professionals installing and configuring it understand how to secure the wireless network. You must have a current CWNA credential to take the CWSP exam.

CWDP: Certified Wireless Design Professional   The CWDP certification exam (CWDP-302) is a professional-level career certification for networkers who are already CWNA certified and have a thorough understanding of RF technologies and applications of 802.11 networks. The CWDP curriculum prepares WLAN professionals to properly design wireless LANs for different applications to perform optimally in different environments. You must have a current CWNA credential to take the CWDP exam.

CWAP: Certified Wireless Analysis Professional   The CWAP certification exam (CWAP-402) is a professional-level career certification for networkers who are already CWNA certified and have a thorough understanding of RF technologies and applications of 802.11 networks. The CWAP curriculum prepares WLAN professionals to analyze, troubleshoot, and optimize any wireless LAN. You must have a current CWNA credential to take the CWAP exam.

CWNE: Certified Wireless Network Expert   The CWNE certification is the highest-level certification in the CWNP program. By successfully completing the CWNE requirements, you will have demonstrated that you have the most advanced skills available in today's wireless LAN market. The CWNE certification requires CWNA, CWAP, CWDP, and CWAP certifications. To earn the CWNE certification, a rigorous application must also be submitted and approved by the CWNE Board of Advisors. A minimum of three years of verifiable, documented, full-time professional work experience related to enterprise Wi-Fi networks is required. CWNE applicants must also submit three endorsements from people familiar with the applicant's enterprise Wi-Fi work history.

CWNT: Certified Wireless Network Trainer   Certified Wireless Network Trainers are qualified instructors certified by the CWNP program to deliver CWNP training courses to IT professionals. CWNTs are technical and instructional experts in wireless technologies, products, and solutions. To ensure a superior learning experience, CWNP Education Partners are required to use CWNTs when delivering training using official CWNP courseware. More information about becoming a CWNT is available on the CWNP website.

How to Become a CWNA

To become a CWNA, you must do the following two things: Agree that you have read and will abide by the terms and conditions of the CWNP Confidentiality Agreement and pass the CWNA certification test.

A copy of the CWNP Confidentiality Agreement can be found online at the CWNP website.

When you sit to take the test, you will be required to accept this confidentiality agreement before you can continue with the test. After you have agreed, you will be able to continue with the test. If you pass the test, you are then a CWNA.

The information for the exam is as follows:

Exam name: Wireless LAN Administrator

Exam number: CWNA-107

Cost: $200 (in US dollars)

Duration: 90 minutes

Questions: 60

Passing score: 70 percent (80 percent for instructors)

Available languages: English

Availability: Register at Pearson VUE (www.vue.com/cwnp)

When you schedule the exam, you will receive instructions regarding appointment and cancellation procedures, ID requirements, and information about the testing center location. In addition, you will receive a registration and payment confirmation letter. Exams can be scheduled weeks in advance, or in some cases, even as late as the same day. Exam vouchers can also be purchased at the CWNP website.

After you have successfully passed the CWNA exam, the CWNP program will award you a certification that is good for three years. To recertify, you will need to pass the current CWNA exam, the CWSP exam, the CWDP exam, or the CWAP exam that is current at that time. If the information you provided to the testing center is correct, you will receive an email from CWNP recognizing your accomplishment and providing you with a CWNP certification number.

Who Should Buy This Book?

If you want to acquire a solid foundation in wireless networking and your goal is to prepare for the exam, this book is for you. You will find clear explanations of the concepts you need to grasp and plenty of help to achieve the high level of professional competency you need in order to succeed.

If you want to become certified as a CWNA, this book is definitely what you need. However, if you just want to attempt to pass the exam without really understanding wireless, this Study Guide is not for you. It is written for people who want to acquire hands-on skills and in-depth knowledge of wireless networking. Many people purchase this book as a general reference guide for 802.11 technology.

How to Use This Book and the Online Resources

We have included several testing features in the book and online resources. These tools will help you retain vital exam content as well as prepare you to sit for the actual exam.

Before You Begin   At the beginning of the book (right after this introduction) is an assessment test that you can use to check your readiness for the exam. Take this test before you start reading the book; it will help you determine the areas that you may need to brush up on. The answers to the assessment test appear on a separate page after the last question of the test. Each answer includes an explanation and a note telling you the chapter in which the material appears.

Chapter Review Questions   To test your knowledge as you progress through the book, there are review questions at the end of each chapter. As you finish each chapter, answer the review questions and then check your answers. You can go back and reread the section that deals with each question you answered incorrectly to ensure that you answer correctly the next time you are tested on the material.

Sybex Online Learning Environment

The Sybex Online Learning Environment for this book includes flashcards, a test engine, and a glossary. To start using these to study for the CWNA exam go to www.wiley.com/go/ sybextestprep, register your book to receive your unique PIN, then one you have the PIN, return to www.wiley.com/go/sybextestprep and register a new account or add this book to an existing account.

Test Engine   The test engine includes three bonus practice exams. You can use them as if you were taking the exam to rate your progress toward being ready. The test engine also includes all of the end-of-chapter review questions and the pre-book assessment questions. You can study by chapter or you can have the test engine mix and match questions from multiple chapters or the bonus exams. The test engine also comes with a practice mode (where you can see hints) as well as an exam mode (like the real test).

Flashcards   These are short questions and answers like you probably used in school but online.

Glossary   A electronic list of key terms and their definitions.

Additional Online Resources

Labs and Exercises   Several chapters in this book have labs that use resources you can download from the book's website (www.wiley.com/go/cwnasg). These labs and exercises provide you with a broader learning experience by supplying hands-on experience and step-by-step problem solving. Some of the included hands-on materials you can download are an **RF signal simulator** to help you learn the radio frequency fundamentals taught in Chapter 3 and **PCAP frame captures** to reinforce what you learn about 802.11 wireless frames in chapters 9 and 17.

White Papers   Several chapters in this book reference wireless networking white papers that are available from the referenced websites. These white papers serve as additional reference material for preparing for the CWNA exam.

Getting Help Online

We hope your experience with the Sybex Online Learning Environment and the additional online resources is smooth. But if you have any issues with the online materials or the book itself, please start by reporting your issue to our 24x7 technical support team at support.wiley.com. They have live online chat as well as email options.

CWNA Exam (CWNA-107) Objectives

The CWNA exam measures your understanding of the fundamentals of RF behavior, your ability to describe the features and functions of wireless LAN components, and your knowledge of the skills needed to install, configure, and troubleshoot wireless LAN hardware peripherals and protocols.

The skills and knowledge measured by this exam were derived from a survey of wireless networking experts and professionals. The results of this survey were used in weighing the subject areas and ensuring that the weighting is representative of the relative importance of the content.

The following chart provides the breakdown of the exam, showing you the weight of each section:

Subject Area	% of Exam
Radio frequency (RF) technologies	15%
WLAN regulations and standards	10%
WLAN protocols and devices	20%
WLAN network architecture	20%
WLAN network security	10%
RF validation	10%
WLAN troubleshooting	15%
Total	**100%**

Radio Frequency (RF) Technologies – 15%

1.1 Define and explain the basic characteristics of RF and RF behavior

1.1.1 Wavelength, frequency, amplitude, phase, sine waves

1.1.2 RF propagation and coverage

1.1.3 Reflection, refraction, diffraction, and scattering

1.1.4 Multipath and RF interference

1.1.5 Gain and loss

1.1.6 Amplification

1.4 Explain and apply the functionality of RF antennas and antenna systems and the mounting options and antenna accessories available

1.4.1 Omni-directional antennas

1.4.2 Semi-directional antennas

1.4.3 Highly directional antennas

1.4.4 Sectorized antennas and antenna arrays

1.4.5 Reading antenna charts for different antenna types

1.4.6 Pole/mast mount

1.4.7 Ceiling mount

1.4.8 Wall mount

1.4.9 Indoor vs. outdoor mounting

1.4.10 RF cables, connectors, and splitters

1.4.11 Amplifiers and attenuators

1.4.12 Lightning arrestors and grounding rods/wires

1.4.13 Towers, safety equipment, and related concerns

WLAN Regulations and Standards – 10%

2.1 Explain the roles of WLAN and networking industry organizations

2.1.1 IEEE

2.1.2 Wi-Fi Alliance

2.1.3 IETF

2.1.4 Regulatory domains and agencies

2.2 Explain the IEEE standard creation process including working groups, naming conventions, drafts, and ratification

2.3 Explain and apply the various Physical Layer (PHY) solutions of the IEEE 802.11-2016 standard as amended including supported channel widths, spatial streams, data rates, and supported modulation types

2.3.1 DSSS – 802.11

2.3.2 HR-DSSS – 802.11b

2.3.3 OFDM – 802.11a

2.3.4 ERP – 802.11g

2.3.5 HT – 802.11n

2.3.6 DMG – 802.11ad

2.3.7 VHT – 802.11ac

3.5 Explain 802.11 channel access methods

3.5.1 DCF

3.5.2 EDCA

3.5.3 RTS/CTS

3.5.4 CTS-to-Self

3.5.5 NAV

3.5.6 Physical carrier sense and virtual carrier sense

3.5.7 Channel width operations

3.5.8 HT operation modes

3.5.9 VHT operating mode field

3.5.10 HT and VHT protection mechanisms

3.5.11 Power save modes

3.6 Describe features of, select, and install WLAN infrastructure devices

3.6.1 Autonomous access points (APs)

3.6.2 Controller-based APs

3.6.3 Cloud-based APs

3.6.4 Distributed APs

3.6.5 Management systems

3.6.6 Mesh APs and routers

3.6.7 WLAN controllers

3.6.8 Remote office controllers and/or APs

3.6.9 PoE injectors and PoE-enabled Ethernet switches

3.6.10 WLAN bridges

3.6.11 Home WLAN routers

3.7 Identify the features, purpose, and use of the following WLAN client devices and adapters

3.7.1 USB adapters

3.7.2 PCI, Mini-PCI, Mini-PCIe, and Half Mini-PCIe cards

3.7.3 Laptops, tablets, and mobile phones

3.7.4 802.11 VoIP handsets

3.7.5 Specialty devices (handheld scanners, push-to-talk, IoT)

3.7.6 Configure Windows, Linux, Chrome OS, and macOS clients

WLAN Network Architecture – 20%

4.1 Identify technology roles for which WLAN solutions are appropriate and describe the typical use of WLAN solutions in those roles

4.1.1 Corporate data access and end-user mobility

4.1.2 Enterprise network extension

4.1.3 WLAN bridging

4.1.4 Last-mile data delivery – Wireless ISP

4.1.5 Small Office/Home Office (SOHO) use

4.1.6 Mobile offices

4.1.7 Educational/classroom use

4.1.8 Industrial

4.1.9 Healthcare

4.1.10 Hotspots

4.1.11 Hospitality

4.1.12 Conference/convention/arena/stadium and large high density deployments

4.1.13 Transportation networks (trains, planes, automobiles)

4.1.14 Law enforcement networks

4.2 Describe and implement Power over Ethernet (PoE)

4.2.1 IEEE 802.3-2015, Clause 33, including 802.3af-2003 and 802.3at-2009

4.2.2 Power source equipment

4.2.3 Powered device

4.2.4 Midspan and endpoint PSEs

4.2.5 Power levels

4.2.6 Power budgets and powered port density

4.3 Define and describe controller-based, distributed, cloud-based, and controller-less WLAN architectures

4.3.1 Core, Distribution, and Access layer forwarding

4.3.2 Centralized data forwarding

4.3.3 Distributed data forwarding

4.3.4 Control, Management, and Data planes

4.3.5 Scalability and availability solutions

4.3.6 Intra- and Inter-controller STA roaming handoffs (OKC and FT)

5.4 Explain and use secure management protocols

5.4.1 HTTPS

5.4.2 SNMPv3

5.4.3 SSH2

5.4.4 VPN

RF Validation – 10%

6.1 Explain the importance of and the process of a post-implementation validation survey

6.1.1 Verify design requirements

6.1.1.1 Coverage

6.1.1.2 Capacity

6.1.1.3 Throughput

6.1.1.4 Roaming

6.1.1.5 Delay

6.1.1.6 Jitter

6.1.1.7 Connectivity

6.1.1.8 Aesthetics

6.1.2 Document actual WLAN implementation results

6.2 Locate and identify sources of RF interference

6.2.1 WLAN devices

6.2.1.1 Co-channel interference (CCI)

6.2.1.2 Adjacent channel interference (ACI)

6.2.2 Non-Wi-Fi devices

6.2.2.1 Airtime utilization

6.2.2.2 Frequencies used

6.2.3 Interference solutions

6.2.4 Spectrum analysis

6.3 Perform application testing to validate WLAN performance

6.3.1 Network and service availability

6.3.2 VoIP testing

6.3.3 Real-time application testing

6.3.4 Throughput testing

6.3.5 Load testing

6.4 Understand and use the basic features of validation tools

6.4.1 Throughput testers (iPerf, TamoSoft Throughput Tester, etc.)

6.4.2 Wireless design software (Ekahau Site Survey, iBwave Wi-Fi, AirMagnet Survey Pro, TamoSoft Survey)

6.4.3 Protocol analyzers

6.4.4 Spectrum analyzers

WLAN Troubleshooting – 15%

7.1 Define and apply industry and vendor recommended troubleshooting processes to resolve common 802.11 wireless networking problems

7.1.1 Identify the problem

7.1.2 Discover the scale of the problem

7.1.3 Define possible causes

7.1.4 Narrow to the most likely cause

7.1.5 Create a plan of action or escalate the problem

7.1.6 Perform corrective actions

7.1.7 Verify the solution

7.1.8 Document the results

7.2 Describe and apply common troubleshooting tools used in WLANs

7.2.1 Protocol analyzer

7.2.2 Spectrum analyzer

7.2.3 Centralized management consoles

7.2.4 WLAN monitoring solutions

7.3 Identify and explain how to solve the following WLAN implementation challenges using features available in enterprise class WLAN equipment and troubleshooting tools

7.3.1 System throughput

7.3.2 CCI and ACI

7.3.3 RF noise and noise floor

7.3.4 RF interference

7.3.5 Hidden nodes

7.3.6 Insufficient PoE power

7.3.7 Lack of coverage

7.4 Troubleshoot common connectivity problems in WLANs (both WLAN connectivity and network connectivity for wireless clients)

7.4.1 No signal or weak signal

7.4.2 Security configuration mismatch

7.4.3 Improper AP configuration

7.4.4 Improper client configuration

7.4.5 Faulty drivers/firmware

7.4.6 Hardware failure

7.4.7 DHCP issues

7.4.8 Captive portal issues

CWNA Exam Terminology

The CWNP program uses specific terminology when phrasing the questions on any of the CWNP exams. The terminology used most often mirrors the same language that is used by the Wi-Fi Alliance and in the IEEE 802.11-2016 standard. The most current IEEE version of the 802.11 standard is the IEEE 802.11-2016 document, which includes all the amendments that have been ratified prior to the document's publication. Standards bodies such as the IEEE often create several amendments to a standard before "rolling up" the ratified amendments (finalized or approved versions) into a new standard.

 To properly prepare for the CWNA exam, any test candidate should become 100 percent familiar with the terminology used by the CWNP program. This book defines and covers all terminology, including acronyms, terms, and definitions.

CWNP Authorized Materials Use Policy

CWNP does not condone the use of unauthorized training materials, aka brain dumps. Individuals who utilize such materials to pass CWNP exams will have their certifications revoked. In an effort to more clearly communicate CWNP's policy on use of unauthorized study materials, CWNP directs all certification candidates to the CWNP Candidate Conduct Policy, which is available on the CWNP website. Please review this policy before beginning the study process for any CWNP exam. Candidates will be required to state that they understand and have abided by this policy at the time of exam delivery.

Tips for Taking the CWNA Exam

Here are some general tips for taking your exam successfully:

- Bring two forms of ID with you. One must be a photo ID, such as a driver's license. The other can be a major credit card or a passport. Both forms must include a signature.

- Arrive early at the exam center so that you can relax and review your study materials, particularly tables and lists of exam-related information.

- Read the questions carefully.

- Do not be tempted to jump to an early conclusion. Make sure you know exactly what the question is asking.

- There will be questions with multiple correct responses.

- When there is more than one correct answer, a message at the bottom of the screen will prompt you to either "choose two" or "choose all that apply." Be sure to read the messages displayed to know how many correct answers you must choose.

- When answering multiple-choice questions you are not sure about, use a process of elimination to get rid of the obviously incorrect answers first. Doing so will improve your odds if you need to make an educated guess.

- Do not spend too much time on one question.

- This is a form-based test; however, you cannot move backward through the exam. You must answer the current question before you can move to the next question, and after you have moved to the next question, you cannot go back and change your answer on a previous question.

- Keep track of your time.

- Because this is a 90-minute test consisting of 60 questions, you have an average of 90 seconds to answer each question. You can spend as much or as little time on any one question, but when 90 minutes is up, the test is over. Check your progress. After 45 minutes, you should have answered at least 30 questions. If you have not, do not panic. You will simply need to answer the remaining questions at a faster pace. If on average you can answer each of the remaining 30 questions 4 seconds quicker, you will recover 2 minutes. Again, do not panic; just pace yourself.

- For the latest pricing on the exams and updates to the registration procedures, visit CWNP's website at www.cwnp.com.

CWNA: Certified Wireless Network Administrator Exam (CWNA-107) Objectives

The *CWNA: Certified Wireless Network Administrator Study Guide, Fifth Edition* was written to cover every CWNA-107 exam objective at a level appropriate to its exam weighting. The following sections provide a breakdown of this book's exam coverage, showing you the weight of each section and listing the chapter where each objective or subobjective is covered.

Subject area	% of exam
Radio frequency (RF) technologies	15%
WLAN regulations and standards	10%
WLAN protocols and devices	20%
WLAN network architecture	20%
WLAN network security	10%
RF validation	10%
WLAN troubleshooting	15%
Total	**100%**

Radio Frequency (RF) Technologies

Objective		Chapter
1.1	RF and RF Behavior	3

Wavelength, frequency, amplitude, phase, sine waves
RF propagation and coverage
Reflection, refraction, diffraction, and scattering
Multipath and RF interference
Gain and loss
Amplification
Attenuation
Absorption
Voltage Standing Wave Ratio (VSWR)
Return Loss
Free Space Path Loss (FSPL)
Delay Spread
Modulation (ASK and PSK)

Objective		Chapter
1.2	RF Mathematics and measurement	1, 4
	Watt and milliwatt	
	Decibel (dB)	
	dBm, dBd, and dBi	
	Noise floor	
	SNR and SINR	
	RSSI	
	Signal metric conversions	
	System operating margin (SOM), fade margin, and link budget calculations	
	Intentional radiator compared with equivalent isotropically radiated power (EIRP)	
1.3	RF Signal and Antenna Concepts	5, 10
	RF and physical line of sight and Fresnel zone clearance	
	Beamwidths	
	Azimuth and Elevation charts	
	Passive gain vs. active gain	
	Isotropic radiator	
	Polarization	
	Antenna diversity types	
	Radio chains	
	Spatial multiplexing (SM)	
	Transmit beamforming (TxBF)	
	Maximal ratio combining (MRC)	
	MIMO and MU-MIMO	
1.4	RF Antennas and Accessories	5
	Omni-directional antennas	
	Semi-directional antennas	
	Highly directional antennas	
	Sectorized antennas and antenna arrays	
	Reading antenna charts for different antenna types	
	Pole/mast mount	
	Ceiling mount	
	Wall mount	

Objective		Chapter
	Indoor vs. outdoor mounting	
	RF cables, connectors, and splitters	
	Amplifiers and attenuators	
	Lightning arrestors and grounding rods/wires	
	Towers, safety equipment, and related concerns	

WLAN Regulations and Standards

Objective		Chapter
2.1	WLAN and networking industry organizations	1
	IEEE	
	Wi-Fi Alliance	
	IETF	
	Regulatory domains and agencies	
2.2	IEEE standard creation process	2
2.3	Physical Layer (PHY) solutions	2, 6, 10, 19
	DSSS 0 802.11	
	HR-DSSS - 802.11b	
	OFDM - 802.11a	
	ERP - 802.11g	
	HT - 802.11n	
	DMB - 802.11ad	
	VHT - 802.11ac	
	TVHT - 802.11af	
	SIG - 802.11ah	
2.4	WLAN functional concepts	6, 10, 13, 19
	Modulation and coding	
	Co-location interference	
	Channel centers and widths (all PHYs)	
	Primary channels	
	Adjacent overlapping and non-overlapping channels	
	Throughput vs. data rate	
	Bandwidth	
	Communication resilience	

Objective		Chapter
2.5	OSI model layers affected by 802.11-2016 standard	1, 2, 8, 15
2.6	802.11 frequency bands	2, 6
2.7	Regulatory domain requirements	1, 2, 6, 13
	Available channels	
	Output power constraints	
	Dynamic frequency selection (DFS)	
	Transmit power control (TPC)	
2.8	WLAN use case scenarios	7, 11, 13, 20
	Wireless LAN (WLAN) – BSS and ESS	
	Wireless PAN (WPAN)	
	Wireless bridging	
	Wireless ad hoc (IBSS)	
	Wireless mesh (MBSS)	

WLAN Protocols and Devices

Objective		Chapter
3.1	802.11 wireless service set components	7, 8, 9, 15
	Stations (STAs)	
	Basic service set (BSS)	
	Basic service area (BSA)	
	SSID	
	BSSID	
	Extended service set (ESS)	
	Ad hoc mode and IBSS	
	Infrastructure mode	
	Distribution system (DS)	
	Distribution system media (DSM)	
	Roaming (Layer 1 and Layer 2)	

Objective		Chapter
3.6	WLAN infrastructure devices	11, 12

Autonomous access points (APs)
Controller-based APs
Cloud-based APs
Distributed APs
Management systems
Mesh APs and routers
WLAN controllers
Remote office controllers and/or APs
PoE injectors and PoE-enabled Ethernet switches
WLAN bridges
Home WLAN routers

3.7	Features, purpose, and use of WLAN client devices	11

USB adapters
PCI, Mini-PCI, Mini-PCIe, and Half Mini-PCIe cards
Laptops, tablets, and mobile phones
802.11 VoIP handsets
Specialty devices (handheld scanners, push-to-talk, IoT)
Configure Windows, Linux, Chrome OS, and macOS clients

WLAN Network Architecture

Objective		Chapter
4.1	WLAN technology roles, solutions, and typical use	14, 20

Corporate data access and end-user mobility
Enterprise network extension
WLAN bridging
Last-mile data delivery – Wireless ISP
Small Office/Home Office (SOHO) use
Mobile offices
Educational/classroom use

Objective		Chapter
	Industrial	
	Healthcare	
	Hotspots	
	Hospitality	
	Conference/convention/arena/stadium and large high density deployments	
	Transportation networks (trains, planes, automobiles)	
	Law enforcement networks	
4.2	Power over Ethernet (PoE)	12
	IEEE 802.3-2015, Clause 33, including 802.3af-2003 and 802.3at-2009	
	Power source equipment	
	Powered device	
	Midspan and endpoint PSEs	
	Power levels	
	Power budgets and powered port density	
4.3	Controller-based, distributed, cloud-based, and controller-less WLAN architectures	1, 11, 13, 15
	Core, Distribution, and Access layer forwarding	
	Centralized data forwarding	
	Distributed data forwarding	
	Control, Management, and Data planes	
	Scalability and availability solutions	
	Intra- and Inter-controller STA roaming handoffs (OKC and FT)	
	Advantages and limitations of each technology	
	Tunneling, QoS, and VLANs	
4.4	Multiple channel architecture (MCA) network model and single channel architecture (SCA) model	11, 13, 15
	BSSID and ESS configuration	
	Channel selection	
	AP placement	
	Co-channel and adjacent channel interference	
	Cell sizing (output power, antenna selection)	

WLAN Network Security

Objective		Chapter
	MAC filtering	
	Improper use of WPA (TKIP/RC4)	
	Open System authentication	
	Wi-Fi Protected Setup (WPS)	
5.2	Effective security mechanisms for enterprise WLANs	16, 17
	WPA2 (CCMP/AES)	
	WPA2-Personal	
	WPA2-Enterprise	
	802.1X/EAP framework	
	RADIUS servers	
	EAP methods	
	Effective pre-shared key (PSK) and passphrase usage	
	Per-user PSK (PPSK)	
5.3	Common security enhancements and tools used in WLANs	2, 15, 16, 18
	Captive portals	
	BYOD and guest networks	
	Protected management frames	
	Fast secure roaming methods	
	Wireless intrusion prevention system (WIPS)	
	Protocol and spectrum analyzers	
5.4	Secure management protocols	15, 17
	HTTPS	
	SNMPv3	
	SSH2	
	VPN	

RF Validation

Objective		Chapter
6.1	Post-implementation validation survey	13, 14
	Verify design requirements	
	Coverage	

WLAN Troubleshooting

Assessment Test

1. At which layers of the OSI model does 802.11 technology operate? (Choose all that apply.)
   A. Data-Link
   B. Network
   C. Physical
   D. Presentation
   E. Transport

2. Which Wi-Fi Alliance certification defines the mechanism for conserving battery life that is critical for handheld devices such as bar code scanners and VoWiFi phones?
   A. WPA2-Enterprise
   B. WPA2-Personal
   C. WMM-PS
   D. WMM-SA
   E. CWG-RF

3. Which of these frequencies has the longest wavelength?
   A. 750 KHz
   B. 2.4 GHz
   C. 252 GHz
   D. 2.4 MHz

4. Which of these terms can best be used to compare the relationship between two radio waves that share the same frequency?
   A. Multipath
   B. Multiplexing
   C. Phase
   D. Spread spectrum

5. A bridge transmits at 10 mW. The cable to the antenna produces a loss of 3 dB, and the antenna produces a gain of 20 dBi. What is the EIRP?
   A. 25 mW
   B. 27 mW
   C. 4 mW
   D. 1,300 mW
   E. 500 mW

6. dBi is an expression of what type of measurement?
   A. Access point gain
   B. Received power

   C. Transmitted power

   D. Antenna gain

   E. Effective output

7. What are some possible effects of voltage standing wave ratio (VSWR)? (Choose all that apply.)

   A. Increased amplitude

   B. Decreased signal strength

   C. Transmitter failure

   D. Erratic amplitude

   E. Out-of-phase signals

8. When installing a higher-gain omnidirectional antenna, which of the following occurs? (Choose two.)

   A. The horizontal coverage increases.

   B. The horizontal coverage decreases.

   C. The vertical coverage increases.

   D. The vertical coverage decreases.

9. 802.11ac VHT radios are backward compatible with which IEEE 802.11 radios? (Choose two)

   A. 802.11 legacy (FHSS) radios

   B. 802.11g (ERP) radios

   C. 802.11 legacy (DSSS) radios

   D. 802.11b (HR-DSSS) radios

   E. 802.11a (OFDM) radios

   F. 2.4 GHz 802.11n (HT) radios

   G. 5 GHz 802.11n (HT) radios

   H. None of the above

10. Which IEEE 802.11 amendment specifies the use of up to eight spatial streams of modulated data bits?

   A. IEEE 802.11n

   B. IEEE 802.11g

   C. IEEE 802.11ac

   D. IEEE 802.11s

   E. IEEE 802.11w

11. Which of the following measures the difference between the power of the primary RF signal compared against the sum of the power of the RF interference and background noise?

   A. Noise ratio

   B. SNR

   C. SINR

   D. BER

   E. DFS

12. What signal characteristics are common in spread spectrum and OFDM-based signaling methods? (Choose two.)

   A. Narrow bandwidth

   B. Low power

   C. High power

   D. Wide bandwidth

13. A service set identifier is often synonymous with which of the following?

   A. IBSS

   B. ESSID

   C. BSSID

   D. BSS

14. Which ESS design scenario is required by the IEEE 802.11-2016 standard?

   A. Two or more access points with overlapping coverage cells

   B. Two or more access points with overlapping disjointed coverage cells

   C. One access point with a single BSA

   D. Two basic service sets interconnected by a distribution system medium (DSM)

   E. None of the above

15. Which CSMA/CA conditions must be met before an 802.11 radio can transmit? (Choose all that apply.)

   A. The NAV timer must be equal to zero.

   B. The random backoff timer must have expired.

   C. The CCA must be idle.

   D. The proper interframe space must have occurred.

   E. The access point must be in PCF mode.

16. Beacon management frames contain which of the following information? (Choose all that apply.)

    A. Channel information

    B. Destination IP address

    C. Basic data rate

    D. Traffic indication map (TIM)

    E. Vendor proprietary information

    F. Time stamp

17. Rebekah McAdams was hired to perform a wireless packet analysis of your network. While performing the analysis, she noticed that many of the data frames were preceded by an RTS frame followed by a CTS frame. What could cause this to occur? (Choose all that apply.)

    A. Because of high RF noise levels, some of the stations have automatically enabled RTS/CTS.

    B. An AP was manually configured with a low RTS/CTS threshold.

    C. A nearby cell phone is causing some of the nodes to enable a protection mechanism.

    D. Legacy 802.11b clients are connected to an 802.11g AP.

18. What is another name for an 802.11 data frame that is also known as a PSDU?

    A. PPDU

    B. MSDU

    C. MPDU

    D. BPDU

19. Which WLAN device uses dynamic layer 2 routing protocols?

    A. WLAN switch

    B. WLAN controller

    C. WLAN router

    D. WLAN mesh access point

20. What term best describes the bulk of the data generated on the Internet being created by sensors, monitors, and machines?

    A. Wearables

    B. Cloud-enabled networking (CEN)

    C. Cloud-based networking (CBN)

    D. Software as a service (SaaS)

    E. Internet of Things (IoT)

21. Which technology subdivides a channel, allowing parallel transmissions of smaller frames to multiple users to occur simultaneously?

    A. OFDMA

    B. OFDM

    C. Channel blocking

    D. Sub-channelization

    E. RTS/CTS

22. What term best describes how Wi-Fi can be used to identify customer behavior and shopping trends?

    A. Radio analytics

    B. Customer analytics

    C. Retail analytics

    D. 802.11 analytics

23. The hidden node problem occurs when one client station's transmissions are not heard by some of the other client stations in the coverage area of a basic service set (BSS). What are some of the consequences of the hidden node problem? (Choose all that apply.)

    A. Retransmissions

    B. Intersymbol interference (ISI)

    C. Collisions

    D. Increased throughput

    E. Decreased throughput

24. What are some potential causes of layer 2 retransmissions? (Choose all that apply.)

    A. RF interference

    B. Low signal-to-noise ratio (SNR)

    C. Dual-frequency transmissions

    D. Fade margin

    E. Multiplexing

25. Which of these solutions would be considered strong WLAN security?

    A. SSID cloaking

    B. MAC filtering

    C. WEP

    D. Shared Key authentication

    E. CCMP/AES

    F. TKIP

26. Which security standard defines port-based access control?

  A. IEEE 802.11x

  B. IEEE 802.3b

  C. IEEE 802.11i

  D. IEEE 802.1X

  E. IEEE 802.11s

27. What is the best tool for detecting an RF jamming denial-of-service attack?

  A. Time-domain analysis software

  B. Protocol analyzer

  C. Spectrum analyzer

  D. Predictive modeling software

  E. Oscilloscope

28. Which of these attacks can be detected by a wireless intrusion detection system (WIDS)? (Choose all that apply.)

  A. Deauthentication spoofing

  B. MAC spoofing

  C. Rogue ad hoc network

  D. Association flood

  E. Rogue AP

29. You have been hired by the US-based XYZ Company to conduct a wireless site survey. Which government agencies need to be informed before a tower that exceeds 200 feet above ground level is installed? (Choose all that apply.)

  A. RF regulatory authority

  B. Local municipality

  C. Fire department

  D. Tax authority

  E. Aviation authority

30. You have been hired by the ABC Corporation to conduct an indoor site survey. What information will be in the final site survey report? (Choose two.)

  A. Security analysis

  B. Coverage analysis

  C. Spectrum analysis

  D. Routing analysis

  E. Switching analysis

31. Name a potential source of RF interference in the 5 GHz U-NII band.

    A. Cordless phones

    B. AM radio

    C. FM radio

    D. Microwave ovens

    E. Bluetooth

32. Which of these measurements are taken for indoor coverage analysis? (Choose all that apply.)

    A. Received signal strength

    B. Signal-to-noise ratio

    C. Noise level

    D. Path loss

    E. Packet loss

33. What is the number one cause of layer 2 retransmissions?

    A. Low SNR

    B. Hidden node

    C. Adjacent cell interference

    D. RF interference

34. What must a powered device (PD) do to be considered PoE compliant (IEEE 802.3-2015, Clause 33)? (Choose all that apply.)

    A. Be able to accept power in either of two ways (through the data lines or unused pairs).

    B. Reply with a classification signature.

    C. Reply with a 35-ohm detection signature.

    D. Reply with a 25-ohm detection signature.

    E. Receive 30 watts of power from the power sourcing equipment.

35. An 802.11n (HT) network can operate on which frequency bands? (Choose all that apply.)

    A. 902–928 MHz

    B. 2.4–2.4835 GHz

    C. 5.15–5.25 GHz

    D. 5.47–5.725 GHz

36. What are some of the methods used to reduce MAC layer overhead, as defined by the 802.11n-2009 amendment? (Choose all that apply.)

    A. A-MSDU

    B. A-MPDU

    C. MCS

    D. PPDU

    E. MSDU

37. How many modulation and coding schemes (MCSs) are defined by the 802.11ac-2013 amendment?

   A. 10

   B. 100

   C. 7

   D. 77

   E. 22

38. Which capabilities defined by the 802.11n-2009 amendment are no longer defined by the 802.11ac-2013 amendment? (Choose all that apply.)

   A. Equal modulation

   B. Unequal modulation

   C. RIFS

   D. SIFS

   E. 40 MHz channels

39. What can be delivered over-the-air to WLAN mobile devices, such as tablets and smartphones, when a mobile device management (MDM) solution is deployed?

   A. Configuration settings

   B. Applications

   C. Certificates

   D. Web clips

   E. All of the above

40. WLAN vendors have begun to offer the capability for guest users to log in to a guest WLAN with preexisting social media credentials, such as Facebook or Twitter usernames and passwords. Which authorization framework can be used for social media logins to WLAN guest networks?

   A. Kerberos

   B. RADIUS

   C. 802.1X/EAP

   D. OAuth

   E. TACACS

Answers to the Assessment Test

1. **A and C.** The IEEE 802.11-2016 standard defines communication mechanisms at only the Physical layer and MAC sublayer of the Data-Link layer of the OSI model. For more information, see Chapter 1.

2. **C.** WMM-PS helps conserve battery power for devices using Wi-Fi radios by managing the time the client devices spend in sleep mode. Conserving battery life is critical for handheld devices such as bar code scanners and VoWiFi phones. To take advantage of power-saving capabilities, both the device and the access point must support WMM-Power Save. For more information, see Chapter 9.

3. **A.** A 750 KHz signal has an approximate wavelength of 1,312 feet, or 400 meters. A 252 GHz signal has an approximate wavelength of less than 0.05 inches, or 1.2 millimeters. Remember, the higher the frequency of a signal, the smaller the wavelength property of an electromagnetic signal. For more information, see Chapter 3.

4. **C.** Phase involves the positioning of the amplitude crests and troughs of two waveforms. For more information, see Chapter 3.

5. **E.** The 10 mW of power is decreased by 3 dB, or divided by 2, giving 5 mW. This is then increased by 20 dBi, or multiplied by 10 twice, giving 500 mW. For more information, see Chapter 4.

6. **D.** Theoretically, an isotropic radiator can radiate an equal signal in all directions. An antenna cannot do this because of construction limitations. However, antennas are often referred to as isotropic radiators because they radiate RF energy. The gain, or increase, of power from an antenna when compared to what an isotropic radiator would generate is known as decibels isotropic (dBi). Another way of phrasing this is decibel gain referenced to an isotropic radiator, or change in power relative to an antenna. dBi is a measurement of antenna gain. For more information, see Chapter 4.

7. **B, C, and D.** Reflected voltage caused by an impedance mismatch may cause a degradation of amplitude, erratic signal strength, or even the worst-case scenario of transmitter burnout. See Chapter 5 for more information.

8. **A and D.** When the gain of an omnidirectional antenna is increased, the vertical coverage area decreases while the horizontal coverage area is increased. See Chapter 5 for more information.

9. **E and G.** 802.11ac (VHT) radios transmit in the 5 GHz U-NII bands and are not compatible with 2.4 GHz radios, such as 802.11 legacy (FHSS) radios, 802.11 legacy (DSSS) radios, 802.11b (HR-DSSS) radios, 802.11g (ERP) radios, or 802.11n radios, which transmit in the 2.4 GHz ISM frequency band. 802.11ac (VHT) radios are backward compatible with 5 GHz 802.11n (HT) radios and 802.11a (OFDM) radios. For more information, see Chapter 6.

10. C. The 802.11ac-2013 amendment defines the use of 256-QAM modulation, up to eight spatial streams, multiuser MIMO, 20 MHz channels, 40 MHz channels, 80 MHz channels, and 160 MHz channels. 802.11 MIMO technology and 40 MHz channels debuted with the ratification of the 802.11n-2009 amendment. For more information, see Chapter 2 and Chapter 10.

11. C. Signal-to-interference-plus-noise (SINR) relates the primary RF signal to both interference and noise. While the noise level tends not to fluctuate much, interference from other devices is likely to be more common and frequent. For more information, see Chapter 4.

12. B and D. Both spread spectrum and OFDM signals utilize bandwidth that is wider than what is required to carry the data and has low transmission power requirements. See Chapter 6 for more information.

13. B. The logical network name of a wireless LAN is often called an ESSID (extended service set identifier) and is essentially synonymous with SSID (service set identifier), which is another term for a logical network name in the most common deployments of a WLAN. For more information, see Chapter 7.

14. E. The scenarios described in options A, B, C, and D are all examples of how an extended service set may be deployed. The IEEE 802.11-2016 standard defines an extended service set (ESS) as "a set of one or more interconnected basic service sets." However, the IEEE 802.11-2016 standard does not mandate any of the examples given in the options. For more information, see Chapter 7.

15. A, B, C, and D. Carrier Sense Multiple Access with Collision Avoidance (CSMA/CA) is a medium access method that utilizes multiple checks and balances to try to minimize collisions. These checks and balances can also be thought of as several lines of defense. The various lines of defense are put in place to hopefully ensure that only one radio is transmitting while all other radios are listening. The four lines of defense include the network allocation vector, the random backoff timer, the clear channel assessment, and interframe spacing. For more information, see Chapter 8.

16. A, C, D, E, and F. From the list of choices, the only information not contained in the beacon management frame is the destination IP address. The body of all 802.11 management frames contains only layer 2 information; therefore, IP information is not included in the frame. Other information that is included in a beacon includes security and QoS parameters. For more information, see Chapter 9.

17. B and D. AP radios can be manually configured to use RTS/CTS for all transmissions. This is usually done to diagnose hidden node problems or to prevent hidden node problems when installing point-to-multipoint networks. 802.11g or 802.11n nodes may have enabled RTS/CTS as their protection mechanism. For more information, see Chapter 9.

18. C. The technical name for an 802.11 data frame is MAC protocol data unit (MPDU). An MPDU contains a layer 2 header, a frame body, and a trailer, which is a 32-bit CRC known as the frame check sequence (FCS). Inside the frame body of an MPDU is a MAC service data unit (MSDU), which contains data from the LLC and layers 3–7. For more information, see Chapter 9.

19. D. WLAN mesh access points create a self-forming WLAN mesh network that automatically connects access points at installation and dynamically updates routes as more clients are added. Most WLAN mesh networks use dynamic layer 2 routing protocols with metrics such as RSSI, SNR, and client load. For more information, see Chapter 11.

20. E. Over the years, most of the data generated on the Internet has been created by human beings. The theory of Internet of Things (IoT) is that in the future, the bulk of the data generated on the Internet might be created by sensors, monitors, and machines. 802.11 radio NICs used as client devices have begun to show up in many types of machines and devices. For more information, see Chapter 11.

21. A. Orthogonal frequency-division multiple access (OFDMA) is a technology that can be found in the proposed 802.11ax draft amendment. It allows 20 MHz channels to be partitioned into as many as 9 smaller channels, providing for multiple-user transmissions. For more information, see Chapter 19.

22. C. To further support and understand customers and their behaviors, retail analytics products are being installed to monitor customer movement and behavior. Strategically placed access points or sensor devices listen for probe frames from Wi-Fi-enabled smartphones. MAC addresses are used to identify each unique device, and signal strength is used to monitor and track the location of the shopper. Retail analytics can identify the path the shopper took while walking through the store, along with the time spent in different areas of the store. This information can be used to identify shopping patterns and to analyze the effectiveness of in-store displays and advertisements. For more information, see Chapter 20.

23. A, C, and E. The stations that cannot hear the hidden node may transmit at the same time that the hidden node is transmitting. This will result in continuous transmission collisions in a half-duplex medium. Collisions will corrupt the frames, and they will need to be retransmitted. Any time retransmissions are necessary, more overhead is added to the medium, resulting in decreased throughput. Intersymbol interference is a result of multipath, not the hidden node problem. For more information, see Chapter 15.

24. A and B. Layer 2 retransmissions can be caused by many different variables in a WLAN environment. Multipath, RF interference, hidden nodes, adjacent cell interference, and low signal-to-noise ratio (SNR) are all possible causes of layer 2 retransmissions. For more information, see Chapter 15.

25. E. Although you can hide your SSID to cloak the identity of your wireless network from script kiddies and non-hackers, it should be clearly understood that SSID cloaking is by no means an end-all wireless security solution. Because of spoofing and because of all the administrative work involved, MAC filtering is not considered a reliable means of security for wireless enterprise networks. WEP and Shared Key authentication are legacy 802.11 security solutions. CCMP/AES is defined as the default encryption type by the IEEE 802.11i security amendment. Cracking the AES cipher would take the lifetime of the sun using the tools that are available today. For more information, see Chapter 17.

26. D. The IEEE 802.1X standard is not specifically a wireless standard and often is mistakenly referred to as IEEE 802.11x. The IEEE 802.1X standard is a port-based access control standard. IEEE 802.1X provides an authorization framework that allows or disallows traffic to pass through a port and thereby access network resources. For more information, see Chapter 17.

27. C. The only tool that will absolutely identify an interfering signal is a spectrum analyzer. A spectrum analyzer is a layer 1 frequency domain tool that can detect any RF signal in the frequency range that is being scanned. Some WLAN vendors offer low-grade spectrum analysis as a built-in feature of their access points. For more information, see Chapter 16.

28. A, B, C, D, and E. 802.11 wireless intrusion detection systems may be able to monitor for as many as 100 or more attacks. Any layer 2 DoS attack and spoofing attack and most rogue devices can be detected. For more information, see Chapter 16.

29. A, B, and E. In the United States, if any tower exceeds a height of 200 feet above ground level (AGL), you must contact both the FCC and FAA, which are communications and aviation regulatory authorities, respectively. Other countries will have similar height restrictions, and the proper RF regulatory authority and aviation authority must be contacted to find out the details. Local municipalities may have construction regulations or height restrictions, and a permit may be required. For more information, see Chapter 14.

30. B and C. The final site survey report, known as the deliverable, will contain spectrum analysis information identifying potential sources of interference. Coverage analysis will also define RF cell boundaries. The final report also contains recommended access point place-ment, configuration settings, and antenna orientation. Capacity planning is considered to be mandatory when designing a WLAN; however, application throughput testing is often an optional analysis report included in the final survey report. Security, switching, and routing analysis are not included in a site survey report. For more information, see Chapter 14.

31. A. Some cordless phones transmit in the 5 GHz U-NII-3 band and are a potential source of RF interference, Bluetooth devices transmit in the 2.4 GHz frequency space. FM and AM radios transmit in licensed frequencies. For more information, see Chapter 14.

32. A, B, C, and E. RF coverage cell measurements that are taken during an indoor passive site survey include received signal strength, noise levels, signal-to-noise ratio (SNR), and data rates. Packet loss can be an additional measurement recorded during an active manual site survey. Packet loss is a calculation needed for an outdoor wireless bridging survey. For more information, see Chapter 14.

33. D. All the answers are possible causes of layer 2 retransmissions; however, RF interference is the main reason for layer 2 frame retransmissions to occur. WLAN performance is negatively impacted if the retransmission rate exceeds 10 percent. For more information, see Chapter 10.

34. A and D. For a powered device (PD) such as an access point to be considered compliant with the IEEE 802.3-2015, Clause 33 PoE standard, the device must be able to receive power through the data lines or the unused twisted pairs of an Ethernet cable. The PD must also reply to the power-sourcing equipment (PSE) with a 25-ohm detection signature. The PD may reply with a classification signature, but it is optional. The current PoE standard allows for a maximum draw of 12.95 watts by the PD from the power-sourcing equipment. For more information, see Chapter 12.

35. B, C, and D. High throughput (HT) technology is defined by the IEEE 802.11n-2009 amendment and is not frequency-dependent. 802.11n (HT) can operate in the 2.4 GHz ISM band as well as all the 5 GHz U-NII frequency bands. For more information, see Chapter 10.

36. A and B. The 802.11n-2009 amendment introduced two new methods of frame aggregation to help reduce the overhead. Frame aggregation is a method of combining multiple frames into a single frame transmission. The first method of frame aggregation is known as an aggregate MAC service data unit (A-MSDU). The second method of frame aggregation is known as an aggregate MAC protocol data unit (A-MPDU). For more information, see Chapter 10.

37. A. The 802.11n-2009 amendment defined 77 modulation and coding schemes (MCSs). The 802.11ac-2013 amendment lowers the number to 10. The 802.11ac data rates are determined by the number of spatial streams, the channel width, the guard interval, and which one of the 10 MCSs is used. For more information, see Chapter 10.

38. B and C. Reduced interframe spacing (RIFS), unequal modulation, Greenfield mode, and implicit beamforming are 802.11n capabilities that are no longer defined with the advent of the 802.11ac amendment. For more information, see Chapter 10.

39. E. Mobile device management (MDM) solutions can be used for both a company-issued device (CID) and a bring your own device (BYOD), which is owned by an employee. MDM solutions offer the capability of over-the-air installation and distribution of security certificates, web clips, applications, and configuration settings. For more information, see Chapter 18.

40. D. The OAuth 2.0 authorization framework enables a third-party application to obtain limited access to an HTTP service and is often used for social login for Wi-Fi guest networks. For more information, see Chapter 18.

Chapter

1

Overview of Wireless Standards, Organizations, and Fundamentals

---

IN THIS CHAPTER, YOU WILL LEARN ABOUT THE FOLLOWING:

✓ **History of wireless local area networks**

✓ **Standards organizations**

- Federal Communications Commission
- International Telecommunication Union Radiocommunication Sector
- Institute of Electrical and Electronics Engineers
- Internet Engineering Task Force
- Wi-Fi Alliance
- International Organization for Standardization

✓ **Core, distribution, and access**

✓ **Communications fundamentals**

- Communication terminology
- Carrier signals
- Keying methods

Wireless local area network (WLAN) technology has a long history that dates back to the 1970s, with roots as far back as the 19th century. This chapter will start with a brief history of WLAN technology. Learning a new technology can seem like a daunting task. There are so many new acronyms, abbreviations, terms, and ideas to become familiar with. One of the keys to learning any subject is to learn the basics. Whether you are learning to drive a car, fly an airplane, or install a wireless computer network, there are basic rules, principles, and concepts that, once learned, provide the building blocks for the rest of your education.

The Institute of Electrical and Electronics Engineers (IEEE) 802.11 technology, more commonly referred to as Wi-Fi, is a standard technology for providing local area network (LAN) communications using radio frequencies (RFs). The IEEE designates the 802.11-2016 standard as a guideline to provide operational parameters for WLANs. Numerous standards organizations and regulatory bodies help govern and direct wireless technologies and the related industry. Having some knowledge of these various organizations can provide you with insight as to how IEEE 802.11 functions and sometimes even how and why the standards have evolved the way they have.

As you become more knowledgeable about wireless networking, you may want or need to read some of the standards documents that are created by the different organizations. Along with the information about the standards bodies, this chapter includes a brief overview of their documents.

In addition to reviewing the various standards organizations that guide and regulate Wi-Fi, this chapter discusses where WLAN technology fits in with basic networking design fundamentals. Finally, this chapter reviews some fundamentals of communications and data keying that are not part of the CWNA exam but that may help you better understand wireless communications.

History of Wireless Local Area Networks

In the 19th century, numerous inventors and scientists, including Michael Faraday, James Clerk Maxwell, Heinrich Rudolf Hertz, Nikola Tesla, David Edward Hughes, Thomas Edison, and Guglielmo Marconi, began to experiment with wireless communications. These innovators discovered and created many theories about the concepts of electrical magnetic *radio frequency (RF)*.

Wireless networking technology was first used by the U.S. military during World War II to transmit data over an RF medium using classified encryption technology to send battle plans across enemy lines. The spread spectrum radio technologies often used in today's WLANs were also originally patented during the era of World War II, although they were not implemented until almost two decades later.

In 1970, the University of Hawaii developed the first wireless network, called ALOHAnet, to wirelessly communicate data between the Hawaiian Islands. The network used a LAN communication Open Systems Interconnection (OSI) layer 2 protocol called ALOHA on a wireless shared medium in the 400 MHz frequency range. The technology used in ALOHAnet is often credited as a building block for the Medium Access Control (MAC) technologies of Carrier Sense Multiple Access with Collision Detection (CSMA/CD) used in Ethernet and Carrier Sense Multiple Access with Collision Avoidance (CSMA/CA) used in 802.11 radios. You will learn more about CSMA/CA in Chapter 8, "802.11 Medium Access."

In the 1990s, commercial networking vendors began to produce low-speed wireless data networking products, most of which operated in the 900 MHz frequency band. The Institute of Electrical and Electronics Engineers (IEEE) began to discuss standardizing WLAN technologies in 1991. In 1997, the IEEE ratified the original 802.11 standard that is the foundation of the WLAN technologies you will be learning about in this book.

This legacy 802.11 technology was deployed between 1997 and 1999 mostly in warehousing and manufacturing environments for the use of low-speed data collection with wireless barcode scanners. In 1999, the IEEE defined higher data speeds with the 802.11b amendment. The introduction of data rates as high as 11 Mbps, along with price reductions, ignited the sales of wireless home networking routers in the small office, home office (SOHO) marketplace. Home users soon became accustomed to wireless networking in their homes and began to demand that their employers also provide wireless networking capabilities in the workplace. After initial resistance to 802.11 technology, small companies, medium-sized businesses, and corporations began to realize the value of deploying 802.11 wireless networking in their enterprises.

If you ask the average user about their 802.11 wireless network, they may give you a strange look. The name that people often recognize for the technology is *Wi-Fi*. Wi-Fi is a marketing term, recognized worldwide by millions of people as referring to 802.11 wireless networking.

What Does the Term *Wi-Fi* Mean?

Many people mistakenly assume that *Wi-Fi* is an acronym for the phrase *wireless fidelity* (much like *hi-fi* is short for *high fidelity*), but Wi-Fi is simply a brand name used to market 802.11 WLAN technology. Ambiguity in IEEE framework standards for wireless communications allowed manufacturers to interpret the 802.11 standard in different ways. As a result, multiple vendors could have IEEE 802.11–compliant devices that did not interoperate with each other. The organization Wireless Ethernet Compatibility Alliance (WECA) was created to further define the IEEE standard in such a way as to force interoperability between vendors. WECA, now known as the Wi-Fi Alliance, chose the term *Wi-Fi* as a marketing brand. The Wi-Fi Alliance champions enforcing interoperability among wireless devices. To be Wi-Fi compliant, vendors must send their products to a Wi-Fi Alliance test lab, which thoroughly tests compliance to the Wi-Fi certifications. More information about the origins of the term Wi-Fi can be found online at Wi-Fi Net News:

```
https://wifinetnews.com/archives/2005/11/wi-fi_stands_fornothing_and_
everything.html
```

Wi-Fi radios are used for numerous enterprise applications and can also be found in laptops, smartphones, cameras, televisions, printers, and many other consumer devices. According to the Wi-Fi Alliance, the billionth Wi-Fi chipset was sold in 2009. According to the technology research firm, 650 Group, (www.650.group.com), semiconductor shipments of Wi-Fi radios surpassed three billion units in 2017. As shown in Figure 1.1, shipments of Wi-Fi radios will continue to rise further into the billions per year. In a survey conducted by the Wi-Fi Alliance, 68 percent of Wi-Fi users would rather give up chocolate than do without Wi-Fi. Since the original standard was created in 1997, 802.11 technology has grown to enormous proportions; Wi-Fi has become part of our worldwide communications culture. A recent report from Telecom Advisory Services estimates that technologies that rely on unlicensed spectrum add $222 billion dollars per year to the U.S. economy. More than $91 billion dollars can be attributed to Wi-Fi.

FIGURE 1.1   Growth of the Wi-Fi industry

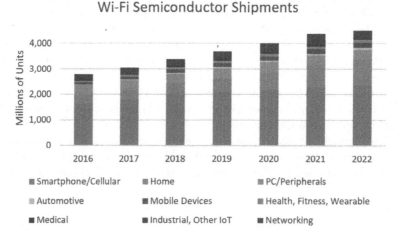

Courtesy of 650 Group

Standards Organizations

Each standards organization discussed in this chapter helps to guide a different aspect of the wireless networking industry.

The International Telecommunication Union Radiocommunication Sector (ITU-R) and local entities such as the Federal Communications Commission (FCC) set the rules for what users can do with a radio transmitter. These organizations manage and regulate frequencies, power levels, and transmission methods. They also work together to help guide the growth and expansion that is being demanded by wireless users.

The Institute of Electrical and Electronics Engineers (IEEE) creates standards for compatibility and coexistence between networking equipment. The IEEE standards must adhere to the rules of the communications organizations, such as the FCC.

The Internet Engineering Task Force (IETF) is responsible for creating Internet standards. Many of these standards are integrated into the wireless networking and security protocols and standards.

The Wi-Fi Alliance performs certification testing to ensure that wireless networking equipment conforms to the 802.11 WLAN communication guidelines, which are similar to the IEEE 802.11-2016 standard.

The International Organization for Standardization (ISO) created the Open Systems Interconnection (OSI) model, which is an architectural model for data communications.

The following sections discuss each of these organizations in greater detail.

Federal Communications Commission

To put it simply, the *Federal Communications Commission (FCC)* regulates communications within the United States as well as communications to and from the United States. Established by the Communications Act of 1934, the FCC is responsible for regulating interstate and international communications by radio, television, wire, satellite, and cable. The task of the FCC in wireless networking is to regulate the radio signals that are used for wireless networking. The FCC has jurisdiction over the 50 states, the District of Columbia, and U.S. possessions. Most countries have governing bodies that function similarly to the FCC.

The FCC and the respective controlling agencies in other countries typically regulate two categories of wireless communications: licensed spectrum and unlicensed spectrum. The difference is that unlicensed users do not have to go through the license application procedures before they can install a wireless system. Both licensed and unlicensed communications are typically regulated in the following five areas:

- Frequency
- Bandwidth
- Maximum power of the intentional radiator (IR)
- Maximum equivalent isotropically radiated power (EIRP)
- Use (indoor and/or outdoor)
- Spectrum sharing rules

 Real World Scenario

What Are the Advantages and Disadvantages of Using an Unlicensed Frequency?

As stated earlier, licensed frequencies require an approved license application, and the financial costs are typically very high. One main advantage of an unlicensed frequency is that permission to transmit on the frequency is free. Although there are no financial costs, you still must abide by transmission regulations and other restrictions. In other words, transmitting in an unlicensed frequency may be free, but there still are rules.

> The main disadvantage to transmitting in an unlicensed frequency band is that anyone else can also transmit in that same frequency space. Unlicensed frequency bands are often very crowded; therefore, transmissions from other individuals can cause interference with your transmissions. If someone else is interfering with your transmissions, you have no legal recourse as long as the other individual is abiding by the rules and regulations of the unlicensed frequency.

Essentially, the FCC and other regulatory bodies set the rules for what the user can do regarding RF transmissions. From there, the standards organizations create the standards to work within these guidelines. These organizations work together to help meet the demands of the fast-growing wireless industry.

The FCC rules are published in the Code of Federal Regulations (CFR). The CFR is divided into 50 titles, which are updated yearly. The title that is relevant to wireless networking is Title 47, *Telecommunications*. Title 47 is divided into many parts; Part 15, "Radio Frequency Devices," is where you will find the rules and regulations regarding wireless networking related to 802.11. Part 15 is further broken down into subparts and sections. A complete reference will look like this example: 47CFR15.3.

International Telecommunication Union Radiocommunication Sector

A global hierarchy exists for management of the RF spectrum worldwide. The United Nations has tasked the *International Telecommunication Union Radiocommunication Sector (ITU-R)* with global spectrum management. The ITU-R strives to ensure interference-free communications on land, sea, and in the skies. The ITU-R maintains a database of worldwide frequency assignments through five administrative regions.

The five administrative regions are broken down as follows:

Region A: The Americas   Inter-American Telecommunication Commission (CITEL)

www.citel.oas.org

Region B: Western Europe   European Conference of Postal and Telecommunications Administrations (CEPT)

www.cept.org

Region C: Eastern Europe and Northern Asia   Regional Commonwealth in the field of Communications (RCC)

www.en.rcc.org.ru

Region D: Africa   African Telecommunications Union (ATU)

www.atu-uat.org

Region E: Asia and Australasia   Asia-Pacific Telecommunity (APT)

www.aptsec.org

In addition to the five administrative regions, the ITU-R defines three radio regulatory regions. These three regions are defined geographically, as shown in the following list. You should check an official ITU-R map to identify the exact boundaries of each region.

- **Region 1:** Europe, Middle East, and Africa
- **Region 2:** Americas
- **Region 3:** Asia and Oceania

The ITU-R radio regulation documents are part of an international treaty governing the use of spectrum. Within each of these regions, the ITU-R allocates and allots frequency bands and radio channels that are allowed to be used, along with the conditions regarding their use. Within each region, local government RF regulatory bodies, such as the following, manage the RF spectrum for their respective countries:

Australia   Australian Communications and Media Authority (ACMA)

www.acma.gov.au

Japan   Association of Radio Industries and Businesses (ARIB)

www.arib.or.jp

United States   Federal Communications Commission (FCC)

www.fcc.gov

It is important to understand that communications are regulated differently in many regions and countries. For example, European RF regulations are very different from the regulations used in North America. When deploying a WLAN, please take the time to learn about rules and policies of the local *regulatory domain authority*. However, since the rules vary around the globe, it is beyond the capabilities of this book to reference the different regulations. Additionally, the CWNA exam will not reference the RF regulations of the FCC or those specific to any other country.

   More information about the ITU-R can be found at www.itu.int/ITU-R.

Institute of Electrical and Electronics Engineers

The *Institute of Electrical and Electronics Engineers*, commonly known as the *IEEE*, is a global professional society with more than 420,000 members in 160 countries. The IEEE's mission is to "foster technological innovation and excellence for the benefit of humanity." To networking professionals, that means creating the standards that we use to communicate.

The IEEE is probably best known for its LAN standards, the IEEE 802 project.

   The 802 project is one of many IEEE projects; however, it is the only IEEE project addressed in this book.

IEEE projects are subdivided into working groups to develop standards that address specific problems or needs. For instance, the IEEE 802.3 working group was responsible for the creation of a standard for Ethernet, and the IEEE 802.11 working group was responsible for creating the WLAN standard. The numbers are assigned as the groups are formed, so the 11 assigned to the wireless group indicates that it was the 11th working group formed under the IEEE 802 project.

As the need arises to revise existing standards created by the working groups, task groups are formed. These task groups are assigned a sequential single letter (multiple letters are assigned if all single letters have been used) that is added to the end of the standard number (for example, 802.11a, 802.11g, and 802.3at). Some letters are not assigned. For example *o* and *l* are not assigned to prevent confusion with the numbers 0 and 1. Other letters may not be assigned to task groups to prevent confusion with other standards. For example, 802.11x has not been assigned because it can be easily confused with the 802.1X standard, and because 802.11*x* has become a common casual reference to the 802.11 family of standards.

You can find more information about the IEEE at www.ieee.org.

It is important to remember that the IEEE standards, like many other standards, are written documents describing how technical processes and equipment should function. Unfortunately, this often allows for different interpretations when the standard is being implemented, so it is common for early products to be incompatible between vendors, as was the case with some of the early 802.11 products.

The history of the 802.11 standard and amendments is covered extensively in Chapter 2, "IEEE 802.11 Standards." The CWNA exam (CWNA-107) is based on the most recently published version of the standard, 802.11-2016. The 802.11-2016 standard can be downloaded from:

http://standards.ieee.org/about/get/802/802.11.html

Internet Engineering Task Force

The *Internet Engineering Task Force*, commonly known as the *IETF*, is an international community of people in the networking industry whose goal is to make the Internet work better. The mission of the IETF, as defined by the organization in a document known as RFC 3935, is "to produce high quality, relevant technical and engineering documents that influence the way people design, use, and manage the Internet in such a way as to make the Internet work better. These documents include protocol standards, best current practices, and informational documents of various kinds." The IETF has no membership fees, and anyone may register for and attend an IETF meeting.

The IETF is one of five main groups that are part of the Internet Society (ISOC). The ISOC groups include the following:

- Internet Engineering Task Force (IETF)
- Internet Architecture Board (IAB)

- Internet Corporation for Assigned Names and Numbers (ICANN)
- Internet Engineering Steering Group (IESG)
- Internet Research Task Force (IRTF)

The IETF is broken into eight subject matter areas: Applications, General, Internet, Operations and Management, Real-Time Applications and Infrastructure, Routing, Security, and Transport. Figure 1.2 shows the hierarchy of the ISOC and a breakdown of the IETF subject matter areas.

FIGURE 1.2 ISOC hierarchy

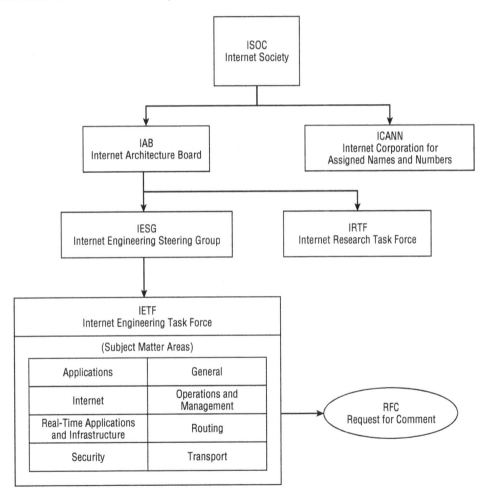

The IESG provides technical management of the activities of the IETF and the Internet standards process. The IETF is made up of a large number of groups, each addressing specific topics. An IETF working group is created by the IESG and is given a specific charter or specific topic to address. There is no formal voting process for the working groups. Decisions in working groups are made by rough consensus, or basically a general sense of agreement among the working group.

The results of a working group are usually the creation of a document known as a *Request for Comments (RFC)*. Contrary to its name, an RFC is not actually a request for comments but a statement or definition. Most RFCs describe network protocols, services, or policies and may evolve into an Internet standard. RFCs are numbered sequentially, and once a number is assigned, it is never reused. RFCs may be updated or supplemented by higher-numbered RFCs. As an example, Mobile IPv4 was described in RFC 3344 in 2002. This document was updated in RFC 4721. In 2012, RFC 5944 made RFC 3344 obsolete. At the top of the RFC document, it states whether the RFC is updated by another RFC and also if it makes any other RFCs obsolete.

Not all RFCs are standards. Each RFC is given a status, relative to its relationship with the Internet standardization process: Informational, Experimental, Standards Track, or Historic. If it is a Standards Track RFC, it could be a Proposed Standard, Draft Standard, or Internet Standard. When an RFC becomes a standard, it still keeps its RFC number, but it is also given an "STD *xxxx*" label. The relationship between the STD numbers and the RFC numbers is not one-to-one. STD numbers identify protocols, whereas RFC numbers identify documents.

Many of the protocol standards, best current practices, and informational documents produced by the IETF affect WLAN security. In Chapter 17, "802.11 Network Security Architecture," you will learn about some of the varieties of the Extensible Authentication Protocol (EAP), which is defined by the IETF RFC 3748.

Wi-Fi Alliance

The *Wi-Fi Alliance* is a global, nonprofit industry association of more than 550 member companies devoted to promoting the growth of WLANs. One of the primary tasks of the Wi-Fi Alliance is to market the Wi-Fi brand and raise consumer awareness of new 802.11 technologies as they become available. Because of the Wi-Fi Alliance's overwhelming marketing success, the majority of the worldwide Wi-Fi users are likely to recognize the Wi-Fi logo, as shown in Figure 1.3.

FIGURE 1.3   Wi-Fi logo

The Wi-Fi Alliance's main task is to ensure the interoperability of WLAN products by providing certification testing. During the early days of the 802.11 standard, the Wi-Fi Alliance further defined some of the ambiguous standards requirements and provided a set of guidelines to ensure compatibility between different vendors. This is still done to help simplify the complexity of the standards and to ensure compatibility. As shown in Figure 1.4, products that pass the Wi-Fi certification process receive a Wi-Fi Interoperability Certificate that provides detailed information about the individual product's Wi-Fi certifications.

FIGURE 1.4 Wi-Fi Interoperability Certificate

Wi-Fi CERTIFIED™ Interoperability Certificate
This certificate lists the features that have successfully
completed Wi-Fi Alliance interoperability testing.
Learn more: www.wi-fi.org/certification/programs

Certification ID: WFAXXXYYY	Page 1 of 3

Date of Last Certification	January 01, 2013
Company	ABC Design
Product	Widget Series 123
Model Number	5678
Product Identifier(s)	AB-CDE-FG (SKU), 123456789 (UPC), AB123 (EAN), 131313TYTY (Other)
Category	Media Adapter
Hardware Version	Product: 11, Wi-Fi Component: 11
Firmware Version	Product: 11, Wi-Fi Component: 11
Operating System	Windows 8
Frequency Band(s)	2.4 GHz, 5 GHz, 60 GHz

Summary of Certifications

CLASSIFICATION	PROGRAM
Connectivity	Wi-Fi CERTIFIED™ a, b, g, n, ac
	Wi-Fi CERTIFIED WiGig™
	WPA™ - Enterprise, Personal
	WPA2™ - Enterprise, Personal
	Wi-Fi Direct®
Optimization	TDLS
	Wi-Fi Agile Multiband
	Wi-Fi Optimized Connectivity
	Wi-Fi TimeSync™
	Wi-Fi Vantage™
	WMM®
	WMM®-Admission Control
	WMM®-Power Save
Access	IBSS with Wi-Fi Protected Setup™
	Passpoint™
	Wi-Fi Protected Setup™
Applications & Services	Miracast® - Display, Source
	Voice-Enterprise
	Voice-Personal
	Wi-Fi Aware™
	Wi-Fi Location
RF Coexistence	CWG-RF Test

FIGURE 1.4 Wi-Fi Interoperability Certificate *(continued)*

 Wi-Fi CERTIFIED™ Interoperability Certificate

Certification ID: WFAXXXYYY   Page 2 of 3

Security

WPA™ – Enterprise, Personal
WPA2™ – Enterprise, Personal
 EAP Type(s)
  – EAP-TLS
  – EAP-TTLS/MSCHAPv2
  – PEAP0/EAP-MSCHAPv2
  – PEAP1/EAP-GTC
  – EAP-SIM
  – EAP-AKA
  – EAP-AKA Prime
  – EAP-FAST
 Protected Management Frames

Wi-Fi CERTIFIED™ a

Wi-Fi CERTIFIED™ b

Wi-Fi CERTIFIED™ g

Wi-Fi CERTIFIED™ n

2.4 GHz, 5 GHz
X Spatial Streams 2.4 GHz
X Spatial Stream 5 GHz
Short Guard Interval 20 MHz
Short Guard Interval 40 MHz
Greenfield Preamble
TX A-MPDU
STBC Receive
STBC Transmit
40 MHz operation in 2.4 GHz, with coexistence mechanisms
40 MHz operation in 5 GHz
HT Duplication Mode (MCS 32)
OBSS on Extension Channel
RIFS Transmit
STAUT Power Management

Wi-Fi CERTIFIED™ ac

X Spatial Stream 5 GHz
Rx MCS 8 (256-QAM)
Rx MCS 8-9 (256-QAM)
Tx STBC 2x1
Rx STBC 2x1
Rx A-MPDU of A-MSDU
Tx SU beamformer
Tx SU beamformee
Low Density Parity Check coding
Tx DL MU-MIMO
Rx DL MU-MIMO
Rx 160 MHz operation
Extended 5 GHz Channel Support
RTS with BW Signaling

Wi-Fi CERTIFIED WiGig™

Optional n Features

2.4 GHz, 5 GHz
X Spatial Streams 2.4 GHz
X Spatial Stream 5 GHz
Short Guard Interval 20 MHz
Short Guard Interval 40 MHz
Greenfield Preamble
TX A-MPDU
STBC Receive
STBC Transmit
40 MHz operation in 2.4 GHz, with coexistence mechanisms
40 MHz operation in 5 GHz
HT Duplication Mode (MCS 32)
OBSS on Extension Channel
RIFS Transmit
STAUT Power Management

Wi-Fi Direct®

2.4 GHz, 5 GHz, 60 GHz
Wi-Fi Direct Send – Transmit, Receive
Wi-Fi Direct for DLNA® – Transmit, Receive
Miracast – Display, Source
Wi-Fi Direct Print – Transmit, Receive
Wi-Fi Direct Toolkit

TDLS

Wi-Fi Agile Multiband

Steer to Cellular Data
Fast Transition: WPA2-Enterprise <or> WPA2-Personal

Wi-Fi Optimized Connectivity

Fast Initial Link Setup Shared Key Authentication
Higher Layer Protocol Encapsulation
Estimated Service Parameters - (AP only)

Wi-Fi TimeSync™

Wi-Fi Vantage™

Release 1 or Release 2

WMM®

WMM®-Admission Control

WMM®-Power Save

IBSS with Wi-Fi Protected Setup™

Passpoint™

Online Signup (OSU) and Policy Provisioning
Open Mobile Alliance™ Device Management (OMA DM)
Wi-Fi Network Icon

The Wi-Fi Alliance, originally named the Wireless Ethernet Compatibility Alliance (WECA), was founded in August 1999. The name was changed to the Wi-Fi Alliance in October 2002.

The Wi-Fi Alliance has certified more than 35,000 Wi-Fi products for interoperability since testing began in April 2000. Multiple Wi-Fi CERTIFIED programs exist, covering basic connectivity, security, quality of service (QoS), and more. Testing of vendor Wi-Fi products is performed in independent authorized test laboratories in eight countries. A listing of these testing laboratories can be found on the Wi-Fi Alliance's website. The guidelines for interoperability for each Wi-Fi CERTIFIED program are usually based on key components and functions that are defined in the IEEE 802.11-2016 standard and various 802.11 amendments. In fact, many of the same engineers who belong to 802.11 task groups are also contributing members of the Wi-Fi Alliance. However, it is important to understand that the IEEE and the Wi-Fi Alliance are two separate organizations. The IEEE 802.11 task group defines the WLAN standards, and the Wi-Fi Alliance defines interoperability certification programs.

The Wi-Fi Alliance certifies 802.11a, b, g, n, and/or ac interoperability to ensure that the essential wireless data transmission works as expected. Each device is tested according to its capabilities. Table 1.1 lists the five different core Wi-Fi transmission technologies along with the frequencies and maximum data rate that each is capable of. Each certified product is required to support one frequency band as a minimum, but it can support both.

TABLE 1.1   Five generations of Wi-Fi

Wi-Fi Technology	Frequency Band	Maximum Data Rate
802.11a	5 GHz	54 Mbps
802.11b	2.4 GHz	11 Mbps
802.11g	2.4 GHz	54 Mbps
802.11n	2.4 GHz, 5 GHz, 2.4 or 5 GHz (selectable), or 2.4 and 5 GHz (concurrent)	600 Mbps
802.11ac	5 GHz	6933.3 Mbps

In the following sections, the Wi-Fi CERTIFIED programs are discussed in the context of different categories:

Connectivity

Wi-Fi CERTIFIED b/g   The Wi-Fi Alliance certifies backward compatibility with legacy 802.11b/g devices that operate in the 2.4 GHz frequency band. Discussion about the operational capabilities of legacy 802.11b/g radios can be found in subsequent chapters.

Wi-Fi CERTIFIED a   The Wi-Fi Alliance certifies backward compatibility with legacy 802.11a radios that transmit in 5 GHz frequency band. Discussion about the operational capabilities of legacy 802.11a radios can be found in subsequent chapters.

Wi-Fi CERTIFIED n   The Wi-Fi Alliance certifies the operational capabilities for 802.11n radios for both the 2.4 GHz and 5 GHz frequency bands. 802.11n introduced PHY and MAC layer enhancements to achieve higher data rates. 802.11n requires *multiple-input-multiple output (MIMO)* radio systems that are backward compatible with 802.11a/b/g technology. A deep discussion about 802.11n can be found in Chapter 10, "MIMO Technology: HT and VHT."

Wi-Fi CERTIFIED ac   The Wi-Fi Alliance certifies the operational capabilities for 802.11ac radios for the 5 GHz frequency band. 802.11ac technology introduced further PHY and MAC layer enhancements to achieve higher data rates beyond 802.11n. 802.11ac radios are backward compatible with 802.11a/n radios. A deep discussion about 802.11ac can be found in Chapter 10.

Wi-Fi Direct   *Wi-Fi Direct* enables Wi-Fi devices to connect directly without the use of an access point (AP), making it easier to print, share, sync, and display. Wi-Fi Direct is ideal for mobile phones, cameras, printers, PCs, and gaming devices needing to establish a one-to-one connection, or even connecting a small group of devices. Wi-Fi Direct is simple to configure (in some cases as easy as pressing a button), provides the same performance and range as other Wi-Fi CERTIFIED devices, and is secured using WPA2 security. Wi-Fi Direct implements technology based on the Wi-Fi Peer-to-Peer technical specification.

Wi-Fi CERTIFIED WiGig   The WiGig certification program is based on technology originally defined in the 802.11ad amendment for *directional multi-gigabit (DMG)* radios that transmit the 60 GHz frequency band. Multi-band Wi-Fi CERTIFIED WiGig devices can seamlessly transfer between the 2.4, 5, or 60 GHz bands. Operational specifications are defined in the Wi-Fi Alliance 60 GHz technical specification. WiGig uses wider channels in 60 GHz to transmit data efficiently at multi-gigabit per second speeds and with low latency at distances of up to 10 meters. WiGig use cases include wireless docking stations, HD video streaming, and other bandwidth-intensive applications.

Security

Wi-Fi Protected Access 2   The Wi-Fi Protected Access 2 (WPA2) certification is based on *robust security network (RSN)* capabilities, security mechanisms that were originally defined in the IEEE 802.11i amendment. All Wi-Fi WPA2–certified devices must support CCMP/AES dynamic encryption methods. The Wi-Fi Alliance specifies two methods for user and device authorization for WLANs. *WPA2 Enterprise* requires support for 802.1X port-based access control security for enterprise deployments. *WPA2-Personal* uses a less complex passphrase method intended for SOHO environments. On January 8th, 2018, the Wi-Fi Alliance announced Wi-Fi Protected Access 3, also known as WPA3, which defines enhancements to the existing WPA2 security capabilities for 802.11 radios. Four new capabilities for both personal and enterprise networks have been announced. You will find a more detailed discussion of WPA2 and WPA3 security in Chapter 17, "802.11 Network Security Architecture."

Extensible Authentication Protocol   WPA2 Enterprise radios support 802.1X security, which requires many components to approve access for users and devices to a WLAN. Enterprise devices must support *Extensible Authentication Protocol (EAP)*, the authentication protocol used within an 802.1X authorization framework. The Wi-Fi Alliance certification program tests for many variants of EAP, including EAP-TLS, EAP-TTLS, EAP-PEAP, and others. More information about 802.1X and EAP can be found in Chapter 17.

WPA2 with Protected Management Frames   This certification is based upon the IEEE 802.11w-2009 amendment. It is the *management frame protection (MFP)* amendment, with a goal of delivering certain types of management frames in a secure manner. The intent is to prevent spoofing of certain types of 802.11 management frames and prevent common layer 2 denial-of-service (DoS) attacks.

Access

Passpoint   *Passpoint* is designed to revolutionize the end-user experience when connecting to Wi-Fi hotspots. This is done by automatically identifying the hotspot provider and connecting to it, automatically authenticating the user to the network using Extensible Authentication Protocol (EAP), and providing secure transmission using WPA2-Enterprise encryption. Passpoint is based on the *Hotspot 2.0* technical specification. More detailed information about Passpoint and Hotspot can be found in Chapter 18, "Bring Your Own Device (BYOD) and Guest Access."

Wi-Fi Protected Setup   *Wi-Fi Protected Setup (WPS)* defines simplified and automatic WPA and WPA2 security configurations for home and small-business users. Users can easily configure a network with security protection by using either near field communication (NFC), a personal identification number (PIN), or a button located on the AP and the client device. WPS technology is defined in the Wi-Fi Simple Configuration technical specification.

IBSS with Wi-Fi Protected Setup   IBSS with Wi-Fi Protected Setup provides easy configuration and strong security for ad-hoc (peer-to-peer) Wi-Fi networks. This is designed for mobile products and devices that have a limited user interface, such as smartphones, cameras, and media players. Features include easy push button or PIN setup, task-oriented short-term connections, and dynamic networks that can be established anywhere.

Applications and Services

Voice-Enterprise   *Voice-Enterprise* offers enhanced support for voice applications in enterprise Wi-Fi networks. Enterprise-grade voice equipment must provide consistently good voice quality under all network load conditions and coexist with data traffic. Many of the mechanisms defined by the IEEE 802.11k, 802.11r and 802.11v amendments are also defined by the Voice-Enterprise certification. Both access point and client devices must support prioritization using Wi-Fi Multimedia (WMM), with voice traffic being placed in the highest-priority queue (Access Category Voice, AC_VO). Voice-Enterprise equipment must also support seamless roaming between APs, WPA2-Enterprise security, optimization of power through the WMM-Power Save mechanism, and traffic management through WMM-Admission Control.

Voice-Personal   *Voice-Personal* offers enhanced support for voice applications in residential and small-business Wi-Fi networks. These networks include one AP, mixed voice and data traffic from multiple devices (such as phones, PCs, printers, and other consumer electronic devices), and support for up to four concurrent phone calls. Both the AP and the client device must be certified to achieve performance matching the certification metrics.

Miracast   *Miracast* seamlessly integrates the display of high-resolution streaming video content between devices. Wireless links are used to replace wired connections. Devices are designed to identify and connect with each other, manage their connections, and optimize the transmission of video content. Miracast is based on the Wi-Fi Display technical specification. The Miracast certification program is for any video-capable device, such as cameras, televisions, projectors, tablets, and smartphones. Paired Miracast devices can stream high-definition (HD) content or mirror displays via a peer-to-peer Wi-Fi connection.

Wi-Fi Aware   Wi-Fi Aware-enabled devices use power-efficient discovery of nearby services or information before making a connection. The *neighbor awareness networking (NAN)* technical specification defines mechanisms for WLAN devices to synchronize channel and time information to allow for the discovery of services. *Wi-Fi Aware* does not require the existence of a WLAN infrastructure, and discovery occurs in the background, even in crowded user environments. Prior to establishing a connection, users can find other nearby users for the purposes of sharing media, local information, and gaming opponents.

Wi-Fi Location   *Wi-Fi Location* is based on the *Fine Timing Measurement (FTM)* protocol defined in the IEEE 802.11-2016 standard. Wi-Fi Location-enabled devices and networks can provide devices with highly accurate indoor location information via the Wi-Fi network without the need for an overlay infrastructure such as iBeacons or a real-time locating system (RTLS). Application and OS developers can create location-based applications and services. Some of the potential uses include asset management, geo-fencing, and hyperlocal marketing.

Optimization

Wi-Fi Multimedia   *Wi-Fi Multimedia (WMM)* is based on the QoS mechanisms that were originally defined in the IEEE 802.11e amendment. WMM enables Wi-Fi networks to prioritize traffic generated by different applications. In a network where WMM is supported by both the AP and the client device, traffic generated by time-sensitive applications, such as voice or video, can be prioritized for transmission on the half-duplex RF medium. WMM certification is mandatory for all core certified devices that support 802.11n and 802.11ac. WMM certification is optional for core certified devices that support 802.11 a, b, or g. WMM mechanisms are discussed in greater detail in Chapter 9, "802.11 MAC."

WMM-Power Save   *WMM-Power Save (WMM-PS)* helps conserve battery power for devices using Wi-Fi radios by managing the time the client device spends in sleep mode. Conserving battery life is critical for handheld devices, such as barcode scanners and voice over Wi-Fi (VoWiFi) phones. To take advantage of power-saving capabilities, both the device and the access point must support WMM-PS. Chapter 9 discusses WMM-PS and legacy power-saving mechanisms in greater detail.

WMM-Admission Control   *WMM-Admission Control (WMM-AC)* allows Wi-Fi networks to manage network traffic based upon channel conditions, network traffic load, and type of traffic (voice, video, best effort data, or background data). The access point allows only the traffic that it can support to connect to the network, based upon the available network resources. WMM-AC uses call admission control (CAC) mechanisms to prevent oversubscription of voice calls through an 802.11 access point.

Wi-Fi CERTIFIED TDLS   The IEEE 802.11z-2010 amendment defines a *Tunneled Direct Link Setup (TDLS)* security protocol. The Wi-Fi Alliance also introduced Wi-Fi CERTIFIED TDLS as a certification program for devices using TDLS to connect directly to one another after they have joined a traditional Wi-Fi network. This allows consumer devices such as TVs, gaming devices, smartphones, cameras, and printers to directly and securely communicate with each other while remaining connected to an access point.

Wi-Fi TimeSync   *Wi-Fi CERTIFIED TimeSync* enables sub-microsecond clock synchronization between multiple devices, aiding precise service coordination and accurate representation of audio, video, or data. The technology supports in-room multichannel audio and video capabilities. Uses for Wi-Fi TimeSync technology include home theater systems, recording studios, camera systems, and more. The technology is based on the IEEE 802.11-2016 Timing Measurement (TM) and Fine Timing Measurement (FTM) protocols.

Wi-Fi Vantage   A growing trend in the Wi-Fi industry is for a managed service provider (MSP) to offer "wireless as a service." Many telecom carrier companies offer MSP services that oversee Wi-Fi operations in airports, stadiums, schools, office buildings, retail and hotel locations, and other venues. Wi-Fi Certified Vantage is a certification program with the goal of elevating the experience for users in managed Wi-Fi networks.

RF Coexistence

Converged Wireless Group-RF Profile   Converged Wireless Group-RF Profile (CWG-RF) was developed jointly by the Wi-Fi Alliance and the Cellular Telecommunications and Internet Association (CTIA), now known as The Wireless Association. CWG-RF defines performance and tests metrics for Wi-Fi and cellular radios in a converged handset to help ensure that both technologies perform well in the presence of the other. Although this test program is not an element of Wi-Fi certification, completion of the testing is mandatory for Wi-Fi enabled handsets

Additional Capabilities

Wi-Fi Home Design   *Wi-Fi Home Design* is a certification program from the Wi-Fi Alliance that enables new home builders to offer turnkey, high-performance Wi-Fi in newly built homes. Home builders can join the Wi-Fi Alliance, follow design guidelines, and certify their design offerings.

Future Certifications

As 802.11 technologies evolve, new Wi-Fi CERTIFIED programs will be defined by the Wi-Fi Alliance. For example, *Wi-Fi HaLow* is a certification designated for low-power devices operating in frequencies below 1 GHz and with greater range. Wi-Fi HaLow is based on the IEEE 802.11ah amendment intended for *Internet of Things (IoT)* devices. The roadmap of the Wi-Fi Alliance also includes updated versions of existing certifications as well as certifications that may be targeted for Wi-Fi operations specific to vertical industries.

Much like the IEEE 802.11 standard and amendments, technology defined in new Wi-Fi Alliance certification programs often takes many years before there is widespread adoption in the marketplace.

Wi-Fi Alliance and Wi-Fi CERTIFIED

Learn more about the Wi-Fi Alliance at www.wi-fi.org. The Wi-Fi Alliance website contains many articles, FAQs, and white papers describing the organization along with additional information about the certification programs. The Wi-Fi Alliance technical white papers are recommended extra reading when preparing for the CWNA exam. The Wi-Fi Alliance also maintains a searchable database of certified Wi-Fi products. Go to www.wi-fi.org/product-finder to verify the certification status of any Wi-Fi device.

International Organization for Standardization

The *International Organization for Standardization*, commonly known as the *ISO*, is a global, nongovernmental organization that identifies business, government, and society needs and develops standards in partnership with the sectors that will put them to use. The ISO is responsible for the creation of the Open Systems Interconnection (OSI) model, which has been a standard reference for data communications between computers since the late 1970s.

Why Is It ISO and Not IOS?

ISO is not a mistyped acronym. It is a word derived from the Greek word *isos*, meaning *equal*. Because acronyms can be different from country to country, based on varying translations, the ISO decided to use a word instead of an acronym for its name. With this in mind, it is easy to see why a standards organization would give itself a name that means *equal*.

The OSI model is the cornerstone of data communications, and learning to understand it is one of the most important and fundamental tasks a person in the networking industry can undertake. Figure 1.5 shows the seven layers of the OSI model.

FIGURE 1.5   The seven layers of the OSI model

OSI Model

Layer 7	Application
Layer 6	Presentation
Layer 5	Session
Layer 4	Transport
Layer 3	Network
Layer 2	Data-Link
Layer 1	Physical

LLC
- - - - - - - -
MAC

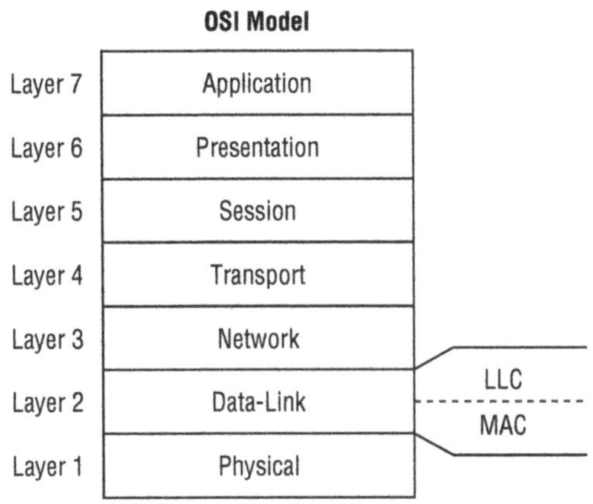

The IEEE 802.11-2016 standard defines communication mechanisms only at the Physical layer and MAC sublayer of the Data-Link layer of the OSI model. How 802.11 technology is used at these two OSI layers is discussed in detail throughout this book.

You should have a working knowledge of the OSI model for both this book and the CWNA exam. Make sure you understand the seven layers of the OSI model and how communications take place at the different layers. If you are not comfortable with the concepts of the OSI model, spend some time reviewing it on the Internet or from a good networking fundamentals book prior to taking the CWNA test. More information about the ISO can be found at www.iso.org.

Core, Distribution, and Access

If you have ever taken a networking class or read a book about network design, you have probably heard the terms *core*, *distribution*, and *access* when referring to networking architecture. Proper network design is imperative no matter what type of network topology is used. The core of the network is the high-speed backbone or the superhighway of the network. The goal of the core is to carry large amounts of information between key data centers or distribution areas, just as superhighways connect cities and metropolitan areas.

The core layer does not route traffic or manipulate packets but rather performs high-speed switching. Redundant solutions are usually designed at the core layer to ensure the fast and reliable delivery of packets. The distribution layer of the network routes or directs traffic toward the smaller clusters of nodes or neighborhoods of the network.

The distribution layer routes traffic between virtual LANs (VLANs) and subnets. The distribution layer is akin to the state and county roads that provide medium travel speeds and distribute the traffic within the city or metropolitan area.

The access layer of the network is responsible for slower delivery of the traffic directly to the end user or end node. The access layer mimics the local roads and neighborhood streets that are used to reach your final address. The access layer ensures the final delivery of packets to the end user. Remember that speed is a relative concept.

Because of traffic load and throughput demands, speed and throughput capabilities increase as data moves from the access layer to the core layer. The additional speed and throughput tends to also mean higher cost.

Just as it would not be practical to build a superhighway so that traffic could travel between your neighborhood and the local school, it would not be practical or efficient to build a two-lane road as the main thoroughfare to connect two large cities, such as New York and Boston. These same principles apply to network design. Each of the network layers—core, distribution, and access—is designed to provide a specific function and capability to the network. It is important to understand how wireless networking fits into this network design model.

Wireless networking can be implemented as either point-to-point or point-to-multipoint solutions. Most wireless networks are used to provide network access to the individual client stations and are designed as point-to-multipoint networks. This type of implementation is designed and installed on the access layer, providing connectivity to the end user. 802.11 wireless networking is most often implemented at the access layer with WLAN clients communicating through strategically deployed access points.

Wireless bridge links are typically used to provide connectivity between buildings, in the same way that county or state roads provide distribution of traffic between neighborhoods. The purpose of wireless bridging is to connect two separate, wired networks wirelessly. Routing data traffic between networks is usually associated with the distribution layer. Wireless bridge links cannot usually meet the speed or distance requirements of the core layer, but they can be very effective at the distribution layer. An 802.11 bridge link is an example of wireless technology being implemented at the distribution layer.

Although wireless is not typically associated with the core layer, you must remember that speed and distance requirements vary greatly between large and small companies, and that one person's distribution layer could be another person's core layer. Very small companies may even implement wireless for all end-user networking devices, forgoing any wired devices except for the connection to the Internet. Higher-bandwidth proprietary wireless bridges and some 802.11 mesh network deployments could be considered an implementation of wireless at the core layer.

Logical Planes of Telecommunication

Telecommunication networks are often defined by three logical planes of operation: management, control, and data. The *management plane* exists for the monitoring and administration of a telecommunications network. The *control plane* is characterized as the intelligence of a network. The *data plane* carries the network user traffic. In an 802.11 environment, these three logical planes of operation function differently depending on the type of WLAN architecture and the WLAN vendor. In Chapter 11, "WLAN Architecture," you will learn about the progressive evolution of enterprise WLAN architecture and where these three logical planes operate.

Communications Fundamentals

Although the CWNA certification is considered to be one of the entry-level certifications in the Certified Wireless Network Professional (CWNP) wireless certification program, it is by no means an entry-level certification in the computing industry. Most of the candidates for the CWNA certificate have experience in other areas of information technology. However, the background and experience of these candidates varies greatly.

Unlike professions for which knowledge and expertise is learned through years of structured training, most computer professionals have followed their own path of education and training.

When people are responsible for their own education, they typically will gain the skills and knowledge that are directly related to their interests or their job. The more fundamental knowledge is often ignored because it is not directly relevant to the tasks at hand. Later, as their knowledge increases and they become more technically proficient, people realize that they need to learn about some of the fundamentals.

Many people in the computer industry understand that, in data communications, bits are transmitted across wires or waves. They even understand that some type of voltage change or wave fluctuation is used to distinguish the bits. When pressed, however, many of these same people have no idea what is actually happening with the electrical signals or the waves.

The following sections review some fundamental communications principles that directly and indirectly relate to wireless communications. Understanding these concepts will help you to better understand what is happening with wireless communications and to more easily recognize and identify the terms used in this profession.

Communication Terminology

We shall now review a few basic networking terms that are often misunderstood: *simplex*, *half-duplex*, and *full-duplex*. These are three dialog methods that are used for communications between people and also between computer equipment.

Simplex   In simplex communications, one device is capable of only transmitting, and the other device is capable of only receiving. FM radio is an example of simplex communications. Simplex communications are rarely used on computer networks.

Half-Duplex   In half-duplex communications, both devices are capable of transmitting and receiving; however, only one device can transmit at a time. Walkie-talkies, or two-way radios, are examples of half-duplex devices. All RF communications by nature are half-duplex, although recent research at Stanford University claims that full-duplex RF communications are possible with transceivers that might be able to cancel self-interference. IEEE 802.11 wireless networks use half-duplex communications.

Full-Duplex   In full-duplex communications, both devices are capable of transmitting and receiving at the same time. A telephone conversation is an example of a full-duplex communication. Most IEEE 802.3 equipment is capable of full-duplex communications. Currently, the only way to accomplish full-duplex communications in a wireless environment is to have a two-channel bidirectional setup, where all transmissions on one channel are transmitted from device A to device B, while all transmissions on the other channel are received on device A from device B. Both device A and device B use two separate radios on different channels.

Understanding Carrier Signals

Because data ultimately consists of bits, the transmitter needs a way of sending both 0s and 1s to transmit data from one location to another. An AC or DC signal by itself does not perform this task. However, if a signal fluctuates or is altered, even slightly, the signal can be interpreted so that data can be properly sent and received. This modified signal is now capable of distinguishing between 0s and 1s and is referred to as a *carrier signal*. The method of adjusting the signal to create the carrier signal is called *modulation*.

Three components of a wave that can fluctuate or be modified to create a carrier signal are amplitude, frequency, and phase.

 This chapter reviews the basics of waves as they relate to the principles of data transmission. Chapter 3, "Radio Frequency Fundamentals," covers radio waves in much greater detail.

All radio-based communications use some form of modulation to transmit data. To encode the data in a signal sent by AM/FM radios, mobile telephones, and satellite television, some type of modulation is performed on the radio signal that is being transmitted. The average person typically is not concerned with how the signal is modulated, only that the device functions as expected. To become a better wireless network administrator, however, it is useful to have a better understanding of what is actually happening when two stations communicate. The rest of this chapter provides an introduction to waves as a basis for understanding carrier signals and data encoding and introduces you to the fundamentals of encoding data.

Amplitude and Wavelength

RF communication starts when radio waves are generated from an RF transmitter and picked up, or "heard," by a receiver at another location. RF waves are similar to the waves that you see in an ocean or lake. Waves are made up of two main components: wavelength and amplitude (see Figure 1.6).

FIGURE 1.6   The wavelength and amplitude of a wave

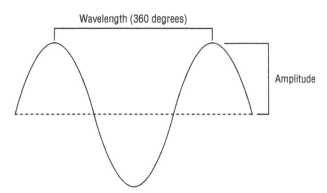

Amplitude   *Amplitude* is the height, force, or power of the wave. If you were standing in the ocean as the waves came to shore, you would feel the force of a larger wave much more than you would a smaller wave. Transmitters do the same thing, but with radio waves. Smaller waves are not as noticeable as bigger waves. A bigger wave generates a much larger electrical signal picked up by the receiving antenna. The receiver can then distinguish between highs and lows.

Wavelength   *Wavelength* is the distance between similar points on two back-to-back waves. When measuring a wave, the wavelength is typically measured from the peak of a wave to the peak of the next wave. Amplitude and wavelength are both properties of waves.

Frequency

Frequency describes a behavior of waves. Waves travel away from the source that generates them. How fast the waves travel, or more specifically, how many waves are generated over a 1-second period of time, is known as frequency. If you were to sit on a pier and count how often a wave hits it, you could tell someone how frequently the waves were coming to shore. Think of radio waves in the same way; however, radio waves travel much faster than the waves in the ocean. If you were to try to count the radio waves that are used in wireless networking, in the time it would take for one wave of water to hit the pier, several billion radio waves would have also hit the pier.

Phase

Phase is a relative term. It is the relationship between two waves with the same frequency. To determine phase, a wavelength is divided into 360 pieces, referred to as *degrees* (see Figure 1.7). If you think of these degrees as starting times, then if one wave begins at the 0 degree point and another wave begins at the 90 degree point, these waves are considered to be 90 degrees out of phase.

FIGURE 1.7 Two waves that are identical but 90 degrees out of phase with each other

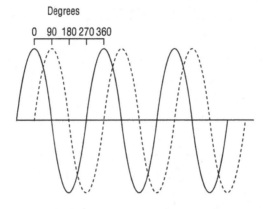

In an ideal world, waves would be created and transmitted from one station and received perfectly intact at another station. Unfortunately, RF communications do not occur in an ideal world. There are many sources of interference and many obstacles that will affect the wave in its travels to the receiving station. Chapter 3 introduces you to some of the outside influences that can affect the integrity of a wave and your ability to communicate between two stations.

Time and Phase

Suppose you have two stopped watches and both are set to noon. At noon you start the first watch, and then you start the second watch one hour later. The second watch is one hour behind the first watch. As time goes by, your second watch will continue to be one hour behind. Both watches will maintain a 24-hour day, but they will be out of sync with each other. Waves that are out of phase behave similarly. Two waves that are out of phase are essentially two waves that have been started at two different times. Both waves will complete full 360-degree cycles, but they will do it out of phase, or out of sync with each other.

Understanding Keying Methods

When data is sent, a signal is transmitted from the transceiver. In order for the data to be transmitted, the signal must be manipulated so that the receiving station has a way of distinguishing 0s and 1s. This method of manipulating a signal so that it can represent multiple pieces of data is known as a *keying method*. A keying method is what changes a signal into a carrier signal. It provides the signal with the ability to encode data so that it can be communicated or transported.

Three types of keying methods are reviewed in the following sections: amplitude-shift keying (ASK), frequency-shift keying (FSK), and phase-shift keying (PSK). These keying methods are also referred to as *modulation techniques*. Keying methods use the following two different techniques to represent data:

Current State   With current state techniques, the current value (the current state) of the signal is used to distinguish between 0s and 1s. The use of the word *current* in this context does not refer to current as in voltage but rather to current as in the present time. Current state techniques will designate a specific or current value to indicate a binary 0 and another value to indicate a binary 1. At a specific point in time, it is the value of the signal that determines the binary value. For example, you can represent 0s and 1s by using an ordinary door. Once a minute you can check to see whether the door is open or closed. If the door is open, it represents a 0, and if the door is closed, it represents a 1. The current state of the door, open or closed, is what determines 0s or 1s.

State Transition   With state transition techniques, the change (or transition) of the signal is used to distinguish between 0s and 1s. State transition techniques may represent a 0 by a change in a wave's phase at a specific time, whereas a 1 would be represented by no change in a wave's phase at a specific time. At a specific point in time, it is the presence of a change or the lack of presence of a change that determines the binary value. The upcoming section "Phase-Shift Keying" provides examples of this in detail, but a door can be used again to provide a simple example. Once a minute you check the door. In this case, if the door is moving (opening or closing), it represents a 0, and if the door is still (either open or closed), it represents a 1. In this example, the state of transition (moving or not moving) is what determines 0s or 1s.

Amplitude-Shift Keying

Amplitude-shift keying (ASK) varies the amplitude, or height, of a signal to represent the binary data. ASK is a current state technique, where one level of amplitude can represent a 0 bit and another level of amplitude can represent a 1 bit. Figure 1.8 shows how a wave can modulate an ASCII letter *K* by using amplitude-shift keying. The larger amplitude wave is interpreted as a binary 1, and the smaller amplitude wave is interpreted as a binary 0.

FIGURE 1.8   An example of amplitude-shift keying (ASCII code of an uppercase *K*)

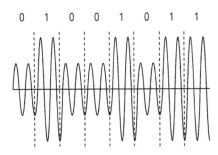

This shifting of amplitude determines the data that is being transmitted. The way the receiving station performs this task is to first divide the signal being received into periods of time known as *symbol periods*. The receiving station then samples or examines the wave during this symbol period to determine the amplitude of the wave. Depending on the value of the wave's amplitude, the receiving station can determine the binary value.

As you will learn later in this book, wireless signals can be unpredictable and also subjected to interference from many sources. When noise or interference occurs, it usually affects the amplitude of a signal. Because a change in amplitude due to noise could cause the receiving station to misinterpret the value of the data, ASK has to be used cautiously.

Frequency-Shift Keying

Frequency-shift keying (FSK) varies the frequency of the signal to represent the binary data. FSK is a current state technique, where one frequency can represent a 0 bit and another frequency can represent a 1 bit (see Figure 1.9). This shifting of frequency determines the data that is being transmitted. When the receiving station samples the signal during the symbol period, it determines the frequency of the wave, and depending on the value of the frequency, the station can determine the binary value.

FIGURE 1.9   An example of frequency-shift keying (ASCII code of an uppercase *K*)

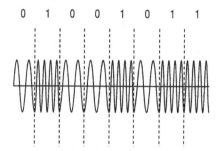

Figure 1.9 shows how a wave can modulate an ASCII letter *K* by using frequency-shift keying. The faster frequency wave is interpreted as a binary 1, and the slower frequency wave is interpreted as a binary 0.

FSK is used in some of the legacy deployments of 802.11 wireless networks. With the demand for faster communications, FSK techniques would require more expensive technology to support faster speeds, making it less practical.

Why Have I Not Heard about Keying Methods Before?

You might not realize it, but you *have* heard about keying methods before. AM/FM radio uses amplitude modulation (AM) and frequency modulation (FM) to transmit the radio stations that you listen to at home or in your automobile. The radio station modulates the voice and music into its transmission signal, and your home or car radio demodulates it.

Phase-Shift Keying

Phase-shift keying (PSK) varies the phase of the signal to represent the binary data. PSK can be a state transition technique, where the change of phase can represent a 0 bit and the lack of a phase change can represent a 1 bit, or vice versa. This shifting of phase determines the data that is being transmitted. PSK can also be a current state technique, where the value of the phase can represent a 0 bit or a 1 bit. When the receiving station samples the signal during the symbol period, it determines the phase of the wave and the status of the bit.

Figure 1.10 shows how a wave can modulate an ASCII letter *K* by using phase-shift keying. A phase change at the beginning of the symbol period is interpreted as a binary 1, and the lack of a phase change at the beginning of the symbol period is interpreted as a binary 0.

FIGURE 1.10   An example of phase-shift keying (ASCII code of an uppercase *K*)

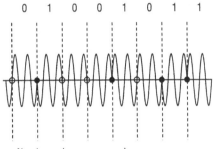

PSK technology is used extensively for radio transmissions as defined by the 802.11-2016 standard. Typically, the receiving station samples the signal during the symbol period, compares the phase of the current sample with the previous sample, and determines the difference. This degree of difference, or *differential*, is used to determine the bit value.

More advanced versions of PSK can encode multiple bits per symbol. Instead of using two phases to represent the binary values, you can use four phases. Each of the four phases is capable of representing two binary values (00, 01, 10, or 11) instead of one (0 or 1), thus shortening the transmission time. When more than two phases are used, this is referred to as *multiple phase-shift keying (MPSK)*. Figure 1.11 shows how a wave can modulate an ASCII letter *K* by using an MPSK method. Four possible phase changes can be monitored, with each phase change now able to be interpreted as 2 bits of data instead of just 1. Notice that there are fewer symbol times in this figure than there are in Figure 1.9.

FIGURE 1.11   An example of multiple phase-shift keying (ASCII code of an uppercase *K*)

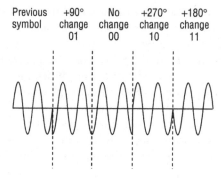

| Previous symbol | +90° change 01 | No change 00 | +270° change 10 | +180° change 11 |

Where Can I Learn More about 802.11 Technology and the Wi-Fi Industry?

Reading this book from cover to cover is a great way to start understanding Wi-Fi technology. Because of the rapidly changing nature of 802.11 WLAN technologies, the authors of this book recommend the following additional resources:

Wi-Fi Alliance   As previously mentioned, the Wi-Fi Alliance is the marketing voice of the Wi-Fi industry and maintains all the industry's certifications. The Wi-Fi Alliance website, www.wi-fi.org, is an excellent resource.

CWNP   The Certified Wireless Networking Professional program maintains learning resources, including user forums and a WLAN white paper database. The website, www.cwnp.com, is also the best source of information about all the vendor-neutral CWNP wireless networking certifications.

WLAN Vendor Websites   Although the CWNA exam and this book take a vendor-neutral approach about 802.11 education, the various WLAN vendor websites are often excellent resources for information about specific Wi-Fi networking solutions. Many of the major WLAN vendors are mentioned throughout this book, and a listing of most of the major WLAN vendor websites can be found in Chapter 20, "WLAN Deployment and Vertical Markets."

Wi-Fi Blogs   In recent years, numerous personal blogs about the subject of Wi-Fi have sprung up all over the Internet. One great example is the Revolution Wi-Fi blog, written by CWNE #84, Andrew vonNagy, at:

    http://revolutionwifi.blogspot.com

Another excellent blog is the Wirednot blog, written by Wi-Fi expert Lee Badman, CWNE #200, at:

    wirednot.wordpress.com

Glenn Cate, CWNE #181, maintains an extensive Wi-Fi blogroll with links to most of the WLAN industry's commercial and personal blogs:

https://gcatewifi.wordpress.com

WLAN Technical Conferences   Different organizations offer technical conferences that are relevant to the WLAN industry. One of the most popular conferences is the wireless-LAN Professionals Conference, held multiple times every year in different regions of the world. You can learn more about the wirelessLAN Professionals Conference at:

www.wlanpros.com

Twitter   The WLAN technical community is very active on Twitter. Wi-Fi industry experts use the social media platform as a way to publically communicate and share information. Lee Badman, CWNE #200, maintains a Wi-Fi question-of-the-day on Twitter via #WIFIQ. You can follow Lee on Twitter: @wirednot. You can also follow this book's co-authors David Coleman and David Westcott on Twitter: @mistermultipath and @davidwestcott, respectively.

Summary

This chapter explained the history of wireless networking and the roles and responsibilities of the following key organizations involved with the wireless networking industry:

- FCC and other regulatory domain authorities
- IEEE
- IETF
- Wi-Fi Alliance

To provide a basic understanding of the relationship between networking fundamentals and 802.11 technologies, we discussed the following concepts:

- OSI model
- Core, distribution, and access

To provide a basic knowledge of how wireless stations transmit and receive data, we introduced some of the components of waves and modulation:

- Carrier signals
- Amplitude
- Wavelength
- Frequency
- Phase
- Keying methods, including ASK, FSK, and PSK

When you are troubleshooting RF communications, having a solid knowledge of waves and modulation techniques can help you understand the fundamental issues behind communications problems and help lead you to a solution.

Exam Essentials

Know the four industry organizations.   Understand the roles and responsibilities of the regulatory domain authorities, the IEEE, the IETF, and the Wi-Fi Alliance.

Understand core, distribution, and access.   Know where 802.11 technology is deployed in fundamental network design.

Explain the difference between simplex, half-duplex, and full-duplex communications.   Know that RF communications, including 802.11 radio communications, are half-duplex.

Understand wavelength, frequency, amplitude, and phase.   Know the definitions of each RF characteristic.

Understand the concepts of modulation.   ASK, FSK, and PSK are three carrier signal modulation techniques.

Review Questions

1. 802.11 technology is typically deployed at which fundamental layer of network architecture?

   A. Core

   B. Distribution

   C. Access

   D. Network

2. Which organization is responsible for enforcing maximum transmit power rules in an unlicensed frequency band?

   A. IEEE

   B. Wi-Fi Alliance

   C. ISO

   D. IETF

   E. None of the above

3. 802.11 wireless bridge links are typically associated with which network architecture layer?

   A. Core

   B. Distribution

   C. Access

   D. Network

4. The 802.11-2016 standard was created by which organization?

   A. IEEE

   B. OSI

   C. ISO

   D. Wi-Fi Alliance

   E. FCC

5. What organization ensures interoperability of WLAN products?

   A. IEEE

   B. ITU-R

   C. ISO

   D. Wi-Fi Alliance

   E. FCC

6. What type of signal is required to carry data?

    A. Communications signal

    B. Data signal

    C. Carrier signal

    D. Binary signal

    E. Digital signal

7. Which keying method is most susceptible to interference from noise?

    A. FSK

    B. ASK

    C. PSK

    D. DSK

8. Which sublayer of the OSI model's Data-Link layer is used for communication between 802.11 radios?

    A. LLC

    B. PLCP

    C. MAC

    D. PMD

9. While performing some research, Janie comes across a reference to a document titled RFC 3935. Which of the following organization's website would be best to further research this document?

    A. IEEE

    B. Wi-Fi Alliance

    C. WECA

    D. FCC

    E. IETF

10. The Wi-Fi Alliance is responsible for which of the following certification programs?

    A. 802.11i

    B. WEP

    C. 802.11-2016

    D. WMM

11. Which wave properties can be modulated to encode data? (Choose all that apply.)

    A. Amplitude

    B. Frequency

    C. Phase

    D. Wavelength

12. The IEEE 802.11-2016 standard defines communication mechanisms at which layers of the OSI model? (Choose all that apply.)

- **A.** Network
- **B.** Physical
- **C.** Transport
- **D.** Application
- **E.** Data-Link
- **F.** Session

13. The height or power of a wave is known as what?

- **A.** Phase
- **B.** Frequency
- **C.** Amplitude
- **D.** Wavelength

14. What are the communication differences between Wi-Fi Direct and Wi-Fi CERTIFIED TDLS devices? (Choose all that apply.)

- **A.** Wi-Fi CERTIFIED TDLS devices never associate to an AP.
- **B.** Wi-Fi Direct devices can communicate with each other without associating to an AP.
- **C.** Wi-Fi CERTIFIED TDLS devices remain associated to an AP while communicating directly with each other.
- **D.** Wi-Fi Direct devices must associate with an AP before they can communicate with each other.

15. What Wi-Fi Alliance certifications are required before a Wi-Fi radio can be certified as Voice-Enterprise compliant? (Choose all that apply.)

- **A.** WMM-Power Save
- **B.** Wi-Fi Direct
- **C.** WPA2-Enterprise
- **D.** Voice-Personal
- **E.** WMM-Admission Control

16. Which of the following wireless communications parameters and usage are typically governed by a local regulatory authority? (Choose all that apply.)

- **A.** Frequency
- **B.** Bandwidth
- **C.** Maximum transmit power
- **D.** Maximum EIRP
- **E.** Indoor/outdoor usage

17. What type of communications do 802.11 radios use to transmit and receive?

   A. Simplex

   B. Half-duplex

   C. Full-duplex

   D. Echo-duplex

18. A wave is divided into degrees. How many degrees make up a complete wave?

   A. 100

   B. 180

   C. 212

   D. 360

19. What are the advantages of using unlicensed frequency bands for RF transmissions? (Choose all that apply.)

   A. There are no governmental regulations.

   B. There is no additional financial cost.

   C. Anyone can use the frequency band.

   D. There are no rules.

20. The OSI model consists of how many layers?

   A. Four

   B. Six

   C. Seven

   D. Nine

Chapter

2

IEEE 802.11 Standard and Amendments

IN THIS CHAPTER, YOU WILL LEARN ABOUT THE FOLLOWING:

✓ Original IEEE 802.11 standard

- IEEE 802.11-2016 ratified amendments
- 802.11a-1999
- 802.11b-1999
- 802.11d-2001
- 802.11e-2005
- 802.11g-2003
- 802.11h-2003
- 802.11i-2004
- 802.11j-2004
- 802.11k-2008
- 802.11n-2009
- 802.11p-2010
- 802.11r-2008
- 802.11s-2011
- 802.11u-2011
- 802.11v-2011
- 802.11w-2009
- 802.11y-2008
- 802.11z-2010
- 802.11aa-2012
- 802.11ac-2013
- 802.11ad-2012

- 802.11ae-2012
- 802.11af-2014

✓ **Post 802.11-2016 ratified amendments**

- 802.11ah
- 802.11ai

✓ **IEEE 802.11 draft amendments**

- 802.11aj
- 802.11ak
- 802.11aq
- 802.11ax
- 802.11ay
- 802.11az
- 802.11ba

✓ **Defunct amendments**

- 802.11F
- 802.11T

✓ **IEEE Task Group m**

As discussed in Chapter 1, "Overview of Wireless Standards, Organizations, and Fundamentals," the Institute of Electrical and Electronics Engineers (IEEE) is the professional society that creates and maintains standards that we use for communications, such as the 802.3 Ethernet standard for wired networking. The IEEE has assigned working groups for several wireless communication standards. For example, the 802.15 working group is responsible for personal area network (PAN) communications using radio frequencies, such as Bluetooth. Another example is the 802.16 standard, which is overseen by the Broadband Wireless Access Standards working group; this technology is often referred to as WiMAX. The focus of this book is the technology as defined by the IEEE 802.11 standard, which provides for local area network (LAN) communications using radio frequencies (RFs).

The 802.11 working group has about 400 active members from more than 200 wireless companies. It consists of standing committees, study groups, and numerous *task groups*. For example, the Standing Committee—Publicity (PSC) is in charge of finding means to better publicize the 802.11 standard. An 802.11 Study Group (SG) is authorized by the executive committee (EC) and is expected to have a short lifespan, typically less than six months. A study group is in charge of investigating the possibility of putting new features and capabilities into the 802.11 standard.

IEEE 802.11: More about the Working Group and 2016 Standard

You can find a quick guide to the IEEE 802.11 working group at:

`http://www.ieee802.org/11/QuickGuide_IEEE_802_WG_and_Activities.htm`

The 802.11-2016 standard and ratified amendments can be accessed from:

`https://ieeexplore.ieee.org/browse/standards/get-program/page`

Some of the standards and ratified amendment documents are free, and others (particularly recently ratified documents) are available for a fee.

Various 802.11 task groups are in charge of revising and amending the original standard that was developed by the MAC task group (MAC) and the PHY task group (PHY). Each group is assigned a letter from the alphabet, and it is common to hear the term *802.11 alphabet soup* when referring to all the amendments created by the multiple 802.11 task groups. When task groups are formed, they are assigned the next highest available letter in the alphabet, although the amendments may not necessarily be ratified in the same order. Quite a few of the 802.11 task group projects have been completed, and amendments to the original standard have been ratified. Other 802.11 task group projects still remain active and exist as draft amendments.

In this chapter, we discuss the original 802.11 standard, the ratified amendments (many of which were incorporated into the 802.11-2007 standard, the 802.11-2012 standard, and the current 802.11-2016 standard), and the draft amendments of various 802.11 task groups.

Original IEEE 802.11 Standard

The original 802.11 standard was published in June 1997 as IEEE Std 802.11-1997, and it is often referred to as 802.11 Prime because it was the first WLAN standard. The standard was revised in 1999, reaffirmed in 2003, and published as IEEE Std 802.11-1999 (R2003). On March 8, 2007, another iteration of the standard was approved, IEEE Std 802.11-2007, and on March 29, 2012, the IEEE Std 802.11-2012 standard was approved. The most recent revision of the standard, IEEE 802.11-2016, was approved on December 7, 2016.

The IEEE specifically defines 802.11 technologies at the Physical layer and the MAC sublayer of the Data-Link layer. By design, the 802.11 standard does not address the upper layers of the OSI model, although there are interactions between the 802.11 MAC layer and the upper layers for parameters such as quality of service (QoS). The PHY task group worked in conjunction with the MAC task group to define the original 802.11 standard. The PHY task group defined three original Physical layer specifications:

Infrared   *Infrared (IR)* technology uses a light-based medium. Although an infrared medium was indeed defined in the original 802.11 standard, it has since been deprecated and removed from the 802.11-2016 standard.

Frequency-Hopping Spread-Spectrum   Radio frequency signals can be defined as narrowband signals or as spread-spectrum signals. An RF signal is considered *spread-spectrum* when the bandwidth is wider than what is required to carry the data. *Frequency-hopping spread-spectrum (FHSS)* is a spread-spectrum technology that was first patented during World War II. Frequency-hopping 802.11 has been deprecated and removed from the 802.11-2016 standard.

Direct-Sequence Spread-Spectrum   *Direct-sequence spread-spectrum (DSSS)* is another spread-spectrum technology that uses fixed channels. DSSS 802.11 radios are known as *Clause 15 devices.*

What Is an IEEE Clause?

The IEEE standards are very organized, structured documents. A standards document is hierarchically structured, with each section numbered (such as 7.3.2.4). The highest level (7) is referred to as a *clause,* with the lower-level sections (3.2.4) referred to as *subclauses.* As an amendment is created, the sections in the amendment are numbered relative to the latest version of the standard, even though the amendment is a separate document. When a standard and its amendments are rolled into a new version of the standard, as was done with IEEE Std 802.11-2016, the clauses and subclauses of all the individual documents are unique, enabling the standard and amendment documents to be combined without having to change any of the section (clause/subclause) numbers. In 2016, the IEEE revised the standard again and rolled into it a group of 5 amendments. Over the years, as new amendments were ratified, some amendments took longer or shorter times to ratify than other amendments. Therefore, the order of some of the clauses was not chronological. Although this is not a requirement, some of the clauses were reordered and renumbered in the IEEE Std 802.11-2012 so that clauses are listed chronologically, based upon when they were ratified. In addition to consolidating some ratified amendments, the IEEE Std 802.11-2016 deleted some outdated clauses and rearranged some of the others. Whenever any clauses are referenced in this book, we will use the current number scheme, defined in the 2016 revised standard. This book references clauses so that you have familiarity with them and understand where to go if you want to learn more about the technology. Note, however, that you will *not* be tested on clause numbers in the CWNA exam (CWNA-107).

As defined by 802.11 Prime, the original frequency space in which 802.11 radios were allowed to transmit was the license-free 2.4 GHz *industrial, scientific, and medical (ISM)* band. DSSS 802.11 radios could transmit within channels subdivided from the entire 2.4 GHz to 2.4835 GHz ISM band. The IEEE was more restrictive for FHSS radios, which were permitted to transmit on 1 MHz subcarriers in the 2.402 GHz to 2.480 GHz range of the 2.4 GHz ISM band.

You will probably never work with any of the legacy radios defined in 802.11 Prime, because the technology is over 20 years old and has been replaced in working WLAN environments. Originally, WLAN vendors had the choice of manufacturing either FHSS radios or DSSS radios. The majority of legacy WLAN deployments used frequency hopping, but some DSSS solutions were available as well. It should also be noted that any references to FHSS radios has been decremented from the current 802.11-2016 standard.

What about the speeds? Data rates defined by the original 802.11 standard were 1 Mbps and 2 Mbps, regardless of which spread-spectrum technology was used. A *data rate* is the number of bits per second the Physical layer carries during a transmission, normally stated as a number of millions of bits per second (Mbps). Keep in mind that data rate is the *speed*, not actual *throughput*. Because of medium access methods and communications overhead, aggregate throughput is typically around half of the available data rate speed.

IEEE 802.11-2016 Ratified Amendments

In the years that followed the publishing of the original 802.11 standard, new task groups were assembled to address potential enhancements to the standard. As of this writing, almost 30 amendments have been ratified and published by the distinctive task groups. In 2007, the IEEE consolidated 8 ratified amendments along with the original standard, creating a single document that was published as *IEEE Std 802.11-2007*. This revision also included corrections, clarifications, and enhancements.

In 2012, the IEEE consolidated 10 ratified amendments into the IEEE Std 802.11-2007 standard, creating a single document that was published as *IEEE Std 802.11-2012*. In addition to consolidating the ratified amendments and making corrections, clarifications, and enhancements to the document, the IEEE reviewed all of the clauses and annexes chronologically. Some of the clauses and annexes were rearranged and renumbered so that they were listed in the order that they were ratified.

Most recently, in 2016, the IEEE consolidated 5 ratified amendments into the IEEE Std 802.11-2012 standard, creating the most recent iteration of the standard, IEEE Std 802.11-2016. In addition to consolidating the ratified amendments and making corrections, some obsolete clauses were removed from the document and some clauses were renumbered.

CWNA Exam Terminology

In 2016, the IEEE consolidated the 2012 standard along with the ratified amendments into a single document that is now published as the *802.11-2016 standard*. Technically, any of the amendments that have been consolidated into an updated standard no longer exist because they have been rolled up into a single document. However, the Wi-Fi Alliance and most WLAN professionals still refer to many of the ratified amendments by name.

Early versions of the CWNA exam did not refer to any of the 802.11 amendments by name and tested you only on the technologies used by each amendment. For example, 802.11b is a ratified amendment that is part of the 802.11-2012 standard. The technology that was originally defined by the 802.11b amendment is called *High-Rate DSSS (HR-DSSS)*. Although the name *802.11b* effectively remains the more commonly used marketing term, older versions of the CWNA exam used only the technical term *HR-DSSS* instead of the more common term *802.11b*. The current version of the CWNA exam (CWNA-107) uses the more common 802.11 amendment terminology, such as 802.11b.

For the CWNA exam (CWNA-107), you should still understand the differences between the different technologies and how each one works. A good grasp of which technologies are defined by each of the amendments will be helpful for your career.

802.11a-1999

During the same year that the 802.11b amendment was approved, another important amendment was also ratified and published as *IEEE Std 802.11a-1999*. The engineers in the Task Group a (TGa) set out to define how 802.11 technologies would operate in the 5 GHz frequency space using an RF technology called *orthogonal frequency-division multiplexing (OFDM)*. 802.11a radios could transmit in three different 100 MHz unlicensed frequency bands in the 5 GHz range. These three bands are called the *Unlicensed National Information Infrastructure (U-NII)* frequency bands. A total of 12 channels were available in the original three U-NII bands. All aspects of the 802.11a ratified amendment can now be found in Clause 17 of the 802.11-2016 standard.

The 2.4 GHz ISM band is a much more crowded frequency space than the 5 GHz U-NII bands. Bluetooth devices, microwave ovens, cordless phones, and numerous other devices all operate in the 2.4 GHz ISM band and are potential sources of interference. In addition, the sheer number of 2.4 GHz WLAN deployments is a problem, especially in environments such as multitenant office buildings.

A big advantage of using 5 GHz WLAN equipment is that the U-NII bands are less crowded. As time passed, the three original U-NII bands also started to become crowded. Regulatory bodies such as the FCC opened up more frequency space in the 5 GHz range, and the IEEE addressed this in the 802.11h amendment. The FCC has also proposed even more 5 GHz spectrum be made available in the near future. Greater detail about all of the 5 GHz U-NII bands can be found in Chapter 6 "Wireless Networks and Spread Spectrum Technologies."

Legacy 802.11a radios initially could transmit in the 12 channels of the U-NII-1, U-NII-2, and U-NII-3 bands; however, the 5 GHz frequency range and channels used by 802.11a radios are dependent on the RF regulatory body of individual countries. The amendment was mostly about the introduction of OFDM technology that provided better higher rates.

NOTE   You will find further discussion about both the ISM and U-NII bands in Chapter 6.

802.11a radios operating in the 5 GHz U-NII bands are classified as *Clause 17 devices*. As defined by the 802.11a amendment, these devices are required to support data rates of 6, 12, and 24 Mbps, with a maximum of 54 Mbps. With the use of a technology called orthogonal frequency-division multiplexing (OFDM), data rates of 6, 9, 12, 18, 24, 36, 48, and 54 Mbps are supported. OFDM is also discussed in Chapter 6.

It should be noted that 802.11a radios cannot communicate with 802.11 legacy, 802.11b, or 802.11g radios for two reasons. First, 802.11a radios use a different RF technology than 802.11 legacy or 802.11b devices. Second, 802.11a devices transmit in the 5 GHz U-NII bands, whereas the 802.11/802.11b/802.11g devices operate in the 2.4 GHz ISM band. The good news is that 802.11a can coexist in the same physical space with 802.11, 802.11b, or 802.11g devices because these devices transmit in separate frequency ranges.

When 802.11a was first ratified, it took almost two years before devices were readily available. When 802.11a devices did become available, the radio chipsets using OFDM were quite expensive. Because of these two factors, widespread deployment of 5 GHz WLANs in the enterprise was rare. Eventually the chipsets become affordable, and the use of 5 GHz frequency bands has grown considerably over the years. WLAN vendors developed dual-frequency access points (APs) with both 2.4 and 5 GHz radios. Most laptops manufactured since 2007 supported dual-frequency radios. The majority of enterprise wireless deployments run both 2.4 GHz and 5 GHz 802.11 wireless networks simultaneously.

802.11b-1999

Although the Wi-Fi consumer market continued to grow at a tremendous rate, 802.11b-compatible WLAN equipment gave the industry the first needed huge shot in the arm. In 1999, the IEEE Task Group b (TGb) published IEEE Std 802.11b-1999, which was later amended and corrected as IEEE Std 802.11b-1999/Cor1-2001. All aspects of the 802.11b ratified amendment can now be found in Clause 16 of the 802.11-2016 standard.

The Physical layer medium that was defined by 802.11b is *High-Rate DSSS (HR-DSSS)*. The frequency space in which 802.11b radio cards can operate is the unlicensed 2.4 GHz to 2.4835 GHz ISM band.

The TGb's main goal was to achieve higher data rates within the 2.4 GHz ISM band. 802.11b radio devices accomplished this feat by using a different spreading/coding technique called *complementary code keying (CCK)* and modulation methods using the phase properties of the RF signal. 802.11 devices used a spreading technique called the *Barker code*. The end result is that 802.11b radio devices supported data rates of 1, 2, 5.5, and 11 Mbps. 802.11b systems are backward compatible with the 802.11 DSSS data rates of 1 Mbps and 2 Mbps. The transmission data rates of 5.5 Mbps and 11 Mbps are known as HR-DSSS. Once again, understand that the supported data rates refer to available bandwidth, not aggregate throughput. 802.11b radios were not backward compatible with legacy 802.11 FHSS radios, because the different spread-spectrum technologies cannot communicate with each other. An optional technology called *Packet Binary Convolutional coding (PBCC)* was removed from the IEEE Std 802.11-2016 standard.

802.11d-2001

The original 802.11 standard was written for compliance with the regulatory domains of the United States, Japan, Canada, and Europe. Regulations in other countries might define different limits on allowed frequencies and transmit power. The 802.11d amendment, which was published as IEEE Std 802.11d-2001, added requirements and definitions necessary to allow 802.11 WLAN equipment to operate in areas not served by the original standard.

Country code information is delivered in fields inside two wireless frames called *beacons* and *probe responses*. This information is then used by 802.11d-compliant devices to ensure that they are abiding by a particular country's frequency and power rules. Figure 2.1 shows an AP configured for use in Mongolia and a capture of a beacon frame containing the country code, frequency, and power information.

FIGURE 2.1   802.11d settings

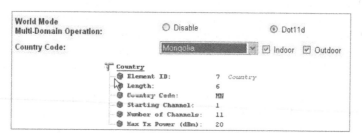

All aspects of the 802.11d ratified amendment can now be found in the 802.11-2016 standard.

> A detailed discussion of beacons, probes, and other wireless frames can be found in Chapter 9, "802.11 MAC."

802.11e-2005

The original 802.11 standard did not define adequate *quality of service (QoS)* procedures for the use of time-sensitive applications such as *Voice over Wi-Fi*. Voice over Wi-Fi is also known as *Voice over Wireless LAN (VoWLAN)*. The terminology used by most vendors and the CWNP program is Voice over Wi-Fi (VoWiFi). Application traffic such as voice, audio, and video has a lower tolerance for latency and jitter and requires priority before standard application data traffic. The 802.11e amendment defines the layer 2 MAC methods needed to meet the QoS requirements for time-sensitive applications over IEEE 802.11 WLANs.

The original 802.11 standard defined two methods in which an 802.11 radio card may gain control of the half-duplex medium. The default method, *Distributed Coordination Function (DCF)*, is a contention-based method determining who gets to transmit on the wireless medium next. The original standard also defined another medium access control method, called *Point Coordination Function (PCF)*, where the access point briefly takes control of the medium and polls the clients. It should be noted that the PCF medium access method was never adopted by WLAN vendors and is considered obsolete.

> Chapter 8, "802.11 Medium Access," describes the DCF and PCF methods of medium access in greater detail.

The 802.11e amendment defines enhanced medium access methods to support QoS requirements. *Hybrid Coordination Function (HCF)* is an additional coordination function that is applied in an 802.11e QoS wireless network. HCF has two access mechanisms to provide QoS. *Enhanced Distributed Channel Access (EDCA)* is an extension to DCF. The EDCA medium access method provides for the "prioritization of frames" based on upper-layer protocols. Application traffic, such as voice or video, is transmitted in a timely fashion on the 802.11 wireless medium, meeting the necessary latency requirements.

Hybrid Coordination Function Controlled Channel Access (HCCA) is an extension of PCF. HCCA gives the access point the ability to provide for "prioritization of stations." In other words, certain client stations will be given a chance to transmit before others. Much like PCF, the HCCA medium access method defined by 802.11e has never been adopted by WLAN vendors.

The Wi-Fi Alliance also has a certification known as *Wi-Fi Multimedia (WMM)*. The WMM certification defines many components of 802.11e and defines traffic prioritization in four access categories with varying degrees of importance. Most aspects of the 802.11e ratified QoS amendment can now be found in Clause 10 of the 802.11-2016 standard.

> **Chapter 8 covers 802.11e and WMM in greater detail.**

802.11g-2003

Another amendment that generated a lot of excitement in the Wi-Fi marketplace was published as IEEE Std 802.11g-2003. 802.11g radios used a new technology called *Extended Rate Physical (ERP)* but were still meant to transmit in the 2.4 GHz to 2.4835 GHz ISM frequency band. All aspects of the 802.11g ratified amendment can now be found in Clause 18 of the 802.11-2016 standard.

The main goal of the Task Group g (TGg) was to enhance the 802.11b Physical layer to achieve greater bandwidth yet remain compatible with the 802.11 MAC sublayer. Two mandatory and two optional ERP Physical layers (PHYs) were defined by the 802.11g amendment.

The mandatory PHYs are ERP-OFDM and ERP-DSSS/CCK. To achieve the higher data rates, a PHY technology called *Extended Rate Physical OFDM (ERP-OFDM)* was mandated. Data rates of 6, 9, 12, 18, 24, 36, 48, and 54 Mbps are possible using this technology, although once again the IEEE required only the data rates of 6, 12, and 24 Mbps. To maintain backward compatibility with 802.11 (DSSS only) and 802.11b networks, a PHY technology called *Extended Rate Physical DSSS (ERP-DSSS/CCK)* was used with support for the data rates of 1, 2, 5.5, and 11 Mbps.

What Is the Difference between ERP-DSSS/CCK, DSSS, and HR-DSSS?

From a technical viewpoint, there is no difference between ERP-DSSS/CCK and DSSS and HR-DSSS. A key point of the 802.11g amendment was to maintain backward compatibility with older 802.11 (DSSS only) and 802.11b radios, while at the same time achieving higher data rates. 802.11g devices (Clause 18 radios) use ERP-OFDM for the higher data rates. ERP-DSSS/CCK is effectively the same technology as DSSS that is used by legacy 802.11 devices (Clause 15 radios) and HR-DSSS that is used by 802.11b devices (Clause 16 radios). Mandated support for ERP-DSSS/CCK allows for backward compatibility with older 802.11 (DSSS only) and 802.11b (HR-DSSS) radios. The technology is explained further in Chapter 6.

The 802.11g ratified amendment also defined two optional PHYs called *ERP-PBCC* and *DSSS-OFDM*. Both of these were removed from the latest standard, IEEE Std 802.11-2016.

What Is the Difference between OFDM and ERP-OFDM?

From a technical viewpoint, there is no difference between OFDM and ERP-OFDM. The only difference is the transmit frequency. OFDM refers to 802.11a devices (Clause 17 radios) that transmit in the 5 GHz U-NII-1, U-NII-2, and U-NII-3 frequency bands. ERP-OFDM refers to 802.11g devices (Clause 18 radios) that transmit in the 2.4 GHz ISM frequency band. The technology is explained further in Chapter 6.

The ratification of the 802.11g amendment triggered monumental sales of Wi-Fi gear in the small office, home office (SOHO), and enterprise markets because of both the higher data rates and the backward compatibility with older equipment.

As mentioned earlier in this chapter, different spread-spectrum technologies cannot communicate with each other, yet the 802.11g amendment mandated support for both ERP-DSSS/CCK and ERP-OFDM. In other words, ERP-OFDM and ERP-DSSS/CCK technologies can coexist, yet they cannot speak to each other. Therefore, the 802.11g amendment called for a *protection mechanism* that allows the two technologies to coexist. The goal of the ERP protection mechanism was to prevent older 802.11b HR-DSSS or 802.11 DSSS radio cards from transmitting at the same time as 802.11g (ERP) radios. Table 2.1 shows a brief overview and comparison of 802.11, 802.11b, 802.11g, and 802.11a.

TABLE 2.1   Original 802.11 amendments comparison

	802.11 legacy	802.11b	802.11g	802.11a
Frequency	2.4 GHz ISM band	2.4 GHz ISM band	2.4 GHz ISM band	5 GHz U-NII-1, U-NII-2, and U-NII-3 bands
Spread-spectrum technology	FHSS or DSSS	HR-DSSS	ERP: ERP-OFDM and ERP-DSSS/CCK are mandatory.	OFDM
Data rates	1, 2 Mbps	DSSS: 1, 2 Mbps HR-DSSS: 5.5 and 11 Mbps	ERP-DSSS/CCK: 1, 2, 5.5, and 11 Mbps	6, 12, and 24 Mbps are mandatory.
			ERP-OFDM: 6, 12, and 24 Mbps are mandatory.	Also supported are 9, 18, 36, 48, and 54 Mbps.

TABLE 2.1   Original 802.11 amendments comparison *(continued)*

	802.11 legacy	802.11b	802.11g	802.11a
			Also supported are 9, 18, 36, 48, and 54 Mbps.	
Backward compatibility	N/A	802.11 DSSS only	802.11b HR-DSSS and 802.11 DSSS	None
Ratified	1997	1999	2003	1999

802.11h-2003

Published as IEEE Std 802.11h-2003, this amendment defined mechanisms for *dynamic frequency selection (DFS)* and *transmit power control (TPC)*. It was originally proposed to satisfy regulatory requirements for operation in the 5 GHz band in Europe and to detect and avoid interference with 5 GHz satellite and radar systems. These same regulatory requirements have also been adopted by the FCC in the United States. The main purpose of DFS and TPC is to provide services where 5 GHz 802.11 radio transmissions will not cause interference with 5 GHz satellite and radar transmissions.

The 802.11h amendment also introduced the capability for 802.11 radios to transmit in a new frequency band, called *U-NII-2 Extended*, with 11 more channels in some regulatory domains. The 802.11h amendment effectively is an extension of the 802.11a amendment. OFDM transmission technology is used in all of the U-NII bands. The radar detection and avoidance technologies of DFS and TPC are defined by the IEEE. However, the RF regulatory organizations in each country still define the RF regulations. In the United States and Europe, radar detection and avoidance is required in both the U-NII-2 and U-NII-2 Extended bands.

DFS is used for spectrum management of 5 GHz channels by OFDM radio devices. The European Radiocommunications Committee (ERC) and the FCC mandate that radio cards operating in the 5 GHz band implement a mechanism to avoid interference with radar systems. DFS is essentially radar-detection and radar-interference avoidance technology. The DFS service is used to meet these regulatory requirements.

The dynamic frequency selection (DFS) service provides for the following:

- An AP will allow client stations to associate based on the supported channel of the access point. The term *associate* means that a station has become a member of the AP's wireless network.

- An AP can quiet a channel to test for the presence of radar.

- An AP may test a channel for the presence of radar before using the channel.

- An AP can detect radar on the current channel and other channels.

- An AP can cease operations after radar detection to avoid interference.
- When interference is detected, the AP may choose a different channel to transmit on and inform all the associated stations.

TPC is used to regulate the power levels used by OFDM radio cards in the 5 GHz frequency bands. The ERC mandates that radio cards operating in the 5 GHz band use TPC to abide by a maximum regulatory transmit power and are able to alleviate transmission power to avoid interference. The TPC service is used to meet the regulatory transmission power requirements.

The transmit power control (TPC) service provides for the following:

- Client stations can associate with an AP based on their transmit power.
- APs and client stations abide by the maximum transmit power levels permitted on a channel, as permitted by regulations.
- An AP can specify the transmit power of any or all stations that are associated with the AP.
- An AP can change transmission power on stations based on factors of the physical RF environment, such as path loss.

The information used by both DFS and TPC is exchanged between client stations and APs inside of management frames. The 802.11h amendment effectively introduced two major enhancements: more frequency space, with the introduction of the U-NII-2 Extended band, and radar avoidance and detection technologies. Some aspects of the 802.11h ratified amendment can now be found in Clause 11.8 and Clause 11.9 of the 802.11-2016 standard.

It should be noted that DFS technology is most often used for radar avoidance as opposed to TPC. Careful consideration should be given when planning a 5 GHz WLAN with DFS channels enabled. A deeper discussion of DFS operations can be found in Chapter 13, "WLAN Design Concepts."

802.11i-2004

From 1997 to 2004, not much was defined in terms of security in the original 802.11 standard. Three key components of any wireless security solution are data privacy (encryption), data integrity (protection from modification), and authentication (identity verification). For seven years, the only defined method of encryption in an 802.11 network was the use of 64-bit static encryption called *Wired Equivalent Privacy (WEP)*.

WEP encryption has long been cracked and is not considered an acceptable means of providing data privacy. The original 802.11 standard defined two methods of authentication. The default method is *Open System authentication*, which effectively allows access to all users regardless of identity. Another defined method is called *Shared Key authentication*, which opens up a whole new can of worms and potential security risks.

The 802.11i amendment, which was ratified and published as IEEE Std 802.11i-2004, defined stronger encryption and better authentication methods. The 802.11i amendment defined a *robust security network (RSN)*. The intended goal of an RSN was to better hide the data flying through the air while at the same time placing a bigger guard at the front door. The 802.11i security amendment is without a doubt one of the most important enhancements to the original 802.11 standard because of the seriousness of properly protecting a wireless network. The major security enhancements addressed in 802.11i are as follows:

Data Privacy   Confidentiality needs have been addressed in 802.11i with the use of a stronger encryption method called *Counter Mode with Cipher Block Chaining Message Authentication Code Protocol (CCMP)*, which uses the *Advanced Encryption Standard (AES)* algorithm. The encryption method is often abbreviated as CCMP/AES, AES CCMP, or often just CCMP. The 802.11i amendment also defined an optional encryption method known as *Temporal Key Integrity Protocol (TKIP)*, which uses the ARC4 stream cipher algorithm and is basically an enhancement of WEP encryption.

Data Integrity   All of the WLAN encryption methods defined by the IEEE employ data-integrity mechanisms to ensure that the encrypted data has not been modified. WEP uses a data-integrity method called the *integrity check value (ICV)*. TKIP uses a method known as the *message integrity check (MIC)*. CCMP uses a much stronger MIC as well as other mechanisms for data integrity. Finally, in the trailer of all 802.11 frames is a 32-bit CRC known as the *frame check sequence (FCS)* that protects the entire body of the 802.11 frame.

Authentication   802.11i defines two methods of authentication using either an *IEEE 802.1X* authorization framework or *preshared keys (PSKs)*. An 802.1X solution requires the use of an *Extensible Authentication Protocol (EAP)*, although the 802.11i amendment does not specify which EAP method to use.

Robust Security Network   A *robust security network (RSN)* defines the entire method of establishing authentication, negotiating security associations, and dynamically generating encryption keys for client stations and access points.

The Wi-Fi Alliance also has a certification known as *Wi-Fi Protected Access 2 (WPA2)*, which is a mirror of the IEEE 802.11i security amendment. WPA version 1 was considered a preview of 802.11i, whereas WPA version 2 is fully compliant with 802.11i. All aspects of the 802.11i ratified security amendment can now be found in Clause 12 of the 802.11-2016 standard.

Wi-Fi security is the top priority when deploying any WLAN, and that is why there is another valued certification called Certified Wireless Security Professional (CWSP). At least 10 percent of the CWNA test will involve questions regarding Wi-Fi security. Therefore, wireless security topics—such as 802.1X, EAP, AES CCMP, TKIP, WPA, and others—are described in more detail in Chapter 17, "802.11 Network Security Architecture," and Chapter 16, "Wireless Attacks, Intrusion Monitoring, and Policy."

802.11j-2004

The main goal set out by the IEEE Task Group j (TGj) was to obtain Japanese regulatory approval by enhancing the 802.11 MAC and 802.11a PHY to additionally operate in Japanese 4.9 GHz and 5 GHz bands. Not all WLAN vendors support this band. The 802.11j amendment was approved and published as IEEE Std 802.11j-2004.

In Japan, 802.11a radio cards could transmit in the lower U-NII band at 5.15 GHz to 5.25 GHz as well as a Japanese licensed/unlicensed frequency space of 4.9 GHz to 5.091 GHz.

802.11a radio cards use OFDM technology and are required to support channel spacing of 20 MHz. When 20 MHz channel spacing is used, data rates of 6, 9, 12, 18, 24, 36, 48, and 54 Mbps are possible using OFDM technology. Japan also has the option of using OFDM channel spacing of 10 MHz, which results in available bandwidth data rates of 3, 4.5, 6, 9, 12, 18, 24, and 27 Mbps. The data rates of 3, 6, and 12 Mbps are mandatory when using 10 MHz channel spacing.

802.11k-2008

The goal of the 802.11 Task Group k (TGk) was to provide a means of radio resource measurement (RRM). The 802.11k-2008 amendment called for measurable client statistical information in the form of requests and reports for the Physical layer 1 and the MAC sublayer of the Data-Link layer 2. 802.11k defined mechanisms in which client station resource data is gathered and processed by an access point or *WLAN controller*. (WLAN controllers are covered in Chapter 11, "WLAN Architecture." For now, think of a WLAN controller as a core device that manages many access points.) In some instances, the client may also request information from an access point or WLAN controller. The following are some of the key radio resource measurements defined under 802.11k:

Transmit Power Control   The 802.11h amendment defined the use of transmit power control (TPC) for the 5 GHz band to reduce interference. Under 802.11k, TPC will also be used in other frequency bands and in areas governed by other regulatory agencies.

Client Statistics   Physical layer information, such as signal-to-noise ratio, signal strength, and data rates, can all be reported back to the access point or WLAN controller. MAC information, such as frame transmissions, retries, and errors, may all be reported back to the access point or WLAN controller as well.

Channel Statistics   Clients may gather noise-floor information based on any RF energy in the background of the channel and report this information back to the access point. Channel-load information may also be collected and sent to the AP. The access point or WLAN controller may use this information for channel management decisions.

Neighbor Reports   802.11k gave client stations the ability to learn from access points or WLAN controllers about other access points where the client stations might potentially roam. AP neighbor report information is shared among WLAN devices to improve roaming efficiency.

Using proprietary methods, a client station keeps a table of known access points and makes decisions on when to roam to another access point. Most client stations make a roaming decision based on the received amplitude of known access points. In other words, a client station decides to roam based on its individual perspective of the RF environment. 802.11k mechanisms provide a client station with additional information about the existing RF environment.

As defined by 802.11k, a client station will request information about neighbor access points on other channels from an access point or WLAN controller. The current AP or WLAN controller will then process that information and generate a *neighbor report* detailing available access points from best to worst. Before a station roams, it will request the neighbor report from the current AP or controller and then decide whether to roam to one of the access points on the neighbor report. Neighbor reports effectively give a client station more information about the RF environment from other existing radios. With the additional information, a client station should make a more informed roaming decision.

802.11n-2009

An event that had a major impact on the Wi-Fi marketplace was the ratification of the 802.11n-2009 amendment. Since 2004, the 802.11 Task Group n (TGn) worked on improvements to the 802.11 standard to provide for greater throughput. Some of the IEEE 802.11 amendments in the past have addressed bandwidth data rates in the 2.4 GHz frequency band. However, the specific objective of the 802.11n-2009 amendment was to increase the throughput in both the 2.4 GHz and 5 GHz frequency bands. The 802.11n-2009 amendment defined a new operation known as *high throughput (HT)*, which provides PHY and MAC enhancements to support data rates of up to 600 Mbps and therefore aggregate throughput above 100 Mbps.

HT Clause 19 radios use *multiple-input, multiple-output (MIMO)* technology in unison with OFDM technology. MIMO uses multiple receiving and transmitting antennas and actually capitalizes on the effects of multipath, as opposed to compensating for or eliminating them. The beneficial consequences of using MIMO are increased throughput and even greater range. 802.11n radios are also backward compatible with legacy 802.11a/b/g radios.

Chapter 10, "MIMO Technology: HT and VHT," discusses 802.11n and MIMO technology in great detail.

802.11p-2010

The mission of the 802.11 Task Group p (TGp) was to define enhancements to the 802.11 standard to support Intelligent Transportation System (ITS) applications. Data exchanges between high-speed vehicles are possible in the licensed ITS band of 5.9 GHz. Additionally, communications between vehicles and roadside infrastructure are supported in the 5 GHz bands, specifically the 5.850 GHz–5.925 GHz band within North America.

Communications may be possible at speeds of up to 200 kilometers per hour (124 mph) and within a range of 1,000 meters (3,281 feet). Very short latencies will also be needed, as some applications must guarantee data delivery within 4 to 50 milliseconds.

802.11p is also known as *Wireless Access in Vehicular Environments (WAVE)* and is a possible foundation for a US Department of Transportation project called *Dedicated Short Range Communications (DSRC)*. The DSRC project envisions a nationwide vehicle and roadside communication network utilizing applications such as vehicle safety services, traffic jam alerts, toll collections, vehicle collision avoidance, and adaptive traffic light control. In Europe, the ETSI Intelligent Transport System (ITS) is based on IEEE 802.11 and 802.11p technology. This standard is designed to provide vehicle to vehicle and vehicle to infrastructure communication. 802.11p will also be applicable to marine and rail communications.

802.11r-2008

The 802.11r-2008 amendment is known as the *fast basic service set transition (FT)* amendment. The technology is more often referred to as *fast-secure roaming* because it defines faster handoffs when roaming occurs between cells in a WLAN using the strong security defined by a robust secure network (RSN). Be aware that there are multiple types of fast-secure roaming that are implemented by different vendors. These include CCKM, PKC, OKC, and fast session resumption. Some vendors support 802.11r, while others do not. 802.11r was proposed primarily because of the time constraints of applications such as VoWiFi. Average time delays of hundreds of milliseconds occur when a client station roams from one access point to another access point.

Roaming can be especially troublesome when using a WPA-Enterprise or WPA2-Enterprise security solution, which requires the use of a RADIUS server for 802.1X/EAP authentication and often takes 700 milliseconds or greater for the client to authenticate. VoWiFi requires a handoff of 100 milliseconds or less to avoid a degradation of the quality of the call or, even worse, a loss of connection.

Under 802.11r, a client station is able to establish a QoS stream and set up a security association with a new access point in an efficient manner that allows bypassing 802.1X/EAP authentication when roaming to a new access point. The client station is able to achieve these tasks either over the wire via the original access point or through the air. Eventually, the client station will complete the roaming process and move to the new access point.

Tactical enterprise deployments of this technology are extremely important for providing more secure communications for VoWiFi. The details of this technology are heavily tested on the CWSP exam.

802.11s-2011

The 802.11s-2011 amendment was ratified in July 2011. 802.11 access points typically act as portal devices to a *distribution system (DS)* that is usually a wired 802.3 Ethernet medium. The 802.11-2016 standard, however, does not mandate that the distribution system use a wired medium. Access points can therefore act as portal devices to a *wireless distribution system (WDS)*. The 802.11s amendment proposes the use of a protocol for adaptive, autoconfiguring systems that support broadcast, multicast, and unicast traffic over a multihop mesh WDS.

The 802.11 Task Group s (TGs) set forth the pursuit of standardizing *mesh networking* using the IEEE 802.11 MAC/PHY layers. The 802.11s amendment defined the use of mesh points (MPs), which are 802.11 QoS stations that support mesh services. A mesh point is capable of using a mandatory mesh routing protocol called *Hybrid Wireless Mesh Protocol (HWMP)* that uses a default path selection metric. Vendors may also use proprietary mesh routing protocols and metrics. As depicted in Figure 2.2, a *mesh access point (MAP)* is a device that provides both mesh functionalities and AP functionalities simultaneously. A *mesh point portal (MPP)* is a device that acts as a gateway to one or more external networks, such as an 802.3 wired backbone.

FIGURE 2.2 Mesh points, mesh APs, and mesh portal

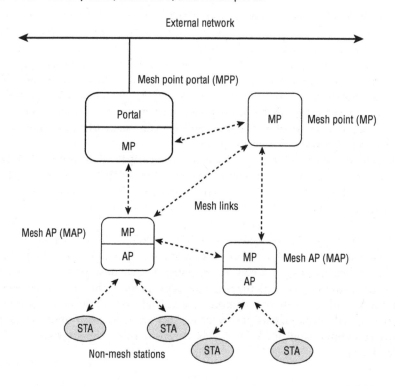

 Further discussion on distribution systems (DSs) and wireless distribution systems (WDSs) can be found in Chapter 7, "Wireless LAN Topologies." You will learn more about 802.11 mesh networking in Chapter 11.

802.11u-2011

The primary objective of the 802.11 Task Group u (TGu) was to address interworking issues between an IEEE 802.11 access network and any external network to which it is connected. A common approach is needed to integrate IEEE 802.11 access networks with external networks in a generic and standardized manner. 802.11u is also often referred to as *Wireless Interworking with External Networks (WIEN)*.

The 802.11u-2011 amendment, ratified in February 2011, defined functions and procedures for aiding network discovery and selection by STAs, information transfer from external networks using QoS mapping, and a general mechanism for the provision of emergency services.

The 802.11u-2011 amendment is the basis for the Wi-Fi Alliance's Hotspot 2.0 specification and its Passpoint certification. This standard and certification is designed to provide seamless roaming for wireless devices between your Wi-Fi network and other partner networks, similar to how cellular telephone networks provide roaming. Passpoint and Hotspot 2.0 are discussed in great detail in Chapter 18, "Bring Your Own Device (BYOD) and Guest Access."

802.11v-2011

The 802.11v-2011 amendment was ratified in February 2011. While 802.11k defines methods of retrieving information from client stations, 802.11v provides for an exchange of information that can potentially ease the configuration of client stations wirelessly from a central point of management. 802.11v-2011 defines *wireless network management (WNM)*, which gives 802.11 stations the ability to exchange information for the purpose of improving the overall performance of the wireless network. Access points and client stations use WNM protocols to exchange operational data so that each station is aware of the network conditions, allowing stations to be more cognizant of the topology and state of the network.

In addition to providing information on network conditions, WNM protocols define mechanisms in which WLAN devices can exchange location information, provide support for the multiple BSSID capability, and offer a new WNM-Sleep mode, in which a client station can sleep for long periods of time without receiving frames from the AP.

Some of the 802.11v mechanisms are defined by the Wi-Fi Alliance as optional mechanisms in the Voice-Enterprise certification.

802.11w-2009

A common type of attack on an 802.11 WLAN is a denial-of-service (DoS) attack. There are a multitude of DoS attacks that can be launched against a wireless network; however, a very common DoS attack occurs at layer 2 using 802.11 management frames. Currently, it is simple for an attacker to edit deauthentication or disassociation frames and then retransmit the frames into the air, effectively shutting down the wireless network.

The goal of the IEEE Task Group w (TGw) was to provide a way of delivering management frames in a secure manner, therefore preventing the management frames from being able to be spoofed. The 802.11w-2009 amendment provided protection for unicast, broadcast, and multicast management frames.

These 802.11w frames are referred to as *robust management frames*. Robust management frames can be protected by the management frame protection service and include disassociation, deauthentication, and robust action frames. Action frames are used to request a station to take action on behalf of another station, and not all action frames are robust.

When unicast management frames are protected, frame protection is achieved by using CCMP. Broadcast and multicast frames are protected using the *Broadcast/Multicast Integrity Protocol (BIP)*. BIP provides data integrity and replay protection using AES-128 in Cipher-Based Message Authentication Code (CMAC) mode. It should be noted that the 802.11w amendment will not put an end to all layer 2 DoS attacks.

You will find a discussion about both layer 1 and layer 2 DoS attacks in Chapter 16, "Wireless Attacks, Intrusion Monitoring, and Policy."

802.11y-2008

Although 802.11 devices mostly operate in unlicensed frequencies, they can also operate on frequencies that are licensed by national regulatory bodies.

The objective of the IEEE Task Group y (TGy) was to standardize the mechanisms required to allow high-powered, shared 802.11 operations with other non-802.11 devices in the 3650 MHz–3700 MHz licensed band in the United States. It should be noted that the mechanisms defined by the 802.11y-2008 amendment can be used in other countries and in other licensed frequencies.

The licensed 3650 MHz to 3700 MHz band requires content-based protocol (CBP) mechanisms to avoid interference between devices. The medium contention method, CSMA/CA (which is used by Wi-Fi radios), can normally accommodate this requirement. However, when standard CSMA/CA methods are not sufficient, the 802.11y-2008 amendment defines *dynamic STA enablement (DSE)* procedures. 802.11 radios broadcast their actual location as a unique identifier in order to help resolve interference with non-802.11 radios in the same frequency.

802.11z-2010

The purpose of IEEE Task Group z (TGz) was to establish and standardize a *direct link setup (DLS)* mechanism to allow operation with non-DLS-capable access points. In most WLAN environments, all frame exchanges between client stations that are associated to the same access point must pass through the access point. DLS allows client stations to bypass the access point and communicate with direct frame exchanges. Some of the earlier amendments have defined DLS communications. The 802.11z-2010 amendment defined enhancements to DLS communications. It should be noted that DLS communications have yet to be used by enterprise WLAN vendors.

802.11aa-2012

The 802.11aa amendment specifies QoS enhancements to the 802.11 Media Access Control (MAC) for robust audio and video streaming for both consumer and enterprise applications. 802.11aa provides improved management, increased link reliability, and increased application performance. The amendment defines *Groupcast with Retries (GCR)*, a flexible service to improve the delivery of group addressed frames. GCR can be provided in an infrastructure BSS by the AP to its associated STAs or in a mesh BSS by a mesh STA and its peer mesh STAs.

802.11ac-2013

The 802.11ac-2013 amendment defines very high throughput (VHT) enhancements below 6 GHz. The technology is used only in the 5 GHz frequency bands where 802.11a/n radios already operate. 802.11ac takes advantage of the greater spectrum space that the 5 GHz U-NII bands can provide. The 2.4 GHz ISM band cannot provide the needed frequency space that would be able to take full advantage of 802.11ac technology. To take full advantage of 802.11ac, even more spectrum would be preferred in 5 GHz. The 802.11ac amendment defines a maximum data rate of 6933.3 Mbps. 802.11ac provides gigabit speeds using the following four major enhancements:

Wider Channels   802.11n introduced the capability of 40 MHz channels, which effectively doubled the data rates. 802.11ac supports channel widths of 20 MHz, 40 MHz, 80 MHz, and 160 MHz channels. This is the main reason that enterprise 802.11ac radios would not be able to operate in the 2.4 GHz ISM band.

New Modulation   802.11ac provides the capability to use 256-QAM modulation, which can provide at least a 30 percent increase in speed over previous modulation methods. 256-QAM modulation requires a very high signal-to-noise (SNR) ratio to be effective.

More Spatial Streams   According to the standard, 802.11ac radios could be built to transmit and receive up to eight spatial streams. In reality, the first couple of generations of 802.11ac chipsets only support up to four spatial streams.

Improved MIMO and Beamforming   While 802.11n defined the use of single-user MIMO radios, very high throughput (VHT) introduces the use of *multi-user MIMO (MU-MIMO)* technology. An access point with MU-MIMO capability could transmit a signal to multiple client stations on the same channel simultaneously, if the client stations support MU-MIMO and are in different physical areas. 802.11ac can utilize explicit beamforming.

The 802.11ac technology has been delivered in two generations of chipsets, often referred to as *waves*. The first wave of 802.11ac chipsets took
advantage of 256-QAM modulation and up to 80 MHz–channels. Most of
the AP hardware uses 3x3:3 radios. The second wave of 802.11ac chipsets
are typically able to use MU-MIMO and up to 160 MHz channels. It should
be understood that the wave terminology is strictly a marketing term when
discussing generations of 802.11ac radios. The AP hardware also supports
4x4:4 radios. Chapter 10 focuses heavily on 802.11ac-2013 and the underlying technology.

802.11ad-2012

The 802.11ad amendment defines performance enhancements using the much higher
unlicensed frequency band of 60 GHz and a transmission method known as *directional
multi-gigabit (DMG)*. The higher frequency range is big enough to support data rates of up
to 7 Gbps. The downside is that 60 GHz will have significantly less effective range than a
5 GHz signal and be limited to line-of-sight communications, as the high frequency signal
will have difficulty penetrating walls.

The 60 GHz Wi-Fi technology has the potential to be used for wireless docking, wireless
displays, wired equivalent data transfers, and streaming of uncompressed video. To provide
seamless transition when roaming between the 60 GHz frequency band and legacy
2.4 GHz or 5 GHz bands, a "fast session transfer" feature was added to the specification.

The DMG technology also required the adoption of a new encryption mechanism. There
was concern that the current CCMP encryption methods may not be able to properly process the higher anticipated data rates. CCMP uses two chained-together AES cryptographic
modes to process 128-bit blocks of data. The 128-bit blocks of data must also be processed
"in order" from the first AES cryptographic mode to the second mode.

The 802.11ad amendment specifies the use of the *Galois/Counter Mode Protocol
(GCMP)*, which also uses AES cryptography. However, GCMP calculations can be run in
parallel and are computationally less intensive than the cryptographic operations of CCMP.

802.11ae-2012

The 802.11ae amendment specifies enhancements to QoS management. A quality-of-
service management frame (QMF) service can be enabled, allowing some of the management frames to be transmitted using a QoS access category that is different from the access
category assigned to voice traffic. This can improve the quality of service of other traffic
streams.

802.11af-2014

The 802.11af amendment allows the use of wireless in the *TV white space (TVWS)* frequencies between 54 MHz and 790 MHz. This technology is sometimes referred to as

White-Fi or *Super Wi-Fi*, but we recommend that you shy away from using these terms, as this technology is not affiliated with the Wi-Fi Alliance, which is the trademark holder of the term Wi-Fi.

In different regions or TV marketplaces, not all of the available TV channels are used by licensed stations. TVWS is the range of TV frequencies that are not used by any licensed station in a specific area. 802.11af-based radios will have to verify what frequencies are available and make sure that they do not cause interference. To achieve this, the 802.11af AP will first need to determine its location, likely through the use of GPS technology. Then the radio device will need to interact with a geographic database to determine the available channels for that given time and location.

The Physical layer is based on the OFDM technology used in 802.11ac, using smaller channel widths than 802.11ac along with a maximum of four spatial streams. This new PHY is known as *television very high throughput (TVHT)*, and is designed to support the narrow TV channels that are made available by TVWS.

The low-bandwidth frequencies that are used mean lower data rates than 802.11a/b/g/n/ac technology. Maximum transmission speed is 26.7 Mbps or 35.6 Mbps, depending on the width of the channel, which is determined by the regulatory domain. Channel width is between 6 MHz and 8 MHz, and up to 4 channels can be bonded together. 802.11af radios can also support up to 4 spatial streams. Using 4 channels and 4 spatial streams, 802.11af has a maximum data rate of about 426 Mbps or 568 Mbps, depending on the regulatory domain. Although the lower TVWS frequencies mean lower data rates, the lower frequencies will provide longer-distance transmissions, along with better penetration through obstructions such as foliage and buildings. This greater distance could result in coverage that is more pervasive, providing contiguous roaming in outdoor office parks, campuses, or public community networks. Another anticipated use is to provide broadband Internet services to rural areas.

It is important to note that the IEEE 802.22-2011 standard, along with at least one other standard that is in development, also specifies wireless communications using the TV white space frequencies. This may cause coexistence problems in the future between these competing technologies. Also, the existence of multiple technologies in the same frequency space may splinter product development and acceptance.

What Happened to the Wireless Gigabit Alliance?

The Wireless Gigabit Alliance (WiGig) was formed to promote wireless communications among consumer electronics, handheld devices, and PCs using the readily available, unlicensed 60 GHz spectrum. On January 3, 2013, it was announced that the activities of the WiGig Alliance would be consolidated into the Wi-Fi Alliance. Since then, the Wi-Fi Alliance has been actively working toward WiGig branding and product certification testing.

Post 802.11-2016 Ratified Amendments

Since the 802.11-2016 document was published, other amendments have been ratified to define further enhancements to 802.11 technology. These amendments include 802.11ah and 802.11ai, as discussed in the following sections.

802.11ah-2016

The 802.11ah amendment defines the use of Wi-Fi in frequencies below 1 GHz. The *Wi-Fi HaLow* certification from the Wi-Fi Alliance is based on mechanisms defined in the IEEE 802.11ah amendment. The lower frequencies will mean lower data rates but longer distances. A likely use for 802.11h will be sensor networks along with backhaul for sensor networks and with extended range Wi-Fi, such as smart homes, automobiles, healthcare, industrial, retail, and agriculture. This internetworking of devices is known as *Internet-of-Things (IoT)* or *machine-to-machine (M2M)* communications.

The available frequencies will vary between countries. For example, the 902–928 MHz unlicensed ISM frequencies are available in the United States, whereas the 863–868 MHz frequencies would likely be available in Europe, and the 755–787 frequencies would likely be available in China.

802.11ai-2016

The goal of the 802.11ai amendment is to provide a *fast initial link setup (FILS)*. It is designed to address the challenges that exist in high-density environments where large numbers of mobile users are continually joining and disconnecting from an extended service set. The amendment is designed to improve user connectivity in high-density environments, such as airports, sports stadiums, arenas, and shopping malls.

FILS is especially important for ensuring that robust security network association (RSNA) links are not degraded as clients roam.

IEEE 802.11 Draft Amendments

What does the future hold in store for us with 802.11 wireless networking? The draft amendments are a looking glass into the enhancements and capabilities that might be available in the near future for 802.11 wireless networking devices. Even greater throughput as well as operations on higher and lower frequencies await us on the wireless horizon.

It is important to remember that draft amendments are proposals that have yet to be ratified. Although some vendors may already be selling products that have some of the capabilities described in the following sections, these features are still considered proprietary. Even though a vendor might be marketing these pre-ratified capabilities, there is no guarantee that their current product will work with future products that are certified as compliant with the forthcoming ratified amendment.

 The CWNA exam (CWNA-107) currently covers all of the technologies defined by the 802.11-2016 standard, as well as any amendments ratified since 2016. *You will not be tested on the draft amendments.* Even though you will not be tested on these amendments, we believe it is important for you to be introduced to the technologies that are being planned and developed, as they will likely change 802.11 wireless networking in the future.

The remaining pages of this chapter provide a glimpse into the future of more advanced and sophisticated Wi-Fi products that could bring this technology to even greater heights.

 Once again, please remember that because these IEEE amendments are still draft documents, they will likely be different from the final, ratified amendments.

802.11aj

The 802.11aj draft amendment is to provide modifications to the IEEE 802.11ad-2012 amendment's PHY and MAC layer to provide support for operating in the Chinese Milli-Meter Wave (CMMW) frequency bands. The CMMW frequency bands are 59–64 GHz. The amendment will also provide modifications to the IEEE 802.11ad-2012 amendment's PHY and MAC layer to provide support for operating in the Chinese 45 GHz frequency band.

802.11ak

The 802.11ak draft amendment is also referred to as *General Link (GLK)*. The task group is exploring enhancement to 802.11 links for use in bridged networks. These bridged networks will be evaluated as potential support for home entertainment systems, industrial control devices, and other products that have both 802.11 wireless and 802.3 wired capabilities. GLK aims to simplify the use of 802.11 between access points and wireless stations, allowing the stations to provide bridging services.

802.11aq

The 802.11aq pre-association service discovery task group is working to develop an 802.11 amendment that enables delivery of network service information prior to the association of stations on an 802.11 network. This amendment hopes to be able to allow advertisement of services to stations prior to the stations' actual association to the network.

802.11ax

The 802.11ax draft amendment, also known as the *high efficiency (HE)* WLAN amendment, is expected to be the next big PHY enhancement to the 802.11 standard. 802.11ax will operate in both the 2.4 GHz and 5 GHz bands. In addition to increasing client throughput, it is also being designed to provide support for more users and higher density environments. Chapter 19, "802.11ax-High Efficiency (HE)," is dedicated to 802.11ax. Although earlier amendments defined methods to achieve higher data rates, 802.11ax uses PHY and MAC layer enhancements for better traffic management of the existing WLAN medium. A key component of 802.11ax is *orthogonal frequency-division multiple access (OFDMA)* technology. OFDMA is a multi-user version of the popular orthogonal frequency-division multiplexing (OFDM) digital modulation scheme. Multiple access is achieved in OFDMA by assigning subsets of subcarriers to individual clients. This allows simultaneous low data rate transmission to/from multiple users.

You will not be tested on 802.11ax technology in the CWNA exam (CWNA-107). However, 802.11ax technology will be arriving in the marketplace in 2018. Furthermore, the technology introduces substantial changes as to how AP radios and client radios communicate. Because this technology is preeminent, the authors of this book decided to include Chapter 19, "802.11ax - High Efficiency (HE)," as an introductory primer to 802.11ax.

802.11ay

The 802.11ay draft amendment is an improvement of the 802.11ad amendment, providing faster speeds and longer range. It is expected to have a maximum data rate of 176 Gbps. 802.11ad uses maximum channel widths of 2.16 GHz. 802.11ay will provide channel bonding of up to four channels, while also adding MIMO, with a maximum of 4 spatial-streams. 256-QAM modulation is also expected to be added. 802.11ay will operate in the unlicensed bands above 45 GHz, and it will provide backward compatibility with 802.11ad. It is expected to be approved some time in 2019. The main uses of 802.11y appear to be DisplayPort, HDMI and USB connectivity, along with connectivity for TV and monitor displays.

802.11az

One of the goals of 802.11az is to improve physical location tracking and positioning of 802.11 devices. Better accuracy could be used with smart building applications and tracking IoT devices. Another goal of this amendment is to increase the energy efficiency of the network.

802.11ba

This amendment is expected to define an energy-efficient data-reception mode, referred to as *wake-up radio(WUR)*. The objective is to increase battery life of devices, such as battery powered IoT devices, without increasing network latency or slowing down performance.

Defunct Amendments

The next two amendments were never ratified and are considered dead in the water. However, the subject matter (roaming and performance testing) of the two amendments is important and therefore we will discuss them in this book.

802.11F

The IEEE Task Group F (TGF) published IEEE Std 802.11F-2003 as a recommended practice in 2003. The amendment was never ratified and was withdrawn in February 2006.

 The use of an uppercase letter designation for an IEEE task group, like that in IEEE Task Group F, indicates that this amendment (F) is considered a recommended practice and not part of the 802.11-2016 standard.

The original published 802.11 standard mandated that vendor APs support *roaming*. A mechanism is needed to allow client stations that are already communicating through one AP to be able to jump from the coverage area of the original AP and continue communications through a new AP. A perfect analogy is the roaming that occurs when using a cell phone. When you are talking on a cell phone while inside a moving vehicle, your phone will roam between cellular towers to allow for seamless communications and hopefully an uninterrupted conversation. Seamless roaming allows for mobility, which is the heart and soul of true wireless networking and connectivity, as depicted in Figure 2.3.

FIGURE 2.3 Seamless roaming

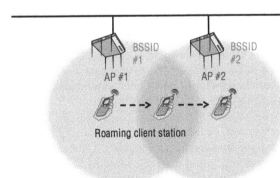

Will Seamless Roaming Work If I Mix and Match Different Vendors' Access Points?

The real-world answer is no. 802.11F was intended to address roaming interoperability between *autonomous access points* from different vendors. The 802.11F amendment was initially only a recommended practice and was eventually withdrawn entirely by the IEEE. WLAN vendors want customers to purchase only the brand of AP that the vendor sells, not the competition's brand of AP. It is the "recommended practice" of this book not to mix different vendors' access points on the same wired network segment. Roaming is discussed in further detail in Chapter 7, "Wireless LAN Topologies," Chapter 9, "802.11 MAC," and Chapter 15, "WLAN Troubleshooting."

Although the 802.11 standard calls for the support of roaming, it fails to dictate how roaming should actually transpire. The IEEE initially intended for vendors to have flexibility in implementing proprietary AP-to-AP roaming mechanisms. The 802.11F amendment was an attempt to standardize how roaming mechanisms work behind the scenes on the distribution system medium, which is typically an 802.3 Ethernet network using TCP/IP networking protocols. 802.11F addressed "vendor interoperability" for AP-to-AP roaming. The final result was a recommended practice to use the *Inter-Access Point Protocol (IAPP)*. IAPP uses announcement and handover processes that result in APs informing other APs about roamed clients as well as delivery of buffered packets. Because the 802.11F amendment was never ratified, the use of IAPP is basically nonexistent.

802.11T

The original goal of the IEEE 802.11 Task Group T (TGT) was to develop performance metrics, measurement methods, and test conditions to measure the performance of 802.11 wireless networking equipment.

 The uppercase *T* in the name *IEEE 802.11T* indicates that this amendment was considered a recommended practice and not a standard. The 802.11T amendment was never ratified and has been dropped.

The 802.11T proposed amendment was also called *Wireless Performance Prediction (WPP)*. Its final objective was consistent and universally accepted WLAN measurement practices. These 802.11 performance benchmarks and methods could be used by independent test labs, manufacturers, and even end users.

Are Throughput Results the Same among Vendors?

Multiple factors can affect throughput in a wireless network, including the physical environment, range, and type of encryption. Another factor that can affect throughput is simply the vendor radio device that is being used for transmissions. Even though the 802.11-2016 standard clearly defines frequency bandwidths, data rate speeds, and medium access methods, throughput results vary widely from vendor to vendor. A throughput performance test using two radio cards from one vendor will most often yield very different results than the same throughput performance test using two radio cards from another vendor. Typically, you will see better throughput results when sticking with one vendor as opposed to mixing vendor equipment. However, sometimes mixing vendor equipment will produce the unexplained consequence of increased throughput. Although standardized 802.11T metrics were never adopted, the Wi-Fi Alliance defines its own metrics for vendor-neutral lab tests for all of the Wi-Fi Alliance certifications.

IEEE Task Group m

The IEEE Task Group m (TGm) started an initiative in 1999 for internal maintenance of the 802.11 standard's technical documentation. 802.11m is often referred to as *802.11 housekeeping* because of its mission of clarifying and correcting the 802.11 standard. Unless you are a member of TGm, this amendment is of little significance. However, this task group also is responsible for "rolling up" ratified amendments into a published document. The following list shows the revisions of the standard that have been created over the years, along with the task group responsible for them.

IEEE Std 802.11-2007	802.11 TGma
IEEE Std 802.11-2012	802.11 TGmb
IEEE Std 802.11-2016	802.11 TGmc
Next 802.11 consolidation	802.11 TGmd

Neither 802.11l nor 802.11o amendments exist, because they are considered typologically problematic. The 802.11ab amendment was skipped to avoid confusion with devices that use both 802.11a and 802.11b PHY technologies, which are often called 802.11a/b devices. The 802.11ag amendment was skipped to avoid confusion with devices that use both 802.11a and 802.11g PHY technologies, which are called 802.11a/g devices. Also, it should be noted that there is no amendment with the name of 802.11x. The term *802.11x* sometimes is used to refer to all the 802.11 standards. The IEEE 802.1X standard, which is a port-based access control standard, is often incorrectly called 802.11x.

Summary

This chapter covered the original 802.11 standard, the amendments consolidated into the 802.11-2007 standard, the 802.11-2012 standard, and the 802.11-2016 standard, as well as 802.11 amendments ratified since the consolidation of the 802.11-2016 standard. This chapter also discussed possible future enhancements. We covered the following:

- All the defined PHY and MAC layer requirements of the original 802.11 Prime standard
- All the approved enhancements to the 802.11 standard in the form of ratified amendments, including higher data rates, different spread-spectrum technologies, quality of service, and security
- Future capabilities and improvements as proposed in the 802.11 draft documents

Although proprietary Wi-Fi solutions exist and will continue to exist in the foreseeable future, standardization brings stability to the marketplace. The 802.11-2016 standard and all the future enhanced supplements provide a much needed foundation for vendors, network administrators, and end users.

The CWNA exam will test your knowledge of the 802.11-2016 standard and all the related technologies.

Exam Essentials

Know the defined spread-spectrum technologies of the original 802.11 standard and the subsequent 802.11-2007, 802.11-2012, and 802.11-2016 standards. Although the original 802.11 standard defined infrared, FHSS, and DSSS, later amendments that are now incorporated in the 802.11-2016 standard also define HR-DSSS, OFDM, ERP, HT, and VHT.

Remember both the required data rates and supported data rates of each PHY. DSSS and FHSS require and support data rates of 1 and 2 Mbps. Other PHYs offer a wider support for data rates. For example, OFDM and ERP-OFDM support data rates of 6, 9, 12, 18, 24, 36, 48, and 54 Mbps, but only the rates of 6, 12, and 24 Mbps are mandatory. With the introduction of 802.11n, it is important to understand the concept of modulation coding schemes (MCSs), which is also defined in 802.11ac. Please understand that data rates are transmission speeds, not aggregate throughput.

Know the frequency bands used by each PHY as defined by the 802.11-2016 standard. 802.11a and 802.11ac equipment operate in the 5 GHz U-NII bands. DSSS, FHSS, HR-DSSS, and ERP (802.11g) devices transmit and receive in the 2.4 GHz ISM band. Understand that 802.11n devices transmit in either the 2.4 GHz or 5 GHz frequency bands.

Define transmit power control and dynamic frequency selection. TPC and DFS are mandated for use in the 5 GHz band. Both technologies are used as a means to avoid interference with radar transmissions.

Explain the defined wireless security standards, both pre-802.11i and post-802.11i. Before the passage of 802.11i, WEP and TKIP were defined. The 802.11i amendment called for the use of CCMP/AES for encryption. For authentication, 802.11i defines either an 802.1X/EAP solution or the use of PSK authentication.

Review Questions

1. An ERP (802.11g) network mandates support for which two spread-spectrum technologies?

   A. ERP-OFDM

   B. FHSS

   C. ERP-PBCC

   D. ERP-DSSS/CCK

   E. CSMA/CA

2. Which amendment defines performance enhancements using the much higher unlicensed frequency band of 60 GHz, and a transmission method known as directional multi-gigabit (DMG)?

   A. 802.11ac

   B. 802.11ad

   C. 802.11ay

   D. 802.11q

   E. 802.11z

3. Which types of devices were defined in the original 802.11 standard? (Choose all that apply.)

   A. OFDM

   B. DSSS

   C. HR-DSSS

   D. IR

   E. FHSS

   F. ERP

4. Which 802.11 amendment defines wireless mesh networking mechanisms?

   A. 802.11n

   B. 802.11u

   C. 802.11s

   D. 802.11v

   E. 802.11k

5. A robust security network (RSN) requires the use of which security mechanisms? (Choose all that apply.)

   A. 802.11x

   B. WEP

   C. IPsec

   D. CCMP/AES

   E. CKIP

   F. 802.1X

6. An 802.11ac radio card can transmit on the _____ frequency and uses _____ spread-spectrum technology.

   A. 5 MHz, OFDM

   B. 2.4 GHz, HR-DSSS

   C. 2.4 GHz, ERP-OFDM

   D. 5 GHz, VHT

   E. 5 GHz, DSSS

7. What are the required data rates of an OFDM station?

   A. 3, 6, and 12 Mbps

   B. 6, 9, 12, 18, 24, 36, 48, and 54 Mbps

   C. 6, 12, 24, and 54 Mbps

   D. 6, 12, and 24 Mbps

   E. 1, 2, 5.5, and 11 Mbps

8. When implementing an 802.1X/EAP RSN network with a VoWiFi solution, what is needed to avoid latency issues during roaming?

   A. Inter-Access Point Protocol

   B. Fast BSS transition

   C. Distributed Coordination Function

   D. Roaming Coordination Function

   E. Lightweight APs

9. Which new technologies debuted in the 802.11ac-2013 amendment? (Choose all that apply.)

   A. MIMO

   B. MU-MIMO

   C. 256-QAM

   D. 40 MHz channels

   E. 80 MHz channels

10. What is the primary reason that OFDM (802.11a) radios cannot communicate with ERP (802.11g) radios?

    A. 802.11a uses OFDM, and 802.11g uses ERP.

    B. 802.11a uses DSSS, and 802.11g uses OFDM.

    C. 802.11a uses OFDM, and 802.11g uses CCK.

    D. 802.11a operates at 5 GHz, and 802.11g operates at 2.4 GHz.

    E. 802.11a requires dynamic frequency selection, and 802.11g does not.

11. Which two technologies are used to prevent 802.11 radios from interfering with radar and satellite transmissions at 5 GHz?

    A. Dynamic frequency selection

    B. Enhanced Distributed Channel Access

    C. Direct-sequence spread-spectrum

    D. Temporal Key Integrity Protocol

    E. Transmit power control

12. Which 802.11 amendments provide for throughput of 1 Gbps or higher? (Choose all that apply.)

    A. 802.11aa

    B. 802.11ab

    C. 802.11ac

    D. 802.11ad

    E. 802.11ae

    F. 802.11af

13. As defined by the 802.11-2016 standard, which equipment is compatible? (Choose all that apply.)

    A. ERP and HR-DSSS

    B. HR-DSSS and FHSS

    C. VHT and OFDM

    D. 802.11h and 802.11a

    E. HR-DSSS and DSSS

14. What is the maximum data rate defined by the 802.11ac amendment?

    A. 54 Mbps

    B. 1300 Mbps

    C. 3466.7 Mbps

    D. 6933.3 Mbps

    E. 60 Gbps

15. What are the security options available as defined in the original IEEE Std 802.11-1999 (R2003)? (Choose all that apply.)

    A. CCMP/AES

    B. Open System authentication

    C. Preshared keys

    D. Shared Key authentication

    E. WEP

    F. TKIP

16. The 802.11u-2011 amendment is also known as what?

    A. WIEN (Wireless Interworking with External Networks)

    B. WLAN (wireless local area networking)

    C. WPP (Wireless Performance Prediction)

    D. WAVE (Wireless Access in Vehicular Environments)

    E. WAP (Wireless Access Protocol)

17. The 802.11-2016 standard defines which two technologies for quality of service (QoS) in a WLAN?

    A. EDCA

    B. PCF

    C. Hybrid Coordination Function Controlled Channel Access

    D. VoIP

    E. Distributed Coordination Function

    F. VoWiFi

18. The 802.11h amendment (now part of the 802.11-2016 standard) introduced which two major changes for 5 GHz radios?

    A. TPC

    B. IAPP

    C. DFS

    D. DMG

    E. FHSS

19. The 802.11n amendment defined which PHY?

    A. HR-DSSS

    B. FHSS

    C. OFDM

    D. PBCC

    E. HT

    F. VHT

20. Which layers of the OSI model are referenced in the 802.11 standard? (Choose all that apply.)

    A. Application

    B. Data-Link

    C. Presentation

    D. Physical

    E. Transport

    F. Network

Chapter 3

Radio Frequency Fundamentals

IN THIS CHAPTER, YOU WILL LEARN ABOUT THE FOLLOWING:

✓ **What is a radio frequency signal**

✓ **Radio frequency characteristics**

- Wavelength
- Frequency
- Amplitude
- Phase

✓ **Radio frequency behaviors**

- Wave propagation
- Absorption
- Reflection
- Scattering
- Refraction
- Diffraction
- Loss (attenuation)
- Free space path loss
- Multipath
- Gain (amplification)

In addition to understanding the OSI model and basic networking concepts, you must broaden your understanding of many other networking technologies in order to properly design, deploy, and administer an 802.11 wireless network. For instance, when administering an Ethernet network, you typically need a comprehension of TCP/IP, bridging, switching, and routing. The skills to manage an Ethernet network will also aid you as a wireless LAN (WLAN) administrator because most 802.11 wireless networks act as "portals" into wired networks. The IEEE defines the 802.11 communications at the Physical layer and the MAC sublayer of the Data-Link layer.

To fully understand the 802.11 technology, you need to have a clear concept of how wireless works at the first layer of the OSI model; and at the heart of the Physical layer is *radio frequency (RF)* communications.

In a wired LAN, the signal is confined neatly inside the wire, and the resulting behaviors are anticipated. However, just the opposite is true for a WLAN. Although the laws of physics apply, RF signals move through the air in a sometimes unpredictable manner. Because RF signals are not saddled inside an Ethernet wire, you should always try to envision a WLAN as an "ever-changing" network.

Does this mean that you must be an RF engineer from Stanford University to perform a WLAN site survey or monitor a Wi-Fi network? Of course not. But if you have a good grasp of the RF characteristics and behaviors defined in this chapter, your skills as a wireless network administrator will be ahead of the curve. Why does a wireless network perform differently in an auditorium full of people than it does inside an empty auditorium? Why does the range of a 5 GHz radio transmitter seem shorter than the range of a 2.4 GHz radio transmitter? These are the types of questions that can be answered with some basic knowledge of how RF signals work and perform.

Wired communications travel across what is known as *bounded medium*. Bounded medium contains or confines the signal (although small amounts of signal leakage does occur). Wireless communications travel across what is known as *unbounded medium*. Unbounded medium does not contain the signal, which radiates into the surrounding environment in all directions (unless restricted or redirected by some outside influence).

In this chapter, we first define what an RF signal does, and then we discuss both the properties and the behaviors of RF.

What Is a Radio Frequency Signal?

This book is by no means intended to be a comprehensive guide to the laws of physics, which is the science of motion and matter. However, a basic understanding of some of the concepts of physics as they relate to radio frequency (RF) is important for even an entry-level wireless networking professional.

The *electromagnetic (EM) spectrum*, which is usually simply referred to as *spectrum*, is the range of all possible electromagnetic radiation. This radiation exists as self-propagating electromagnetic waves that can move through matter or space. Examples of electromagnetic waves include gamma rays, X-rays, visible light, and radio waves. Radio waves are electromagnetic waves occurring on the radio frequency portion of the electromagnetic spectrum, as pictured in Figure 3.1.

FIGURE 3.1   Electromagnetic spectrum

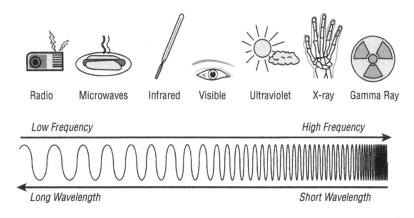

An RF signal starts out as an electrical *alternating current (AC)* signal that is originally generated by a transmitter. This AC signal is sent through a copper conductor (typically a coaxial cable) and radiated out of an antenna element in the form of an electromagnetic wave. This electromagnetic wave is the wireless signal. Changes of electron flow in an antenna, otherwise known as *current*, produce changes in the electromagnetic fields around the antenna.

An alternating current is an electrical current with a magnitude and direction that varies cyclically, as opposed to direct current, the direction of which stays in a constant form. The shape and form of the AC signal—defined as the *waveform*—is what is known as a *sine wave*, as shown in Figure 3.2. Sine wave patterns can also be seen in light, sound, and the ocean. The fluctuation of voltage in an AC current is known as cycling, or *oscillation*.

FIGURE 3.2   A sine wave

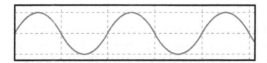

An RF electromagnetic signal radiates away from the antenna in a continuous pattern that is governed by certain properties, such as wavelength, frequency, amplitude, and phase. Additionally, electromagnetic signals can travel through mediums of different materials or travel in a perfect vacuum. When an RF signal travels through a vacuum, it moves at the speed of light, which is 299,792,458 meters per second, or 186,282 miles per second.

 To simplify mathematical calculations that use the speed of light, it is common to approximate the value by rounding it up to 300,000,000 meters per second or rounding it down to 186,000 miles per second. Any references to the speed of light in this book will use the approximate values.

RF electromagnetic signals travel using a variety or combination of movement behaviors. These movement behaviors are referred to as *propagation behaviors*. We discuss some of these propagation behaviors—including absorption, reflection, scattering, refraction, diffraction, amplification, and attenuation—later in this chapter.

Radio Frequency Characteristics

The following characteristics, defined by the laws of physics, exist in every RF signal:

- Wavelength
- Frequency
- Amplitude
- Phase

You will look at each of these in more detail in the following sections.

Wavelength

As stated earlier, an RF signal is an alternating current (AC) that continuously changes between a positive and negative voltage. An oscillation, or cycle, of this alternating current is defined as a single change from up to down to up or as a change from positive to negative to positive.

A *wavelength* is the distance between the two successive crests (peaks) or two successive troughs (valleys) of a wave pattern, as pictured in Figure 3.3. In simpler words, a wavelength is the distance that a single cycle of an RF signal actually travels.

FIGURE 3.3   Wavelength

Distance (360 degrees)

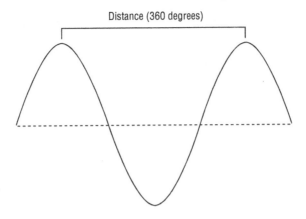

Although the physical length of waves can be different, which you will learn about shortly, waves have similar relative properties and are measured the same. Each wave is measured from peak to peak, and each has a trough between the two peaks. A relative measurement, known as *degrees*, is used to refer to different points along this span. As shown in Figure 3.3, the distance of a complete wave, from peak to peak, is 360 degrees. The trough is at 180 degrees. The first point where the wave crosses the horizontal line is at 90 degrees, and the second point where it crosses the horizontal line is at 270 degrees. You will not need to deal with degrees, except for understanding the concept of phase later in this chapter.

The Greek symbol λ (lambda) represents wavelength. Frequency is usually denoted by the Latin letter *f*. The Latin letter *c* represents the speed of light in a vacuum. This is derived from *celeritas*, the Latin word meaning speed.

It is very important to understand that there is an inverse relationship between wavelength and frequency. The three components of this inverse relationship are frequency (f, measured in hertz, or Hz), wavelength (λ, measured in meters, or m), and the speed of light (c, which is a constant value of 300,000,000 m/sec). The following reference formulas illustrate the relationship: $\lambda = c/f$ and $f = c/\lambda$. A simplified explanation is that the higher the frequency of an RF signal, the smaller the wavelength of that signal. The larger the wavelength of an RF signal, the lower the frequency of that signal.

AM radio stations operate at much lower frequencies than WLAN 802.11 radios, while satellite radio transmissions occur at much higher frequencies than WLAN radios. For instance, radio station WSB-AM in Atlanta broadcasts at 750 KHz and has a wavelength of 1,312 feet, or 400 meters. That is quite a distance for one single cycle of an RF signal to travel. In contrast, some radio navigation satellites operate at a very high frequency, near 252 GHz, and a single cycle of the satellite's signal has a wavelength of less than 0.05 inches, or 1.2 millimeters. Figure 3.4 illustrates a comparison of these two extremely different types of RF signals.

FIGURE 3.4   750 KHz wavelength and 252 GHz wavelength

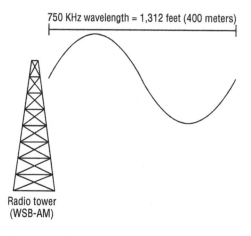

750 KHz wavelength = 1,312 feet (400 meters)

Radio tower
(WSB-AM)

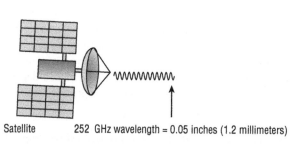

Satellite   252 GHz wavelength = 0.05 inches (1.2 millimeters)

As RF signals travel through space and matter, they lose signal strength (attenuate). It is often thought that a higher frequency electromagnetic signal with a smaller wavelength will attenuate faster than a lower frequency signal with a larger wavelength. In reality, the frequency and wavelength properties of an RF signal do not cause attenuation. Distance is the main cause of attenuation. All antennas have an effective area for receiving power, known as the *aperture*. The amount of RF energy that can be captured by the aperture of an antenna is smaller with higher frequency antennas. Although wavelength and frequency do not cause attenuation, the perception is that higher frequency signals with smaller wavelengths attenuate faster than signals with a larger wavelength. Theoretically, in a vacuum, electromagnetic signals will travel forever. However, as a signal travels through our atmosphere, the signal will attenuate to amplitudes below the receive sensitivity threshold of a receiving radio. Essentially, the signal will arrive at the receiver, but it will be too weak to be detected.

The perception is that the higher frequency signal with a smaller wavelength will not travel as far as the lower frequency signal with a larger wavelength. The reality is that the amount of energy that can be captured by the aperture of a high frequency antenna is smaller than the amount of RF energy that can be captured by a low frequency antenna. A good analogy to a receiving radio would be the human ear. The next time you hear a car coming down the street with loud music, notice that the first thing you hear will be the bass (lower frequencies). This practical example demonstrates that the lower frequency signals with the larger wavelength will be heard from a greater distance than the higher frequency signal with the smaller wavelength.

The majority of WLAN radios operate in either the 2.4 GHz frequency range or the 5 GHz range. In Figure 3.5, you see a comparison of a single cycle of the two waves generated by different frequency WLAN radios.

FIGURE 3.5 2.45 GHz wavelength and 5.775 GHz wavelength

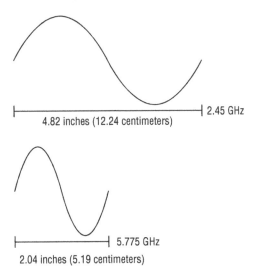

4.82 inches (12.24 centimeters) — 2.45 GHz

5.775 GHz
2.04 inches (5.19 centimeters)

Higher frequency signals will generally attenuate faster than lower frequency signals as they pass through various physical mediums, such as brick walls. This is important for a wireless engineer to know for two reasons. First, the coverage distance is dependent on the attenuation through the air (referred to as free space path loss, discussed later in this chapter). Second, the higher the frequency, typically the less the signal will penetrate through obstructions. For example, a 2.4 GHz signal will pass through walls, windows, and doors with greater amplitude than a 5 GHz signal. Think of how much farther you can receive an AM station's signal (lower frequency) versus an FM station's signal (higher frequency).

 Note that the length of a 2.45 GHz wave is about 4.8 inches, or 12 centimeters. The length of a 5.775 GHz wave is a distance of only about 2 inches, or 5 centimeters.

As you can see in Figure 3.4 and Figure 3.5, the wavelengths of the different frequency signals are different because, although each signal cycles only one time, the waves travel dissimilar distances. In Figure 3.6, you see the formulas for calculating wavelength distance in either inches or centimeters.

FIGURE 3.6   Wavelength formulas

Wavelength (inches) = 11.811/frequency (GHz)

Wavelength (centimeters) = 30/frequency (GHz)

 Throughout this study guide, you will be presented with various formulas. You will not need to know these formulas for the CWNA certification exam. The formulas are in this study guide to demonstrate concepts and to be used as reference material.

🌐 **Real World Scenario**

How Does the Wavelength of a Signal Concern Me?

It is often thought that a higher frequency electromagnetic signal with a smaller wavelength will attenuate faster than a lower frequency signal with a larger wavelength. In reality, the frequency and wavelength properties of an RF signal do not cause attenuation. Distance is the main cause of attenuation. All antennas have an effective area for receiving power, known as the aperture. The amount of RF energy that can be captured by the aperture of an antenna is smaller with higher frequency antennas. Although wavelength and frequency do not cause attenuation, the perception is that higher frequency signals with smaller wavelengths attenuate faster than signals with a larger wavelength. When all other aspects of the wireless link are similar, Wi-Fi equipment using 5 GHz radios will have a shorter range and a smaller coverage area than Wi-Fi equipment using 2.4 GHz radios.

Part of the design of a WLAN includes what is called a *site survey*. A major aspect of the site survey is to validate zones, or cells, of usable received signal coverage in your facilities. If single radio access points are being used, the 2.4 GHz access points can typically provide greater RF footprints (coverage area) for client stations than the higher frequency equipment. More 5 GHz access points would have to be installed to provide the same coverage that can be achieved by a lesser number of 2.4 GHz access points. The penetration of these signals will also reduce coverage for 5 GHz more than it will for 2.4 GHz. Most enterprise Wi-Fi vendors sell dual-frequency access points (APs) with both 2.4 GHz and 5 GHz radios. Site survey planning and coverage analysis for dual-frequency APs should initially be based on the higher frequency 5 GHz signal, which effectively provides a smaller coverage area. As a side note, WLAN design encompasses much more than coverage planning. Designing for client capacity and airtime consumption are just as important as coverage design. These design practices are covered in great detail in Chapter 13, "WLAN Design Concepts."

Frequency

As previously mentioned, an RF signal cycles in an alternating current in the form of an electromagnetic wave. You also know that the distance traveled in one signal cycle is the wavelength. But what about how often an RF signal cycles in a certain time period?

Frequency is the number of times a specified event occurs within a specified time interval. A standard measurement of frequency is *hertz (Hz)*, which was named after the German physicist Heinrich Rudolf Hertz. An event that occurs once in 1 second has a frequency of 1 Hz. An event that occurs 325 times in 1 second is measured as 325 Hz. The frequency at which electromagnetic waves cycle is also measured in hertz. Thus, the number of times an RF signal cycles in 1 second is the frequency of that signal, as pictured in Figure 3.7.

FIGURE 3.7   Frequency

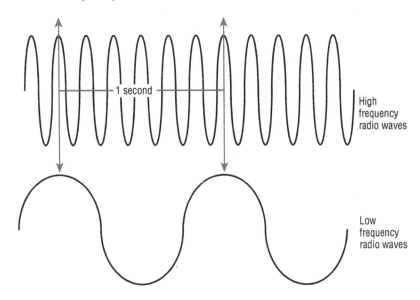

Different metric prefixes can be applied to the hertz (Hz) measurement of radio frequencies to make working with very large frequencies easier:

1 hertz (Hz) = 1 cycle per second

1 kilohertz (KHz) = 1,000 cycles per second

1 megahertz (MHz) = 1,000,000 (million) cycles per second

1 gigahertz (GHz) = 1,000,000,000 (billion) cycles per second

So, when we are talking about 2.4 GHz WLAN radios, the RF signal is oscillating 2.4 billion times per second!

Inverse Relationship

Remember that there is an inverse relationship between wavelength and frequency. The three components of this inverse relationship are frequency (f, measured in hertz, or Hz), wavelength (λ, measured in meters, or m), and the speed of light (c, which is a constant value of 300,000,000 m/sec). The following reference formulas illustrate the relationship: λ = c/f and f = c/ λ. A simplified explanation is that the higher the frequency of an RF signal, the shorter the wavelength will be of that signal. The longer the wavelength of an RF signal, the lower the frequency will be of that signal.

Amplitude

Another very important property of an RF signal is the *amplitude*, which can be characterized simply as the signal's strength, or power. When speaking about wireless transmissions, this is often referenced as how loud or strong the signal is. *Amplitude* can be defined as the maximum displacement of a continuous wave. With RF signals, the amplitude corresponds to the electrical field of the wave. When you look at an RF signal using an oscilloscope, the amplitude is represented by the positive crests and negative troughs of the sine wave.

In Figure 3.8, you can see that λ represents wavelength and *a* represents amplitude. The first signal's crests and troughs have more magnitude; thus, the signal has more amplitude. The second signal's crests and troughs have decreased magnitude, and therefore the signal has less amplitude.

FIGURE 3.8   Amplitude

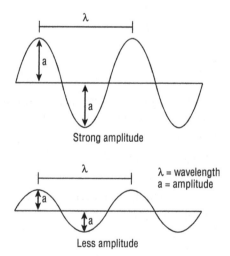

Strong amplitude

λ = wavelength
a = amplitude

Less amplitude

Although the signal strength (amplitude) is different, the frequency and wavelength of the signal remains constant. A variety of factors can cause an RF signal to lose amplitude, otherwise known as *attenuation*, which we discuss later in this chapter, in the section "Loss (Attenuation)."

When discussing signal strength in a WLAN, amplitude is usually referred to as either transmit amplitude or received amplitude. *Transmit amplitude* is typically defined as the amount of initial amplitude that leaves the radio transmitter. For example, if you configure an access point to transmit at 15 milliwatts (mW), that is the transmit amplitude. Cables and connectors will attenuate the transmit amplitude, while most antennas will amplify the transmit amplitude. When a radio receives an RF signal, the received signal strength is most often referred to as *received amplitude*. An RF signal strength measurement taken during a validation site survey is an example of received amplitude.

Different types of RF technologies require varying degrees of transmit amplitude. AM radio stations may transmit narrow band signals with as much power as 50,000 watts (W). The radios used in most indoor 802.11 access points have a transmit power range between 1 mW and 100 mW. You will learn later that Wi-Fi radios can receive and demodulate signals with amplitudes as low as billionths of a milliwatt.

Phase

Phase is not a property of just one RF signal but instead involves the relationship between two or more signals that share the same frequency. The phase involves the relationship between the position of the amplitude crests and troughs of two waveforms.

Phase can be measured in distance, time, or degrees. If the peaks of two signals with the same frequency are in exact alignment at the same time, they are said to be *in phase*. Conversely, if the peaks of two signals with the same frequency are not in exact alignment at the same time, they are said to be *out of phase*. Figure 3.9 illustrates this concept.

What is important to understand is the effect that phase has on amplitude when a radio receives multiple signals. Signals that have 0 (zero) degree phase separation actually combine their amplitude, which results in a received signal of much greater signal strength, potentially as much as twice the amplitude. If two RF signals are 180 degrees out of phase (the peak of one signal is in exact alignment with the trough of the second signal), they cancel each other out, and the effective received signal strength is null. Phase separation has a cumulative effect. Depending on the amount of phase separation of two signals, the received signal strength may be either increased or diminished. The phase difference between two signals is very important to understanding the effects of an RF phenomenon known as multipath, which is discussed later in this chapter.

FIGURE 3.9 Phase relationships

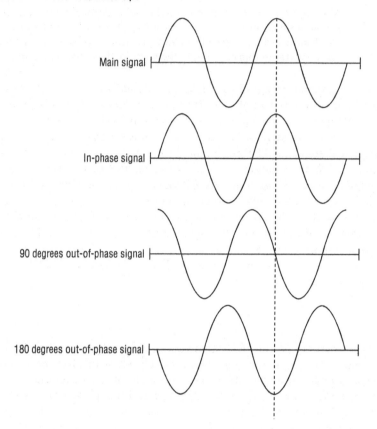

NOTE On the book's web page, which you can find by going to www.wiley .com/go/cwnasg, is a freeware Windows-based program called EMANIM. Toward the end of this chapter, you will use this program to execute Exercise 3.1, which is a lab that demonstrates the changes in amplitude due to phase relationships of RF signals.

Radio Frequency Behaviors

As an RF signal travels through the air and other mediums, it can move and behave in different manners. RF propagation behaviors include absorption, reflection, scattering, refraction, diffraction, free space path loss, multipath, attenuation, and gain. The following sections describe these behaviors.

Wave Propagation

Now that you have learned about some of the various characteristics of an RF signal, it is important to understand the way an RF signal behaves as it moves away from an antenna. As previously stated, electromagnetic waves can move through a perfect vacuum or pass through materials of different mediums. The way in which the RF waves move— known as *wave propagation*—can vary drastically depending on the materials in the signal's path; for example, drywall will have a much different effect on an RF signal than metal or concrete.

What happens to an RF signal between two locations is a direct result of how the signal propagates. When we use the term *propagate*, try to envision an RF signal broadening or spreading as it travels farther away from the antenna. An excellent analogy is shown in Figure 3.10, which depicts an earthquake. Note the concentric seismic rings that propagate away from the epicenter of the earthquake. Near the epicenter the waves are strong and concentrated, but as the seismic waves move away from the epicenter, the waves broaden and weaken. RF waves behave in much the same fashion. The manner in which a wireless signal moves is often referred to as *propagation behavior*.

FIGURE 3.10   Propagation analogy

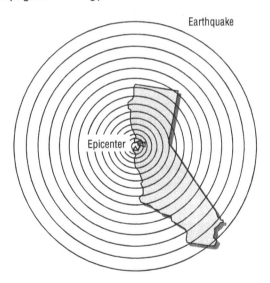

As a WLAN engineer, you should understand RF propagation behaviors for making sure that access points are deployed in the proper location, for making sure the proper type of antenna is chosen, and for monitoring the performance of the wireless network.

Absorption

The most common RF behavior is *absorption*. If a signal does not bounce off an object, move around an object, or pass through an object, then 100 percent absorption has occurred. Most materials will absorb some amount of an RF signal to varying degrees.

Brick and concrete walls will absorb a signal significantly, whereas drywall will absorb a signal to a lesser degree. A 2.4 GHz signal will be 1/16 the original power after propagating through a concrete wall. That same signal will only lose 1/2 the original power after passing though drywall material. Water is another example of a medium that can absorb a signal to a large extent. Absorption is a leading cause of attenuation (loss), which is discussed later in this chapter. The amplitude of an RF signal is directly affected by how much RF energy is absorbed. Even objects with large water content can absorb signals, such as paper, cardboard, and fish tanks.

 Real World Scenario

User Density

Mr. Barrett performed a wireless site survey at an airport terminal. He determined how many access points were required and their proper placement so that he would have the necessary RF coverage. Ten days later, during a snowstorm, the terminal was crammed with people who were delayed due to the weather. During these delays, the signal strength and quality of the WLAN was less than desirable in many areas of the terminal. What happened? Human bodies!

An average adult body is 50 to 65 percent water. Water causes absorption, which results in attenuation. User density is an important factor when designing a wireless network. One reason is the effects of absorption. Another reason is capacity planning for airtime consumption, which we discuss in Chapter 13, "WLAN Design Concepts."

Reflection

One of the most important RF propagation behaviors to be aware of is reflection. When a wave hits a smooth object that is larger than the wave itself, depending on the medium, the wave may bounce in another direction. This behavior is categorized as *reflection*. An analogous situation could be a child bouncing a ball off a sidewalk and the ball changing direction. Figure 3.11 depicts another analogy, a laser beam pointed at a single small mirror. Depending on the angle of the mirror, the laser beam bounces or reflects off in a different direction. RF signals can reflect in the same manner, depending on the objects or materials the signals encounter.

FIGURE 3.11   Reflection analogy

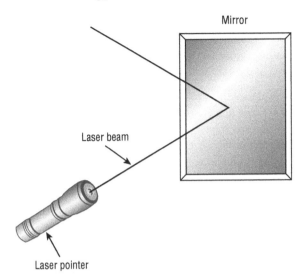

There are two major types of reflections: *sky wave reflection* and *microwave reflection*. Sky wave reflection can occur in frequencies below 1 GHz, where the signal has a very large wavelength. The signal bounces off the surface of the charged particles of the ionosphere in the earth's atmosphere. This is why you can be in Charlotte, North Carolina, and listen to radio station WLS-AM from Chicago on a clear night.

Microwave signals, however, exist between 1 GHz and 300 GHz. Because they are higher frequency signals, they have much smaller wavelengths, thus the term *microwave*. Microwaves can bounce off smaller objects like a metal door. Microwave reflection is what we are concerned about in Wi-Fi environments. In an outdoor environment, microwaves can reflect off large objects and smooth surfaces, such as buildings, roads, bodies of water, and even the earth's surface. In an indoor environment, microwaves reflect off smooth surfaces, such as doors, walls, and file cabinets. Anything made of metal will absolutely cause reflection. Other materials, such as glass and concrete, may cause reflection as well.

What Is the Impact of Reflection?

Reflection can be the cause of serious performance problems in a legacy 802.11a/b/g WLAN. As a wave radiates from an antenna, it broadens and disperses. If portions of this wave are reflected, new wave fronts will appear from the reflection points. If these multiple waves all reach the receiver, the multiple reflected signals cause an effect called *multipath*.

Multipath can degrade the strength and quality of the received signal or even cause data corruption or cancelled signals. (Further discussion of multipath occurs later in this chapter. Hardware solutions to compensate for the negative effects of multipath in this environment, such as directional antennas and antenna diversity, are discussed in Chapter 5, "Radio Frequency Signal and Antenna Concepts.")

Reflection and multipath were often considered primary enemies when deploying legacy 802.11a/b/g radios. 802.11n and 802.11ac radios utilize *multiple-input multiple-output (MIMO)* antennas and advanced digital signal processing (DSP) techniques to take advantage of multipath. MIMO technology is covered extensively in Chapter 10, "MIMO Technology: HT and VHT."

Scattering

Did you know that the color of the sky is blue because the molecules of the atmosphere are smaller than the wavelength of light? This blue sky phenomenon is known as *Rayleigh scattering* (named after the 19th-century British physicist John William Strutt, Lord Rayleigh). The shorter blue wavelength light is absorbed by the gases in the atmosphere and radiated in all directions. This is an example of an RF propagation behavior called *scattering*, sometimes called *scatter*.

Scattering can most easily be described as multiple reflections. These multiple reflections occur when the electromagnetic signal's wavelength is larger than pieces of whatever medium the signal is reflecting from or passing through.

Scattering can happen in two ways. The first type of scattering is on a lower level and has a lesser effect on the signal quality and strength. This type of scattering may manifest itself when the RF signal moves through a substance and the individual electromagnetic waves are reflected off the minute particles within the medium. Smog in our atmosphere and sandstorms in the desert can cause this type of scattering.

The second type of scattering occurs when an RF signal encounters some type of uneven surface and is reflected into multiple directions. Chain link fences, wire mesh in stucco walls or old plaster walls, tree foliage, and rocky terrain commonly cause this type of scattering. When striking the uneven surface, the main signal dissipates into multiple reflected signals, which can cause substantial signal downgrade and may even cause a loss of the received signal.

Figure 3.12 shows a flashlight being shined against a disco mirror ball. Note how the main signal beam is completely displaced into multiple reflected beams with less amplitude and into many different directions.

FIGURE 3.12   Scattering analogy

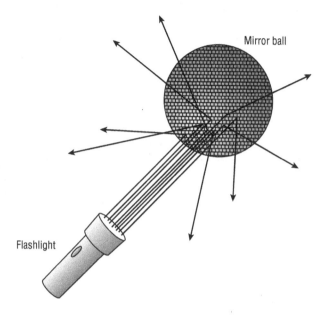

Mirror ball

Flashlight

Refraction

In addition to RF signals being absorbed or bounced (via reflection or scattering), if certain conditions exist, an RF signal can actually be bent, in a behavior known as *refraction*. A straightforward definition of refraction is the bending of an RF signal as it passes through a medium with a different density, thus causing the direction of the wave to change. RF refraction most commonly occurs as a result of atmospheric conditions.

> When you are dealing with long-distance outdoor bridge links, an instance of refractivity change that might be a concern is what is known as the *k-factor*. A k-factor of 1 means there is no bending. A k-factor of less than 1, such as 2/3, represents the signal bending away from the earth. A k-factor of more than 1 represents bending toward the earth. Normal atmospheric conditions have a k-factor of 4/3, which is bending slightly toward the curvature of the earth.

The three most common causes of refraction are water vapor, changes in air temperature, and changes in air pressure. In an outdoor environment, RF signals typically refract slightly back down toward the earth's surface. However, changes in the atmosphere may cause the signal to bend away from the earth. In long-distance outdoor wireless bridge links, refraction can be an issue. An RF signal may also refract through certain types of glass and other materials that are found in an indoor environment. Figure 3.13 shows two examples of refraction.

FIGURE 3.13 Refraction

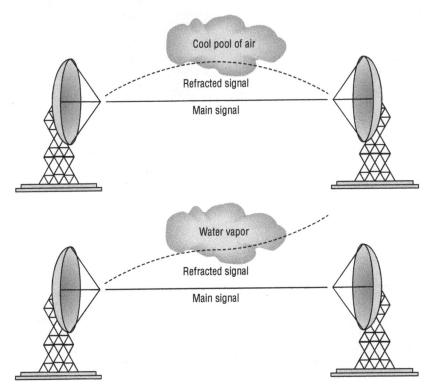

Diffraction

Not to be confused with refraction, another RF propagation behavior exists that also bends the RF signal; it is called *diffraction*. Diffraction is the bending of an RF signal around an object (whereas refraction, as you recall, is the bending of a signal as it passes through a medium). Diffraction is the bending and the spreading of an RF signal when it encounters an obstruction. The conditions that must be met for diffraction to occur depend entirely on the shape, size, and material of the obstructing object, as well as the exact characteristics of the RF signal, such as polarization, phase, and amplitude.

Typically, diffraction is caused by some sort of partial blockage of the RF signal, such as a small hill or a building that sits between a transmitting radio and a receiver. The waves that encounter the obstruction bend around the object, taking a longer and different path. The waves that did not encounter the object do not bend and maintain the shorter and original path. The analogy depicted in Figure 3.14 is a rock sitting in the middle of a river. Most of the current maintains the original flow; however, some of the current that encounters the rock will reflect off the rock and some will diffract around the rock.

FIGURE 3.14 Diffraction analogy

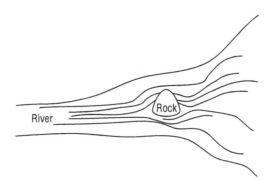

Sitting directly behind the obstruction is an area known as the *RF shadow*. Depending on the change in direction of the diffracted signals, the area of the RF shadow can become a dead zone of coverage or still possibly receive degraded signals. The concept of RF shadows is important when selecting antenna locations. Mounting an AP to a beam or other wall structure can create a virtual RF blind spot, in the same way that mounting a pole will create a shadow from a light mounted to it.

Loss (Attenuation)

Loss, also known as *attenuation*, is best described as the decrease of amplitude or signal strength. A signal may lose strength when transmitted on a wire or in the air. On the wired portion of the communications (RF cable), the AC electrical signal will lose strength because of the electrical impedance of the coaxial cabling and other components, such as connectors.

 In Chapter 5, we discuss impedance, which is the measurement of opposition to the AC current. You will also learn about impedance mismatches, which can create signal loss on the wired side.

After the RF signal is radiated into the air via the antenna, the signal will attenuate due to absorption, distance, or possibly the negative effects of multipath. You already know that as an RF signal passes through different mediums, the signal can be absorbed into the medium, which in turn causes a loss of amplitude. Different materials typically yield different attenuation results. A 2.4 GHz RF signal that passes through drywall will attenuate 3 decibels (dB) and lose half of the original amplitude. A 2.4 GHz signal that is absorbed through a concrete wall will attenuate 12 dB, which is 16 times less amplitude than the original signal. As discussed earlier, water is a major source of absorption as well as dense materials such as cinder blocks, all of which lead to attenuation.

Although the term "loss" may have a negative connotation, attenuation is not always undesirable. In Chapter 13, "WLAN Design Concepts," you will learn that using the attenuation characteristics of walls for signal isolation can actually be helpful. Using the natural RF characteristics of an indoor environment to achieve a better WLAN design is an important concept.

EXERCISE 3.1

Visual Demonstration of Absorption

In this exercise, you will use a program called EMANIM to view the attenuation effect of materials due to absorption. EMANIM is a free program found on the book's page at www.wiley.com/go/cwnasg. This is a special version of EMANIM that was developed specifically for this book and that contains an extra set of menu choices.

1. Download and install the EMANIM program by double-clicking emanim_setup.exe.

2. From the main EMANIM menu, click Phenomenon.

3. Click Sybex CWNA Study Guide.

4. Click Exercise E.

   When a radio wave crosses matter, the matter absorbs part of the wave. As a result, the amplitude of the wave decreases. The extinction coefficient determines how much of the wave is absorbed by unit length of material.

5. Vary the length of the material and the extinction coefficient for Wave 1 to see how it affects the absorption.

Both loss and gain can be gauged in a relative measurement of change in power called decibels (dB), which is discussed extensively in Chapter 4, "Radio Frequency Components, Measurements, and Mathematics." Table 3.1 shows different attenuation values for several materials.

TABLE 3.1   Attenuation comparison of materials

Material	2.4 GHz
Elevator shaft	–30 dB
Concrete wall	–12 dB
Wood door	–3 dB

Material	2.4 GHz
Nontinted glass windows	−3 dB
Drywall	−3 dB
Drywall (hollow)	−2 dB
Cubicle wall	−1 dB

Table 3.1 is meant as a reference chart and is not information that will be covered on the CWNA exam. Actual measurements may vary from site to site, depending on specific environmental factors.

It is important to understand that an RF signal will also lose amplitude merely as a function of distance due to free space path loss. Also, reflection propagation behaviors can produce the negative effects of multipath and, as a result, cause attenuation in signal strength.

Free Space Path Loss

Because of the laws of physics, an electromagnetic signal will attenuate as it travels, despite the lack of attenuation caused by obstructions, absorption, reflection, diffraction, and so on. *Free space path loss (FSPL)* is the loss of signal strength caused by the natural broadening of the waves, often referred to as *beam divergence*. RF signal energy spreads over larger areas as the signal travels farther away from an antenna, and as a result, the strength of the signal attenuates.

One way to illustrate free space path loss is to use a balloon analogy. Before a balloon is filled with air, it remains small but has a dense rubber thickness. After the balloon is inflated and has grown and spread in size, the rubber becomes very thin. RF signals lose strength in much the same manner. Luckily, this loss in signal strength is logarithmic and not linear; thus, the amplitude does not decrease as much in a second segment of equal length as it decreases in the first segment. A 2.4 GHz signal will change in power by about 80 dB after 100 meters but will lessen only another 6 dB in the next 100 meters.

Here are the formulas to calculate free space path loss:

$$FSPL = 36.6 + 20\log_{10}(f) + 20\log_{10}(D)$$

FSPL = path loss in dB

f = frequency in MHz

D = distance in miles between antennas

$$FSPL = 32.44 + 20\log_{10}(f) + 20\log_{10}(D)$$

FSPL = path loss in dB

f = frequency in MHz

D = distance in kilometers between antennas

 NOTE Free space path loss formulas are provided as a reference and are not included on the CWNA exam. Many online calculators for FSPL and other RF calculators can also be found with a simple web search.

An even simpler way to estimate free space path loss (FSPL) is called the *6 dB rule*. (Remember for now that decibels are a measure of gain or loss; further details of dB are covered extensively in Chapter 4.) The 6 dB rule states that doubling the distance will result in a loss of amplitude of 6 dB. Table 3.2 shows estimated path loss and confirms the 6 dB rule.

TABLE 3.2 Attenuation due to free space path loss

Distance (meters)	Attenuation (dB)	
	2.4 GHz	**5 GHz**
1	40	46.4
10	60	66.4
100	80	86.4
1000	100.0	106.4
2000	106.1	112.4
4000	112.1	118.5
8000	118.1	124.5

🌐 Real World Scenario

Why Is Free Space Path Loss Important?

All radio devices have what is known as a *receive sensitivity level*. The radio receiver can properly interpret and receive a signal down to a certain fixed amplitude threshold. If a radio receives a signal above its amplitude threshold, the signal is powerful enough for the radio to sense and interpret the signal. For example, if you were to whisper a secret to someone, you'd need to make sure that you whisper loud enough for them to hear and understand it.

If the amplitude of a received signal is below the radio's receive sensitivity threshold, the radio can no longer properly sense and interpret the signal. The concept of free space path loss also applies to road trips in your car. When you are in a car listening to an AM radio station, eventually you will drive out of range and the radio will no longer be able to receive and process the music.

In addition to the radio being able to receive and interpret a signal, the received signal must be not only strong enough to be heard but also strong enough to be heard above any RF background noise, typically referred to as the *noise floor*. The signal must be louder than any background noise. In the example of whispering a secret to someone, if you were whispering the secret while an ambulance was driving past with the siren blasting, even though you were whispering loud enough for the person to hear you, the noise from the siren would be too loud for the person to distinguish what you were saying.

When designing both indoor WLANs and outdoor wireless bridge links, you must make sure that the RF signal will not attenuate below the receive sensitivity level of your WLAN radio simply because of free space path loss, and you must make sure that the signal does not attenuate near or below the noise floor. You typically achieve this goal indoors by performing a site survey. An outdoor bridge link requires a series of calculations called a *link budget*. (Site surveys are covered in Chapter 14, "Site Survey and Validation" and link budgets are covered in Chapter 4.)

Multipath

Multipath is a propagation phenomenon that results in two or more paths of a signal arriving at a receiving antenna at the same time or within nanoseconds of each other. Because of the natural broadening of the waves, the propagation behaviors of reflection, scattering, diffraction, and refraction will occur differently in dissimilar environments. When a signal encounters an object, it may reflect, scatter, refract, or diffract. These propagation behaviors can all result in multiple paths of the same signal.

In an indoor environment, reflected signals and echoes can be caused by long hallways, walls, desks, floors, file cabinets, and numerous other obstructions. Indoor environments with large amounts of metal surfaces—such as airplane hangars, warehouses, and factories—are notoriously high-multipath environments because of all the reflective surfaces. The propagation behavior of reflection is typically the main cause of high-multipath environments.

In an outdoor environment, multipath can be caused by a flat road, a large body of water, a building, or atmospheric conditions. Therefore, we have signals bouncing and bending in many different directions. The principal signal will still travel to the receiving antenna, but some of the bouncing and bent signals may also find their way to the receiving antenna via different paths. In other words, multiple paths of the RF signal arrive at the receiver, as shown in Figure 3.15.

FIGURE 3.15 Multipath

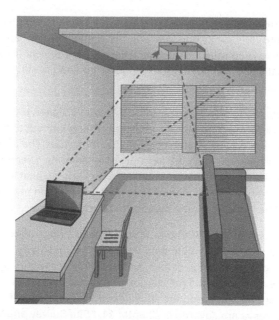

It usually takes a bit longer for reflected signals to arrive at the receiving antenna because they must travel a longer distance than the principal signal. The time differential between these signals can be measured in billionths of a second (nanoseconds). The time differential between these multiple paths is known as the *delay spread*. You will learn later in this book that certain spread spectrum technologies are more tolerant than others when it comes to delay spread.

So, what exactly happens when multipath presents itself? In the days of legacy analog television signal transmissions, multipath caused a visible ghost effect with a faded duplicate to the right of the main image. With modern digital TV transmissions, multipath can manifest itself as pixilation, freezes, or in the worst cases, total loss of picture from data corruption. With RF signals, the effects of multipath can be either constructive or destructive. Quite often they are destructive. Because of the differences in phase of the multiple paths, the combined signal will often attenuate, amplify, or become corrupted. These effects are sometimes called *Rayleigh fading*, another phenomenon named after British physicist Lord Rayleigh.

The four possible results of multipath are as follows:

Upfade   This is increased signal strength. When the multiple RF signal paths arrive at the receiver at the same time and are in phase or partially out of phase with the primary wave, the result is an increase in signal strength (amplitude). Smaller phase differences of between

0 and 120 degrees will cause *upfade*. Please understand, however, that the final received signal can never be stronger than the original transmitted signal because of free space path loss. Upfade is an example of constructive multipath.

Downfade   This is decreased signal strength. When the multiple RF signal paths arrive at the receiver at the same time and are out of phase with the primary wave, the result is a decrease in signal strength (amplitude). Phase differences of between 121 and 179 degrees will cause *downfade*. Decreased amplitude as a result of multipath would be considered destructive multipath.

Nulling   This is signal cancellation. When the multiple RF signal paths arrive at the receiver at the same time and are 180 degrees out of phase with the primary wave, the result will be *nulling*. Nulling is the complete cancellation of the RF signal. A complete cancellation of the signal is obviously destructive.

Data Corruption   Because of the difference in time between the primary signal and the reflected signals (known as the delay spread), along with the fact that there may be multiple reflected signals, the receiver can have problems demodulating the RF signal's information. The delay spread time differential can cause bits to overlap with each other, and the end result is corrupted data. This type of multipath interference is often known as *intersymbol interference (ISI)*. Data corruption is the most common occurrence of destructive multipath.

The bad news is that high-multipath environments can result in data corruption because of intersymbol interference caused by the delay spread. The good news is that the receiving station will detect the errors through an 802.11-defined cyclic redundancy check (CRC) because the checksum will not calculate accurately. The 802.11 standard requires that most unicast frames be acknowledged by the receiving station with an acknowledgment (ACK) frame; otherwise, the transmitting station will have to retransmit the frame. The receiver will not acknowledge a frame that has failed the CRC. Therefore, unfortunately, the frame must be retransmitted, but this is better than it being misinterpreted.

Layer 2 retransmissions negatively affect the overall throughput of any 802.11 WLAN and can also affect the delivery of time-sensitive packets of applications, such as VoIP. In Chapter 15, "WLAN Troubleshooting," we discuss the multiple causes of layer 2 retransmissions and how to troubleshoot and minimize them. Multipath is one of the main causes of layer 2 retransmissions that negatively affect the throughput and latency of a legacy 802.11a/b/g WLAN.

So, how is a hapless WLAN engineer supposed to deal with destructive multipath issues? Multipath can be a serious problem when working with legacy 802.11a/b/g equipment. The use of directional antennas will often reduce the number of reflections, and antenna diversity can also be used to compensate for the negative effects of multipath. Sometimes, reducing transmit power or using a lower-gain antenna can solve the problem as long as there is enough signal to provide connectivity to the remote end. In this chapter, we have mainly focused on the destructive effects that multipath has on legacy 802.11a/b/g radio transmissions. Multipath has a constructive effect with 802.11n and 802.11ac radio transmissions that utilize multiple-input multiple-output (MIMO) antenna diversity and *maximal ratio combining (MRC)* signal-processing techniques.

In the past, data corruption of legacy 802.11a/b/g transmissions caused by multipath had to be dealt with, and using unidirectional antennas to cut down on reflections was commonplace in high-multipath indoor environments. Now that MIMO technology used by 802.11n and 802.11ac radios is commonplace, multipath is now our friend, and using unidirectional antennas is rarely needed indoors. However, unidirectional MIMO patch antennas can still be used indoors to provide sectored coverage in high-density user environments.

EXERCISE 3.2

Visual Demonstration of Multipath and Phase

In this exercise, you will use a program called EMANIM to view the effect on amplitude due to various phases of two signals arriving at the same time.

1. From the book's page at www.wiley.com/go/cwnasg, download and install the EMANIM program by double-clicking emanim_setup.exe.

2. From the main EMANIM menu, click Phenomenon.

3. Click Sybex CWNA Study Guide.

4. Click Exercise A.

    Two identical, vertically polarized waves are superposed. (You might not see both of them because they cover each other.) The result is a wave having double the amplitude of the component waves.

5. Click Exercise B.

    Two identical, 70-degree out-of-phase waves are superposed. The result is a wave with an increased amplitude over the component waves.

6. Click Exercise C.

    Two identical, 140-degree out-of-phase waves are superposed. The result is a wave with a decreased amplitude over the component waves.

7. Click on Exercise D.

    Two identical, vertically polarized waves are superposed. The result is a cancellation of the two waves.

Gain (Amplification)

Gain, also known as *amplification*, can best be described as the increase of amplitude, or signal strength. The two types of gain are known as active gain and passive gain. A signal's amplitude can be boosted by the use of external devices.

Active gain is usually caused by the transceiver or the use of an amplifier on the wire that connects the transceiver to the antenna. Many transceivers are capable of transmitting at different power levels, with the higher power levels creating a stronger or amplified signal. An amplifier is usually bidirectional, meaning that it increases the AC voltage both inbound and outbound. Active gain devices require the use of an external power source.

Passive gain is accomplished by focusing the RF signal with the use of an antenna. Antennas are passive devices that do not require an external power source. Instead, the internal workings of an antenna focus the signal more powerfully in one direction than another. An increase in signal amplitude is the result of either active gain prior to the signal reaching the antenna or passive gain focusing the signal radiating from the antenna.

 The proper use of antennas is covered extensively in Chapter 5.

Two very different tools can be used to measure the amplitude of a signal at a given point. The first, a frequency-domain tool, can be used to measure amplitude in a finite frequency spectrum. The frequency-domain tool used by WLAN engineers is called a *spectrum analyzer.* The second tool, a time-domain tool, can be used to measure how a signal's amplitude changes over time. The conventional name for a time-domain tool is an *oscilloscope.* Figure 3.16 shows how both of these tools can be used to display amplitude. It should be noted that spectrum analyzers are often used by WLAN engineers during site surveys. An oscilloscope is rarely if ever used when deploying a WLAN; however, oscilloscopes are used by RF engineers in laboratory test environments.

FIGURE 3.16   RF signal measurement tools

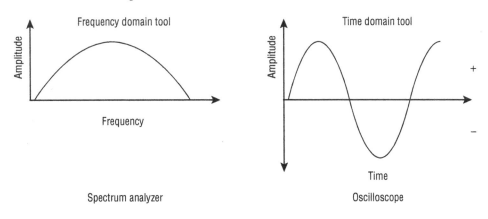

Summary

This chapter covered the meat and potatoes, the basics, of radio frequency signals. To properly design and administer a WLAN network, it is essential to have a thorough understanding of the following principles of RF properties and RF behaviors:

- Electromagnetic waves and how they are generated
- The relationship between wavelength, frequency, and the speed of light
- Signal strength and the various ways in which a signal can either attenuate or amplify
- The importance of the relationship between two or more signals
- How a signal moves by bending, bouncing, or absorbing in some manner

When troubleshooting an Ethernet network, the best place to start is always at layer 1, the Physical layer. WLAN troubleshooting should also begin at the Physical layer. Learning the RF fundamentals that exist at layer 1 is an essential step in proper wireless network administration.

Exam Essentials

Understand wavelength, frequency, amplitude, and phase.   Know the definition of each RF characteristic and how each can affect WLAN design.

Remember all the RF propagation behaviors.   Be able to explain the differences between each RF behavior (such as reflection, diffraction, scattering, and so on) and the various mediums that are associated with each behavior.

Understand what causes attenuation.   Loss can occur either on the wire or in the air. Absorption, free space path loss, and multipath downfade are all causes of attenuation.

Define free space path loss.   Despite the lack of any obstructions, electromagnetic waves attenuate in a logarithmic manner as they travel away from the transmitter.

Remember the four possible results of multipath and their relationship to phase. Multipath may cause downfade, upfade, nulling, and data corruption. Understand that the effects of multipath can be either destructive or constructive.

Know the results of intersymbol interference and delay spread.   The time differential between a primary signal and reflected signals may cause corrupted bits and affect throughput and latency due to layer 2 retransmissions.

Explain the difference between active gain and passive gain.   Transceivers and RF amplifiers are active devices, whereas antennas are passive devices.

Explain the difference between transmit amplitude and received amplitude.   Transmit amplitude is typically defined as the amount of initial amplitude that leaves the radio transmitter. When a radio receives an RF signal, the received signal strength is most often referred to as received amplitude.

Review Questions

1. What are some results of multipath interference? (Choose all that apply.)

    A. Scattering delay

    B. Upfade

    C. Excessive retransmissions

    D. Absorption

2. What term best defines the linear distance traveled in one positive-to-negative-to-positive oscillation of an electromagnetic signal?

    A. Crest

    B. Frequency

    C. Trough

    D. Wavelength

3. Which of the following statements are true about amplification? (Choose all that apply.)

    A. All antennas require an outside power source.

    B. RF amplifiers require an outside power source.

    C. Antennas are passive gain amplifiers that focus the energy of a signal.

    D. RF amplifiers passively increase signal strength by focusing the AC current of the signal.

4. A standard measurement of frequency is called what?

    A. Hertz

    B. Milliwatt

    C. Nanosecond

    D. Decibel

    E. K-factor

5. When an RF signal bends around an object, this propagation behavior is known as what?

    A. Stratification

    B. Refraction

    C. Scattering

    D. Diffraction

    E. Attenuation

6. When the multiple RF signals arrive at a receiver at the same time and are _____ with the primary wave, the result can be _____ of the primary signal.

    A. out of phase, scattering

    B. in phase, intersymbol interference

    C. in phase, attenuation

    D. 180 degrees out of phase, amplification

    E. in phase, cancellation

    F. 180 degrees out of phase, cancellation

7. Which of the following statements are true? (Choose all that apply.)

    A. When upfade occurs, the final received signal will be stronger than the original transmitted signal.

    B. When downfade occurs, the final received signal will never be stronger than the original transmitted signal.

    C. When upfade occurs, the final received signal will never be stronger than the original transmitted signal.

    D. When downfade occurs, the final received signal will be stronger than the original transmitted signal.

8. What is the frequency of an RF signal that cycles 2.4 million times per second?

    A. 2.4 hertz

    B. 2.4 MHz

    C. 2.4 GHz

    D. 2.4 kilohertz

    E. 2.4 KHz

9. What is the best example of a time-domain tool that could be used by an RF engineer?

    A. Oscilloscope

    B. Spectroscope

    C. Spectrum analyzer

    D. Refractivity gastroscope

10. What are some objects or materials that are common causes of reflection? (Choose all that apply.)

    A. Metal

    B. Trees

    C. Asphalt road

    D. Lake

    E. Carpet floors

11. Which of these propagation behaviors can result in multipath? (Choose all that apply.)

    A. Refraction

    B. Diffraction

    C. Reflection

    D. Scattering

    E. None of the above

12. Which behavior can be described as an RF signal encountering a chain link fence, causing the signal to bounce into multiple directions?

A. Diffraction

B. Scattering

C. Reflection

D. Refraction

E. Multiplexing

13. Which 802.11 radio technologies are most impacted by the destructive effects of multipath? (Choose all that apply.)

A. 802.11a

B. 802.11b

C. 802.11g

D. 802.11n

E. 802.11ac

F. 802.11i

14. Which of the following can cause refraction of an RF signal traveling through it? (Choose all that apply.)

A. Shift in air temperature

B. Change in air pressure

C. Humidity

D. Smog

E. Wind

F. Lightning

15. Which of the following statements about free space path loss are true? (Choose all that apply.)

A. RF signals will attenuate as they travel, despite the lack of attenuation caused by obstructions.

B. Path loss occurs at a constant linear rate.

C. Attenuation is caused by obstructions.

D. Path loss occurs at a logarithmic rate.

16. What term is used to describe the time differential between a primary signal and a reflected signal arriving at a receiver?

A. Path delay

B. Spread spectrum

C. Multipath

D. Delay spread

17. What is an example of a frequency-domain tool that could be used by an RF engineer?

    A. Oscilloscope

    B. Spectroscope

    C. Spectrum analyzer

    D. Refractivity gastroscope

18. Using knowledge of RF characteristics and behaviors, which two options should a WLAN engineer be most concerned about when doing an indoor site survey? (Choose the best two answers.)

    A. Concrete walls

    B. Indoor temperature

    C. Wood-lath plaster walls

    D. Drywall

19. Which three properties are interrelated?

    A. Frequency, wavelength, and the speed of light

    B. Frequency, amplitude, and the speed of light

    C. Frequency, phase, and amplitude

    D. Amplitude, phase, and the speed of sound

20. Which RF behavior best describes a signal striking a medium and bending in a different direction?

    A. Refraction

    B. Scattering

    C. Diffusion

    D. Diffraction

    E. Microwave reflection

Chapter

4

Radio Frequency Components, Measurements, and Mathematics

IN THIS CHAPTER, YOU WILL LEARN ABOUT THE FOLLOWING:

✓ **Components of RF communications**

- Transmitter
- Antenna
- Receiver
- Intentional radiator
- Equivalent isotropically radiated power (EIRP)

✓ **Units of power and comparison**

- Watt (W)
- Milliwatt (mW)
- Decibel (dB)
- Decibels relative to an isotropic radiator (dBi)
- Decibels relative to a half-wave dipole antenna (dBd)
- Decibels relative to 1 milliwatt (dBm)
- Inverse square law

✓ **RF mathematics**

  ■ Rule of 10s and 3s

✓ **Noise floor**

✓ **Signal-to-noise ratio (SNR)**

✓ **Signal-to-interference-plus-noise ratio (SINR)**

✓ **Received signal strength indicator (RSSI)**

✓ **Link budget**

✓ **Fade margin/system operating margin**

To put it simply, data communication is the transferring of information between computers. No matter what form of communication is being used, many components are required to achieve a successful transfer. Before examining individual components, let us initially keep things simple and look at the three basic requirements for successful communications:

- Two or more devices want to communicate.

- There must be a medium, a means, or a method for them to use to communicate.

- There must be a set of rules for them to use when they communicate. (This is covered in Chapter 8, "802.11 Medium Access.")

These three basic requirements are the same for all forms of communication, whether a group of people are having a conversation at a dinner party, two computers are transmitting data via a dial-up modem, or many computers are communicating via a wireless network.

The existence of a computer network essentially implies that the first requirement is met. If we did not have two or more devices that wanted to share data, we would not need to create the network in the first place. The CWNA certification program also assumes this and is therefore rarely if ever concerned specifically with the data itself. It is assumed that we have data; our concern is to transmit and receive it.

This chapter focuses on the second requirement: the medium, means, or method of communication. We cover the components of radio frequency (RF), which make up what we refer to as the medium for wireless communications. Here we are concerned with the transmission of the RF signal and the role of each device and component along the transmission path. We also show how each device or component affects the transmission.

In Chapter 3, "Radio Frequency Fundamentals," you learned that there are many RF behaviors that affect the signal as it leaves the transmitter and travels toward the receiver. As the signal moves through the various radio system components and then propagates through the air, the signal's amplitude changes. Some components increase the power of the signal (gain), whereas other components decrease the power (loss). In this chapter, you will learn how to quantify and measure the power of the waves and calculate how the waves are affected by both internal and external influences. Using these calculations, you will be able to determine whether you will have the means to communicate between devices.

RF Components

Many components contribute to the successful transmission and reception of an RF signal. Figure 4.1 shows the key components that are covered in the following sections. In addition to knowing the function of the components, it is important to understand how the strength of the signal is specifically affected by each of the components.

FIGURE 4.1   RF components

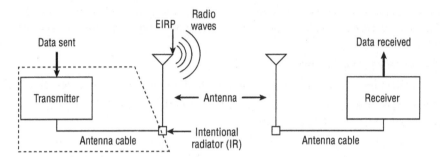

Later in this chapter, when we discuss RF mathematics, we will show you how to calculate the effect that each of the components has on the signal.

Transmitter

The *transmitter* is the initial component in the creation of the wireless medium. The computer hands off the data to the transmitter, and it is the transmitter's job to begin the RF communication.

In Chapter 1, "Overview of Wireless Standards, Organizations, and Fundamentals," you learned about carrier signals and modulation methods. When the transmitter receives the data, it begins generating an alternating current (AC) signal. This AC signal determines the frequency of the transmission. For example, for a 2.4 GHz signal, the AC signal oscillates around 2.4 billion times per second, whereas for a 5 GHz signal, the AC signal oscillates around 5 billion times per second. This oscillation determines the frequency of the radio wave.

    The exact frequencies used are covered in Chapter 6, "Wireless Networks and Spread Spectrum Technologies."

The transmitter takes the data provided and modifies the AC signal by using a modulation technique to encode the data into the signal. This modulated AC signal is now a carrier signal, containing (or carrying) the data to be transmitted. The carrier signal is then transported either directly to the antenna or through a cable to the antenna.

In addition to generating a signal at a specific frequency, the transmitter is responsible for determining the original transmission amplitude, or what is more commonly referred to as the *power level*, of the transmitter. The higher the amplitude of the wave, the more powerful the wave is and the farther it can be received. The power levels that the transmitter is allowed to generate are determined by the local regulatory domain authorities, such as the Federal Communications Commission (FCC) in the United States.

Although we are explaining the transmitter and receiver separately in this chapter, and although functionally they are different components, typically they are one device that is referred to as a *transceiver* (transmitter/receiver). Typical wireless devices that have transceivers built into them are access points, bridges, and client adapters.

Antenna

An *antenna* provides two functions in a communication system. When connected to the transmitter, it collects the AC signal that it receives from the transmitter and directs, or radiates, the RF waves away from the antenna in a pattern specific to the antenna type. When connected to the receiver, the antenna takes the RF waves that it receives through the air and directs the AC signal to the receiver. The receiver converts the AC signal to bits and bytes. As you will see later in this chapter, the signal that is received is much less than the signal that is generated. This signal loss is analogous to two people trying to talk to each other from opposite ends of a football field. Because of distance alone (free space), the yelling from one end of the field may be heard as barely louder than a whisper on the other end.

The RF transmission of an antenna is usually compared or referenced to an isotropic radiator. An *isotropic radiator* is a *point source* that radiates signal equally in all directions. The sun is probably one of the best examples of an isotropic radiator. It generates equal amounts of energy in all directions. Unfortunately, it is not possible to manufacture an antenna that is a perfect isotropic radiator. The structure of the antenna itself influences the output of the antenna, similar to the way the structure of a light bulb affects the bulb's ability to emit light equally in all directions.

There are two ways to increase the power output from an antenna. The first is to generate more power at the transmitter, as stated in the previous section. The other is to direct, or focus, the RF signal that is radiating from the antenna. This is similar to how you can focus light from a flashlight. If you remove the lens from the flashlight, the bulb is typically not very bright and radiates in almost all directions. To make the light brighter, you could use more powerful batteries, or you could put the lens back on. The lens is not actually creating more light; it is merely focusing the light that was radiating in all different directions into a narrow area. Some antennas radiate waves similar to how the bulb without the lens does, whereas others radiate focused waves similar to how the flashlight with the lens does.

In Chapter 5, "Radio Frequency Signal and Antenna Concepts," you will learn about the types of antennas and how to properly and most effectively use them.

Receiver

The *receiver* is the final component in the wireless medium. The receiver takes the carrier signal that is received from the antenna and translates the modulated signals into 1s and 0s. It then takes this data and passes it to the computer to be processed. The job of the receiver is not always an easy one. The signal that is received is a much less powerful signal than what was transmitted because of the distance it has traveled and the effects of free space path loss (FSPL). The signal is also often unintentionally altered due to interference from other RF sources and multipath.

Intentional Radiator

The FCC Code of Federal Regulations (CFR) Part 15 defines an *intentional radiator (IR)* as "a device that intentionally generates and emits radio frequency energy by radiation or induction." Basically, it is something that is specifically designed to generate RF, as opposed to something that generates RF as a by-product of its main function, such as a motor that incidentally generates RF noise.

Regulatory bodies such as the FCC limit the amount of power that is allowed to be generated by an IR. The IR consists of all the components from the transmitter to the antenna but not including the antenna, as shown in Figure 4.1. The power output of the IR is thus the sum of all the components from the transmitter to the antenna (again not including the antenna). The components making up the IR include the transmitter, all cables and connectors, and any other equipment (grounding connectors, lightning arrestors, amplifiers, attenuators, and so forth) between the transmitter and the antenna. The power of the IR is measured at the connector that provides the input to the antenna. Because this is the point where the IR is measured and regulated, we often refer to this point alone as the IR. This power level is typically measured in milliwatts (mW) or decibels relative to 1 milliwatt (dBm). Using the flashlight analogy, the IR is all the components up to the lightbulb socket but not the bulb and lens. This is the raw power, or signal, and now the bulb and lens can focus the signal.

Equivalent Isotropically Radiated Power

Equivalent isotropically radiated power (EIRP) is the highest RF signal strength that is transmitted from a particular antenna. To understand this better, consider our flashlight example. Let us assume that the bulb without the lens generates 1 watt of power. When you put the lens on the flashlight, it focuses that 1 watt of light. If you were to look at the light now, it would appear much brighter. If you were to measure the brightest point of the light that was being generated by the flashlight, because of the effects of the lens it may be equal to the brightness of an 8-watt bulb. So by focusing the light, you are able to make the equivalent isotropically radiated power of the focused bulb equal to 8 watts.

It is important for you to know that you can find other references to EIRP as *equivalent isotropic radiated power* and *effective isotropic radiated power*. The use of EIRP in this book is consistent with the FCC definition, "equivalent isotropically radiated power, the product of the power supplied to the antenna and the antenna gain in a given direction relative to an isotropic antenna." Even though the terms that the initials stand for at times may differ, the definition of EIRP is consistent.

As you learned earlier in this chapter, antennas are capable of focusing, or directing, RF energy. This focusing capability can make the effective output of the antenna much greater than the signal entering the antenna. Because of this ability to amplify the output of the RF signal, regulatory bodies such as the FCC limit the amount of EIRP from an antenna.

In the next section of this chapter, you will learn how to calculate how much power is being provided to the antenna (IR) and how much power is coming out of the antenna (EIRP).

 Real World Scenario

Why Are IR and EIRP Measurements Important?

As you learned in Chapter 1, the regulatory domain authority in an individual country or region is responsible for maximum transmit power regulations. The FCC and other regulatory domain authorities usually define maximum power output for the intentional radiator (IR) and a maximum equivalent isotropically radiated power (EIRP) that radiates from the antenna. In laymen's terms, the FCC regulates the maximum amount of power that goes into an antenna and the maximum amount of power that comes out of an antenna.

You will need to know the definitions of IR and EIRP measurements. However, the CWNA exam (CWNA-107) will not test you on any power regulations, because they vary from country to country. It is advisable to educate yourself about the maximum transmit power regulations of the country where you plan on deploying a WLAN so that no violations occur. The transmit power of most indoor WLAN radios varies in a range between 1 mW and 100 mW. Therefore, you usually do not need to concern yourself with power regulations when deploying indoor WLAN equipment. However, knowledge of power regulations is important for outdoor WLAN deployments.

Units of Power and Comparison

When an 802.11 wireless network is designed, two key components are coverage and performance. A good understanding of RF power, comparison, and RF mathematics can be very helpful during the network design phase.

In the following sections, we will introduce you to an assortment of *units of power* and *units of comparison*. It is important to know and understand the various types of units of measurement and how they relate to each other. Some of the numbers that you will be working with will represent actual units of power, and others will represent relative units of comparison. Actual units of power are ones that represent a known or set value.

To say that a man is 6 feet tall is an example of an actual measurement. Since the man's height is a known value, in this case 6 feet, you know exactly how tall he is. Relative units are comparative values comparing one item to a similar type of item. For example, if you wanted to tell someone how tall the man's wife is by using comparative units of measurement, you could say that she is five-sixths his height. You now have a comparative measurement: If you know the actual height of either one, you can then determine how tall the other is.

Comparative units of measurement are useful when working with units of power. As you will see later in this chapter, we can use these comparative units of power to compare the area that one access point can cover vs. another access point. Using simple mathematics, we can determine things such as how many watts are needed to double the distance of a signal from an access point.

Units of power are used to measure transmission amplitude and received amplitude. In other words, units of transmit or received power measurements are *absolute power* measurements. Units of comparison are often used to measure how much gain or loss occurs because of the introduction of cabling or an antenna. Units of comparison are also used to represent a difference in power from point A to point B. In other words, units of comparison are measurements of a *change in power*.

Here is a list of the units of power, followed by another list of the units of comparison, all of which are covered in the following sections:

Units of power (absolute)

- watt (W)
- milliwatt (mW)
- decibels relative to 1 milliwatt (dBm)

Units of comparison (relative)

- decibel (dB)
- decibels relative to an isotropic radiator (dBi)
- decibels relative to a half-wave dipole antenna (dBd)

Watt

A *watt (W)* is the basic unit of power, named after James Watt, an 18th-century Scottish inventor. One watt is equal to 1 ampere (amp) of current flowing at 1 volt. To give a better explanation of a watt, we will use a modification of the classic water analogy.

Many of you are probably familiar with a piece of equipment known as a power washer. If you are not familiar with it, it is a machine that connects to a water source, such as a garden hose, and enables you to direct a stream of high-pressure water at an object, with the premise that the fast-moving water will clean the object. The success of a power washer is based on two components: the pressure applied to the water and the volume of water used over a period of time, also known as flow. These two components provide the power of the water stream. If you increase the pressure, you will increase the power of the stream. If you increase the flow of the water, you will also increase the power of the stream. The power of the stream is equal to the pressure times the flow.

A watt is very similar to the output of the power washer. Instead of the pressure generated by the machine, electrical systems have voltage. Instead of water flow, electrical systems have current, which is measured in amps. So the amount of watts generated is equal to the volts times the amps.

Milliwatt

A *milliwatt (mW)* is also a unit of power. To put it simply, a milliwatt is 1/1,000 of a watt. The reason you need to be concerned with milliwatts is because most of the indoor 802.11 equipment that you will be using transmits at power levels between 1 mW and 100 mW. Remember that the transmit power level of a radio will be attenuated by any cabling and will be amplified by the antenna. Although regulatory bodies such as the FCC may allow intentional radiator (IR) power output of as much as 1 watt, only rarely in point-to-point communications, such as in building-to-building bridge links, would you use 802.11 equipment with more than 300 mW of transmit power.

 Real World Scenario

What Do a Wi-Fi Vendor's Transmit Power Settings Represent?

All Wi-Fi vendors offer the capability to adjust the transmit power settings of an access point. A typical AP radio will usually have transmit power capabilities of 1 mW to 100 mW. However, not every Wi-Fi vendor will represent transmit power values the same way. The transmit power settings of most vendors represent the IR, whereas the transmit power settings of other vendors might actually be the EIRP instead. Furthermore, Wi-Fi vendors might also indicate the transmit amplitude in either mW or dBm—for example, 32 mW or +15 dBm—yet some might simply indicate transmit power in the form of a percentage value, such as 32 percent. You will need to refer to your specific Wi-Fi vendor's deployment guide to fully understand the transmit amplitude value.

Decibel

The first thing you should know about the *decibel (dB)* is that it is a unit of comparison, not a unit of power. Therefore, it is used to represent a difference between two values. In other words, a dB is a relative expression and a measurement of change in power. In wireless networking, decibels are often used either to compare the power of two transmitters or, more often, to compare the difference or loss between the EIRP output of a transmitter's antenna and the amount of power received by the receiver's antenna.

Decibel is derived from the term *bel*. Employees at Bell Telephone Laboratories needed a way to represent power losses on telephone lines as power ratios. They defined a bel as the ratio of 10 to 1 between the power of two sounds. Let us look at an example: An access point transmits data at 100 mW. Laptop1 receives the signal from the AP at a power level of 10 mW, and laptop2 receives the signal from the AP at a power level of 1 mW. The difference between the signal from the access point (100 mW) to laptop1 (10 mW) is 100:10, or a 10:1 ratio, or 1 bel. The difference between the signal from laptop1 (10 mW) to laptop2 (1 mW) is also a 10:1 ratio, or 1 bel. So the power difference between the access point and laptop2 is 2 bels.

Bels can be looked at mathematically by using logarithms. Not everyone understands or remembers logarithms, so we will review them. First, we need to look at raising a number to a power. If you take 10 and raise it to the third power ($10^3 = y$), what you are actually doing is multiplying three 10s ($10 \times 10 \times 10$). If you do the math, you will calculate that y is equal to 1,000. So the solution is $10^3 = 1,000$. When calculating logarithms, you change the formula to $10^y = 1,000$. Here you are trying to figure out what power 10 needs to be raised to in order to get to 1,000. You know in this example that the answer is 3. You can also write this equation as $y = \log_{10}(1,000)$ or $y = \log_{10}1,000$. So the complete equation is $3 = \log_{10}(1,000)$. Here are some examples of power and log formulas:

$10^1 = 10$	$\log_{10}(10) = 1$
$10^2 = 100$	$\log_{10}(100) = 2$
$10^3 = 1,000$	$\log_{10}(1,000) = 3$
$10^4 = 10,000$	$\log_{10}(10,000) = 4$

Now let us go back and calculate the bels from the access point to the laptop2 example by using logarithms. Remember that bels are used to calculate the ratio between two powers. Let us refer to the power of the access point as P_{AP} and the power of laptop2 as P_{L2}. So the formula for this example would be $y = \log_{10}(P_{AP}/P_{L2})$. If you plug in the power values, the formula becomes $y = \log_{10}(100/1)$, or $y = \log_{10}(100)$. So this equation is asking, 10 raised to what power equals 100? The answer is 2 bels ($10^2 = 100$).

Okay, this is supposed to be a section about decibels, but so far we have covered just bels. In certain environments, bels are not exact enough, which is why we use decibels instead. A decibel is equal to 1/10 of a bel. To calculate decibels, all you need to do is multiply bels by 10. So the formulas for bels and decibels are as follows:

$$\text{bels} = \log_{10}(P_1/P_2)$$

$$\text{decibels} = 10 \times \log_{10}(P_1/P_2)$$

Now let us go back and calculate the decibels for the example of the access point to laptop2. The formula now is $y = 10 \times \log_{10}(P_{AP}/P_{L2})$. If you plug in the power values, the formula becomes $y = 10 \times \log_{10}(100/1)$, or $y = 10 \times \log_{10}(100)$. So the answer is +20 decibels. +20 decibels is the equivalent of +2 bels.

> You do not need to know how to calculate logarithms for the CWNA exam. These examples are here only to give you some basic understanding of what they are and how to calculate them. Later in this chapter, you will learn how to calculate decibels without using logarithms.

Now that you have learned about decibels, you are probably still wondering why you cannot just work with milliwatts. You can if you want, but because power changes are calculated using logarithmic formulas, the differences between values can become extremely large and more difficult to deal with. It is easier to say that a 100 mW signal decreased by 70 decibels than to say that it decreased to 0.00001 milliwatts. Because of the scale of the numbers, you can see why decibels can be easier to work with.

 Real World Scenario

Why Should You Use Decibels?

As you learned in Chapter 3, many behaviors can adversely affect a wave. One of the behaviors that you learned about was free space path loss.

If a 2.4 GHz access point is transmitting at 100 mW, and a laptop is 100 meters (0.1 kilometer) away from the access point, the laptop is receiving only about 0.000001 milliwatts of power. The difference between the numbers 100 and 0.000001 is so large that it does not have much relevance to someone looking at it. Additionally, it would be easy for someone to accidentally leave out a zero when writing or typing 0.00001 (as we just did).

If you use the FSPL formula to calculate the decibel loss for this scenario, the formula would be as follows:

$$decibels = 32.4 + 20\log_{10}(2,400) + 20\log_{10}(0.1)$$

The answer is a loss of 80.004 dB, which is approximately 80 decibels of loss. This number is easier to work with and less likely to be mis-written or mis-typed.

Decibels relative to an isotropic radiator (dBi)

Earlier in this chapter, we compared an antenna to an isotropic radiator. Theoretically, an isotropic radiator can radiate an equal signal in all directions. An antenna cannot do this, because of construction limitations. In other instances, you do not want an antenna to radiate in all directions, because you want to focus the signal of the antenna in a particular direction. Whichever the case may be, it is important to be able to calculate the radiating power of the antenna so that you can determine how strong a signal is at a certain distance from the antenna. You may also want to compare the output of one antenna to that of another.

The gain, or increase, of power from an antenna when compared to what an isotropic radiator would generate is known as *decibels isotropic (dBi)*. Another way of phrasing this is *decibel gain referenced to an isotropic radiator* or *change in power relative to an antenna*. Since antennas are measured in gain, not power, you can conclude that dBi is a relative measurement, not an absolute power measurement. dBi is simply a measurement of antenna gain. The dBi value is measured at the strongest point, or the focus point, of the antenna signal. Because antennas always focus their energy more in one direction than another, the dBi value of an antenna is always a positive gain, not a loss. There are, however, antennas with a dBi value of 0, which are often referred to as *no-gain*, or *unity-gain*, antennas.

A common antenna used on access points is the half-wave dipole antenna. The half-wave dipole antenna is a small, typically rubber or plastic encased, general-purpose omnidirectional antenna. A 2.4 GHz half-wave dipole antenna has a dBi value of 2.14.

Any time you see *dBi*, think *antenna gain*.

Decibels relative to a half-wave dipole antenna (dBd)

The antenna industry uses two dB scales to describe the gain of antennas. The first scale, which you just learned about, is dBi, which is used to describe the gain of an antenna relative to a theoretical isotropic antenna. The other scale used to describe antenna gain is *decibels dipole (dBd)*, or *decibel gain relative to a dipole antenna*. So a dBd value is the increase in gain of an antenna when it is compared to the signal of a dipole antenna. As you will learn in Chapter 5, dipole antennas are also omnidirectional antennas. Therefore, a dBd value is a measurement of omnidirectional antenna gain, not unidirectional antenna gain. Because dipole antennas are measured in gain, not power, you can also conclude that dBd is a relative measurement, not a power measurement.

The definition of dBd seems simple enough, but what happens when you want to compare two antennas and one is represented with dBi and the other with dBd? This is actually quite simple. A standard dipole antenna has a dBi value of 2.14. If an antenna has a value of 3 dBd, this means that it is 3 dB greater than a dipole antenna. Because the value of a dipole antenna is 2.14 dBi, all you need to do is add 3 to 2.14. So a 3 dBd antenna is equal to a 5.14 dBi antenna.

Do not forget that dB, dBi, and dBd are comparative, or relative, measurements and not units of power.

> ### Real World Scenario
>
> #### The Real Scoop on dBd
>
> When working with 802.11 equipment, it is not likely that you will have an antenna with a dBd value. 802.11 antennas typically are measured using dBi. On the rare occasion that you do run into an antenna measured with dBd, just add 2.14 to the dBd value and you will know the antenna's dBi value.

Decibels relative to 1 milliwatt (dBm)

Earlier when you read about bels and decibels, you learned that they measured differences or ratios between two signals. Regardless of the type of power that was being transmitted, all you really knew was that the one signal was greater or less than the other by a particular number of bels or decibels. dBm also provides a comparison, but instead of comparing a signal to another signal, it is used to compare a signal to 1 milliwatt of power. *dBm* means *decibels relative to 1 milliwatt*. So what you are doing is setting dBm to 0 (zero) and equating that to 1 milliwatt of power. Because dBm is a measurement that is compared to a known value, 1 milliwatt, it is actually a measure of absolute power. Because decibels (relative) are referenced to 1 milliwatt (absolute), think of a dBm as an absolute assessment that measures change of power referenced to 1 milliwatt. You can now state that 0 dBm is equal to 1 milliwatt. Using the formula $dBm = 10 \times \log_{10}(P_{mW})$, you can determine that 100 mW of power is equal to +20 dBm.

If you happen to have the dBm value of a device and want to calculate the corresponding milliwatt value, you can do that too. The formula is $P_{mW} = 10^{(dBm \div 10)}$.

Remember that 1 milliwatt is the reference point and that 0 dBm is equal to 1 mW. Any absolute power measurement of +dBm indicates amplitude greater than 1 mW. Any absolute power measurement of –dBm indicates amplitude less than 1 mW. For example, we stated earlier that the transmission amplitude of most 802.11 radios usually ranges from 1 mW to 100 mW. A transmission amplitude of 100 mW is equal to +20 dBm. Because of FSPL, received signals will always measure below 1 mW. A very strong received signal is –40 dBm, which is the equivalent of 0.0001 mW (1/10,000th of 1 milliwatt).

It might seem a little ridiculous to have to deal with both milliwatts and dBm. If milliwatts are a valid measurement of power, why not just use them? Why do you have to, or want to, also use dBm? These are good questions that are asked often by students. One reason is simply that dBm absolute measurements are often easier to grasp than measurements in the millionths and billionths of a single milliwatt. Most 802.11 radios can interpret received signals from –30 dBm (1/1,000th of 1 mW) to as low as –100 dBm (1/10 of a billionth of 1 mW). The human brain can grasp –100 dBm much easier than 0.0000000001 milliwatts. During a site survey, WLAN engineers will always determine coverage zones by recording the received signal strength in –dBm values.

Another very practical reason to use dBm can be shown using the FSPL formula again. Following are two FSPL equations. The first equation calculates the decibel loss of a 2.4 GHz signal at 100 meters (0.1 kilometer) from the RF source, and the second calculates the decibel loss of a 2.4 GHz signal at 200 meters (0.2 kilometer) from the RF source:

$$FSPL = 32.4 + 20\log_{10}(2{,}400) + 20\log_{10}(0.1) = 80.00422 \text{ dB}$$

$$FSPL = 32.4 + 20\log_{10}(2{,}400) + 20\log_{10}(0.2) = 86.02482 \text{ dB}$$

In this example, by doubling the distance from the RF source, the signal decreased by about 6 dB. If you double the distance between the transmitter and the receiver, the received signal will decrease by 6 dB. No matter what numbers are chosen, if the distance is doubled, the decibel loss will be 6 dB. This rule also implies that if you increase the amplitude by 6 dB, the usable distance will double. This *6 dB rule* is very useful for comparing cell sizes or estimating the coverage of a transmitter. The 6 dB rule is also useful for understanding antenna gain, because every 6 dB of extra antenna gain will double the usable distance of an RF signal. Remember, if you were working with milliwatts, this rule would not be relevant. By converting milliwatts to dBm, you have a more practical way to compare signals.

Remember the *6 dB rule:* +6 dB doubles the distance of the usable signal; −6 dB halves the distance of the usable signal.

Using dBm also makes it easy to calculate the effects of antenna gain on a signal. If a transmitter generates a +20 dBm signal and the antenna adds 5 dBi of gain to the signal, then the power that is radiating from the antenna (EIRP) is equal to the sum of the two numbers, which is +25 dBm.

Inverse Square Law

You just learned about the 6 dB rule, which states that a +6 dB change in signal will double the usable distance of a signal, and a −6 dB change in signal will halve the usable distance of a signal. This rule and these numbers are based on the *inverse square law*, originally developed by Isaac Newton.

This law states that the change in power is equal to 1 divided by the square of the change in distance. In other words, as the distance from the source of a signal doubles, the energy is spread out over four times the area, resulting in one-fourth of the original intensity of the signal.

This means that if you are receiving a signal at a certain power level and a certain distance (D) and you double the distance (change in distance = 2), the new power level will change by $1/(2)^2$. To use this principle to calculate the EIRP at a specific distance, the formula is P/(4 × pi × r²), where P equals the initial EIRP power and r equals the original (reference) distance.

Let us also review the formula for free space path loss:

$$FSPL = 36.6 + 20\log_{10}(F) + 20\log_{10}(D)$$

FSPL = path loss in dB

F = frequency in MHz

D = distance in miles between antennas

$$FSPL = 32.4 + 20\log_{10}(F) + 20\log_{10}(D)$$

FSPL = path loss in dB

F = frequency in MHz

D = distance in kilometers between antennas

The concept of FSPL is also based on Newton's inverse square law. The main variable for the inverse square law is simply distance. The FSPL formula is also based on distance but includes another variable: frequency.

RF Mathematics

When the topic of RF mathematics is discussed, most people cringe and panic because they expect formulas that have logarithms in them. Fear not. You are about to learn RF math without having to use logarithms. If you want to refresh yourself on some of your math skills before going through this section, review the following:

- Addition and subtraction using the numbers 3 and 10
- Multiplication and division using the numbers 2 and 10

No, we are not kidding. If you know how to add and subtract using 3 and 10 and how to multiply and divide using 2 and 10, you have all of the math skills you need to perform RF math. Read on, and we will teach you how.

Rule of 10s and 3s

Before you fully delve into the *rule of 10s and 3s*, it is important to know that this rule may not give you the exact same answers that you would get if you used the logarithmic formulas. The rule of 10s and 3s provides approximate values, not necessarily exact values. If you are an engineer creating a product that must conform to RF regulatory guidelines, you will need to use logarithms to calculate the exact values. However, if you are a network designer planning a network for your company, you will find that the rule of 10s and 3s will provide you with the numbers and accuracy you need to properly plan your network.

This section will teach you the essential calculations. All the calculations will be based on the following four rules of the 10s and 3s:

- For every 3 dB of gain (relative), double the absolute power (mW).
- For every 3 dB of loss (relative), halve the absolute power (mW).
- For every 10 dB of gain (relative), multiply the absolute power (mW) by a factor of 10.
- For every 10 dB of loss (relative), divide the absolute power (mW) by a factor of 10.

For example, if your access point is configured to transmit at 100 mW and the antenna is rated for 3 dBi of passive gain, the amount of power that will radiate out of the antenna (EIRP) will be 200 mW. Following the rule that you just learned, you will see that the 3 dB of gain from the antenna caused the 100 mW signal from the access point to double. Conversely, if your access point is configured to transmit at 100 mW and is attached to a cable that introduces 3 dB of loss, the amount of absolute amplitude at the end of the cable will be 50 mW. Here you can see that the 3 dB of loss from the cable caused the 100 mW signal from the access point to be halved.

In another example, if your access point is configured to transmit at 40 mW and the antenna is rated for 10 dBi of passive gain, the amount of power that radiates out of the antenna (EIRP) will be 400 mW. Here you can see that the 10 dB of gain from the antenna caused the 40 mW signal from the access point to increase by a factor of 10. Conversely, if your access point is configured to transmit at 40 mW and is attached to a cable that introduces 10 dB of loss, the amount of absolute amplitude at the end of the cable will be 4 mW. Here you can see that the 10 dB of loss from the cable caused the 40 mW signal from the access point to be decreased by a factor of 10.

If you remember these rules, you will be able to quickly perform RF calculations. After reviewing these rules, see Exercise 4.1, which will take you through a step-by-step procedure for using the rule of 10s and 3s. As you work through the step-by-step procedure, remember that dBm is a unit of power and that dB is a unit of change. dB is a value of change that can be applied to dBm. If you have a +10 dBm signal and it increases by 3 dB, you can add these two numbers together to get a result of +13 dBm signal.

EXERCISE 4.1

Step-by-Step Use of the Rule of 10s and 3s

1. On a sheet of paper, create two columns. The header of the first column should be **dBm**, and the header of the second column should be **mW**.

dBm   mW

2. Next to the dBm header, place a + sign and a − sign, and next to the mW header place a × sign and a ÷ sign.

   These will help you to remember that any math performed on the dBm column is addition or subtraction and any math performed on the mW column is multiplication or division.

   ```
   +                               ×
   −                               ÷
        dBm           mW
   ```

3. To the left of the + and − signs, write the numbers **3** and **10**, and to the right of the × and ÷ signs, write the numbers **2** and **10**.

   Any addition or subtraction to the dBM column can be performed using only the numbers 3 and 10. Any multiplication or division to the mW column can be performed using only the numbers 2 and 10.

   ```
   3   +                       ×   2
   10  −                       ÷   10
        dBm           mW
   ```

4. If there is a + on the left, there needs to be an × on the right. If there is a − on the left, there needs to be a ÷ on the right.

5. If you are adding or subtracting a 3 on the left, you must be multiplying or dividing by a 2 on the right. If you are adding or subtracting a 10 on the left, you must be multiplying or dividing by a 10 on the right.

6. The last thing you need to do is to put a **0** under the dBm column and a **1** under the mW column.

   Remember that the definition of dBm is *decibels relative to 1 milliwatt*. So now the chart shows that 0 dBm is equal to 1 milliwatt.

   ```
   3   +                       ×   2
   10  −                       ÷   10
        dBm           mW
         0             1
   ```

Before we continue with an example, it is important to emphasize that a change of ±3 dB equates to a doubling or halving of the power, no matter what power measurement is being used. In our usage of the rule of 10s and 3s, we are dealing with milliwatts because that is the typical transmission amplitude measurement used by 802.11 equipment. However, it is important to remember that a +3 dB increase means a doubling of the power regardless of the power scale used. So a +3 dB increase of 1.21 gigawatts of power would result in 2.42 gigawatts of power.

An animated explanation of the rule of 10s and 3s, along with examples, has been created using Microsoft PowerPoint and can be downloaded from this book's online resource area (www.wiley.com/go/cwnasg). The names of the PowerPoint files are as follows:

- 10s and 3s Template.ppt
- Rule of 10s and 3s Example 1.ppt
- Rule of 10s and 3s Example 2.ppt
- Rule of 10s and 3s Example 3.ppt
- Rule of 10s and 3s Example 4.ppt

EXERCISE 4.2

Rule of 10s and 3s Example

You have a wireless bridge that generates a 100 mW signal. The bridge is connected to an antenna via a cable that creates –3 dB of signal loss. The antenna provides 10 dBi of signal gain. In this example, calculate the IR and EIRP values.

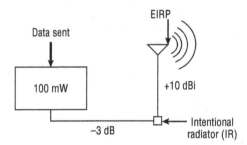

As a reminder, and as shown in the graphic, the IR is the signal up to but not including the antenna, and the EIRP is the signal radiating from the antenna.

1. The first step is to determine whether by using 10 or 2, and × or ÷, you can go from 1 mW to 100 mW.

   It is not too difficult to realize that multiplying 1 by 10 twice will give you 100. So the bridge is generating 100 mW, or +20 dBm, of power.

		dBm	mW	
3	+			× 2
10	–			÷ 10
		0	1	
+ 10		10	10	× 10
+ 10		20	100	× 10

2. Next you have the antenna cable, which is introducing –3 dB of loss to the signal. After you calculate the effect of the –3 dB loss, you know the value of the IR. You can represent the IR as either +17 dBm or 50 mW.

	3 +					×	2
	10 –		dBm	mW		÷	10
			0	1			
		+ 10	10	10	× 10		
		+ 10	20	100	× 10		
		– 3	17	50	÷ 2		

3. All that is left is to calculate the increase of the signal due to the gain from the antenna. Because the gain is 10 dBi, you add 10 to the dBm column and multiply the mW column by 10. This gives you an EIRP of +27 dBm, or 500 mW.

	3 +					×	2
	10 –		dBm	mW		÷	10
			0	1			
		+ 10	10	10	× 10		
		+ 10	20	100	× 10		
		– 3	17	50	÷ 2		
		+ 10	27	500	× 10		

The numbers chosen in the example were straightforward, using the values that are part of the template. In the real world, however, this will not be the case. Using a little creativity, you can calculate gain or loss for any integer. Unfortunately, the rule of 10s and 3s does not work for fractional or decimal numbers. For those numbers, you need to use the logarithmic formula.

dB gain or loss is cumulative. If, for example, you had three sections of cable connecting the transceiver to the antenna and each section of cable provided 2 dB of loss, all three cables would create 6 dB of loss. Using the rule of 10s and 3s, subtracting 6 dBs is equal to subtracting 3 dBs twice. Decibels are very flexible. As long as you come up with the total that you need, they do not care how you do it.

Table 4.1 shows how to calculate all integer dB loss and gain from –10 to +10 by using combinations of just 10s and 3s. Take a moment to look at these values and you will realize that with a little creativity, you can calculate the loss or gain of any integer.

TABLE 4.1   dB loss and gain (–10 through +10)

Loss or Gain (dB)	Combination of 10s and 3s
–10	–10
–9	–3 –3 –3

TABLE 4.1   dB loss and gain (−10 through +10) *(continued)*

Loss or Gain (dB)	Combination of 10s and 3s
−8	−10 −10 +3 +3 +3 +3
−7	−10 +3
−6	−3 −3
−5	−10 −10 +3 +3 +3 +3 +3
−4	−10 +3 +3
−3	−3
−2	−3 −3 −3 −3 +10
−1	−10 +3 +3 +3
+1	+10 −3 −3 −3
+2	+3 +3 +3 +3 −10
+3	+3
+4	+10 −3 −3
+5	+10 +10 −3 −3 −3 −3 −3
+6	+3 +3
+7	+10 −3
+8	+10 +10 −3 −3 −3 −3
+9	+3 +3 +3
+10	+10

RF Math Summary

It is important to remember that the bottom line is that you are trying to calculate the power at different points in the RF system and the effects caused by gain or loss. If you want to perform the RF math calculations by using the logarithmic formulas, here they are:

$$dBm = 10 \times \log_{10}(P_{mW})$$

$$mW = 10^{(dBm \div 10)}$$

If you want to use the rule of 10s and 3s, just remember these four simple tasks and you will not have a problem:

- 3 dB gain = mW × 2
- 3 dB loss = mW ÷ 2
- 10 dB gain = mW × 10
- 10 dB loss = mW ÷ 10

Table 4.2 provides a quick reference guide comparing the absolute power measurements of milliwatts to the absolute power dBm values.

TABLE 4.2   dBm and mW conversions

dBm	mW	Power Level
+36 dBm	4,000 mW	4 watts
+30 dBm	1,000 mW	1 watt
+20 dBm	100 mW	1/10th of 1 watt
+10 dBm	10 mW	1/100th of 1 watt
0 dBm	1 mW	1/1,000th of 1 watt
−10 dBm	0.1 mW	1/10th of 1 milliwatt
−20 dBm	0.01 mW	1/100th of 1 milliwatt
−30 dBm	0.001 mW	1/1,000th of 1 milliwatt
−40 dBm	0.0001 mW	1/10,000th of 1 milliwatt
−50 dBm	0.00001 mW	1/100,000th of 1 milliwatt
−60 dBm	0.000001 mW	1 millionth of 1 milliwatt
−70 dBm	0.0000001 mW	1 ten-millionth of 1 milliwatt
−80 dBm	0.00000001 mW	1 hundred-millionth of 1 milliwatt
−90 dBm	0.000000001 mW	1 billionth of 1 milliwatt

Noise Floor

The *noise floor* is the ambient or background level of radio energy on a specific channel. This background energy can include modulated or encoded bits coming from nearby 802.11 transmitting radios or unmodulated energy coming from non-802.11 devices, such as microwave ovens, Bluetooth devices, portable telephones, and so on. Anything electromagnetic has the potential of raising the amplitude of the noise floor on a specific channel.

The amplitude of the noise floor, which is sometimes simply referred to as "background noise," varies in different environments. For example, the noise floor of a 2.4 GHz industrial, scientific, and medical (ISM) channel might be about –100 dBm in a typical environment. However, a noisier RF environment, such as a manufacturing plant, might have a noise floor of –90 dBm because of the electrical machinery operating within the plant. It should also be noted that the noise floor of 5 GHz channels is almost always lower than the noise floor of 2.4 GHz channels because the 5 GHz frequency bands are less crowded.

Signal-to-Noise Ratio

Many Wi-Fi vendors define signal quality as the *signal-to-noise ratio (SNR)*. As shown in Figure 4.2, the SNR is the difference in decibels between the received signal and the background noise level (noise floor), not actually a ratio. For example, if a radio receives a signal of –85 dBm and the noise floor is measured at –100 dBm, the difference between the received signal and the background noise is 15 dB. The SNR is 15 dB.

FIGURE 4.2   Signal-to-noise ratio

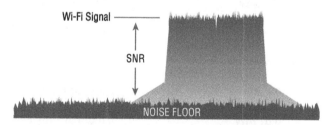

Data transmissions can become corrupted with a very low SNR. If the amplitude of the noise floor is too close to the amplitude of the received signal, data corruption will likely occur and result in layer 2 retransmissions. The retransmissions will negatively affect both throughput and latency. An SNR of 25 dB or greater is considered good signal quality, and an SNR of 10 dB or lower is considered very poor signal quality.

Signal-to-Interference-Plus-Noise Ratio

For years SNR has been a standard measurement for Wi-Fi networks. Over the past few years, the term *signal-to-interference-plus-noise (SINR)* ratio has appeared and is being used by vendors. SINR is the difference between the power of the primary RF signal, compared against the sum of the power of the RF interference and background noise. This difference is measured in decibels.

SNR tends to be a value that is interpreted and looked at over time, since the RF level of background noise tends to be somewhat consistent over time. SINR, however, relates the primary RF signal to both interference and noise. While the noise level tends not to fluctuate much, interference from other devices is likely to be more common and frequent. Since interference can occur more frequently, SINR is a better indicator of what is happening at a specific time.

Received Signal Strength Indicator

Receive sensitivity refers to the power level of an RF signal required to be successfully received by the receiver radio. The lower the power level that the receiver can successfully process, the better the receive sensitivity. Think of this as if you were at a hockey game. There is an ambient level of noise that exists from everything around you. There is a certain volume that you have to speak at for your neighbor to hear you. That level is the receiver sensitivity. It is the weakest signal that the transceiver can decode under normal circumstances. With that said, if the noise in a particular area is louder than normal, then the minimum level you have to yell gets louder.

In WLAN equipment, receive sensitivity is usually defined as a function of network speed. Wi-Fi vendors will usually specify their receive sensitivity thresholds at various data rates; an example vendor specification for a 2.4 GHz radio is listed in Table 4.3. For any given receiver, more power is required by the receiver radio to support the higher data rates. Different speeds use different modulation techniques and encoding methods, and the higher data rates use encoding methods that are more susceptible to corruption. The lower data rates use modulation-encoding methods that are less susceptible to corruption.

TABLE 4.3   Receive sensitivity thresholds (vendor example)

Data Rate	Received Signal Amplitude
MCS7	−77 dBm
MCS6	−78 dBm
MCS5	−80 dBm
MCS4	−85 dBm
MCS3	−88 dBm
MCS2	−90 dBm
MCS1	−90 dBm

TABLE 4.3   Receive sensitivity thresholds (vendor example) *(continued)*

Data Rate	Received Signal Amplitude
MCS0	–90 dBm
54 Mbps	–79 dBm
48 Mbps	–80 dBm
36 Mbps	–85 dBm
24 Mbps	–87 dBm
18 Mbps	–90 dBm
12 Mbps	–91 dBm
9 Mbps	–91 dBm
6 Mbps	–91 dBm

The 802.11-2016 standard defines the *received signal strength indicator (RSSI)* as a relative metric used by 802.11 radios to measure signal strength (amplitude). The 802.11 RSSI measurement parameter can have a value from 0 to 255. The RSSI value is designed to be used by the WLAN hardware manufacturer as a relative measurement of the RF signal strength that is received by an 802.11 radio. RSSI metrics are typically mapped to receive sensitivity thresholds expressed in absolute dBm values, as shown in Table 4.4. For example, an RSSI metric of 30 might represent –30 dBm of received signal amplitude. The RSSI metric of 0 might be mapped to –110 dBm of received signal amplitude. Another vendor might use an RSSI metric of 255 to represent –30 dBm of received signal amplitude and 0 to represent –100 dBm of received signal amplitude.

TABLE 4.4   Received signal strength indicator (RSSI) metrics (vendor example)

RSSI	Receive Sensitivity Threshold	Signal Strength (%)	Signal-to-Noise Ratio	Signal Quality (%)
30	–30 dBm	100%	70 dB	100%
25	–41 dBm	90%	60 dB	100%
20	–52 dBm	80%	43 dB	90%
21	–52 dBm	80%	40 dB	80%

RSSI	Receive Sensitivity Threshold	Signal Strength (%)	Signal-to-Noise Ratio	Signal Quality (%)
15	–63 dBm	60%	33 dB	50%
10	–75 dBm	40%	25 dB	35%
5	–89 dBm	10%	10 dB	5%
0	–110 dBm	0%	0 dB	0%

The 802.11-2016 standard defines another metric, called *signal quality (SQ)*, which is a measure of pseudonoise (PN) code correlation quality received by a radio. In simpler terms, the signal quality could be a measurement of what might affect coding techniques, which relates to the transmission speed. You will learn about coding techniques in Chapter 6. Anything that might increase the bit error rate (BER), such as a low SNR, might be indicated by SQ metrics.

Information parameters from both RSSI and SQ metrics can be passed along from the PHY layer to the MAC sublayer. Some SQ parameters might also be used in conjunction with RSSI as part of a clear channel assessment (CCA) scheme.

Although SQ metrics and RSSI metrics are technically separate measurements, most Wi-Fi vendors refer to both together as simply *RSSI metrics*. For the purposes of this book, whenever we refer to RSSI metrics, we are referring to both SQ and RSSI metrics.

According to the 802.11-2016 standard, "the RSSI is a measure of the RF energy received. Mapping of the RSSI values to actual received power is implementation dependent." In other words, WLAN vendors can define RSSI metrics in a proprietary manner. The actual range of the RSSI value is from 0 to a maximum value (less than or equal to 255) that each vendor can choose on its own (known as *RSSI_Max*). Many vendors publish their implementation of RSSI values in product documents and/or on their website. Some WLAN vendors do not publish their RSSI metrics. Because the implementation of RSSI metrics is proprietary, two problems exist when trying to compare RSSI values between different manufacturers' wireless cards. The first problem is that the manufacturers may have chosen two different values as the RSSI_Max. WLAN vendor A may have chosen a scale from 0 to 100, whereas WLAN vendor B may have chosen a scale from 0 to 60. Because of the difference in scale, WLAN vendor A may indicate a signal with an RSSI value of 25, whereas vendor B may indicate that same signal with a different RSSI value of 15. Also, the radio card manufactured by WLAN vendor A uses more RSSI metrics and is probably more sensitive when evaluating signal quality and SNR.

The second problem with RSSI is that manufacturers could take their range of RSSI values and compare them to a different range of values. WLAN vendor A may take its

100-number scale and relate it to dBm values of –110 dBm to –10 dBm, whereas WLAN vendor B may take its 60-number scale and relate it to dBm values of –95 dBm to –35 dBm. Not only do we have different numbering schemes, we also have different ranges of values.

Although the way in which Wi-Fi vendors implement RSSI may be proprietary, most vendors are alike in that they use RSSI thresholds for very important mechanisms, such as roaming and dynamic rate switching. During the *roaming* process, clients make the decision to move from one access point to the next. RSSI thresholds are key factors for clients when they initiate the roaming handoff. RSSI thresholds are also used by vendors to implement *dynamic rate switching (DRS)*, which is a process used by 802.11 radios to shift between data rates. Roaming is discussed in several chapters of this book, and DRS is discussed in greater detail in Chapter 13, "WLAN Design Concepts."

Because comparing RSSI between vendors can be difficult due to the potential for using values from different numbering scales, many network-monitoring programs convert the RSSI values to percentages, thereby creating a common comparison.

To calculate RSSI percentage, the software will compare the actual signal against the RSSI_Max value, which is part of the IEEE 802.11 standard. Most vendors use 0 as the base value for their calculations. From there, they must simply divide the RSSI received value by the RSSI_Max, as shown in the following formula:

$$\text{RSSI/RSSI\_Max} = \text{RSSI percentage}$$

From the previous example in this section, vendor A had a scale of 0 to 100, and an RSSI value of 25, whereas vendor B had a scale of 0 to 60, and an RSSI of 15. The RSSI percentage would be the same for both vendors: 25%.

 Real World Scenario

Can an 802.11 Network Card Truly Measure the Noise Floor?

It should be understood that the earlier 802.11 wireless network interface cards (NIC) were not spectrum analyzers, and though they could transmit and receive data at a prodigious rate, they could not see raw ambient RF signals. Since the only things getting past the NIC's encoding filter were bits, all the information reported by the NIC had to come from the bits they received. If you turned on a microwave oven near a wireless NIC, no data bits were being generated by the microwave, so the NIC would always report a noise variable of zero. In the absence of encoded RF signals coming from other 802.11 devices, the noise variable could not be used to report the noise floor. The only device that could truly measure non-encoded RF energy was a *spectrum analyzer*.

We know that you may have seen many screens generated by your various 802.11 devices that displayed signal (from the RSSI variable) and another value displayed as signal-to-noise ratio (SNR), showing the comparison between the RSSI and the noise floor. The developers of the wireless NICs knew that the RF folks out there "lived, breathed, and died" by signal, noise, and signal-to-noise ratio data.

WLAN professionals demanded a noise variable in order to perform signal calculations, so various Wi-Fi vendor organizations came up with unique ways to guess the noise floor. Because 802.11 wireless NICs could only process bits, they needed to come up with algorithms to calculate a noise variable based on the bits going through the NIC.

As with RSSI measurements, different manufacturers of 802.11 equipment calculated noise in different ways. Some vendors flatly refused to make up a number for noise only based on bits. Other vendors developed sophisticated algorithms for calculating noise.

More recently, 802.11 chip manufacturers figured out how to turn off the encoding filters and use the RF signals coming through the antenna to become rudimentary spectrum analyzers. However, this is in lieu of being an 802.11 NIC capable of processing data. Typically, these newer chips could be either a lightweight spectrum analyzer or a Wi-Fi card processing data, but usually not both at the same time, since the front-end filter would identify an 802.11 signal and pass it on to the 802.11 protocol stack, not the spectrum analyzer. Some APs can operate in what is sometimes termed "hybrid" mode. These APs can perform both 802.11 and spectrum analysis functions at the same time, although there is often degradation in WLAN performance. Additionally, some WLAN vendors offer APs with an integrated spectrum analyzer chipset that operates independent of the WLAN radio. So what is the best tool to accurately measure the noise floor in any environment? A high-quality spectrum analyzer. A high-quality portable spectrum analyzer uses a spectrum analyzer chipset capable of measuring non-encoded RF energy, and the portability of it makes it the best tool to measure the true noise floor. Keep in mind, however, that what a spectrum analyzer sees as the noise floor may be different from the interpretation of the noise floor by an 802.11 radio in a WLAN client or access point.

If you would like to learn more about the differences between 802.11 NICs and spectrum analyzers, read *CWAP: Certified Wireless Analysis Professional Official Study Guide* (Sybex, 2011).

Link Budget

When radio communications are deployed, a *link budget* is the sum of all the planned and expected gains and losses from the transmitting radio, through the RF medium, to the receiver radio. The purpose of link budget calculations is to guarantee that the final received signal amplitude is above the receiver sensitivity threshold of the receiver radio.

Link budget calculations include original transmit gain, passive antenna gain, and active gain from RF amplifiers. All gain must be accounted for—including RF amplifiers and antennas—and all losses must be accounted for—including attenuators, FSPL, and insertion loss. Any hardware device installed in a radio system adds a certain amount of signal attenuation, called *insertion loss*. Cabling is rated for dB loss per 100 feet, and connectors typically add about 0.5 dB of insertion loss.

You have already learned that RF also attenuates as it travels through free space. Figure 4.3 depicts a point-to-point wireless bridge link and shows that loss occurs as the signal moves through various RF components, as well as the signal loss caused by FSPL.

FIGURE 4.3   Link budget components

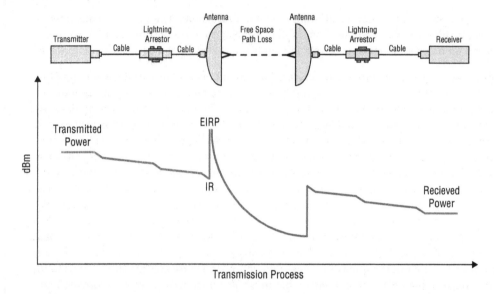

At the top of the figure are the components that make up a point-to-point link, while the lower section of the figure represents the increase or decrease of the RF signal as the signal moves from the transmitter to the receiver. The transmitter begins the process by generating a signal of a specific value. This signal decreases as it travels to the transmitting antenna. The cables, connectors, and the lightning arrestor all attenuate or decrease the signal. The antenna focuses the transmitted signal, producing gain, which increases the signal. The biggest loss of signal of any RF transmission is caused by free space path loss (FSPL) as the signal travels to the receiving antenna. This antenna focuses the received signal, producing gain, increasing the signal. The signal then decreases again as it travels through the cables, connectors, and the lightning arrestor, until it ultimately arrives at the receiver. This figure is a representation and is not drawn to scale.

Let us look at the link budget calculations of a 2.4 GHz point-to-point wireless bridge link, as depicted in Figure 4.4 and Table 4.5. In this case, the two antennas are 2 kilometers apart, and the original transmission is +10 dBm. Notice the amount of insertion loss caused by each RF component, such as the cabling and the lightning arrestors. The antennas passively amplify the signal, and the signal attenuates as it travels through free space. The final received signal at the receiver end of the bridge link is –51.5 dBm.

FIGURE 4.4   Point-to-point link budget gain and loss

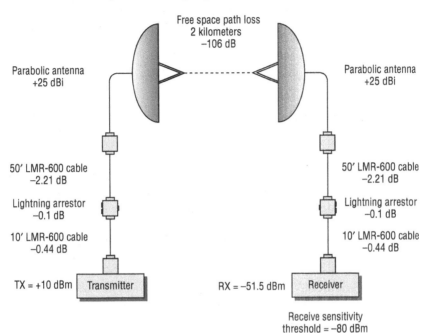

TABLE 4.5   Link budget calculations

Component	Gain or Loss	Signal Strength
Transceiver (original transmission signal)		+10 dBm
10′ LMR-600 cable	−0.44 dB	+9.56 dBm
Lightning arrestor	−0.1 dB	+9.46 dBm
50′ LMR-600 cable	−2.21 dB	+7.25 dBm
Parabolic antenna	+25 dBi	+32.25 dBm
Free space path loss	−106 dB	−73.75 dBm
Parabolic antenna	+25 dBi	−48.75 dBm
50′ LMR-600 cable	−2.21 dB	−50.96 dBm

TABLE 4.5 Link budget calculations *(continued)*

Component	Gain or Loss	Signal Strength
Lightning arrestor	–0.1 dB	–51.06 dBm
10' LMR-600 cable	–0.44 dB	–51.5 dBm
Receiver (final received signal)		–51.5 dBm

Let us assume that the receive sensitivity threshold of the receiver radio is –80 dBm. Any signal received with amplitude above –80 dBm can be understood by the receiver radio, whereas any amplitude below –80 dBm cannot be understood. The link budget calculations determined that the final received signal is –51.5 dBm, which is well above the receive sensitivity threshold of –80 dBm. There is a 28.5 dB buffer between the final received signal and the receive sensitivity threshold. The 28.5 dB buffer that was determined during link budget calculations is known as the *fade margin*, which is discussed in the next section.

You may be wondering why these numbers are negative when up until now most of the dBm numbers you have worked with have been positive. Figure 4.5 shows a simple summary of the gains and losses in an office environment. Until now you have worked primarily with calculating the IR and EIRP. It is the effect of FSPL that makes the values negative, as you will see in the calculations based on Figure 4.5. In this example, the received signal is the sum of all components, which is as follows:

$$+20 \text{ dBm} + 5 \text{ dBi} - 73.98 \text{ dB} + 2.14 \text{ dBi} = -46.84 \text{ dBm}$$

FIGURE 4.5 Office link budget gain and loss

Final signal at the RX = –46.84 dBm

Although the initial transmission amplitude will almost always be above 0 dBm (1 mW), the final received signal amplitude will always be well below 0 dBm (1 mW) because of FSPL.

Fade Margin/System Operating Margin

Fade margin is a level of desired signal above what is required. A good way to explain fade margin is to think of it as a comfort zone. If a receiver has a receive sensitivity of –80 dBm, a transmission will be successful as long as the signal received is greater than –80 dBm. The problem is that the signal being received fluctuates because of many outside influences, such as interference and weather conditions. To accommodate for the fluctuation, it is a common practice to plan for a 10 dB to 25 dB buffer above the receive sensitivity threshold of a radio used in a bridge link. The 10 dB to 25 dB buffer above the receive sensitivity threshold is the fade margin.

Let us say that a receiver has a sensitivity of –80 dBm, and a signal is typically received at –76 dBm. Under normal circumstances, this communication is successful. However, because of outside influences, the signal may fluctuate by ±10 dB. This means that most of the time, the communication is successful, but on those occasions that the signal has fluctuated to –86 dBm, the communication will be unsuccessful. By adding a fade margin of 20 dB in your link budget calculations, you are now stating that for your needs, the receive sensitivity is –60 dBm, and you will plan your network so that the received signal is greater than –60 dBm. If the received signal fluctuates, you have already built in some padding—in this case, 20 dB.

Refer to Figure 4.4. If you required a fade margin of 10 dB above the receive sensitivity of –80 dBm, the amount of signal required for the link would be –70 dBm. Since the signal is calculated to be received at –51.5 dBm, you would have a successful communication. This signal would even be adequate if you chose a fade margin of 20 dB.

Because RF communications can be affected by many outside influences, it is common to have a fade margin to provide a level of link reliability. By increasing the fade margin, you are essentially increasing the reliability of the link. Think of the fade margin as the buffer or margin of error for received signals that is used when designing and planning an RF system. After the RF link has been installed, it is important to measure the link to see how much buffer or padding there actually is. This functional measurement is known as the *system operating margin (SOM)*. The SOM is the difference between the actual received signal and the signal necessary for reliable communications.

 Real World Scenario

When Are Fade Margin Calculations Needed?

Whenever an outdoor WLAN bridge link is designed, link budget and fade margin calculations will be an absolute requirement. For example, an RF engineer may perform link budget calculations for a 2-mile point-to-point bridge link and determine that the final received signal is 5 dB above the receive sensitivity threshold of a radio at one end of a bridge link. It would seem that RF communications will be just fine; however, because of downfade caused by multipath and weather conditions, a fade margin buffer is needed. A torrential downpour can attenuate a signal as much as 0.08 dB per mile (0.05 dB per kilometer) in both the 2.4 GHz and 5 GHz frequency ranges. Over long-distance bridge links, a fade margin of 25 dB is usually recommended to compensate for attenuation due to changes in RF behaviors, such as multipath, and due to changes in weather conditions, such as rain, fog, or snow.

When deploying a WLAN indoors where high multipath or high noise floor conditions exist, the best practice is to plan for a fade margin of about 5 dB above the vendor's recommended receive sensitivity amplitude. For example, a –70 dBm or stronger signal falls above the RSSI threshold for the higher data rates for most WLAN vendor radios. During the indoor site survey, RF measurements of –70 dBm will often be used to determine coverage areas for higher data rates. In a noisy environment, RF measurements of –65 dBm utilizing a 5 dB fade margin is a recommended best practice.

EXERCISE 4.3

Link Budget and Fade Margin

In this exercise, you will use a Microsoft Excel file to calculate a link budget and fade margin. You will need Excel installed on your computer.

1. From the book's online resource area (www.wiley.com/go/cwnasg), download the file LinkBudget.xls to your desktop, and then open it.

2. In row 10, enter a link distance of **25** kilometers.

    Note that the path loss due to a 25-kilometer link is now 128 dB in the 2.4 GHz frequency.

3. In row 20, enter **128** for path loss in dB.

4. In row 23, change the radio receiver sensitivity to –**80** dBm.

    Notice that the final received signal is now –69 dBm, and the fade margin is only 11 dB.

5. Try increasing the "radio transmitter output power" to see how the connection would fare and to determine how much power would be needed to ensure a fade margin of 20 dB. You can also change the other components, such as antenna gain and cable loss, to ensure a fade margin of 20 dB.

Summary

This chapter covered the following six key areas of RF communications:

- RF components
- RF measurements
- RF mathematics
- RSSI thresholds
- Link budgets
- Fade margins

It is important to understand how each RF component affects the output of the transceiver. Whenever a component is added, removed, or modified, the output of the RF communications is changed. You need to understand these changes and make sure that the system conforms to regulatory standards. The following RF components were covered in this chapter:

- Transmitter

- Receiver

- Antenna

- Isotropic radiator

- Intentional radiator (IR)

- Equivalent isotropically radiated power (EIRP)

In addition to understanding the components and their effects on the transmitted signal, you must know the different units of power and comparison that are used to measure the output and the changes to the RF communications:

- Units of power

  - Watt

  - Milliwatt

  - dBm

- Units of comparison

  - dB

  - dBi

  - dBd

After you become familiar with the RF components and their effects on RF communications, and you know the different units of power and comparison, you need to understand how to perform the actual calculations and determine whether your RF communication will be successful. It is important to know how to perform the calculations and some of the terms and concepts involved with making sure that the RF link will work properly. These concepts and terms are as follows:

- Rule of 10s and 3s

- Noise floor

- Signal-to-noise ratio (SNR)

- Signal-to-interference-plus-noise ratio (SINR)

- Receive sensitivity

- Received signal strength indicator (RSSI)

- Link budget

- System operating margin (SOM)

- Fade margin

Exam Essentials

Understand the RF components.   Know the function of each of the components and which components add gain and which components add loss.

Understand the units of power and comparison.   Make sure you are comfortable with the difference between units of power (absolute) and units of comparison (relative). Know all the units of power and comparison, what they measure, and how they are used.

Be able to perform simple RF mathematics.   No logarithms will be on the test; however, you must know how to use the rule of 10s and 3s. You need to be able to calculate a result based on a scenario, power value, or comparative change.

Understand the practical uses of RF mathematics.   When all is said and done, the ultimate question is, will the RF communication work? This is where an understanding of RSSI, SOM, fade margin, and link budget is important.

Be able to explain the importance of measuring the SNR and the noise floor.   Understand that the ambient background level of radio energy on a specific channel can corrupt 802.11 data transmissions. Understand that the only device that can truly measure unmodulated RF energy is a spectrum analyzer.

Define RSSI.   Understand that radios use RSSI metrics to interpret signal strength and quality. 802.11 radios use RSSI metrics for decisions such as roaming and dynamic rate switching.

Understand the necessity of a link budget and fade margin.   A link budget is the sum of all gains and losses from the transmitting radio, through the RF medium, to the receiver radio. The purpose of link budget calculations is to guarantee that the final received signal amplitude is above the receiver sensitivity threshold of the receiver radio. Fade margin is a level of desired signal above what is required.

Review Questions

1. Which of the following is a better indicator of what outside influences are affecting an RF signal at a specific moment in time?

   A. RSSI

   B. SNR

   C. EIRP

   D. SINR

2. A point source that radiates RF signal equally in all directions is known as what?

   A. Omnidirectional signal generator

   B. Omnidirectional antenna

   C. Intentional radiator

   D. Nondirectional transmitter

   E. Isotropic radiator

3. When calculating the link budget and system operating margin of a point-to-point outdoor WLAN bridge link, which factors should be taken into account? (Choose all that apply.)

   A. Distance

   B. Receive sensitivity

   C. Transmit amplitude

   D. Antenna height

   E. Cable loss

   F. Frequency

4. The sum of all the components from the transmitter to the antenna, not including the antenna, is known as what? (Choose two.)

   A. IR

   B. Isotropic radiator

   C. EIRP

   D. Intentional radiator

5. The highest RF signal strength that is transmitted from an antenna is known as what?

   A. Equivalent isotropically radiated power

   B. Transmit sensitivity

   C. Total emitted power

   D. Antenna radiated power

6. Select the absolute units of power. (Choose all that apply.)

   A. Watt

   B. Milliwatt

   C. Decibel

   D. dBm

   E. Bel

7. Select the units of comparison (relative). (Choose all that apply.)

   A. dBm

   B. dBi

   C. Decibel

   D. dBd

   E. Bel

8. 2 dBd is equal to how many dBi?

   A. 5 dBi

   B. 4.41 dBi

   C. 4.14 dBi

   D. The value cannot be calculated.

9. 23 dBm is equal to how many mW?

   A. 200 mW

   B. 14 mW

   C. 20 mW

   D. 23 mW

   E. 400 mW

10. A wireless bridge is configured to transmit at 100 mW. The antenna cable and connectors produce a 3 dB loss and are connected to a 16 dBi antenna. What is the EIRP?

   A. 20 mW

   B. 30 dBm

   C. 2,000 mW

   D. 36 dBm

   E. 8 W

11. A WLAN transmitter that emits a 400 mW signal is connected to a cable with a 9 dB loss. If the cable is connected to an antenna with 19 dBi of gain, what is the EIRP?

   A. 4 W

   B. 3,000 mW

   C. 3,500 mW

   D. 2 W

12. WLAN vendors use RSSI thresholds to trigger which radio card behaviors? (Choose all that apply.)

    A. Receive sensitivity

    B. Roaming

    C. Retransmissions

    D. Dynamic rate switching

13. Received signal strength indicator (RSSI) metrics are used by 802.11 radios to define which RF characteristics?

    A. Signal strength

    B. Phase

    C. Frequency

    D. Modulation

14. dBi is a measure of what?

    A. The output of the transmitter

    B. The signal increase caused by the antenna

    C. The signal increase of the intentional transmitter

    D. The comparison between an isotropic radiator and the transceiver

    E. The strength of the intentional radiator

15. Which of the following are valid calculations when using the rule of 10s and 3s? (Choose all that apply.)

    A. For every 3 dB of gain (relative), double the absolute power (mW).

    B. For every 10 dB of loss (relative), divide the absolute power (mW) by a factor of 2.

    C. For every 10 dB of loss (absolute), divide the relative power (mW) by a factor of 3.

    D. For every 10 mW of loss (relative), multiply the absolute power (dB) by a factor of 10.

    E. For every 10 dB of loss (relative), halve the absolute power (mW).

    F. For every 10 dB of loss (relative), divide the absolute power (mW) by a factor of 10.

16. A WLAN transmitter that emits a 100 mW signal is connected to a cable with a 3 dB loss. If the cable were connected to an antenna with 7 dBi of gain, what would be the EIRP at the antenna element?

    A. 200 mW

    B. 250 mW

    C. 300 mW

    D. 400 mW

17. In a normal wireless bridged network, the greatest loss of signal is caused by what component?

  A. Receive sensitivity

  B. Antenna cable loss

  C. Lightning arrestor

  D. Free space path loss

18. To double the effective distance of a signal at a specific power level, the EIRP must be increased by how many dBs?

  A. 3 dB

  B. 6 dB

  C. 10 dB

  D. 20 dB

19. During a site survey of a point-to-point link between buildings at a manufacturing plant, the WLAN engineer determines that the noise floor is extremely high because of all the machinery that is operating in the buildings. The engineer is worried about a low SNR and poor performance due to the high noise floor. What is a suggested best practice to deal with this scenario?

  A. Increase the access points' transmission amplitude.

  B. Mount the access points higher.

  C. Double the distance of the AP signal with 6 dBi of antenna gain.

  D. Plan for coverage cells with a 5 dB fade margin.

  E. Increase the transmission amplitude of the client radios.

20. Which value should not be used to compare wireless network cards manufactured by different WLAN vendors?

  A. Receive sensitivity

  B. Transmit power range

  C. Antenna dBi

  D. RSSI

Chapter

5

Radio Frequency Signal and Antenna Concepts

IN THIS CHAPTER, YOU WILL LEARN ABOUT THE FOLLOWING:

✓ Azimuth and elevation charts (antenna radiation envelopes)

✓ Interpreting polar charts

✓ Beamwidth

✓ Antenna types

  ▪ Omnidirectional antennas

  ▪ Semidirectional antennas

  ▪ Highly directional antennas

  ▪ Sector antennas

  ▪ Antenna arrays

✓ Visual line of sight

✓ RF line of sight

✓ Fresnel zone

✓ Earth bulge

✓ Antenna polarization

✓ Antenna diversity

✓ Multiple-input, multiple-output (MIMO)

  ▪ MIMO antennas

✓ **Antenna connection and installation**

- Voltage standing wave ratio (VSWR)
- Signal loss
- Antenna mounting

✓ **Antenna accessories**

- Cables
- Connectors
- Splitters
- Amplifiers
- Attenuators
- Lightning arrestors
- Grounding rods and wires

✓ **Regulatory compliance**

To be able to communicate between two or more transceivers, the radio frequency (RF) signal must be radiated from the antenna of the transmitter with enough power so that it is received and understood by the receiver. The installation of antennas has the greatest ability to affect whether or not the communication is successful. Antenna installation can be as simple as placing an access point in the middle of a small office to provide full coverage for your company, or it can be as complex as installing an assortment of directional antennas, kind of like piecing together a jigsaw puzzle. Do not fear this process; with proper understanding of antennas and how they function, you may find successfully planning for and installing antennas in a wireless network to be a skillful and rewarding task.

This chapter focuses on the categories and types of antennas and the different ways that they can direct an RF signal. Choosing and installing antennas is like choosing and installing lighting in a home. When installing home lighting, you have many choices: table lamps, ceiling lighting, narrow- or wide-beam directional spotlights. Chapter 4, "Radio Frequency Components, Measurements, and Mathematics," introduced you to the concept of antennas focusing an RF signal. In this chapter, you will learn about the various types of antennas, their radiation patterns, and how to use the different antennas in different environments.

You will also learn that even though we often use light to explain RF radiation, differences exist between the ways the two behave. You will learn about aiming and aligning antennas, and you will learn that what you see is not necessarily what you will get.

In addition to learning about antennas, you will learn about the accessories that may be needed for proper antenna installation. In office environments, you may simply need to connect the antenna to the access point. In outdoor installations, you will need special cable and connectors, lightning arrestors, and special mounting brackets. We will introduce you to the components necessary to successfully install an antenna.

To summarize: You will gain the knowledge that will enable you to properly select, install, and align antennas. These skills will help you successfully implement a wireless network, whether it is a point-to-point network between two buildings or a network providing wireless coverage throughout an office building.

Azimuth and Elevation Charts (Antenna Radiation Envelopes)

There are many types of antennas designed for many different purposes, just as there are many types of lights designed for many different purposes. When purchasing lighting for your home, it is easy to compare two lamps by turning them on and looking at the way each disperses the light. Unfortunately, it is not possible to compare antennas in the same way.

An actual side-by-side comparison of antennas requires you to walk around the antenna with an RF meter, take numerous signal measurements, and then plot the measurements, either on the ground or on a piece of paper that represents the environment. Besides the fact that this is a time-consuming task, the results could be skewed by outside influences on the RF signal, such as furniture or other RF signals in the area. To assist potential buyers with their purchasing decision, antenna manufacturers create *azimuth charts* and *elevation charts*, commonly known as radiation patterns, for their antennas. These radiation patterns are created in controlled environments, where the results cannot be skewed by outside influences, and represent the signal pattern that is radiated by a particular model of antenna. These charts are commonly known as *polar charts* or *antenna radiation envelopes*.

In addition to the antenna polar charts, there are numerous companies that offer software that allows you to perform predictive wireless network design. This type of software uses the antenna radiation patterns in conjunction with the RF attenuation properties of a building's structure to create a projected wireless coverage plan.

Figure 5.1 shows the azimuth and elevation charts of an omnidirectional antenna. The azimuth chart, labeled H-plane, shows the top-down view of the radiation pattern of the antenna. Since this is an omnidirectional antenna, as you can see from the azimuth chart, its radiation pattern is almost perfectly circular. The elevation chart, labeled E-plane, shows the side view of the radiation pattern of the antenna. There is no standard that requires the antenna manufacturers to align the degree marks of the chart with the direction that the antenna is facing, so unfortunately it is up to the reader of the chart to understand and interpret it.

Here are a few pointers that will help you interpret the radiation charts:

- In either chart, the antenna is placed at the center of the chart.
- Azimuth chart = H-plane = top-down view
- Elevation chart = E-plane = side view

The outer ring of the chart usually represents the strongest signal of the antenna. The chart does not represent distance or any level of power or strength; it represents only the relationship of power between different points on the chart.

One way to think of the chart is to consider the way a shadow behaves. If you were to move a flashlight closer or farther from your hand, the shadow of your hand would grow larger or smaller. The shadow does not represent the size of the hand. The shadow represents the relative shape of the hand. Whether the shadow is large or small, the shape and pattern of the shadow of the hand is identical. With an antenna, the radiation pattern will grow larger or smaller depending on how much power the antenna receives, but the shape and the relationships represented by the patterns will always stay the same.

Figure 5.2 shows a representation of another omnidirectional antenna. This graphic was generated by a predictive modeling solution from iBwave (www.ibwave.com). On the left side of the graphic is the H-plane and E-plane representation of the antenna. The main image displays a 3-dimensional mesh rendering of the antenna coverage.

FIGURE 5.1   Azimuth and elevation charts

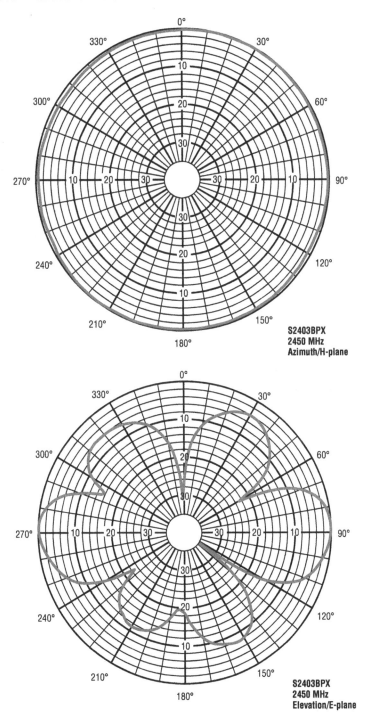

FIGURE 5.2   Omnidirectional antenna: 3-dimensional view

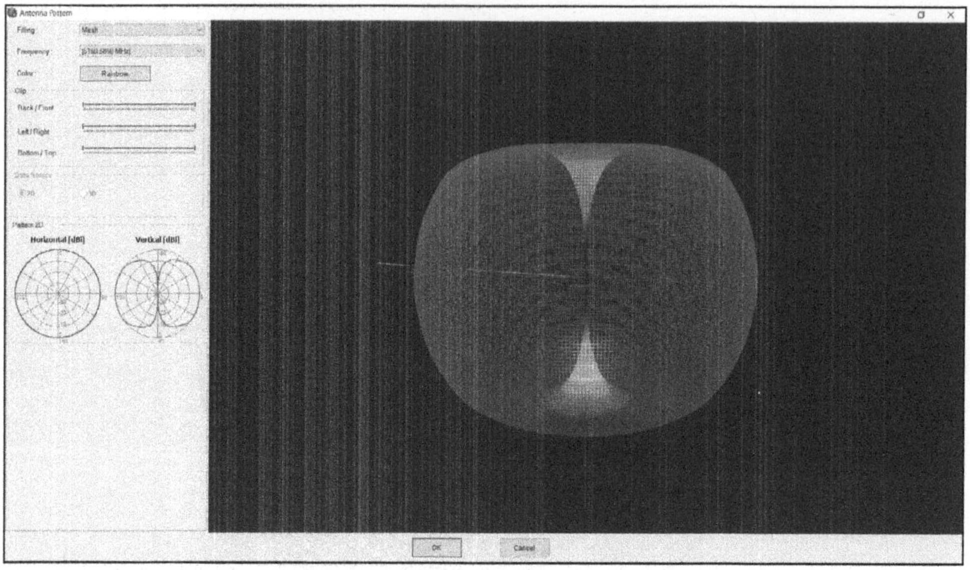

Courtesy of iBwave

Interpreting Polar Charts

As previously mentioned, the antenna azimuth (H-plane) and elevation (E-plane) charts are commonly referred to as polar charts. These charts are often misinterpreted and misread. One of the biggest reasons these charts are misinterpreted is that they represent the decibel (dB) mapping of the antenna coverage. This dB mapping represents the radiation pattern of the antenna; however, it does this using a logarithmic scale instead of a linear scale. Remember that the logarithmic scale is a variable scale, based on exponential values, so the polar chart is actually a visual representation using a variable scale.

Take a look at Figure 5.3. The numbers inside the four boxes in the upper-left corner tell you how long and wide each box is. So, even though visually in our drawing we represented the boxes as the same size, in reality each one is twice as long and wide as the previous one. It is easier to draw the four boxes as the same physical size and just put the number in each box to represent the actual size of the box. In the middle drawing, we drew the boxes showing the relative size of the four boxes.

What if we had more boxes, say 10? By representing each box using the same-sized drawing, it is easier to illustrate the boxes, as shown with the boxes in the lower-left corner. In this example, if we tried to show the actual differences in size, as we did in the middle of the drawing, we could not fit this drawing on the page in the book. In fact, the room that you are in may not have enough space for you to even draw this. Because the scale changes so drastically, it is necessary to not draw the boxes to scale so that we can still represent the information.

FIGURE 5.3   Logarithmic/linear comparison

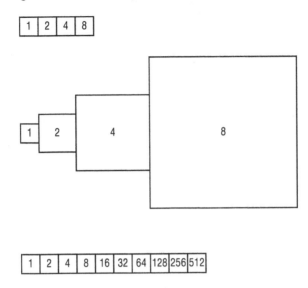

In Chapter 4, you learned about RF math. One of the rules you learned was the *6 dB rule*, which states that a 6 dB decrease of power decreases the effective distance the signal travels by half. A 10 dB decrease of power decreases the effective distance the signal travels by approximately 70 percent. In Figure 5.4, the left polar chart displays the logarithmic representation of the elevation chart of an omnidirectional antenna. This is what you are typically looking at on an antenna brochure or specification sheet. Someone who is untrained in reading these charts would look at the chart and be impressed with how much vertical coverage the antenna provides but would likely be disappointed with the actual coverage. When reading the logarithmic chart, you must remember that for every 10 dB decrease from the peak signal, the actual distance decreases by 70 percent. Each concentric circle on this logarithmic chart represents a change of 5 dB. Figure 5.4 shows the logarithmic pattern of an elevation chart of an omnidirectional antenna along with a linear representation of its coverage. Notice that the first side lobe is about 10 dB weaker than the main lobe. Remember to compare where the lobes are relative to the concentric circles. This 10 dB decrease on the logarithmic chart is equal to a 70 percent decrease in range on the linear chart. Comparing both charts, you see that the side lobes on the logarithmic chart are essentially insignificant when adjusted to the linear chart. As you can see, this omnidirectional antenna has very little vertical coverage.

To give you another comparison, Figure 5.5 shows the logarithmic pattern of the elevation chart of a directional antenna along with a linear representation of the vertical coverage area of this antenna. We rotated the polar chart on its side so that you can better visualize the antenna mounted on the side of a building and aiming at another building.

FIGURE 5.4   Omnidirectional polar chart (E-plane)

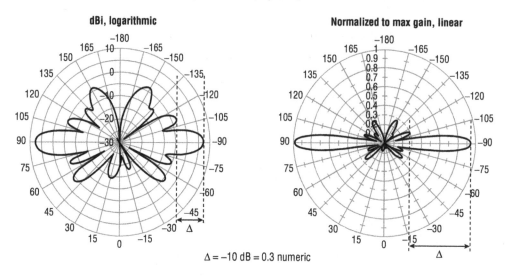

Δ = −10 dB = 0.3 numeric

FIGURE 5.5   Directional polar chart (E-plane)

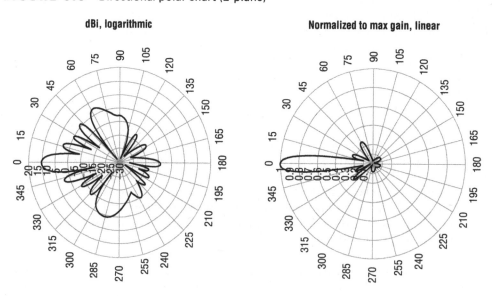

Beamwidth

Many flashlights have adjustable lenses, enabling the user to widen or tighten the concentration of light. RF antennas are capable of focusing the power that is radiating from them, but unlike flashlights, antennas are not adjustable. The user must decide how much focus is desired prior to the purchase of the antenna.

Beamwidth is the measurement of how broad or narrow the focus of an antenna is—and is measured both horizontally and vertically. It is the measurement from the center, or strongest point, of the antenna signal to each of the points along the horizontal and vertical axes, where the signal decreases by half power (–3 dB), as shown in Figure 5.6. These –3 dB points are often referred to as *half-power points*. The distance between the two half-power points on the horizontal axis is measured in degrees, giving the horizontal beamwidth measurement. The distance between the two half-power points on the vertical axis is also measured in degrees, giving the vertical beamwidth measurement.

FIGURE 5.6 Antenna beamwidth

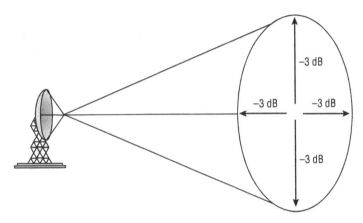

Most of the time when you are deciding which antenna will address your communications needs, you will look at the manufacturer's spec sheets to determine the technical specifications of the antenna. The manufacturer typically includes the numerical values for the horizontal and vertical beamwidths of the antenna. It is important to understand how these numbers are calculated. Figure 5.7 illustrates the process.

1. Determine the scale of the polar chart.

   On this chart, the outer solid line represents the peak signal level. Moving towards the middle of the circle, the next solid line is –10 dB less than the peak signal level, the next one closer to the middle is –20 dB less than the peak signal level, and the final solid line is –30 dB less than the peak signal level. Moving towards the middle of the circle, the dotted lines represent –5 dB, –15 dB, and –25 dB less than the peak signal level.

2. To determine the beamwidth of this antenna, locate the point on the chart where the antenna signal is the strongest.

   In this example, the signal is strongest where the number 1 arrow is pointing.

3. Move along the antenna pattern away from the peak signal (as shown by the two number 2 arrows) until you reach the point where the antenna pattern is 3 dB closer to the center of the diagram (as shown by the two number 3 arrows).

   This is why you needed to know the scale of the chart first.

4. Draw a line from each of these points to the middle of the polar chart (as shown by the dark dotted lines).

5. Measure the distance in degrees between these lines to calculate the beamwidth of the antenna.

   In this example, the beamwidth of this antenna is about 28 degrees.

FIGURE 5.7   Beamwidth calculation

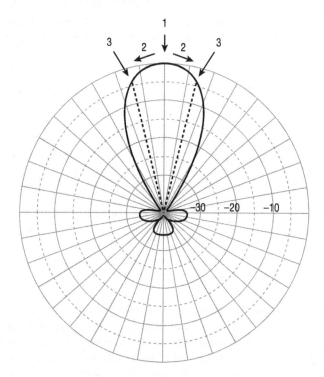

It is important to realize that even though the majority of the RF signal that is generated is focused within the beamwidth of the antenna, a significant amount of signal can still radiate from outside the beamwidth, from what is known as the antenna's side or rear lobes. As you look at the azimuth charts of different antennas, you will notice that some of these side and rear lobes are fairly significant. Although the signal of these lobes is drastically less than the signal of the main beamwidth, they are dependable and, in certain implementations, very functional. When you are aligning point-to-point antennas, it is important that you make sure they are actually aligned to the main lobe, not to a side lobe.

Table 5.1 shows the types of antennas used in 802.11 communications.

Table 5.1 provides reference information that will be useful as you learn about various types of antennas in this chapter.

TABLE 5.1   Antenna beamwidth

Antenna Type	Horizontal Beamwidth (in Degrees)	Vertical Beamwidth (in Degrees)
Omnidirectional	360	7 to 80
Patch/panel	30 to 180	6 to 90
Yagi	30 to 78	14 to 64
Sector	60 to 180	7 to 17
Parabolic dish	4 to 25	4 to 21

Antenna Types

There are three main categories of antennas:

Omnidirectional   *Omnidirectional antennas* radiate RF in a fashion similar to the way a table or floor lamp radiates light. They are designed to provide general coverage horizontally in all directions.

Semidirectional   *Semidirectional antennas* radiate RF in a fashion similar to the way a wall sconce radiates light away from the wall or the way a street lamp shines light down on a street or a parking lot, providing a directional light across a large area.

Highly directional   *Highly directional antennas* radiate RF in a fashion similar to the way a spotlight focuses light on a flag or a sign.

Each type of antenna is designed with a different objective in mind.

It is important to keep in mind that this section is discussing types of antennas, not lighting. Although it is useful to refer to lighting to provide analogies to antennas, it is critical to remember that, unlike lighting, RF signals can travel through solid objects, such as walls and floors.

In addition to antennas acting as radiators and focusing signals that are being transmitted, they focus signals that are received. If you were to walk outside and look up at a star, it would appear fairly dim. If you were to look at that same star through binoculars, it would appear brighter. If you were to use a telescope, it would appear even brighter. Antennas function in a similar way. Not only do they amplify the signal that is being transmitted, they also amplify the signal that is being received. High-gain microphones operate in the same way, enabling you to not only watch the action of your favorite sport on television but to also hear the action.

Antennas or Antennae?

Although it is not a matter of critical importance, many people are often curious whether the plural of *antenna* is *antennas* or *antennae*. The simple answer is both, but the complete answer is, it depends. When *antenna* is used as a biological term, the plural is *antennae*, such as the antennae of a bug. When it is used as an electronics term, the plural is typically *antennas*, such as the antennas on an access point. You should note that this is not always the case, and you may find some regions that use antennas as the plural and others that use antennae. For additional information, please read more at http://grammarist.com/usage/antennae-antennas.

Omnidirectional Antennas

Omnidirectional antennas radiate an RF signal in all directions. The small, rubber-coated *dipole antenna*, often referred to as a *rubber duck* antenna, is the classic example of an omnidirectional antenna and is the default antenna of many access points, although most of the antennas nowadays are encased in plastic instead of rubber. A perfect omnidirectional antenna would radiate an RF signal equally in all directions, like the theoretical isotropic radiator discussed in Chapter 4, "Radio Frequency Components, Measurements, and Mathematics." The closest thing to an isotropic radiator is the omnidirectional dipole antenna.

An easy way to explain the radiation pattern of a typical omnidirectional antenna is to hold your index finger straight up (this represents the antenna) and place a bagel on it as if it were a ring (this represents the RF signal). If you were to slice the bagel in half horizontally, as if you were planning to spread butter on it, the cut surface of the bagel would represent the azimuth chart, or H-plane, of the omnidirectional antenna. If you took another bagel and sliced it vertically instead, essentially cutting the hole you are looking through in half, the cut surface of the bagel would now represent the elevation, or E-plane, of the omnidirectional antenna.

In Chapter 4, you learned that antennas can focus or direct the signal they are transmitting. It is important to know that the higher the dBi or dBd value of an antenna, the more focused the signal. When discussing omnidirectional antennas, it is not uncommon to initially question how it is possible to focus a signal that is radiated in all directions. With higher-gain omnidirectional antennas, the vertical signal is decreased and the horizontal power is increased.

Figure 5.8 shows the elevation view of three theoretical antennas. Notice that the signal of the higher-gain antennas is elongated, or more focused horizontally. The horizontal beamwidth of omnidirectional antennas is always 360 degrees, and the vertical beamwidth ranges from 7 to 80 degrees, depending on the particular antenna.

FIGURE 5.8   Vertical radiation patterns of omnidirectional antennas

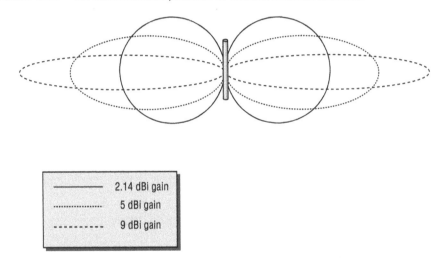

Because of the narrower vertical coverage of the higher-gain omnidirectional antennas, it is important to carefully plan how they are used. Placing one of these higher-gain antennas on the first floor of a building may provide good coverage to the first floor, but because of the narrow vertical coverage, the second and third floors may receive minimal signal. In some installations, you may want this; in others, you may not. Indoor installations typically use low-gain omnidirectional antennas with gain of about 2.14 dBi.

Antennas are most effective when the length of the element is an even fraction (such as 1/4 or 1/2) or a multiple of the wavelength (λ). A 2.4 GHz half-wave dipole antenna (see Figure 5.9) consists of two elements, each 1/4 wave in length (about 1 inch), running in the opposite direction from each other. Higher-gain omnidirectional antennas are typically constructed by stacking multiple dipole antennas on top of each other and are known as *collinear antennas*.

FIGURE 5.9   Half-wave dipole antenna

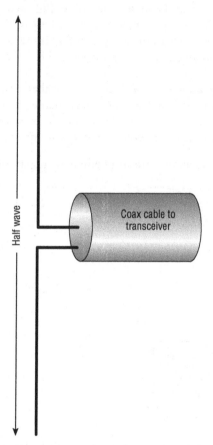

Omnidirectional antennas are typically used in point-to-multipoint environments. The omnidirectional antenna is connected to a device (such as an access point) that is placed at the center of a group of client devices, providing central communication capabilities to the surrounding clients. High-gain omnidirectional antennas can also be used outdoors to connect multiple buildings together in a point-to-multipoint configuration. A central building would have an omnidirectional antenna on its roof, and the surrounding buildings would have directional antennas aimed at the central building. In this configuration, it is important to ensure that the gain of the omnidirectional antenna is high enough to provide the coverage necessary but not so high that the vertical beamwidth is too narrow to provide an adequate signal to the surrounding buildings.

Figure 5.10 shows an installation where the gain is too high. The building to the left will be able to communicate, but the building on the right will likely have problems. To solve the problem shown in Figure 5.10, sector arrays using a down-tilt configuration are used instead of high-gain omnidirectional antennas. Sector antennas are discussed later in this chapter.

FIGURE 5.10   Improperly installed omnidirectional antenna

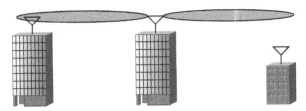

Semidirectional Antennas

Unlike omnidirectional antennas, which radiate an RF signal in all directions, semidirectional antennas are designed to direct a signal in a specific direction. Semidirectional antennas are used for short- to medium-distance communications, with long-distance communications being served by highly directional antennas.

It is common to use semidirectional antennas to provide a network bridge between two buildings in a campus environment or down the street from each other. Longer distances would be served by highly directional antennas.

The following three types of antennas fit into the semidirectional category:

- Patch
- Panel
- Yagi (pronounced *YAH-gee*)

Patch and panel antennas, as shown in Figure 5.11, are more accurately classified or referred to as planar antennas. *Patch* refers to a particular way of designing the radiating elements inside the antenna. Unfortunately, it has become common practice to use the terms *patch antenna* and *panel antenna* interchangeably. If you are unsure of the antenna's specific design, it is better to refer to it as a *planar antenna*.

FIGURE 5.11   The exterior of a patch antenna and the internal antenna element

 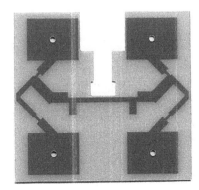

These antennas can be used for outdoor point-to-point communications up to about a mile but are more commonly used as a central device to provide unidirectional coverage from the access point to the clients in an indoor environment. It is common for patch or panel antennas to be connected to access points to provide directional coverage within a building. Planar antennas can be used effectively in libraries, warehouses, and retail stores with long aisles of shelves. Because of the tall, long shelves, omnidirectional antennas often have difficulty providing RF coverage effectively.

In contrast, planar antennas can be placed high on the side walls of the building, aiming through the rows of shelves. The antennas can be alternated between rows, with every other antenna being placed on the opposite wall. Since planar antennas have a horizontal beamwidth of 180 degrees or less, a minimal amount of signal will radiate outside of the building. With the antenna placement alternated and aimed from opposite sides of the building, the RF signal is more likely to radiate down the rows, providing the necessary coverage.

Before the advent of 802.11 MIMO radios, patch and panel antennas were used indoors with legacy 802.11/a/b/g radios to help reduce reflections and hopefully reduce the negative effects of multipath. Semidirectional indoor antennas were often deployed in high multipath environments, such as warehouses or retail stores with a lot of metal racks or shelving. With MIMO technology, patch and panel antennas are no longer needed to reduce multipath, because multipath is constructive with MIMO technology.

802.11n and 802.11ac MIMO patch antennas are still used indoors but for a much different reason. The most common use case for deploying a MIMO patch antenna indoors is a high-density environment. A high-density environment can be described as a small area where numerous Wi-Fi client devices exist. An example might be a school gymnasium or a meeting hall packed with people using multiple Wi-Fi radios. In a high-density scenario, an omnidirectional antenna might not be the best solution for coverage. MIMO patch and panel antennas are often mounted from the ceiling or wall and aimed downward to provide tight "sectors" of coverage. The most common use of indoor MIMO patch antennas is for high-density environments. A discussion about using indoor MIMO patch antennas can be found in Chapter 13, "WLAN Design Concepts."

Yagi-Uda antennas, shown in Figure 5.12, are more commonly known as just *Yagi antennas*. They are typically used for short- to medium-distance point-to-point communications of up to about 2 miles, although high-gain Yagi antennas can be used for longer distances.

Another benefit of semidirectional antennas is that they can be installed high on a wall and tilted downward toward the area to be covered. This cannot be done with an omnidirectional antenna without causing the signal on the other side of the antenna to be tilted upward. Since the only RF signal that radiates from the back of a semidirectional antenna is incidental, the ability to aim it vertically is an additional benefit.

FIGURE 5.12   The exterior of a Yagi antenna and the internal antenna element

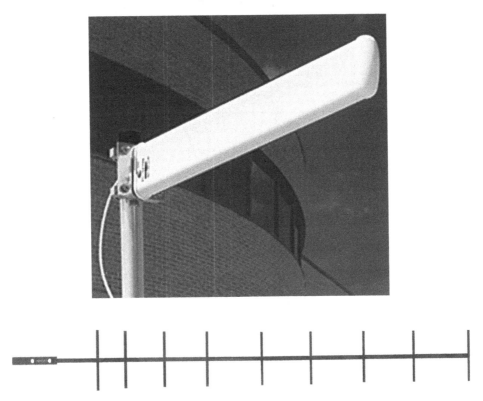

Figure 5.13 shows the radiation patterns of a typical semidirectional panel antenna. Remember that these are actual azimuth and elevation charts from a specific antenna and that every manufacturer and model of antenna will have a slightly different radiation pattern.

Highly Directional Antennas

Highly directional antennas are used strictly for point-to-point communications, typically to provide network bridging between two buildings. They provide the most focused, narrow beamwidth of any of the antenna types.

FIGURE 5.13   Radiation pattern of a typical semidirectional panel antenna

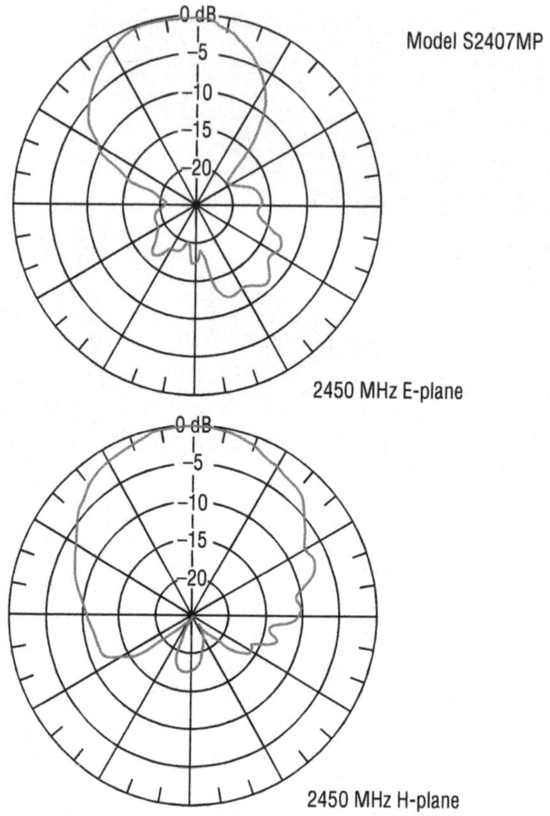

Model S2407MP

2450 MHz E-plane

2450 MHz H-plane

There are two types of highly directional antennas: *parabolic dish antennas* and *grid antennas*:

Parabolic Dish Antennas   The parabolic dish antenna is similar in appearance to the small digital satellite TV antennas that can be seen on the roofs of many houses.

Grid Antennas   As shown in Figure 5.14, the grid antenna resembles the grill of a barbecue, with the edges slightly curved inward. The spacing of the wires on a grid antenna is determined by the wavelength of the frequencies that the antenna is designed for.

Because of the high gain of highly directional antennas, they are ideal for long-distance point-to-point communications.

Because of the long distances and narrow beamwidth, highly directional antennas are affected more by antenna wind loading, which is antenna movement or shifting caused by wind. Even a slight movement of a highly directional antenna can cause the RF beam to be aimed away from the receiving antenna, interrupting or compromising the RF communications. In high-wind environments, grid antennas, because of the spacing between the wires, are less susceptible to wind load and may be a better choice.

FIGURE 5.14   Grid antenna

Image courtesy of Ventev (www.ventevinfra.com).

Another option in high-wind environments is to choose an antenna with a wider beamwidth. In this situation, if the antenna were to shift slightly, the signal would still be received because of its wider coverage area. Keep in mind that a wider beam means less gain. If a solid dish is used, it is highly recommended that a protective cover known as a *radome* be used to help offset some of the effects of the wind. No matter which type of antenna is installed, the quality of the mount and antenna will have a huge effect in reducing wind load.

Sector Antennas

Sector antennas are a special type of high-gain, semidirectional antenna that provide a pie-shaped coverage pattern. These antennas are typically installed in the middle of the area where RF coverage is desired and placed back-to-back with other sector antennas. Individually, each antenna services its own piece of the pie, but as a group, all the pie pieces fit together and provide omnidirectional coverage for the entire area. Combining multiple sector antennas to provide 360 degrees of horizontal coverage is known as a *sectorized array.*

Unlike other semidirectional antennas, a sector antenna generates very little RF signal behind the antenna (*back lobe*) and therefore does not interfere with the other sector antennas it is working with. The horizontal beamwidth of a sector antenna is from 60 to

180 degrees, with a narrow vertical beamwidth of 7 to 17 degrees. Sector antennas typically have a gain of at least 10 dBi.

Installing a group of sector antennas to provide omnidirectional coverage for an area provides many benefits over installing a single omnidirectional antenna:

▪ To begin with, sector antennas can be mounted high over the terrain and tilted slightly downward, with the tilt of each antenna at an angle appropriate for the terrain it is covering. Omnidirectional antennas can also be mounted high over the terrain; however, if an omnidirectional antenna is tilted downward on one side, the other side will be tilted upward.

▪ Since each antenna covers a separate area, each antenna can be connected to a separate transceiver and can transmit and receive independently of the other antennas.

    This provides the capability for all the antennas to be transmitting at the same time, providing much greater throughput. A single omnidirectional antenna is capable of transmitting to only one device at a time.

▪ The last benefit of the sector antennas over a single omnidirectional antenna is that the gain of the sector antennas is much greater than the gain of the omnidirectional antenna, providing a much larger coverage area.

Historically, sector antennas were used extensively for cell phone communications, and are rarely used in Wi-Fi installations. Sector antennas are sometimes used for Wi-Fi as a last-mile outdoor coverage solution by some wireless Internet service providers (WISPs). Outdoor sector antennas are occasionally used in stadium deployments.

 Real World Scenario

Cellular Sector Antennas Are Everywhere

As you walk or drive around your town or city, look for radio communications towers. Many of these towers have what appear to be rings of antennas around them. These rings of antennas are sector antennas. If a tower has more than one grouping or ring around it, then it is likely that multiple cellular carriers are using the same tower.

Antenna Arrays

An *antenna array* is a group of two or more antennas that are integrated together to provide coverage. These antennas operate together to perform what is known as *beamforming*. Beamforming is a method of concentrating RF energy. Concentrating a signal means that the power of the signal will be greater; therefore, the SNR at the receiver should also be greater, providing a better transmission.

There are three different types of beamforming:

- Static beamforming
- Dynamic beamforming
- Transmit beamforming

The following sections explain each of these beamforming methods.

Static Beamforming

Static beamforming is performed by using directional antennas to provide a fixed radiation pattern. Static beamforming uses multiple directional antennas, all clustered together but aimed away from a center point or location. Static beamforming is just another term occasionally used when referring to an indoor sectorized array. Wi-Fi vendor Riverbed/Xirrus manufactures an indoor sectorized array access point solution that uses directional antennas to create multiple beam sectors.

Dynamic Beamforming

Dynamic beamforming focuses the RF energy in a specific direction and in a particular shape. Like static beamforming, the direction and shape of the signal is focused. Unlike static beamforming, however, the radiation pattern of the signal can change on a frame-by-frame basis. This can provide the optimal power and signal for each station. As shown in Figure 5.15, dynamic beamforming uses an *adaptive antenna array* that maneuvers the beam in the direction of a targeted receiver. The technology is often referred to as *smart antenna technology*, or *beamsteering*. Dynamic beamforming capabilities are not available on the client side.

FIGURE 5.15   Dynamic beamforming—adaptive antenna array

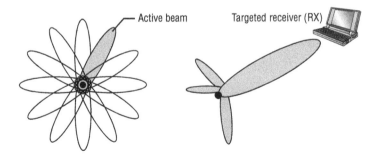

Dynamic beamforming can focus a beam in the direction of an individual client for downstream unicast transmissions between an access point and the targeted client. However, any broadcast frames, such as beacons, are transmitted using an omnidirectional pattern so that the access point can communicate with all nearby client stations in all directions. Note that although Figure 5.15 illustrates the concept, the actual beam is probably more like the signal pattern generated by the antenna shown in Figure 5.13.

Transmit Beamforming

Transmit beamforming (TxBF) is performed by transmitting multiple phase-shifted signals with the hope and intention that they will arrive in-phase at the location where the transmitter believes that the receiver is located. Unlike dynamic beamforming, TxBF does not change the antenna radiation pattern and an actual directional beam does not exist. In truth, transmit beamforming is not really an antenna technology; it is a digital signal–processing technology on the transmitting device that duplicates the transmitted signal on more than one antenna to optimize a combined signal at the client. However, carefully controlling the phase of the signals transmitted from multiple antennas has the effect of improving gain, thus emulating a higher-gain unidirectional antenna. Transmit beamforming is all about adjusting the phase of the transmissions.

The 802.11n amendment defined two types of transmit beamforming: *implicit TxBF* and *explicit TxBF*. Implicit TxBF uses an implicit channel-sounding process to optimize the phase differentials between the transmit chains. Explicit TxBF requires feedback from the stations in order to determine the amount of phase-shift required for each signal. The 802.11ac amendment defines explicit TxBF, requiring the use of channel measurement frames, and both the transmitter and the receiver must support beamforming. 802.11ac will be discussed in greater detail in Chapter 10, "MIMO Technology: HT and VHT."

Visual Line of Sight

When light travels from one point to another, it travels across what is perceived to be an unobstructed straight line, known as the visual *line of sight (LOS)*. For all intents and purposes, it is a straight line, but because of the possibility of light refraction, diffraction, and reflection, there is a slight chance that it is not. If you have been outside on a summer day and looked across a hot parking lot at a stationary object, you may have noticed that, because of the heat rising from the pavement, the object that you were looking at seemed to be moving. This is an example of how visual LOS is sometimes altered slightly. When it comes to RF communications, visual LOS has no bearing on whether the RF transmission is successful.

RF Line of Sight

Point-to-point RF communication also needs to have an unobstructed line of sight between the two antennas. So, the first step for installing a point-to-point system is to make sure that, from the installation point of one of the antennas, you have a clear direct path to the other antenna. Unfortunately, for RF communications to work properly, this is not sufficient. An additional area around the visual LOS needs to remain clear of obstacles and obstructions. This area around the visual LOS is known as the Fresnel zone and is often referred to as *RF line of sight*.

Fresnel Zone

The *Fresnel zone* (pronounced *FRUH-nel*—the *s* is silent) is an imaginary, elongated, football-shaped area (American football) that surrounds the path of the visual LOS between two point-to-point antennas. Figure 5.16 shows an illustration of the Fresnel zone's football-like shape.

FIGURE 5.16   Fresnel zone

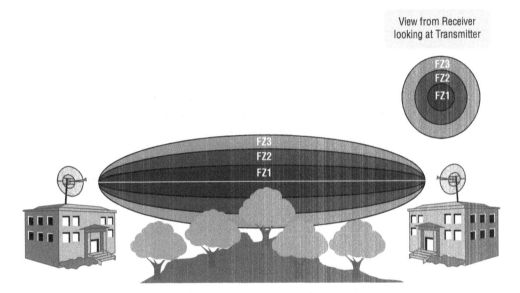

Theoretically, there is an infinite number of Fresnel zones, or concentric ellipsoids (the football shape), that surround the visual LOS. The closest ellipsoid is known as the first Fresnel zone, the next one is the second Fresnel zone, and so on, as depicted in Figure 5.16. For simplicity's sake, and because they are the most relevant for this section, only the first two Fresnel zones are displayed in the figure. The subsequent Fresnel zones have little effect on communications.

If the first Fresnel zone becomes even partly obstructed, the obstruction will negatively influence the integrity of the RF communication. In addition to the obvious reflection and scattering that can occur if there are obstructions between the two antennas, the RF signal can be diffracted, or bent, as it passes an obstruction of the Fresnel zone. This diffraction of the signal decreases the amount of RF energy that is received by the antenna and may even cause the communications link to fail.

Figure 5.17 illustrates a link that is one mile long. The top solid line is a straight line from the center of one antenna to the other. The dotted line shows 60 percent of the bottom half of the first Fresnel zone. The bottom solid line shows the bottom half of the first Fresnel zone. The trees are potential obstructions along the path.

FIGURE 5.17   Fresnel zone clearances of 60 percent and 100 percent

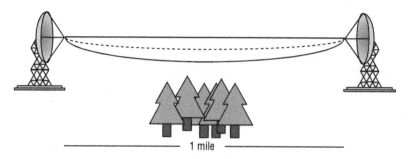

1 mile

Under no circumstances should you allow any object or objects to encroach more than 40 percent into the first Fresnel zone of an outdoor point-to-point bridge link. Anything more than 40 percent is likely to make the communications link unreliable. Even an obstruction of less than 40 percent is likely to impair the performance of the link. Therefore, we recommend that you not allow any obstruction of the first Fresnel zone, particularly in wooded areas, where the growth of trees may obstruct the Fresnel zone in the future.

The typical obstacles that you are likely to encounter are trees and buildings. It is important to periodically visually check your link to make sure that trees have not grown into the Fresnel zone or that buildings have not been constructed that encroach into the Fresnel zone. Do not forget that the Fresnel zone exists below, to the sides of, and above the visual LOS. If the Fresnel zone does become obstructed, you will need to either move the antenna (usually raise it) or remove the obstacle (usually with a chain saw—just kidding).

To determine whether an obstacle is encroaching into the Fresnel zone, you need to be familiar with a few formulas that enable you to calculate its radius. Don't fret; you will not be tested on these formulas.

The first formula enables you to calculate the radius of the first Fresnel zone at the midpoint between the two antennas. This is the point where the Fresnel zone is the largest. This formula is as follows:

$$radius = 72.2 \times \sqrt{[D \div (4 \times F)]}$$

D = distance of the link in miles

F = transmitting frequency in GHz

This is the optimal clearance that you want along the signal path. Although this is the ideal radius, it is not always feasible. Therefore, the next formula will be very useful. It can be used to calculate the radius of the Fresnel zone that will enable you to have 60 percent of the Fresnel zone unobstructed. This is the minimum amount of clearance you need at the midpoint between the antennas. Here is the formula:

$$radius\ (60\%) = 43.3 \times \sqrt{[D \div (4 \times F)]}$$

D = distance of the link in miles

F = transmitting frequency in GHz

Both of these formulas are useful, but they also have major shortcomings. These formulas calculate the radius of the Fresnel zone at the midpoint between the antennas. Since this is the point where the Fresnel zone is the largest, these numbers can be used to determine the minimum height the antennas need to be above the ground. You have to know this number, because if you place the antennas too low, the ground would encroach on the Fresnel zone and cause degradation to the communications. The problem is that if there is a known object somewhere other than the midpoint between the antennas, it is not possible to calculate the radius of the Fresnel zone at that point by using these equations. The following formula can be used to calculate the radius of any Fresnel zone at any point between the two antennas:

$$\text{radius} = 72.2 \times \sqrt{[(N \times d1 \times d2) \div (F \times D)]}$$

N = which Fresnel zone you are calculating (usually 1 or 2)

d1 = distance from one antenna to the location of the obstacle in miles

d2 = distance from the obstacle to the other antenna in miles

D = total distance between the antennas in miles (D = d1 + d2)

F = frequency in GHz

Figure 5.18 shows a point-to-point communications link that is 10 miles long. There is an obstacle (tree) that is 3 miles away from one antenna and 40 feet tall. So, the values and the formula to calculate the radius of the Fresnel zone at a point 3 miles from the antenna are as follows:

N = 1 (for first Fresnel zone)

d1 = 3 miles

d2 = 7 miles

D = 10 miles

F = 2.4 GHz

$$\text{radius at 3 miles} = 72.2 \times \sqrt{[(1 \times 3 \times 7) \div (2.4 \times 10)]}$$

$$\text{radius at 3 miles} = 72.2 \times \sqrt{[21 \div 24]}$$

$$\text{radius at 3 miles} = 67.53 \text{ feet}$$

FIGURE 5.18   Point-to-point communication with potential obstacle

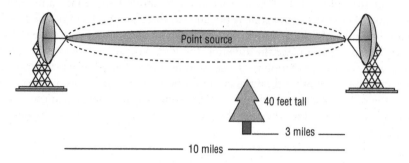

So, if the obstacle is 40 feet tall and the Fresnel zone at that point is 67.53 feet tall, then the antennas need to be mounted at least 108 feet above the ground to have complete clearance (40′ + 67.53′ = 107.53′; we rounded up.) If we are willing to allow the obstruction to encroach up to 40 percent into the Fresnel zone, we need to keep 60 percent of the Fresnel zone clear. So, 60 percent of 67.53 feet is 40.52 feet. The absolute minimum height of the antennas will need to be 81 feet (40′ + 40.52′ = 80.52′; again, we rounded up). In the next section, you will learn that, because of the curvature of the earth, you will need to raise the antennas even higher to compensate for the earth's bulge.

When highly directional antennas are used, the beamwidth of the signal is smaller, causing a more focused signal to be transmitted. Many people think that a smaller beamwidth would decrease the size of the Fresnel zone. This is not the case. The size of the Fresnel zone is a function of the frequency being used and the distance of the link. Since the only variables in the formula are frequency and distance, the size of the Fresnel zone will be the same regardless of the antenna type or beamwidth. The first Fresnel zone is technically the area around the point source, where the waves are in phase with the point source signal. The second Fresnel zone is then the area beyond the first Fresnel zone, where the waves are out of phase with the point source signal. All the odd-numbered Fresnel zones are in phase with the point source signal, and all the even-numbered Fresnel zones are out of phase.

If an RF signal of the same frequency but out of phase with the primary signal intersects the primary signal, the out-of-phase signal will cause degradation or even cancellation of the primary signal. (This was demonstrated in Chapter 3, "Radio Frequency Fundamentals," using the EMANIM software.) One of the ways that an out-of-phase signal can intercept the primary signal is by reflection. It is, therefore, important to consider the second Fresnel zone when evaluating point-to-point communications. If the height of the antennas and the layout of the geography are such that the RF signal from the second Fresnel zone is reflected toward the receiving antenna, degradation of the link can occur. Although this is not a common occurrence, the second Fresnel zone should be considered when planning or troubleshooting the connection, especially in flat, arid terrain, like a desert. You should also be cautious of metal surfaces or calm water along the Fresnel zone.

Please understand that the Fresnel zone is three-dimensional. Can something impede on the Fresnel zone from above? Although trees do not grow from the sky, a point-to-point bridge link could be shot under a railroad trestle or a freeway. In these rare situations, consideration would have to be given to proper clearance of the upper radius of the first Fresnel zone. A more common scenario would be the deployment of point-to-point links in an urban city environment. Very often building-to-building links must be shot between other buildings. In these situations, other buildings have the potential of impeding the side radiuses of the Fresnel zone.

Until now, all the discussion about the Fresnel zone has related to point-to-point communications. The Fresnel zone exists in all RF communications; however, it is in outdoor point-to-point communications where it can cause the most problems. Indoor environments have so many walls and other obstacles where there is already so much reflection, refraction, diffraction, and scattering that the Fresnel zone is not likely to play a big part in the success or failure of the link.

Earth Bulge

When you are installing long-distance point-to-point RF communications, another variable that must be considered is the curvature of the earth, also known as the *earth bulge*. Because the landscape varies throughout the world, it is impossible to specify an exact distance for when the curvature of the earth will affect a communications link. The recommendation is that if the antennas are more than seven miles away from each other, you should take into consideration the earth bulge, because after seven miles, the earth itself begins to impede on the Fresnel zone. The following formula can be used to calculate the additional height that the antennas will need to be raised to compensate for the earth bulge:

$$H = D^2 \div 8$$

H = height of the earth bulge in feet

D = distance between the antennas in miles

You now have all the pieces to estimate how high the antennas need to be installed. Remember, this is an estimate that is being calculated, because it is assumed that the terrain between the two antennas does not vary. You need to know or calculate the following three things:

- The 60 percent radius of the first Fresnel zone
- The height of the earth bulge
- The height of any obstacles that may encroach into the Fresnel zone, and the distance of those obstacles from the antenna

Taking these three pieces and adding them together gives you the following formula, which can be used to calculate the antenna height:

$$H = \text{obstacle height} + \text{earth bulge} + \text{Fresnel zone}$$

$$H = OB + (D^2 \div 8) + \left(43.3 \times \sqrt{D \div (4 \times F)}\right)$$

OB = obstacle height

D = distance of the link in miles

F = transmitting frequency in GHz

Figure 5.19 shows a point-to-point link that spans a distance of 12 miles. In the middle of this link is an office building that is 30 feet tall. A 2.4 GHz signal is being used to communicate between the two towers. Using the formula, we calculate that each of the antennas needs to be installed at least 96.4 feet above the ground:

$$H = 30 + (12^2 \div 8) + \left(43.3 \times \sqrt{[12 \div (4 \times 2.4)]}\right)$$

$$H = 30 + 18 + 48.4$$

$$H = 96.4$$

FIGURE 5.19   Calculating antenna height

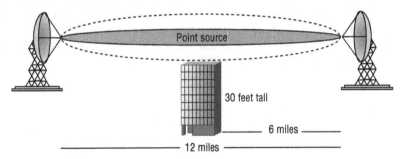

Although these formulas are useful, the good news is that you do not need to know them for the test. Many of these formulas are also available in the form of free online calculators. One such resource is www.everythingrf.com/rf-calculators.

Antenna Polarization

Another consideration when installing antennas is antenna polarization. Although a lesser-known concern, it is extremely important for successful communications. Proper polarization alignment is vital when installing any type of antenna. As waves radiate from an antenna, the amplitude of the waves can oscillate either vertically or horizontally. It is important to have the polarization of the transmitting and receiving antennas oriented the same in order to receive the strongest possible signal. Whether the antennas are installed with horizontal or vertical polarization is usually irrelevant, so long as both antennas are aligned with the same polarization.

When discussing antennas, the proper term is antenna *polarization*, which refers to the alignment or orientation of the waves. The use of the term *polarity* is incorrect. Polarity refers to positive or negative voltage, which has no relevance to antenna orientation.

Polarization is not as important for indoor communications, because the polarization of the RF signal often changes when it is reflected, which is a common occurrence indoors. Most access points use low-gain omnidirectional antennas, which should be mounted from the ceiling with vertical polarization. Laptop manufacturers build antennas into the sides of the monitor. When the laptop monitor is in the upright position, the internal antennas have vertical polarization as well.

When you are aligning a point-to-point or point-to-multipoint bridge, proper polarization is extremely important. If the best received signal level (RSL) you receive when aligning the antennas is 15 dB to 20 dB less than your estimated RSL, there is a good chance you have cross-polarization. If this difference exists on only one side and the other has a higher signal, you are likely aligned to a side lobe.

An excellent video, *Beam Patterns and Polarization of Directional Antennas*, is available for download from the book's online resource area, which you can access at www.wiley.com/go/cwnasg. This 3-minute video explains and demonstrates the effects of antenna side lobes and polarization. The filename of the video is Antenna Properties.wmv.

Antenna Diversity

Wireless networks, especially indoor networks, are prone to multipath signals. To help compensate for the effects of multipath, antenna diversity, also called spatial diversity, is commonly implemented in wireless networking equipment, such as access points. *Antenna diversity* exists when an access point has two or more antennas with a receiver functioning together to minimize the negative effects of multipath.

Because the wavelengths of 802.11 wireless networks are less than 5 inches long, the antennas can be placed very near each other and still allow antenna diversity to be effective. When the access point senses an RF signal, it compares the signal that it is receiving on both antennas and uses whichever antenna has the higher signal strength to receive the frame of data. This sampling is performed on a frame-by-frame basis, choosing whichever antenna has the higher signal strength.

Most pre-802.11n radios use *switched diversity*. When receiving incoming transmissions, switched diversity listens with multiple antennas. Multiple copies of the same signal arrive at the receiver antennas with different amplitudes. The signal with the best amplitude is chosen, and the other signals are ignored. The AP will use one antenna as long as the signal is above a predefined signal level. If the signal degrades below the acceptable level, then the AP will use the signal received on the other antenna.

This method of listening for the best received signal is also known as *receive diversity*. Switched diversity is also used when transmitting, but only one antenna is used. The transmitter will transmit out of the diversity antenna where the best amplitude signal was last heard. The method of transmitting out of the antenna where the last best-received signal was heard is known as *transmit diversity*.

> When an access point has two antenna ports for antenna diversity, the antennas should have identical gain and should be installed in the same location and with the same orientation. You should not be running antenna cables to antennas in opposite directions to try to provide better coverage. Remember, when diversity is used, the transceiver will switch between the antennas; therefore, the antennas need to provide essentially the same coverage. The distance between the antennas should be a factor of the wavelength (1/4, 1/2, 1, 2).

Because the antennas are so close to each other, it is not uncommon to doubt that antenna diversity is actually beneficial. As you may recall from Chapter 4, the strength of an RF signal that is received is often less than 0.00000001 milliwatts. At this level of signal, the slightest difference between the signals that each antenna receives can be significant. Another factor to remember is that the access point is often communicating with multiple client devices at different locations. These clients are not always stationary, which further affects the path of the RF signal.

The access point has to handle transmitting data differently from receiving data. When the access point needs to transmit data back to the client, it has no way of determining which antenna the client would receive from the best. An access point can handle transmitting data by using the antenna that it used most recently to receive the data. You will remember that this is often referred to as *transmit diversity*. Not all access points are equipped with this capability.

There are many kinds of antenna diversity. Laptops with internal cards usually have diversity antennas mounted inside the laptop monitor. Remember that because of the half-duplex nature of the RF medium, when antenna diversity is used, only one antenna is operational at any given time. In other words, a radio card transmitting a frame with one antenna cannot be receiving a frame with the other antenna at the same time.

Multiple-Input, Multiple-Output

Multiple-input, multiple-output (MIMO) is another, more sophisticated form of antenna diversity. Unlike conventional antenna systems, where multipath propagation is an impairment, MIMO (pronounced MY-*moh*) systems take advantage of multipath. MIMO can safely be described as a wireless radio architecture that can receive or transmit using multiple antennas concurrently. Complex signal-processing techniques enable significant enhancements to reliability, range, and throughput in MIMO systems. These techniques send data by using multiple simultaneous RF signals. The receiver then reconstructs the data from those signals.

802.11n and 802.11ac radios use MIMO technology. One of the key goals when installing a MIMO device is to ensure that each of the signals from the different radio chains travel with different signal polarization. This can be done by aligning or orienting the antennas so that the path that each signal travels is at least slightly different. This will help to introduce delay between the different MIMO signals, which will improve the ability for the MIMO receiver to process the different signals. Different types of MIMO antennas are discussed in the next section, and MIMO technology is explored in much further detail in Chapter 10, as it is a key component of 802.11n and 802.11ac.

MIMO Antennas

With the need and desire to increase the throughput and capacity of wireless networks, the installation of 802.11n and 802.11ac access points has become the norm. 802.11n and 802.11ac have become commonplace not only for indoor networks but also for outdoor networks and point-to-point networks. MIMO antenna selection and placement is important in each of these environments.

Indoor MIMO Antennas

There is usually not much decision-making involved regarding the antennas on an indoor MIMO access point. Many of the new enterprise MIMO access points have the antennas integrated into the chassis of the access point, with no antennas protruding from the access point. If the antennas are not integrated into the access point chassis, the MIMO access point likely has three or four omnidirectional antennas directly attached to it. In some cases, the antennas are detachable, allowing you to instead choose higher-gain omnidirectional or indoor MIMO patch antennas.

Outdoor MIMO Antennas

As with the indoor access points, multipath provides a benefit for successful and higher data rate communications with outdoor MIMO devices. This benefit may not be realized if the environment does not have reflective surfaces that induce multipath. Therefore, it is important to try to change the radiation path of the antennas while maintaining the same range and coverage with all the antennas. In an outdoor environment, achieving this goal requires more knowledge and technology than can usually be achieved by leaving the antenna choice and placement up to the designer or installer of the networks. Therefore, many of the access point and antenna manufacturers have designed both omnidirectional and directional MIMO antennas.

To distinguish the different radio chain signals from each other, the directional MIMO antennas incorporate multiple antenna elements within one physical antenna. The antenna will have two, three, or four connectors to connect it to the access point. If the access point that the antenna is being connected to is a multi-radio access point, the access point will have multiple antenna connectors for each radio. It is important to make sure that the cables from the antenna are connected to the antenna jacks for the same radio.

To provide omnidirectional MIMO coverage with multiple radio chain APs, special sets of omnidirectional antennas are available. Each set is made up of an omnidirectional antenna with vertical polarization and a second omnidirectional antenna with horizontal polarization. It is a little strange to use these antennas, because in the past with legacy non-802.11n access points, if two omnidirectional antennas were installed on an access point, it was important to purchase identical antennas. With outdoor MIMO omnidirectional antennas, the antennas are purchased as a set, but they are typically of different lengths and widths because of the different polarization that each antenna has. If you are not familiar with these new antenna pairs, you may think that you were shipped the wrong product, due to the antennas not looking the same. When trying to provide omnidirectional coverage, special single-chassis antennas exist that can be used with MIMO APs. One particular type, known as a *down-tilt antenna*, is made up of multiple antenna elements mounted within one antenna body. The antenna is typically mounted in a high location, is mounted horizontally above the area of coverage, and is faced down at the floor or ground below. The horizontal coverage area is omnidirectional. The vertical coverage behaves like a typical omnidirectional antenna but with more vertical signal/coverage below the antenna than above.

Antenna Connection and Installation

In addition to the physical antenna being a vital component in the wireless network, the installation and connection of the antenna to the wireless transceiver is critical. If the antenna is not properly connected and installed, any benefit that the antenna introduces to the network can be instantly wiped out. Three key components associated with the proper installation of the antenna are voltage standing wave ratio (VSWR), signal loss, and the actual mounting of the antenna.

Voltage Standing Wave Ratio

Voltage standing wave ratio (VSWR) is a measurement of the change in impedances to an AC signal. Voltage standing waves exist because of impedance mismatches or variations between devices in an RF communications system. Impedance is a value of ohms of electrical resistance to an AC signal. A standard unit of measurement of electrical resistance is the ohm, named after German physicist Georg Ohm. When the transmitter generates the AC radio signal, the signal travels along the cable to the antenna. Some of this incident (or forward) energy is reflected back toward the transmitter because of impedance mismatch.

Mismatches may occur anywhere along the signal path but are usually due to abrupt impedance changes between the radio transmitter and cable and between the cable and the antenna. The amount of energy reflected depends on the level of mismatch between the transmitter, cable, and antenna. The ratio between the voltage of the reflected wave and the voltage of the incident wave, at the same point along the cable, is called the *voltage reflection coefficient*, usually designated by the Greek letter rho (ρ).

In an ideal system, where there are no mismatches (the impedance is the same everywhere), all the incident energy will be delivered to the antenna (except for the resistive losses in the cable itself), and there will be no reflected energy. The cable is said to be *matched*; and the voltage reflection coefficient is exactly zero; and the *return loss*, in dB, is infinite. Return loss is essentially the dB difference between the power sent to the antenna and the power reflected back; a higher value is better than a lower value. The combination of incident and reflected waves traveling back and forth along the cable creates a resulting *standing wave* pattern along the length of the line. The standing wave pattern is periodic (it repeats) and exhibits multiple peaks and troughs of voltage, current, and power.

VSWR is a numerical relationship between the measurement of the maximum voltage along the line (what is generated by the transmitter) and the measurement of the minimum voltage along the line (what is received by the antenna). As shown in the equation, VSWR is therefore a ratio of impedance mismatch, with 1:1 (no impedance) being optimal but unobtainable; typical values range from 1.1:1 to as much as 1.5:1. VSWR military specs are 1.1:1.

$$VSWR = V_{max} \div V_{min}$$

When the transmitter, cable, and antenna impedances are matched (that is, there are no standing waves), the voltage along the cable will be constant. This matched cable is also referred to as a *flat line*, because there are no peaks and troughs of voltage along the length of the cable. In this case, VSWR is 1:1. As the degree of mismatch increases, the VSWR increases with a corresponding decrease in the power delivered to the antenna. Table 5.2 shows this effect.

TABLE 5.2   Signal loss caused by VSWR

VSWR	Radiated Power	Lost Power	Return Loss	dB Power Loss
1:1	100%	0%	Infinite	0 dB
1.5:1	96%	4%	14 dB	Nearly 0 dB
2:1	89%	11%	9.5 dB	< 1 dB
6:1	50%	50%	2.9 dB	3 dB

If VSWR is large, this means that a large amount of voltage is being reflected back toward the transmitter. This, of course, means a decrease in power or amplitude (loss) of the signal that is supposed to be transmitted. This loss of forward amplitude is known as *return loss* and can be measured in dB. Additionally, the power that is being reflected back is then directed back into the transmitter. If the transmitter is not protected from excessive reflected power or large voltage peaks, it can overheat and fail. Understand that VSWR may cause decreased signal strength, erratic signal strength, or even transmitter failure.

The first thing that can be done to minimize VSWR is to make sure that the impedance of all the wireless networking equipment is matched. Most wireless networking equipment has an impedance of 50 ohms; however, you should check the manuals to confirm this. When attaching the different components, make sure that all connectors are installed and crimped properly and that they are snugly tightened.

Signal Loss

When connecting an antenna to a transmitter, the main objective is to make sure that as much of the signal that is generated by the transmitter is received by the antenna to be transmitted. To achieve this, it is important to pay particular attention to the cables and connectors that connect the transmitter to the antenna. In the "Antenna Accessories" section later in this chapter, we review the cables, connectors, and many other components that are used when installing antennas. If inferior components are used, or if the components are not installed properly, the access point will most likely function below its optimal capability.

Antenna Mounting

As previously mentioned, proper installation of the antenna is one of the most important tasks to ensure an optimally functioning network. The following are key areas to be concerned with when installing antennas:

- Placement
- Mounting
- Appropriate use and environment
- Orientation and alignment
- Safety
- Maintenance

Placement

The proper placement of an antenna depends on the type of antenna. Omnidirectional antennas typically are placed at the center of the area where you want coverage. Remember that lower-gain omnidirectional antennas provide broader vertical coverage, whereas higher-gain omnidirectional antennas provide less vertical coverage. Be careful not to place high-gain omnidirectional antennas too high above the ground, because the narrow vertical coverage may cause the antenna to provide insufficient signal to clients located on the ground.

When installing directional antennas, make sure you know both the horizontal and vertical beamwidths so that you can properly aim the antennas. Also make sure that you are aware of the amount of gain the antenna is adding to the transmission. If the signal is too strong, it will overshoot the area that you are looking to provide coverage to. This can be a security risk, and you may want to decrease the amount of power that the transceiver is generating to reduce the coverage area, provided that this signal decrease does not compromise the performance of your link. In addition to being a security risk, overshooting your coverage area is considered rude.

If you are installing an outdoor directional antenna, in addition to the concerns regarding the horizontal and vertical beamwidths, make sure that you have correctly calculated the Fresnel zone and mounted the antenna accordingly.

Indoor Mounting Considerations

After deciding where to place the antenna, the next step is to decide how to mount it. There are numerous ways of mounting antennas indoors. Most access points have at least a couple of keyhole-type mounts for hanging the access point off of a couple of screws on a wall. Most enterprise-class access points have mounting kits that allow you to mount the access point to a wall or ceiling. Many of these kits are designed to easily attach directly to the metal rails of a drop ceiling.

Two common concerns are aesthetics and security. Many organizations, particularly those that provide hospitality-oriented services, such as hotels and hospitals, are concerned about the aesthetics of the installation of the antennas. Specialty enclosures and ceiling tiles can help to hide the installation of the access points and antennas. Other organizations, particularly schools, are concerned with securing the access points and antennas from theft or vandalism. An access point can be locked in a secure enclosure, with a short cable connecting it to the antenna. If security is a concern, mounting the antenna high on the wall or ceiling can also minimize unauthorized access.

If access points or antennas are installed below the ceiling, children or teens will often try to jump up and hit the antennas or throw things at them in an attempt to move them. This also needs to be considered when choosing locations to install antennas.

Outdoor Mounting Considerations

Many antennas, especially outdoor antennas, are mounted on masts or towers. It is common to use mounting clamps and U-bolts to attach the antennas to the masts. For mounting directional antennas, specially designed tilt-and-swivel mounting kits are available to make it easier to aim and secure the antenna. If the antenna is being installed in a windy location (and what rooftop or tower is not windy?), make sure that you take into consideration wind load and properly secure the antenna.

Appropriate Use and Environment

Make sure that indoor access points and antennas are not used for outdoor communications. Outdoor access points and antennas are specifically built to withstand the wide range of temperatures they may be exposed to. It is important to make sure that the environment where you are installing the equipment is within the operating temperature range of the access point and antennas. The extreme cold weather of northern Canada may be too cold for some equipment, whereas the extreme heat of the desert in Saudi Arabia may be too hot. Outdoor access points and antennas are also built to stand up to other elements, such as rain, snow, and fog. In addition to installing the proper devices, make sure that the mounts you use are designed for the environment in which you are installing the equipment.

With the expansion of wireless networking, it is becoming more common to not only install wireless devices in harsh environments but also to install them in potentially flammable or combustible environments, such as mines and oil rigs. Installation of access points and antennas in these environments requires special construction of the devices or the installation of the devices in special enclosures.

In the following sections, you will learn about four classification standards. The first two standards designate how a device will stand up to harsh conditions, and the following two standards designate the environments in which a device is allowed to operate. These are just four examples of standards that exist and how they apply to equipment and environments. You will need to do research to determine if there are requirements to which you must (or should) adhere to in your country, region, or environment when installing equipment.

Ingress Protection Rating

The *Ingress Protection Rating* is sometimes referred to as the International Protection Rating and is commonly referred to as the *IP Code* (not to be confused with Internet Protocol, which is part of TCP/IP). The IP Rating system is published by the International Electrotechnical Commission (IEC). The IP Code is represented by the letters *IP* followed by two digits or a digit and one or two letters, such as IP66.

The first digit of the IP Code classifies the degree of protection that the device provides against the intrusion of solid objects, and the second digit classifies the degree of protection that the device provides against the intrusion of water. If no protection is provided for either of these classifications, the digit is replaced with the letter *X*.

The solids digit can be a value between 0 and 6, with protection ranging from no protection (0) up to dust tight (6). The liquids digit can be a value between 0 and 8, including, for example, no protection (0), dripping water (1), water splashing from any direction (4), powerful water jets (6), and immersion greater than one meter (8).

NEMA Enclosure Rating

The *NEMA Enclosure Rating* is published by the United States National Electrical Manufacturers Association (NEMA). The NEMA ratings are similar to the IP ratings, but the NEMA ratings also specify other features, such as corrosion resistance, gasket aging, and construction practices.

The NEMA enclosure types are defined in the NEMA Standards Publication 250-2014, "Enclosures for Electrical Equipment (1000 Volts Maximum)." This document defines the degree of protection from such things as solid foreign objects, such as dirt, dust, lint, and fibers, along with the ingress of water, oil, and coolant. The rating for the NEMA enclosures is in the form of a number or a number followed by a letter, such as Type 2 or Type 12 K.

NEMA enclosures, like the one shown in Figure 5.20, are often needed to protect outdoor APs from weather conditions. Many WLAN vendors also manufacture outdoor APs that are already NEMA rated.

FIGURE 5.20   NEMA enclosure

Image courtesy of Ventev (www.ventevinfra.com).

ATEX Directives

There are two ATEX directives:

ATEX 2014/34/EU   A revision of the earlier ATEX 94/9/EC directive, this went into effect on April 20, 2016. ATEX 2014/34/EU pertains to equipment and protective systems that are intended to be used in potentially explosive atmospheres.

ATEX 137   Also known as directive ATEX 99/92/EC, ATEX 137 pertains to the workplace and is intended to protect and improve the safety and health of workers at risk from explosive atmospheres.

Organizations in the European Union must follow these directives to protect employees. The ATEX directives inherit their name from the French title of the 94/9/EC directive: "Appareils destinés à être utilisés en ATmosphères EXplosibles."

Employers must classify work areas where explosive atmospheres may exist into different zones. Areas can be classified for gas-vapor-mist environments or dust environments. These regulations apply to all equipment, whether mechanical or electrical, and are categorized for mining and surface industries. ATEX enclosure units are often needed in these environments for obvious safety reasons.

National Electrical Code Hazardous Locations

The *National Electrical Code (NEC)* is a standard for the safe installation of electrical equipment and wiring. The document itself is not a legally binding document, but it can and has been adopted by many local and state governments in the United States, thus making it law in those places. A substantial part of the NEC discusses hazardous locations. The NEC classifies hazardous locations by type, condition, and nature. The hazardous location type is defined as follows:

- **Class I:** Gas or vapor
- **Class II:** Dust
- **Class III:** Fibers and flyings

The type is further subdivided by the conditions of the hazardous location:

- **Division 1:** Normal conditions (for example, a typical day at the loading dock)
- **Division 2:** Abnormal conditions (same loading dock, but a container is leaking its contents)

A final classification defines a group for the hazardous substance, based on the nature of the substance. This value is represented by an uppercase letter ranging from *A* through *G*.

Orientation and Alignment

Before installing an antenna, make sure you read the manufacturer's recommendations for mounting it. This suggestion is particularly important when installing directional antennas. Since directional antennas may have different horizontal and vertical beamwidths, and because directional antennas can be installed with different polarization, proper orientation can make the difference between being able to communicate or not:

1. Make sure that the antenna polarization is consistent on both ends of a directional link.

2. Decide on the mounting technique and ensure that it is compatible with the mounting location.

3. Align the antennas. Remember that you need to align both the direction of the antenna and its vertical tilt.

4. Weatherproof the cables and connectors and secure them from movement.

5. Document and photograph each installation of the access point and antennas. This can help you troubleshoot problems in the future and allows you to more easily determine if there has been movement in the installation or antenna alignment.

As previously mentioned, with the transition to 802.11n, 802.11ac, and MIMO, special two-radio chain, outdoor omnidirectional MIMO antennas have been designed to be installed as pairs, with one antenna generating a signal with vertical polarization and the other generating a signal with horizontal polarization.

Safety

We cannot emphasize enough the importance of being careful when installing antennas. Most of the time, the installation of an antenna requires climbing ladders, towers, or rooftops. Gravity and wind have a way of making an installation difficult for both the climber and the people below helping.

Plan the installation before you begin, making sure you have all the tools and equipment that you will need to install the antenna. Unplanned stoppages of the installation and relaying forgotten equipment up and down a ladder add to the risk of injury.

Be careful when working with your antenna or near other antennas. Highly directional antennas are focusing high concentrations of RF energy. This large amount of energy can be dangerous to your health. Do not power on your antenna while you are working on it, and do not stand in front of other antennas that are near where you are installing your antenna. You probably do not know the frequency or power output of these other antenna systems, nor the potential health risks that you might be exposed to.

When installing antennas (or any device) on ceilings, rafters, or masts, make sure they are properly secured. Even a one-pound antenna can be deadly if it falls from the rafters of a warehouse.

If you will be installing antennas as part of your job, we recommend that you take an RF health and safety course. In the United States, these courses will teach you the FCC and the US Department of Labor Occupational Safety and Health Administration (OSHA) regulations and how to be safe and compliant with the standards. Similar courses can be found in many other countries around the world. We suggest looking for courses that are appropriate to your country or region.

If you need an antenna installed on any elevated structure, such as a pole, tower, or even a roof, consider hiring a professional installer. Professional climbers and installers are trained and, in some places, certified to perform these types of installations. In addition to the training, they have the necessary safety equipment and proper insurance for the job.

If you are planning to install wireless equipment as a profession, you should develop a safety policy that is approved by your local occupational safety representative. You should also receive certified training on climbing safety in addition to RF safety training. First aid and CPR training are also highly recommended.

Maintenance

There are two types of maintenance: preventive and diagnostic. When installing an antenna, it is important to prevent problems from occurring in the future. This seems like simple advice, but since antennas are often difficult to get to after they have been installed, it is especially prudent advice. Two key problems that can be minimized with proper preventative measures are wind damage and water damage. When installing the antenna, ensure that all the nuts, bolts, screws, and so on are installed and tightened properly. Also make sure all the cables are properly secured so that they are not thrashed about by the wind.

To help prevent water damage, cold-shrink tubing or coaxial sealant can be used to minimize the risk of water getting into the cable or connectors. Another common method is a combination of electrical tape and mastic, installed in layers to provide a completely watertight installation. If mastic is used, be sure to first tightly wrap the connection with electrical tape before applying the mastic. If the connection ever needs to be disconnected and reattached, it will be virtually impossible to remove the mastic if it has been applied directly to the connector.

WARNING Heat-shrink tubing should not be used, because the cable can be damaged by the heat that is necessary to shrink the wrapping. Silicone also should not be used, because air bubbles can form under the silicone and moisture can collect.

Another cabling technique is the drip loop. To create a drip loop, when a cable is run down to a connector, run the cable down below the connector and then loop it up to the connector, creating a small loop or "U" of cable below the connector. Drip loops are also used when a cable is run into a building or structure. A drip loop prevents water from flowing down the cable and onto a connector or into the hole where a cable enters the building. Any water that is flowing down the cable will continue to the bottom of the loop and then drip off.

Antennas are typically installed and forgotten about until they break. It is advisable to periodically perform a visual inspection of the antenna and, if needed, verify its status with the installation documentation. If the antenna is not easily accessible, a pair of binoculars or a camera with a very high zoom lens can make this a simple task.

Antenna Accessories

In Chapter 4, we introduced the main components of RF communications. There are additional components that are either not as significant or not always installed as part of the communications link. Important specifications for all antenna accessories include frequency response, impedance, VSWR, maximum input power, and insertion loss. We discuss some of these components and accessories in the following sections.

Cables

The improper installation or selection of cables can detrimentally affect the RF communications more than just about any other component or outside influence. It is important to remember this fact when installing antenna cables. The following list addresses some concerns when selecting and installing cables:

- Make sure you select the correct cable.

  The impedance of the cable needs to match the impedance of the antenna and transceiver. If there is an impedance mismatch, the return loss from VSWR will affect the link.

- Make sure the cable you select will support the frequencies that you will be using.

  Typically, cable manufacturers list cutoff frequencies, which are the lowest and highest frequencies that the cable supports. This is often referred to as frequency response. For instance, LMR cable is a popular brand of coaxial cable used in RF communications. LMR-1200 will not work with 5 GHz transmissions. LMR-900 is the highest you can use. However, you can use LMR-1200 for 2.4 GHz operations.

- Cables introduce signal loss into the communications link.

  Cable vendors provide charts and calculators to assist you with determining the amount of signal loss. Figure 5.21 is an attenuation chart for LMR cable produced by Times Microwave Systems. The left side of the chart lists different types of LMR cable. The farther you move down the list, the better the cable is. The better cable is typically thicker, stiffer, more difficult to work with, and, of course, more expensive. The chart shows how much decibel loss the cable will add to the communications link per 100 feet of cable. The column headers list the frequencies that may be used with the cable. For example, 100 feet of LMR-400 cable used on a 2.5 GHz network (2,500 MHz) would decrease the signal by 6.8 dB.

- Attenuation increases with frequency. If you convert from a 2.4 GHz WLAN to a 5 GHz WLAN, the loss caused by the cable will be greater.

- Either purchase the cables precut and preinstalled with the connectors or hire a professional cabler to install the connections (unless you are a professional cabler).

  Improperly installed connectors will add more loss to the communications link, which can nullify the extra money you spend for the better-quality cable. It can also introduce return loss in the cable due to reflections.

Connectors

Many types of connectors are used to connect antennas to 802.11 equipment. Part of the reason for this is that the FCC Report & Order 04-165 requires that amplifiers have either unique connectors or electronic identification systems to prevent the use of noncertified antennas. This requirement was created to prevent people from connecting higher-gain antennas, either intentionally or unintentionally, to a transceiver. An unauthorized high-gain antenna could exceed the maximum equivalent isotropically radiated power (EIRP) that is allowed by the FCC or other regulatory body.

FIGURE 5.21   Coaxial cable attenuation

Times Microwave Systems (Attenuation dB/100 ft)											
LMR Cable\Frequency	30	50	150	220	450	900	1,500	1,800	2,000	2,500	5,800
100A	3.9	5.1	8.9	10.9	15.8	22.8	30.1	33.2	35.2	39.8	64.1
195	2	2.5	4.4	5.4	7.8	11.1	14.5	16	16.9	19	29.9
195UF	2.3	3	5.3	6.4	9.3	13.2	17.3	19	20.1	22.6	35.6
200	1.8	2.3	4	4.8	7	9.9	12.9	14.2	15	16.9	26.4
200UF	2.1	2.7	4.8	5.8	8.3	11.9	15.5	17.1	18	20.2	31.6
240	1.3	1.7	3	3.7	5.3	7.6	9.9	10.9	11.5	12.9	20.4
240UF	1.6	2.1	3.6	4.4	6.3	9.1	11.8	13	13.8	15.5	24.4
300	1.1	1.4	2.4	2.9	4.2	6.1	7.9	8.7	9.2	10.4	16.5
300UF	1.3	1.6	2.9	3.5	5.1	7.3	9.5	10.5	11.1	12.5	19.8
400	0.7	0.9	1.5	1.9	2.7	3.9	5.1	5.7	6	6.8	10.8
400UF	0.8	1.1	1.8	2.2	3.3	4.7	6.2	6.8	7.2	8.1	13
500	0.5	0.7	1.2	1.5	2.2	3.1	4.1	4.6	4.8	5.5	8.9
500UF	0.6	0.8	1.5	1.8	2.6	3.8	5	5.5	5.8	6.6	10.6
600	0.4	0.5	1	1.2	1.7	2.5	3.3	3.7	3.9	4.4	7.3
600UF	0.5	0.7	1.2	1.4	2.1	3	4	4.4	4.7	5.3	8.7
900	0.3	0.4	0.7	0.8	1.2	1.7	2.2	2.5	2.6	3	4.9
1200	0.2	0.3	0.5	0.6	0.9	1.3	1.7	1.9	2	2.3	not supported
1700	0.1	0.2	0.3	0.4	0.6	0.9	1.3	1.4	1.5	1.7	not supported

UF = Ultraflex (more flexible cable)

In response to this regulation, cable manufacturers sell *pigtail* adapter cables. These pigtail cables are usually short segments of cable (typically about 2 feet long) with different connectors on each end. They act as adapters, changing the connector and allowing a different antenna to be used.

Many of the same principles of cables apply to the connectors as well as many of the other accessories. RF connectors need to be of the correct impedance to match the other RF equipment. They also support specific ranges of frequencies. The connectors add signal loss to the RF link, and lower-quality connectors are more likely to cause connection or VSWR problems. RF connectors on average add about 1/2 dB of insertion loss.

Splitters

Splitters are also known as signal splitters, RF splitters, power splitters, and power dividers. A splitter is a connector or cable that divides an RF signal into two or more separate signals. Only in an unusually special or unique situation would you possibly use an RF splitter. When you install a splitter, not only will the signal be degraded because it is being split (known as *through loss*), but also each connector will add its own insertion loss to the signal. There are so many variables and potential problems with this configuration that we recommend this type of installation be attempted only by a very RF-knowledgeable person and only for temporary installations.

A more practical, but again rare, use of a splitter is to monitor the power that is being transmitted. The splitter can be connected to the transceiver and then split to the antenna and a power meter. This approach would enable you to actively monitor the power that is being sent to the antenna.

Amplifiers

An RF *amplifier* takes the signal that is generated by the transceiver, increases it, and sends it to the antenna. Unlike the antenna providing an increase in gain by focusing the signal, an amplifier provides an overall increase in power by adding electrical energy to the signal, which is referred to as *active gain*.

Amplifiers can be purchased as either unidirectional or bidirectional devices. Unidirectional amplifiers perform the amplification in only one direction, either when transmitting or when receiving. Bidirectional amplifiers perform the amplification in both directions.

The amplifier's increase in power is created using one of the following two methods:

Fixed-Gain   With the fixed-gain method, the output of the transceiver is increased by the amount of the amplifier.

Fixed-Output   A fixed-output amplifier does not add to the output of the transceiver. It simply generates a signal equal to the output of the amplifier, regardless of the power generated by the transceiver.

 Adjustable variable-gain amplifiers also exist, but using them is not a recommended practice. The unauthorized adjustment of a variable-gain amplifier may result in either violation of power regulations or insufficient transmission amplitude.

Since most regulatory bodies have a maximum power regulation of 1 watt or less at the intentional radiator (IR), the main purpose of using amplifiers is to compensate for cable loss, as opposed to boosting the signal for range. Therefore, when installing an amplifier, install it as close to the antenna as possible. Because the antenna cable adds loss to the signal, the shorter antenna cable will produce less loss and allow more signal to the antenna.

Additionally, it is important to note that an amplifier increases noise as well as signal strength. It is not uncommon for an amplifier to raise the noise floor by 10 dB or more.

 Amplifiers must typically be certified with the system in use according to regulatory bodies such as the FCC. If an amplifier is added to a wireless network and it has not been certified, then it is illegal. It is far better to further engineer the system than to use an amplifier.

Attenuators

In some situations, it may be necessary to decrease the amount of signal that is radiating from the antenna. In some instances, even the lowest power setting of the transceiver may generate more signal than you want. In this situation, you can add a fixed-loss or a variable-loss *attenuator*. Attenuators are typically small devices about the size of a C-cell battery or smaller, with cable connectors on both sides. Attenuators absorb energy, decreasing the signal as it travels through. Fixed-loss attenuators provide a set amount of dB loss. A variable-loss attenuator has a dial or switch configuration on it that enables you to adjust the amount of energy that is absorbed.

Variable-loss attenuators may be used during outdoor site surveys to simulate loss caused by various grades of cabling and different cable lengths. Another interesting use of a variable attenuator is to test the actual fade margin on a point-to-point link. By gradually increasing the attenuation until there is no more link, you can use that number to determine the actual fade margin of the link.

Lightning Arrestors

The purpose of a *lightning arrestor* is to redirect (shunt) transient currents caused by nearby lightning strikes or ambient static away from your electronic equipment and into the ground. Lightning arrestors are used to protect electronic equipment from the sudden surge of power that a nearby lightning strike or static buildup can cause. You may have noticed the use of the phrase "nearby lightning strike." This wording is used because lightning arrestors are not capable of protecting against a direct lightning strike. Lightning arrestors can typically protect against surges of up to 5,000 amperes at up to 50 volts. The IEEE specifies that lightning arrestors should be capable of redirecting the transient current in less than 8 microseconds. Most lightning arrestors are capable of doing it in less than 2 microseconds.

The lightning arrestor is installed between the transceiver and the antenna. Any devices installed between the lightning arrestor and the antenna will not be protected by the lightning arrestor. Therefore, the lightning arrestor is typically placed closer to the antenna, with all other communications devices (amplifiers, attenuators, etc.) installed between the lightning arrestor and the transceiver. After a lightning arrestor has performed its job by protecting the equipment from an electrical surge, it will have to be replaced, or it may have a replaceable gas discharge tube (like a fuse). Most installations place the lightning arrestor at the egress to the building. Cable grounding kits can be installed near the antenna and at every 100 feet.

Fiber-optic cable can also be used to provide additional lightning protection. A short piece of fiber-optic cable can be inserted into the Ethernet cable that connects the wireless bridge to the rest of the network. Ethernet-to-fiber adapters, known as transceivers, convert the electrical Ethernet signal to a light-based fiber signal and then back to Ethernet. Because fiber-optic cable is constructed of glass and uses light and not electricity to transmit data, it does not conduct electricity. It is important to make sure that the power supply for the adapters is protected as well.

The fiber-optic cable acts as a kind of safety net, should the lightning arrestor fail due to a much higher transient current or even a direct lightning strike. Realize that if there is a direct lightning strike to the antenna, you can plan on replacing all the components, from the fiber-optic cable to the antenna. Furthermore, a direct lightning strike may also arc over the fiber link and still cause damage to equipment on the opposite side of the fiber link. Grounding the RF cables as well can help prevent this from happening.

 Real World Scenario

Not Only Is Lightning Unpredictable, the Results Are Too!

A business in a five-story, 200-year-old brick brownstone in the North End neighborhood of Boston had a lightning strike or a nearby lightning strike. This building was not even one of the tallest buildings in the area, and it was at the bottom of a small hill and surrounded by other similar buildings. An electrical current traveled down the water vent pipe and past a bundle of Ethernet cables. A transient current on the Ethernet cables damaged the transceiver circuits on the Ethernet adapters in the PCs and on the individual ports on the Ethernet hub. About half of the Ethernet devices in the company failed, and about half of the ports on the hub were no longer functioning. Yet all the software recognized the adapters, and all the power and port lights worked flawlessly. The problem appeared to be related to the cabling.

You often will not know that the problem is related to lightning, and the symptoms may be misleading. Testing the lightning arrestors can help with your diagnosis.

Grounding Rods and Wires

When lightning strikes an object, it is looking for the path of least resistance, or more specifically, the path of least impedance. This is where lightning protection and grounding equipment come into play. A grounding system, which is made up of a grounding rod and wires, provides a low-impedance path to the ground. This low-impedance path is installed to encourage the lightning to travel through it instead of through your expensive electronic equipment.

Grounding rods and wires are also used to create what is referred to as a *common ground*. One way of creating a common ground is to drive a copper rod into the ground and connect your electrical and electronic equipment to this rod by using wires or straps (grounding wires). The grounding rod should be fully driven into the ground, leaving enough of the rod accessible to attach the ground wires to it. By creating a common ground, you have created a path of least impedance for all your equipment, should lightning cause an electrical surge.

Regulatory Compliance

In Chapter 4, you learned about the RF components along with the concepts of intentional radiator (IR) and equivalent isotropically radiated power (EIRP). In this chapter, you learned about antennas and the many aspects of antenna operations and installation. Although there are many antenna, cabling, and component options when you are configuring a wireless network, the reality is that you are often limited as to your antenna choices because of regulatory compliance. Although each regulatory body operates independently, there are similarities between how these organizations operate and certify equipment. This section will briefly explain the process in the United States, as regulated by the FCC.

In order for an access point manufacturer to sell its product within a country or region, it must prove that its product operates within the rules of the relevant regulatory domain, such as the FCC. The FCC creates documents that specify the rules that must be followed by the manufacturer, also referred to as the *responsible party* or the *grant holder*. The manufacturer will send its equipment to be tested by the regulatory body or an authorized testing organization, which will perform the compliance testing on the equipment. If the equipment passes the testing, the device will ultimately be issued an ID number and a *Grant of Certification*.

Most people are not familiar with this process and do not realize that when the company sends its product to be tested, what is submitted is the complete system that the manufacturer will market and sell as the product, which includes the intentional radiator (AP), any cabling and connectors, and the antenna or antennas that it wants to be able to use with the AP. Most companies will have their AP certified with a grouping of antennas that provide different gain and beamwidth characteristics.

An intentional radiator can be operated only with an antenna with which it is authorized. The FCC does allow an antenna to be substituted with a different one, provided that two key conditions are met:

- The gain of the new antenna must be the same or lower than the antenna that the system was certified with.

- The new antenna must be of the same type, which means that the antenna must have the same in-band and out-of-band characteristics.

The gain of an antenna is easily identified and provided with most antennas, so the first criteria is relatively simple to meet. However, meeting the second criteria—for example, determining if one antenna is of the same type as another—requires verifying both the in-band and out-of-band characteristics of the antenna. How a new antenna will perform as part of a combined system for out-of-band requirements can be difficult to predict from datasheet information alone. The out-of-band requirements typically include spurious emissions limits over a very wide frequency range (9 KHz to 300 GHz), band edge emissions (a mask-type requirement), limits on harmonics generated, and very low limits on noise in specific restricted bands. Verifying a new antenna against all these "out-of-band" requirements could require testing as extensive as the original certification testing itself. Therefore, if you wanted to replace your antenna with an antenna from a third-party manufacturer, you may need to work with them to ensure that the use of their antenna with your AP is acceptable by the local regulatory domain.

Summary

This chapter focused on RF signal and antenna concepts. The antenna is a key component of successful RF communications. The following four types of antennas are used with 802.11 networks:

- Omnidirectional (dipole, collinear)
- Semidirectional (patch, panel, Yagi)
- Highly directional (parabolic dish, grid)
- Sector

The antenna types produce different signal patterns, which can be viewed on azimuth and elevation charts.

This chapter also reviewed the following key concerns when installing point-to-point communications:

- Visual line of sight
- RF line of sight
- Fresnel zone
- Earth bulge
- Antenna polarization

The final section of this chapter covered VSWR and antenna mounting issues, along with antenna accessories and their roles.

Exam Essentials

Know the different categories and types of antennas, how they radiate signals, and which type of environment they are used in.   Make sure you know the three main categories of antennas and the different types of antennas. Know the similarities and differences among them, and understand when and why you would use one antenna over another. Make sure that you understand azimuth and elevation charts, beamwidth, antenna polarization, and antenna diversity.

Fully understand the Fresnel zone.   Make sure you understand all the issues and variables involved with installing point-to-point communications. You are not required to memorize the Fresnel zone or earth bulge formulas; however, you will need to know the principles regarding these topics and when and why you would use the formulas.

Understand the concerns associated with connecting and installing antennas and antenna accessories.   Every cable, connector, and device between the transceiver and the antenna affects the signal that gets radiated from the antenna. Understand which devices provide gain and which devices provide loss. Understand what VSWR is and which values are good or bad. Know the different antenna accessories, what they do, and why and when you would use them.

Review Questions

1. Which of the following refers to the polar chart of an antenna as viewed from above the antenna? (Choose all that apply)
   A. Horizontal view
   B. Vertical view
   C. H-plane
   D. E-plane
   E. Elevation chart
   F. Azimuth chart

2. The azimuth chart represents a view of an antenna's radiation pattern from which direction?
   A. Top
   B. Side
   C. Front
   D. Both top and side

3. What is the definition of the horizontal beamwidth of an antenna?
   A. The measurement of the angle of the main lobe as represented on the azimuth chart
   B. The distance between the two points on the horizontal axis where the signal decreases by a third. This distance is measured in degrees.
   C. The distance between the two –3 dB power points on the horizontal axis, measured in degrees
   D. The distance between the peak power and the point where the signal decreases by half. This distance is measured in degrees.

4. Which antennas are highly directional? (Choose all that apply.)
   A. Patch
   B. Panel
   C. Parabolic dish
   D. Grid
   E. Sector

5. Semidirectional antennas are often used for which of the following purposes? (Choose all that apply.)
   A. To provide short-distance point-to-point communications
   B. To provide long-distance point-to-point communications

    C. To provide unidirectional coverage from an access point to clients in an indoor environment

    D. To reduce reflections and the negative effects of multipath

6. The Fresnel zone should not be blocked by more than what percentage to maintain a reliable communications link?

    A. 20 percent

    B. 40 percent

    C. 50 percent

    D. 60 percent

7. The size of the Fresnel zone is controlled by which factors? (Choose all that apply.)

    A. Antenna beamwidth

    B. RF line of sight

    C. Distance

    D. Frequency

8. When a long-distance point-to-point link is installed, earth bulge should be considered beyond what distance?

    A. 5 miles

    B. 7 miles

    C. 10 miles

    D. 30 miles

9. A network administrator replaced some coaxial cabling used in an outdoor bridge deployment after water damaged the cabling. After replacing the cabling, the network administrator noticed that the EIRP increased drastically and is possibly violating the maximum EIRP power regulation mandate. What are the possible causes of the increased amplitude? (Choose all that apply.)

    A. The administrator installed a shorter cable.

    B. The administrator installed a lower-grade cable.

    C. The administrator installed a higher-grade cable.

    D. The administrator installed a longer cable.

    E. The administrator used a different-color cable.

10. Which of the following is true for 802.11n and 802.11ac radios?

    A. The transceiver combines the signal from multiple antennas to provide better coverage.

    B. Transceivers can transmit from multiple antennas at the same time.

    C. The transceiver samples multiple antennas and chooses the best received signal from one antenna.

    D. Transceivers can transmit from only one antenna at a time.

11. To establish a four-mile point-to-point bridge link in the 5 GHz U-NII-3 band, what factors should be taken under consideration? (Choose all that apply.)

    A. Fresnel zone with 40 percent or less blockage

    B. Earth bulge calculations

    C. Minimum of 16 dBi of passive gain

    D. Proper choice of semidirectional antennas

    E. Proper choice of highly directional antennas

12. The ratio between the maximum peak voltage and minimum voltage on a line is known as what?

    A. Signal flux

    B. Return loss

    C. VSWR

    D. Signal incidents

13. What are some of the possible negative effects of an impendence mismatch? (Choose all that apply.)

    A. Voltage reflection

    B. Blockage of the Fresnel zone

    C. Erratic signal strength

    D. Decreased signal amplitude

    E. Amplifier/transmitter failure

14. When determining the mounting height of a long-distance point-to-point antenna, which of the following needs to be considered? (Choose all that apply.)

    A. Frequency

    B. Distance

    C. Visual line of sight

    D. Earth bulge

    E. Antenna beamwidth

    F. RF line of sight

15. Which of the following are true about cables? (Choose all that apply.)

    A. They cause impedance on the signal.

    B. They work regardless of the frequency.

    C. Attenuation decreases as frequency increases.

    D. They add loss to the signal.

16. Amplifiers can be purchased with which of the following features? (Choose all that apply.)

    A. Bidirectional amplification

    B. Unidirectional amplification

    C. Fixed gain

    D. Fixed output

17. The signal between the transceiver and the antenna will be reduced by which of the following methods? (Choose all that apply.)

    A. Adding an attenuator

    B. Increasing the length of the cable

    C. Shortening the length of the cable

    D. Using cheaper-quality cable

18. Lightning arrestors will defend against which of the following?

    A. Direct lightning strikes

    B. Power surges

    C. Transient currents

    D. Improper common grounding

19. The radius of the second Fresnel zone is _____. (Choose all that apply.)

    A. The first area where the signal is out of phase with the point source

    B. The first area where the signal is in phase with the point source

    C. Smaller than the first Fresnel zone

    D. Larger than the first Fresnel zone

20. While aligning a directional antenna, you notice that the signal drops as you turn the antenna away from the other antenna, but then it increases a little. This increase in signal is caused by what?

    A. Signal reflection

    B. Frequency harmonic

    C. Side band

    D. Side lobe

Chapter

6

Wireless Networks and Spread Spectrum Technologies

IN THIS CHAPTER, YOU WILL LEARN ABOUT THE FOLLOWING:

✓ **Narrowband and spread spectrum**

- Multipath interference

✓ **Frequency-hopping spread spectrum**

- Hopping sequence
- Dwell time
- Hop time
- Modulation

✓ **Direct-sequence spread spectrum**

- DSSS data encoding
- Modulation

✓ **Orthogonal frequency-division multiplexing**

- Convolutional coding
- Modulation

✓ **Orthogonal frequency-division multiple access**

✓ **Industrial, scientific, and medical bands**

- 900 MHz ISM band
- 2.4 GHz ISM band
- 5.8 GHz ISM band

✓ **Unlicensed National Information Infrastructure bands**

- U-NII-1
- U-NII-2A
- U-NII-2C
- U-NII-3
- Future U-NII bands

✓ **3.6 GHz band**

✓ **4.9 GHz band**

✓ **Future Wi-Fi frequencies**

- Below 1 GHz
- 60 GHz
- TV white space

✓ **2.4 GHz channels**

✓ **5 GHz channels**

- Long-term Evolution in 5 GHz

✓ **Adjacent, nonadjacent, and overlapping channels**

✓ **Throughput vs. bandwidth**

✓ **Communication resilience**

In this chapter, you will learn about the different spread spectrum transmission technologies and frequency ranges that are supported by the 802.11 standard and amendments. You will learn how these frequencies are divided into different channels and some of the proper and improper ways of using the channels. Additionally, you will learn about the various types of spread spectrum technologies. You will also learn about orthogonal frequency-division multiplexing (OFDM) and the similarities and differences between OFDM and spread spectrum.

This chapter contains many references to FCC specifications and regulations. The CWNA exam does not test you on any regulatory domain-specific information. Any FCC references are provided strictly to help you understand the technology better. It is important to realize that similarities often exist between the regulations of different regulatory domains. Therefore, understanding the rules of another country's regulatory domain can help you interpret the rules of your regulatory domain.

Narrowband and Spread Spectrum

There are two primary radio frequency (RF) transmission methods: *narrowband* and *spread spectrum*. A narrowband transmission uses very little bandwidth to transmit the data it is carrying, whereas a spread spectrum transmission uses more bandwidth than is necessary to carry its data. Spread spectrum technology spreads the data to be transmitted across the frequencies that it is using. For example, a narrowband radio might transmit data on 2 MHz of frequency space, whereas a spread spectrum radio might transmit data over a 20 MHz frequency space. Figure 6.1 shows a rudimentary comparison of how a narrowband and spread spectrum signal relate to each other. Because narrowband signals take up a single or very narrow band of frequencies, intentional jamming or unintentional interference of this frequency range is likely to cause disruption in the signal. Because spread spectrum uses a wider range of frequency space, it is typically less susceptible to intentional jamming or unintentional interference from outside sources, unless the interfering signal was also spread across the range of frequencies used by the spread spectrum communications.

FIGURE 6.1   Overlay of narrowband and spread spectrum frequency use

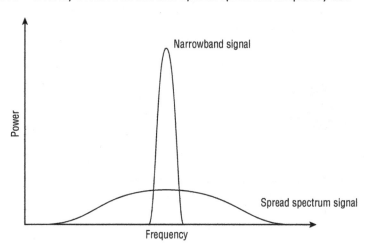

Narrowband signals are typically transmitted using much higher power than spread spectrum signals. Typically, the FCC or other local regulatory bodies require that narrowband transmitters be licensed to minimize the risk of two narrowband transmitters interfering with each other. AM and FM radio stations are examples of narrowband transmitters that are licensed to make sure that two stations in the same or nearby market are not transmitting on the same frequency.

Spread spectrum signals are transmitted using very low power levels.

🌐 Real World Scenario

Who Invented Spread Spectrum?

Spread spectrum was originally patented on August 11, 1942, by actress Hedy Kiesler Markey (Hedy Lamarr) and composer George Antheil. It was originally designed to be a radio guidance system for torpedoes, a purpose for which it was never used. The idea of spread spectrum was ahead of its time. It was not until 1957 that further development on spread spectrum occurred, and in 1962 frequency-hopping spread spectrum was used for the first time between the US ships at the blockade of Cuba during the Cuban Missile Crisis.

If you would like to learn more about the interesting history of spread spectrum, search the Internet for Lamarr and Antheil. There are many websites with articles about these two inventors and even copies of the original patent. Neither inventor made any money from their patent because it expired before the technology was developed.

Multipath Interference

One of the problems that can occur with RF communications is *multipath* interference. Multipath occurs when a reflected signal arrives at the receiving antenna after the primary signal. Figure 6.2 illustrates a signal traveling from the client to the AP. In this illustration, you can see three different signals, each traveling a different path of different distance and duration. This is similar to the way an echo is heard after the original sound.

FIGURE 6.2   Multipath diagram

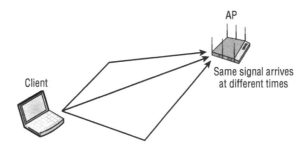

To illustrate multipath further, let us use an example of yelling to a friend across a canyon. Assume you are going to yell, "Hello, how are you?" to your friend. To make sure that your friend understands your message, you might pace your message and yell each word one second after the previous word. If your friend heard the echo (multipath reflection of your voice) a half-second after the main sound arrived, your friend would hear "HELLO hello HOW how ARE are YOU you." (Echoes are represented by lowercase.) Your friend would be able to interpret the message because the echo arrived between the main signals, or the sound of your voice. However, if the echo arrived one second after the main sound, the echo for the word *hello* would arrive at the same time the word *HOW* arrives. With both sounds arriving at the same time, it may be difficult to understand the message. As a result, you may need to ask your friend to repeat the message.

RF data communications behave the same way as the sound example. At the receiver, the delay between the main signal and the reflected signal is known as the *delay spread*. A typical delay spread in an indoor environment can vary from 30 to 270 nanoseconds (ns). If the delay spread is too great, data from the reflected signal may interfere with the same data stream from the main signal; this is referred to as *intersymbol interference (ISI)*. Spread spectrum systems are not as susceptible to ISI because they spread their signals across a range of frequencies. These various frequencies produce different delays in multipath, such that some wavelengths may be affected by ISI whereas others may not. Because of this behavior, spread spectrum signals are typically more tolerant of multipath interference than narrowband signals.

802.11 (DSSS), 802.11b (HR-DSSS), and 802.11g (ERP) are tolerant of delay spread only to a certain extent. 802.11 (DSSS) and 802.11b (HR-DSSS) can tolerate delay spread of up to 500 nanoseconds. Even though the delay spread can be tolerated, performance is much better when the delay spread is lower. A transmitter will drop to a lower data rate when the delay spread increases. Longer symbols are used when transmitting at the lower data rates. When longer symbols are used, longer delays can occur before ISI occurs.

Because of OFDM's greater tolerance of delay spread, an 802.11a/g transmitter can maintain 54 Mbps with a delay spread of up to about 150 nanoseconds. This depends on the 802.11a/g chipset that is being used in the transmitter and receiver. Some chipsets are not as tolerant and switch to a lower data rate at a lower delay spread value.

Prior to 802.11n and 802.11ac MIMO technology, multipath had always been a concern. It was a condition that could drastically affect the performance and throughput of the wireless LAN. With the introduction of MIMO, multipath is actually a condition that can now enhance and increase the performance of the wireless LAN. The enhanced digital signal processing techniques of MIMO devices take advantage of multiple simultaneous transmissions and can actually benefit from the effects of multipath. You will learn more about 802.11n/ac and MIMO in Chapter 10, "MIMO Technology: HT and VHT."

Frequency-Hopping Spread Spectrum

Frequency-hopping spread spectrum (FHSS) was used in the original 802.11 standard and provided 1 and 2 Mbps RF communications using the 2.4 GHz ISM band for legacy radios. The majority of legacy FHSS radios were manufactured between 1997 and 1999. The IEEE specified that in North America, 802.11 FHSS would use 79 MHz of frequencies, from 2.402 GHz to 2.480 GHz.

The IEEE 802.11-2016 standard has deprecated FHSS, completely removing it from the current document. 802.11 FHSS technology will not be tested on the CWNA exam. 802.11 vendors stopped manufacturing 802.11 FHSS adapters and access points a long time ago. Most organizations have long since transitioned from 802.11 FHSS to one of the newer and faster transmission methods. It is still important to understand the basics behind FHSS, as there are other technologies, such as Bluetooth, that use FHSS. Be aware that even though Bluetooth uses FHSS, the number of hops, dwell time, and hopping sequence is very different from 802.11 FHSS. It is also important to note that Bluetooth does operate in the 2.4 GHz ISM band, the same band used by 802.11b/g/n devices.

Generally, the way FHSS works is that it transmits data by using a small frequency carrier space, then hops to another small frequency carrier space and transmits data, then to another frequency, and so on, as illustrated in Figure 6.3. More specifically, FHSS transmits data by using a specific frequency for a set period of time, known as the *dwell time*. When the dwell time expires, the system changes to another frequency and begins to transmit on that frequency for the duration of the dwell time. Each time the dwell time is reached, the system changes to another frequency and continues to transmit.

FIGURE 6.3   FHSS components

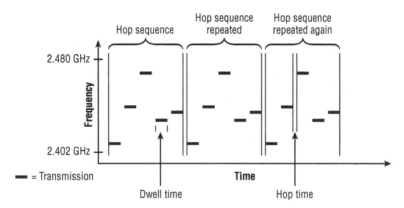

Hopping Sequence

FHSS radios use a predefined *hopping sequence* (also called a *hopping pattern* or *hopping set*) comprising a series of small carrier frequencies, or *hops*. Instead of transmitting on one set channel or finite frequency space, an FHSS radio transmits on a sequence of subchannels called hops. Each time the hop sequence is completed, it is repeated. Figure 6.3 shows a make-believe hopping sequence that consists of five hops.

The original IEEE 802.11 standard mandated that each hop be 1 MHz in size. These individual hops were then arranged in predefined sequences. In North America and most of Europe, the hopping sequences contained at least 75 hops but no more than 79 hops. Other countries had different requirements; for example, France used 35 hops, whereas Spain and Japan used 23 hops in a sequence. For successful transmissions to occur, all FHSS transmitters and receivers must be synchronized on the same carrier hop at the same time. The original 802.11 standard defined hopping sequences that could be configured on an FHSS access point, and the hopping sequence information is delivered to client stations via the beacon management frame.

Dwell Time

Dwell time is a defined amount of time that the FHSS system transmits on a specific frequency before it switches to the next frequency in the hop set. The local regulatory body typically limits the amount of dwell time. For example, the FCC specifies a maximum dwell time of 400 milliseconds (ms) per carrier frequency during any 30-second period of time. Typical dwell times are around 100 ms to 200 ms. The original IEEE 802.11 standard specified that a hopping sequence consist of at least 75 frequencies, 1 MHz wide. Because the standard specified a maximum bandwidth of 79 MHz, the maximum number of hops possible for a hop set would be 79. With an FHSS hop sequence consisting of 75 hops and a dwell time of 400 ms, it would take about 30 seconds to complete the hop sequence. After the hop sequence is complete, it is repeated.

Hop Time

Hop time is not a specified period of time but rather a measurement of the amount of time it takes for the transmitter to change from one frequency to another. Hop time is typically a fairly small number, often about 200 to 300 microseconds (μs). With typical dwell times of 100 to 200 milliseconds (ms), hop times of 200 to 300 μs are insignificant. Insignificant or not, the hop time is essentially wasted time, or overhead, and is the same regardless of the dwell time. The longer the dwell time, the less often the transmitter has to waste time hopping to another frequency, resulting in greater throughput. If the dwell time is shorter, the transmitter has to hop more frequently, thus decreasing throughput.

Modulation

FHSS uses Gaussian frequency-shift keying (GFSK) to encode the data. Two-level GFSK (2GFSK) uses two frequencies to represent a 0 or a 1 bit. Four-level GFSK (4GFSK) uses four frequencies, with each frequency representing 2 bits (00, 01, 10, or 11). Because it takes transmission cycles before the frequency can be determined, the symbol rate (the rate that the data is sent) is only about one or two million symbols per second, a fraction of the 2.4 GHz carrier frequency.

What Is the Significance of the Dwell Time?

Because FHSS transmissions jump inside a frequency range of 79 MHz, a narrowband signal or noise would disrupt only a small range of frequencies and would produce only a minimal amount of throughput loss. Decreasing the dwell time can further reduce the effect of interference. Conversely, because the radio is transmitting data during the dwell time, the longer the dwell time, the greater the throughput.

Direct-Sequence Spread Spectrum

Direct-sequence spread spectrum (DSSS) was originally specified in the primary, or root, 802.11 standard and provides 1 and 2 Mbps RF communications using the 2.4 GHz ISM band. An updated implementation of DSSS (HR-DSSS) was also specified in the 802.11b addendum and provides 5.5 and 11 Mbps RF communications using the same 2.4 GHz ISM band. The 802.11b 5.5 and 11 Mbps speeds are known as *high-rate DSSS (HR-DSSS)*.

Current 2.4 GHz 802.11 devices are backward compatible with legacy 802.11 DSSS devices. This means that an 802.11b device can transmit using DSSS at 1 and 2 Mbps and using HR-DSSS at 5.5 and 11 Mbps.

DSSS 1 and 2 Mbps are specified in Clause 15 of the 802.11-2016 standard. HR-DSSS 5.5 and 11 Mbps are specified in Clause 16 of the 802.11-2016 standard.

Unlike FHSS, where the transmitter jumps between frequencies, DSSS is set to one channel. The data transmitted is spread across the range of frequencies that make up the channel. The process of spreading the data across the channel is known as *data encoding*.

DSSS Data Encoding

In Chapter 3, "Radio Frequency Fundamentals," you learned about the many ways that RF signals can get altered or corrupted. Because 802.11 uses an unbounded medium with a huge potential for RF interference, it had to be designed to be resilient enough that data corruption could be minimized. To achieve this, each bit of data is encoded and transmitted as multiple bits of data.

The task of adding additional, redundant information to the data is known as *processing gain*. In this day and age of data compression, it seems strange that we would use a technology that adds data to our transmission, but by doing so, the communication is more resistant to data corruption. The system converts the 1 bit of data into a series of bits that are referred to as *chips*. To create the chips, a Boolean XOR is performed on the data bit and a fixed-length bit sequence pseudorandom number (PN) code. Using a PN code known as the Barker code, the binary data 1 and 0 are represented by the following chip sequences:

Binary data 1 = 1 0 1 1 0 1 1 1 0 0 0

Binary data 0 = 0 1 0 0 1 0 0 0 1 1 1

This sequence of chips is then spread across a wider frequency space. Although 1 bit of data might need only 2 MHz of frequency space, the 11 chips will require 22 MHz of frequency carrier space. This process of converting a single data bit into a sequence is often called *spreading* or *chipping*. The receiving radio converts, or *de-spreads*, the chip sequence back into a single data bit. When the data is converted to multiple chips and some of the chips are not received properly, the radio will still be able to interpret the data by looking at the chips that were received properly. When the Barker code is used, as many as 9 of the 11 chips can be corrupted, yet the receiving radio will still be able to interpret the sequence and convert them back into a single data bit. This chipping process also makes the communication less likely to be affected by intersymbol interference because it uses more bandwidth.

 After the Barker code is applied to data, a series of 11 bits, referred to as chips, represent the original single bit of data. This series of encoded bits makes up 1 bit of data. To help prevent confusion, it is best to think of and refer to the encoded bits as *chips*.

The Barker code uses an 11-chip PN; however, the length of the code is irrelevant. To help provide the faster speeds of HR-DSSS, another more complex code, *complementary code keying (CCK)*, is utilized. CCK uses an 8-chip PN, along with different PNs for different bit sequences. CCK can encode 4 bits of data with 8 chips (5.5 Mbps) and can encode 8 bits of data with 8 chips (11 Mbps). Although it is interesting to learn about, a thorough understanding of CCK is not required for the CWNA exam.

Modulation

After the data has been encoded using a chipping method, the transmitter needs to modulate the signal to create a carrier signal containing the chips. *Differential binary phase-shift keying (DBPSK)* uses two phase shifts: one that represents a 0 chip and another that represents a 1 chip. To provide faster throughput, *differential quadrature phase-shift keying (DQPSK)* uses four phase shifts, allowing each of the four phase shifts to modulate 2 chips (00, 01, 10, 11) instead of just 1 chip, doubling the speed.

Table 6.1 shows a summary of the data encoding and modulation techniques used by 802.11 and 802.11b.

TABLE 6.1   DSSS and HR-DSSS encoding and modulation overview

	Data Rate (Mbps)	Encoding	Chip Length	Bits Encoded	Modulation
DSSS	1	Barker coding	11	1	DBPSK
DSSS	2	Barker coding	11	1	DQPSK
HR-DSSS	5.5	CCK coding	8	4	DQPSK
HR-DSSS	11	CCK coding	8	8	DQPSK

Orthogonal Frequency-Division Multiplexing

Orthogonal frequency-division multiplexing (OFDM) is one of the most popular communications technologies, used in both wired and wireless communications. The 802.11-2016 standard specifies the use of OFDM at 5 GHz and also specifies the use of ERP-OFDM at 2.4 GHz. OFDM and ERP-OFDM are the same technology used by 802.11a and 802.11g radios, respectively. OFDM is not a spread spectrum technology, even though it has similar properties to spread spectrum, such as low transmit power and using more bandwidth than is required to transmit data. Because of these similarities, OFDM is often referred to as a spread spectrum technology even though technically that reference is incorrect. A 20 MHz

OFDM channel consists of 64 separate, closely and precisely spaced frequencies, often referred to as *subcarriers*, as illustrated in Figure 6.4. OFDM subcarriers are sometimes also referred to as OFDM *tones*.

FIGURE 6.4   802.11a/g Channels and OFDM subcarriers

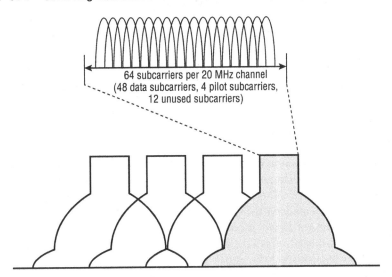

The frequency width of each subcarrier is 312.5 KHz. The subcarriers are also transmitted at lower data rates, but because there are so many subcarriers, overall data rates are higher. Also, because of the lower subcarrier data rates, delay spread is a smaller percentage of the symbol period, which means that ISI is less likely to occur. In other words, OFDM technology is more resistant to the negative effects of multipath than DSSS and FHSS spread spectrum technologies. Figure 6.5 represents four of the subcarriers. One of the subcarriers is highlighted so that you can more easily understand the drawing. Notice that the frequency spacing of the subcarriers has been chosen so that the harmonics overlap and provide cancellation of most of the unwanted signals. The spacing of the subcarriers is orthogonal, so they will not interfere with one another. An ODFM symbol time is 3.2 μs. Subcarrier spacing is equal to the reciprocal of the OFDM symbol time. For example: 1 cycle/.00000032 = 312,500 cycles per second = 312.5 kHz.

FIGURE 6.5   Subcarrier signal overlay

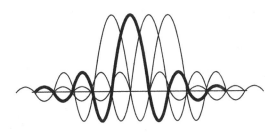

Twelve of the 64 subcarriers in a 20 MHz OFDM channel are unused and serve as guard bands. A *guard band* is an unused part of the radio spectrum between radio bands, for the purpose of preventing interference. Forty-eight of the subcarriers are used to transmit modulated data. The other four subcarriers are known as *pilot carriers* and are used for dynamic calibration between a transmitter and receiver. These four pilot tones are used as references for phase and amplitude by the demodulator, allowing the receiver to synchronize itself as it demodulates the data in the other subcarriers.

In Chapter 10, you will learn that 802.11n/ac radios also use OFDM technology. 802.11n/ac radios also transmit on a 20 MHz channel that consists of 64 total subcarriers, although only eight subcarriers are guard bands. Fifty-two of the subcarriers are used to transmit modulated data, and four subcarriers function as pilot carriers.

Convolutional Coding

To make OFDM more resistant to narrowband interference, a form of error correction known as *convolutional coding* is performed. The 802.11-2016 standard defines the use of convolutional coding as the error-correction method to be used with OFDM technology. It is a *forward error correction (FEC)* that allows the receiving system to detect and repair corrupted bits.

There are many levels of convolutional coding. Convolutional coding uses a ratio between the bits transmitted versus the bits encoded to provide these different levels. The lower the ratio, the less resistant the signal is to interference and the greater the data rate will be. Table 6.2 displays a comparison between the technologies used to create the different data rates of both 802.11a and 802.11g. Notice that the data rates are grouped by pairs based on modulation technique and that the difference between the two speeds is caused by the different levels of convolutional coding. A detailed explanation of convolutional coding is extremely complex and far beyond the knowledge needed for the CWNA exam.

TABLE 6.2 802.11a and 802.11g data rate and modulation comparison

Data Rates (Mbps)	Modulation Method	Coded Bits per Subcarrier	Data Bits per OFDM Symbol	Coded Bits per OFDM Symbol	Coding Rate (Data Bits/ Coded Bits)
6	BPSK	1	24	48	1/2
9	BPSK	1	36	48	3/4
12	QPSK	2	48	96	1/2
18	QPSK	2	72	96	3/4
24	16-QAM	4	96	192	1/2
36	16-QAM	4	144	192	3/4

Data Rates (Mbps)	Modulation Method	Coded Bits per Subcarrier	Data Bits per OFDM Symbol	Coded Bits per OFDM Symbol	Coding Rate (Data Bits/ Coded Bits)
48	64-QAM	6	192	288	2/3
54	64-QAM	6	216	288	3/4

Modulation

OFDM uses binary phase-shift keying (BPSK) and quadrature phase-shift keying (QPSK) phase modulation for the lower ODFM data rates. The higher OFDM data rates use 16-QAM, 64-QAM, and 256-QAM modulation. *Quadrature amplitude modulation (QAM)* combines both phase and amplitude modulation. A *constellation diagram*, also known as a constellation map, is a two-dimensional diagram often used to represent QAM modulation. A constellation diagram is divided into four quadrants, and different locations in each quadrant can be used to represent data bits. Areas on the quadrant relative to the horizontal axis can be used to represent various phase shifts. In Figure 6.6, the first two bits are represented by a shift in phase. Notice that the first two bits are the same across each column. Areas relative to the vertical axis are used to represent amplitude shifts. In this example, the last two bits are represented by a shift in amplitude. Notice that the last two bits are the same across each row. A 16-QAM constellation diagram uses four different phase shifts and four different amplitude shifts to generate a total of 16 4-digit combinations. Each of the 16 different points within the four quadrants are used to represent four data bits. Chapter 10 explains constellation charts in much more detail.

FIGURE 6.6   16-QAM constellation diagram

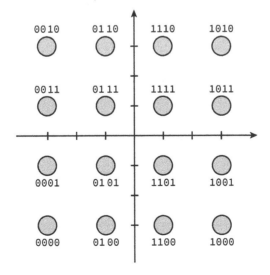

Orthogonal Frequency-Division Multiple Access

Orthogonal frequency-division multiple access (OFDMA) is a multi-user version of the OFDM technology. 802.11a/g/n/ac radios currently use orthogonal frequency-division multiplexing (OFDM) for single-user transmissions on an 802.11 frequency. OFDMA subdivides a channel into smaller frequency allocations, called *resource units (RUs)*. By subdividing the channel, parallel transmissions of smaller frames to multiple users can occur simultaneously. Think of OFDMA as a technology that partitions a channel into smaller sub-channels so that simultaneous multi-user transmissions can occur. For example, a traditional 20 MHz channel might be partitioned into as many as nine smaller channels. Each smaller channel would consist of a set number of OFDM subcarriers. OFDMA technology has been used for cellular communications for many years and will soon be used by Wi-Fi radios, as proposed in the IEEE 802.11ax draft amendment. A detailed explanation about OFDMA can be found in Chapter 19, "802.11ax - High Efficiency (HE)."

Industrial, Scientific, and Medical Bands

The original IEEE 802.11 standard and the subsequent 802.11b, 802.11g, and 802.11n amendments all define communications in the frequency range between 2.4 GHz and 2.4835 GHz. This frequency range is one of three frequency ranges known as the *industrial, scientific, and medical (ISM) bands*. The frequency ranges of the ISM bands are as follows:

- 902 MHz to 928 MHz (26 MHz wide)
- 2.4 GHz to 2.5 GHz (100 MHz wide)
- 5.725 GHz to 5.875 GHz (150 MHz wide)

The ISM bands are defined by the ITU Telecommunication Standardization Sector (ITU-T) in S5.138 and S5.150 of the Radio Regulations. Although the FCC governs the use of the ISM bands defined by the ITU-T in the United States, their usage in other countries may be different because of local regulations. The 900 MHz band is known as the industrial band; the 2.4 GHz band is known as the scientific band; and the 5.8 GHz band is known as the medical band.

Note that all three of these bands are license-free bands, and there are no restrictions on what types of equipment can be used in any of them. For example, a radio used in medical equipment can be used in the 900 MHz industrial band.

900 MHz ISM Band

The 900 MHz ISM band is 26 MHz wide and spans from 902 MHz to 928 MHz. In the past, this band was used for wireless networking; however, most wireless networks now use higher frequencies, which are capable of faster throughput.

Another factor limiting the use of the 900 MHz ISM band is that in many parts of the world, part of the 900 MHz frequency range has already been allocated to the Global System for Mobile Communications (GSM) for use by mobile phones. Although the

900 MHz ISM band is rarely used for networking, many products, such as baby monitors, wireless home telephones, and wireless headphones, use this frequency range.

802.11 radios do not operate in the 900 MHz ISM band, but many older legacy deployments of wireless networking did operate in this band. Some vendors still manufacture non-802.11 wireless networking devices that operate in the 900 MHz ISM band. This is a particularly popular frequency that is used for wireless ISPs because of its superior foliage penetration over the 2.4 GHz and 5 GHz frequency ranges.

2.4 GHz ISM Band

The 2.4 GHz ISM band has historically been the most common band used for wireless networking communications. The 2.4 GHz ISM band is 100 MHz wide and spans from 2.4 GHz to 2.5 GHz. Use of the 2.4 GHz ISM for wireless LANs is defined by the IEEE in the 802.11-2016 standard. With most of the current 802.11 radio chipsets now including 5 GHz capabilities, the usage of the 2.4 GHz ISM band has decreased, although the 2.4 band is still extremely overcrowded. The following 802.11 radios transmit in this band:

- 802.11 (FHSS radios or DSSS radios)
- 802.11b (HR-DSSS radios)
- 802.11g (ERP radios)
- 802.11n (HT radios)

In addition to being used by 802.11 WLAN equipment, the 2.4 GHz ISM band is also used by microwave ovens, cordless home telephones, baby monitors, and wireless video cameras. Other RF technologies, such as Bluetooth and Zigbee, also transmit in the 2.4 GHz band. Bluetooth uses FHSS transmissions, and Zigbee uses DSSS transmissions. The 2.4 GHz ISM band is heavily used, and one of the big disadvantages of using 802.11b/g/n 2.4 GHz radios is the potential for interference.

Please keep in mind that not every country's RF regulatory body will allow for transmissions across the entire 2.4 GHz to 2.5 GHz ISM band. The IEEE 802.11-2016 standard allows for WLAN transmissions in this band across 14 channels. However, each country can determine which channels can be used. A discussion of all the 2.4 GHz channels occurs later in this chapter.

5.8 GHz ISM Band

The 5.8 GHz ISM band is 150 MHz wide and spans from 5.725 GHz to 5.875 GHz. As with the other ISM bands, the 5.8 GHz ISM band is used by many of the same types of consumer products: baby monitors, cordless telephones, and cameras. It is not uncommon for novices to confuse the 5.8 GHz ISM band with the U-NII-3 band, which spans from 5.725 GHz to 5.85 GHz. Both unlicensed bands span the same frequency space; however, the 5.8 GHz ISM band is 25 MHz larger.

The IEEE 802.11a amendment (now part of the 802.11-2016 standard) states that "the OFDM PHY shall operate in the 5 GHz band, as allocated by a regulatory body in its operational region." Most countries allow for OFDM transmissions in channels of the various U-NII bands, which are discussed in this chapter. The United States has also always allowed OFDM transmissions on channel 165, which, until April 2014, resided in the 5.8 GHz ISM band. Historically, channel 165 has been sparsely used. In April 2014, the U-NII-3 band was expanded to include channel 165.

From the perspective of Wi-Fi channels, the 5.8 GHz ISM band is not relevant; however, many of the consumer devices that operate in the 5.8 GHz ISM band can cause RF interference with 802.11 radios that transmit in the U-NII-3 band.

Unlicensed National Information Infrastructure Bands

The IEEE 802.11a amendment designated WLAN transmissions within the frequency space of the three 5 GHz bands, each with four channels. These frequency ranges are known as the *Unlicensed National Information Infrastructure (U-NII) bands.* The 802.11a amendment defined three groupings, or bands, of U-NII frequencies, often known as the lower, middle, and upper U-NII bands. These three bands are typically designated as U-NII-1 (lower), U-NII-2 (middle), and U-NII-3 (upper).

When the 802.11h amendment was ratified, the IEEE designated more frequency space for WLAN transmissions. This frequency space, which consists of 12 additional channels, was originally referred to as U-NII-2 Extended and is now known as the U-NII-2C band.

Wi-Fi radios that currently transmit in the 5 GHz U-NII bands include radios that use the following technologies:

- 802.11a (OFDM radios)
- 802.11n (HT radios)
- 802.11ac (VHT radios)

Keep in mind that not every country's RF regulatory body will allow for transmissions in all these bands. The IEEE 802.11-2016 standard allows for WLAN transmissions in all four of the U-NII bands across 25 channels. However, each country may be different due to different channel and power regulations.

U-NII-1

U-NII-1, the lower U-NII band, is 100 MHz wide and spans from 5.150 GHz to 5.250 GHz. A total of four 20 MHz 802.11 channels reside in the U-NII-1 band. In the past, the U-NII-1 band was restricted by the FCC for indoor use only in the United States. As of April 2014, the FCC lifted this restriction. Prior to 2004, the FCC required that all U-NII-1-capable devices have permanently attached antennas. This meant that any 802.11a device that supported U-NII-1 could not have a detachable antenna, even if the device supported other frequencies or standards.

In 2004, the FCC changed the regulations to allow detachable antennas, provided that the antenna connector is unique. This requirement is similar to the antenna requirements for the other U-NII bands and the 2.4 GHz ISM band. Always remember that the 5 GHz power and transmit regulations are often different in other countries. Take care to ensure that you do not exceed the limitations of your local regulatory body.

U-NII-2A

The original U-NII-2 band is now referred to as the U-NII-2A band. U-NII-2A, the original middle U-NII band, is 100 MHz wide and spans from 5.250 GHz to 5.350 GHz. A total of four 20 MHz 802.11 channels reside in the U-NII-2 band. 802.11 radios that transmit in the U-NII-2 band must support dynamic frequency selection (DFS).

U-NII-2C

The U-NII-2 Extended band is now more often referred to as the U-NII-2C band. The U-NII-2C band is 255 MHz wide and spans from 5.470 GHz to 5.725 GHz. Most 5 GHz 802.11 radios can transmit on a total of eleven 20 MHz 802.11 channels that reside in the U-NII-2 band. However, with the advent of 802.11ac technology, a new channel 144 was added to the U-NII-2C band, for a total of 12 channels. 802.11 radios that transmit in the U-NII-2 band must support dynamic frequency selection (DFS). Operations for WLAN communications were first allowed in this band with the ratification of the 802.11h amendment. Prior to the ratification of this amendment, 5 GHz WLAN communications were allowed in only U-NII-1, U-NII-2A, and U-NII-3.

Dynamic Frequency Selection and Transmit Power Control

In Chapter 2, "IEEE 802.11 Standards," you learned that the 802.11h amendment defined the use of transmit power control (TPC) and dynamic frequency selection (DFS) to avoid interference with radar transmissions. Any 5 GHz WLAN products manufactured in the United States or Canada on or after July 20, 2007 are required to support dynamic frequency selection if they transmit in all the U-NII-2 bands. FCC Rule #15.407(h)(2) requires that WLAN products operating in the U-NII-2 bands support DFS, to protect WLAN communications from interfering with military or weather radar systems. Europe also requires DFS safeguards. DFS is a mechanism that detects the presence of radar signals and dynamically guides a transmitter to switch to another channel. Prior to the start of any transmission, a radio equipped with DFS capability must continually monitor the radio environment for radar pulse transmissions. If a radio determines that a radar signal is present, it must either select another channel to avoid interference with radar or go into a "sleep mode" if no other channel is available. TPC is required to protect the Earth Exploration Satellite Service (EESS). Once again, the local regulatory agencies determine how TPC and DFS restrictions are imposed in any of the U-NII bands.

U-NII-3

U-NII-3, the upper U-NII band, is 125 MHz wide and spans from 5.725 GHz to 5.850 GHz. This band is typically used for outdoor point-to-point communications but can also be used indoors in some countries, including the United States. Many of the countries in Europe do not use the U-NII-3 band for WLAN unlicensed communications. Some European countries allow transmission in the U-NII-3 band with the purchase of an inexpensive license.

In Table 6.3, notice that five 20 MHz 802.11 channels reside in the U-NII-3 band. In April 2014, the FCC expanded the size of the U-NII-3 band from 100 MHz to 125 MHz. Channel 165, formerly in the 5.8 GHz ISM band, is now available as part of the U-NII-3 band.

TABLE 6.3 The 5 GHz U-NII bands

Band	Frequency	Channels
U-NII-1	5.15 GHz–5.25 GHz	4 channels
U-NII-2A	5.25 GHz–5.35 GHz	4 channels
U-NII-2C	5.47 GHz–5.725 GHz	12 channels*
U-NII-3	5.725 GHz–5.85 GHz	5 channels

*With the advent of 802.11ac technology, a new channel 144 has been added to the U-NII-2C band, for a total of 12 channels. Currently, the majority of 5 GHz radios do not yet transmit on channel 144.

Future U-NII Bands

In January 2013, the FCC proposed two new U-NII bands. The first proposed U-NII-2 band was supposed to occupy the frequency space of 5.35 GHz–5.47 GHz and would have provided six 20 MHz channels. However, it looks like the FCC has decided that the U-NII-2B band will not be made available for Wi-Fi use. While the FCC is denying expansion of Wi-Fi into the U-NII-2B band, there is still the possibility for additional frequency expansion at the top end of the 5 GHz band. The U-NII-4 frequency band, 5.85 GHz–5.925 GHz, was reserved decades ago by US and European regulatory bodies to allow *Wireless Access in Vehicular Environments (WAVE)* communications from vehicle-to-vehicle and vehicle-to-roadway. This is the realm of 802.11p, and the band is designated as *dedicated short-range communications (DSRC)*. The automobile industry is seeing significant innovation towards self-driving automobiles and enhanced safety features, such as blind-side monitoring. These technologies may rely heavily on the U-NII-4 band.

Currently, there has been little usage of the U-NII-4 band by the automotive industry. Thus, there is thinking by the FCC that this band can be shared by traditional Wi-Fi users in places where DSRC is not being used. Although the U-NII-4 band offers only 75 MHz of frequency space, it is perfectly positioned to bridge with the U-NII-3 band. Therefore, as shown in Figure 6.7, U-NII-4 offers the potential of four additional 20 MHz channels, two additional 40 MHz channels, one additional 80 MHz, or an additional 160 MHz-wide channel. Although the addition of U-NII-4 appears to be a logical choice to allow for the expansion of Wi-Fi communications, the FCC has ordered additional testing and comment collection on this issue, and no timeline has been publicly announced.

FIGURE 6.7   U-NII bands

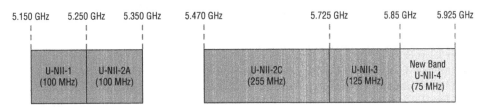

There have also been recent FCC inquiries into the possibility of Wi-Fi communications expanding into the 6 GHz frequency band. The FCC is soliciting comments on allowing unlicensed band devices (such as Wi-Fi radios) to share bandwidth in the 5.925 GH–6.425 GHz and 6.425 GHz–7.125 GHz frequency ranges. For a more detailed explanation about potential frequency space that can be used for Wi-Fi communications, we suggest that you read the blog "New Spectrum Status for FCC Regions," at www.wirelesstrainingsolutions .com/new-spectrum. This blog is authored by Rick Murphy, CWNE #10.

3.6 GHz Band

In 2008, the 802.11y amendment was ratified. This amendment specified the use of the frequency range of 3.65 GHz to 3.7 GHz. This was approved as a licensed band for use in the United States. Unlike other licensed frequencies, the use of this frequency range was non-exclusive and included limitations when used near certain satellite earth stations. Although the project was designed for use in the United States, it was carefully designed to be able to operate in other countries without the need to ratify a new amendment (a process that can take several years to complete). It was designed to operate in any 5 MHz, 10 MHz, or 20 MHz channel. Regulators can make any frequency range available for use.

4.9 GHz Band

The 802.11-2016 standard defines the frequency range of 4.94 GHz to 4.99 GHz in the United States for public safety organizations to use for the protection of life, health, or property. This band is actually a licensed band and is reserved strictly for public safety.

This frequency range has also been approved in other countries, such as Canada and Mexico.

In 2004, the 802.11j amendment was ratified, providing support for the 4.9 GHz to 5.091 GHz frequency range for use in Japan. This amendment was later incorporated in the 802.11-2016 standard.

Because of the proximity of these frequencies to the U-NII-1 band, we are seeing more wireless radios providing support for this band.

Future Wi-Fi Frequencies

The 2.4 GHz ISM band was the original license-free range of frequencies, known as a *frequency band*, that has been used for Wi-Fi communications since 1997. Although 802.11a was ratified in 1999, the use of the 5 GHz U-NII bands really did not start to catch on until about 2006. Wi-Fi use in the 5 GHz frequency bands is now the norm for a number of reasons: The 2.4 GHz band remains overcrowded, and the 5 GHz bands are wider and have more channels. The IEEE continues to look toward other spectrum space for future Wi-Fi communications.

Below 1 GHz

The 802.11ah amendment defines the use of Wi-Fi in frequencies below 1 GHz. The *Wi-Fi HaLow* certification from the Wi-Fi Alliance is based on mechanisms defined in the IEEE 802.11ah amendment. The lower frequencies will mean lower data rates but longer distances. A likely use for 802.11h will be sensor networks along with backhaul for sensor networks and extended range Wi-Fi, such as smart homes, automobiles, health care, industrial, retail, and agriculture. This internetworking of devices is known as the *Internet of Things (IoT)* or *machine-to-machine (M2M)* communications.

The available frequencies will vary between countries. For example, the 902–928 MHz unlicensed ISM frequencies are available in the United States, whereas the 863–868 MHz frequencies would likely be available in Europe, and the 755–787 MHz frequencies would likely be available in China.

60 GHz

As mentioned in Chapter 2, the 802.11ad ratified amendment defines very high throughput (VHT) technology that operates in the unlicensed 60 GHz frequency band. New PHY and MAC layer enhancements have the potential of accomplishing speeds of up to 7 Gbps. Since these ultrahigh frequencies have difficulty penetrating through walls, the technology will most likely be used to provide bandwidth-intensive and short-distance communications indoors, such as high definition (HD) video streaming. The technology will not be backward compatible with other 802.11 technology.

Tri-band radios will have the capability to provide Wi-Fi access in the 2.4 GHz, 5 GHz, and 60 GHz bands. This tri-band capability should provide for seamless handoff between devices in the short coverage area of the 60 GHz band and the greater coverage area of either the 2.4 GHz or 5 GHz band.

TV White Space

As mentioned in Chapter 2, *TV white space (TVWS)* is a term used to describe the use of Wi-Fi technology in the unused television RF spectrum. The TV white space spectrum is the former VHF and UHF frequency bands between 54 and 790 MHz. The 802.11af-2014 amendment defines Wi-Fi operations within these unused frequency ranges. One of the immediate gains will be greater range because the white space frequencies are below 1 GHz.

2.4 GHz Channels

To better understand how legacy 802.11 (DSSS), 802.11b (HR-DSSS), and 802.11g (ERP) radios are used, it is important to know how the IEEE 802.11-2016 standard divides the 2.4 GHz ISM band into 14 separate channels, as listed in Table 6.4. Although the 2.4 GHz ISM band is divided into 14 channels, the FCC or local regulatory body designates which channels are allowed to be used. Table 6.4 also shows a sample of how channel support can vary.

TABLE 6.4   2.4 GHz frequency channel plan

Channel ID	Center Frequency (GHz)	US (FCC)	Canada (IC)	Many European Countries
1	2.412	X	X	X
2	2.417	X	X	X
3	2.422	X	X	X
4	2.427	X	X	X
5	2.432	X	X	X
6	2.437	X	X	X
7	2.442	X	X	X

TABLE 6.4 2.4 GHz frequency channel plan *(continued)*

Channel ID	Center Frequency (GHz)	US (FCC)	Canada (IC)	Many European Countries
8	2.447	X	X	X
9	2.452	X	X	X
10	2.457	X	X	X
11	2.462	X	X	X
12	2.467			X
13	2.472			X
14	2.484			

X = supported channel

Channels are designated by their center frequency. How wide the channel is depends on the technology used by the 802.11 transmitter. When DSSS and HR-DSSS 802.11 radios are transmitting, each channel is 22 MHz wide and is often referenced by the center frequency ± 11 MHz. For example, channel 1 is 2.412 GHz ± 11 MHz, which means that channel 1 spans from 2.401 GHz to 2.423 GHz. It should also be noted that within the 2.4 GHz ISM band, the distance between channel center frequencies is only 5 MHz. Because each channel is 22 MHz wide, and because the separation between center frequencies of each channel is only 5 MHz, the channels will have overlapping frequency space. With the introduction of OFDM in 802.11a, along with its expanded use in 802.11g, 802.11n, and 802.11ac, the frequency width used by an OFDM channel is approximately 20 MHz (as defined by the spectral mask, which you will learn about later in this chapter).

Figure 6.8 shows an overlay of all the channels and how they overlap. Channels 1, 6, and 11 have been highlighted because, as you can see, they are separated from each other by enough frequencies that they do not overlap. In order for two channels to not overlap, they must be separated by at least five channels, or 25 MHz. Channels such as 2 and 9 do not overlap, but when 2 and 9 are selected, there is no additional legal channel that can be chosen that does not overlap either 2 or 9. In the United States and Canada, the only three simultaneously nonoverlapping channels are 1, 6, and 11. In regions where channels 1 through 13 are allowed to be used, there are different combinations of three nonoverlapping channels, although channels 1, 6, and 11 are commonly chosen. Enterprise deployments of three or more access points in the 2.4 GHz ISM band normally use channels 1, 6, and 11, which are all considered nonoverlapping.

FIGURE 6.8   2.4 GHz channel overlay diagram

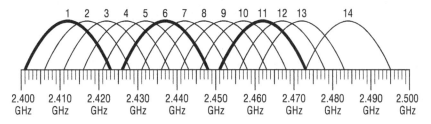

The IEEE 802.11-2016 definitions of nonoverlapping channels in the 2.4 GHz ISM band can be somewhat confusing if not properly explained. Legacy 802.11 (DSSS), 802.11b (HR-DSSS), and 802.11g (ERP) channels all use the same numbering schemes and have the same center frequencies. However, the individual channel's frequency space may overlap. Figure 6.9 shows channels 1, 6, and 11 with 25 MHz of spacing between the center frequencies. These are the most commonly used *nonoverlapping channels* in North America and most of the world for 802.11b/g/n networks.

FIGURE 6.9   802.11/b/g/n center frequencies

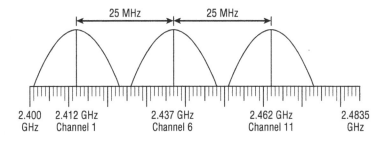

What exactly classifies DSSS or HR-DSSS channels as nonoverlapping? According to the original 802.11 standard, legacy DSSS channels had to have at least 30 MHz of spacing between the center frequencies to be considered nonoverlapping. In a deployment of legacy DSSS equipment using a channel pattern of 1, 6, and 11, the channels were considered overlapping because the center frequencies were only 25 MHz apart. Although DSSS channels 1, 6, and 11 were defined as overlapping, these were still the only three channels used in channel reuse patterns when legacy networks were deployed. This is of little significance anymore because most 2.4 GHz deployments now use 802.11b/g/n technology.

HR-DSSS was introduced under the 802.11b amendment, which states that channels need a minimum of 25 MHz of separation between the center frequencies to be considered nonoverlapping. Therefore, when 802.11b was introduced, channels 1, 6, and 11 were considered nonoverlapping.

The 802.11g amendment, which allows for backward compatibility with 802.11b HR-DSSS, also requires 25 MHz of separation between the center frequencies to be considered nonoverlapping. Under the 802.11g amendment, channels 1, 6, and 11 are also considered nonoverlapping for both ERP-DSSS/CCK and ERP-OFDM.

Although it is very common to represent the RF signal of a particular channel with an arch-type line, this is not a true representation of the signal. For example, in addition to the main *carrier frequency*, or main frequency, sideband carrier frequencies are generated, as shown in Figure 6.10. In this example, the IEEE defines a *transmit spectrum mask*, specifying that the first sideband frequency (–11 MHz to –22 MHz from the center frequency and +11 MHz to +22 MHz from the center frequency) must be at least 30 dB less than the main frequency. The mask also specifies that any additional sideband carrier frequencies (–22 MHz from the center frequency and beyond and +22 MHz from the center frequency and beyond) must be at least 50 dB less than the main frequency.

FIGURE 6.10   IEEE 802.11b transmit spectrum mask

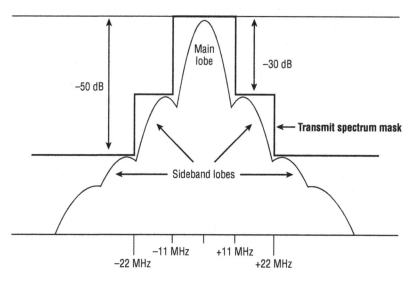

Figure 6.10 illustrates the transmit spectrum mask of an HR-DSSS channel at 2.4 GHz. The transmit spectrum mask is defined to minimize interference between devices on different frequencies. Even though the sideband carrier frequencies are mere whispers of signals compared to the main carrier frequency, even a whisper is noticeable when the person whispering is close to you. This is true for RF devices, too.

Figure 6.11 represents 802.11b RF signals on channels 1, 6, and 11. A signal-level line indicates an arbitrary level of reception by the access point on channel 6. At level 1, meaning the AP on channel 6 receives only the signals above the level 1 line, the signals from channel 1 and channel 11 do not intersect (interfere) with the signals on channel 6. However, at the level 2 line, the signals from channel 1 and channel 11 do intersect (interfere) slightly with the signals on channel 6. At the level 3 line, there is significant interference from the signals from channel 1 and channel 11. Because of the potential for this situation, it is important to separate access points (usually 5 to 10 feet is sufficient) so that interference from sideband frequencies does not occur. This separation is important both horizontally and vertically.

FIGURE 6.11   Sideband carrier frequency interference

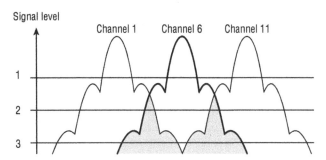

Although most 2.4 GHz transmissions are performed at higher data rates using OFDM, it is still important to understand the older, slower technologies because they are still part of the standard and are still used. In the next section, you will learn about 5 GHz channels and see a transmit spectrum mask for 5 GHz OFDM transmissions. This mask is about 20 MHz wide and is almost *identical* to what is required when transmitting using OFDM in the 2.4 GHz band. Remember that OFDM technology is used by 802.11g and 802.11n radios in the 2.4 GHz band.

5 GHz Channels

802.11a/n and 802.11ac radios transmit in the 5 GHz U-NII bands: U-NII-1, U-NII-2A, U-NII-2C, and U-NII-3. To prevent interference with other possible bands, extra bandwidth is used as a guard band. In the U-NII-1 and U-NII-2 bands, the centers of the outermost channels of each band must be 30 MHz from the band's edge. An extra 20 MHz of bandwidth exits in the U-NII-3 band. The unused bandwidth at the edge of each band is known as a *guard band*. The original three U-NII bands each had four nonoverlapping channels with 20 MHz separation between the center frequencies. A fifth channel was added to U-NII-3. The U-NII-2C band has 12 nonoverlapping channels with 20 MHz of separation between the center frequencies. The U-NII-2C band was an 11-channel band for many years, but an extra channel, 144, was added to the band with the advent of 802.11ac. If you want to calculate the center frequency of a channel, multiply the channel by 5 and then add 5000 to the result—for example, channel 36 times 5 equals 180, then add 5000, for a center frequency of 5180 MHz, or 5.18 GHz.

Figure 6.12 shows the eight U-NII-1 and U-NII-2A channels in the top graphic, the 12 U-NII-2C channels in the center graphic, and the five U-NII-3 channels in the bottom graphic. Channel 36 is highlighted so that it is easier to distinguish a single carrier and its sideband frequencies. The IEEE does not specifically define a channel width; however, the spectral mask of an OFDM channel is approximately 20 MHz.

FIGURE 6.12   U-NII channels

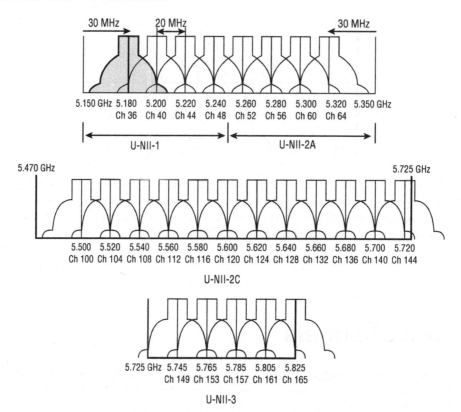

Figure 6.13 depicts a wide overview of all the 5 GHz channels that can currently be used by 802.11 transmitters. A total of twenty-five 20 MHz channels in the 5 GHz U-NII bands can be used when designing a WLAN with a channel reuse pattern. The channels you can use, of course, depend on the regulations of each country. For example, in most of Europe, the U-NII-3 band is still considered a licensed band, meaning that only 20 channels are available for a channel reuse pattern. In the United States, all the channels were available until 2009. Notice in Figure 6.13 that DFS is required in U-NII-2A and U-NII-2C channels. In these bands, 802.11 radios are required to use dynamic frequency selection (DFS) to avoid interference with radar. In 2009, the Federal Aviation Authority (FAA) reported interference with *Terminal Doppler Weather Radar (TDWR)* systems. As a result, the FCC suspended certification of 802.11 devices in the U-NII-2 and U-NII-2E bands that required DFS. Eventually, certification was re-established; however, the rules changed and 802.11 radios were not allowed to transmit in the 5.60 GHz–5.65 GHz frequency space where TDWR operates. As shown in Figure 6.13, channels 120, 124, and 128 reside in the TDWR frequency space and could not be used for many years in the United States; therefore, not all channels were available for a 20 MHz channel reuse. In

2014, the FCC changed the rules and the TDWR frequency space was once again available. It should also be noted that some enterprise WLAN deployments have completely avoided the use of the DFS channels altogether because some client devices simply have not been certified for any of the DFS channels. A more detailed discussion of 5 GHz channel reuse patterns and channel planning appears in Chapter 13, "WLAN Design Concepts."

FIGURE 6.13   U-NII channel overview

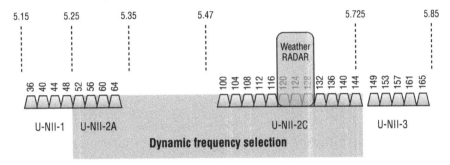

802.11n technology introduced the capability of bonding together two 20 MHz channels to create a larger 40 MHz channel. Channel bonding effectively doubles the frequency bandwidth, meaning double the data rates that can be available to 802.11n radios. 40 MHz channels will be discussed in greater detail in Chapter 10. As shown in Figure 6.14, a total of twelve 40 MHz channels are available to be used in a reuse pattern when deploying an enterprise WLAN. However, in the past, two of the 40 MHz channels have not been used in the United States because they fall within the TDWR band. In Europe, two of the 40 MHz channels cannot be used because they fall within the U-NII-3 band that requires licensing in many countries.

FIGURE 6.14   U-NII 40 MHz, 80 MHz, and 160 MHz channels

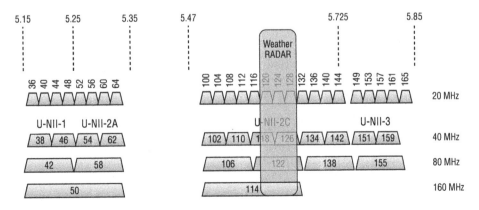

Figure 16.14 also depicts the 80 MHz channels and 160 MHz channels that could possibly be used by 802.11ac radios. In reality, there is not enough frequency space to provide for enough of these channels for proper channel reuse patterns. 802.11ac capabilities, including 80 MHz and 160 MHz channels, are discussed in Chapter 10.

As 802.11ac technology becomes more commonplace, the need for the extra frequency space is even more important. As shown in Figure 6.15, if the U-NII-4 spectrum is indeed made available, channel reuse patterns using larger channels are possible. A total of fourteen 40 MHz channels would be available for 802.11n or 802.11ac radios. A reuse pattern of seven 80 MHz channels could be designed for use by 802.11ac radios. There would even be enough frequency space for three 160 MHz channels. Please keep in mind that the proposed extra spectrum is still not currently available and that rules and regulations about frequency use can vary from country to country.

FIGURE 6.15   Potential 20 MHz, 40 MHz, 80 MHz, and 160 MHz channels

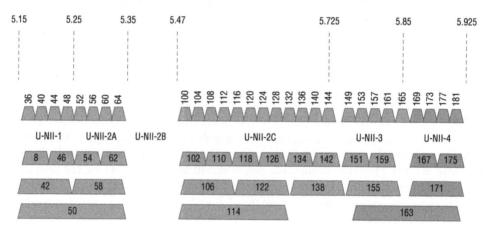

As stated earlier, the IEEE does not specifically define a channel width; however, the spectral mask of an OFDM channel is approximately 20 MHz. The spectral mask of 802.11n or 802.11ac bonded channels is obviously much larger.

Figure 6.16 shows the OFDM spectrum mask. Note that the sideband carrier frequencies do not drop off very quickly; therefore, the sideband frequencies of two adjacent valid channels overlap and are more likely to cause interference. The 802.11a amendment, which originally defined the use of OFDM, required only 20 MHz of separation between the center frequencies for channels to be considered nonoverlapping. All 20 MHz channels in the 5 GHz U-NII bands use OFDM and have 20 MHz of separation between the center frequencies. Therefore, all 5 GHz OFDM channels are considered nonoverlapping by the IEEE. In reality, some sideband carrier frequency overlap exists between any two adjacent 5 GHz channels. Luckily, due to the number of channels and the channel spacing, it is easy to separate adjacent channels and prevent interference with a proper channel reuse plan.

FIGURE 6.16   OFDM spectrum mask

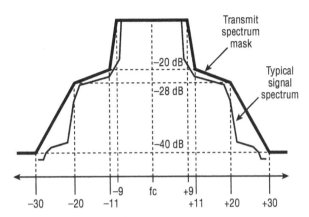

Long-Term Evolution in 5 GHz

Even though this book is about 802.11 wireless technology, there are other RF technologies you should be familiar with, including Bluetooth, Zigbee, and cellular technology. *Long-Term Evolution (LTE)* is a standard for high-speed wireless voice and data communication for cellular devices. The next generation of LTE technology currently generating a lot of press is Fifth-Generation Wireless Systems, also known as *5G*. In contrast to previous generations of LTE technology, 5G uses both unlicensed and licensed frequency spectrum for cellular communications.

You may hear the terms 5G and *unlicensed LTE* used interchangeably. That is wrong. Try to think of 5G as a framework, encapsulating both technology and business use cases, while unlicensed LTE is one of many cellular technologies potentially used in a 5G environment. For the scope of this chapter, the relationship between unlicensed LTE and Wi-Fi is the primary discussion.

One of the main reasons enterprise WLANs are deployed in the unlicensed 5 GHz band is that the unlicensed 2.4 GHz band is already overcrowded. Furthermore, 5 GHz has more frequency space and traditionally has mainly been used for Wi-Fi communications. However, cellular companies have been successfully lobbying for LTE communication in the unlicensed 5 GHz frequency band where Wi-Fi currently operates.

Will unlicensed LTE transmissions in the 5 GHz frequency band interfere with Wi-Fi (and vice versa)? The answer is yes, but the impact depends on several variables. You should first realize that many flavors of LTE are being defined for use in unlicensed spectrum, including the 5 GHz band.

Wi-Fi integrators and designers will have to be aware of the unlicensed LTE capabilities. Remember that this is a very simplified overview; if (when) presented with a challenge to design in the presence of unlicensed LTE, please do your homework. Some of the unlicensed LTE technologies include the following:

LTE-U   LTE-Unlicensed (LTE-U) technology is used in the unlicensed U-NII-1 and U-NII-3 bands in the 5 GHz spectrum. LTE-U uses 20 MHz channels for downlink communication from an LTE-U base station. The uplink and control plane communications remain on the licensed spectrum. The telecom carrier aggregates the unlicensed 5 GHz downlink communications together with the uplink communications on a licensed (400 MHz–3.8 GHz) anchor channel. LTE-U does not implement any "listen before talk" protection mechanisms and will talk over Wi-Fi communications in the same frequency space. LTE-U uses a *Carrier Sense Adaptive Transmission (CSAT)* algorithim. CSAT dynamically chooses a free channel, to avoid interfering with Wi-Fi. However, if no free channel is available, CSAT shares the channel using an on/off duty cycle, which provides 50 percent of the duty cycle for LTE and 50 percent of the duty cycle for Wi-Fi communications on a 5 GHz channel.

LTE-LAA   Another LTE technology intended for unlicensed frequencies is *License Assisted Access (LAA)*, which was originally also defined for downlink only. A new variant called *enhanced Licensed Assisted Access (eLAA)* also provides for uplink capabilities. eLAA uses up to 20 MHz wide channels but can operate bi-directionally for data transfers. eLAA uses both licensed and unlicensed spectrum for data but also uses licensed spectrum for control plane communications. LTE-LAA also uses carrier aggregation for the unlicensed 5 GHz channel together with the licensed anchor channel.

LTE-LAA uses *Listen Before Talk (LBT)* as a carrier sense protection mechanism, somewhat similar to clear channel assessment (CCA) capabilities used by 802.11 radios. An LTE-LAA base station utilizes an RF *energy detect (ED)* threshold to defer LTE transmissions. However, there is no *signal detect (SD)* threshold for specifically detecting 802.11 transmissions. At the time of this writing, LTE-LAA avoids transmission on DFS channels.

MulteFire   Currently, *MulteFire* operates exclusively in the unlicensed 5 GHz band for downlink, uplink, and control plane communications. No licensed anchor channel is necessary and, therefore, no carrier aggregation. MulteFire is intended for small cell deployments and, for all intended purposes, MulteFire could be viewed as a direct competitor to Wi-Fi. Similar to LTE-LAA, MulteFire uses Listen Before Talk (LBT) as a carrier sense protection mechanism. MulteFire also operates on 20 MHz channels.

LTE-WLAN Aggregation   *LTE-WLAN Aggregation (LWA)* provides an alternative to LTE in the 5 GHz spectrum, wherein LTE data transfers remain on a licensed channel, while Wi-Fi communications remain on an unlicensed channel. However, LWA provides for aggregation of the data across both links as a single flow of traffic. A mobile handset that supports both LTE and Wi-Fi could be configured to use both links simultaneously. From a Wi-Fi perspective, the main advantage is that there is no need to share a 5 GHz channel, because the LTE communications remain on licensed spectrum.

Because of the political influence and lobbying by cellular providers and chipset manufacturers, unlicensed LTE in the 5 GHz band is slowly becoming a reality. It remains to be seen which unlicensed LTE technology will become the most prevalent, and different regions of the world might adopt different unlicensed LTE standards. Keep in mind that LTE will have to abide by the same transmit power regulations in the unlicensed 5 GHz band that Wi-Fi might abide by. However, most indoor deployments of Wi-Fi APs use a maximum of 100 mW, whereas an LTE base station might transmit at 1 watt. Despite the defined co-existence techniques, it is fairly safe to assume that there will be interference between LTE and Wi-Fi. As unlicensed LTE moves into the 5 GHz band, new challenges for WLAN design and deployment are inevitable.

Adjacent, Nonadjacent, and Overlapping Channels

In the preceding paragraphs, you learned how the IEEE 802.11-2016 standard defines non-overlapping channels. DSSS (legacy) channels require 30 MHz of separation between the center frequencies to be considered nonoverlapping. HR-DSSS (802.11b) and ERP (802.11g) channels require 25 MHz of separation between the center frequencies to be considered nonoverlapping. And, finally, 5 GHz OFDM channels require 20 MHz of separation between the center frequencies to be considered nonoverlapping. Why are these definitions important? When deploying a WLAN, it is important to have overlapping cell coverage in order for roaming to occur. However, it is just as important for these coverage cells not to have overlapping frequency space. A channel reuse pattern is needed because overlapping frequency space causes degradation in performance. The design aspects of channel reuse patterns are discussed in detail in Chapter 13, "WLAN Design Concepts."

 **Real World Scenario**

What Is the Significance of Adjacent Channels?

Most Wi-Fi vendors use the term *adjacent channel interference* to refer to the degradation of performance resulting from overlapping frequency space that occurs because of an improper channel reuse design. In the WLAN industry, an adjacent channel is considered to be the next or previous numbered channel. For example, channel 3 is adjacent to channel 2. The concept of adjacent channel interference is discussed in detail in Chapter 13.

Throughput vs. Bandwidth

Wireless communication is typically performed within a constrained set of frequencies known as a *frequency band*. This frequency band is the *bandwidth*. Frequency bandwidth does play a part in the eventual throughput of the data, but many other factors also determine throughput. In addition to frequency bandwidth, data encoding, modulation, medium contention, encryption, and many other factors also play a large part in data throughput.

Care should be taken not to confuse frequency bandwidth with data bandwidth. Data encoding and modulation determine data rates, which are sometimes also referred to as data bandwidth. Simply look at the 5 GHz channels and OFDM as an example. OFDM 802.11a radios can transmit at 6, 9, 12, 18, 24, 36, 48, or 54 Mbps, yet the frequency bandwidth for all the U-NII band channels is the same for all these speeds. What changes between all these speeds (data rates) is the modulation and coding technique. The proper term for the changes in speed due to modulation and coding is *data rates*; however, they are also often referred to as *data bandwidth*.

One of the surprising facts when explaining wireless networking to a layperson is the actual throughput that an 802.11 wireless network provides. When novices walk through a computer store and see the packages of 802.11 devices, they likely assume that a device that is labeled as 300 Mbps is going to provide throughput of 300 Mbps. A medium access method known as Carrier Sense Multiple Access with Collision Avoidance (CSMA/CA) attempts to ensure that only one radio device can be transmitting on the medium at any given time. Because of the half-duplex nature of the medium and the overhead generated by CSMA/CA, the actual aggregate throughput is typically 50 percent or less of the data rates for 802.11a/b/g legacy transmissions, and 60–70 percent of the data rates for 802.11n/ac transmissions. In addition to the throughput being affected by the half-duplex nature of 802.11 communications, the throughput is affected differently based on the frequency used. HT and OFDM technologies are used in both the 5 GHz and 2.4 GHz bands. Because of the higher level of RF noise that is typical in the 2.4 GHz ISM band, throughput of 2.4 GHz devices will typically be less than the 5 GHz devices.

It is also very important to understand that the 802.11 RF medium is a *shared* medium, meaning that in any discussion of throughput, it should be thought of as *aggregate throughput*. For example, if a data rate is 54 Mbps, because of CSMA/CA, the aggregate throughput might be about 20 Mbps. If five client stations were all downloading the same file from an FTP server at the same time, the perceived throughput for each client station would be about 4 Mbps under ideal circumstances. When 802.11n and 802.11ac radios are used, the aggregate throughput can be as much as 65 percent of the advertised data rate under ideal conditions The medium contention overhead created by CSMA/CA is typically about 35 percent of the bandwidth. Medium contention overhead is 50 percent or more when using legacy 802.11a/b/g radios.

RTS/CTS (which you will learn about in Chapter 9, "802.11 MAC") can also affect throughput by adding communication overhead.

Variables at almost all layers of the OSI model can affect the throughput of 802.11 communications. It is important to understand the different causes, their effects, and what, if anything, can be done to minimize their effect on overall data throughput.

Communication Resilience

Many of the technologies that have been covered in this chapter either directly or indirectly provide resilience to 802.11 communications. Spread spectrum spreads the data across a range of frequencies, making it less likely for a narrowband RF signal to cause interference.

FHSS is inherently more resilient to narrowband interference than OFDM, and OFDM is more resilient to narrowband interference than DSSS. Spread spectrum technology uses a range of frequencies, which inherently adds resilience because delay spread and ISI will vary between the different frequencies. Additionally, data encoding provides error-recovery methods, helping to reduce the need for retransmission of the data.

Summary

This chapter focused on the technologies that make up wireless networking and spread spectrum. 802.11, 802.11b, 802.11g, and 802.11n radios use the 2.4 GHz ISM band, while 802.11a/n and 802.11ac radios use the 5 GHz U-NII bands. 802.11n HT radios can use both the 2.4 GHz ISM band and the 5 GHz U-NII bands. The following ISM and U-NII bands were discussed in this chapter:

- ISM 902–928 MHz—industrial
- ISM 2.4–2.5 GHz—scientific
- ISM 5.725–5.875 GHz—medical
- U-NII-1 5.150–5.250 GHz
- U-NII-2A 5.250–5.350 GHz
- U-NII-2C 5.470–5.725 GHz
- U-NII-3 5.725–5.85 GHz—upper
- Proposed U-NII-4 5.85–5.925 GHz

    In addition to the ISM and U-NII bands, the following bands were discussed:

- 4.94–4.99 GHz—US public safety
- 4.9–5.091 GHz—Japan
- 60 GHz
- <1 GHz
- TV white space (TVWS)

    Spread spectrum technology was introduced and described in detail, along with OFDM and convolutional coding.

Exam Essentials

Know the technical specifications of all the ISM and U-NII bands.   Make sure that you know all the frequencies, bandwidth uses, and channels.

Know spread spectrum.   Spread spectrum can be complicated and has different flavors. Understand FHSS, DSSS, and OFDM. (Although OFDM is not a spread spectrum technology, it has similar properties, and you have to know it.) Understand how coding and modulation work with spread spectrum and OFDM.

Understand the similarities and differences between the transmission methods discussed in this chapter. There are differences and similarities between many of the topics in this chapter. Carefully compare and understand them. Minor subtleties can be difficult to recognize when you are taking the test.

Review Questions

1. Which technology has a greater tolerance of delay spread? (Choose all that apply.)
    A. DSSS
    B. FHSS
    C. OFDM
    D. HT

2. Which of the following are valid U-NII bands? (Choose all that apply.)
    A. 5.150 GHz–5.250 GHz
    B. 5.470 GHz–5.725 GHz
    C. 5.725 GHz–5.85 GHz
    D. 5.725 GHz–5.875 GHz

3. Which technologies are used by 802.11 radios in the 2.4 GHz ISM band? (Choose all that apply.)
    A. HT
    B. ERP
    C. DSSS
    D. HR-DSSS

4. 802.11n (HT radios) can transmit in which frequency bands? (Choose all that apply.)
    A. 2.4 GHz–2.4835 GHz
    B. 5.47 GHz–5.725 GHz
    C. 902 GHz–928 GHz
    D. 5.15 GHz–5.25 GHz

5. In the U-NII-1 band, what is the center frequency of channel 40?
    A. 5.2 GHz
    B. 5.4 GHz
    C. 5.8 GHz
    D. 5.140 GHz

6. What is the channel and band of a Wi-Fi transmission whose center frequency is 5.300 GHz?
    A. U-NII-1 channel 30
    B. U-NII-1 channel 48
    C. U-NII-2A channel 56
    D. U-NII-2A channel 60

7. When a single-channel OFDM signal is transmitted, what is the approximate frequency width of the transmission?

   A. 20 MHz

   B. 22 MHz

   C. 25 MHz

   D. 40 MHz

   E. 80 MHz

   F. 160 MHz

8. What best describes hop time?

   A. The period of time that the transmitter waits before hopping to the next frequency

   B. The period of time that the standard requires when hopping between frequencies

   C. The period of time that the transmitter takes to hop to the next frequency

   D. The period of time the transmitter takes to hop through all the FHSS frequencies

9. As defined by the IEEE 802.11-2016 standard, how much separation is needed between center frequencies of channels in the U-NII-2C band?

   A. 10 MHz

   B. 20 MHz

   C. 22 MHz

   D. 25 MHz

   E. 30 MHz

10. When deploying a modern 802.11n/ac wireless network with only two access points, which of these 2.4 GHz channel groupings would be considered nonoverlapping? (Choose all that apply.)

    A. Channels 1 and 3

    B. Channels 7 and 10

    C. Channels 3 and 8

    D. Channels 5 and 11

    E. Channels 6 and 10

11. Which U-NII band is currently designated for Wireless Access in Vehicular Environments (WAVE) communications?

    A. U-NII-1

    B. U-NII-2A

    C. U-NII-2B

    D. U-NII-2C

    E. U-NII-3

    F. U-NII-4

12. When data is corrupted by previous data from a reflected signal, this is known as what?

    A. Delay spread

    B. ISI

    C. Forward error creation

    D. Bit crossover

13. Assuming all channels are supported by a 5 GHz access point, how many possible 20 MHz channels can be configured on the access point?

    A. 4

    B. 11

    C. 12

    D. 25

14. Which of these technologies is the most resilient against the negative effects of multipath?

    A. FHSS

    B. DSSS

    C. HR-DSSS

    D. OFDM

15. What is the average amount of aggregate throughput at any data rate when legacy 802.11a/b/g/n/ac radios are transmitting?

    A. 80 percent

    B. 75 percent

    C. 50 percent

    D. 100 percent

16. Which U-NII band proposed by the FCC may provide for 75 MHz of additional spectrum at 5 GHz for Wi-Fi communications?

    A. U-NII-1

    B. U-NII-2A

    C. U-NII-2B

    D. U-NII-2C

    E. U-NII-3

    F. U-NII-4

17. In the United States, 802.11 radios were not allowed to transmit on which range of frequencies in order to avoid interference with Terminal Doppler Weather Radar (TDWR) systems?

    A. 5.15 GHz–5.25 GHz

    B. 5.25 GHz–5.25 GHz

    C. 5.60 GHz–5.65 GHz

    D. 5.85 GHz–5.925 GHz

18. Which modulation types are used by OFDM technology? (Choose all that apply.)

    A. QAM

    B. Phase

    C. Frequency

    D. Hopping

19. The Barker code converts a bit of data into a series of bits that are referred to as what?

    A. Chipset

    B. Chips

    C. Convolutional code

    D. Complementary code

20. A 20 MHz OFDM channel uses how many 312.5 KHz data subcarriers when transmitted by an 802.11a/g radio?

    A. 54

    B. 52

    C. 48

    D. 36

Chapter

7

Wireless LAN Topologies

IN THIS CHAPTER, YOU WILL LEARN ABOUT THE FOLLOWING:

✓ **Wireless networking topologies**

- Wireless wide area network (WWAN)
- Wireless metropolitan area network (WMAN)
- Wireless personal area network (WPAN)
- Wireless local area network (WLAN)

✓ **802.11 stations**

- Client station
- Access point station
- Integration service (IS)
- Distribution system (DS)
- Wireless distribution system (WDS)

✓ **802.11 service sets**

- Service set identifier (SSID)
- Basic service set (BSS)
- BSA should be after BSS
- Basic service set identifier (BSSID)
- Multiple basic service set identifiers
- Extended service set (ESS)
- Independent basic service set (IBSS)
- Personal basic service set (PBSS)
- Mesh basic service set (MBSS)
- QoS basic service set (QBSS)

✓ **802.11 configuration modes**

- Access point modes
- Client station modes

A computer network is a system that provides communications between computers. Computer networks can be configured as peer-to-peer, as client-server, or as clustered central processing units (CPUs) with distributed dumb terminals. A networking *topology* is defined simply as the physical and/or logical layout of nodes in a computer network. Any individual who has taken a networking basics class is already familiar with the bus, ring, star, mesh, and hybrid topologies that are often used in wired networks.

All topologies have advantages and disadvantages. A topology may cover very small areas or exist as a worldwide architecture. Wireless topologies also exist as defined by the physical and logical layout of wireless hardware. Many wireless technologies are available and can be arranged into four major wireless networking topologies. The 802.11-2016 standard defines one specific type of wireless communication. Within the 802.11-2016 standard are different types of topologies, known as *service sets*. Over the years, vendors have also used 802.11 hardware using variations of these topologies to meet specific wireless networking needs. This chapter covers the topologies used by a cross section of radio frequency (RF) technologies and covers 802.11-specific wireless local area network (WLAN) topologies.

Wireless Networking Topologies

Although the main focus of this study guide is 802.11 wireless networking, which is a local area technology, other wireless technologies and standards exist in which wireless communications span either smaller or larger areas of coverage. Examples of other wireless technologies include cellular, Bluetooth, and Zigbee. All these different wireless technologies can be arranged into the following four major wireless topologies:

- Wireless wide area network (WWAN)
- Wireless metropolitan area network (WMAN)
- Wireless personal area network (WPAN)
- Wireless local area network (WLAN)

Additionally, although the 802.11-2016 standard is a WLAN standard, the same technology can sometimes be deployed in different wireless network topologies, as discussed in the following sections.

Wireless Wide Area Network

A wide area network (WAN) provides RF coverage over a vast geographical area. A WAN might traverse an entire state, region, or country or even span worldwide. The best example of a WAN is the Internet. Many private and public corporate WANs consist of hardware infrastructure, such as T1 lines, fiber optics, and routers. Protocols used for wired WAN communications include Frame Relay, ATM, Multiprotocol Label Switching (MPLS), and others.

A *wireless wide area network (WWAN)* also covers broad geographical boundaries but obviously uses a wireless medium instead of a wired medium. WWANs typically use cellular telephone technologies or proprietary licensed wireless bridging technologies. Cellular providers, such as AT&T Mobility, Verizon, and Vodafone, use a variety of competing technologies to carry data. Some examples of these cellular technologies are general packet radio service (GPRS), code division multiple access (CDMA), time division multiple access (TDMA), Long Term Evolution (LTE), and Global System for Mobile Communications (GSM). Data can be carried to a variety of devices, such as smartphones, tablet PCs, and cellular USB modems.

In the past, data rates and bandwidth using these technologies were relatively slow when compared to other wireless technologies, such as 802.11. Much like Wi-Fi, data transfer rates have improved with the progression of multiple generations of cellular technology. Additionally, convergence and co-existence between Wi-Fi technology and cellular technologies is becoming a reality.

Wireless Metropolitan Area Network

A *wireless metropolitan area network (WMAN)* provides RF coverage to a metropolitan area, such as a city and the surrounding suburbs. WMANs have been created for some time by matching different wireless technologies, and recent advancements have made this more practical. One wireless technology that is often associated with a WMAN is defined by the 802.16 standard. This standard defines broadband wireless access and is sometimes referred to as *Worldwide Interoperability for Microwave Access (WiMAX)*. The WiMAX Forum is responsible for compatibility and interoperability testing of wireless broadband equipment, such as 802.16 hardware.

802.16 technology is viewed as a direct competitor to other broadband services, such as DSL and cable. Although 802.16 wireless networking is typically thought of as a last-mile data-delivery solution, the technology might also be used to provide access to users over citywide areas.

More information about the 802.16 standard can be found at http://ieee802.org/16. Learn more about WiMAX at www.wimaxforum.org.

In the past, a lot of press was generated about the possibility of citywide deployments of Wi-Fi networks, giving city residents access to the Internet throughout a metropolitan area. Although 802.11 technology was initially never intended to be used to provide access over such a wide area, many cities had initiatives to achieve this very feat. The equipment that was being used for these large-scale 802.11 deployments was proprietary wireless mesh routers or mesh access points (APs). Many of these cities scrapped their initial plans to deploy 802.11 technology simply because the technology could not scale across an entire city. However, some WLAN vendors have partnered with 4G/LTE service providers and have had success with 802.11 WMAN deployments using as many as 100,000 APs for metro access. Telco service providers have also begun to use 802.11u defined mechanisms to offload cellular data to Wi-Fi networks.

Wireless Personal Area Network

A *wireless personal area network (WPAN)* is a wireless computer network used for communication between computer devices within close proximity of a user. Devices such as laptops, gaming devices, tablet PCs, and smartphones can communicate with each other by using a variety of wireless technologies. WPANs can be used for communication between devices or as portals to higher-level networks, such as local area networks (LANs) and/or the Internet. The most common technologies in WPANs are Bluetooth and infrared. Infrared is a light-based medium, whereas Bluetooth is a radio-frequency medium that uses frequency-hopping spread spectrum (FHSS) technology.

The IEEE 802.15 Working Group focuses on technologies used for WPANs, such as Bluetooth and Zigbee. Zigbee is another RF technology that has the potential of low-cost wireless networking between devices in a WPAN architecture.

> You can find further information about the 802.15 WPAN standards at www.ieee802.org/15. Although originally defined by the IEEE in the 802.15 standard, Bluetooth standards are now overseen by the Bluetooth Special Interest Group (SIG). To learn more about Bluetooth, visit www.bluetooth.com. The Zigbee Alliance provides information about Zigbee technology at www.zigbee.org. To learn more about infrared communications, visit the Infrared Data Association website (www.irda.org).

The best example of 802.11 Wi-Fi radios being used in a WPAN scenario would be as peer-to-peer connections. We provide more information about 802.11 peer-to-peer networking later in this chapter, in the section entitled "Independent Basic Service Set." Apple's AirDrop technology, which works over Bluetooth and Wi-Fi, is another example of a WPAN used to transfer files between computers or tablets.

Wireless Local Area Network

As you learned in earlier chapters, the 802.11-2016 standard is defined as a *wireless local area network (WLAN)* technology. Local area networks provide networking for a building or campus environment. The 802.11 wireless medium is a perfect fit for local area networking simply because of the range and speeds that are defined by the 802.11-2016 standard and future amendments. The majority of 802.11 wireless network deployments are indeed LANs that provide access at businesses and homes.

WLANs typically use multiple 802.11 access points connected by a wired network backbone. In enterprise deployments, WLANs are used to provide end users with access to network resources and network services and a gateway to the Internet. Although 802.11

hardware can be used in other wireless topologies, the majority of Wi-Fi deployments are WLANs, which is how the technology was originally defined by the IEEE 802.11 Working Group. The discussion of WLANs usually refers to 802.11 solutions; however, other proprietary and competing WLAN technologies do exist.

Please note that large corporations can deploy and manage 802.11 WLANs on a global scale. Enterprise Wi-Fi networks with many geographical locations can be managed centrally using a network management server (NMS) and might also be connected via virtual private networks (VPNs). A more in-depth discussion of Wi-Fi management and scaling can be found in Chapter 11, "WLAN Architecture."

802.11 Stations

The main component of an 802.11 wireless network is the radio, which is referred to by the 802.11 standard as a *station (STA)*. The radio can reside inside an access point or be used as a client station. All stations are identified by a unique MAC address. The 802.11-2016 standard specifies architectural services that stations use within various 802.11 topologies. There are three categories of 802.11 services that operate at the 802.11 MAC sublayer:

Station Service   The *station service (SS)* exists in all 802.11 stations, including client stations and access points. The station service provides the following:

- Authentication
- Deauthentication
- Data confidentiality
- MSDU delivery
- Dynamic frequency selection (DFS)
- Transmit power control (TPC)
- Higher layer timer synchronization
- QoS traffic scheduling
- Radio measurement
- Dynamic station enablement (DSE)

Although these station services operate at the MAC sublayer, many are also dependent on information from the Physical layer. For example, DFS-capable radios detect radar pulses over the physical RF medium but use 802.11 MAC frame exchanges for channel switch announcements. The bulk of these station services are discussed in greater detail in various chapters of this book.

Distribution System Service   The *distribution system service (DSS)* operates only within APs and mesh portals. The DSS is used to manage client station associations, reassociations, disassociations, and more. A more detailed discussion of DSS follows later in this chapter.

PBSS Control Point Service   The *PBSS control point service (PCPS)* is defined specifically for 802.11ad radios when operated in a very specific 802.11 topology called a personal basic service set (PBSS), which is discussed later in this chapter. The PCPS handles associations, reassociations, disassociations, and QoS traffic scheduling when the PBSS topology is deployed.

Client Station

Any radio that is not used in an access point is typically referred to as a *client station* or a non-AP station. Client station radios can be used in laptops, tablets, scanners, smartphones, and many other mobile devices. Client stations must contend for the half-duplex RF medium in the same manner that an access point radio contends for the RF medium. When client stations have a layer 2 connection with an access point, they are known as *associated*. Once associated, client stations can take advantage of the portal functionality that an AP provides. Although 802.11 client stations can be stationary, they are assumed to be mobile and may be able to maintain communications when roaming between access points. All client stations support the station services.

Access Point Station

An 802.11 *access point (AP)* station is a radio that functions as a wireless portal from which other client stations can communicate. In general, an AP has all the same functionally as a client station. However, a key difference between an AP station and a client station is the portal functionality. An access point provides a portal functionality allowing associated client stations to communicate via the wireless medium to another physical medium, such as an 802.3 Ethernet network. The technical term for this portal functionality is *distribution system access function (DSAF)*.

As stated earlier, APs also use the distribution system service (DSS) to manage client associations. A good analogy would be the *content-addressable memory (CAM)* tables in a managed switch. Managed wired switches maintain dynamic MAC address tables, known as *CAM tables*, that can direct frames to ports based on the destination MAC address of a frame. Similarly, an AP maintains an *association table* of connected WLAN clients and directs traffic.

Integration Service

The 802.11-2016 standard defines an *integration service (IS)* that enables delivery of MSDUs between the distribution system (DS) and a non-IEEE-802.11 LAN via a portal. A simpler way of defining the integration service is to characterize it as a frame format transfer method. The portal is usually either an access point or a WLAN controller. The payload of a wireless 802.11 data frame is the layer 3–7 information known as the *MAC service data unit (MSDU)*. The eventual destination of this payload is usually to a wired network infrastructure. Because the wired infrastructure is a different physical medium, an 802.11

data frame payload must be effectively transferred into an 802.3 Ethernet frame. For example, a VoWiFi phone sends an 802.11 data frame to a standalone AP. The MSDU payload of the frame is a VoIP packet with a final destination of an IP PBX that resides at the 802.3 network core. The job of the integration service is to remove the 802.11 header and trailer and then encase the MSDU VoIP payload inside an 802.3 frame. The 802.3 frame is then sent on to the Ethernet network. The integration service performs the same actions in reverse when an 802.3 frame payload must be transferred into an 802.11 frame that is eventually transmitted by the access point radio.

It is beyond the scope of the 802.11-2016 standard to define how the integration service operates. Normally, the integration service transfers data frame payloads between an 802.11 and 802.3 medium. However, the integration service could transfer an MSDU between the 802.11 medium and some sort of other medium. If 802.11 user traffic is forwarded at the edge of a network, the integration service exists in an access point. The integration service mechanism normally takes place inside a WLAN controller when 802.11 user traffic is tunneled back to a WLAN controller.

Distribution System

As previously mentioned, a key difference between 802.11 AP radios and client stations is the portal functionality of an AP. This portal capability is also known as a distribution system access function (DSAF). The 802.11-2016 standard defines a *distribution system (DS)* that is used to interconnect a set of basic service sets via integrated LANs to create an extended service set (ESS). Service sets are described in detail later in this chapter. Access points by their very nature are portal devices. Wireless traffic can be destined back onto the wireless medium or forwarded to the integration service. The DS consists of the following two main components:

Distribution System Medium   A logical physical medium used to connect access points is known as a *distribution system medium (DSM)*. The most common example is an 802.3 medium.

Distribution System Service   As discussed earlier, the distribution system service (DSS) is used in APs to manage client station associations, reassociations, and disassociations. The distribution system service also uses the layer 2 addressing of the 802.11 MAC header to eventually forward the layer 3–7 information (MSDU) either to the integration service or to another wireless client station. A full understanding of DSS is beyond the scope of the CWNA exam.

A single access point or multiple access points may be connected to the same distribution system medium. The majority of 802.11 deployments use an AP as a portal into an 802.3 Ethernet backbone, which serves as the distribution system medium. Access points are usually connected to a switched Ethernet network, which often also offers the advantage of supplying power to the APs via Power over Ethernet (PoE).

An access point may also act as a portal device into other wired and wireless mediums. The 802.11-2016 standard by design does not care, nor does it define, onto which medium an access point translates and forwards data. Therefore, an access point can be characterized as a translational bridge between two mediums. The AP translates and forwards data between the 802.11 medium and whatever medium is used by the distribution system medium. Once again, the distribution system medium will almost always be an 802.3 Ethernet network, as shown in Figure 7.1. In the case of a wireless mesh network, the handoff is through a series of wireless devices, with the final destination typically being an 802.3 network.

FIGURE 7.1   Distribution system medium

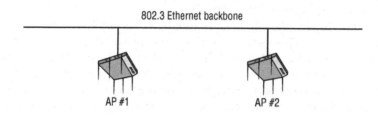

Wireless Distribution System

The 802.11-2016 standard defines a mechanism for wireless communication using a four-MAC-address frame format. The standard describes such a frame format but does not describe how such a mechanism or frame format would be used. This mechanism is known as a *wireless distribution system (WDS)*. Although the DS normally uses a wired Ethernet backbone, it is possible to use a wireless connection instead. A WDS can connect access points together using what is referred to as a *wireless backhaul*. As depicted in Figure 7.2, the most common real-world example of a WDS is when access points function in a mesh deployment to provide both coverage and backhaul. As shown in Figure 7.3, another real-world example of a WDS is an 802.11 outdoor bridge link used to provide wireless back-haul connectivity between two buildings. More detailed discussion about mesh networks and WLAN bridging can be found in multiple chapters in this book.

FIGURE 7.2   Wireless distribution system—mesh backhaul

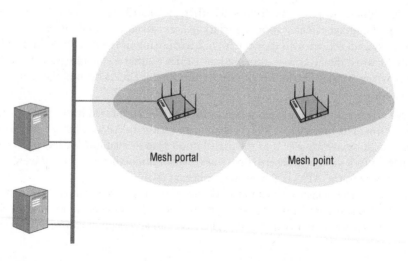

FIGURE 7.3   Wireless distribution system—WLAN bridge backhaul

Which Distribution System Is Most Desirable?

Whenever possible, a wired network will usually be the best option for the distribution system. Because most enterprise deployments already have a wired 802.3 infrastructure in place, integrating a wireless network into an Ethernet network is the most logical solution. A wired distribution system medium does not encounter many of the problems that may affect a WDS, such as physical obstructions and radio frequency interference. A mesh backhaul network is sometimes the better option if cabling is difficult. Additionally, an outdoor 802.11 bridge link may be the only option to interconnect two buildings or structures. If the occasion does arise when a wired network cannot connect access points together, a WDS might be a viable alternative. A more desirable WDS solution utilizes different frequencies and radios for client access and distribution.

802.11 Service Sets

The 802.11-2016 standard defines multiple 802.11 topologies, known as *service sets*, which describe how these radios may be used to communicate with each other. These 802.11 topologies are known as a basic service set (BSS), extended service set (ESS), independent basic service set (IBSS), personal basic service set (PBSS), mesh basic service set (MBSS), and a QoS basic service set (QBSS). In the following sections, we describe all of the components that comprise the various 802.11 service sets.

Service Set Identifier

The *service set identifier (SSID)* is a logical name used to identify an 802.11 wireless network. The SSID wireless network name is comparable to a Windows workgroup name. The radios use this logical name in several different 802.11 frame exchanges. The SSID is a configurable setting on all 802.11 radios, including access points and client stations. The SSID can be made up of as many as 32 characters and is case sensitive. Figure 7.4 shows an SSID configuration of an access point.

FIGURE 7.4 Service set identifier

Wireless Network		
Name (SSID) *	Sybex Wi-Fi	Broadcast SSID Using
Broadcast Name *	Sybex Wi-Fi	☑ WiFi0 Radio (2.4 GHz or 5 GHz)
		☑ WiFi1 Radio (5 GHz only)

Most access points have the ability to cloak an SSID and keep the network name hidden from illegitimate end users. Hiding the SSID is a very weak attempt at security that is not defined by the 802.11-2016 standard. However, it is an option many administrators still mistakenly choose to implement.

In order for clients to roam seamlessly, the APs must advertise the same SSID configured with the same security.

SSID cloaking is discussed in Chapter 17, "802.11 Network Security Architecture."

Basic Service Set

The *basic service set (BSS)* is the cornerstone topology of an 802.11 network. The communicating devices that make up a BSS consist of one AP radio with one or more client stations. Client stations join the AP wireless domain and begin communicating through the AP. Stations that are members of a BSS have a layer 2 connection and are called *associated*. Figure 7.5 depicts a standard basic service set.

FIGURE 7.5 Basic service set

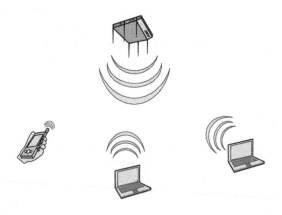

Typically the AP is connected to a distribution system medium, but that is not a requirement of a basic service set. If an AP is serving as a portal to the distribution system, it is likely that the client stations will communicate via the AP with network resources that reside on the DSM. The real-world goal of a BSS is for clients to be able to communicate through the AP to network servers as well as access a gateway to the Internet.

Most communications over a WLAN are client/server-based. However, if 802.11 client stations wish to communicate directly with each other, they will relay their data through the AP. In the typical BSS, peer-to-peer communications from one client station to another client station can occur as long as the traffic is forwarded through the AP. Clients that support *tunneled direct link setup (TDLS)* are the rare exception to this rule. TDLS-capable clients can communicate directly with each other and bypass the AP. TDLS clients will remain associated to the AP and still participate as a member of the BSS. Conversely, most WLAN vendors offer *client isolation* capabilities to block peer-to-peer communications between clients associated to an AP.

You may see other 802.11 terminology used to describe a basic service set. For example, a VHT BSS refers to a basic service set with an 802.11ac access point. A DMG BSS service set refers to a basic service set with an 802.11ad access point.

Basic Service Area

The physical area of coverage provided by an access point in a BSS is known as the *basic service area (BSA)*. Figure 7.6 shows a typical BSA. Client stations can move throughout the coverage area and maintain communications with the AP as long as the received signal between the radios remains above received signal strength indicator (RSSI) thresholds. Client station and AP radios can also shift between concentric zones of variable data rates that exist within the BSA. The process of moving between data rates is known as *dynamic rate switching* and is discussed in Chapter 13, "WLAN Design Concepts."

The size and shape of a BSA depends on many variables, including AP transmit power, antenna gain, receive sensitivity, and physical surroundings. Because environmental and physical surroundings often change, the BSA can often be fluid. When drawing a BSA, it is common to draw a circle around the AP to illustrate the theoretical coverage area. In reality, the real coverage area will have a disproportional shape due to the existing indoor or outdoor environment. You can also make the argument that the effective range of the BSA is from the perspective of any connected client station because all client devices interpret RSSI differently.

FIGURE 7.6   Basic service area

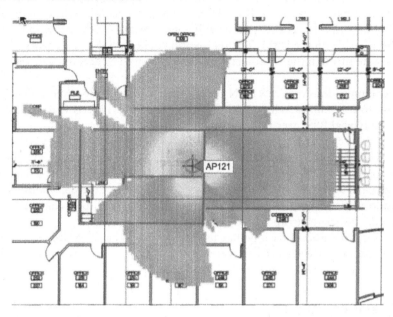

Basic Service Set Identifier

The 48-bit (6-octet) MAC address of an access point's radio is known as the *basic service set identifier (BSSID)*. The simple definition of a BSSID is that it is the MAC address of the radio network interface in an access point. However, the proper definition is that the BSSID address is the layer 2 identifier of each individual BSS.

In the previous section, you learned that a basic service set consists of an AP with one or more stations associated with the AP. If two basic service sets are near each other and both are advertising the same SSID, a client station needs to identify the one BSS from the other. In order for clients to roam seamlessly, the APs must advertise the same SSID configured with the same security. The client station, however, still needs to see a unique layer 2 identifier of each AP in order to roam. The BSSID provides each BSS with a unique identifier, thus the name BSSID. The phrase *BSS transition* refers to the roaming process of the client station moving from one BSS to another BSS.

> Do not confuse the BSSID address with the SSID. The service set identifier (SSID) is the logical WLAN name that is user-configurable, whereas the BSSID is the layer 2 MAC address of a radio provided by the hardware manufacturer.

As shown in Figure 7.7, the BSSID address is found in the MAC header of most 802.11 wireless frames and is used for identification purposes of the basic service set. The BSSID address plays a role in directing 802.11 traffic within the basic service set. Remember, the BSSID address is used as a unique layer 2 identifier of the basic service set and is critical for the roaming process.

FIGURE 7.7   Basic service set identifier

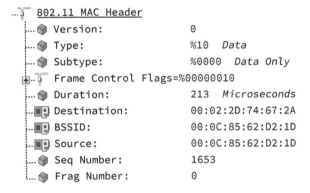

```
...┬ 802.11 MAC Header
   ├...● Version:                         0
   ├...● Type:                            %10   Data
   ├...● Subtype:                         %0000   Data Only
   ├...┬ Frame Control Flags=%00000010
   ├...● Duration:                        213   Microseconds
   ├...▣ Destination:                     00:02:2D:74:67:2A
   ├...▣ BSSID:                           00:0C:85:62:D2:1D
   ├...▣ Source:                          00:0C:85:62:D2:1D
   ├...● Seq Number:                      1653
   └...● Frag Number:                     0
```

Multiple Basic Service Set Identifiers

As you have learned, every WLAN has a logical name (the SSID), and each WLAN BSS has a unique layer 2 identifier (the BSSID). The BSSID can be the physical MAC address of the access point radio; however, multiple BSSIDs may be created for a radio interface using sub-interfaces, as shown in Figure 7.8. The multiple BSSIDs are usually increments of the original MAC address of the AP's radio.

FIGURE 7.8   Incremental BSSID addresses

Name	MAC addr	SSID	Chan(Width)
Wifi1	c413:e204:2f60		48(20MHz)
Wifi1.1	c413:e204:2f64	green	48(20MHz)
Wifi1.2	c413:e204:2f65	blue	48(20MHz)
Wifi1.3	c413:e204:2f66	red	48(20MHz)

If the radio interface of an AP has a MAC address, why would you need multiple BSSIDs and not just use the MAC address of the AP as the layer 2 identifier? The reason is that enterprise WLAN vendors provide a means for APs to support multiple WLANs at the same time.

As depicted in Figure 7.9, multiple WLANs can exist within each AP's coverage area. Each WLAN has a unique logical name (SSID) and a unique layer 2 identifier (BSSID), and each SSID is usually mapped to a unique virtual local area network (VLAN), which is mapped to a unique subnet (layer 3). In other words, multiple layer 2/3 domains can exist within one layer 1 domain. Try to envision multiple basic service sets that are linked to multiple VLANs, yet they all exist within the same coverage area of a single access point.

FIGURE 7.9   Multiple basic service set identifiers (BSSIDs)

SSID	VLAN	Subnet
blue	201	192.168.10.0/24
green	202	192.168.20.0/24
red	203	192.168.30.0/24

SSID	BSSID
blue	C413:e204:2f64
green	C413:e204:2f65
red	C413:e204:2f66

Access Point
radio MAC address: **C413:e204:2f60**

Client A
SSID: blue
IP Address: 192.168.10.25

Client B
SSID: green
IP Address: 192.168.20.40

Client C
SSID: red
IP Address: 192.168.30.77

🌐 Real World Scenario

Will Multiple SSIDs and BSSIDs Affect the Performance?

The simple answer is yes, if there are too many SSIDs transmitted from the same source AP radio. When creating multiple SSIDS, the existence of multiple basic service sets (BSSs) results in excessive amounts of MAC layer overhead. Many WLAN vendors support the transmission of as many as 16 SSIDs and effectively 16 BSSs from a single AP radio. While identified by a unique BSSID, each basic service set will have its own set of beacons, probe responses, and other management and control frame overhead. If a single access point radio is transmitting 16 beacons at an interval of 100ms each, an extreme amount of extra MAC layer overhead is created and will result is serious performance issues. Because of this potential degradation in performance, most WLAN design engineers recommend the broadcasting of no more than three or four SSIDs.

Extended Service Set

While a BSS might be considered the cornerstone 802.11 topology, an *extended service set (ESS)* 802.11 topology is analogous to an entire stone building. An extended service set is two or more identically configured basic service sets connected by a distribution system medium. Usually an extended service set is a collection of multiple access points and their associated client stations, all united by a single DSM. The *extended service area (ESA)* is the coverage area of the ESS in which all clients can communicate and roam. The most common example of an ESS has access points with overlapping coverage cells, as shown in Figure 7.10. The purpose behind an ESS with overlapping coverage cells is to provide seamless roaming to the client stations. Coverage overlap is really duplicate coverage from the perspective of a Wi-Fi client station and is discussed in greater detail in Chapter 13.

FIGURE 7.10 Extended service set, seamless roaming

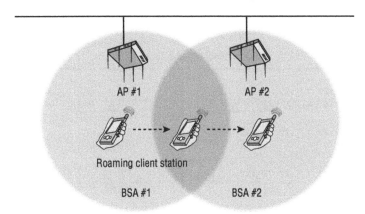

Although seamless roaming is usually a key aspect of WLAN design, there is no requirement for an ESS to guarantee uninterrupted communications. For example, an ESS can utilize multiple access points with nonoverlapping coverage cells, as shown in Figure 7.11. In this scenario, a client station that leaves the basic service area (BSA) of the first access point will lose connectivity. The client station will later reestablish connectivity as it moves into the coverage cell of the second access point. This method of station mobility between disjointed cells is sometimes referred to as *nomadic roaming*.

FIGURE 7.11 Extended service set, nomadic roaming

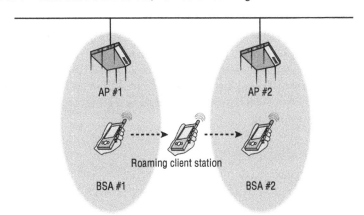

Note that both examples of the previously mentioned extended service sets share a distribution system. As stated earlier in this chapter, the distribution system medium is usually an 802.3 Ethernet network; however, the DS may use another type of medium. In an extended service set, the access points all share the same SSID name. The logical network name of an ESS is often called an *extended service set identifier (ESSID)*. The terminology of ESSID and SSID is synonymous. As Figure 7.12 illustrates, access points in an ESS where roaming is required must all share the same logical name (SSID) and security configuration settings, but must have unique layer 2 identifiers (BSSIDs) for each unique BSA coverage cell.

FIGURE 7.12 SSID and BSSIDs within an ESS

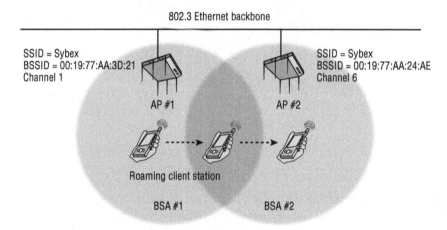

Independent Basic Service Set

The third service set topology defined by the 802.11 standard is an *independent basic service set (IBSS)*. The radios that make up an IBSS network consist solely of client stations (STAs), and no access point is deployed. An IBSS network that consists of just two STAs is analogous to a wired crossover cable. An IBSS can, however, have multiple client stations in one physical area communicating in an ad hoc fashion. Figure 7.13 depicts four client stations communicating with each other in a peer-to-peer fashion.

All of the stations transmit frames to each other directly and do not route their frames from one client to another. All client station frame exchanges in an IBSS are peer-to-peer. All stations in an IBSS must contend for the half-duplex medium, and at any given time only one STA can be transmitting.

FIGURE 7.13 Independent basic service set

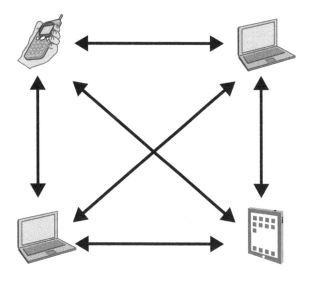

> The independent basic service set has two other names. Wi-Fi vendors often refer to an IBSS as either a *peer-to-peer network* or an *ad hoc network.*

In order for IBSS communications to succeed, all stations must be transmitting on the same frequency channel. Furthermore, this entire set of standalone wireless stations connected together as a group must share the same SSID WLAN name. Another caveat of an IBSS is that a BSSID address is created. Earlier in this chapter, we defined a BSSID as the MAC address of the radio inside an access point. So, how can an independent basic service set have a BSSID if no access point is used in the IBSS topology? The first station that starts up in an IBSS randomly generates a BSSID in the MAC address format. This randomly generated BSSID is a virtual MAC address and is used for layer 2 identification purposes within the IBSS.

Personal Basic Service Set

Similar to the IBSS, a *personal basic service set (PBSS)* is an 802.11 WLAN topology in which 802.11ad stations communicate directly with each other. A PBSS can be established only by *directional multi-gigabit (DMG)* radios that transmit the 60 GHz frequency band. Similar to an IBSS, there is no centralized access point that functions as a portal to a distribution system medium, such as a wired 802.3 Ethernet network. In contrast to an IBSS, one client assumes the role of the *PBSS control point (PCP)*. The PCP client uses DMG Beacon and Announce frames to provide for synchronized medium contention between all clients participating within the PBSS.

As previously stated, a PBSS can be established only by 802.11ad-compliant radios that transmit the 60 GHz frequency. It should be noted that DMG radios can also communicate with other DMG radios via a BSS or IBSS topology.

Mesh Basic Service Set

The 802.11 standard has long defined BSS, ESS, and IBSS service sets. The 802.11-2016 standard also defines a service set for an 802.11 mesh topology. When access points support mesh functions, they may be deployed where wired network access is not possible. The mesh functions are used to provide wireless distribution of network traffic, and the set of APs that provide mesh distribution form a *mesh basic service set (MBSS)*. An MBSS requires features that are not necessary in a BSS, ESS, or IBSS because the purpose of an MBSS is different from the other topologies. As shown in Figure 7.14, one or more mesh APs will typically be connected to the wired infrastructure. Any mesh AP connected to the upstream wired medium is known as a *mesh portal*. The 802.11 technical terminology for a mesh portal station is *mesh gate*, so it is sometimes referred to as a mesh gateway. Any other mesh APs that are not connected to the wired network will form wireless backhaul connections back to the mesh portals to reach the wired network. Mesh APs that are not connected to the upstream wired infrastructure are known as *mesh points*, or MPs. The backhaul connection between a mesh point and a mesh portal is considered to be a wireless distribution system (WDS). Client stations that are associated to the mesh points have their traffic forwarded through the wireless backhaul. Usually the MBSS uses the 5 GHz radios for backhaul communications.

FIGURE 7.14   Mesh basic service set

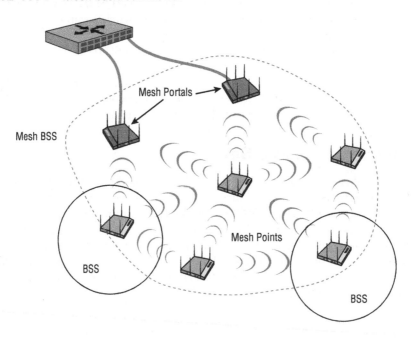

The mesh nodes in an MBSS function much like routers in a network, because their goal is to discover neighbor mesh stations, identify possible and best connections back to the portal, form neighbor links, and share link information. Keep in mind that 802.11 frame exchanges are layer 2 operations; therefore, mesh routing of 802.11 traffic is based on MAC addresses and not on IP addresses. A *hybrid wireless mesh protocol (HWMP)* is defined as the default path selection protocol for an MBSS. HWMP is both proactive and reactive and is effectively a dynamic layer 2 routing protocol. Note that WLAN vendors have offered mesh capabilities for many years using proprietary layer 2 mesh protocols. For competitive reasons, the standard version of HWRP is not supported by enterprise WLAN vendors. They continue to use their own dynamic layer 2 mesh mechanisms, utilizing metrics such as RSSI, SNR, client load, and hop counts to determine the best path for the backhaul traffic. The WLAN infrastructure used to provide WLAN mesh networking has gone through several generations, which are discussed in Chapter 11, "WLAN Architecture."

QoS Basic Service Set

Quality of service (QoS) mechanisms can be implemented within all of the 802.11 service sets. The QoS enhancements are available to QoS STAs associated with a QoS access point in a QoS BSS. QoS stations may also belong to the same QoS IBSS. Older radios that do not support QoS mechanisms are known as *non-QoS STAs* and *non-QoS APs*. Chapter 8, "802.11 Medium Access," discusses QoS mechanisms in greater detail. QoS mechanisms are a requirement for the Wi-Fi Multimedia (WMM) certification, which is strictly enforced by the Wi-Fi Alliance. Any 802.11 enterprise access point manufactured in the past 10 years supports WMM QoS mechanisms by default. Therefore, each basic service set in most enterprise deployments is considered to be a *QoS basic service set (QBSS)*.

 Real World Scenario

Vendor Considerations When Deploying and Integrating 802.11 WLAN Infrastructure

When deploying 802.11 infrastructure, the recommended practice is to purchase the equipment from one vendor. A bridge from vendor A is not likely to work with a bridge from vendor B. A mesh point from vendor A most likely will not communicate with a mesh portal from vendor B. Another example of unlikely interoperability is roaming handoffs between different vendor APs. Client stations may not be able to roam effectively when using a mix of different WLAN vendor access points.

The main purpose of an 802.11 AP is to act as a portal to a wired network infrastructure. Although 802.11 technology operates at layers 1 and 2, there are always higher layer design considerations. All WLAN vendors have different strategies on how to integrate into a preexisting wired network infrastructure. For that reason, the normal best practice is to stick with one enterprise WLAN vendor when deploying and integrating 802.11 infrastructure.

802.11 Configuration Modes

While the 802.11-2016 standard defines all radios as stations (STAs), an access point radio and a client station radio can each be configured in a number of ways. The default configuration of an AP radio is to allow it to operate inside a basic service set (BSS) as a portal device to a wired network infrastructure. However, an AP can be configured to function in other operational modes. Client stations can be configured to participate in either a BSS or an IBSS 802.11 service set.

Access Point Modes

The default configuration of some WLAN vendor access points is known as *root mode*. The main purpose of an AP is to serve as a portal to a distribution system. The normal default setting of an AP is root mode, which allows the AP to transfer data back and forth between the DS and the 802.11 wireless medium. Not all vendors have the same names for this mode of operation. For example, many Wi-Fi vendors use the term *AP mode* or *access mode* instead of root mode.

The default root configuration of an AP radio allows it to operate as part of a BSS. There are, however, other operational modes in which an AP may be configured:

Mesh Mode   The AP radio operates as a wireless backhaul radio for a mesh environment. Depending on the vendor, the backhaul radio may also allow for client access. Mesh mode is sometimes also referred to as *repeater mode*.

Sensor Mode   The AP radio is converted into a sensor radio, allowing the AP to integrate into a wireless intrusion detection system (WIDS) architecture. An AP in sensor mode is in a continuous listening state while scanning between multiple channels. Sensor mode is also often referred to as *monitor mode* or *scanner mode*.

Bridge Mode   The AP radio is converted into a wireless bridge. This typically adds extra MAC-layer intelligence to the device and gives the AP the capability to learn and maintain tables about MAC addresses from the wired side of the network.

Workgroup Bridge Mode   The AP radio is transformed into a workgroup bridge, providing wireless backhaul for connected 802.3 wired clients.

AP as a Client Mode   The AP radio functions as a client device that can then associate to other APs. This operational mode is sometimes used for troubleshooting purposes.

The 802.11-2016 standard does not define these AP operational modes; therefore, every WLAN vendor will have different capabilities. These modes of operation are "radio configuration modes" and may be able to be applied to a 2.4 GHz radio in an AP, a 5 GHz radio in an AP, or both radios. WLAN vendors often use different terminology for the various available configuration modes. You can see an example on one vendor's AP configurable modes in Figure 7.15.

FIGURE 7.15 Access point configuration modes

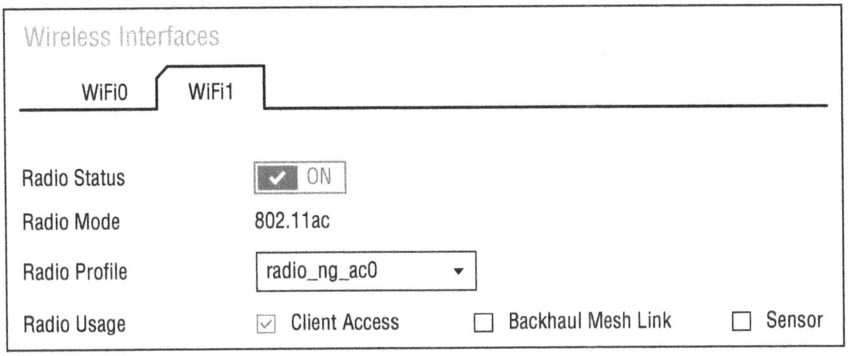

Wireless Interfaces

| WiFi0 | WiFi1 |

Radio Status	✔ ON
Radio Mode	802.11ac
Radio Profile	radio_ng_ac0 ▼
Radio Usage	☑ Client Access ☐ Backhaul Mesh Link ☐ Sensor

Client Station Modes

A client station may operate in one of two states, as shown in Figure 7.16. The default mode for an 802.11 client radio is typically *infrastructure mode*. When running in infrastructure mode, the client station will allow communication via an access point. Infrastructure mode allows for a client station to participate in a basic service set or an extended service set. Clients that are configured in this mode may communicate, via the AP, with other wireless client stations within a BSS. This client mode is often not apparent because client devices have this operational mode enabled by default so that clients can easily discover APs.

FIGURE 7.16 Client station configuration modes

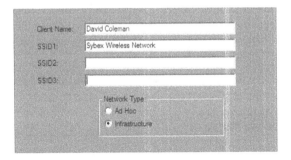

Clients may also communicate through the AP with other networking devices that exist on the distribution system, such as servers or wired desktops.

The second client station mode is called *ad hoc mode*. Other client vendors may refer to this as *peer-to-peer mode*. 802.11 client stations set to ad hoc mode participate in an IBSS topology and do not communicate via an access point. All station transmissions and frame exchanges are peer-to-peer. Many client stations, such as tablets and smartphones, may not have a configuration mode for ad hoc communications.

Summary

This chapter covered the major types of generic wireless topologies as well as the topologies specific to 802.11 wireless networking:

- The four wireless architectures that can be used by many different wireless technologies
- The 802.11 service sets, as defined by the 802.11-2016 standard, and the various aspects and purposes defined for each service set
- Operational configuration modes of both access points and client stations

As a wireless network administrator, you should have a full understanding of the defined 802.11 service sets and how they operate. Administrators typically oversee the design and management of an 802.11 ESS, but there is a good chance that they will also deploy 802.11 radios using a variety of operational modes.

Exam Essentials

Know the four major types of wireless topologies.   Understand the differences between a WWAN, WLAN, WPAN, and WMAN.

Explain the various 802.11 service sets.   Be able to fully expound on all the components, purposes, and differences of a basic service set, an extended service set, an independent basic service set, a personal basic service set, a QoS basic service set, and a mesh basic service set. Understand how the 802.11 radios interact with each other in each service set.

Identify the various ways in which an 802.11 radio can be used.   Understand that the 802.11 standard expects a radio to be used either as a client station or inside an access point. Also understand that an 802.11 radio can be used for other purposes, such as bridging, mesh, and so on.

Explain the purpose of the distribution system.   Know that the DS consists of two pieces: distribution system services (DSS) and the distribution system medium (DSM). Understand that the medium used by the DS can be any type of medium. Explain the functions of a wireless distribution system (WDS).

Define SSID, BSSID, and ESSID.   Be able to explain the differences or similarities of all three of these addresses and the function of each.

Describe the various ways in which an ESS can be implemented and the purpose behind each design.   Explain the three ways in which the coverage cells of the ESS access points can be designed and the purpose behind each design.

Explain access point and client station configuration modes.   Remember all the configuration modes of both an AP and a client station.

Review Questions

1. The logical name of an 802.11 wireless network is known as which type of address? (Choose all that apply.)

   A. BSSID

   B. MAC address

   C. IP address

   D. SSID

   E. ESSID

2. Which two 802.11 topologies require the use of an AP?

   A. WPAN

   B. IBSS

   C. BSS

   D. PBSS

   E. ESS

3. The 802.11 standard defines which medium to be used in a distribution system?

   A. 802.3

   B. 802.15

   C. 802.5 token ring

   D. Ethernet

   E. None of the above

4. Which option is a wireless computer topology used for communication between computer devices within close proximity of a person?

   A. WWAN

   B. WMAN

   C. WLAN

   D. WPAN

5. Which 802.11 service set may allow for client roaming?

   A. ESS

   B. BSS

   C. IBSS

   D. PBSS

6. What factors might affect the size of a basic service area (BSA) of an AP? (Choose all that apply.)

   A. Antenna gain

   B. CSMA/CA

   C. Transmission power

   D. Indoor/outdoor surroundings

   E. Distribution system

7. What is the default configuration mode that allows an AP radio to operate in a BSS?

   A. Scanner

   B. Repeater

   C. Root

   D. Access

   E. Nonroot

8. Which terms describe an 802.11 topology involving STAs but no access points? (Choose all that apply.)

   A. BSS

   B. Ad hoc

   C. DSSS

   D. Infrastructure

   E. IBSS

   F. Peer-to-peer

9. STAs operating in the default infrastructure mode may communicate in which of the following scenarios? (Choose all that apply.)

   A. 802.11 frame exchanges with other STAs via an AP

   B. 802.11 frame exchanges with an AP in scanner mode

   C. 802.11 frame peer-to-peer exchanges directly with other STAs

   D. Frame exchanges with network devices on the DSM

   E. All of the above

10. Which of the following are included in the topologies defined by the 802.11-2016 standard? (Choose all that apply.)

    A. DSSS

    B. ESS

    C. BSS

    D. IBSS

    E. FHSS

    F. PBSS

11. Which wireless topology provides citywide wireless coverage?

    A. WMAN

    B. WLAN

    C. WPAN

    D. WAN

    E. WWAN

12. At which layer of the OSI model is a BSSID address used?

    A. Physical

    B. Network

    C. Session

    D. Data-Link

    E. Application

13. The BSSID address can be found in which topologies? (Choose all that apply.)

    A. FHSS

    B. IBSS

    C. ESS

    D. HR-DSSS

    E. BSS

14. Which 802.11 service set defines mechanisms for mesh networking?

    A. BSS

    B. PBSS

    C. ESS

    D. MBSS

    E. IBSS

15. Which 802.11 service set is defined specifically for directional multi-gigabit (DMG) radios?

    A. BSS

    B. ESS

    C. IBSS

    D. PBSS

    E. MBSS

16. The 802.11-2016 standard specifies architectural services that stations use within various 802.11 topologies. Which service is used by both client stations and AP stations?

    A. Station service

    B. Distribution service

    C. PBSS control point service

    D. Integration service

    E. Bus service

17. A network consisting of clients and two or more APs with the same SSID connected by an 802.3 Ethernet backbone is one example of which 802.11 topology? (Choose all that apply.)

    A. Public basic service set

    B. Basic service set

    C. Extended service set

    D. Independent basic service set

    E. Ethernet service set

18. Which term best describes two access points communicating with each other wirelessly while also allowing clients to communicate through the access points?

    A. WDS

    B. DS

    C. DSS

    D. DSSS

    E. DSM

19. What components make up a distribution system? (Choose all that apply.)

    A. HR-DSSS

    B. DSS

    C. DSM

    D. DSSS

    E. QBSS

20. What type of wireless topology is defined by the 802.11-2016 standard?

    A. WAN

    B. WLAN

    C. WWAN

    D. WMAN

    E. WPAN

Chapter

8

802.11 Medium Access

IN THIS CHAPTER, YOU WILL LEARN ABOUT THE FOLLOWING:

✓ **CSMA/CA vs. CSMA/CD**

- Collision detection

✓ **Distributed Coordination Function (DCF)**

- Physical carrier sense
- Virtual carrier sense
- Pseudo-random backoff timer
- Interframe space

✓ **Point Coordination Function (PCF)**

✓ **Hybrid Coordination Function (HCF)**

- Enhanced Distributed Channel Access (EDCA)
- HCF Controlled Channel Access (HCCA)

✓ **Wi-Fi Multimedia (WMM)**

✓ **Airtime Fairness**

One of the difficulties we had in writing this chapter was that in order for you to understand how a wireless station gains access to the medium, we have to teach more than what is needed for the CWNA exam. The details are needed to grasp the concepts; however, it is the concepts that you will be tested on. If you find the details of this chapter interesting, then after reading this book, you should consider reading CWAP Certified Wireless Analysis Professional Official Study Guide: Exam PW0-270, by David A. Westcott, David D. Coleman, et al. (Sybex, 2011), which gets into the nitty-gritty details of 802.11 communications. If you decide to take the CWAP exam, at that time you will need to know details far beyond what we have included in this chapter. But for now, take the details for what they are: a foundation for helping you understand the overall process of how a wireless station gains access to the half-duplex medium.

CSMA/CA vs. CSMA/CD

Network communication requires a set of rules to provide controlled and efficient access to the network medium. *Medium access control (MAC)* is the generic term used when discussing the concept of access. There are many ways of providing medium access. The early mainframes used polling, which sequentially checked each terminal to see whether there was data to be processed. Later, token-passing and contention methods were used to provide access to the medium. Two forms of contention that are heavily used in today's networks are *Carrier Sense Multiple Access with Collision Detection (CSMA/CD)* and *Carrier Sense Multiple Access with Collision Avoidance (CSMA/CA)*.

CSMA/CD is well known and is used for 802.3 Ethernet networks. 802.11 WLANs utilize the lesser-known CSMA/CA for medium access. Stations using either access method must first listen to see whether any other device is transmitting; otherwise, the station must wait until the medium is available. The difference between CSMA/CD and CSMA/CA exists at the point when a client wants to transmit and no other clients are presently transmitting.

A CSMA/CD wired node first checks whether another node is transmitting. If no other wired node is transmitting on the Ethernet medium, the node sends the first bit of information. If no collision is detected, the node continues to send the other bits of information while continuously checking whether a collision has been detected. If a collision is detected, the wired node calculates a random amount of time to wait before starting the process again. 802.11 wireless radios are not capable of transmitting and receiving at the same time, so they are not capable of detecting a collision during their transmission. For this reason, 802.11 wireless networking uses CSMA/CA instead of CSMA/CD to try to avoid collisions.

When a CSMA/CA station has determined that no other stations are transmitting, the 802.11 radio will choose a random backoff value. The station will then wait an additional period of time, based on that backoff value, before transmitting. During this time, the station continues to monitor to make sure that no other stations begin transmitting. Because of the half-duplex nature of the RF medium, it is necessary to ensure that at any given time only one 802.11 radio has control of the medium. CSMA/CA is a process used to ensure that only one 802.11 radio is transmitting at a time. Is this process perfect? Absolutely not! Collisions still occur when two or more radios transmit at the same time.

The IEEE 802.11-2016 standard defines a function called *Distributed Coordination Function (DCF)* that allows for automatic medium sharing between compatible PHYs through the use of the CSMA/CA protocol. DCF also leverages the use of ACK frames as a delivery-verification method. CSMA/CA utilizes multiple checks and balances to try to minimize collisions. These checks and balances can also be thought of as several lines of defense. The various lines of defense are put in place to once again hopefully ensure that only one radio is transmitting while all other radios are listening. CSMA/CA minimizes the risk of collisions without excessive overhead. Additionally, the 802.11-2016 standard encompasses a Hybrid Coordination Function (HCF) that specifies advanced *quality-of-service (QoS)* methods.

This entire process is covered in more detail in the next section of this chapter.

CSMA/CA Overview

Carrier sense determines whether the medium is busy. *Multiple access* ensures that every radio gets a fair shot at the medium (but only one at a time). *Collision avoidance* means only one radio gets access to the medium at any given time, hopefully avoiding collisions.

Collision Detection

In the previous section, we mentioned that 802.11 radios cannot transmit and receive at the same time and therefore cannot detect collisions. So, if they cannot detect a collision, how do they know whether one occurred? The answer is simple. As shown in Figure 8.1, every time an 802.11 radio transmits a unicast frame, if the frame is received properly, the 802.11 radio that received the frame will reply with an *acknowledgment (ACK)* frame. The ACK frame is a method of delivery verification of unicast frames. 802.11n and 802.11ac radios make use of frame aggregation, which groups multiple unicast frames together. The delivery of aggregated frames is verified using a *Block ACK*.

FIGURE 8.1 Unicast acknowledgment

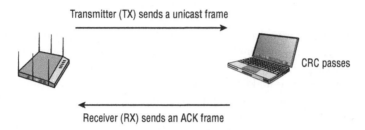

Transmitter (TX) sends a unicast frame

CRC passes

Receiver (RX) sends an ACK frame

The majority of unicast 802.11 frames must be acknowledged. Broadcast and multicast frames do not require an acknowledgment. If any portion of a unicast frame is corrupted, the *cyclic redundancy check (CRC)* will fail and the receiving 802.11 radio will not send an ACK frame to the transmitting 802.11 radio. If an ACK frame is not received by the original transmitting radio, the unicast frame is not acknowledged and will have to be retransmitted.

This process does not specifically determine whether a collision occurs—in other words, there is no collision detection. However, if an ACK frame is not received by the original radio, there is collision assumption. Think of the ACK frame as a method of delivery verification for unicast 802.11 frames. If no proof of delivery is provided, the original radio assumes there was a delivery failure and retransmits the frame.

Distributed Coordination Function

Distributed Coordination Function (DCF) is the fundamental access method of 802.11 communications, and the CSMA/CA process is the foundation of the DCF. With the addition of the 802.11e amendment, which is now part of the 802.11-2016 standard, an enhanced coordination function known as *Hybrid Coordination Function (HCF)* builds further on DCF access methods. In the following sections, you will learn about some of the components that are part of the CSMA/CA process. Here are the four main components of the CSMA/CA protocol, as defined by DCF:

- Physical carrier sense

- Virtual carrier sense

- Pseudo-random backoff timer

- Interframe spaces

802.11 radios use carrier sense mechanisms to determine whether the wireless medium is busy. Think of it like listening for a busy signal when you call someone on the phone. There are two ways that a carrier sense is performed: physical carrier sense and virtual carrier sense. 802.11 radios also contend for the medium using a pseudo-random algorithm and a backoff timer prior to transmission. Interframe spaces are also used to further provide priority levels for access to the wireless medium.

Think of these four components as checks and balances that work together at the same time to ensure that only one 802.11 radio is transmitting on the half-duplex medium. These four components will be explained separately, but it is important to understand that all four mechanisms are functioning at the same time.

Physical Carrier Sense

CSMA/CA utilizes a line of defense to ensure that a station does not transmit while another is already transmitting: The 802.11-2016 standard defines a *physical carrier sense* mechanism to determine if the radio frequency (RF) medium is busy.

Physical carrier sense is performed constantly by all stations that are not transmitting or receiving. When a station performs a physical carrier sense, it is actually listening to the channel to see whether any other RF transmissions are occupying the channel.

Physical carrier sense has two purposes:

- The first purpose is to determine whether a frame transmission is inbound for a station to receive. If the medium is busy, the radio will attempt to synchronize with the transmission.

- The second purpose is to determine whether the medium is busy before transmitting. The medium must be clear before a station can transmit.

To achieve these two physical carrier sense goals, 802.11 radios use a *clear channel assessment (CCA)* to appraise the RF medium. The CCA involves listening for RF transmissions at the Physical layer. 802.11 radios use two separate CCA thresholds when listening to the RF medium. As shown in Figure 8.2, the *signal detect (SD)* threshold is used to identify any 802.11 preamble transmissions from another transmitting 802.11 radio. The preamble is a component of the Physical layer header of 802.11 frame transmissions. The preamble is used for synchronization between transmitting and receiving 802.11 radios. The SD threshold is sometimes referred to as the *preamble carrier sense threshold*. The signal detect (SD) threshold is statistically around 4 dB signal-to-noise ratio (SNR) for most 802.11 radios to detect and decode an 802.11 preamble. In other words, an 802.11 radio can usually decode any incoming 802.11 preamble transmissions at a received signal of about 4 dB above the noise floor.

FIGURE 8.2   Clear channel assessment (CCA)

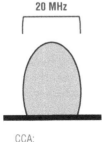

20 MHz

Signal Detect (SD) threshold is statistically a 4 dB signal-to-noise ratio (SNR) to detect an 802.11 preamble.

Energy Detect (ED) threshold is 20 dB above the signal detect threshold.

CCA:
SD = 4 dB SNR
ED = SD + 20 dB

The *energy detect (ED)* threshold is used to detect any other type of RF transmissions during the clear channel assessment (CCA). Remember that the 2.4 GHz and 5 GHz bands are license-free bands and other non 802.11 RF transmissions may occupy a channel. As shown in Figure 8.2, the ED threshold is 20 dB higher than the signal detect threshold. For example, if the noise floor of channel 36 were at –95 dBm, the SD threshold for detecting 802.11 transmissions would be around –91 dBm, and the ED threshold for detecting other RF transmissions would be –71 dBm. If the noise floor of channel 40 were at –100 dBm, the SD threshold for detecting 802.11 transmissions would be around –96 dBm, and the ED threshold for detecting other RF transmissions would be –76 dBm.

Approximately, 4 microseconds is needed for both the signal and energy detect assessments during the CCA. Think of the signal detect as a method of detecting and deferring for 802.11 radio transmissions. Think of the energy detect as a method of detecting and deferring for any signals from non-802.11 transmitters. As shown in Table 8.1, both thresholds are used together during the CCA to determine whether the medium is busy and therefore defer transmissions.

TABLE 8.1   Clear channel assessment thresholds

Signal Detect (SD)	Energy Detect (ED)	Transmit or Defer
Idle	Idle	OK to transmit
Busy	Idle	Defer and begin demodulating OFDM symbol
Idle	Busy	Defer one OFDM slot time

The definition of both of these CCA thresholds is somewhat vague in the 802.11-2016 standard, which has often resulted in a misunderstanding of the actual threshold values. The interpretation of these thresholds by WLAN manufacturers of 802.11 client and AP radios will often differ. To complicate matters further, please remember that the receive sensitivity capabilities between radios can vary widely. Because of the difference in receive sensitivity, the perception of the noise floor can be quite different between 802.11 radios. Therefore, the two CCA thresholds may also vary due to differences in radio receive sensitivity. The CCA thresholds discussed in this section are based on transmissions on a 20 MHz channel. In Chapter 10, you will learn that 802.11n and 802.11ac radios have the capability to transmit on larger channels by bonding together multiple 20 MHz channels. For example, an 802.11n/ac radio can transmit and receive on a 40 MHz channel using a bonded primary and secondary channel. Please understand that the CCA thresholds between primary and secondary channels are also different, which will be discussed in later chapters.

Virtual Carrier Sense

As pictured in Figure 8.3, one of the fields in the MAC header of an 802.11 frame is the *Duration/ID field*. When a client transmits a unicast frame, the Duration/ID field contains a value from 0 to 32,767. The Duration/ID value represents the time, in microseconds, that is required to transmit an active frame exchange process so that other radios do not interrupt the process. In the example shown in Figure 8.4, the client that is transmitting the data frame calculates how long it will take to receive an ACK frame and includes that length of time in the Duration/ID field in the MAC header of the transmitted unicast data frame. The value of the Duration/ID field in the MAC header of the ACK frame that follows is 0 (zero). The duration value in any frame is always about the amount of time needed to complete a frame exchange between two radios. To summarize, the value of the Duration/ID field indicates how long the RF medium will be busy during a frame exchange, before another station can contend for the medium.

FIGURE 8.3   Duration/ID field

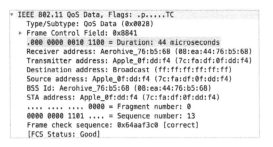

FIGURE 8.4   Duration value of SIFS + ACK

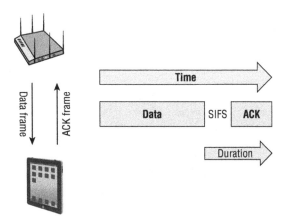

The majority of the time, the Duration/ID field contains a Duration value that is used to reset other stations' network allocation vector (NAV) timers. In the rare case of a PS-Poll frame, the Duration/ID is used as an ID value of a client station using legacy power management. Power management is discussed in Chapter 9, "802.11 MAC."

Virtual carrier sense uses a timer mechanism known as the *network allocation vector (NAV)*. The NAV timer maintains a prediction of future traffic on the medium based on Duration value information seen in a previous frame transmission. When an 802.11 radio is not transmitting, it is listening. As depicted in Figure 8.5, when the listening radio hears a frame transmission from another station, it looks at the header of the frame and determines whether the Duration/ID field contains a Duration value or an ID value. If the field contains a Duration value, the listening station will set its NAV timer to this value. The listening station will then use the NAV as a countdown timer, knowing that the RF medium should be busy until the countdown reaches 0.

FIGURE 8.5 Virtual carrier sense

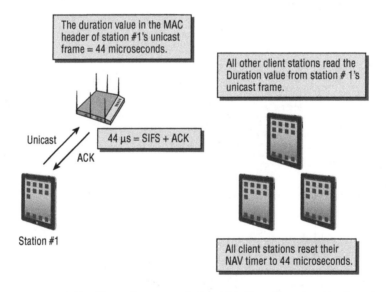

This process essentially allows the transmitting 802.11 radio to notify the other stations that the medium will be busy for a period of time (Duration/ID value). The stations that are not transmitting listen and hear the Duration/ID, set a countdown timer (NAV), and wait until their timer hits 0 before they can contend for the medium and eventually transmit on the medium. A station cannot contend for the medium until its NAV timer is 0, nor can a station transmit on the medium if the NAV timer is set to a nonzero value. As stated earlier, CSMA/CA uses several lines of defense to prevent collisions, and the NAV timer is often considered one line of defense. Because the Duration/ID value inside an 802.11 MAC header is used to set the NAV timer, virtual carrier sense is a layer 2 carrier sense mechanism.

It is important to understand that both virtual carrier sense and physical carrier sense are always happening at the same time. Virtual carrier sense is a layer 2 line of defense, whereas physical carrier sense is a layer 1 line of defense. If one line of defense fails, hopefully the other will prevent collisions from occurring.

Pseudo-Random Backoff Timer

An 802.11 station may contend for the medium during a window of time known as the *backoff time*. At this point in the CSMA/CA process, the station selects a random backoff value using a pseudo-random backoff algorithm.

The station chooses a random number from a range called a *contention window (CW)* value. After the random number is chosen, the number is multiplied by the *slot time* value. This starts a *pseudo-random backoff timer*. Please do not confuse the backoff timer with the NAV timer. As mentioned earlier, the NAV timer is a virtual carrier sense mechanism used to reserve the medium for further transmissions. The pseudo-random backoff timer is the final timer used by a station before it transmits. The station's backoff timer begins to count down ticks of a clock known as *slots*. When the backoff timer is equal to zero, the client can reassess the channel and, if it is clear, begin transmitting.

Slot time sizes are dependent on the Physical layer specification (PHY) in use. For example, legacy 802.1b radios use an HR-DSSS PHY that defines a slot time of 20 μs. 802.11a/g/n/ac radios all use OFDM and a slot time of 9 μs. During the OFDM slot time, 4 μs is needed for signal detect (SD) and energy detect (ED) carrier sense, and 4 μs is needed to listen for an OFDM symbol.

If no medium activity occurs during a particular slot time, then the backoff timer is decremented by a slot time. If the physical or virtual carrier sense mechanisms sense a busy medium, the backoff timer decrement is suspended, and the backoff timer value is maintained. When the medium is idle for a duration of a DIFS, AIFS, or EIFS period, the backoff process resumes and continues the countdown from where it left off. When the backoff timer reaches zero, transmission commences. The following example is a simple review of the entire backoff process:

- An OFDM station selects a random number from a contention window of 0–15. For this example, the number chosen is 4.

- The station multiplies the random number of 4 by a slot time of 9 μs.

- The random backoff timer has a value of 36 μs (4 slots).

- For every slot time during which there is no medium activity, the backoff timer is decremented by a slot time.

- The station decrements the backoff timer until the timer is zero.

- The station performs a final CCA and transmits if the medium is clear.

- If the medium is not clear, the client defers one slot time, assesses the medium with another CCA, and then transmits if the medium is clear.

The whole point of the backoff procedure is that all 802.11 radios get a chance to transmit on the RF medium; however, a pseudo-random process is needed to ensure that they all take turns. A good analogy would be to write down the numbers 0–15 on 16 pieces of paper and put all the pieces of paper in a hat. Then four people would each choose one piece of paper from the hat. The person with the lowest number would get to transmit on the medium first.

Figure 8.6 illustrates the process in a different way. Assume three 802.11ac clients on channel 40 all want to transmit at the same time. The length of each station's backoff timer is entirely random based on the contention widow. Client station #1 might be assigned a CW value of 6, multiplied by the 9 μs slot time, for a backoff timer of 54 μs. Client station #2 might be assigned a CW value of 5, multiplied by the 9 μs slot time, for a backoff timer of 45 μs. Client station #3 might be assigned a CW value of 4, multiplied by the 9 μs slot time, for a backoff timer of 36 μs. All three stations would then begin to decrement their 9 μs slots of their backoff timers. The station #3 backoff timer would be decremented first and would be the first to transmit on the medium.

FIGURE 8.6 Pseudo-random backoff timer

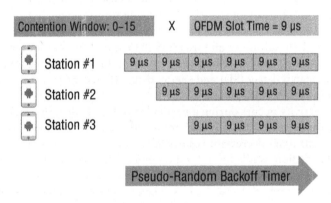

Please remember that the RF medium is half-duplex and that only one radio at a time can transmit. Therefore, all 802.11 radios must contend for the medium, including any access point radios. Within any basic service set (BSS), the AP is always the busiest transmitter; however, the AP radio is given no special priority and must also contend for the medium along with all the clients that might be associated to the AP.

The random backoff timer is another line of defense and helps minimize the likelihood of two stations trying to communicate at the same time, although it does not fully prevent this from occurring. If a station does not receive an ACK, it starts the carrier sense process over again. What if a frame transmission is corrupted and a retransmission is necessary? Unsuccessful transmissions cause the CW size to increase exponentially up to a maximum value, as shown in Figure 8.7. In other words, stations must contend for the medium for any retransmissions; however, the odds are not necessarily as good with each successive retransmission. This ensures better stability in high-capacity conditions.

FIGURE 8.7 Example of exponential increase of the contention window (CW)

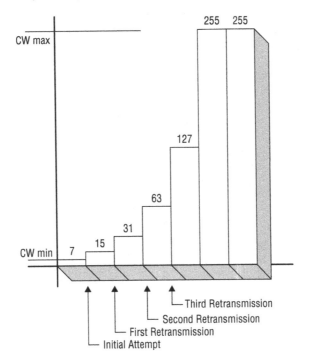

Interframe Space

Interframe space (IFS) is a period of time that exists between transmissions of wireless frames. There are ten types of interframe spaces. Following is a partial list, from shortest to longest:

- Reduced interframe space (RIFS)—highest priority
- Short interframe space (SIFS)—second highest priority
- PCF interframe space (PIFS)—middle priority
- DCF interframe space (DIFS)—lowest priority
- Arbitration interframe space (AIFS)—used by QoS stations
- Extended interframe space (EIFS)—used after receipt of corrupted frames

The actual length of time of each interframe space varies depending on the transmission speed of the network. Interframe spaces are one line of defense used by CSMA/CA to ensure that only certain types of 802.11 frames are transmitted following certain interframe spaces. For example, only ACK frames, block ACK frames, data frames, and clear-to-send (CTS) frames can follow a SIFS. The two most common interframe spaces used are the SIFS and the DIFS. As pictured in Figure 8.8, the ACK frame is the highest-priority frame, and the use of a SIFS ensures that it will be transmitted first, before any other type of 802.11 frame. Most other 802.11 frames follow a longer period of time, called a DIFS. Stations use SIFS to maintain control of the medium during a frame-exchange sequence. Other stations cannot gain access to the medium during the sequence because they must wait for the longer DIFS period of time.

FIGURE 8.8 SIFS and DIFS

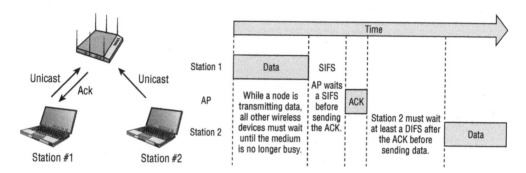

Interframe spaces are all about which type of 802.11 traffic is allowed next. Interframe spacing also acts as a backup mechanism to virtual carrier sense, which was discussed earlier in this chapter.

Point Coordination Function

In addition to DCF, the IEEE 802.11-2016 standard defines an additional, optional medium access method known as *Point Coordination Function (PCF)*. This access method is a form of polling. The AP performs the function of the *point coordinator (PC)*. Because an AP is taking the role of the point coordinator, the PCF medium access method will work in only a basic service set (BSS). PCF cannot be utilized in an ad hoc network because no AP exists in an independent basic service set (IBSS). Because polling is performed from a central device, PCF provides managed access to the medium.

In order for PCF to be used, both the AP and the station must support it. If PCF is enabled, DCF will still function. The AP will alternate between PCF mode and DCF mode. While the AP is functioning in PCF mode, that time is known as the *contention-free period (CFP)*. During the contention-free period, the AP polls only clients in PCF mode about their intention to send data. This is a method of prioritizing clients. While the AP is functioning in DCF mode, that time is known as the *contention period (CP)*.

More information about PCF can be found in clause 10.4 of the 802.11-2016 standard document, which you can download from the IEEE website:

`https://ieeexplore.ieee.org/document/7786995/`

As we stated earlier, PCF is an optional access method, and as of this writing, we do not know of any WLAN vendor that has implemented PCF. As a matter of fact, the IEEE now considers PCF to be obsolete and will most likely remove all references to PCF in later revisions of the 802.11 standard. You will not be tested on PCF in the CWNA exam.

Hybrid Coordination Function

The 802.11e quality-of-service amendment added a new coordination function to 802.11 medium contention, known as *Hybrid Coordination Function (HCF)*. The 802.11e amendment and HCF have since been incorporated into the 802.11-2016 standard. HCF combines capabilities from both DCF and PCF and adds enhancements to them to create two channel access methods: Enhanced Distributed Channel Access (EDCA) and HCF Controlled Channel Access (HCCA).

The DCF and PCF medium-contention mechanisms discussed earlier allow for an 802.11 radio to transmit a single frame. After transmitting a frame, the 802.11 station must contend for the medium again before transmitting another frame. HCF defines the ability for an 802.11 radio to send multiple frames when transmitting on the RF medium. When an HCF-compliant radio contends for the medium, it receives an allotted amount of time to send frames. This period of time is called a *transmit opportunity (TXOP)*. During this TXOP, an 802.11 radio may send multiple frames in what is called a *frame burst*. A *short interframe space (SIFS)* is used between each frame to ensure that no other radios transmit during the frame burst.

Enhanced Distributed Channel Access

Enhanced Distributed Channel Access (EDCA) is a wireless medium access method that provides differentiated access that directs traffic to four access-category QoS priority queues. EDCA is an extension of DCF. The EDCA medium access method prioritizes traffic using priority tags that are identical to 802.1D priority tags. Priority tags provide a mechanism for implementing QoS at the MAC level.

Different classes of service are available, represented in a 3-bit user priority (UP) field in an IEEE 802.1Q header added to an Ethernet frame. 802.1D enables priority queuing (enabling some Ethernet frames to be forwarded ahead of others within a switched Ethernet network). Figure 8.9 depicts 802.1D priority tags from the Ethernet side that are used to direct traffic to access-category queues.

FIGURE 8.9   EDCA and 802.1D priority tags

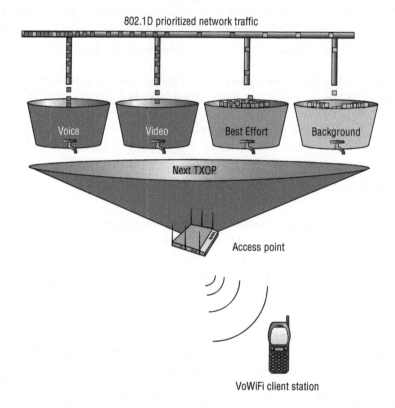

EDCA defines four access categories, based on the eight UPs. The four access categories, from lowest priority to highest priority, are AC_BK (Background), AC_BE (Best Effort), AC_VI (Video), and AC_VO (Voice). For each access category, an enhanced version of DCF known as *Enhanced Distributed Channel Access Function (EDCAF)* is used to contend for a TXOP. Frames with the highest-priority access category have the lowest backoff values and therefore are more likely to get a TXOP. The specific details of this process are beyond the scope of the CWNA exam.

HCF Controlled Channel Access

HCF Controlled Channel Access (HCCA) is a wireless medium access method that uses a QoS-aware centralized coordinator known as a *hybrid coordinator (HC)*, which operates differently from the point coordinator in a PCF network. The HC is built into the AP and has a higher priority of access to the wireless medium. Using this higher priority level, it can allocate TXOPs to itself and other stations to provide a limited-duration controlled access phase (CAP), providing contention-free transfer of QoS data. The specific details of this process are beyond the scope of the CWNA exam. As with PCF, as of this writing, we do not know of any vendor that has implemented HCCA.

Wi-Fi Multimedia

Prior to the adoption of the 802.11e amendment, no adequate QoS procedures had been defined for the use of time-sensitive applications, such as *Voice over Wi-Fi (VoWiFi)*. Application traffic, such as voice, audio, and video, has a lower tolerance for latency and jitter and requires priority before standard data traffic. The 802.11e amendment defined the layer 2 MAC methods needed to meet the QoS requirements for time-sensitive applications over IEEE 802.11 wireless LANs. The Wi-Fi Alliance introduced the *Wi-Fi Multimedia (WMM)* certification as a partial mirror of 802.11e amendment.

Because WMM is based on EDCA mechanisms, 802.1D priority tags from the Ethernet side are used to direct traffic to four access-category priority queues. The WMM certification provides for traffic prioritization via four access categories, as described in Table 8.2.

TABLE 8.2   Wi-Fi multimedia access categories

Access Category	Description	802.1D Tags
WMM Voice priority	This is the highest priority. It allows multiple and concurrent VoIP calls with low latency and toll voice quality.	7, 6
WMM Video priority	This supports prioritized video traffic before other data traffic. A single 802.11g or 802.11a channel can support three to four SDTV video streams or one HDTV video stream.	5, 4
WMM Best Effort priority	This is traffic from applications or devices, such as Internet browsing, that cannot provide QoS capabilities, such as legacy devices. This traffic is not as sensitive to latency but is affected by long delays.	0, 3
WMM Background priority	This is low-priority traffic that does not have strict throughput or latency requirements. This traffic includes file transfers and print jobs.	2, 1

The whole point of WMM is to prioritize different classes of application traffic during the medium-contention process. As shown in Figure 8.10, in the voice access category has better odds when contending for the medium during the backoff process. For voice traffic, a minimum wait time of a SIFS plus two slots is required and then a contention window of 0–3 slots before transmitting on the medium. Best effort traffic must wait a minimum time of a SIFS and three slots and then the contention window is 0–15 slots. The contention process is still entirely pseudo-random; however, the odds are better for the voice traffic.

FIGURE 8.10   WMM access category timing

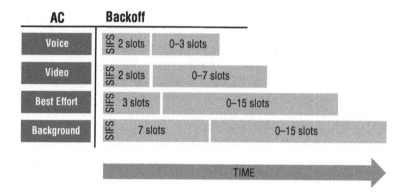

The Wi-Fi Alliance also defined *WMM-PS (Power Save)*, which uses 802.11e power-saving mechanisms to increase the battery life of client devices. You can find more information about power management in Chapter 9.

Another Wi-Fi Alliance certification is *WMM-Admission Control*, which defines the use of management frames for the signaling between an AP and a client station. Client stations can request to send a *traffic stream (TS)* of frames of a particular WMM access category. A traffic stream can be unidirectional or bidirectional. An AP will evaluate a request frame from a client station against the network load and channel conditions. If the AP can accommodate the request, it accepts the request and grants the client station the medium time for a traffic stream. If the request is rejected, the client device is not allowed to initiate the requested traffic stream and may decide to delay the traffic stream, associate with a different AP, or establish a best-effort traffic stream outside the operation of WMM-Admission Control. WMM-Admission Control improves the performance for time-sensitive data, such as video and voice. WMM-Admission Control improves the reliability of applications in progress by preventing the oversubscription of bandwidth.

Important Wi-Fi Alliance White Papers

The Wi-Fi Alliance has two white papers that we recommend you read to learn more about WMM. Both white papers are available for download at the Wi-Fi Alliance website: www.wi-fi.org.

- *Wi-Fi CERTIFIED for WMM—Support for Multimedia Applications with Quality of Service in Wi-Fi Networks*

- *WMM Power Save for Mobile and Portable Wi-Fi CERTIFIED Devices*

Airtime Fairness

One of the important features of 802.11 is its ability to support many different data rates. This allows older technologies to still communicate alongside newer devices, along with enabling devices to maintain communications by shifting to slower data rates as they move away from an access point. The ability to use these slower data rates is paramount to 802.11 communications; however, it can also be a huge hindrance to the overall performance of the network and to individual devices operating at faster data rates.

Since 802.11 is contention-based, each radio must contend for its turn to communicate, then transmit, and then go back to the contention process. As each radio takes its turn transmitting, the other 802.11 radios must wait. If the transmitting radio is using a fast data rate, the other radios do not have to wait long. If the transmitting radio is using a slow data rate, the other radios will have to wait a much longer period of time. When 802.11 radios transmit at very low data rates, such as 1 Mbps and 2 Mbps, effectively they cause medium-contention overhead for higher data rate transmitters due to the long wait time while the slower devices are transmitting.

To try to understand this, look at Figure 8.11. The top portion of the figure illustrates the normal operation of two stations, each sending eight frames. One station is sending eight frames at a higher data rate, and the other station is sending eight frames at a lower data rate. If a high-speed and a low-speed device coexist in the same WLAN, they have to share or contend for the time to transmit. In other words, both stations will statistically get an equal number of times to access the RF medium even though one of the stations is capable of transmitting at a higher rate and requires much less airtime to transmit the same amount of data. Because there is no priority given to the station with the higher data rate, both stations finish transmitting their eight frames over the same period of time.

FIGURE 8.11   Airtime fairness example

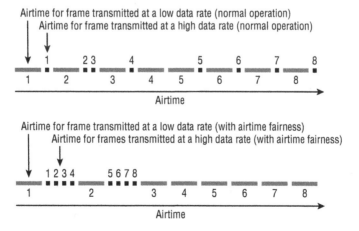

Instead of allocating equal access to the network between devices, the goal of *airtime fairness* is to allocate equal time, as opposed to equal opportunity. Airtime fairness can provide better time management of the RF medium. In the bottom half of Figure 8.11, airtime fairness is enabled; you can see that the station with the higher data rate transmission is given priority before the station with the lower data rates. Effectively, this is a much better use of transmission time because the higher data rate station does not have to remain idle waiting during the lower data rate transmission. Notice that the faster station transmitted all eight frames in a much shorter time period, and the slower rate station still sent all eight frames in about the same period as before. Airtime fairness effectively achieves better time management of the medium by cutting down on wait times. The net result is better performance, higher capacity, and more throughput over the Wi-Fi network.

Currently, no 802.11 standards or amendments define airtime fairness or how to implement it, nor are vendors required to implement it. Most vendors use airtime fairness mechanisms only for downstream transmissions from an AP to an associated client. Airtime fairness mechanisms are normally used for prioritizing the higher data rate downstream transmissions from an AP over the lower data rate downstream transmissions from an AP. At least one vendor also makes claims of upstream airtime fairness capability. Any implementation of airtime fairness is a proprietary solution developed by each WLAN vendor. No matter how each vendor implements its solution, the underlying goal is essentially the same: to prevent slower devices from bogging down the rest of the network.

Although each WLAN vendor takes its own approach to implementing airtime fairness, it is typical for them to analyze the downstream client traffic and assign different weighting, based on such characteristics as current throughput, client data rates, SSID, PHY type, and other variables. Algorithms are used to process this information and determine the number of opportunities for each client's downstream transmissions. If implemented properly, airtime fairness makes better use of the medium by providing preferential access for higher data rate transmissions.

Summary

This chapter focused on 802.11 medium access. Every station has the right to communicate, and the management of access to the wireless medium is controlled through medium access control (MAC). We discussed the difference between CSMA/CD and CSMA/CA as contention methods. CSMA/CA uses a pseudo-random contention method called Distributed Coordination Function (DCF). DCF uses four lines of defense to ensure that only one 802.11 radio is transmitting on the half-duplex medium.

The 802.11e quality-of-service amendment added a new coordination function to 802.11 medium contention, known as Hybrid Coordination Function (HCF). The Wi-Fi Multimedia (WMM) certification was introduced by the Wi-Fi Alliance as a partial mirror of the 802.11e amendment. WMM is designed to meet the QoS requirements for time-sensitive applications, such as audio, video, and voice, over IEEE 802.11.

Airtime fairness was introduced as a way for vendors to provide faster devices with preferential access to the medium when operating alongside devices that are transmitting at slower data rates.

Exam Essentials

Understand the similarities and differences between CSMA/CA and CSMA/CD. Understand both access methods and know what makes them similar and what makes them different.

Define the four checks and balances of CSMA/CA and DCF. Understand that virtual carrier sense, physical carrier sense, interframe spacing, and the pseudo-random backoff timer all work together to ensure that only one 802.11 radio is transmitting on the half-duplex medium.

Define virtual carrier sense and physical carrier sense. Understand the purpose and basic mechanisms of the two carrier senses.

Define HCF quality-of-service mechanisms. Hybrid Coordination Function defines the use of TXOPs and access categories in EDCA as well as the use of TXOPs and polling during HCCA.

Understand the Wi-Fi Multimedia certification. WMM is designed to provide quality-of-service capabilities to 802.11 wireless networks. WMM is a partial mirror of the 802.11e amendment. WMM currently provides for traffic priority via four access categories.

Understand the importance of airtime fairness and what it does. Airtime fairness provides devices operating at faster data rates with preferential access to the medium. This preferential treatment provides all devices with equal access, resulting in all devices equally sharing the available transmission bandwidth.

Review Questions

1. Which medium contention and access method is used as the foundation of 802.11 Distributed Coordination Function (DCF)? (Choose all that apply.)

   A. Carrier Sense Multiple Access with Collision Detection (CSMA/CD)

   B. Carrier Sense Multiple Access with Collision Avoidance (CSMA/CA)

   C. Token passing

   D. Demand priority

2. 802.11 collision detection is achieved using which technology?

   A. Network allocation vector (NAV)

   B. Clear channel assessment (CCA)

   C. Duration/ID value

   D. Receiving an ACK from the destination station

   E. Positive collision detection cannot be determined.

3. ACK and CTS frames follow which interframe space?

   A. EIFS

   B. DIFS

   C. PIFS

   D. SIFS

   E. AIFS

4. The carrier sense portion of CSMA/CA is performed by using which of the following methods? (Choose all that apply.)

   A. Contention window

   B. Backoff timer

   C. Channel sense window

   D. Clear channel assessment

   E. NAV timer

5. After the station has performed the carrier sense and determined that no other devices are transmitting for a period of a DIFS interval, what is the next step for the station?

   A. Wait the necessary number of slot times before transmitting if a random backoff value has already been selected.

   B. Begin transmitting.

   C. Select a random backoff value.

   D. Begin the random backoff timer.

6. Physical carrier sense uses which two thresholds during the clear channel assessment to determine if the medium is busy?

   A. RF detect

   B. Signal detect

   C. Transmission detect

   D. Energy detect

   E. Random detect

7. Which of the following terms are affiliated with the virtual carrier sense mechanism? (Choose all that apply.)

   A. Contention window

   B. Network allocation vector

   C. Random backoff time

   D. Duration/ID field

8. The goal of allocating equal time as opposed to equal opportunity is known as what?

   A. Access fairness

   B. Opportunistic medium access

   C. CSMA/CA

   D. Airtime fairness

9. CSMA/CA and DCF define which mechanisms that attempt to ensure that only one 802.11 radio can transmit on the half-duplex RF medium? (Choose all that apply.)

   A. Pseudo-random backoff timer

   B. Virtual carrier sense

   C. Collision detection

   D. Physical carrier sense

   E. Interframe spacing

10. The Wi-Fi Alliance certification called Wi-Fi Multimedia (WMM) is based on which wireless medium access method defined by the 802.11-2016 standard?

    A. DCF

    B. PCF

    C. EDCA

    D. HCCA

    E. HSRP

11. Hybrid Coordination Function (HCF) defines what allotted period of time in which a station can transmit multiple frames?

    A. Block acknowledgment

    B. Polling

    C. Virtual carrier sense

    D. Physical carrier sense

    E. TXOP

12. WMM is based on EDCA and provides for traffic prioritization via which of the following access categories? (Choose all that apply.)

    A. WMM Voice priority

    B. WMM Video priority

    C. WMM Audio priority

    D. WMM Best Effort priority

    E. WMM Background priority

13. As defined by WMM, which type of application traffic has the highest priority for transmission on the half-duplex RF medium?

    A. Best Effort

    B. Video

    C. Voice

    D. Background

14. What information that comes from the wired network is used to assign traffic into access categories on a WLAN controller?

    A. Duration/ID

    B. 802.1D priority tags

    C. Destination MAC address

    D. Source MAC address

15. What are the two reasons that 802.11 radios use physical carrier sense? (Choose all that apply.)

    A. To synchronize incoming transmissions

    B. To synchronize outgoing transmissions

    C. To reset the NAV

    D. To start the random backoff timer

    E. To assess the RF medium

16. Which carrier sense method is used to detect and decode 802.11 transmissions?

    A. Network allocation vector

    B. Signal detect

    C. Energy detect

    D. Virtual carrier sense

17. Which field in the MAC header of an 802.11 frame resets the NAV timer for all listening 802.11 stations?

    A. NAV

    B. Frame control

    C. Duration/ID

    D. Sequence number

    E. Strictly ordered bit

18. The EDCA medium access method provides for the prioritization of traffic via priority queues that are matched to eight 802.1D priority tags. What are the EDCA priority queues called?

    A. TXOP

    B. Access categories

    C. Priority levels

    D. Priority bits

    E. PT

19. ACKs are required for which of the following frames?

    A. Unicast

    B. Broadcast

    C. Multicast

    D. Anycast

20. Which two components of the pseudo-random backoff algorithm are used to create the pseudo-random backoff timer?

    A. Contention window

    B. Network allocation vector

    C. Duration/ID

    D. Slot time

Chapter

9

802.11 MAC

IN THIS CHAPTER, YOU WILL LEARN ABOUT THE FOLLOWING:

✓ **Packets, frames, and bits**

✓ **Data-Link layer**
- MAC service data unit
- MAC protocol data unit

✓ **Physical layer**
- PLCP service data unit
- PLCP protocol data unit

✓ **802.11 and 802.3 interoperability**

✓ **802.11 MAC header**
- Frame Control field
- Duration/ID field
- MAC layer addressing
- Sequence Control field
- QoS Control field
- HT Control field

✓ **802.11 frame body**

✓ **802.11 trailer**

✓ **802.11 state machine**

✓ **Management frames**
- Beacon
- Authentication
- Association
- Reassociation

- Disassociation
- Deauthentication
- Action frame

✓ **Control frames**

- ACK frame
- Block acknowledgment
- PS-Poll
- RTS/CTS
- CTS-to-Self
- Protection mechanisms

✓ **Data frames**

- QoS and non-QoS data frames
- Non-data carrying frames

✓ **Power management**

- Legacy power management
- WMM-Power Save and U-APSD
- MIMO power management

This chapter presents all the components of the 802.11 MAC frame format. We discuss how upper-layer information is encapsulated within an 802.11 frame format. We discuss in detail the 802.11 MAC header and MAC addressing. We cover the three major 802.11 frame types and a majority of the 802.11 frame subtypes. We discuss the 802.11 state machine, which defines how stations discover, join, and leave a basic service set (BSS). Finally, we discuss legacy 802.11 power management and enhanced WMM-PS power management, which are methods used to save battery life.

Packets, Frames, and Bits

When learning about any technology, at times you need to step back and focus on the basics. If you have ever flown an airplane, you know that it is important, when things get difficult, to refocus on the number one priority, the main objective—and that is to fly the airplane. Navigation and communications are secondary to flying the airplane. When dealing with any complex technology, it is easy to forget the main objective; this is as true with 802.11 communications as it is with flying. With 802.11 communications, the main objective is to transfer user data from one computing device to another.

As data is processed in a computer and prepared to be transferred from one computer to another, it starts at the upper layers of the OSI model and moves down until it reaches the Physical layer, where it is ultimately transferred to the other devices. Initially, a user may want to transfer a word processing document from their computer to a shared network disk on another computer. This document will start at the Application layer and work its way down to the Physical layer, get transmitted to the other computer, and then work its way back up the layers of the OSI model to the Application layer on the other computer.

As data travels down the OSI model for the purpose of being transmitted, each layer adds header information to that data. This enables the data to be reassembled when it is received by the other computer. At the Network layer, an IP header is added to the data that came from layers 4–7. A layer 3 IP *packet*, or datagram, encapsulates the data from the higher layers. At the Data-Link layer, a MAC header is added and the IP packet is encapsulated inside a *frame*. Ultimately, when the frame reaches the Physical layer, a PHY header with more information is added to the frame.

Data is eventually transmitted as individual bits at the Physical layer. A *bit* is a binary digit, taking a value of either 0 or 1. Binary digits are a basic unit of communication in digital computing. A *byte* of information consists of 8 bits. An *octet* is another name for a byte of data. The CWNA exam uses the terms octet and byte interchangeably.

In this chapter, we discuss how upper-layer information moves down the OSI model through the Data-Link and Physical layers from an 802.11 perspective.

Data-Link Layer

The 802.11 *Data-Link layer* is divided into two sublayers. The upper portion is the IEEE 802.2 *Logical Link Control (LLC)* sublayer, which is identical for all 802-based networks, although it is not used by all IEEE 802 networks. The bottom portion of the Data-Link layer is the *Media Access Control (MAC) sublayer*. The 802.11 standard defines operations at the MAC sublayer.

MAC Service Data Unit

When the Network layer (layer 3) sends data to the Data-Link layer, that data is handed off to the LLC and becomes known as the *MAC service data unit (MSDU)*. The MSDU contains data from the LLC and layers 3–7. A simple definition of the MSDU is that it is the data payload that contains the IP packet plus some LLC data. Later in this chapter, you will learn about the three major 802.11 frame types. 802.11 management and control frames do not carry upper-layer information. Only 802.11 data frames carry an MSDU payload in the frame body. The 802.11-2016 standard states that the maximum size of the MSDU is 2,304 bytes. The maximum frame body size is determined by the maximum MSDU size (2,304 octets) plus any overhead from encryption.

The ratification of the 802.11n-2009 amendment introduced aggregate MSDU (A-MSDU). With A-MSDU, the maximum frame body size is determined by the maximum A-MSDU size of 3,839 or 7,935 octets, depending on the STA's capability, plus any overhead from encryption. You will learn more about A-MSDUs in Chapter 10, "MIMO Technology: HT and VHT."

MAC Protocol Data Unit

When the LLC sublayer sends the MSDU to the MAC sublayer, the MAC header information is added to the MSDU to identify it. The MSDU is now encapsulated in a *MAC protocol data unit (MPDU)*. A simple definition of an 802.11 MPDU is that it is an 802.11 frame. As shown in Figure 9.1, an 802.11 MPDU consists of the following three basic components:

MAC Header   Frame control information, duration information, MAC addressing, sequence control, QoS control information, and HT control information are all found in the MAC header. The 802.11 MAC header is discussed in more detail later in this chapter.

Frame Body   The frame body component can vary in size and contains information that is different depending on the frame type and frame subtype. The MSDU upper-layer payload is encapsulated in the frame body. The MSDU layer 3–7 payload is protected when using encryption.

Frame Check Sequence   The frame check sequence (FCS) comprises a 32-bit *cyclic redundancy check (CRC)* that is used to validate the integrity of received frames.

FIGURE 9.1   802.11 MPDU

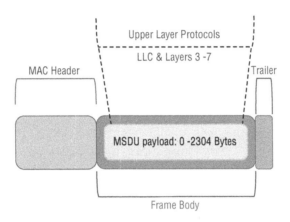

At this point, the frame is ready to be passed onto the Physical layer, which will then further prepare the frame for transmission.

Physical Layer

Similar to the way the Data-Link layer is divided into two sublayers, the *Physical layer* is also divided into two sublayers. The upper portion of the Physical layer is known as the *Physical Layer Convergence Procedure (PLCP)* sublayer, and the lower portion is known as the *Physical Medium Dependent (PMD)* sublayer. The PLCP prepares the frame for transmission by taking the frame from the MAC sublayer and creating the PLCP protocol data unit (PPDU). The PMD sublayer then modulates and transmits the data as bits.

PLCP Service Data Unit

The *PLCP service data unit (PSDU)* is a view of the MPDU from the Physical layer. The MAC layer refers to the frame as the MPDU, whereas the Physical layer refers to this same frame as the PSDU. The only difference is from which layer of the OSI model you are looking at the frame.

PLCP Protocol Data Unit

When the PLCP receives the PSDU, it then prepares the PSDU to be transmitted and creates the *PLCP protocol data unit (PPDU)*. The PLCP adds a preamble and PHY header to the PSDU. The preamble is used for synchronization between transmitting and receiving 802.11 radios. It is beyond the scope of this book and the CWNA exam to discuss all the details of the preamble and PHY header. When the PPDU is created, the PMD sublayer takes the PPDU and modulates the data bits and begins transmitting.

Figure 9.2 depicts a flowchart that shows the upper-layer information moving between the Data-Link and Physical layers.

FIGURE 9.2   Data-Link and Physical layers

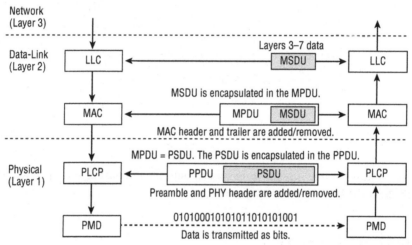

802.11 and 802.3 Interoperability

As you learned in Chapter 7, "Wireless LAN Topologies," the 802.11-2016 standard defines an *integration service (IS)* that enables delivery of MSDUs between the distribution system (DS) and a non-IEEE-802.11 local area network (LAN), via a portal. A simpler way of defining the integration service is to characterize it as a frame format transfer method. The portal is usually either an access point or a WLAN controller. As mentioned earlier, the payload of a wireless 802.11 data frame is the upper layer 3–7 information known as the MSDU. The eventual destination of this payload usually resides on a wired network infrastructure. Because the wired infrastructure is a different physical medium, an 802.11 data frame payload (MSDU) must be effectively transferred into an 802.3 Ethernet frame. For example, a VoWiFi phone transmits an 802.11 data frame to an access point. The MSDU payload of the frame is the VoIP packet with a final destination of a PBX server residing on the wired network. The job of the integration service is to first remove the 802.11 header and trailer and then encase the MSDU VoIP payload inside an 802.3 Ethernet frame. Normally, the integration service transfers frame payloads between an 802.11 medium and an 802.3 medium. However, the IS could transfer an MSDU between the 802.11 medium and some sort of other medium. All the IEEE 802 frame formats share similar characteristics, including the 802.11 frame format. Because the frames are similar, it makes it easier to translate the frames as they move from the 802.11 wireless network to the 802.3 wired network, and vice versa.

One of the differences between 802.3 Ethernet and 802.11 wireless frames is the frame size. 802.3 frames have a maximum size of 1,518 bytes with a maximum data payload of 1,500 bytes. If the 802.3 frames are 802.1Q tagged for VLANs and user priority, the maximum size of the 802.3 frame is 1,522 bytes with a data payload of 1,504 bytes. As you have just learned, 802.11 frames are capable of transporting frames with an MSDU payload of 2,304 bytes of upper-layer data. This means that as the data moves between the wireless network and the wired network, the AP may receive a data frame that is too large for the wired network. This is rarely a problem thanks to the TCP/IP protocol suite. TCP/IP, the most common communications protocol used on networks, typically has an IP *maximum transmission unit (MTU)* size of 1,500 bytes. IP packets are usually 1,500 bytes based on the MTUs. When the IP packets are passed down to 802.11, even though the maximum size of the MSDU is 2,304 bytes, the size will be limited to the 1,500 bytes of the IP packets.

Ethernet frames with more than 1,500 bytes of payload are called *jumbo frames* and usually carry a payload of up to 9,000 bytes. Many Gigabit Ethernet switches and Gigabit Ethernet network interface cards can support jumbo frames. 802.11 WLANs do not support jumbo frames; however, there is no need because of 802.11 frame aggregation. In Chapter 10, "MIMO Technology: HT and VHT," you will learn about more efficient methods of MSDU payload delivery via both A-MSDU and A-MPDU frame aggregation. Please note that the MTU settings of Ethernet ports of some WLAN controllers and APs must be configured for 9,000 bytes to support jumbo frames outbound on the wired network.

802.11 MAC Header

Every 802.11 frame contains a MAC header that contains layer 2 information. The layer 2 information is not encrypted and is always visible when viewed with a protocol analyzer. As shown in Figure 9.3, the 802.11 MAC header has nine major fields, four of which are used for addressing. The remaining fields include the Frame Control field, the Duration/ID field, the Sequence Control field, the QoS Control field, and the HT Control field. It is beyond the scope of the CWNA exam to explain in great detail the purpose of every field and subfield in the 802.11 MAC header; however, we will proceed with a high-level discussion about some of these fields.

FIGURE 9.3   802.11 MAC header

Octets: 2	2	6	0 or 6	0 or 6	0 or 2	0 or 6	0 or 2	0 or 4	variable	4
Frame Control	Duration /ID	Address 1	Address 2	Address 3	Sequence Control	Address 4	QoS Control	HT Control	Frame Body	FCS

MAC header

Frame Control Field

The first two bytes of the MAC header consists of 11 subfields within the *Frame Control field*. These subfields include Protocol Version, Type, Subtype, To DS, From DS, More Fragments, Retry, Power Management, More Data, Protected Frame, and +HTC/Order. We will now discuss some of these subfields, which are illustrated in Figure 9.4

FIGURE 9.4   Frame Control field

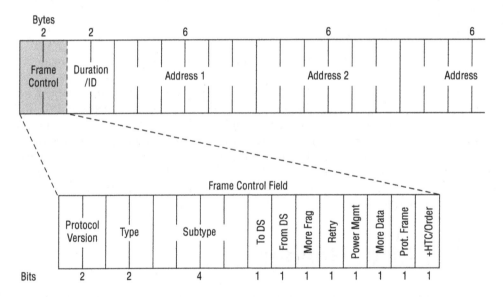

The *Protocol Version field* is a consistent 2-bit field that is always placed at the beginning of all 802.11 MAC headers. This field is simply used to indicate which protocol version of 802.11 technology is being used by the frame. All 802.11 frames have the Protocol Version field always set to 0. All other values are reserved. In other words, there is currently only one version of 802.11 technology. The IEEE could possibly in the future define another version of 802.11 technology that would not be backward compatible with the current version 0.

Unlike many wired network standards, such as IEEE 802.3, which uses a single data frame type, the IEEE 802.11-2016 standard defines three major frame types: *management*, *control*, and *data*. These 802.11 frame types are further subdivided into multiple subtypes. In reality, a fourth major 802.11 frame type, *extension frames*, is defined for use with *directional multi-gigabit (DMG)* 802.11ad radios that operate in 60 GHz. The discussion of 802.11 extension frames is beyond the scope of the CWNA exam. The Type field of the 802.11 header identifies whether the frame is a management, control, data, or extension frame. As shown in Table 9.1, the 2-bit Type field identifies whether the frame is a control, data, management, or extension frame. A value of 00 means the type is a management frame; a value of 01 indicates a control frame; a value of 10 indicates a data frame; and a value of 11 indicates an extension frame.

TABLE 9.1 802.11 frame types

Bits	Frame Type	Purpose
00	Management	Used to discover APs and to join a BSS
01	Control	Used to acknowledge successful transmissions and reserve the wireless medium
10	Data	Used to carry an upper-layer MSDU payload
11	Extension	A new, flexible frame format, currently used only with 802.11ad

Do not confuse 802.11 management, control, and data frames with the three telecommunications planes of the same name. A discussion of the management, control, and data planes as related to WLAN network architectural operations can be found in Chapter 11, "WLAN Architecture."

The *Type field* and *Subtype field* are used together to identify the function of the frame. Because there are many different kinds of management, control, and data frames, a 4-bit Subtype field is needed. For example, Figure 9.5 shows a portion of a frame capture of a management frame. The Subtype field indicates that the frame is a beacon management frame.

FIGURE 9.5 Type and Subtype fields

The *Retry field* is a single significant bit of information found in all 802.11 MAC headers. The Retry field comprises a single bit of the Frame Control field and is perhaps one of the most important fields in the MAC header. If the Retry bit has a value of 0, an original transmission of the frame is occurring. If the Retry bit is set to a value of 1 in either a management or data frame, the transmitting radio is indicating that the frame being sent is a retransmission. Figure 9.6 shows the position of the Retry field in the MAC header.

FIGURE 9.6   Retry field

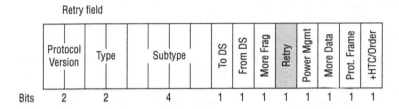

As discussed many times in this book, every time an 802.11 radio transmits a unicast frame, if the frame is received properly and the cyclic redundancy check (CRC) of the FCS passes, the 802.11 radio that received the frame will reply with an acknowledgment (ACK) frame. If the ACK is received, the original station knows that the frame transfer was successful. If any portion of a unicast frame is corrupted, the CRC will fail, and the receiving 802.11 radio will not send an ACK frame to the transmitting 802.11 radio. If an ACK frame is not received by the original transmitting radio, the unicast frame is not acknowledged and will have to be retransmitted. The Retry bit is an indication that the frame being transmitted is a retransmission as opposed to an original frame transmission. Any good 802.11 protocol analyzer can compile layer 2 retransmission rates by observing management and data frames that have the Retry field set to a value of 1.

The *Protected Frame field* is a single bit and is used to indicate whether the MSDU payload of a data frame is encrypted. The Protected Frame field is a subfield of the Frame Control field. When the Protected Frame field is set to a value of 1 in a data frame, the MSDU payload of the data frame is indeed encrypted. The Protected Frame field does not indicate which type of encryption is being used; it indicates only that the MSDU payload of the frame body is encrypted. The encryption can be Wired Equivalent Privacy (WEP), Temporal Key Integrity Protocol (TKIP), or Counter Mode with Cipher Block Chaining Message Authentication Code Protocol (CCMP).

Duration/ID Field

A very important field that was discussed earlier in the book is the *Duration/ID field*. As you learned in Chapter 8, "802.11 Medium Access," the duration value in the MAC header of a transmitting station is used to reset the NAV timer of other listening stations.

To review, virtual carrier-sense uses a timer mechanism known as the *network allocation vector (NAV)*. The NAV timer maintains a prediction of future traffic on the medium based on Duration value information seen in a previous frame transmission. When an 802.11 radio is not transmitting, it is listening. When the listening radio hears a frame transmission from another station, it looks at the header of the frame and determines whether the Duration/ID field contains a Duration value or an ID value. If the field contains a Duration value, the listening station will set its NAV timer to this value. The listening station will then use the NAV as a countdown timer, knowing that the RF medium should be busy until the countdown reaches 0.

This field is almost always used for duration value information for the virtual carrier-sense purposes just mentioned. However, a second way in which the Duration/ID field is used is during the legacy power-management process. During this process, clients will use the field in a PS-Poll control frame as an identifier to an AP during the power-management process. A more detailed discussion about power management follows later in this chapter.

MAC Layer Addressing

Much like in an 802.3 Ethernet frame, the header of an 802.11 frame contains MAC addresses. A MAC address is one of the following two types:

Individual Address   Individual addresses are assigned to unique stations on the network (also known as a *unicast address)*.

Group Address   A multiple-destination address (group address) could be used by one or more stations on a network. There are two kinds of group addresses:

 Multicast-Group Address   An address used by an upper-layer entity to define a logical group of stations is known as a *multicast-group address.*

 Broadcast Address   A group address that indicates all stations that belong to the network is known as a *broadcast address.* A broadcast address, all bits with a value of 1, defines all stations on a local area network. In hexadecimal, the broadcast address would be FF:FF:FF:FF:FF:FF.

Although there are similarities, the MAC addressing used by 802.11 frames is much more complex than Ethernet frames. 802.3 frames have only a source address (SA) and destination address (DA) in the layer 2 header. As shown earlier in Figure 9.3, 802.11 frames have up to four address fields in the MAC header. 802.11 frames typically use only three of the MAC address fields. However, an 802.11 frame sent within a wireless distribution system (WDS) requires all four MAC addresses. Certain frames may not contain some of the address fields. Even though the number of address fields is different, both 802.3 and 802.11 identify a source address and a destination address and use the same MAC address format. The first three octets are known as the *organizationally unique identifier (OUI)*, and the last three octets are known as the *extension identifier.*

As shown in Figure 9.7, there are four 802.11 MAC address fields: Address 1, Address 2, Address 3, and Address 4. Depending on how the To DS and From DS fields are used, the definition of each of the four MAC address fields will change. The five possible definitions are as follows:

Source Address (SA) The MAC address of the original sending station is known as the *source address (SA)*. The source address can originate from either a wireless station or the wired network.

Destination Address (DA) The MAC address that is the final destination of the layer 2 frame is known as the *destination address (DA)*. The final destination may be a wireless station or a destination on the wired network, such as a server.

Transmitter Address (TA) The MAC address of an 802.11 radio that is transmitting the frame onto the half-duplex 802.11 medium is known as the *transmitter address (TA)*.

Receiver Address (RA) The MAC address of the 802.11 radio that is intended to receive the incoming transmission from the transmitting station is known as the *receiver address (RA)*.

Basic Service Set Identifier (BSSID) This is the MAC address that is the layer 2 identifier of the basic service set (BSS). The *basic service set identifier (BSSID)* is the MAC address of the AP's radio or is derived from the MAC address of the AP's radio if multiple basic service sets exist.

FIGURE 9.7 802.11 MAC addressing

Bits: 2	2	4	1	1	1	1	1	1	1	1
Protocol Version	Type	Subtype	To DS	From DS	More Frag	Retry	Power Mgmt	More Data	Prot. Frame	+HTC/ Order

Frame Control field

To DS	From DS	Address 1	Address 2	Address 3	Address 4
0	0	RA = DA	TA = SA	BSSID	N/A
0	1	RA = DA	TA = BSSID	SA	N/A
1	0	RA = BSSID	TA = SA	DA	N/A
1	1	RA	TA	DA	SA

- SA = MAC address of the original sender (wired or wireless)
- DA = MAC address of the final destination (wired or wireless)
- TA = MAC address of the transmitting 802.11 radio
- RA = MAC address of the receiving 802.11 radio
- BSSID = L2 identifier of the basic service set (BSS)

The *To DS field* and the *From DS field* are each 1 bit and are used in combination to change the meaning of the four MAC addresses in an 802.11 header. These two bits also indicate the flow of the 802.11 data frames between a WLAN environment and the distribution system (DS). The DS is normally an Ethernet wired network. Depending on how the To DS and From DS fields are used together with the four MAC addresses, the definition of each field will change. One constant, however, is that the Address 1 field will always be the receiver address (RA) but may have a second definition as well. Address 2 will always be the transmitter address (TA) but also may have a second definition. Address 3 is normally used for additional MAC address information. Address 4 is used only in the case of a WDS.

There are four possible combinations of these two bits. The first combination of the To DS and From DS bits is as follows:

To DS = 0

From DS = 0

When both bits are set to 0, several different scenarios can exist. The most common scenario is that these are management or control frames. Management and control frames do not have an MSDU payload, so their final destination is never the distribution system (DS). Management and control frames exist only at the MAC sublayer and therefore have no need to be translated by the integration service (IS) and never are sent to the wired network. Figure 9.8 depicts the MAC addressing used for a probe request management frame sent by a client to an AP. The third address field carries additional information and is used to identify the BSSID. Address fields 1 and 3 have the same values because the AP is both the RA and BSSID. Figure 9.9 depicts the MAC addressing used for a probe response management frame sent by an AP to a client. Address fields 1 and 3 have the same values because the AP is both the TA and BSSID.

FIGURE 9.8   To DS:0 From DS:0 (Probe request)

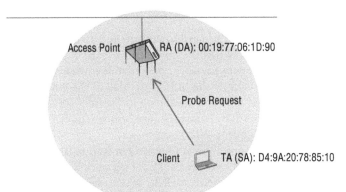

Access Point — RA (DA): 00:19:77:06:1D:90

Probe Request

Client — TA (SA): D4:9A:20:78:85:10

To DS: 0 From DS: 0

Address #1: RA (DA): 00:19:77:06:1D:90
Address #2: TA (SA): D4:9A:20:78:85:10
Address #3: BSSID: 00:19:77:06:1D:90

FIGURE 9.9   To DS:0 From DS:0 (Probe response)

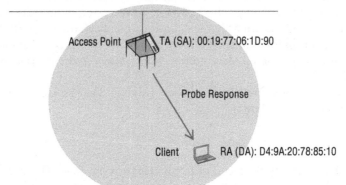

To DS: 0 From DS: 0
Address #1: RA (DA): D4:9A:20:78:85:10
Address #2: TA (SA): 00:19:77:06:1D:90
Address #3: BSSID: 00:19:77:06:1D:90

Another scenario when both DS bits are set to 0 is a direct data frame transfer from one STA to another STA within an independent basic service set (IBSS), more commonly known as an ad hoc network. The third scenario involves what is known as a station-to-station link (STSL), which involves a data frame being sent directly from one client station to another client station that belongs to the same BSS, thereby bypassing the AP.

The second combination of the To DS and From DS bits is as follows:

To DS = 0

From DS = 1

The To DS and From DS bits can be used to indicate the direction and flow of 802.11 data frames within a typical BSS. When the To DS bit is set to 0 and the From DS bit is set to 1, it indicates that an 802.11 data frame is being sent downlink from an access point to a client station. The original source of the MSDU payload of the 802.11 data frame is an address that exists on the wired network. As shown in Figure 9.10, an example would be a DHCP server that resides on the 802.3 network forwarding a DHCP lease offer through an AP with the final destination being an 802.11 client station. The address of the access point radio is 00:19:77:06:1D:90, and the address of the client station is D4:9A:20:78:85:10. The address of the DHCP server that resides on the 802.3 network is 00:0A:E4:DA:92:F7. The Address 1 field is always the receiver address (RA), which is the client station and the final destination address (DA). The Address 2 field is always the transmitter address (TA) and is the access point that is also the BSSID. The Address 3 field carries additional information and is used to identify the source address (SA) of the DHCP server that exists on the 802.3 medium.

FIGURE 9.10   To DS:1 From DS:0 – Downlink traffic

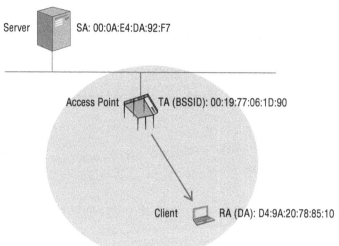

To DS: 0 From DS: 1
Address #1: RA (DA): D4:9A:20:78:85:10
Address #2: TA (BSSID): 00:19:77:06:1D:90
Address #3: SA: 00:0A:E4:DA:92:F7

The third combination of the To DS and From DS bits is as follows:

To DS = 1

From DS = 0

When the To DS bit is set to 1 and the From DS bit is set to 0, it indicates that an 802.11 data frame is being sent uplink from a client station to an access point. In most cases, the final destination of the MSDU payload of the data frame is an address that exists on the wired network. As shown in Figure 9.11, an example of this scenario would be a client station sending a DHCP request packet through an AP to a DHCP server that resides on the 802.3 network. The address of the DHCP server that resides on the wired network is 00:0A:E4:DA:92:F7. The Address 1 field is always the receiver address (RA), which is the access point radio and the BSSID. The Address 2 field is always the transmitter address (TA) and is the client station that is also the source address (SA). The Address 3 field carries additional information and is used to identify the destination address (DA) of the DHCP server that exists on the 802.3 medium.

FIGURE 9.11 To DS:1 From DS:0 – Uplink traffic

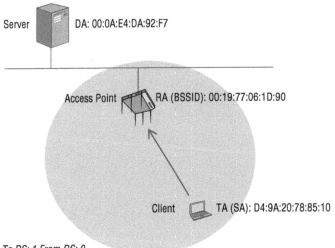

To DS: 1 From DS: 0
Address #1: RA (BSSID): 00:19:77:06:1D:90
Address #2: TA (SA): D4:9A:20:78:85:10
Address #3: DA: 00:0A:E4:DA:92:F7

The fourth combination of the To DS and From DS bits is as follows:

To DS = 1

From DS = 1

When the To DS bit and the From DS bit are both set to 1, this is the only time that a data frame uses the four-address format. Although the standard does not define procedures for using this format, WLAN vendors often implement what is known as a *wireless distribution system (WDS)*. Examples of a WDS include WLAN bridges and mesh networks. In these WDS scenarios, a data frame is being sent across a second wireless medium before eventually being forwarded to a wired medium. When the To DS and From DS fields are both set to a value of 1, a WDS is being used, and four addresses are needed.

Figure 9.12 shows an example of an 802.11 mesh 5 GHz backhaul link between a mesh point and a mesh portal. A client station that is associated to the 2.4 GHz radio of the mesh point wants to send a frame to a server that resides on the 802.3 backbone. When the frame is forwarded over the 5 GHz wireless backhaul, the To DS and From DS bits are both set to 1, and four addresses are needed. The Address 1 field is always the receiver address (RA), which in this case is the 5 GHz radio of the mesh portal. The Address 2 field always is the transmitter address (TA), which in this example is the 5 GHz radio of the mesh point. The Address 3 field holds the destination address (DA), which is the server on the wired network. The Address 4 field is the source address, which is the client station that is associated to the 2.4 GHz radio of the mesh point. From this example, you can see why four addresses would be needed across the mesh backhaul, which is the WDS.

FIGURE 9.12   To DS:1 From DS:1 – Mesh backhaul

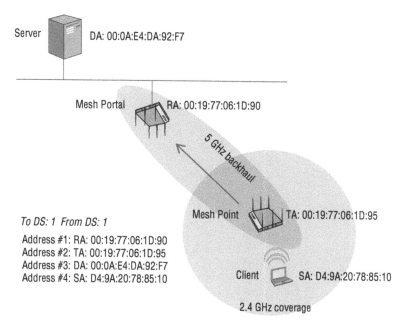

Figure 9.13 shows an example of an 802.11 point-to-point bridge link between two buildings. A frame needs to be sent from a wired server in building 1 to a wired desktop in building 2. The Address 1 field is always the receiver address (RA), which in this case is the WLAN bridge in building 2. The Address 2 field always is the transmitter address (TA), which in this example is the WLAN bridge from building 1. The Address 3 field holds the destination address (DA), which is the desktop in building 2, and the Address 4 field is the source address, which is the server from building 1. From this example, you can see why four addresses would be needed across the WLAN bridge link, which is the WDS.

Sequence Control Field

The *Sequence Control field* is a 16-bit field comprising two subfields and is used when 802.11 MSDUs are fragmented. The 802.11-2016 standard allows for fragmentation of frames. *Fragmentation* breaks an 802.11 frame into smaller pieces known as fragments, adds header information to each fragment, and transmits each fragment individually. All 802.11 APs and some client stations can be configured with a fragmentation threshold. If the fragmentation threshold is set at 300 bytes, any MSDU larger than 300 bytes will be fragmented. Figure 9.14, depicts a 1,200-byte MSDU with a sequence number of 542. Based on a threshold of 300 bytes, Figure 9.14 displays the fragmentation from the bottom upward because the fragments move down the OSI stack. The information shown in the Sequence Control field is also needed for a receiver radio to reassemble the fragments.

FIGURE 9.13 To DS:1 From DS:1 – WLAN bridge link

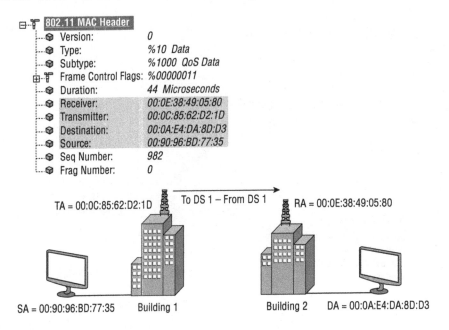

FIGURE 9.14 Fragmentation

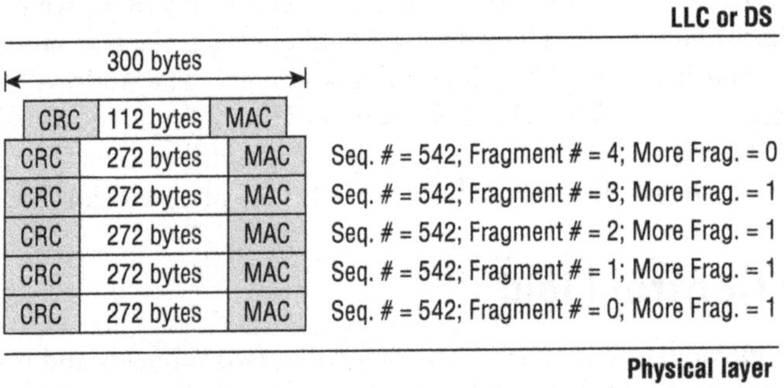

Although the same amount of actual data is being transmitted, each fragment requires its own header, and the transmission of each fragment is followed by a short interframe space (SIFS) and an ACK. In a properly functioning 802.11 network, smaller fragments will actually decrease data throughput because of the MAC sublayer overhead of the additional header, SIFS, and ACK of each fragment. On the other hand, if the network is experiencing a large amount of data corruption, lowering the 802.11 fragmentation setting may improve data throughput. Fragments are always sent in what is known as a *fragment burst*. Fragmentation was sometimes used in legacy 802.11/b/g networks but is no longer needed with 802.11n/ac networks that support frame aggregation and Block ACKs. Because fragmentation is rarely used, a deep discussion of fragmentation is beyond the scope of the CWNA exam.

The transmission of a fragment is treated the same way as the transmission of a frame. Therefore, every fragment must participate in the CSMA/CA medium access and must be followed by an ACK. If a fragment is not followed by an ACK, it will be retransmitted.

QoS Control Field

The *QoS Control field* is a 16-bit field that identifies the quality-of-service (QoS) parameters of a data frame. Note that not all data frames contain a QoS Control field. The QoS Control field is used only in the MAC header of QoS data frames. As you learned in Chapter 8, on a wired 802.3 Ethernet network, different classes of service are available, represented in a 3-bit User Priority field in an IEEE 802.1Q header added to an Ethernet frame. 802.1D enables priority queuing (enabling some Ethernet frames to be forwarded ahead of others within a switched Ethernet network). These 802.1D service classes are mapped to *Wi-Fi Multimedia (WMM)* access categories. WMM provides for 802.11 traffic prioritization in four access categories: voice, video, best effort, and background. The QoS Control field is sometime referred to as the WMM QoS Control field because the QoS Control field effectively indicates the WMM class of service of the QoS data frame.

HT Control Field

The HT Control field is used for link adaptation, transmit beamforming (TxBF), and other advanced capabilities of 802.11n/ac transmitters and receivers. The HT Control field is used only in management frames and QoS data frames when the +HTC/Order subfield of the Frame Control field is set to a value of 1. A full explanation of the HT Control field is beyond the scope of the CWNA exam.

Although we have extensively covered the 802.11 MAC header, it is beyond the scope of the CWNA exam to explain the purpose of every field and subfield of the 802.11 MAC header as well as the purpose of all the fixed fields and information elements used in the numerous types of 802.11 management, control, and data frames. For an in-depth look at the 802.11 frame format, we suggest you read *CWAP Certified Wireless Analysis Professional Official Study Guide: Exam PW0-270* (Sybex, 2011).

802.11 Frame Body

As you have already learned, there are three major 802.11 frame types: management, control, and data. It should be noted that not all three frame types carry the same type of payload in the frame body. As a matter of fact, control frames do not even have a body.

Another name for an 802.11 management frame is a *management MAC protocol data unit (MMPDU)*. Management frames have a MAC header, a frame body, and a trailer; however, management frames do not carry any upper-layer information. There is no MSDU encapsulated in the MMPDU frame body, which carries only layer 2 information fields and information elements. *Information fields* are fixed-length mandatory fields in the body of a management frame. *Information elements* are variable in length and are optional. One example of an information element would be the *RSN information element*, which contains information about the type of authentication and encryption being used within a BSS. The payload in an MMPDU frame body is not encrypted. We will discuss the different subtypes of 802.11 management frames later in this chapter.

Control frames are used to clear the channel, acquire the channel, and provide unicast frame acknowledgments. They contain only header information and a trailer. Control frames do not have a frame body. We will also discuss the different subtypes of 802.11 control frames later in this chapter.

Only 802.11 data frames carry an upper-layer MSDU payload in the frame body. When encryption is used, the MSDU payload is protected. Please note that certain subtypes of data frames, such as the null function frame, do not have a frame body. We will also discuss the various subtypes of 802.11 data frames later in this chapter.

802.11 Trailer

The main purpose of the 802.11 trailer is to carry data integrity check information for the entire frame. Found in every 802.11 trailer is the *frame check sequence (FCS)*, also known as the *FCS field*, which contains a 32-bit cyclic redundancy check (CRC) that is used to validate the integrity of received frames. As shown in Figure 9.15, the FCS is calculated over all the fields of the MAC header and the Frame Body field. These are referred to as the *calculation fields*.

FIGURE 9.15   Frame check sequence

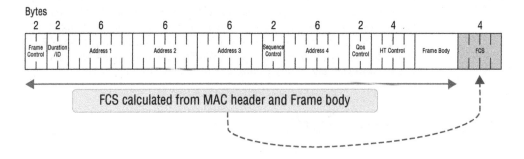

The FCS is calculated using the following standard generator polynomial of degree 32:

$$G(x) = x^{32} + x^{26} + x^{23} + x^{22} + x^{16} + x^{12} + x^{11} + x^{10} + x^8 + x^7 + x^5 + x^4 + x^2 + x + 1$$

You will need to fully understand this formula for the CWNA exam. (Just kidding!) What you will absolutely need to understand for the exam is what happens if the CRC fails or passes when a unicast frame is received by an 802.11 station. As mentioned earlier in this chapter, every time an 802.11 radio transmits a unicast frame, if the frame is received properly and the cyclic redundancy check (CRC) of the FCS passes, the 802.11 radio that received the frame will reply with an acknowledgment (ACK) frame. If the ACK is received, the original station knows that the frame transfer was successful. All unicast 802.11 frames must be acknowledged. Broadcast and multicast frames do not require an acknowledgment.

If any portion of a unicast frame is corrupted, the CRC will fail, and the receiving 802.11 radio will not send an ACK frame to the transmitting 802.11 radio. If an ACK frame is not received by the original transmitting radio, the unicast frame is not acknowledged and will have to be retransmitted.

802.11 State Machine

The 802.11-2016 standard defines four states of client connectivity. These four states are often referred to as the *802.11 state machine*. 802.11 management frame communications are used between a client station and an AP as a client transitions between the four states towards established layer 2 connectivity. These four states are as follows:

- **State 1:** Initial start state, unauthenticated and unassociated
- **State 2:** Authenticated, not associated
- **State 3:** Authenticated and associated (pending RSN authentication)
- **State 4:** Authenticated and associated

The purpose of the 802.11 state machine is for clients and an AP to be able to discover each other and establish a secure relationship, with the final goal of the client joining the basic service set (BSS). If no security is used, only three states are needed. In most cases, PSK or 802.1X/EAP authentication is required, and all four states will then occur.

Figure 9.16 illustrates the management frame exchanges that occur between these states. In the next section, we will discuss in great detail all the management frames that are used between a client station and an AP when joining or leaving a BSS.

FIGURE 9.16   802.11 state machine

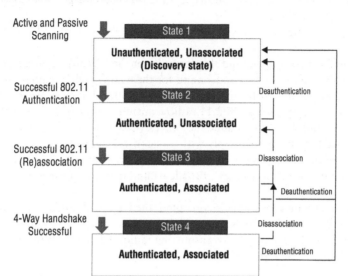

Management Frames

Within any BSS, a large percentage of the WLAN traffic consists of 802.11 *management frames*. Management frames are used by wireless stations to join and leave the basic service set (BSS). They are not necessary on wired networks, since physically connecting or disconnecting the network cable performs this function. However, because wireless networking is an unbounded medium, it is necessary for the wireless station to first find a compatible WLAN, then authenticate to the WLAN (assuming they are allowed to connect), and then associate with the WLAN (typically with an AP) to gain access to the wired network (the distribution system). In most cases, RSN security is also required.

Another name for an 802.11 management frame is *management MAC protocol data unit (MMPDU)*. Management frames do not carry any upper-layer information. There is no MSDU encapsulated in the MMPDU frame body, which carries only layer 2 information fields and information elements. Information fields are fixed-length fields in the body of a management frame. Information elements are variable in length.

Following is a list of all 14 management frame subtypes, as defined by the 802.11 standard and ratified amendments:

- Association request
- Association response
- Reassociation request
- Reassociation response

- Probe request
- Probe response
- Beacon
- Announcement traffic indication message (ATIM)
- Disassociation
- Authentication
- Deauthentication
- Action
- Action No ACK
- Timing advertisement

We will now discuss the most commonly used 802.11 management frames.

Beacon

One of the most important 802.11 frame types is the *beacon*, also referred to as the *beacon management frame*. Beacons are essentially the heartbeat of the wireless network. The AP of a basic service set broadcasts the beacons while the clients listen for the beacon frames. Client stations transmit beacons only when participating in an independent basic service set (IBSS), also known as ad hoc mode. Each beacon contains a time stamp, which client stations use to keep their clocks synchronized with the AP. Because so much of successful wireless communications is based on timing, it is imperative that all stations be in sync with each other. By performing Exercise 9.1, you will be able to inspect the contents of a beacon frame using a wireless packet analyzer. Table 9.2 includes a partial list of the information that can be found inside the body of a beacon frame.

TABLE 9.2 Beacon frame contents

Information Type	Description
Time Stamp	Synchronization information
Spread Spectrum Parameter Sets	FHSS-, DSSS-, HR-DSSS-, ERP-, OFDM-, HT-, or VHT-specific information
SSID	Logical WLAN name
Data Rates	Basic and supported rates
Service Set capabilities	Extra BSS or IBSS parameters
Channel information	Channel used by the AP or IBSS

TABLE 9.2 Beacon frame contents *(continued)*

Information Type	Description
Traffic Indication Map (TIM)	A field used during the power-save process
BSS Load	A field defined by 802.11e that is a good indicator of channel utilization
QoS capabilities	Quality-of-service and Enhanced Distributed Channel Access (EDCA) information
Robust Security Network (RSN) capabilities	TKIP or CCMP cipher information and authentication method
HT and VHT capabilities	802.11n and 802.11ac capabilities
Vendor Proprietary information	Vendor-unique or vendor-specific information

The beacon frame contains all the necessary information for a client station to learn about the parameters of the basic service set before joining the BSS. Beacons are transmitted at a targeted time of every 102.4 milliseconds, which means an AP transmits the beacon about 10 times per second. This interval can be configured on APs, but it cannot be disabled. Some WLAN design guides recommend raising the *beacon interval* as a means of reducing overhead. In most cases, raising the beacon interval is a very bad idea because it may negatively impact client connectivity. The AP uses the beacon frame to advertise to the client stations all the configured capabilities of any BSS that the AP provides.

If an AP has been configured for multiple SSIDs, the AP will transmit beacon frames for each SSID. The overhead consequences of transmitting multiple beacons will be discussed in great detail in Chapter 13, "WLAN Design Concepts."

EXERCISE 9.1

Viewing Beacon Frames

1. To perform this exercise, you need to first download the CWNA-CH9.PCAPNG file from the book's online resource area, which can be accessed at www.wiley.com/go/cwnasg.

2. After the file is downloaded, you will need packet analysis software to open the file. If you do not already have a packet analyzer installed on your computer, you can download Wireshark from www.wireshark.org.

3. Using the packet analyzer, open the CWNA-CH9.PCAPNG file. Most packet analyzers display a list of capture frames in the upper section of the screen, with each frame numbered sequentially in the first column.

4. Click one of the first 10 frames. All of these frames are beacon frames.

5. After selecting one of the beacon frames, in the lower section of the screen, browse through the information found inside the beacon frame body. You can expand a section by clicking on the plus sign next to the section.

Passive Scanning

In order for a station to be able to connect to an AP, it must first discover an AP. A station discovers an AP by either listening for an AP (passive scanning) or searching for an AP (active scanning). Client stations exist in state 1 of the 802.11 state machine during these discovery phases. In *passive scanning*, the client station listens for the beacon frames that are continuously being sent by the APs, as shown in Figure 9.17. As mentioned previously, beacons are sent at a targeted time of every 102.4 milliseconds. In busy WLAN environments, the exact transmission time will vary slightly due to medium contention of all stations in the BSS, including the AP.

FIGURE 9.17 Passive scanning

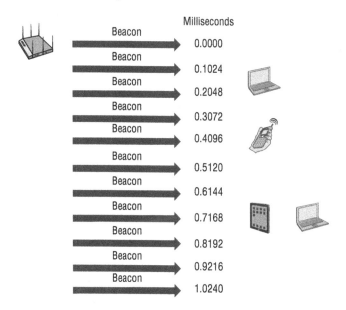

The client station will listen for the beacons that contain the same SSID that has been preconfigured in the client station's software utility. Passive scanning provides an initial means for a client to learn about all the BSS capabilities that the AP supports. When the station hears a beacon, it can then attempt to connect to that WLAN using subsequent management frames. If the client station hears beacons from multiple APs with the same SSID, it will determine which AP has the best signal, and it will attempt to connect to that AP.

Also, a station can use either or both methods of scanning to discover existing WLANs. When an independent basic service set is deployed, all the stations in ad hoc mode take turns transmitting the beacons since there is no AP. Passive scanning occurs in an ad hoc environment, just as it does in a basic service set.

Active Scanning

Discovering a WLAN by scanning all possible channels and listening to beacons is not an efficient method for a client to find all APs on all channels. To enhance this discovery process, client stations also use what is called *active scanning*. In addition to passively scanning for APs, client stations will actively scan for them. In active scanning, the client station transmits management frames known as *probe requests*. The probe request frame also contains information about the client station capabilities that can initially be shared with an AP. Some of the client information found in the probe request frame includes supported data rates, HT/VHT capabilities, SSID parameters, and more.

These probe requests can contain the SSID of the specific WLAN that the client station is looking for or can look for any SSID. A client station that is looking for any available SSID sends a probe request with the SSID field set to null. A probe request with specific SSID information is known as a *directed probe request*. A probe request without SSID information is known as a *null probe request*. Another term sometimes used for null probe requests is wildcard SSID.

If a directed probe request is sent, all APs that support that specific SSID and hear the request should reply by sending a *probe response*. The information that is contained inside the body of a probe response frame is the same information that can be found in a beacon frame, with the exception of the traffic indication map (TIM). Just like the beacon frame, the probe response frame contains all the necessary information for a client station to learn about the parameters of the basic service set before joining the BSS.

If a null probe request is sent, all APs that hear the request will reply by sending a probe response. As shown in Figure 9.18, a client sends a null probe request on channel 36, and three APs with three different SSIDs will all respond.

FIGURE 9.18   Active scanning—null probe request

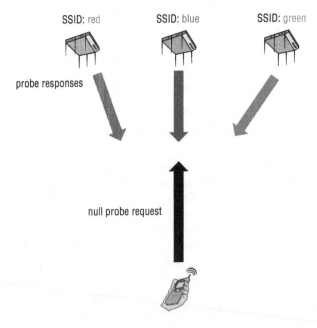

If a directed probe request is sent, only APs configured with the same SSID will reply. As depicted in Figure 9.19, a client sends a directed probe request with an SSID of blue, and the only AP that responds is the AP that also supports the blue SSID.

FIGURE 9.19   Active scanning—directed probe request

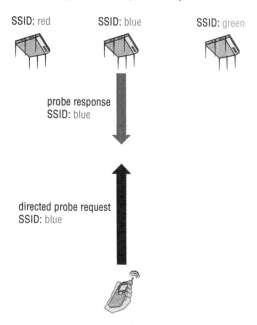

SSID: red   SSID: blue   SSID: green

probe response
SSID: blue

directed probe request
SSID: blue

One drawback to passive scanning is that beacon management frames are broadcast only on the same channel as the AP. In contrast, active scanning uses probe request frames that are sent out across all available channels by the client station. If a client station receives probe responses from multiple APs, signal strength and quality characteristics are typically used by the client station to determine which AP has the best signal and thus which AP to connect to. As shown in Figure 9.20, a client station will sequentially send probe requests on each of the supported channels. In fact, it is common for a client station that is already associated to an AP and transmitting data to go off-channel and continue to send probe requests every few seconds across other channels. The main purpose of off-channel probing is so that a client station can find other APs to potentially roam to. By continuing to actively scan and send probe requests across multiple channels, a client station can maintain and update a list of known APs. If a client station needs to roam, it can typically do so faster and more efficiently.

How often a client station goes off-channel for active scanning purposes is proprietary and depends on the client device drivers. For example, an 802.11 radio in a mobile device, such as a smartphone or tablet, will probably send probe requests across all channels more frequently than an 802.11 radio in a laptop. Some client devices have the ability to adjust the probing rate. By performing Exercise 9.2, you will be able to look at probe request and probe response frames.

FIGURE 9.20   Probe requests—multiple channels

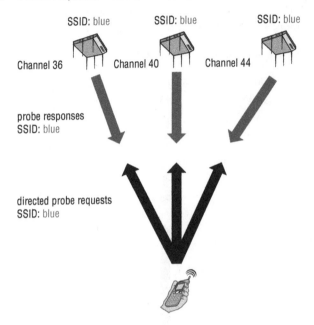

Understanding Probe Requests and Probe Responses

1. To perform this exercise, you need to first download the CWNA-CH9.PCAPNG file from the book's online resource area, which can be accessed at www.wiley.com/go/cwnasg.

2. After the file is downloaded, you will need packet analysis software to open the file. If you do not already have a packet analyzer installed on your computer, you can download Wireshark from www.wireshark.org.

3. Using the packet analyzer, open the CWNA-CH9.PCAPNG file. Most packet analyzers display a list of capture frames in the upper section of the screen, with each frame numbered sequentially in the first column.

4. Scroll down the list of frames and click frame #13684, which is a probe request.

5. In the lower section of the screen, look at the SSID field in the frame body and notice that this is a directed probe request.

6. Click frame #13685, which is a probe response.

7. In the lower section of the screen, browse through the information found inside the frame body and notice that the information is similar to a beacon frame.

8. Click frame #429, which is a probe request. Look at the SSID field in the frame body and notice that this is a null probe request, since it does not contain an SSID value.

9. Click frames #430, #432, #434, #436, and #438. Notice that there are five probe responses to the null probe request. Each probe response has a different SSID.

Authentication

Authentication is the first of two steps required to connect to the 802.11 basic service set. Both authentication and association must occur, in that order, before an 802.11 client can pass traffic through the AP to another device on the network.

Authentication is a process that is often misunderstood. When many people think of authentication, they think of what is commonly referred to as network authentication—entering a username and password in order to get access to the network. In this chapter, we are referring to 802.11 authentication. When an 802.3 device needs to communicate with other devices, the first step is to plug the Ethernet cable into the wall jack. When this cable is plugged in, the client creates a physical link to the wired switch and is now able to start transmitting frames. When an 802.11 device needs to communicate, it must first authenticate with the AP, or with the other stations if it is configured for ad hoc mode. This authentication is not much more of a task than plugging the Ethernet cable into the wall jack. The 802.11 authentication merely establishes an initial connection between the client and the AP. Think of this as authenticating that both of the devices are valid 802.11 devices.

Once a client station has discovered an AP by either active or passive scanning, the client station will use 802.11 authentication management frames to proceed to state 2 of the 802.11 state machines.

The original 802.11 standard defined two different methods of authentication: Open System authentication and Shared Key authentication. Shared Key authentication uses Wired Equivalent Privacy (WEP) to authenticate client stations and requires that a static WEP key be configured on both the station and the access point. Because WEP is an outdated security method, Shared Key authentication is simply not used anymore. Shared Key authentication is briefly described in Chapter 17, "802.11 Network Security Architecture."

Open System Authentication

Open System authentication provides authentication without performing any type of client verification. It is essentially an exchange of hellos between the client and the AP. It is considered a null authentication because no exchange or verification of identity takes place between the devices. Open System authentication occurs with an exchange of frames between the client and the AP, as shown in Exercise 9.3. Once a client station has exchanged authentication management frames with an AP, the client has moved to state 2 of the 802.11 state machine.

Because of its simplicity, Open System authentication is also used in conjunction with more advanced network security authentication methods, such as PSK authentication and 802.1X/EAP.

Using Open System Authentication

1. To perform this exercise, you need to first download the CWNA-CH9.PCAPNG file from the book's online resource area, which can be accessed at www.wiley.com/go/cwnasg.

2. After the file is downloaded, you will need packet analysis software to open the file. If you do not already have a packet analyzer installed on your computer, you can download Wireshark from www.wireshark.org.

3. Using the packet analyzer, open the CWNA-CH9.PCAPNG file. Most packet analyzers display a list of capture frames in the upper section of the screen, with each frame numbered sequentially in the first column.

4. Scroll down the list of frames and click frame #871, which is an authentication request.

5. In the lower section of the screen, look at the 802.11 MAC header and note the source address and destination address.

6. Click frame #873, which is an authentication response. Look at the 802.11 MAC header and note that the source address is the AP's BSSID and that the destination address is the MAC address of the client that sent the authentication request. Look at the frame body and note that authentication was successful.

Association

After the station has authenticated with the AP, the next step is for it to associate with the AP. When a client station associates, it becomes a member of a basic service set (BSS). *Association* means that the client station has established layer 2 connectivity with the AP and joined the BSS. The client station sends an association request management frame to the AP, seeking permission to join the BSS. The AP sends an association response management frame to the client, either granting or denying permission to join the BSS. These frames are used by the client station to move towards state 3 of the 802.11 state machine. In the body of the association response frame is an association identifier (AID), a unique association number given to every associated client. You will learn later in this chapter that the AID is used during power management. In Exercise 9.4, you will see that the association request and response frames are also used as a final capabilities notification between the AP and the client station.

Understanding Association

1. To perform this exercise, you need to first download the CWNA-CH9.PCAPNG file from the book's online resource area, which can be accessed at www.wiley.com/go/cwnasg.

2. After the file is downloaded, you will need packet analysis software to open the file. If you do not already have a packet analyzer installed on your computer, you can download Wireshark from www.wireshark.org.

3. Using the packet analyzer, open the CWNA-CH9.PCAPNG file. Most packet analyzers display a list of capture frames in the upper section of the screen, with each frame numbered sequentially in the first column.

4. Scroll down the list of frames and click frame #875, which is an association request. Look at the frame body.

5. Click frame #877, which is the association response. Look at the frame body and note that the association was successful and that the client received an AID number.

If no RSN security is used, once a client station has completed association, the device has reached state 3 of the 802.11 state machine and has joined the BSS. At this point, the client station can move beyond layer 2, request an IP address, and begin upper-layer communications. Figure 9.21 illustrates all the frame exchanges needed between a client station and an AP for the client station to reach state 3 and join the BSS.

FIGURE 9.21 Joining the BSS

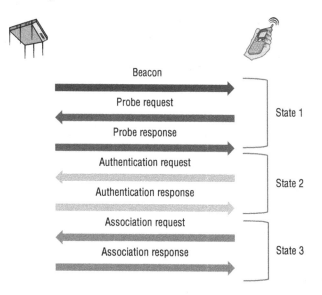

What about state 4 of the 802.11 state machine? If PSK authentication or 802.1X/ EAP is configured on the AP, the client station has still not joined the BSS. The client is associated; however, RSN authentication is pending. If PSK authentication is used, the client and AP must have a matching WPA2 passphrase. Additionally, another frame exchange, called the 4-Way Handshake, must also occur to create dynamic encryption keys for both radios.

If 802.1X/EAP authentication is used, a series of EAP authentication frames will be exchanged between the client and a RADIUS server to validate the client's security credentials. The 4-Way Handshake exchange also occurs after 802.1X/EAP to create dynamic encryption keys for both radios.

Once either RSN authentication method has completed and after the 4-Way Handshake exchange has created the encryption keys, the client station has reached state 4 of the 802.11 state machine and is a member of the BSS. At this point, the client station can move beyond layer 2, request an IP address, and begin upper-layer communications. A detailed explanation about the 4-Way Handshake can be found in Chapter 17.

Basic and Supported Rates

As you have learned in earlier chapters, the 802.11-2016 standard defines supported rates for various RF technologies. For example, HR-DSSS (802.11b) radios are capable of supporting data rates of 1, 2, 5.5, and 11 Mbps. ERP (802.11g) radios are capable of supporting the HR-DSSS data rates but are also capable of supporting ERP-OFDM rates of 6, 9, 12, 18, 24, 36, 48, and 54 Mbps.

Specific data rates can be configured for any AP as *required rates*. The 802.11-2016 standard defines required rates as *basic rates*. Please understand that an AP will transmit all management frames at the lowest configured basic rate. Data frames can be transmitted at much higher supported data rates.

In order for a client station to successfully associate with an AP, the station must be capable of communicating by using the configured basic rates that the AP requires. If the client station is not capable of communicating with all of the basic rates, the client station will not be able to associate with the AP and will not be allowed to join the BSS.

In addition to the basic rates, the AP defines a set of supported rates. This set of supported rates is advertised by the AP in the beacon frame and is also in some of the other management frames. The supported rates are data rates that the AP offers to a client station, but the client station does not have to support all of them.

Reassociation

When a client station decides to roam to a new AP, it will send a *reassociation* request frame to the new AP. Reassociation frames are used by a client station to transition from an original BSS to a new BSS. A reassociation request frame is effectively a roaming request sent from a client station to a target AP.

Reassociation occurs after the client and the AP move through the following steps:

1. In the first step, the client station sends a reassociation request frame to the new AP.

   As shown in Figure 9.22, the reassociation request frame includes the BSSID (MAC address) of the AP's radio it is currently connected to. (We will refer to this as the original AP.)

FIGURE 9.22 Reassociation process

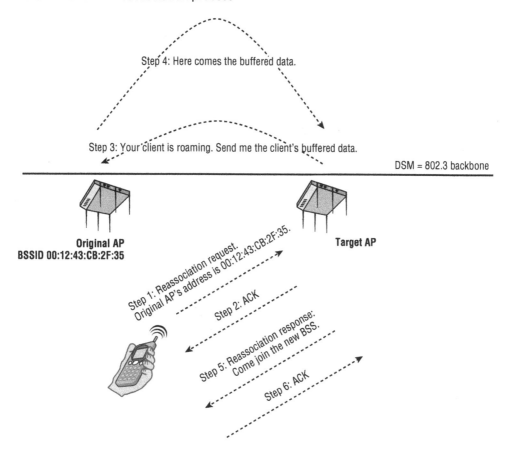

2. The new AP then replies to the station with an ACK.

3. The new AP attempts to communicate with the original AP by using the distribution system medium (DSM).

The new AP attempts to notify the original AP about the roaming client and requests that the original AP forward any buffered data. Please remember that any communications between APs via the DSM are not defined by the 802.11-2016 standard and are proprietary. In a controller-based WLAN solution, the inter-AP communications might occur within the controller. In a non-controller architecture, APs will communicate with each other at the edge of the network.

4. If this communication is successful, the original AP will use the distribution system medium to forward any buffered data to the new AP.

5. The new AP sends a reassociation response frame to the roaming client via the wireless medium.

6. The client sends an ACK to the new AP, confirming that it received the reassociation response and intends to roam. In most cases, WPA2 security will be in place and there will be one final step that is not depicted in Figure 9.22. The client and target AP will proceed with a 4-Way Handshake frame exchange to generate unique encryption keys between the two radios.

If the reassociation is not successful, the client will retain its connection to the original AP and either continue to communicate with it or attempt to roam to another AP. In Exercise 9.5, you can look at the reassociation request and reassociation response frames.

EXERCISE 9.5

Understanding Reassociation

1. To perform this exercise, you need to first download the CWNA-CH9.PCAPNG file from the book's online resource area, which can be accessed at www.wiley.com/go/cwnasg.

2. After the file is downloaded, you will need packet analysis software to open the file. If you do not already have a packet analyzer installed on your computer, you can download Wireshark from www.wireshark.org.

3. Using the packet analyzer, open the CWNA-CH9.PCAPNG file. Most packet analyzers display a list of capture frames in the upper section of the screen, with each frame numbered sequentially in the first column.

4. Scroll down the list of frames and click frame #7626, which is a reassociation request. Look at the frame body and note the address of the current AP.

5. Click frame #7628, which is the reassociation response. Look at the frame body and note that the reassociation was successful and that the client received an AID number.

Disassociation

Disassociation is a notification, not a request. If a station wants to disassociate from an AP, or an AP wants to disassociate from stations, either device can send a disassociation frame. This is a polite way of terminating the association. A client will do so when you shut down the operating system. An AP might do so if it is being disconnected from the network for maintenance. A disassociation frame sent by an AP sends a client from state 3 or 4 back to state 2 of the 802.11 state machine. Every disassociation frame carries a reason code as to why disassociation is occurring. For example, an AP might send a disassociation frame with reason code 4 to a client that was been inactive. All the possible reason codes can be found in section 9.4.17 of the 802.11-2016 standard.

Deauthentication

Like disassociation, a *deauthentication* frame is a notification, not a request. If a station wants to deauthenticate from an AP, or if an AP wants to deauthenticate from stations, either device can send a deauthentication frame. Because authentication is a prerequisite for association, a deauthentication frame will automatically cause a disassociation to occur.

A deauthentication frame sent by an AP sends a client from states 2, 3, or 4 all the way back to state 1 of the 802.11 state machine. Deauthentication frames effectively force a client station to start over in order to find and join a BSS. Every deauthentication frame carries a reason code as to why deauthentication is occurring. For example, an AP might send a deauthentication frame with reason code 23 to a client that fails 802.1X/EAP authentication and force the client back to state 1. All the possible reason codes can be found in section 9.4.17 of the 802.11-2016 standard.

Action Frame

An *action* frame is a type of management frame used to trigger specific actions in a BSS. Action frames can be sent by access points or client stations. The action frame provides information and direction for what to do. Action frames were first introduced in 802.11h because the subtype for management frames had been exhausted. An action frame is sometimes referred to as a "management frame that can do anything." As new 802.11 technologies evolve, there is a need for new management frames to carry information and trigger specific actions. Instead of creating new management frames, an action frame can get the job done. Figure 9.23 shows the action frame structure.

FIGURE 9.23   Action frame structure

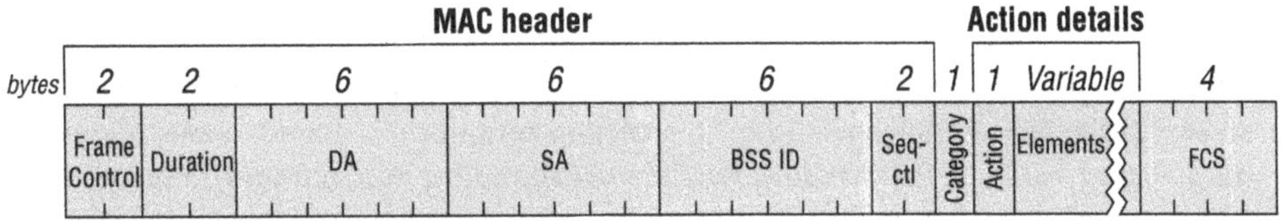

The action frame body contains the following three sections:

- **Category:** Describes the action frame type. Category allows you to know which family the action frame belongs to and which protocol introduced it

- **Action:** The action to perform. It is usually a number. You need to know the category to understand which action is called.

- **Elements:** Adds additional information specific to the action

A complete list of all the current action frames can be found in section 9.6 of the 802.11-2016 standard. One example of how action frames are used is as a *channel switch announcement (CSA)* from an AP transmitting on *a dynamic frequency selection (DFS)* channel. If radar is detected on the current DFS frequency, the AP will inform all associated client stations to move to another channel. Action frames are also used as *transmit power control (TPC)* request and report frames. An AP can tell associated client stations that also support TPC to adjust their transmit power levels to match the AP power levels. A deeper discussion of DFS channels and TPC mechanisms can be found in Chapter 13.

Another example of an action frame is *neighbor report* requests and responses that 802.11k–compliant radios can use. As shown in Exercise 9.6, client stations use neighbor report information to gain information from the associated AP about potential roaming neighbors. As defined by the 802.11k-2008 amendment, the neighbor report information assists the fast roaming process by providing a method for the client to request the associated AP to measure and report about neighboring APs available within the same mobility domain. This can speed up the client scanning process by informing the client device of nearby APs to which it may roam. The neighbor report information is typically delivered through a request/report frame exchange inside 802.11 action frames.

EXERCISE 9.6

Viewing Action Frames

1. To perform this exercise, you need to first download the ACTION.PCAPNG file from the book's online resource area, which can be accessed at www.wiley.com/go/cwnasg.

2. After the file is downloaded, you will need packet analysis software to open the file. If you do not already have a packet analyzer installed on your computer, you can download Wireshark from www.wireshark.org.

3. Using the packet analyzer, open the ACTION.PCAPNG file. Most packet analyzers display a list of capture frames in the upper section of the screen, with each frame numbered sequentially in the first column.

4. Scroll down the list of frames and click packet #103, which is an 802.11 action frame transmitted by an Apple iOS client device. Typically, in the lower section of the screen is the packet details window. This section contains annotated details about the selected frame. In this window, locate and expand the action frame. In the body of the action frame, expand the tagged parameters and notice that this action frame is being used as a neighbor report request. The client is asking the AP if the AP has information about any neighboring APs.

5. Click packet #105, which is an 802.11 action frame transmitted by the AP. In the body of the action frame, expand the fixed parameters and notice that this action frame is being used as a neighbor report response. In the body of the action frame, expand the tagged parameters and view the neighbor report about the AP with a BSSID of 08:ea:44:76:b5:68, which is transmitting on channel 48.

Control Frames

802.11 *control frames* assist with the delivery of the data frames and are transmitted at one of the basic rates. Control frames are also used to clear the channel, acquire the channel, and provide unicast frame acknowledgments. As previously mentioned, control frames have only a MAC header and a trailer; they do not have a frame body. Information found in the MAC header is sufficient for accomplishing the tasks defined for 802.11 control frames.

Following is a list of all 12 control frame subtypes, as currently defined by the 802.11-2016 standard:

- Beamforming report poll
- VHT NDP announcement
- Control frame extension
- Control wrapper
- Block ACK request (BAR)
- Block ACK (BlockAck)
- Power save-poll (PS-Poll)
- Request-to-send (RTS)
- Clear-to-send (CTS)
- Acknowledgment (ACK)
- Contention Free-End (CF-End)
- CF-End + CF-ACK

We will now discuss the most commonly used 802.11 control frames.

ACK Frame

The *ACK frame* is one of the 12 control frames and one of the key components of the 802.11 CSMA/CA medium access control method. Since 802.11 is a wireless medium that cannot guarantee successful data transmission, the only way for a station to know that a frame it transmitted was properly received is for the receiving station to notify the transmitting station. This notification is performed using an ACK.

The ACK is a simple frame consisting of 14 octets of information, as depicted in Figure 9.24. When a station receives a unicast frame, it waits for a short period of time, known as a *short interframe space (SIFS)*. The receiving station copies the MAC address of the transmitting station from the data frame and places it in the Receiver Address (RA) field of the ACK frame. As you will see in Exercise 9.7, the receiving station then replies by transmitting the ACK. If all goes well, the station that sent the unicast frame receives the ACK with its MAC address in the RA field and now knows that the frame was received and was not corrupted. The delivery of every unicast frame must be verified; otherwise, a retransmission must take place. The ACK frame is a very important control frame because it is used for delivery verification of all 802.11 unicast frames. If a collision occurs or if any portion of a unicast frame is corrupted, the cyclic redundancy check (CRC) will fail, and the receiving 802.11 radio will not return an ACK frame to the transmitting 802.11 radio.

FIGURE 9.24   ACK control frame

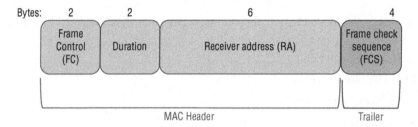

Every unicast frame must be followed by an ACK frame. If for any reason the unicast frame is corrupted, the 32-bit CRC known as the frame check sequence (FCS) will fail and the receiving station will not send an ACK. If a unicast frame is not followed by an ACK, it is retransmitted. With a few rare exceptions, broadcast and multicast frames do not require acknowledgment.

EXERCISE 9.7

Understanding Acknowledgment

1. To perform this exercise, you need to first download the CWNA-CH9.PCAPNG file from the book's online resource area, which can be accessed at www.wiley.com/go/cwnasg.

2. After the file is downloaded, you will need packet analysis software to open the file. If you do not already have a packet analyzer installed on your computer, you can download Wireshark from www.wireshark.org.

3. Using the packet analyzer, open the CWNA-CH9.PCAPNG file. Most packet analyzers display a list of capture frames in the upper section of the screen, with each frame numbered sequentially in the first column.

4. Scroll down the list of frames and click frame #29073, which is a data frame.

5. Click frame #29074, which is an ACK frame.

6. Observe the frame exchanges between frame #29073 and frame #29088. Notice that all the unicast frames are being acknowledged by the receiving station.

Block Acknowledgment

The 802.11e amendment introduced a *Block acknowledgment (BA)* mechanism that is now defined in the 802.11-2016 standard. A Block ACK improves channel efficiency by aggregating several acknowledgments into one single acknowledgment frame. There are two types of Block ACK mechanisms: immediate and delayed.

- The immediate Block ACK is designed for use with low-latency traffic.
- The delayed Block ACK is more suitable for latency-tolerant traffic.

For the purposes of this book, we will discuss only the immediate Block ACK. As pictured in Figure 9.25, an originator station sends a block of QoS data frames to a recipient station. The originator requests acknowledgment of all the QoS data frames by sending a Block ACK request (BAR) frame. Instead of acknowledging each unicast frame independently, the block of QoS data frames are all acknowledged by a single Block ACK. A bitmap in the Block ACK frame is used to indicate the status of all the received data frames. If only one of the frames is corrupted, only that frame will need to be retransmitted. The use of a Block ACK instead of a traditional ACK is a more efficient method that cuts down on medium contention overhead. Block ACKs were initially defined to be used with a "frame burst," as shown in Figure 9.25. However, Block ACKs are more commonly used with A-MPDU frame aggregation. Please see Chapter 10 for more details.

FIGURE 9.25   Immediate Block ACK

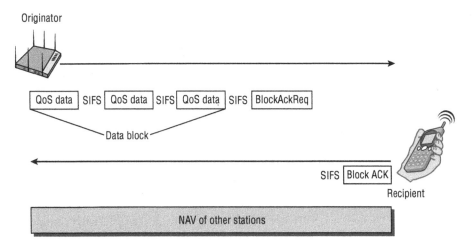

PS-Poll

When legacy power management is in use, the *PS-Poll* frame is an 802.11 control frame used by client stations to request that an AP send buffered traffic for the client station. Clients using legacy power management will send a PS-Poll frame to the access point to request that the AP send the buffered unicast frame to the station. Inside the PS-Poll frame, the Duration/ID field is used as an association ID (AID) value. In other words, the station will identify itself to the AP and request the buffered unicast frame. The Duration/ID field is now used strictly as an identifier and is not being used for duration or resetting NAV timers. Only power-save poll (PS-Poll) frames use this field as the AID. The legacy power-management process using PS-Poll frames is discussed in greater detail later in this chapter.

RTS/CTS

In order for a client station to participate in a BSS, it must be able to communicate with the AP. This is straightforward and logical; however, it is possible for the client station to be able to communicate with the AP but not be able to hear or be heard by any of the other client stations. This can be a problem because, as you may recall, a station performs collision avoidance by setting its NAV when it hears another station transmitting (virtual carrier sense) and by listening for RF (physical carrier sense). If a station cannot hear the other stations, or if it cannot be heard by the other stations, there is a greater likelihood that a collision will occur. *Request-to-send/clear-to-send (RTS/CTS)* is a mechanism that performs a NAV distribution and helps prevent collisions from occurring. This NAV distribution reserves the medium prior to the transmission of the data frame.

Let us look at the RTS/CTS from a slightly more technical perspective. This will be a basic explanation, as an in-depth explanation is beyond the scope of the exam. When a station uses RTS/CTS, every time the station wants to transmit a frame, it must perform an RTS/CTS exchange prior to the normal data transmissions. When the transmitting station goes to transmit a data frame, it first sends an RTS frame. The duration value of the RTS frame resets the NAV timers of all listening stations so that they must wait until the CTS, Data, and ACK frames have been transmitted. The receiving station, the AP, then sends a CTS, which is also used for NAV distribution. The duration value of the CTS frame resets the NAV timer of all listening stations so that they must wait until the Data and ACK frames have been transmitted.

As you can see in Figure 9.26, the duration value of the RTS frame represents the time, in microseconds, that is required to transmit the CTS/Data/ACK exchange plus three SIFS intervals. The duration value of the CTS frame represents the time, in microseconds, that is required to transmit the Data/ACK exchange plus two SIFS intervals. If any station did not hear the RTS, it should hear the CTS. When a station hears either the RTS or the CTS, it will set its NAV to the value provided. At this point, all stations in the BSS should have their NAV set, and the stations should wait until the entire data exchange is complete. Figure 9.27 depicts an RTS/CTS exchange between a client station and an AP.

RTS/CTS is used primarily in two situations. It can be used when a hidden node exists (this is covered in Chapter 15, "WLAN Troubleshooting"), or it can be used automatically as a protection mechanism when different technologies, such as 802.11b/g/n, coexist in the same basic service set. Figure 9.27 depicts the RTS/CTS frame exchange.

CTS-to-Self

CTS-to-Self is also used automatically as a protection mechanism when different technologies, such as 802.11b/g/n, coexist in the same basic service set. One of the benefits of using CTS-to-Self over RTS/CTS as a protection mechanism is that the throughput will be higher because fewer frames are being sent.

When a station using CTS-to-Self wants to transmit data, it performs a NAV distribution by sending a CTS frame. This CTS notifies all other stations that they must wait until the Data/ACK frame exchange has completed. Any station that hears the CTS will set their NAV to the value provided.

FIGURE 9.26   RTS/CTS duration values

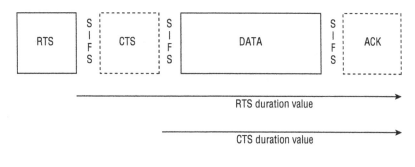

RTS duration value

CTS duration value

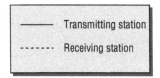

Transmitting station

------ Receiving station

FIGURE 9.27   RTS/CTS frame exchange

RTS duration = CTS/Data/ACK exchange
CTS duration = Data/ACK exchange
DATA duration = ACK
ACK duration = 0 (exchange is over)

Station C

Station 3 does not hear the RTS but does hear the CTS and resets the NAV timer for the Data/ACK exchange.

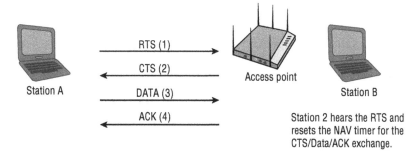

Station A

RTS (1)

CTS (2)

DATA (3)

ACK (4)

Access point

Station B

Station 2 hears the RTS and resets the NAV timer for the CTS/Data/ACK exchange.

CTS-to-Self is better suited for use by an AP as opposed to client stations. It is important to make sure that all stations hear the CTS to reserve the medium, which is most likely to occur if it is being sent by an AP. If a client station were to use CTS-to-Self, there is a chance that another client station on the opposite side of the BSS might be too far away from the CTS-to-Self and would not realize that the medium is busy. Even though this is true, from our experience, it appears that most client stations use CTS-to-Self to reserve the medium instead of RTS/CTS. CTS-to-Self is used because of the decreased overhead when compared with RTS/CTS. Some vendors allow the user to select whether the client station uses RTS/CTS or CTS-to-Self when in protected mode.

Protection Mechanisms

When 802.11g technology debuted in 2006, the 802.11 standard had to provide for a way for both DSSS and OFDM technologies to coexist in the same 2.4 GHz RF environment. The technical term for 802.11g technology is *Extended Rate Physical (ERP)*. An ERP (802.11g) radio can effectively communicate using either OFDM or HR-DSSS transmissions. However, older 802.11 or 802.11b radios only communicate using DSSS or HR-DSSS transmissions. An ERP protection mechanism is defined so that DSSS or HR-DSSS transmissions do not happen when two 802.11g radios are communicating using OFDM.

A good analogy would be verbal languages. Think of an 802.11g radio as speaking both English and Spanish, whereas an 802.11b radio can speak only English. An ERP protection mechanism is defined so that when Spanish is spoken between 802.11g radios, the 802.11b radios will not interrupt the conversation because the 802.11b radios speak only English. The protection mechanism uses RTS/CTS or CTS-to-Self.

In Chapter 8, you learned that one of the ways of preventing collisions is for the stations to set a countdown timer known as the network allocation vector (NAV). This notification is known as *NAV distribution*. NAV distribution is done through the Duration/ID field that is part of the data frame. When a station transmits a data frame, the listening stations use the Duration/ID field to set their NAV timers. Unfortunately, this is not inherently possible in a mixed environment. If an 802.11g device were to transmit a data frame, 802.11b devices would not be able to interpret the data frame or the Duration/ID value because the 802.11b HR-DSSS devices are not capable of understanding 802.11g ERP-OFDM transmissions. The 802.11b devices would not set their NAV timers and could incorrectly believe that the medium is available. To prevent this from happening, the 802.11g ERP stations switch into what is known as *protected mode*.

As shown in Figure 9.28 and Figure 9.29, when an 802.11g device wants to transmit data, it will first perform a NAV distribution by transmitting a *request-to-send/clear-to-send (RTS/CTS)* exchange with the AP or by transmitting a CTS-to-Self using a data rate and modulation method that the 802.11b HR-DSSS stations can understand. The RTS/CTS or CTS-to-Self will hopefully be heard and understood by all of the 802.11b and 802.11g stations. The RTS/CTS or CTS-to-Self will contain a Duration/ID value that all the listening stations will use to set their NAV timers. To put it simply, using a language that all stations can understand, the ERP (802.11g) device notifies all the stations to reset their

NAV values. After the RTS/CTS or CTS-to-Self has been used to reserve the medium, the 802.11g station can transmit a data frame by using OFDM modulation without worrying about collisions with 802.11b HR-DSSS or legacy 802.11 DSSS stations.

FIGURE 9.28   Protection mechanism—RTS/CTS

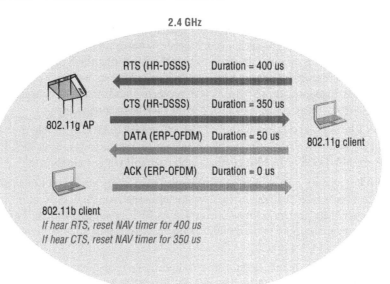

FIGURE 9.29   Protection mechanism—CTS-to-Self

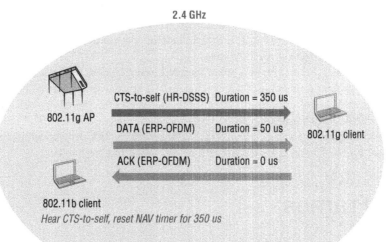

Within an ERP basic service set, the HR-DSSS (802.11b) and legacy 802.11 DSSS stations are known as *non-ERP stations*. The purpose of the protection mechanism is so that ERP stations (802.11g) can coexist with non-ERP stations (802.11b and 802.11 legacy) within the same BSS. This allows the ERP stations to use the higher ERP-OFDM data rates to transmit and receive data yet still maintain backward compatibility with the older legacy non-ERP stations.

So what exactly triggers the ERP protection mechanism? When an ERP (802.11g) AP decides to enable the use of a protection mechanism, it needs to notify all the ERP (802.11g) stations in the BSS that protection is required. It accomplishes this by setting the NonERP_Present bit in beacons and probe response frames, which will notify the ERP stations that protected mode is required. There is an assortment of reasons why protected mode may be enabled. The following are three scenarios that can trigger protection in an ERP basic service set:

- If a non-ERP STA associates with an ERP AP, the ERP AP will enable the NonERP_Present bit in its own beacons, enabling protection mechanisms in its BSS. In other words, an HR-DSSS (802.11b) client association will trigger protection.

- If an ERP AP hears a beacon from an AP where the supported data rates contain only 802.11b or 802.11 DSSS rates, it will enable the NonERP_Present bit in its own beacons, enabling protection mechanisms in its BSS. In simpler terms, if an 802.11g AP hears a beacon frame from an 802.11 or 802.11b AP or ad hoc client, a protection mechanism will be triggered.

- If an ERP AP hears a management frame (other than a probe request) where the supported rate includes only 802.11 or 802.11b rates, the NonERP_Present bit may be set to 1.

To summarize: The ERP protection mechanism is used so that legacy 802.11b and 802.11g stations can coexist in the same BSS by using either the RTS/CTS or CTS-to-Self protection mechanism. In Chapter 13, "WLAN Design Concepts," you will learn that eliminating legacy 802.11b clients from any enterprise WLAN is a highly recommended practice. Disabling the 802.11b data rates on an AP will reduce airtime consumption and increase performance of the WLAN. If 802.11b clients do not belong to the BSS, there is no need for RTS/CTS protection mechanisms.

If eliminating 802.11b clients is a recommended best practice, why does this chapter cover the ERP protection mechanism in such detail? The answer is that the ERP protection mechanism is also the foundation for coexistence between 802.11n/ac devices and earlier legacy devices. RTS/CTS and CTS-to-Self are once again needed when 802.11n/ac stations are communicating in the same BSS as 802.11a/b/g stations. HT protection mechanisms that also use RTS/CTS and CTS-to-Self will be discussed in Chapter 10.

Data Frames

Most 802.11 *data frames* carry the actual data that is passed down from the higher-layer protocols. The layer 3–7 MSDU payload is normally encrypted for data privacy reasons. However, some 802.11 data frames carry no MSDU payload at all but do have a specific

MAC control purpose within a BSS. Any data frames that do not carry an MSDU payload are not encrypted, because a layer 3–7 data payload does not exist.

There are a total of 15 data frame subtypes. The two most common data frames are the *data* subtype (usually referred to as the *simple data frame)* and the *QoS data* subtype. The difference between the two is that QoS data frames carry class of service information in the QoS Control field. Simple data frames are sometimes also referred to as *non-QoS data frames.*

Both data frames have MSDU upper-layer information encapsulated in the frame body. For data privacy reasons, the MSDU data payload is usually encrypted. After the payload has been decrypted, the integration service that resides in access points and WLAN controllers takes the MSDU payload of the data frame and transfers the MSDU into 802.3 Ethernet frames.

The maximum frame body size of an 802.11 simple data frame or a QoS data frame is determined by the maximum MSDU size (2,304 bytes) plus any overhead from encryption. WEP encryption adds 8 bytes of overhead to the frame body of an 802.11 data frame. TKIP encryption adds 20 bytes of overhead to the frame body of an 802.11 data frame. CCMP encryption adds 16 bytes of overhead to the frame body of an 802.11 data frame.

In reality, most of the 15 data frame subtypes do not really exist. In Chapter 8, you learned about two 802.11 medium access control methods: Point Coordination Function (PCF) and HCF Controlled Channel Access (HCCA). Both access methods define mechanisms were the AP controls the medium via polling. As of this writing, we do not know of any WLAN vendor that supports either PCF or HCCA. Therefore, 11 of the 802.11 data frames defined by the 802.11-2016 standard exist only on paper. In the following list of all 15 of data frame subtypes, we have indicated all the PCF or HCCA data frames that really are never used. The first four data frame subtypes that are listed are the only ones to be concerned with. The 15 data frame subtypes are as follows:

- Data (simple data frame)
- Null (no data)
- QoS Data
- QoS Null (no data)
- Data + CF-ACK [PCF only]
- Data + CF-Poll [PCF only]
- Data + CF-ACK + CF-Poll [PCF only]
- CF-ACK (no data) [PCF only]
- CF-ACK + CF-Poll (no data) [PCF only]
- CF-Poll (no data) [PCF only]
- QoS Data + CF-ACK [HCCA only]
- QoS Data + CF-Poll [HCCA only]
- QoS Data + CF-ACK + CF-Poll [HCCA only]
- QoS CF-Poll (no data) [HCCA only]
- QoS CF-ACK + CF-Poll (no data) [HCCA only]

QoS and Non-QoS Data Frames

QoS mechanisms are a requirement for the Wi-Fi Multimedia (WMM) certification; this is strictly enforced by the Wi-Fi Alliance. Any 802.11 enterprise access point and most WLAN clients manufactured in the last 10 years support WMM QoS mechanisms by default. Therefore, each basic service set in most enterprise deployments is considered to be a *quality of service basic service set (QBSS)*, and most modern-day radios are considered to be QoS stations. QoS stations are capable of transmitting both QoS data frames and non-QoS data frames. As shown in Table 9.3, it is not uncommon to have a wireless network that consists of both QoS and non-QoS stations. In this type of mixed environment, it is likely that QoS devices will transmit both QoS data frames and non-QoS data frames, depending on the capabilities of the receiving station.

Whenever a non-QoS device is involved in the communication, either as the transmitting station or as the receiving station, a non-QoS data frame must be used. Broadcast frames are transmitted by default as non-QoS frames, unless the transmitting station knows that all the stations in the basic service set (BSS) are QoS capable. Like the broadcast frames, multicast frames are transmitted by default as non-QoS frames, unless the transmitting station knows that all the stations in the basic service set that are members of the multicast group are QoS capable, in which case the multicast frame will be a QoS frame.

TABLE 9.3   QoS and non-QoS transmissions

Transmitting Station	Receiving Station	Data Frame Subtype Used
Non-QoS station	Non-QoS station	Simple data frame (non-QoS)
Non-QoS station	QoS station	Simple data frame (non-QoS)
QoS station	QoS station	QoS data frame
QoS station	Non-QoS station	Simple data frame
Non-QoS station	Broadcast	Simple data frame
Non-QoS station	Multicast	Simple data frame
QoS station	Broadcast	Simple data frame, unless the transmitting station knows that all stations in the BSS are QoS capable, in which case a QoS data frame would be used
QoS station	Multicast	Simple data frame, unless the transmitting station knows that all stations in the BSS that are members of the multicast group are QoS capable, in which case a QoS data frame would be used

Non-Data Carrying Frames

As strange as it may sound, some 802.11 data frames do not actually carry any data. The Null and QoS Null frames are both non-data carrying frames. Both of these 802.11 data frame types have a header and a trailer but do not have a frame body that transports an MSDU payload. These frames are sometimes referred to as *null function frames* because the payload is null, yet the frames still serve a purpose. Client stations use the null function frames to inform the AP of changes in power-save status by changing the Power Management bit. When a client station decides to go off-channel for active scanning purposes, the client station will send a null function frame to the AP with the *Power Management bit* set to 1. As demonstrated in Exercise 9.8, when the Power Management bit is set to 1, the AP buffers all of that client's 802.11 frames. When the client station returns to the AP's channel, the station sends another null function frame with the Power Management bit set to 0. The AP then transmits the client's buffered frames. Some vendors also use the null function frame to implement proprietary power-management methods.

EXERCISE 9.8

Using Data Frames

1. To perform this exercise, you need to first download the CWNA-CH9.PCAPNG file from the book's online resource area, which can be accessed at www.wiley.com/go/cwnasg.

2. After the file is downloaded, you will need packet analysis software to open the file. If you do not already have a packet analyzer installed on your computer, you can download Wireshark from www.wireshark.org.

3. Using the packet analyzer, open the CWNA-CH9.PCAPNG file. Most packet analyzers display a list of capture frames in the upper section of the screen, with each frame numbered sequentially in the first column.

4. Scroll down the list of frames and click frame #97903, which is an unencrypted simple data frame. Look at the frame body and notice the upper-layer information, such as IP addresses and TCP port. This information is visible because no encryption is being used.

5. Click frame #34019, which is a QoS data frame. Look at the 802.11 MAC header and notice the QoS Control field. Notice the priority level is set to Best Effort.

6. Click frame #9507, which is a null function frame. Look at the 802.11 MAC header. Look in the Frame Control field and note that the Power Management bit is set to 1. The AP will now buffer the client's traffic.

Power Management

One of the main uses of wireless networking is to provide mobility for the client station. Client mobility goes hand in hand with battery-operated client stations. When battery-operated devices are used, one of the biggest concerns is how long the battery will last until it needs to be recharged. To increase the battery time, a bigger, longer-lasting battery can be used or power consumption can be decreased. The 802.11 standard includes a power-management feature that can be enabled to help increase battery life. Battery life is extremely important for smartphones, tablets, handheld scanners, and VoWiFi phones. The battery life of mobile devices usually needs to last at least one 8-hour work shift. The two legacy power-management modes supported by the 802.11 standard are active mode and power-save mode. 802.11 power-management methods have also been enhanced by both the ratified 802.11e-2005 amendment and the ratified 802.11n-2009 amendment. Further power management enhancements are defined by the 802.11ac-2013 amendment and the 802.11ax draft amendment.

Legacy Power Management

Active mode is a legacy power-management mode used by very old 802.11 stations. When a station is operating in active mode, the wireless station is always ready to transmit or receive data. Active mode provides no battery conservation. In the MAC header of an 802.11 frame, the Power Management field is 1 bit in length and is used to indicate the power-management mode of the station. A value of 0 indicates that the station is in active mode. Stations running in active mode will achieve higher throughput than stations running in power-save mode, but the battery life will typically be much shorter.

Stations that are always connected to a power source should be configured to use active mode.

Power-save mode is an optional mode for 802.11 stations. When a client station is operating in power-save mode, it will shut down some of the transceiver components for a period of time to conserve power. The wireless radio basically takes a short nap. The station indicates that it is using power-save mode by changing the value of the Power Management bit to 1. When the Power Management bit is set to 1, the AP is informed that the client station is using power management, and the AP buffers all of that client's 802.11 frames.

Traffic Indication Map

If a station is part of a basic service set, it will notify the AP that it is enabling power-save mode by changing the Power Management field to 1. When the AP receives a frame from a station with this bit set to 1, the AP knows that the station is in power-save mode. If the AP then receives any data that is destined for the station in power-save mode, the AP will store the information in a buffer. Any time a station associates to an AP, the station receives an *association identifier (AID)*. The AP uses this AID to keep track of the stations that are associated and the members of the BSS. If the AP is buffering data for a station in power-save mode, when the AP transmits its next beacon, the AID of the station will be seen in a field of the beacon frame known as the *traffic indication map (TIM)*. The TIM field is a list of all stations that have undelivered data buffered on the AP, waiting to be delivered. Every beacon will include the AID of the station until the data is delivered.

After the station notifies the AP that it is in power-save mode, the station shuts down part of its transceiver to conserve energy. A station can be in one of two states, either awake or doze:

- During the awake state, the client station can receive frames and transmit frames.
- During the doze state, the client station cannot receive or transmit any frames and operates in a very low power state to conserve power.

Because beacons are transmitted at a consistent predetermined interval, known as the *target beacon transmission time (TBTT)*, all stations know when beacons will occur. The station will remain asleep for a short period of time and awaken in time to hear a beacon frame. The station does not have to awaken for every beacon. To conserve more power, the station can sleep for a longer period of time and then awaken in time to hear an upcoming beacon. How often the client station awakens is based on a client variable called the *listen interval* and is usually vendor specific.

As shown in Figure 9.30, when the station receives the beacon, it checks to see whether its AID is set in the TIM, indicating that a buffered unicast frame waits. If so, the station will remain awake and will send a PS-Poll control frame to the AP. Inside the PS-Poll frame, the Duration/ID field is used as an association ID (AID) value. In other words, the station will identify itself to the AP and request the buffered unicast frame. When the AP receives the PS-Poll frame, it will send the buffered unicast frame to the station. The station will stay awake while the AP transmits the buffered unicast frame. When the AP sends the data to the station, the station needs to know when all the buffered unicast data has been received so that it can go back to sleep. Each unicast frame contains a 1-bit field called the More Data field. When the station receives a buffered unicast frame with the More Data field set to 1, the station knows that it cannot go back to sleep yet because there is some more buffered data that it has not yet received. When the More Data field is set to 1, the station knows that it needs to send another PS-Poll frame and wait to receive the next buffered unicast frame.

After all the buffered unicast frames have been sent, the More Data field in the last buffered frame will be set to 0, indicating that there is currently no more buffered data, and the station will go back to sleep. The AP will set the value of the station's AID bit to 0, and when the next TBTT arrives, the AP will send a beacon. The station will remain asleep for a short period of time and again awaken in time to hear a beacon frame. When the station receives the beacon, it will again check to see whether its AID is set in the TIM. Assuming that there are no buffered unicast frames awaiting this station, the station's AID will not be set to 1 in the TIM, and the station can simply go back to sleep until it is time to wake up and check again.

FIGURE 9.30   Legacy power management

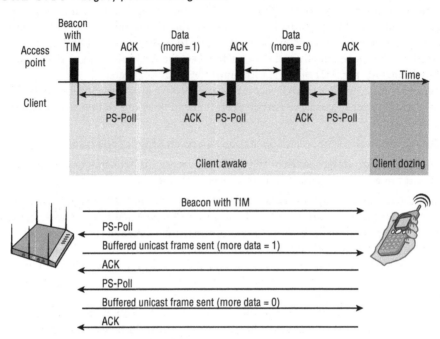

Delivery Traffic Indication Map

In addition to unicast traffic, network traffic includes multicast and broadcast traffic. Because multicast and broadcast traffic is directed to all stations, the BSS needs to provide a way to make sure that all stations are awake to receive these frames. A *delivery traffic indication map (DTIM)* is used to ensure that all stations using power management are awake when multicast or broadcast traffic is sent. DTIM is a special type of TIM. A TIM or DTIM is transmitted as part of every beacon.

A configurable setting on the AP called the *DTIM interval* determines how often a DTIM beacon is transmitted. A DTIM interval of 3 means that every third beacon is a DTIM beacon, whereas a DTIM interval of 1 means that every beacon is a DTIM beacon. Every beacon contains DTIM information that informs the stations when the next DTIM will occur. A DTIM value of 0 indicates that the current TIM is a DTIM. All stations will wake up in time to receive the beacon with the DTIM. If the AP has multicast or broadcast traffic to be sent, it will transmit the beacon with the DTIM and then immediately send the multicast or broadcast data.

After the multicast or broadcast data is transmitted, if a station's AID was in the DTIM, the station will remain awake and will send a PS-Poll frame and proceed with retrieving its buffered unicast traffic from the AP. If a station did not see its AID in the DTIM, or if its AID was set to 0, then the station can go back to sleep.

The DTIM interval is important for any application that uses multicasting. For example, many VoWiFi vendors support *push-to-talk* capabilities that send VoIP traffic to a multicast address. A misconfigured DTIM interval would cause performance issues during a push-to-talk multicast.

WMM-Power Save and U-APSD

The main focus of the 802.11e amendment, which is now part of the 802.11-2016 standard, is quality of service. However, the IEEE 802.11e amendment also introduced an enhanced power-management method called *automatic power save delivery (APSD)*. The two APSD methods that are defined are *scheduled automatic power save delivery (S-APSD)* and *unscheduled automatic power save delivery (U-APSD)*. The S-APSD power-management method is beyond the scope of this book. The Wi-Fi Alliance's *WMM-Power Save (WMM-PS)* certification is based on U-APSD. WMM-PS is an enhancement over the legacy power-saving mechanisms already discussed. The goal of WMM-PS is to have client devices spend more time in a doze state and consume less power. WMM-PS is also designed to minimize latency for time-sensitive applications, such as voice, during the power-management process.

The legacy power-management methods have several limitations. As shown earlier in Figure 9.30, a client using legacy power management must first wait for a beacon with a TIM before the client can request buffered unicast frames. The client must also send a unique PS-Poll frame to the AP to request every single buffered unicast frame. This ping-pong power-management method increases the latency of time-sensitive applications, such as voice. The clients must also stay awake during the ping-pong process, which results in reduced battery life. In addition, the amount of time that the clients spend dozing is determined by the vendor's driver and not by the application traffic.

WMM-PS uses a trigger mechanism to receive buffered unicast traffic based on WMM access categories. You learned in Chapter 8 that 802.1D priority tags from the Ethernet side are used to direct traffic to four different WMM access-category priority queues. The access-category queues are voice, video, best effort, and background. As shown in Figure 9.31, the client station sends a trigger frame related to a WMM access category to inform the AP that the client is awake and ready to download any frames that the AP may have buffered for that access category. The trigger frame can also be an 802.11 data frame, thus eliminating the need for a separate PS-Poll frame. The AP will then send an ACK to the client and proceed to send a frame burst of buffered application traffic during a transmit opportunity (TXOP).

The advantages of this enhanced power-management method include the following:

- Applications now control the power-save behavior by setting doze periods and sending trigger frames. VoWiFi phones will obviously send triggers to the AP frequently during voice calls, whereas a laptop radio using a data application will have a longer doze period.

- The trigger and delivery method eliminates the need for PS-Poll frames.

- The client can request to download buffered traffic and does not have to wait for a beacon frame.

- All the downlink application traffic is sent in a faster frame burst during the AP's TXOP.

FIGURE 9.31 WMM-PS

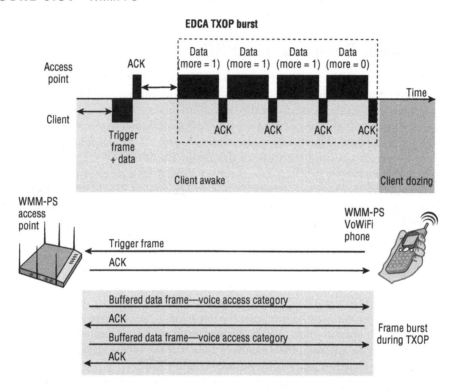

A couple of conditions have to be met in order for a Wi-Fi client to use the enhanced WMM-PS mechanisms:

- The client must be Wi-Fi CERTIFIED for WMM-PS.
- The AP must be Wi-Fi CERTIFIED for WMM-PS.

It should be noted that applications that do not support WMM-PS can still coexist with WMM-Power Save–enabled applications. The data from the other applications will be delivered with legacy power-save methods.

MIMO Power Management

The ratified 802.11n-2009 amendment defined two power-management methods that can be used with multiple-input, multiple-output (MIMO) radios. The first method is called *spatial multiplexing power save (SM power save).* The purpose of SM power save is to enable a MIMO 802.11n/ac device to power down all but one of its radio chains. A more detailed discussion about MIMO power-management methods is presented in Chapter 10. In Chapter 19, "802.11ax: High Efficiency (HE)," you will also learn about additional proposed power-management enhancements that will help conserve battery life for IoT devices.

Summary

This chapter covered the following key areas of the MAC architecture:

- 802.11 frame format
- Major 802.11 frame types
- 802.11 frame subtypes
- 802.11 state machine
- Protection mechanisms
- Power management

It is important to understand the makeup of the three major 802.11 frame types and the purpose of each individual 802.11 frame and how they are used in scanning, authentication, association, and other MAC processes. You should understand the need for an ERP protection mechanism, without which mixed-mode networks would not be able to function. Both RTS/CTS and CTS-to-Self provide ERP (802.11g) and HT (802.11n) protection mechanisms.

To help manage battery life, power management can be configured on a wireless station. Active mode provides no battery conservation of any kind, whereas power-save mode can be invaluable for increasing the battery life of laptop and handheld computing devices. WMM and 802.11n have also enhanced power-management capabilities. We discussed the following power-management pieces in this chapter:

- Traffic indication map (TIM)
- Delivery traffic indication map (DTIM)
- WMM-Power Save (WMM-PS)

Exam Essentials

Explain the differences between a PPDU, a PSDU, an MPDU, and an MSDU. Understand at which layer of the OSI model each data unit operates and what each data unit comprises.

Understand the 802.11 frame format. Describe the key components of the 802.11 MAC header. Be able to explain 802.11 MAC addressing. Understand that the 802.11 trailer is used for data integrity.

Know the three major 802.11 frame types. Make sure you know the function of the management, control, and data frames. Know what makes the major frame types different. Data frames contain an MSDU, whereas management and control frames do not. Understand the purpose of each individual frame subtype covered in this chapter.

Know the media access control (MAC) process and all the frames that are used during this process. Understand the function of each of the following: active scanning, passive scanning, beacons, probe requests, probe responses, authentication, association, reassociation, disassociation, and deauthentication.

Know the importance of the ACK frame for determining that a unicast frame was received and uncorrupted.   Understand that after a unicast frame is transmitted, there is a short interframe space (SIFS) and then the receiving station replies by transmitting an ACK. If this process is completed successfully, the transmitting station knows the frame was received and was not corrupted. Understand that Block ACKs are used with frames bursts and A-MPDU frame aggregation.

Understand the importance of ERP protection mechanisms and how they function.   Protected mode allows 2.4 GHz HT (802.11n), ERP (802.11g), HR-DSSS (802.11b), and legacy DSSS devices to coexist within the same BSS. Protected mode can be provided by RTS/CTS or CTS-to-Self. CTS-to-Self is strictly a protection mechanism, but RTS/CTS can also be manually configured and used to identify or prevent hidden nodes.

Understand all the technologies that make up power management.   Power management can be enabled to decrease power usage and increase battery life. Understand how buffered unicast traffic is received in a different way from buffered broadcast and multicast traffic. Understand the power-management enhancements defined by WMM-PS.

Review Questions

1. Which of the following 802.11 frames carry an MSDU payload that may eventually be transferred by the integration service into an 802.3 Ethernet frame?

    A. 802.11 management frames

    B. 802.11 control frames

    C. 802.11 data frames

    D. 802.11 action frames

    E. 802.11 association frames

2. Which of the following contains only LLC data and the IP packet but does not include any layer 2 802.11 data?

    A. MPDU

    B. PPDU

    C. PSDU

    D. MSDU

    E. MMPDU

3. Based on the 802.11 frame capture shown here, which type of networking communication is occurring?

802.11 MAC Header	
Version:	0
Type:	%10 Data
Subtype:	%1000 QoS Data
Frame Control Flags=	%00000011
Duration:	44 Microseconds
Receiver:	00:0E:38:49:05:80
Transmitter:	00:0C:85:62:D2:1D
Destination:	00:0A:E4:DA:8D:DS
Source:	00:90:96:BD:77:35
Seq Number:	982
Frag Number:	0

    A. AP to client station

    B. Client station to server

    C. Client station to AP

    D. Server to client station

    E. AP to AP

4. The presence of what type of transmissions can trigger the protection mechanism within an ERP basic service set? (Choose all that apply.)

   A. Association of an HR-DSSS client

   B. Association of an ERP-OFDM client

   C. HR-DSSS beacon frame

   D. ERP beacon frame with the NonERP_Present bit set to 1

   E. Association of an FHSS client

5. Which of the following information is included in a probe response frame? (Choose all that apply.)

   A. Channel information

   B. Supported data rates

   C. Basic data rates

   D. SSID

   E. Traffic indication map

6. Which of the following statements are true about beacon management frames? (Choose all that apply.)

   A. Beacons can be disabled to hide the network from intruders.

   B. Time-stamp information is used by the clients to synchronize their clocks.

   C. In a BSS, clients share the responsibility of transmitting the beacons.

   D. Beacons can contain vendor-proprietary information.

7. Which of the following statements regarding the four MAC address fields in the 802.11 MAC header are accurate? (Choose all that apply.)

   A. Address 2 is always the transmitter address (TA).

   B. Address 3 is always the transmitter address (TA).

   C. Address 1 is always the basic service set identifier (BSSID).

   D. Address 1 is always the receiver address (RA).

   E. Address 3 is always the basic service set identifier (BSSID).

   F. Address 2 is always the receiver address (RA).

8. When a station sends an RTS, the Duration/ID field notifies the other stations that they must set their NAV timers to which of the following values?

   A. 213 microseconds

   B. The time necessary to transmit the Data and ACK frames

   C. The time necessary to transmit the CTS frame

   D. The time necessary to transmit the CTS, Data, and ACK frames

9. How does a client station indicate that it is using power-save mode?

   A. It transmits a frame to the AP with the Sleep field set to 1.

   B. It transmits a frame to the AP with the Power Management field set to 1.

   C. Using DTIM, the AP determines when the client station uses power-save mode.

   D. It does not need to, because power-save mode is the default.

10. What would cause an 802.11 station to retransmit a unicast frame? (Choose all that apply.)

   A. The transmitted unicast frame was corrupted.

   B. The ACK frame from the receiver was corrupted.

   C. The receiver's buffer was full.

   D. The transmitting station's buffer was full.

11. If a station is in power-save mode, how does it know that the AP has buffered unicast frames waiting for it?

   A. By examining the PS-Poll frame

   B. By examining the TIM field

   C. When it receives an ATIM

   D. When the Power Management bit is set to 1

   E. By examining the DTIM interval

12. When is an 802.11ac AP transmitting on 5 GHz required by the IEEE 802.11-2016 standard to respond to probe request frames from nearby client stations? (Choose all that apply.)

   A. When the probe request frames contain a null SSID value

   B. When the probe request client is also an 802.11ac radio

   C. When the probe request is encrypted

   D. When the probe request has the Power Management bit set to 1

   E. When the probe request frames contain the correct SSID value

13. Which of the following statements about scanning are true? (Choose all that apply.)

   A. There are two types of scanning: passive and active.

   B. Stations must transmit probe requests in order to learn about local APs.

   C. The 802.11 standard allows APs to ignore probe requests for security reasons.

   D. It is common for stations to continue to send probe requests after being associated to an AP.

14. Given that an 802.11 MAC header can have as many as four MAC addresses, which types of addresses are not found in an 802.3 MAC header? (Choose all that apply.)

   A. SA

   B. BSSID

   C. DA

   D. RA

   E. TA

15. When a client station is first powered on, what is the order of frames generated by the client station and AP?

    A. Probe request/response, association request/response, authentication request/response

    B. Probe request/response, authentication request/response, association request/response

    C. Association request/response, authentication request/response, probe request/response

    D. Authentication request/response, association request/response, probe request/response

16. WLAN users have recently complained about gaps in audio and problems with the push-to-talk capabilities with the ACME Company's VoWiFi phones. What could be the cause of this problem?

    A. A misconfigured TIM setting

    B. A misconfigured DTIM setting

    C. A misconfigured ATIM setting

    D. A misconfigured BTIM setting

17. The WLAN help desk gets a call that suddenly all the legacy 802.11b wireless barcode scanners cannot connect to any of the 802.11n APs. However, all the 802.11g/n clients can still connect. What are the possible causes of this problem? (Choose all that apply.)

    A. The WLAN administrator disabled the 1, 2, 5.5, and 11 Mbps data rates.

    B. The WLAN administrator disabled the 6 and 9 Mbps data rates.

    C. The WLAN administrator enabled the 6 and 9 Mbps data rates as basic rates.

    D. The WLAN administrator configured all the APs on channel 6.

18. Which of these 802.11 data frames carry an MSDU payload? (Choose all that apply.)

    A. Non-QoS data frame

    B. QoS data frame

    C. Null frame

    D. QoS null frame

19. What are some examples of how an 802.11 action frame can be used? (Choose all that apply.)

    A. An action frame can function as a probe request.

    B. An action frame can function as a neighbor report request.

    C. An action frame can function as a probe response.

    D. An action frame can function as a channel switch announcement.

    E. An action frame can function as a beacon.

20. What can you conclude about this frame based on the frame capture graphic shown? (Choose all that apply.)

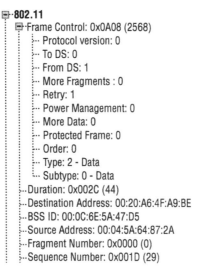

```
⊟ 802.11
    ⊟ Frame Control: 0x0A08 (2568)
        ⋮·· Protocol version: 0
        ⋮·· To DS: 0
        ⋮·· From DS: 1
        ⋮·· More Fragments : 0
        ⋮·· Retry: 1
        ⋮·· Power Management: 0
        ⋮·· More Data: 0
        ⋮·· Protected Frame: 0
        ⋮·· Order: 0
        ⋮·· Type: 2 - Data
        ⋯· Subtype: 0 - Data
    ⋮·· Duration: 0x002C (44)
    ⋮·· Destination Address: 00:20:A6:4F:A9:BE
    ⋮·· BSS ID: 00:0C:6E:5A:47:D5
    ⋮·· Source Address: 00:04:5A:64:87:2A
    ⋮·· Fragment Number: 0x0000 (0)
    ⋮·· Sequence Number: 0x001D (29)
```

A. This is a unicast frame.

B. This is a multicast frame.

C. This is a broadcast frame.

D. This a transmission between a mesh point AP and a mesh portal AP.

E. This frame is encrypted.

F. The frame check sequence (FCS) of the previous attempt of the same frame failed at the receiving station.

Chapter

10

MIMO Technology: HT and VHT

IN THIS CHAPTER, YOU WILL LEARN ABOUT THE FOLLOWING:

✓ **MIMO**

- Radio chains
- Spatial multiplexing (SM)
- MIMO diversity
- Space-time block coding (STBC)
- Cyclic shift diversity (CSD)
- Transmit beamforming (TxBF)
- 802.11ac explicit beamforming

✓ **Multi-user MIMO**

- Multi-user beamforming

✓ **Channels**

- 20 MHz channels
- 40 MHz channels
- Forty MHz Intolerant
- 80 MHz and 160 MHz channels

✓ **Guard interval (GI)**

✓ **256-QAM modulation**

✓ **802.11n/ac PPDUs**

- Non-HT
- HT Mixed
- VHT

✓ **802.11n/ac MAC**

- A-MSDU
- A-MPDU
- Block acknowledgment
- Power management
- Modulation and coding scheme
- 802.11ac data rates

✓ **HT/VHT protection mechanisms**

- HT protection modes (0–3)

✓ **Wi-Fi Alliance Certification**

In this chapter, we discuss both MIMO-based Wi-Fi technologies: high throughput (HT), originally defined by the 802.11n-2009 amendment, and very high throughput (VHT), originally defined by the 802.11ac-2013 amendment. Both technologies provide PHY and MAC enhancements and increased data rates.

The original main objective of the 802.11n amendment was to increase the data rates and the throughput in both the 2.4 GHz and 5 GHz frequency bands. The 802.11n amendment defines an operation mode known as *high throughput (HT)*, which provides PHY and MAC enhancements to provide for transmission rates potentially as high as 600 Mbps. The 802.11ac amendment defined a new operation mode, known as *very high throughput (VHT)*. VHT operates only in the 5 GHz U-NII bands and provides PHY and MAC enhancements that allow for transmission rates potentially as high as 6933 Mbps.

802.11n introduced a whole new approach to the Physical layer, using a technology called *multiple-input, multiple-output (MIMO)*, which requires the use of multiple radios and antennas. As you learned in earlier chapters, multipath is an RF behavior that can cause performance degradation in legacy 802.11a/b/g WLANs. 802.11n and 802.11ac radios use MIMO technology, which takes advantage of multipath to increase throughput as well as range.

Besides the use of MIMO technology, HT and VHT mechanisms provide for enhanced throughput using other methods. We will discuss the use of wider channels that provide greater frequency bandwidth. Enhancements to the MAC sublayer also provide for greater throughput with the use of frame aggregation. The 802.11e amendment defined enhancements to power management, and later, the 802.11n amendment also provided for new power-management techniques.

802.11ac expanded and in some cases simplified many of the technologies of 802.11n, while also introducing a new technology known as *multi-user MIMO (MU-MIMO)*. Although MU-MIMO has received fanfare for its technology and capabilities, very few clients support MU-MIMO; WLANs rarely benefit from MU-MIMO. Therefore, for the most part 802.11ac is an enhancement or extension of 802.11n, expanding on the capabilities of 802.11n, providing faster Wi-Fi. Table 10.1 provides a summary of the differences between 802.11n and 802.11ac.

TABLE 10.1 Comparison of 802.11n and 802.11ac

Technology	802.11n - HT	802.11ac - VHT
Frequency	2.4 GHz and 5 GHz	5 GHz only
Modulation	BPSK, QPSK, 16-QAM, 64-QAM	BPSK, QPSK, 16-QAM, 64-QAM, 256-QAM

TABLE 10.1   Comparison of 802.11n and 802.11ac *(continued)*

Technology	802.11n - HT	802.11ac - VHT
Channel widths	20 MHz, 40 MHz	20 MHz, 40 MHz, 80 MHz, 160 MHz
Spatial streams	Up to four	Up to eight on APs, up to four on clients
Short guard interval support	Yes	Yes
Beamforming	Multiple types, both implicit and explicit; not typically implemented	Explicit beamforming with null data packets (NDPs)
Number of modulation and coding schemes (MCSs)	77	10
Support for A-MSDU and A-MPDU	Yes	Yes, all frames transmitted as A-MPDU
MIMO support	Single-user MIMO	Single-user MIMO and multiuser MIMO (MU-MIMO)
Maximum number of simultaneous user transmissions	One	Four
Maximum data rate	600 Mbps	6933 Mbps

Finally, we discuss the various modes of operation for HT and VHT networks, and how their radio transmissions can coexist in the same WLAN environment with radios that use the legacy technologies we have discussed throughout this book. The technologies that are implemented to deploy HT and VHT networks are so complex that entire books are dedicated to them. In this chapter, we cover the key components of HT and VHT, along with the knowledge needed to properly prepare you for the CWNA exam.

MIMO

The heart and soul of the 802.11n and 802.11ac amendments exists at the Physical (PHY) layer, with the use of a technology known as multiple-input, multiple-output (MIMO). MIMO requires the use of multiple radios and antennas, called *radio chains*, which are defined later in this chapter. MIMO radios transmit multiple radio signals at the same time to take advantage of multipath.

In traditional 802.11 environments, the phenomenon of multipath has long caused problems. *Multipath* is a propagation phenomenon that results in two or more paths of the same signal arriving at a receiving antenna at the same time or within nanoseconds of each other. Due to the natural broadening of the waves, the propagation behaviors of reflection, scattering, diffraction, and refraction will occur. A signal may reflect off an object or may scatter, refract, or diffract. These propagation behaviors can each result in multiple paths of the same signal. As you learned in Chapter 3, "Radio Frequency Fundamentals," the negative effects of multipath can include loss of amplitude and data corruption. MIMO systems, however, take advantage of multipath and, believe it or not, multipath then becomes your friend.

In a typical indoor environment, multiple RF signals sent by a MIMO radio will take multiple paths to reach the MIMO receivers. For example, as shown in Figure 10.1, multiple copies of the three original signals will be received by multiple antennas. The MIMO receiver will then use advanced *digital signal processing (DSP)* techniques to sort out the originally transmitted signals. A high multipath environment actually helps a MIMO receiver differentiate between the unique data streams carried on the multiple RF signals. As a matter of fact, if multiple signals sent by a MIMO transmitter all arrive simultaneously at the receiver, the signals can cancel each other and the performance is basically the same as a non-MIMO system.

FIGURE 10.1 MIMO operation and multipath

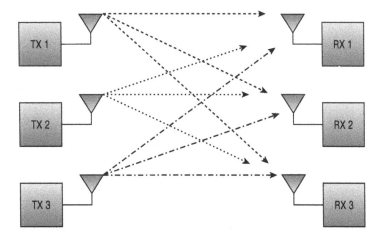

Transmitting multiple streams of data with a method called *spatial multiplexing (SM)* provides for greater throughput and takes advantage of the old enemy known as multipath. MIMO systems can also use multiple antennas to provide for better transmit and receive diversity, which can increase range and reliability. There are various transmit and receive diversity techniques. Space-time block coding (STBC) and cyclic shift diversity (CSD) are transmit diversity techniques where the same transmit data is sent out of multiple antennas. STBC communication is possible only between MIMO devices. CSD diversity signals can be received by either 802.11n or legacy devices. Transmit beamforming (TxBF) is a technique where the same signal is transmitted over multiple antennas, and the antennas act like a phased array. Maximal ratio combining (MRC) is a type of receive diversity technique where multiple received signals are combined, thus improving sensitivity. Spatial multiplexing and diversity techniques are explained in greater detail in the following sections.

Radio Chains

Legacy 802.11 radios transmit and receive RF signals by using a *single-input, single-output (SISO)* system. SISO systems use a single radio chain. A *radio chain* is defined as a single radio and all its supporting architecture, including mixers, amplifiers, and analog/digital converters.

A MIMO system consists of multiple radio chains, with each radio chain having its own antenna. A MIMO system is characterized by the number of transmitters and receivers used by the multiple radio chains. For example, a 2×3 MIMO system would consist of three radio chains with two transmitters and three receivers. A 3×3 MIMO system would use three radio chains with three transmitters and three receivers. In a MIMO system, the first number always references the transmitters (TX), and the second number references the receivers (RX).

Figure 10.2 illustrates both 2×3 and 3×3 MIMO systems. Please note that both systems utilize three radio chains; however, the 3×3 system has three transmitters, whereas the 2×3 system has only two transmitters.

FIGURE 10.2   2×3 and 3×3 MIMO

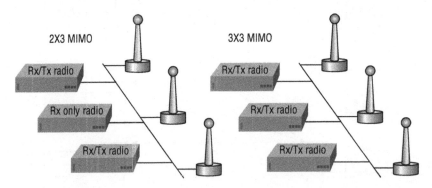

The use of multiple transmitters in a MIMO system provides for the transmission of more data via spatial multiplexing. The use of multiple receivers increases signal-to-noise ratio (SNR) because of advanced MIMO antenna diversity. Both of these benefits are discussed in greater detail in the following sections. The 802.11n standard allows for MIMO systems up to 4×4 using four radio chains. Each radio chain requires power. A 2×2 MIMO system would require much less of a power draw than a 4×4 MIMO system. 802.11n radios can have up to 4 radios chains, and 802.11ac access point radios can have up to 8 radio chains.

Spatial Multiplexing

You have already learned that MIMO radios will transmit multiple signals. A MIMO radio also has the ability to send independent unique data streams. Each independent data stream is known as a *spatial stream*, and each unique stream can contain data that is different from the other streams transmitted by one or more of the other radio chains. Each stream will also travel a different path, because there is at least a half-wavelength of space between the multiple transmitting antennas. The fact that the multiple streams follow different paths to the receiver because of the space between the transmitting antennas is known as *spatial diversity*. Sending multiple independent streams of unique data using spatial diversity is often also referred to as *spatial multiplexing (SM)* or *spatial diversity multiplexing (SDM)*.

The benefit of sending multiple unique data streams is that throughput is drastically increased. If a MIMO access point sends two unique data streams to a MIMO client station that receives both streams, the throughput is effectively doubled. If a MIMO access point sends three unique data streams to a MIMO client station that receives all three streams, the throughput is effectively tripled.

Do not confuse the independent unique streams of data with the number of transmitters. In fact, when referring to MIMO radios, it is important to also reference how many unique streams of data are sent and received by MIMO radios. Most Wi-Fi vendors use a three-number syntax when describing MIMO radio capabilities. In a MIMO system, the first number always references the transmitters (TX), and the second number references the receivers (RX). The third number represents how many unique streams of data can be sent or received.

For example, a 3×3:2 MIMO system would use three transmitters and three receivers, but only two unique data streams are utilized. A 3×3:3 MIMO system would use three transmitters and three receivers with three unique data streams.

Figure 10.3 depicts a 3×3:3 MIMO AP transmitting three independent streams of unique data to a 3×3:3 MIMO client.

FIGURE 10.3 Multiple spatial streams

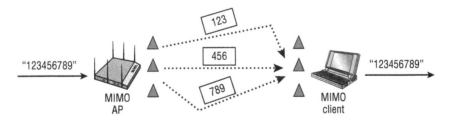

It is important to understand that not all 802.11n or 802.11ac radios have the same MIMO capabilities. Many earlier 802.11n access points deployed 3×3:2 instead of 3×3:3 MIMO radios. Many WLAN vendors also offer less expensive 802.11n or 802.11ac access points using 2×2:2 MIMO radios. On the client side, a variety of combinations have existed, with many laptops using 3×3:2 or 3×3:3 MIMO radios. Some handheld devices still only use the legacy SISO capabilities found in 802.11a/b/g. However, older mobile devices, such as smartphones and tablets, deploy 802.11n radios with 1×1:1 capabilities that effectively function as a SISO radio with some 802.11n capabilities. A 1×1:1 radio will not offer the full advantages of using multiple spatial streams; however, some of the other 802.11n PHY and MAC enhancements can still be employed.

If good RF conditions exist, when a 3×3:3 access point and a 3×3:3 client device are communicating with each other, three spatial streams can be used for unicast transmissions. However, when a 3×3:3 access point and a 2×2:2 client device are communicating with each other, only two spatial streams will be used for unicast transmissions. When a client radio joins a basic service set (BSS), the access point is advised about the client radio's MIMO capabilities.

The 802.11n amendment does allow for the use of up to a 4×4:4 MIMO system. The majority of enterprise 802.1n access points have traditionally been 2×2:2 or 3×3:3. The majority of enterprise WLAN vendors APs are dual-band. WLAN vendors now offer dual-band APs with a 4×4:4 802.11ac radio for 5 GHz and 4×4:4 802.11n radio for 2.4 GHz. There is a wide variety of MIMO capabilities among 802.11n/ac client devices. Laptops usually have either 2×2:2 or 3×3:3 radios, while most mobile devices, such as smartphones and tablets, now have 2×2:2 MIMO radios. As previously mentioned, many older 802.11n tablets and smartphones used 1×1:1 MIMO radios. Mobile devices typically only use either 1×1:1 or 2×2:2 MIMO radios because additional radio chains would drain the battery life of the mobile devices quicker. In the past, the majority of mobile devices originally transmitted only on the 2.4 GHz band; however, most smartphone and tablet vendors now typically offer dual-frequency radios that transmit on both the 2.4 and 5 GHz frequency bands.

According to the 802.11n amendment, multiple spatial streams can be sent with the same (equal) modulation, or they can be sent using different (unequal) modulation. For example, a 3×3:3 MIMO radio can transmit three data streams using the same 64-quadrature amplitude modulation (QAM) technique. Another example is a 3×3:3 MIMO radio transmitting two streams by using 64-QAM and the third stream using quadrature phase-shift keying (QPSK) modulation because of a higher noise floor. A 3×3:3 MIMO system using *equal modulation* would accomplish greater throughput than a 3×3:3 MIMO system using *unequal modulation*. Although unequal modulation is theoretically and technically possible, WLAN vendors have never implemented unequal modulation with 802.11n radios. The 802.11ac amendment eliminated support for this feature.

MIMO Diversity

If you cover one of your ears with your hand, will you hear better or worse with a single ear? Obviously, you will hear better with two ears. Do you think you would be able to hear more clearly if you had three or four ears instead of just two? Do you think you would be able to hear sounds from greater distances if you had three or four ears instead of just two? Yes, a human being would hear more clearly and with greater range if equipped with more than two ears. MIMO systems employ advanced antenna diversity capabilities that are analogous to having multiple ears.

Antenna diversity often is mistaken for the spatial multiplexing capabilities that are utilized by MIMO. Antenna diversity (both receive and transmit) is a method of using multiple antennas to survive the negative effects of multipath. As you just learned, MIMO takes advantage of multipath with spatial multiplexing to increase data capacity. Simple *antenna diversity* is a method of compensating for multipath, as opposed to utilizing multipath. Multipath produces multiple copies of the same signal that arrive at the receiver with different amplitudes.

In Chapter 5, "Radio Frequency Signal and Antenna Concepts," you learned about traditional antenna diversity, which consists of one radio with two antennas. Most SISO radios use *switched diversity*. When receiving RF signals, switched diversity systems listen with multiple antennas. Multiple copies of the same signal arrive at the receiver antennas with different amplitudes. The signal with the best amplitude is chosen, and the other signals are ignored. Switched diversity is also used when transmitting, but only one antenna is used. The transmitter will transmit out of the diversity antenna where the best amplitude signal was last heard.

As the distance between a transmitter and receiver increases, the received signal amplitude decreases to levels closer to the noise floor. As the signal-to-noise ratio (SNR) diminishes, the odds of data corruption grow. Listening with two antennas increases the odds of hearing at least one signal without corrupted data. Now, imagine if you had three or four antennas listening for the best received signal by using switched diversity. The probabilistic odds of hearing signals with stronger amplitudes and uncorrupted data have increased even more. The increased probability of hearing at least one uncorrupted signal in a switched diversity system using three or four antennas often results in increased range.

When receive diversity is used, the signals may also be linearly combined by using a signal processing technique called *maximal ratio combining (MRC)*. MRC algorithms are used to combine multiple received signals by looking at each unique signal and optimally combining the signals in a method that is additive as opposed to destructive. MIMO systems using MRC will effectively raise the SNR level of the received signal. As shown in Figure 10.4, maximal ratio combining is useful when a non-MIMO radio transmits to a MIMO receiver and multipath occurs. The MRC algorithm focuses on the signal with the highest SNR level; however, it may still combine information from the noisier signals. The end result is that less data corruption occurs because a better estimate of the original data has been reconstructed.

FIGURE 10.4   Maximal ratio combining

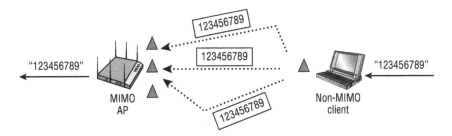

MRC uses a receive-combining function that assesses the phase and SNR of each incoming signal. Each received signal is phase-shifted so that they can be combined. The amplitude of the incoming signals is also modified to focus on the signal with the best SNR.

Space-Time Block Coding

Space-time block coding (STBC) is a method where the same information is transmitted on two or more antennas; however, the number of antennas must be even. It is a type of transmit diversity. STBC can be used when the number of radio chains exceeds the number of spatial streams. By sending copies of the same information on multiple antennas, the actual rate of the data transmitted does not increase as transmit antennas are added. STBC does, however, increase the receiver's ability to detect signals at a lower SNR than otherwise would be possible. The receive sensitivity of the radio system improves. STBC and cyclic shift diversity (CSD) are transmit diversity techniques where the same transmit data is sent out of multiple antennas. STBC communication is possible only between MIMO devices. CSD diversity signals can be received by either MIMO or legacy SISO devices.

Cyclic Shift Diversity

Cyclic shift diversity (CSD) is another transmit diversity technique specified in 802.11n and 802.11ac. Unlike STBC, a signal from a transmitter that uses CSD can be received by legacy 802.11g and 802.11a devices. For mixed mode deployments, where 802.11n coexists with 802.11g and 802.11a devices, there is a need to have a way of transmitting the symbols in the legacy OFDM preamble over multiple transmit antennas. CSD is used and a cyclic delay is applied to each of the transmitted signals. The delays are calculated to minimize the correlation between the multiple signals. A conventional legacy system would treat the multiple received signals as multipath versions of the same signal. The cyclic delay is chosen to be within the limits of the guard interval (GI) so that it does not cause excessive intersymbol interference (ISI). A MIMO system has no problem using the multiple signals to improve the overall SNR of the preamble. The details of how CSD works will not be part of the CWNA exam. CSD is one of the finer and least discussed features of 802.11n/ac but nonetheless still important to equipment vendor radio designers.

Transmit Beamforming

The 802.11n amendment also proposed an optional PHY capability, called *transmit beamforming (TxBF)*, which uses phase adjustments. Transmit beamforming can be used when there are more transmitting antennas than there are spatial data streams.

Transmit beamforming is a method that allows a MIMO transmitter using multiple antennas to adjust the phase and amplitude of the outgoing transmissions in a coordinated method. When multiple copies of the same signal are sent to a receiver, the signals will usually arrive out of phase with each other. If the transmitter (TX) knows about the RF characteristic of the receiver's location, the phase of the multiple signals sent by a MIMO transmitter can be adjusted. When the multiple signals arrive at the receiver, they are in phase, resulting in constructive multipath instead of the destructive multipath caused by out-of-phase signals. Carefully controlling the phase of the signals transmitted from multiple antennas has the effect of emulating a directional antenna.

Because transmit beamforming results in constructive multipath communication, the result is a higher signal-to-noise ratio and greater received amplitude. Therefore, transmit beamforming will result in greater range for individual clients communicating with an access point. Transmit beamforming will also result in higher throughput because of the higher SNR, which allows for the use of more complex modulation methods that can encode more data bits. The higher SNR also results in fewer layer 2 retransmissions.

Transmit beamforming could be used together with spatial multiplexing (SM); however, the number of spatial streams is constrained by the number of receiving antennas. For example, a 4×4:4 MIMO radio might be transmitting to a 2×2:2 MIMO radio, which can receive only two spatial streams. The 4×4:4 MIMO radio will send only two spatial streams but might also use the other antennas to form beams that are more focused to the receiving 2×2:2 MIMO receiver. In practice, transmit beamforming will probably be used when spatial multiplexing is not the best option. As shown in Figure 10.5, when utilizing transmit beamforming, the transmitter will not be sending multiple unique spatial streams but will instead be sending multiple streams of the same data with the phase adjusted for each RF signal.

FIGURE 10.5   Transmit beamforming data

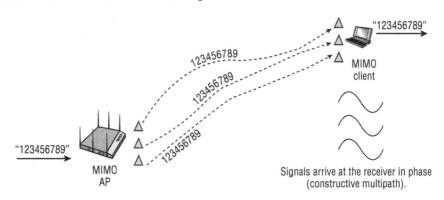

Transmitters that use beamforming will try to adjust the phase of the signals based on feedback from the receiver by using *sounding frames*. The transmitter is considered the *beamformer*, while the receiver is considered the *beamformee*. The beamformer and the beamformee work together to educate each other about the characteristics of the MIMO channel. This exchange of sounding frames is used to measure the RF channel and create a computative assessment on how to better steer RF energy to a receiver. The assessment is known as a *steering matrix*.

Transmit beamforming relies on *implicit feedback* or *explicit feedback* from both the transmitter and receiver. Any frame can be used as a sounding frame. Null function data frames can be used if another frame is not used. When using implicit feedback, the beam-former sends a sounding frame and then receives long training symbols transmitted by the beamformee, which allows the MIMO channel between the beamformee and beamformer to be estimated by the beamformer. In other words, there is no direct feedback from the beamformee, and thus the beamformer creates the steering matrix. A good analogy for implicit feedback is sonar. Sonar is a method in which submarines use sound propagation underwater to detect other vessels. A submarine sends out a sound wave, and based on the characteristics of the returning sound wave, the crew can determine the type of vessel that might be in the path of the submarine. However, there is no direct explicit feedback from the vessel to the submarine.

Much more information can be exchanged between two HT radios if they are both capable of explicit feedback. When using explicit feedback, the beamformee makes a direct estimate of the channel from training symbols sent to it by the beamformer. The beamfor-mee takes that information and sends additional feedback back to the beamformer. In other words, the beamformee creates the steering matrix. The beamformer then transmits based on the feedback from the beamformee. It should be noted that explicit beamforming was never adopted by WLAN vendors for 802.11n radios. One WLAN vendor did adopt some implicit beamforming capabilities in some of their 802.11n APs. However, the 802.11ac amendment defines only explicit beamforming, and WLAN vendors have built the capability into 802.11ac APs. Explicit beamforming, as used with 802.11ac, is explained in the sec-tions to follow.

802.11ac Explicit Beamforming

In this section, you will learn about explicit beamforming used by 802.11ac radios and how it is used with multiuser MIMO (MU-MIMO). To perform beamforming, the multiple radio chains in the AP transmit the same information through different antennas. The APs time their transmissions so that the waves of all the antennas arrive at the receiving radio at the same time and in phase with each other. This should result in a signal increase of approximately 3 decibels. This increase in signal strength can move the communications between the radios to a higher data rate. The increase will not likely be sufficient to affect the higher 256-QAM data rates (or the lower data rates), but it will affect communications in the middle data rate ranges.

Just like 802.11n, 802.11ac calls the transmitting radio the beamformer and the receiv-ing radio the beamformee. Beamforming can be adjusted on a frame-by-frame basis, so for one transmission the AP can be the beamformer and for another the client can be the beamformer. An AP can send a beamformed transmission to a station, and if a station sup-ports multiple radio chains, it can send beamformed frames to the AP.

The 802.11n amendment defined multiple beamforming methods. However, 802.11ac uses only explicit beamforming and requires support by both the transmitter and receiver in order for beamforming to be used. Explicit beamforming uses an interactive calibration process to identify how to perform the transmission using the multiple radio chains. This process is known as *channel sounding*.

To begin the process, the beamformer transmits a null data packet (NDP) announcement frame, which notifies the beamformee of the intent to send a beamformed transmission. The beamformer then follows this with an NDP frame. The beamformee processes each OFDM subcarrier and creates feedback information. The feedback contains information regarding power and the phase shift between each pair of transmit and receive antennas. This information is used to create a feedback matrix, which is then compressed and sent back to the beamformer. Figure 10.6 shows this exchange of frames. The beamformer uses the feedback matrix to calculate a steering matrix, which is used to direct the data transmission to the beamformee.

FIGURE 10.6   Single-user beamforming sounding process

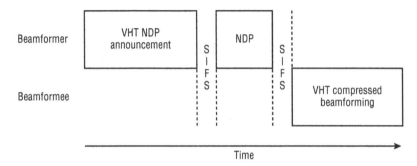

Although this section explained the beamforming process, it really explained what is known as *single-user beamforming*. The next section will introduce you to multiuser MIMO (MU-MIMO). After explaining MU-MIMO, we will continue to further explain beamforming, specifically multiuser beamforming.

Multi-User MIMO

Prior to 802.11ac, an 802.11 AP was able to communicate with only one device at a time. When an AP made a transmission, it was addressed to a single client device. With 802.11ac, it is possible for an 802.11ac AP to communicate with up to four devices simultaneously using MU-MIMO technology. However, not all 802.11ac radios support MU-MIMO, and support only the single-user communications of MIMO. Additionally, 802.11n radios support only the single-user communications of MIMO and do not support MU-MIMO. To distinguish the standard MIMO technology that was introduced with 802.11n from MU-MIMO, we will refer to it as *single-user MIMO (SU-MIMO)*.

Both 802.11n and 802.11ac APs are capable of transmitting multiple streams of data. Due to technology costs and battery consumption, however, many of the most common and popular client devices used on wireless networks are capable of transmitting only a single stream of data. This means that when an access point is communicating with a wireless tablet or other handheld device, much of the potential of the technology is not being utilized.

The goal of MU-MIMO is to use as many spatial streams as possible, whether the transmission is with one client using four spatial streams or with four clients using one spatial stream each. MU-MIMO is supported only for downstream transmissions from an 802.11ac AP to 802.11ac clients. Figure 10.7 shows an AP that is capable of transmitting four spatial streams downlink. In this illustration, the AP is using two spatial streams to transmit to a laptop, a third stream to transmit to a tablet, and a fourth stream to transmit to an 802.11ac-capable smartphone.

FIGURE 10.7   Multiuser MIMO

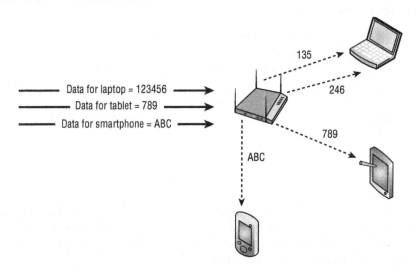

A five-number syntax is sometimes used when describing MU-MIMO radio capabilities. In a MU-MIMO system, the first number always references the transmitters (TX), and the second number references the receivers (RX). The third number represents how many unique single-user (SU) streams of data can be sent or received. The fourth number references how many multiple user (MU) streams can be transmitted. A fifth number is used to represent a MU-MIMO group or how many MU-MIMO clients are receiving transmissions at the same time. For example, a 4×4:4:3:3 802.11ac AP could transmit up to four spatial streams to a single user when operating in SU-MIMO mode. However, only three spatial streams could be used for MU-MIMO transmissions to as many as three MU-MIMO capable clients, with each client receiving a single stream. What if an 802.11ac AP was operating as a 4×4:4:3:2 MU-MIMO AP. Only two clients would belong to the MU-MIMO group. Since three spatial streams are available for MU-MIMO transmissions, 1 spatial stream would be destined for one client and 2 spatial streams for the other client that belonged to the MU-MIMO group.

So how would this work if there were 20 MU-MIMO clients associated to the 802.11ac AP? The AP would make the decision on which clients received the downlink MU-MIMO transmissions and which clients are assigned to the MU-MIMO client group. For example, three clients could receive spatial streams simultaneously in the first downlink transmission and then three different clients would receive spatial streams simultaneously in the next downlink transmission.

It should be understood that 802.11ac MU-MIMO transmissions are only downlink from an 802.11ac MU-MIMO AP to multiple 802.11ac MU-MIMO-capable clients. Uplink MU-MIMO does not exist, although it has been proposed in the 802.11ax draft amendment. Beamforming is a critical part of MU-MIMO. The previous section of this chapter explained explicit beamforming. In the following section, we explain why explicit beamforming is necessary for MU-MIMO to function.

Multi-User Beamforming

Earlier in this chapter, we explained how 802.11ac performs explicit beamforming. We also explained the principles of MU-MIMO. It is now time to discuss these two technologies together and the importance of beamforming to the success of MU-MIMO.

With single-user MIMO, beamforming uses phase adjustments of an RF signal to a client to increase the signal strength—and hopefully allow the AP and client to communicate using a higher data rate. With MU-MIMO, the task of beamforming is not just performed for transmitting to a single client; it's performed for transmitting to up to four clients at a time.

To begin the MU-MIMO beamforming process, the AP performs a channel sounding procedure, similar but more complex than with SU-MIMO. To begin the process, the AP transmits a *null data packet (NDP)* announcement frame, notifying multiple beamformees of the intent to send a beamformed transmission. The AP then follows this with an NDP frame. As with beamforming to a single user, each beamformee processes each OFDM subcarrier and creates feedback information, creating a compressed *feedback matrix*. The first beamformee responds to the AP with its compressed feedback matrix. The AP then polls each additional beamformee sequentially using Beamforming Report Poll frames. Figure 10.8 illustrates this process.

FIGURE 10.8 Multiuser beamforming sounding process

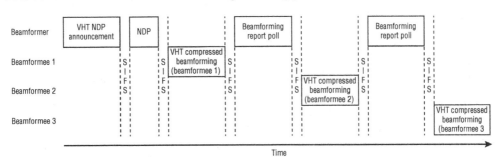

The AP then uses the feedback matrix from each of the beamformees to create a single *steering matrix*. The steering matrix defines transmit parameters for communications between each of the antennas on the AP and each of the antennas on each of the client devices, as illustrated in Figure 10.9.

FIGURE 10.9   Beamformed transmissions in a MU-MIMO environment

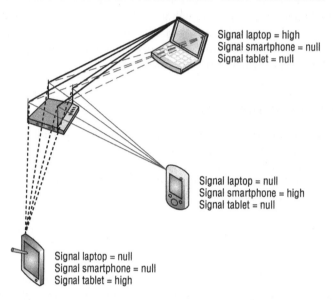

Signal laptop = high
Signal smartphone = null
Signal tablet = null

Signal laptop = null
Signal smartphone = high
Signal tablet = null

Signal laptop = null
Signal smartphone = null
Signal tablet = high

It is important to remember that in Figure 10.9, the AP is sending 16 transmissions, 4 from each antenna. Of those 16 transmissions, the receiving antenna needs to be able to distinguish and interpret the signal that is directed toward it while trying to ignore the other 12 transmissions. The signals beamformed for a specific client are timed to arrive at the same time and in sync, effectively producing a stronger signal. The beamformees that are too close to each other could experience inter-user interference from signals directed toward other users. Ideally, the users are physically separated enough from each other, and the beamformed signal for the intended user device is strong, while that signal received by the other users is lower. Figure 10.9 illustrates how the different signals should be recognized by the user. If the user devices are separated enough, then the beamformed signal to the intended user should be strong and the signals received by the other users should be *null* or weak.

After the AP transmits the multiuser frame, each client station must acknowledge its frame. As stated earlier, MU-MIMO is performed only from the AP to the client, so the acknowledgments must be single-user transmissions. Since every 802.11ac frame is an A-MPDU frame, the delivery of all the individual MPDUs is verified with a Block ACK. When a Block ACK is required, the originator of the frame—in this case, the AP—sends a *Block Acknowledgment Request (BAR)* frame to the receiver, which replies with a Block ACK. Since this is a MU-MIMO frame, the AP sends a BAR frame to a user, waits for the Block ACK from that user, and then sequentially repeats the process with the other users. Figure 10.10 illustrates this sequence.

FIGURE 10.10   MU-MIMO block acknowledgments

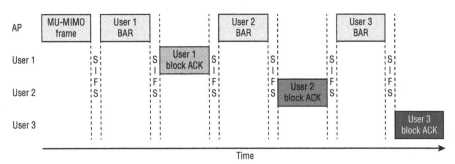

Downlink MU-MIMO capabilities were introduced in the second generation of 802.11ac access points; however, widespread use of MU-MIMO technology is rare. Although MU-MIMO sounds great on paper, real-world implementation is not practical for the following reasons:

- Very few MU-MIMO-capable 802.11ac clients exist in the current marketplace, and the technology is rarely used in the enterprise. The clients must also support explicit beamforming.

- MU-MIMO requires spatial diversity; therefore, physical distance between the clients is necessary. Most modern-day enterprise deployments of Wi-Fi involve a high density of users, which is not conducive for MU-MIMO conditions.

- Because MU-MIMO requires spatial diversity, a sizable distance between the clients and the AP is necessary. Most modern-day enterprise deployments of Wi-Fi involve a high density of users, which is not conducive for MU-MIMO conditions.

- MU-MIMO requires transmit beamforming (TxBF), which requires sounding frames. The sounding frames add excessive overhead, especially when the bulk of data frames are small. The overhead from the sounding frames usually negates any performance gained from an 802.11ac AP transmitting downstream simultaneously to multiple 802.11ac clients.

It should be noted that the 802.11ax amendment has also proposed upstream MU-MIMO communications. More information can be found in Chapter 19, "802.11ax: High Efficiency."

Channels

In previous chapters, you learned that the 802.11a amendment defined the capabilities of radios using *orthogonal frequency-division multiplexing (OFDM)* technology in the 5 GHz U-NII bands. 802.11g defined the capabilities of radios using ERP-OFDM, which is effectively the same technology, except that transmissions occur in the 2.4 GHz ISM band. The 802.11n amendment also defines the use of OFDM channels. However, key differences exist for 802.11n (HT) radios. As mentioned earlier in this chapter, 802.11n (HT) radios can operate in either frequency.

You have already learned that MIMO radios use spatial multiplexing to send multiple independent streams of unique data. Spatial multiplexing is one method of increasing the throughput. The OFDM channels used by MIMO radios use more subcarriers, and there is also an option to bond channels together. The greater frequency bandwidth provided by the OFDM channels used by 802.11n (HT) and 802.11ac (VHT) radios can also provide for higher data rates and more potential throughput.

20 MHz Channels

As you learned in Chapter 6, "Wireless Networks and Spread Spectrum Technologies," 802.11a and 802.11g radios use 20 MHz OFDM channels. As shown in Figure 10.11, each channel consists of 64 subcarriers. Forty-eight of the subcarriers transmit data, while four of the subcarriers are used as pilot tones for dynamic calibration between the transmitter and receiver. The remaining subcarriers are not used. OFDM technology also employs the use of convolutional coding and forward error correction.

FIGURE 10.11   20 MHz non-HT (802.11a/g) channel

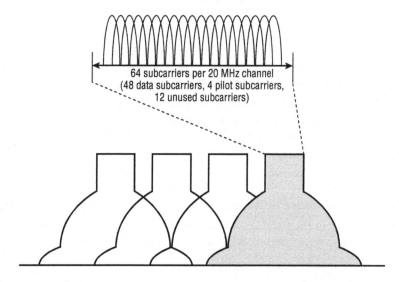

802.11n (HT) and 802.11ac (VHT) radios also use the same OFDM technology. The 20 MHz channels used by HT and VHT radios have four extra data subcarriers than a non-HT OFDM channel. As a result, the HT 20 MHz channel with a single spatial stream can provide greater aggregate throughput for the same frequency space. As shown in Figure 10.12, an HT/VHT 20 MHz OFDM channel also has 64 subcarriers. However, 52 of the subcarriers transmit data, while 4 of the subcarriers are used as pilot tones for dynamic calibration between the transmitter and receiver. In other words, although some unused subcarriers still exist, an HT or VHT radio makes use of four additional subcarriers for data transmissions. Using these four additional subcarriers is a more efficient use of the available frequency space in the 20 MHz channels.

FIGURE 10.12 20 MHz HT or VHT channel

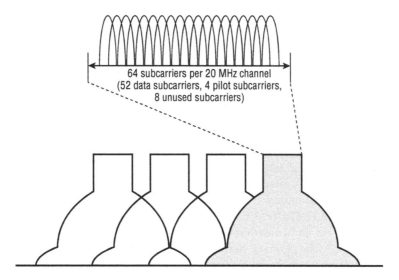

40 MHz Channels

802.11n and 802.11ac radios also have the capability of using 40 MHz OFDM channels. As shown in Figure 10.13, the 40 MHz channels use 128 OFDM subcarriers; 108 of the subcarriers transmit data, whereas 6 of the subcarriers are used as pilot tones for dynamic calibration between the transmitter and receiver. The remaining subcarriers are not used. A 40 MHz channel effectively doubles the frequency bandwidth available for data transmissions.

FIGURE 10.13 40 MHz HT or VHT channel

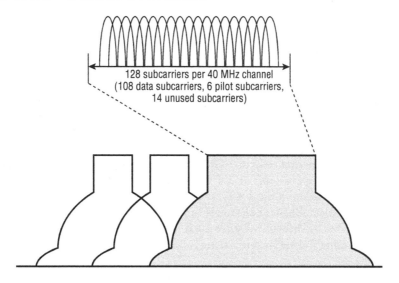

The 40 MHz channels used by HT and VHT radios are essentially two 20 MHz OFDM channels that are bonded together. Each 40 MHz channel consists of a primary and secondary 20 MHz channel. The primary and secondary 20 MHz channels must be adjacent 20 MHz channels in the frequency in which they operate. As shown in Figure 10.14, the two 20 MHz channels used to form a 40 MHz channel are designated as primary and secondary and are indicated by two fields in the body of certain 802.11 management frames. The primary field indicates the number of the primary channel. With 802.11n 40 MHz channels, a positive or negative offset indicates whether the secondary channel is one channel above or one channel below the primary channel. 802.11ac (VHT) does not reference any channel offsets but instead references the center frequency of the 40 MHz channel. However, WLAN vendors do not specify a center frequency when configuring a 40 MHz channel on an 802.11ac access point. Instead, a 20 MHz channel number is selected and that 20 MHz channel functions as the primary channel. Primary and secondary channels are used together only for data frame transmissions between an 802.11n/ac AP and 802.11n/ac client. For backward compatibility, all 802.11 management and control frames are transmitted only on the primary channel. Additionally, only the primary channel is used for data transmissions between an 802.11n/ac AP and legacy 802.11a/g clients.

FIGURE 10.14   Channel bonding

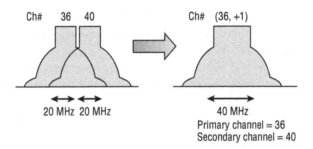

Channel reuse patterns using 40 MHz channels at 5 GHz are feasible because of all the frequency bandwidth available in the 5 GHz U-NII bands. The use of 40 MHz channels in the 5 GHz frequency bands makes perfect sense because there are many more 20 MHz channels that can be bonded together in various pairs, as shown in Figure 10.15.

Deploying 40 MHz channels at 2.4 GHz unfortunately does not scale with a multiple-channel reuse pattern. 40 MHz channels are an issue only with 802.11n radios, since 802.11ac radios operate only in 5 GHz. As you learned in earlier chapters, although 14 channels are available at 2.4 GHz, there are only three nonoverlapping 20 MHz channels available in the 2.4 GHz ISM band. When the smaller channels are bonded together to form 40 MHz channels in the 2.4 GHz ISM band, any two 40 MHz channels will overlap, as shown in Figure 10.16. In other words, only one 40 MHz channel can be used at 2.4 GHz, and the possibility of a channel reuse pattern is essentially impossible.

FIGURE 10.15   Channel bonding—5 GHz U-NII bands

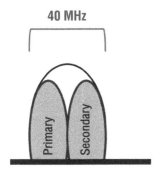

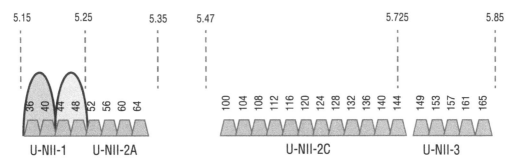

FIGURE 10.16   Channel bonding—2.4 GHz ISM band

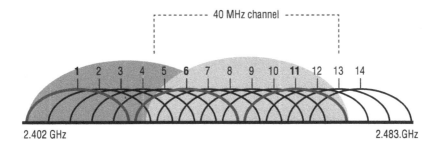

Forty MHz Intolerant

As you just learned, only one nonoverlapping 40 MHz channel can be deployed in the 2.4 GHz band; therefore, a channel reuse pattern using multiple 40 MHz channels in 2.4 GHz is impossible. However, it is still possible to turn on channel bonding in the 2.4 GHz band. A 2.4 GHz 802.11n access point transmitting on a 40 MHz channel will interfere with other nearby APs that have been deployed using a standard 20 MHz channel reuse pattern of 1, 6, and 11. By default, 802.11n clients and APs should use 20 MHz channels when transmitting in the 2.4 GHz band. They can also advertise that they are *Forty MHz Intolerant* using various 802.11n management frames. Any 802.11n AP using a 40 MHz channel will be forced to switch back to using only 20 MHz channels if they receive the frames from nearby 802.11n 2.4 GHz stations that are intolerant.

Effectively, Forty MHz Intolerant operations are a protection against your next-door neighbor who might deploy a 40 MHz channel and interfere with your 2.4 GHz 20 MHz channels. Enterprise WLAN access points should have 20 MHz channels as the default setting at 2.4 GHz. It should be noted that the Forty MHz Intolerant operations are meant for 2.4 GHz only and are not permitted in 5 GHz.

80 MHz and 160 MHz Channels

802.11ac introduced two new channel widths: 80 MHz and 160 MHz. Just as a 40 MHz channel is created by combining two 20 MHz channels, an 80 MHz channel combines four 20 MHz channels. As shown in Figure 10.17, an 80 MHz channel consists of 256 subcarriers, of which 234 are used to transmit data, 8 are used as pilot carriers, and the remaining 14 are unused.

FIGURE 10.17   80 MHz VHT (802.11ac) channel

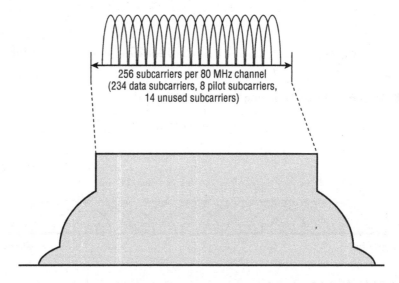

256 subcarriers per 80 MHz channel
(234 data subcarriers, 8 pilot subcarriers,
14 unused subcarriers)

The second channel width that was introduced with 802.11ac is a 160 MHz channel. As you might deduce, the 160 MHz channel is made up of two 80 MHz channels; however, the two 80 MHz channels do not have to be adjacent. If the channels are adjacent, then it is referred to as a 160 MHz channel. If they are not adjacent, then it is referred to as an 80+80 MHz channel. Since these channels can be adjacent or separated, they are treated as two individual 80 MHz channels, and you do not gain any unused subcarriers between the channels. Therefore, a 160 MHz channel is simply two 80 MHz channels and consists of

512 subcarriers, with 468 used to transmit data, 16 used as pilot carriers, and the remaining 28 are unused.

Figure 10.18 shows all the various combinations of 20 MHz, 40 MHz, 80 MHz, and contiguous 160 MHz channels in the 5 GHz U-NII bands. Please note that this figure also depicts the proposed U-NII-4 channels, which are not yet available for Wi-Fi communications. Even though 80 MHz and 160 MHz channels are available with 802.11ac radios, they should not be used in enterprise. In Chapter 13, "WLAN Design Concepts," you will learn that a 20 MHz channel reuse design is still the preferred method. A 40 MHz channel reuse design can also work in the enterprise with careful planning. 80 MHz and 160 MHz channel implementations do not scale in an enterprise WLAN.

FIGURE 10.18   20, 40, 80, and 160 MHz channels

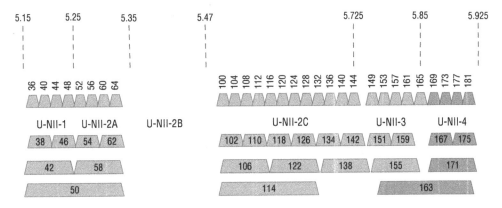

Guard Interval

For digital signals, data is modulated onto the carrier signal in bits or in collections of bits, called *symbols*. When 802.11a/g radios transmit at 54 Mbps, each OFDM symbol contains 288 bits; 216 of these bits are data, and 72 of the bits are error-correction bits. All the data bits of an OFDM symbol are transmitted across the 48 data subcarriers of a 20 MHz non-HT channel.

802.11a/g radios use an 800-nanosecond *guard interval (GI)* between OFDM symbols. The guard interval is a period of time between symbols that accommodates the late arrival of symbols over long paths. In a multipath environment, symbols travel different paths, so some symbols arrive later. A "new" symbol may arrive at a receiver before a "late" symbol has been completely received. This is known as *intersymbol interference (ISI)* and can result in data corruption.

In earlier chapters, we discussed ISI and delay spread. The *delay spread* is the time differential between multiple paths of the same signal. Normal delay spread is from

50 nanoseconds to 100 nanoseconds, and a maximum delay spread is about 200 nanoseconds. The guard interval should be two to four times the length of the delay spread. Think of the guard interval as a buffer for the delay spread. The normal guard interval is an 800-nanosecond buffer between symbol transmissions; however, in most indoor environments, a 400-nanosecond buffer will provide enough separation between transmissions. As shown in Figure 10.19, a guard interval will compensate for the delay spread and help prevent intersymbol interference. If the guard interval is too short, intersymbol interference can still occur.

FIGURE 10.19 Guard interval

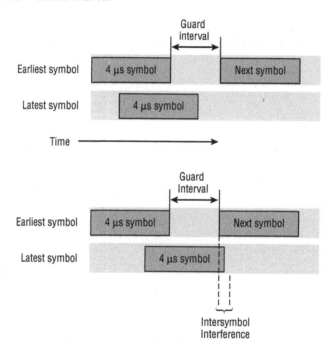

HT/VHT technology introduced the capability of a 400-nanosecond guard interval, sometimes referred to as the *short guard interval*. 802.11n/ac radios also use an 800-nanosecond guard interval; however, a shorter 400-nanosecond guard interval is optional. A shorter guard interval results in a shorter symbol time, which has the effect of increasing data rates by about 10 percent. If the optional, shorter 400-nanosecond guard interval is used with an 802.11n/ac radio, throughput will increase; however, the odds of an intersymbol interference occurrence increase. If intersymbol interference does indeed occur because of the shorter GI, the result is data corruption. If data corruption occurs, layer 2 retransmissions will increase and the throughput will be adversely affected. If throughput goes down because of a shorter GI setting, the guard interval setting of 800 nanoseconds should be used instead. In most indoor environments, the short guard interval of 400 nanoseconds is preferred. The long guard interval may be needed indoors in high multipath environments. The long guard interval is normally used outdoors.

256-QAM Modulation

In Chapter 1, "Overview of Wireless Standards, Organizations, and Fundamentals," we explained how waves are manipulated, or modulated, in order to carry data. The chapter described how amplitude, frequency, or phase could be varied to represent a single bit of data or even multiple bits of data. Over the years, newer and faster modulation methods have been incorporated into 802.11 Physical layer (PHY) technologies. With the introduction of each new and faster PHY, a newer modulation method, capable of encoding more bits, was also introduced and thus increased the effective speed and performance of the network. It is important to remember that even as new transmission and modulation methods are introduced, the older and slower methods are still supported and used.

As a client moves away from an access point and the signal level decreases, dynamic rate switching causes the client to shift to a slower data rate to maintain a connection. Even though we tend to highlight the latest and greatest technologies that are introduced with the most recent standard or amendment, the older and slower technologies are still key and necessary components of any infrastructure. This section describes 256-QAM, which was introduced with the 802.11ac amendment. (QAM is the acronym for quadrature amplitude modulation and is pronounced "kwam," which rhymes with "Tom.") The following is a list of modulation methods that are used with 802.11 networks:

DBPSK—Differential binary phase-shift keying

DQPSK—Differential quadrature phase-shift keying

BPSK—Binary phase-shift keying

QPSK—Quadrature phase-shift keying

16-QAM—16 quadrature amplitude modulation

64-QAM—64 quadrature amplitude modulation

256-QAM—256 quadrature amplitude modulation

256-QAM is an evolutionary upgrade that was introduced with 802.11ac. The 802.11a amendment introduced 64-QAM modulation. 64-QAM identifies 64 unique values. 64-QAM essentially performs a phase shift that can differentiate eight different levels and also performs an amplitude shift, which can also differentiate eight different levels. Combine the two of them and the system has the ability to identify the 64 unique values. Having 64 distinct values provides the ability for each value to represent 6 bits (2^6 = 64). QAM is often represented by symbols displayed in a constellation chart, as shown in Figure 10.20. Each dot represents a unique symbol—a different grouping of 6 bits.

FIGURE 10.20   64-QAM constellation chart

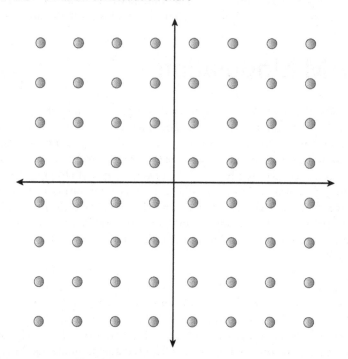

With 802.11ac, a new modulation method was introduced, 256-QAM. 256-QAM identifies 256 unique values, using 16 different levels of phase shift and 16 different levels of amplitude shift. Because there are 256 distinct values, each value is able to represent 8 bits ($2^8 = 256$). Figure 10.21 shows the constellation chart for 256-QAM.

Now that we have provided a basic explanation of 64-QAM and 256-QAM, we need to delve a little deeper so that you understand what is happening and how the two differ. When a 64-QAM radio transmits data, it modifies the amplitude and phase of the wave and then transmits it. The receiving radio must then take the signal and identify the amplitude and phase modifications that were made to identify which of the 64 symbols was transmitted. This is not always easy, because noise and interference can make it difficult to identify the values of the transmitted signal.

As an analogy, consider an archer shooting arrows at a target. Suppose that the target is a 2-meter-square board with evenly spaced 1-inch dots in rows and columns of eight that stands on the roof of a building, which we will call the target roof. Ten meters away from the target roof is another roof where an Olympic archer is standing. We will call this the shooting roof. In perfect conditions, our Olympic archer never misses. However, the space between these two buildings is unpredictably windy, not only from side to side, but also with updrafts and downdrafts. From the shooting roof, we ask the archer to shoot an arrow at a specific dot on the target. Since the winds are so unpredictable, the archer does not make any adjustments or corrections, only aims for the chosen dot and hopes that the winds do not push the arrow too far off target.

FIGURE 10.21   256-QAM constellation chart

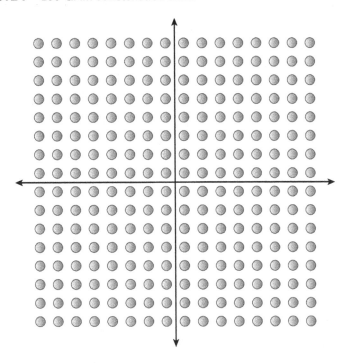

When the arrow hits the target, a person on the target roof looks at the location of the arrow and, using a ruler, measures the distance of the arrow from the nearest dots and attempts to identify which dot the archer was shooting at. As an example, Figure 10.22 shows four dots and the location where the arrow hit. In this figure, the person would identify the upper-right dot as the one that the archer was aiming for. At this short distance, unless the wind was incredibly strong, the archer should be able to accurately shoot the arrow on or near the correct dot, and the person at the target should be able to properly identify which dot the archer was aiming for.

If we incrementally move the archer farther away from the target, the wind will interfere more with the flight of the arrow, possibly making it drift farther from the dot that the archer was aiming for. The farther away from the target, the less successful the archer will be. 64-QAM behaves similarly. The transmitting radio modulates the signal and transmits it. The amplitude and phase adjustments are exact, and a perfectly modulated signal is generated by the radio. Noise, interference, and signal attenuation alter the signal so that when it is received, it has been modified. The receiver maps the signal on the constellation diagram and calculates the error vector to identify the constellation point, which equates to the data that was transmitted. *Error vector magnitude (EVM)* is a measure used to quantify the performance of a radio receiver or transmitter in regards to modulation accuracy. With QAM modulation, EVM is a measure of how far a received signal is from a constellation point.

FIGURE 10.22   Example target

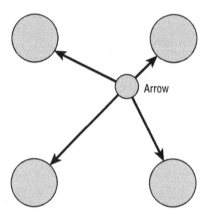

Arrow

So, now that you have a general idea of how 64-QAM behaves, how does it relate to 256-QAM? That is fairly straightforward. In our archer analogy, instead of having 64 dots on the target, we use the same size target but place 256 dots on it. This means that there is less room for error, and the effect of the wind on the arrow is much more critical. In the same way, 256-QAM is more sensitive to noise and interference. Because of this, 802.11ac receiver performance requires about 5 dB of additional gain as compared to 64-QAM.

256-QAM is used for the highest modulation coding sets. To achieve these higher data rates, higher signal-to-noise ratios are needed. This also means that the clients need to be close to the AP in order to achieve these data rates. Since a 256-QAM signal can transmit 8 bits per subcarrier, compared with the 6 bits that were transmitted with 64-QAM, a speed increase of 33 percent is achieved solely by deploying this feature.

As mentioned earlier, 802.11ac is supported only in the 5 GHz bands, leaving 802.11n as the fastest technology supported in 2.4 GHz. The fastest modulation officially supported by the 802.11n amendment is 64-QAM. With 802.11ac dual-band radios, even though 256-QAM is not part of the 802.11n standard, the technology is already integrated into the 802.11ac radio chipsets. Therefore, many vendors are enabling this technology on the 2.4 GHz radios, unofficially increasing the fastest 2.4 GHz data rates by about 33 percent.

Although 802.11ac is for 5 GHz radios only, some WLAN vendors offer support for 256-QAM for the 2.4 GHz 802.11n radio of an access point. TurboQAM is a Broadcom marketing term for non-standard 256-QAM support on a 2.4 GHz radio. For this capability to work, clients would also have to support 256-QAM on their 2.4 GHz radios. Additionally, achieving the necessary SNR for 256-QAM is normally going to be a challenge due to the very high noise floor that usually exists in the 2.4 GHz band.

802.11n/ac PPDUs

In earlier chapters, you learned that a MAC service data unit (MSDU) is the layer 3–7 payload of an 802.11 data frame. You also learned that a MAC protocol data unit (MPDU) is a technical name for an entire 802.11 frame. An MPDU consists of a layer 2 header, body, and trailer.

When an MPDU (802.11 frame) is sent down from layer 2 to the Physical layer, a preamble and PHY header are added to the MPDU. This creates what is called a *Physical Layer Convergence Procedure protocol data unit (PPDU).* The details of the PHY preamble and header are well beyond the scope of the CWNA exam. The main purpose of the preamble is to use bits to synchronize transmissions at the Physical layer between two 802.11 radios. The main purpose of the PHY header is to use a signal field to indicate how long it will take to transmit the 802.11 frame (MPDU) and to notify the receiver of the MCS (data rate) that is being used to transmit the MPDU. The 802.11-2016 standard defines the use of multiple PPDU structures and preambles. Both 802.11n and 802.11ac introduced new PHY headers.

Non-HT

The first PPDU format is called *non-HT* and is often also referred to as a legacy format because it was originally defined by Clause 17 of the 802.11-2016 standard for OFDM transmissions. As shown in Figure 10.23, the non-HT PPDU consists of a preamble that uses legacy short and long training symbols for synchronization. An OFDM symbol consists of 12 bits. The header contains the signal field, which indicates the time needed to transmit the payload of the non-HT PPDU, which, of course, is the MPDU (802.11 frame). Support for the non-HT legacy format is mandatory for 802.11n radios, and transmissions can occur in only 20 MHz channels. The non-HT format effectively is the same format used by legacy 802.11a and 802.11g radios.

FIGURE 10.23 PPDU formats

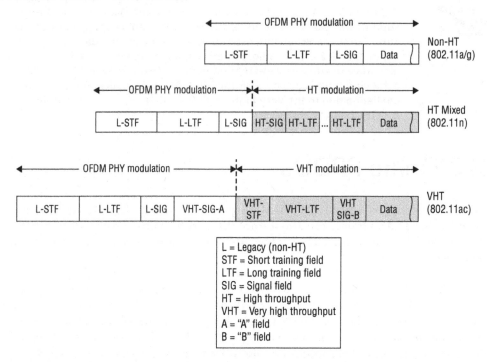

L = Legacy (non-HT)
STF = Short training field
LTF = Long training field
SIG = Signal field
HT = High throughput
VHT = Very high throughput
A = "A" field
B = "B" field

HT Mixed

The second PPDU format defined in the 802.11-2016 standard is the *HT Mixed* format. As shown in Figure 10.23, the beginning of the preamble contains the non-HT training symbols and legacy signal field that can be decoded by legacy 802.11a and 802.11g radios. The rest of the HT Mixed preamble and header cannot be decoded by legacy 802.11a/g devices. HT information includes the HT-SIG and HT training symbols.

The HT Signal (HT-SIG) contains information about the MCS, frame length, 20 MHz or 40 MHz channel size, frame aggregation, guard interval, and STBC. The HT Short Training Field (HT-STF) and HT Long Training Field (HT-LTF) are used for synchronization between MIMO radios.

Non-802.11n receivers will not be able to read the frame, but the length field in the legacy section of the header will allow them to know how long the medium is going to be busy, and they will therefore stay silent without having to do an energy detect at each cycle. The HT Mixed format supports both HT and legacy 802.11a/g OFDM radios. The HT Mixed format is also considered mandatory, and transmissions can occur in both 20 MHz and 40 MHz channels. When a 40 MHz channel is used, all broadcast traffic must be sent on a legacy 20 MHz channel so as to maintain interoperability with the 802.11a/g non-HT clients. Also, any transmissions to and from the non-HT clients will have to use a legacy 20 MHz channel.

VHT

The final PPDU format defined by the 802.11-2016 standard is the VHT format for 802.11ac radios. As shown in Figure 10.23, the preamble is compatible with both legacy 802.11a/g radios and 802.11n radios. The non-VHT portion of the PHY header can be understood by legacy 802.11a/n devices while the VHT portion of the PHY header can only be understood by 802.11ac radios. The VHT portion of the PHY header can be used to indicate whether or not SU-MIMO or MU-MIMO transmissions are occurring.

802.11n/ac MAC

So far, we have discussed all the enhancements to the Physical layer that MIMO radios use to achieve greater bandwidth and throughput. The 802.11-2016 standard also addresses enhancements to the MAC sublayer of the Data-Link layer to increase throughput and improve power management. Medium contention overhead is addressed by using two methods of frame aggregation. Enhancements are also addressed using block acknowledgments to limit the amount of fixed MAC overhead. Three methods of power management are defined for MIMO radios.

A-MSDU

As you can see in Figure 10.24, every time a unicast 802.11 frame is transmitted, a certain amount of fixed overhead exists as a result of the PHY header, MAC header, MAC trailer, interframe spacing, and acknowledgment frame. Medium contention overhead also exists because of the time required when each frame must contend for the medium.

FIGURE 10.24   802.11 unicast frame overhead

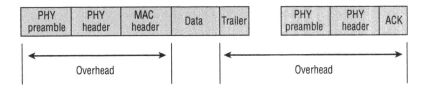

The 802.11n amendment introduced two methods of frame aggregation to help reduce the overhead. *Frame aggregation* is a method of combining multiple frames into a single frame transmission. The fixed MAC layer overhead is reduced, and overhead caused by the random backoff timer during medium contention is also minimized.

The first method of frame aggregation is known as *aggregate MAC service data unit (A-MSDU)*. As you learned in earlier chapters, the MSDU is the layer 3–7 payload of a data frame. As Figure 10.25 shows, multiple MSDUs can be aggregated into a single frame transmission.

FIGURE 10.25 A-MSDU

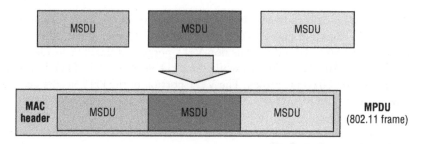

A MIMO access point using A-MSDU aggregation would receive multiple 802.3 frames, remove the 802.3 headers and trailers, and then wrap the multiple MSDU payloads into a single 802.11 frame for transmission. The aggregated MSDUs will have a single wireless receiver when wrapped together in a single frame. All the MSDUs within an A-MSDU are encrypted together as a single encrypted payload. It should be noted, however, that the individual MSDUs must all be of the same 802.11e QoS access category. Voice MSDUs cannot be mixed with Best Effort or Video MSDUs inside the same aggregated frame. Many of the initial 802.11n chipsets implemented A-MSDU.

A-MPDU

The second method of frame aggregation is known as *aggregate MAC protocol data unit (A-MPDU)*. As you learned in earlier chapters, the MPDU is an entire 802.11 frame, including the MAC header, body, and trailer. As shown in Figure 10.26, multiple MPDUs can be aggregated into a single PPDU transmission. Notice in Figure 10.26 that the A-MPDU comprises multiple MPDUs and is prepended with a PHY header.

FIGURE 10.26 A-MPDU

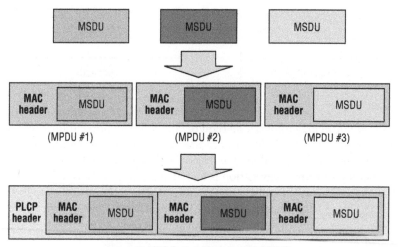

Aggregate MAC Protocol Data Unit (A-MPDU)

The individual MPDUs within an A-MPDU must all have the same receiver address. Unlike A-MSDU, the data payload of each MPDU is encrypted and decrypted separately. Much like MSDU aggregation, individual MPDUs must all be of the same 802.11e QoS access category. Voice MPDUs cannot be mixed with Best Effort or Video MPDUs inside the same aggregated frame. Please note that MPDU aggregation has more overhead than MSDU aggregation because each MPDU has an individual MAC header and trailer. However, A-MPDU utilizes Block ACKs for delivery verification, whereas A-MSDU uses ACKs for delivery verification. An A-MPDU frame reduces the per-frame overhead and requires only a single Block ACK. CRC errors can be detected in the individual MPDU frames, and therefore an entire A-MPDU does not need to be retransmitted, only the individual MPDU that is corrupted. Therefore, A-MPDU is less susceptible to noise than A-MSDU. For this reason, the majority of WLAN vendors choose to use A-MPDU in the second generation of 802.11n chipsets.

The 802.11ac amendment defined only the use of A-MPDU aggregation. All 802.11ac frames are transmitted using the aggregate MAC protocol data unit (A-MPDU) format, even if only a single frame is being transmitted. Although A-MPDU is required for 802.11ac transmission, it should be noted that A-MSDU and A-MPDU can be used together. The payload of an A-MPDU transmission can be multiple A-MSDUs.

Block Acknowledgment

As you learned in earlier chapters, all 802.11 unicast frames must be followed by an ACK frame for delivery verification purposes. Multicast and broadcast frames are not acknowledged. An A-MSDU contains multiple MSDUs all wrapped in a single frame with one MAC header and one destination. Therefore, only normal acknowledgments are required when using A-MSDU aggregation. However, an A-MPDU contains multiple MPDUs, each with its own unique MAC header. Each of the individual MPDUs must be acknowledged; this is accomplished by using a Block ACK frame. Block ACKs were first introduced by the 802.11e amendment as a method of acknowledging multiple individual 802.11 frames during a *frame burst*. Block acknowledgments are also needed to cover the multiple MPDUs that are aggregated inside a single A-MPDU transmission.

As shown in Figure 10.27, when A-MSDU aggregation is used, a standard 802.11 ACK frame is used to verify the delivery of the A-MSDU transmission. The star in both Figures 10.27 and 10.28 represents a single corrupted bit. If any portion of the A-MSDU frame is corrupted, however, the receiver will not respond with an ACK frame, and the entire A-MSDU frame must be retransmitted. As depicted in Figure 10.28, with A-MPDU aggregation, if one of the MPDUs is corrupted, the Block ACK will inform the transmitter which MPDU is corrupted. Only the corrupted MPDU must be retransmitted, as opposed to the entire A-MPDU. The A-MPDU method is more efficient than A-MSDU because of the use of Block ACKs and less retransmission overhead. This is the main reason that A-MPDU is required for 802.11ac radios.

FIGURE 10.27   A-MSDU, ACKs, and retransmissions

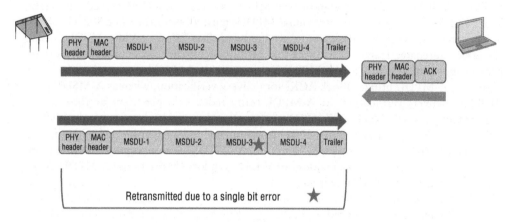

Retransmitted due to a single bit error ★

FIGURE 10.28   A-MPDU, Block ACKs, and retransmissions

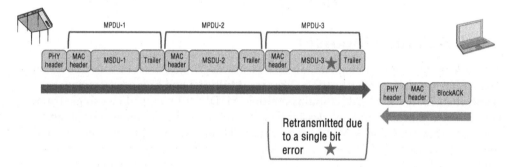

Retransmitted due to a single bit error ★

Power Management

As the 802.11 standard is amended, power-management capabilities continue to be enhanced. The 802.11e QoS amendment introduced *unscheduled automatic power save delivery (U-APSD)*, which is the mechanism used by WMM Power Save (WMM-PS). The 802.11n amendment introduced two power-management mechanisms that are used by 802.11n (HT) radios. The 802.11n power-management mechanisms are meant as supplements to WMM-PS when MIMO radios are used.

802.11n/ac radios still support a basic power-save mode, which is based on the original 802.11 power-management mechanisms. Access points buffer frames for stations in basic power-save mode. The stations wake up when delivery traffic indication message (DTIM) beacons are broadcast and the stations download their buffered frames.

The first power-management method introduced with 802.11n is called *spatial multiplexing power save (SM power save)*. The purpose of SM power save is to allow a MIMO 802.11n/ac device to power down all but one of its radios. For example, a 4×4 MIMO device with four radio chains would power down three of the four radios, thus conserving power. SM power save defines two methods of operation: static and dynamic.

When static SM power save is utilized, a MIMO client station powers down all the client's radios except for one single radio. Effectively, the MIMO client station is now the equivalent of a SISO radio that is capable of sending and receiving only one spatial stream. The client uses an SM power save action frame to inform the access point that the MIMO client is using only one radio and is capable of receiving only one spatial stream from the AP. The SM power save action frame is also used to tell the AP that the client station has powered up all of its radios and now is capable of transmitting and receiving multiple spatial streams once again.

When dynamic SM power save is utilized, the MIMO client can also power down all but one of the client's radios but can power up the radios again much more rapidly. The client station disables all but one of the radios after a frame exchange. An access point can trigger the client to wake up the sleeping radios by sending a request-to-send (RTS) frame. The client station receives the RTS frame, powers up the sleeping radios, and sends a clear-to-send (CTS) frame back to the access point. The client can now once again transmit and receive multiple spatial streams. The client uses an SM power save action frame to inform the AP of the client's dynamic power save state.

The second power-management method introduced with 802.11n is *power save multi-poll (PSMP)*. PSMP is an extension of automatic power save delivery (APSD) that was defined by the 802.11e amendment. Unscheduled PSMP (U-PSMP) is similar to U-APSD and uses the same delivery-enabled and trigger-enabled mechanisms. Scheduled PSMP (S-PSMP) is also similar to S-APSD and is an effective method for streaming data and other scheduled transmissions. S-PSMP uses a frame, PSMP Action Frame, to schedule downlink and uplink transmissions. PSMP downlink transmission time (DTT) is the time scheduled for the AP to transmit to its connected stations, and PSMP uplink transmission time (UTT) is the time scheduled for the stations to transmit to the AP.

VHT TXOP power save is another power-management technique that was introduced as part of the 802.11ac amendment. If a client sees that a *transmit opportunity (TXOP)* is allocated to another client, VHT TXOP power save allows the client to turn off its radio during the duration of the transmission. A TXOP can occur over several frames, which could allow the client to doze for an extended period of time. The AP must make sure to remember that the client is unavailable during the time, and not attempt to send any frames to the dozing client.

Modulation and Coding Scheme

802.11n data rates are defined with a *modulation and coding scheme (MCS) matrix*. Non-HT radios that used OFDM technology (802.11a/g) use data rates of 6 Mbps to 54 Mbps, based on the modulation and coding method that is used. HT radios, however, define data rates based on numerous factors, including modulation, coding method, the number of spatial streams, channel size, and guard interval. Each modulation and coding scheme (MCS) is a variation of these multiple factors. Seventy-seven modulation and coding schemes (MCSs) exist for both 20 MHz HT channels and 40 MHz HT channels. Eight modulation and coding schemes are *mandatory* for 20 MHz HT channels, as shown in Table 10.2. The eight mandatory MCSs for 20 MHz HT channels are comparable to basic (required) rates.

TABLE 10.2   MCS—20 MHz HT channel, one spatial stream

MCS Index	Modulation	Spatial Streams	Data Rates	
			800 ns GI	**400 ns GI**
0	BPSK	1	6.5 Mbps	7.2 Mbps
1	QPSK	1	13.0 Mbps	14.4 Mbps
2	QPSK	1	19.5 Mbps	21.7 Mbps
3	16-QAM	1	26.0 Mbps	28.9 Mbps
4	16-QAM	1	39.0 Mbps	43.3 Mbps
5	64-QAM	1	52.0 Mbps	57.8 Mbps
6	64-QAM	1	58.5 Mbps	65.0 Mbps
7	64-QAM	1	65.0 Mbps	72.2 Mbps

As you can see in Table 10.2, the modulation type, the guard interval, and the number of spatial streams all determine the eventual data rate. Table 10.3 describes the modulation and coding schemes for a 40 MHz channel using four spatial streams.

TABLE 10.3   MCS—40 MHz HT channel, four spatial streams

MCS Index	Modulation	Spatial Streams	Data Rates	
			800 ns GI	**400 ns GI**
24	BPSK	4	54.0 Mbps	60.0 Mbps
25	QPSK	4	108.0 Mbps	120.0 Mbps
26	QPSK	4	162.0 Mbps	180.0 Mbps
27	16-QAM	4	216.0 Mbps	240.0 Mbps
28	16-QAM	4	324.0 Mbps	360.0 Mbps

MCS Index	Modulation	Spatial Streams	Data Rates	
			800 ns GI	**400 ns GI**
29	64-QAM	4	432.0 Mbps	480.0 Mbps
30	64-QAM	4	486.0 Mbps	540.0 Mbps
31	64-QAM	4	540.0 Mbps	600.0 Mbps

802.11n (HT) defined 77 different modulation and coding schemes (MCSs). HT radios defined MCSs based on numerous factors, including modulation, coding method, the number of spatial streams, channel size, and guard interval. 802.11n also defined MCSs that allowed unequal modulation, which is the use of different modulation and coding schemes at the same time on different spatial streams. 802.11ac (VHT) simplified this by defining only 10 MCS options, as shown in Table 10.4.

TABLE 10.4 VHT MCS, modulation, code rate, and data rate

VHT MCS Value	Modulation	Code Rate (R)	20 MHz Data Rate (Mbps)
0	BPSK	1/2	7.2
1	QPSK	1/2	14.4
2	QPSK	3/4	21.7
3	16-QAM	1/2	28.9
4	16-QAM	3/4	43.3
5	64-QAM	2/3	57.8
6	64-QAM	3/4	65.0
7	64-QAM	5/6	72.2
8	256-QAM	3/4	86.7
9	256-QAM	5/6	96.3*

*MCS 9 is supported only for 40 MHz, 80 MHz, and 160 MHz channels, not for 20 MHz channels.

The first eight modulation and coding schemes are mandatory; however, most vendors support the last two, which provide 256-QAM modulation. The Code Rate (R) column shows the error-correcting code used by each MCS. Error-correcting codes add redundant information to assist with error recovery. The code rate is represented by a fraction. The first number (numerator) represents the quantity of user data bits, relative to the number of bits on the channel (denominator)—the higher the code rate, the more data transmitted and the less redundancy provided. The last column represents the maximum achievable data rate for each MCS. The data rate is based on a 20 MHz channel, a single spatial stream, and a short guard interval (400 ns).

802.11ac Data Rates

There is not one specific enhancement that provides 802.11ac with its faster data rates but a combination of enhancements. This section will review the key components involved with increased performance and explain how 802.11ac can boast theoretical data rates of up to 6933 Mbps.

The first enhancement toward the increased data rates of 802.11ac is 256-QAM. This is incorporated into MCS 8 and MCS 9. Table 10.5 shows the maximum data rate for each MCS operating on a single spatial stream and a 20 MHz channel and using a 400 ns short guard interval. Due to technical and practical reasons, some MCS values are not supported with certain channel width and spatial stream combinations. There are 10 such instances. MCS 6 is not supported for an 80 MHz channel when using three or seven spatial streams. MCS 9 has the most exceptions. It will not work with a 20 MHz channel when using one, two, four, five, seven, or eight spatial streams. It will not work with an 80 MHz channel with six spatial streams, and it will not work with a 160 MHz channel with three spatial streams.

TABLE 10.5 802.11ac data rate factors

MCS	20 MHz Data Rate	Spatial Stream Multiplier	Channel Width Multiplier
0	7.2	× 1 (1 streams)	× 1.0 (20 MHz)
1	14.4	× 2 (2 streams)	× 2.1 (40 MHz)
2	21.7	× 3 (3 streams)	× 4.5 (80 MHz)
3	28.9	× 4 (4 streams)	× 9.0 (160 MHz)
4	43.3	× 5 (5 streams)	
5	57.8	× 6 (6 streams)	

MCS	20 MHz Data Rate	Spatial Stream Multiplier	Channel Width Multiplier
6	65.0	× 7 (7 streams)	
7	72.2	× 8 (8 streams)	
8	86.7		
9*	96.3		

*MCS 9 is supported only for 40, 80, and 160 channels, not for 20 MHz channels.

As shown in Table 10.5, even though there are only 10 MCSs for VHT, there are two other variables that determine the data rate. Each MCS can use up to eight spatial streams and four different channel widths. Each spatial stream is a transmission capable of the data rate provided by the MCS that it is using to transmit. Calculating the data rate increase is simply a matter of multiplying the 20 MHz data rate by the number of spatial streams, as shown in the Spatial Stream Multiplier column of Table 10.5.

The last variable to factor into the increase in data rates is the channel width. Earlier in this chapter we explained that when combining channels, not only did we increase the throughput due to doubling of the channel, but we also gained a little more channel space from the area between the two bonded channels. Therefore, the increase for a 40 MHz channel is 2.1 times, and the increase for an 80 MHz channel is 4.5 times. Since a 160 MHz channel consists of two 80 MHz channels, either side by side or separated, there is no additional gain. The multiplier for a 160 MHz channel is simply twice that of the 80 MHz channel, 9 times. Table 10.6 shows the maximum data rate across each of the channel widths for each MCS when operating on a single spatial stream and using a 400 ns short guard interval.

TABLE 10.6 Maximum data rates (Mbps) - VHT

MCS	20 MHz	40 MHz	80 MHz	160 MHz
0	7.2	15.0	32.5	65.0
1	14.4	30.0	65.0	130.0
2	21.7	45.0	97.5	195.0
3	28.9	60.0	130.0	260.0
4	43.3	90.0	195.0	390.0

TABLE 10.6 Maximum data rates (Mbps) - VHT *(continued)*

MCS	20 MHz	40 MHz	80 MHz	160 MHz
5	57.8	120.0	260.0	520.0
6	65.0	135.0	292.5	585.0
7	72.2	150.0	325.0	650.0
8	86.7	180.0	390.0	780.0
9	96.3*	200.0	433.3	866.7

*MCS 9 is supported only for 40, 80, and 160 channels, not for 20 MHz channels.

Confused yet? Although 802.11ac defines only 10 MCSs, there are still more than 300 possible VHT data rates dependent on the variables on guard interval, spatial streams, and channel width. And as previously mentioned, there are 77 MCSs for 802.11n that determine HT data rates. You are not expected to remember all the possible data rate combinations for the CWNA exam. However, a good reference for 802.11n/ac data rates is available at www.mcsindex.com.

HT/VHT Protection Mechanisms

In earlier chapters, you learned about the protection mechanisms used in an ERP (802.11g) network. RTS/CTS and CTS-to-Self mechanisms are used to ensure that 802.11b HR-DSSS clients do not transmit when ERP-OFDM transmissions are occurring. The 802.11n/ac radios require backward compatibility with 802.11a and 802.1b/g radios. Therefore, protection mechanisms are also needed for 802.11n/ac radios to co-exist with 802.11a/b/g radios. The 802.11n amendment originally defined *HT protection modes* that enable RTS/CTS and CTS-to-Self to protect 802.11n/ac data frame transmissions. Please note that the rules for HT protection modes also apply for any 802.11ac (VHT) radios, not just 802.11n (HT) radios.

HT Protection Modes (0–3)

To ensure backward compatibility with older 802.11a/b/g radios, 802.11n/ac access points may signal to other 802.11n/ac stations when to use one of four HT protection modes. As previously mentioned, HT protection is also use for VHT radios. A field in the beacon frame, the HT Protection field, has four possible settings of 0–3. Much like an ERP

(802.11g) access point, the protection modes may change dynamically, depending on devices that are nearby or associated to the 802.11n/ac access point. The protection mechanisms that are used are RTS/CTS, CTS-to-Self, Dual-CTS, or other protection methods. The four modes are as follows:

Mode 0—Greenfield (No Protection) Mode   This mode is referred to as *Greenfield* because only HT radios are in use. All the HT client stations must also have the same operational capabilities. If the HT basic service set is a 20 MHz BSS, all the stations must be 20 MHz capable. If the HT basic service set is a 20/40 MHz BSS, all the stations must be 20/40 capable. If these conditions are met, there is no need for protection.

Mode 1—HT Nonmember Protection Mode   In this mode, all the stations in the BSS must be HT stations. Protection mechanisms kick in when a non-HT client station or a non-HT access point is heard that is not a member of the BSS. For example, an HT AP and stations may be transmitting on a 40 MHz HT channel. A non-HT 802.11a access point or client station is detected to be transmitting in a 20 MHz space that interferes with either the primary or secondary channel of the 40 MHz HT channel.

Mode 2—HT 20 MHz Protection Mode   In this mode, all the stations in the BSS must be HT stations and are associated to a 20/40 MHz access point. If a 20 MHz–only HT station associates to the 20/40 MHz AP, protection must be used. In other words, the 20/40-capable HT stations must use protection when transmitting on a 40 MHz channel in order to prevent the 20 MHz–only HT stations from transmitting at the same time.

Mode 3—Non-HT Mixed Mode   This protection mode is used when one or more non-HT stations are associated to the HT access point. The HT basic service set can be either 20 MHz or 20/40 MHz capable. If any 802.11a/b/g radios associate to the BSS, protection will be used. Mode 3 will probably be the most commonly used protection mode because most basic service sets will most likely have legacy 802.11a/b/g devices as members.

Wi-Fi Alliance Certification

The Wi-Fi Alliance maintains a vendor certification program for 802.11n called *Wi-Fi CERTIFIED n*, along with a certification for 802.11ac called *Wi-Fi CERTIFIED ac*. Products are tested for both mandatory and optional baseline capabilities, as described in Table 10.7. All certified products must also support both Wi-Fi Multimedia (WMM) quality-of-service (QoS) mechanisms and WPA/WPA2 security mechanisms. Most enterprise-grade APs that are 802.11ac capable have two radios: a 2.4 GHz radio and a 5 GHz radio. Since 802.11ac operates only in the 5 GHz frequencies, the 5 GHz radio would be tested using Wi-Fi CERTIFIED ac requirements, and the 2.4 GHz radio would be tested using Wi-Fi CERTIFIED n requirements. Note that 802.11ac devices will be tested to ensure interoperability with earlier technologies that operate in the 5 GHz frequencies, which includes 802.11n and 802.11a.

TABLE 10.7   Wi-Fi CERTIFIED n baseline requirements

Feature	Explanation	Type
Support for two spatial streams	Access points are required to transmit and receive at least two spatial streams. Client stations are required to transmit and receive at least one spatial stream.	Mandatory
Support for three spatial streams	Access points and client stations capable of transmitting and receiving three spatial streams	Optional (tested if implemented)
Support for A-MPDU and A-MSDU in receive mode; support for A-MPDU in transmit mode	Required for all devices. Reduces MAC layer overhead.	Receive mode mandatory
Support for Block ACK	Required for all devices. Sends a single Block ACK frame to acknowledge multiple received frames.	Mandatory
2.4 GHz operation	Devices can be 2.4 GHz only, 5 GHz only, or dual-band. For this reason, both frequency bands are listed as optional.	Optional (tested if implemented)
5 GHz operation	Devices can be 2.4 GHz only, 5 GHz only, or dual-band. For this reason, both frequency bands are listed as optional.	Optional (tested if implemented)
Concurrent operation in 2.4 GHz and 5 GHz bands	This mode is tested for APs only. APs capable of operating in both bands are certified as "concurrent dual-band."	Optional (tested if implemented)
40 MHz channels in the 5 GHz band	Bonding of two adjacent 20 MHz channels to create a single 40 MHz channel. Provides twice the frequency bandwidth.	Optional (tested if implemented)

Feature	Explanation	Type
20/40 MHz coexistence mechanisms in the 2.4 GHz band	If an AP supports 40 MHz channels in the 2.4 GHz band, coexistence mechanisms are required. The default 2.4 GHz channel size is 20 GHz.	Optional (tested if implemented)
Greenfield preamble	The Greenfield preamble cannot be interpreted by legacy stations. The Greenfield preamble improves efficiency of the 802.11n networks with no legacy devices.	Optional (tested if implemented)
Short guard interval (short GI), 20 and 40 MHz	The short GI is 400 nanoseconds; the traditional GI is 800 nanoseconds. Improves data rates by 10 percent.	Optional (tested if implemented)
Space-time block coding (STBC)	Improves reception by encoding data streams in blocks across multiple antennas. Access points can be certified for STBC.	Optional (tested if implemented)
HT Duplicate mode	Allows an AP to send the same data simultaneously on each 20 MHz channel within a bonded 40 MHz channel.	Optional (tested if implemented)

In June 2013, prior to the ratification of the 802.11ac amendment, the Wi-Fi Alliance published its vendor certification program for 802.11ac, Wi-Fi CERTIFIED ac. 802.11ac products are tested for both the mandatory and optional baseline capabilities listed in Table 10.8. As with Wi-Fi CERTIFIED n products, Wi-Fi CERTIFIED ac products must support both Wi-Fi Multimedia (WMM) quality-of-service mechanisms and WPA2/WPA2 security mechanisms. Unlike Wi-Fi CERTIFIED n devices, Wi-Fi CERTIFIED ac devices do not operate in both the 2.4 GHz and 5 GHz frequency bands. Wi-Fi CERTIFIED ac devices operate only in the 5 GHz frequency band. As previously mentioned, this is due to the limited frequency range available in the 2.4 GHz ISM band. Therefore, they only need to be backward compatible with 5 GHz 802.11a/n certified devices.

TABLE 10.8 Wi-Fi CERTIFIED ac baseline requirements

Feature	Mandatory	Optional
Channel width	20, 40, 80 MHz	80+80, 160 MHz
Modulation and coding scheme	MCS 0–7	MCS 8, 9
Spatial streams	One for clients, two for APs	Two to eight
Guard Interval	Both Long (800 nanoseconds) and Short (400 nanoseconds)	
Beamforming feedback		Respond to beamforming sounding
Space-time block coding (STBC)		Transmit and receive STBC
Low-density parity check (LDPC)		Transmit and receive LDPC
Multiuser MIMO		Up to four spatial streams per client, using the same MCS

Summary

In this chapter, you learned the history of the 802.11n and 802.11ac amendments and how the Wi-Fi Alliance certifies interoperability. We also discussed all the methods used by HT and VHT radios to increase throughput and range at the Physical layer. In addition to PHY enhancements, these radios utilize MAC layer mechanisms to enhance throughput and power management. Finally, we discussed the different modes of operation that are used for protection mechanisms and co-existence with older legacy 802.11a/b/g technologies. Since the 802.11ac standard is designed to operate only in the 5 GHz frequency range, dual-radio 802.11ac access points will continue to support 802.11n in the 2.4 GHz frequency range for some time to come. If you are interested in learning more about 802.11ac, we recommend that you read *802.11ac: A Survival Guide*, by Matthew Gast (O'Reilly Media, 2013). Matthew has done an excellent job of explaining the core technologies that make up 802.11ac, in an in-depth and concise format.

Exam Essentials

Define the differences between MIMO and SISO. Understand that SISO devices use only one radio chain, whereas MIMO systems use multiple radio chains.

Define the differences between 802.11n and 802.11ac. Understand how 802.11ac is similar and different from 802.11n. Know which 802.11ac technologies are evolutionary and which are revolutionary. Explain why 802.11ac is being implemented only in the 5 GHz band.

Understand spatial multiplexing. Describe how SM takes advantage of multipath and sends multiple spatial streams, resulting in increased throughput.

Explain the difference between SU-MIMO and MU-MIMO. Explain how many spatial streams are supported by 802.11ac along with the additional resources necessary to implement more spatial streams. Explain the technological differences between sending a SU-MIMO signal and a MU-MIMO.

Explain MU-MIMO. Explain the MU-MIMO process and the conditions under which it will be most successful. Explain how beamforming makes this possible. Explain the requirements for adding more spatial streams. Explain how QoS is implemented in a MU-MIMO environment.

Describe explicit beamforming. Describe the communications between the AP and the client to perform explicit beamforming. Describe the benefits of explicit beamforming.

Understand 20 MHz, 40 MHz, 80 MHz, and 160 MHz channels. Understand the differences between 20 MHz, 40 MHz, 80 MHz, and 160 MHz channels. Explain how 160 MHz channels are actually two 80 MHz channels. Explain how 802.11ac radios will dynamically switch to narrower channels if the wider channel is not available. Describe the importance of primary channel selection for each channel width.

Explain the guard interval. Describe how the guard interval compensates for intersymbol interference. Discuss the use of both 800- and 400-nanosecond GIs.

Understand modulation and coding schemes. Explain how modulation and coding schemes (MCSs) are used to define data rates. Explain all the variables that can affect the data rates.

Explain the HT/VHT PPDU formats. Describe the differences between non-HT legacy, HT Mixed, and VHT PPDU.

Understand HT MAC enhancements. Explain how frame aggregation is used to increase throughput at the MAC sublayer. Define the new power-management methods used by HT/VHT radios.

Explain the HT protection modes. Describe the differences between protection modes 0–3. Understand that these protection modes are used for both HT and VHT radios.

Understand 64-QAM and 256-QAM. Explain how 256-QAM is similar to and different from 64-QAM. Describe the significance of the constellation chart and the pros and cons of the denser 256-QAM.

Review Questions

1. Which of the following modulation methods are supported with 802.11ac? (Choose all that apply.)

   A. BPSK

   B. BASK

   C. 32-QAM

   D. 64-QAM

   E. 256-QAM

2. How can a MIMO system increase throughput at the Physical layer? (Choose all that apply.)

   A. Spatial multiplexing

   B. A-MPDU

   C. Transmit beamforming

   D. 40 MHz channels

3. Which new power-management method defined by the 802.11n amendment conserves power by powering down all but one radio?

   A. A-MPDU

   B. Power Save protection

   C. PSMP

   D. SM power save

   E. PS mode

4. The guard interval is used as a buffer to compensate for which type of interference?

   A. Co-channel interference

   B. Adjacent cell interference

   C. RF interference

   D. HT interference

   E. Intersymbol interference

5. Which of the following channel widths are supported in 802.11ac? (Choose all that apply.)

   A. 20 MHz

   B. 40 MHz

   C. 80 MHz

   D. 80+80 MHz

   E. 160 MHz

6. What could an 802.11n (HT) radio use to increase throughput at the MAC sublayer of the Data-Link layer? (Choose all that apply.)

   A. A-MSDU

   B. A-MPDU

   C. Guard interval

   D. Block ACKs

7. Which of the following technologies is part of explicit beamforming? (Choose all that apply.)

   A. Channel sounding

   B. Feedback matrix

   C. Sounding matrix

   D. Steering matrix

   E. Null data packet

   F. Channel matrix

8. A 3×3:2 MIMO radio can transmit and receive how many unique streams of data?

   A. Two

   B. Three

   C. Four

   D. Three equal and four unequal streams

   E. None—the streams are not unique data

9. Name a capability not defined for A-MPDU.

   A. Multiple QoS access categories

   B. Independent data payload encryption

   C. Individual MPDUs having the same receiver address

   D. MPDU aggregation

10. Which HT protection modes would allow only for the association of 802.11a/g clients to an 802.11ac access point? (Choose all that apply.)

    A. Mode 0—Greenfield mode

    B. Mode 1—HT nonmember protection mode

    C. Mode 2—HT 20 MHz protection mode

    D. Mode 3—HT Mixed mode

11. Which of these capabilities are considered mandatory for an 802.11n access point, as defined by the Wi-Fi Alliance's vendor certification program called Wi-Fi CERTIFIED n? (Choose all that apply.)

    A. Three spatial streams in receive mode

    B. WPA/WPA2

    C. WMM

    D. Two spatial streams in transmit mode

    E. 2.4 GHz–40 MHz channels

12. MIMO radios use which mechanisms for transmit diversity? (Choose all that apply.)

    A. Maximal ratio combining (MRC)

    B. Direct-sequence spread spectrum (DSSS)

    C. Space-time block coding (STBC)

    D. Cyclic shift diversity (CSD)

13. 802.11n (HT) radios are backward compatible with which of the following types of 802.11 radios? (Choose all that apply.)

    A. 802.11b radios (HR-DSSS)

    B. 802.11a radios (OFDM)

    C. 802.11 legacy radios (FHSS)

    D. 802.11g radios (ERP)

14. How many modulation and coding schemes are defined in 802.11ac?

    A. 8

    B. 10

    C. 64

    D. 77

    E. 256

15. Which 802.11ac MCS range defines all of the MCSs that are mandatory?

    A. MCS 0–2

    B. MCS 0–4

    C. MCS 0–6

    D. MCS 0–7

    E. MCS 0–8

    F. MCS 0–9

16. A WLAN consultant has recommend that a new 802.11n/ac network use only 40 MHz channels in the 5 GHz U-NII bands. Why would he recommend 40 MHz channels be used only in 5 GHz and not in 2.4 GHz?

    A. HT/VHT radios do not require DFS and TPC in the 5 GHz bands.

    B. HT/VHT radios get better range using TxBF in the 5 GHz bands.

    C. 40 MHz channels do not scale in the 2.4 GHz ISM band.

    D. 5 GHz VHT radios are less expensive than 2.4 GHz HT radios.

17. 802.11ac (VHT) radios are backward compatible with which of the following types of 802.11 technology? (Choose all that apply.)

    A. 802.11b (HR-DSSS)

    B. 802.11a (OFDM)

    C. 802.11g (ERP)

    D. 802.11n (HT)

18. Which frequencies are defined for 802.11n (HT) radio transmissions? (Choose all that apply.)

    A. 902–928 MHz

    B. 2.4–2.4835 GHz

    C. 5.15–5.25 GHz

    D. 5.25–5.35 MHz

19. Which PHY layer mechanism might be used to increase throughput for an HT/VHT radio in a clean RF environment with minimal reflections and low multipath?

    A. Maximal ratio combining

    B. 400-nanosecond guard interval

    C. Switched diversity

    D. Spatial multiplexing

    E. Spatial diversity

20. The 802.11ac amendment defines a maximum of how many spatial streams for a client?

    A. One

    B. Two

    C. Four

    D. Eight

Chapter

11

WLAN Architecture

IN THIS CHAPTER, YOU WILL LEARN ABOUT THE FOLLOWING:

✓ **WLAN client devices**

- 802.11 radio form factors
- 802.11 radio chipsets
- Client utilities

✓ **Management, Control, and Data planes**

- Management plane
- Control plane
- Data plane

✓ **WLAN architecture**

- Autonomous WLAN architecture
- Centralized network management systems
- Centralized WLAN architecture
- Distributed WLAN architecture
- Hybrid WLAN architecture

✓ **Specialty WLAN infrastructure**

- Enterprise WLAN routers
- WLAN mesh access points
- WLAN bridges
- WLAN array
- Real-time location systems
- VoWiFi

✓ **Cloud networking**

✓ **Infrastructure management**

- Protocols for management

✓ **Application programming interface**

- Transport and data formats
- WLAN APIs
- Common applications

In Chapter 7, "Wireless LAN Topologies," we discussed the various 802.11 WLAN topologies. You learned that both client and access point stations can be arranged in 802.11 service sets to provide wireless access to another medium. In this chapter, we discuss the multiple devices that can be used in 802.11 topologies. Many choices exist for client station radio cards that can be used in desktops, laptops, smartphones, tablets, and so on.

We also discuss the three logical planes of network operation and where they apply in a WLAN. This chapter provides an overview of the many different WLAN architectures that are available today. We also explore the progression of WLAN infrastructure devices over the years. We also cover the purpose of many WLAN specialty devices that exist in today's Wi-Fi marketplace.

WLAN Client Devices

The main hardware in a Wi-Fi *network interface card (NIC)* is a half-duplex radio transceiver, which can exist in many hardware formats and chipsets. All Wi-Fi client NICs require a special driver to interface with the operating system, as well as software utilities to interface with the end user. Laptop Wi-Fi radios can work with Windows, Linux, ChromeOS, and macOS, although they require a different driver and client software for each operating system. The drivers for many manufacturers' radios may already be included in the operating system, but often newer radios require or can benefit from an updated driver installation. Many vendors will provide an online automated method to update drivers; however, some may require that the driver be installed manually in the operating system. First-generation Wi-Fi radio drivers are often buggy. An administrator or user should always ensure that the most current generation of drivers is installed. A large percentage of Wi-Fi issues are resolved by simply upgrading WLAN client drivers.

With a software interface, the end user can configure a NIC to participate in a WLAN by using configuration settings that pertain to identification, security, and performance. These client utilities may be the manufacturer's own software utility or an incorporated software interface built into the operating system.

Next, we discuss the various radio NIC formats, the chipsets that are used, and software client utilities.

802.11 Radio Form Factors

802.11 radios are used in both client NICs and access points. The following sections focus mainly on how Wi-Fi radios can be used as client devices. 802.11 radios are manufactured in many *form factors*, meaning the NIC comes in different shapes and sizes. Many Wi-Fi radio form factors, such as USB, are meant to be used as add-on external devices, although the majority of Wi-Fi devices now use internal or integrated form factors.

External Wi-Fi Radios

When 802.11 WLANs were first deployed, the only option you had when purchasing an 802.11 client NIC was a standard PC Card adapter, which was a peripheral for laptop computers. The PC Card form factor was developed by the Personal Computer Memory Card International Association (PCMCIA). Three legacy *PCMCIA* adapters, also known as PC cards, are shown in Figure 11.1. The PCMCIA radio card could be used in any laptop or handheld device that had a PC card slot. Most PC cards had only internal integrated antennas, whereas others had both integrated antennas and external connectors. Laptops are no longer manufactured with PC card slots, and PCMCIA radios have become obsolete.

FIGURE 11.1   PCMCIA adapter/PC card

Courtesy of Cisco Systems, Inc. Unauthorized use not permitted.

Eventually, other radio form factors hit the marketplace, including the *ExpressCard* format. ExpressCard was a hardware standard that replaced PCMCIA cards. Most laptop manufacturers replaced PCMCIA slots with the smaller ExpressCard slots.

Secure Digital (SD) and *CompactFlash (CF)* were two peripheral radio form factors that were originally used with handheld personal digital assistants (PDAs). These radios typically required very low power and were smaller than the size of a matchbook. The use of the SD and CF formats with handheld devices quickly became obsolete because handheld devices integrated embedded form factor 802.11 radios directly into their products.

We have discussed a few Wi-Fi radio form factors that can be used as external radios with laptops and other mobile devices. However, *Universal Serial Bus (USB)* 802.11 radios remain the most popular choice for external Wi-Fi radios because almost all computers have USB ports. USB technology provides simplicity of setup and does not require an external power source. 802.11 USB radios exist either in the form of a small dongle device (see Figure 11.2) or as an external wired USB device with a separate USB cable connector. The dongle devices are compact and portable for use with a laptop computer, and the external devices can be connected to a desktop computer with a USB extension cable and placed on top of a desk for better reception.

FIGURE 11.2 802.11 USB radio

802.11n/ac radios are available in both USB 2.0 and USB 3.0 form factors and can operate in both the 2.4 GHz and 5 GHz frequency bands. Be aware that there are some disadvantages when using a USB radio form factor. USB 2.0 technology defines data transfers of only up to 480 Mbps, which will limit the available 802.11 data rates. USB 3.0 technology defines potential data transfers of up to 5 Gbps. USB 3.0 Wi-Fi radios can therefore take advantage of higher 802.11n/ac data rates. Please be aware that the circuitry in some USB 3.0 devices has been known to cause RF interference in the 2.4 GHz band. USB 3.0 devices of various types have been shown to raise the noise floor 5–20 dB, which can cause serious performance issues with internal 802.11 radios in a laptop.

Internal Wi-Fi Radios

For many years, external Wi-Fi radios were the norm because laptops did not have internal Wi-Fi radio capabilities. Laptops and other mobile devices now include internal Wi-Fi radios. An internal radio format that was initially used was the *Mini PCI*. The Mini PCI was a variation of the Peripheral Component Interconnect (PCI) bus technology and was designed for use mainly in laptops. A Mini PCI radio was often used inside access points and was also the main type of radio used by manufacturers as the internal 802.11 wireless adapter inside laptops. The next generation bus technology form factor is the smaller *Mini PCI Express* and even smaller *Half Mini PCI Express*. It is almost impossible to buy a new laptop today that does not have an internal Mini PCI or Mini PCI Express radio, as shown in Figure 11.3. A Mini PCI or Mini PCI Express radio card typically is installed from the bottom of the laptop and is connected to small antennas that are mounted along the edges of the laptop's monitor.

FIGURE 11.3 Mini PCI and Mini PCI Express radios

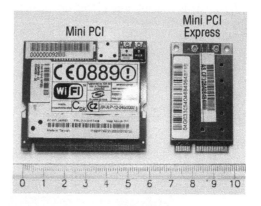

Real World Scenario

Advantages of Using an External USB Radio with a Laptop

Although Mini PCI, Mini PCI Express, and Half Mini PCI Express radios are removable from some laptops, there is no guarantee that any of these form factors will work in another vendor's laptop. One advantage of using USB Wi-Fi adapters is that they can be moved and used in different laptops. Additionally, a laptop with an older 802.11 internal radio can be instantly upgraded to newer 802.11 technology at a low cost using a USB radio. Also, WLAN engineers usually use a USB radio when running 802.11 protocol analyzer software and/or site survey software applications. These applications often require a special driver for the 802.11 radio that will overwrite and/or conflict with the radio's original driver. Using an independent and external Wi-Fi radio for troubleshooting and site surveys is a common practice so that the driver of the internal Wi-Fi radio remains intact.

Mobile Devices

We have mainly discussed the various types of 802.11 radio NIC formats that are used with laptops. 802.11 radios are also used in many other types of handheld devices, such as smartphones, tablets, bar code scanners, and VoWiFi phones. Bar code scanners, such as the Honeywell mobile device pictured in Figure 11.4, have made use of 802.11 radios for many years.

FIGURE 11.4   Bar code scanner

Courtesy of Honeywell

Although older handheld devices did use some of the previously mentioned form factors, manufacturers of most handheld devices use an embedded form factor 802.11 radio (usually a single chip form factor that is embedded into the device's motherboard). Figure 11.5 shows a single chip Broadcom Wi-Fi radio that is found inside some models of the Apple iPhone. Almost all mobile devices such as smartphones and tablets use a single chip form factor that is embedded on the device's motherboard. The embedded radios often use a combo chipset for both Wi-Fi and Bluetooth radios.

FIGURE 11.5   Embedded 802.11 radio

For many years, most people thought of only using their laptop for Wi-Fi connectivity. With the advent of smartphones and tablets, there has been a handheld client population explosion of mobile devices. In recent years, the number of mobile devices connecting to enterprise WLANs has exceeded the number of laptops connecting to the same enterprise WLANs. Technology research firm, 650 Group, (www.650group.com), estimates that by 2025, the number of smartphones, tablets, PCs and peripherals in use will reach over 12 billion units worldwide.

Users now expect Wi-Fi connectivity with numerous mobile devices in addition to their laptops. Because of the proliferation of personal mobile devices, a *bring your own device (BYOD)* policy is often needed to define how employees' personal devices may access the corporate WLAN. A *mobile device management (MDM)* solution might also be needed for onboarding both personal mobile devices and company-issued devices onto the WLAN. BYOD strategies and MDM solutions are discussed in great detail in Chapter 18, "Bring Your Own Device (BYOD) and Guest Access."

Wearables

Another big technology trend has been wearable computers, simply known as *wearables*. A wearable computing device is worn on a person's body and/or clothing. Wearables are meant to provide a constant interaction between a person and a computer, and the wearable becomes an extension of a user's body or mind. Although the concept of wearable computers is not new, wearables with embedded Wi-Fi radios have begun to find their way into the marketplace. Examples of wearable computers include smart watches, wristbands, exercise sensors, and glasses.

Much as with smartphones and tablets, users may want to connect to the company WLAN using their personal wearable computer devices. New challenges lie ahead for how IT administrators will manage the onboarding and access policies of wearables to the corporate WLAN. Additionally, wearables have the potential for numerous applications in enterprise verticals, such as healthcare and retail. Silicon Valley-based research firm, 650 Group, projects that the number of wearable devices shipped will rise from 1 billion in 2017 to over 5 billion in 2025.

Internet of Things

When speaking about RFID devices, the phrase *Internet of Things (IoT)* is usually credited to Kevin Ashton:

> www.rfidjournal.com/articles/view?4986

Over the years, most of the data generated on the Internet has been created by human beings. The theory of Internet of Things is that in the future, the bulk of the data generated on the Internet might be created by sensors, monitors, and machines. It should be noted that 802.11 radio NICs used as client devices have begun to show up in many types of machines and solutions. Wi-Fi radios already exist in gaming devices, stereo systems, and video cameras. Appliance manufacturers are putting Wi-Fi NICs in washing machines, refrigerators, and automobiles. The use of Wi-Fi radios in sensor and monitoring devices, as well as RFID, has many applications in numerous enterprise vertical markets.

Technology research firm, 650 Group, estimates that by 2025, the number of wirelessly connected IoT devices will be 53 billion units worldwide, far exceeding the expected 28 billion number of PCs, tablets, smartphones and other connected personal devices. Could this be the beginning of the self-aware Skynet predicted by the Terminator movies? All kidding aside, a large portion of IoT devices will most likely connect to the Internet with a Wi-Fi radio. Once again, new challenges lie ahead; IT administrators must manage the onboarding, access, and security policies of IoT devices connecting to the corporate WLAN.

The bulk of IoT devices with an 802.11 radio currently transmit in the 2.4 GHz frequency band only. Please understand that not all IoT devices use Wi-Fi radios. IoT devices may use other RF technology, such as Bluetooth or Zigbee. IoT devices may also have an Ethernet networking interface in addition to the RF interfaces.

 **Real World Scenario**

How Do I Know What Kind of Radio Is in My Laptop or Mobile Device?

Often, a laptop or mobile device manufacturer will list the radio model in the specification sheet for the laptop or mobile device. However, some manufacturers may not list detailed radio specifications and capabilities. What if you want to find out if the radio is a 1x1:1 MIMO radio or maybe a 3x3:3 MIMO radio? Does the radio support 40 MHz channels or only 20 MHz channels? On laptops, you might find some of the radio's capabilities by simply looking at the radio drivers from within the OS. Another method of identifying the Wi-Fi radio in your device is by the FCC ID. In the United States, all Wi-Fi radios must be certified by the Federal Communications Commission (FCC) government agency. The FCC maintains a searchable equipment authorization database at transition.fcc.gov/oet/ea/fccid. You can enter the FCC ID of your device into the database search engine and find documentation and pictures submitted by the manufacturer to the FCC. The FCC database is very useful in helping to identity Wi-Fi radio models and specifications if the information is not available on the manufacturer's website.

Client Device Capabilities

In Chapter 13, "WLAN Design Concepts," we will discuss the importance of understanding the capabilities of the clients you have deployed in an enterprise environment. We will also discuss the importance of upgrading your WLAN client population with newer 802.11 technology when you also upgrade your access points. Most businesses and corporations can eliminate many of the client connectivity and performance problems by simply upgrading company-owned client devices before updating the WLAN infrastructure. Sadly, the opposite is often more common, with companies spending many hundreds of thousands of dollars on technology upgrades with new access points while still deploying legacy clients.

Always remember that all 802.11 client radios do not have the same capabilities. Legacy 802.11b radios have a maximum data rate of 11 Mbps, and legacy 802.11a/g radios have a maximum data rate of 54 Mbps. Laptop, smartphone, and tablet manufacturers now ship their products with 802.11n/ac radios, which are capable of much higher data rates. Be aware, however, that even modern client devices may not have the same capabilities. Some higher-end laptops may have 3x3:3 MIMO radios, but the bulk of laptops have 2x2:2 MIMO radios. Additionally, most smartphones and tablets now have 2x2:2 radios, but many older 802.11n mobile devices were 1x1:1.

The first several generations of tablet PCs and smartphones had 1x1:1 radios that operated only in the 2.4 GHz frequency band. Most modern clients are dual-frequency with 2x2:2 MIMO radios that operate in both the 2.4 GHz and 5 GHZ frequency bands. Also, the majority of new clients will usually support 40 MHz channels. However, many other 802.11 technologies, such as 802.11k, 802.11r, and 802.11v, may not be supported, even in new client devices.

IoT devices with an 802.11 radio usually operate only in the 2.4 GHz frequency band and very often may employ older 802.11g chipset technology to keep costs down.

802.11 Radio Chipsets

A group of integrated circuits designed to work together is often marketed as a *chipset*. Many 802.11 chipset manufacturers exist and sell their chipset technology to the various radio manufacturers and WLAN vendors. Legacy chipsets will obviously not support all the same features as newer chipset technologies. For example, a legacy chipset may support only 802.11a/b/g technology, whereas newer chipsets will support 802.11n/ac technology.

Some chipsets may support the ability to transmit only on the 2.4 GHz ISM band; other chipsets can transmit on either the 2.4 GHz or 5 GHz unlicensed frequencies. Chipsets that support both frequencies are used in 802.11a/b/g/n/ac client radios. The chipset manufacturers incorporate newer 802.11 technologies as they develop. Many proprietary technologies turn up in the individual chipsets, and some of these technologies will become part of the standard in future 802.11 amendments.

 You can find detailed information about some of the most widely used Wi-Fi chipsets and radio manufacturers at the following URLs: www.qca.qualcomm.com, www.broadcom.com, and www.intel.com.

Client Utilities

An end user must have the ability to configure a wireless client NIC. Therefore, a software interface is needed in the form of *client utilities*. Much like a driver is the interface between a radio NIC and an operating system, the Wi-Fi client utility is effectively the software interface between the radio NIC and you. The software interface usually has the ability to create multiple connection profiles. One profile may be used to connect to the wireless network at work, another to connect at home, and a third to connect at a hotspot.

Configuration settings for a client utility typically include the service set identifier (SSID), transmit power, WPA/WPA2 security settings, WMM quality-of-service capabilities, and power-management settings. Another technical term often used for WLAN client utilities is *supplicant*. The supplicant terminology is most often used when discussing 802.1X/EAP security. As mentioned in Chapter 7, some client NICs can also be configured for either infrastructure mode or ad hoc mode. Most good client utilities have some sort of statistical information display, along with some sort of received signal strength indicator (RSSI) measurement tool. Some client utilities also allow for the adjustment of client roaming thresholds.

The following three major types, or categories, of client utilities exist:

- Integrated operating system client utilities
- Vendor-specific client utilities
- Third-party client utilities

The software interface that is most widely used to configure a Wi-Fi radio is usually the integrated operating system Wi-Fi client utilities. Laptop users will most likely use the Wi-Fi NIC configuration interface that is a part of the OS running on the laptop. The client software utilities are different depending on the OS of the laptop being used. The capabilities of the Wi-Fi client utilities also vary between different versions of operating systems. For example, the Windows 8 client utility is different from the Windows 10 client utility. Older macOS client utilities are different from the macOS 10.13 (High Sierra) client utility. Figure 11.6 shows the Windows 10 Wi-Fi client utility. Some OSs, such as the macOS 10.13, offer Wi-Fi diagnostic tools and signal strength indicators, as shown in Figure 11.7.

FIGURE 11.6   Integrated OS client utility for Windows 10

FIGURE 11.7   Wireless diagnostic tool for macOS 10.13

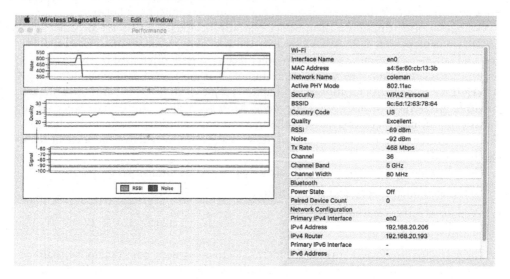

The operating systems of handheld devices usually also include some sort of Wi-Fi client utility. Figure 11.8 shows the client interface found in the Apple iOS 11.0, which runs on iPads and iPhones.

FIGURE 11.8   Integrated OS client utility for iOS 11.0

Vendor-specific software client utilities are sometimes available for use instead of an integrated operating system software interface. SOHO client utilities are usually simplistic in nature and are designed for ease of use for the average home user. The majority of vendor-specific software utilities are for peripheral device WLAN radios. The use of vendor-specific client utilities has decreased dramatically in recent years as the use of peripheral Wi-Fi radios has also declined. Enterprise-grade vendor client utilities provide the software interface for the more expensive enterprise-grade vendor cards. Typically, the enterprise-class utilities support more configuration features and have better statistical tools. Figure 11.9 shows the Intel PROSet wireless client interface that can be used on Windows-based laptops with an Intel Wi-Fi radio.

FIGURE 11.9   Enterprise-class client utility

The last type of software interface for an 802.11 radio card is a third-party client utility, such as SecureW2's Enterprise Client for Windows, pictured in Figure 11.10. Much like any integrated OS client software, a third-party WLAN supplicant will work with radio cards from different vendors, making administrative support much easier. In the past, third-party client utilities often brought the advantage of supporting many different EAP types, giving a WLAN administrator a wider range of security choices. The main disadvantage of third-party client utilities is that they usually cost extra money. Because integrated client utilities have improved over the years, the use of third-party Wi-Fi client utilities has declined.

FIGURE 11.10 Third-party client utility

Management, Control, and Data Planes

Telecommunication networks are often defined as three logical planes of operation:

Management Plane   The *management plane* is defined by administrative network management, administration, and monitoring. An example of the management plane would be any network-management solution that can be used to monitor routers and switches and other wired network infrastructure. A centralized network-management server can be used to push both configuration settings and firmware upgrades to network devices.

Control Plane   The *control plane* consists of control or signaling information and is often defined as network intelligence or protocols. Dynamic layer 3 routing protocols, such as OSPF or BGP, used to forward data would be an example of control plane intelligence found in routers. Content addressable memory (CAM) tables and Spanning Tree Protocol (STP) are control plane mechanisms used by layer 2 switches for data forwarding.

Data Plane   The *data plane*, also known as the *user plane*, is the location in a network where user traffic is actually forwarded. An individual router where IP packets are forwarded is an example of the data plane. An individual switch forwarding an 802.3 Ethernet frame is an example of the data plane.

In an 802.11 environment, these three logical planes of operation function differently depending on the type of WLAN architecture and the WLAN vendor. For example, in a legacy autonomous AP environment, all three planes of operation existed in each standalone access point (although the control plane mechanisms were minimal). When WLAN controller solutions were first introduced in 2002, all three planes of operation were shifted into a centralized device. In modern deployments, the planes of operation may be divided between access points, WLAN controllers, and/or a wireless network management system (WNMS).

 Do not confuse the management, control, and data planes with 802.11 MAC frame types. In this chapter, the discussion of management, control, and data planes is related to WLAN network architectural operations.

Management Plane

The functions of the *management plane* within an 802.11 WLAN are as follows:

WLAN Configuration   Examples include the configuration of SSID, security, WMM, channel, and power settings.

WLAN Monitoring and Reporting   Monitoring of layer 2 statistics, such as ACKs, client associations, reassociations, and data rates, occurs in the management plane. Examples of upper-layer monitoring and reporting include application visibility, IP connectivity, TCP throughput, latency statistics, and stateful firewall sessions.

WLAN Firmware Management   The ability to upgrade access points and other WLAN devices with the latest vendor operational code is included here.

Control Plane

The *control plane* is often defined by protocols that provide the intelligence and interaction between equipment in a network. Here are a few examples of control plane intelligence:

Adaptive RF   Coordinated channel and power settings for multiple access points are provided by the control plane. The majority of WLAN vendors implement some type of *adaptive RF* capability. Adaptive RF is also referred to by the more technical term *radio resource management (RRM)*.

Roaming Mechanisms   The control plane also provides support for roaming handoffs between access points. Capabilities may include layer 3 roaming, maintaining stateful firewall sessions of clients, and forwarding of buffered packets. Fast secure roaming mechanisms, such as opportunistic key caching (OKC) and fast BSS transition (FT), may also be used to forward master encryption keys between access points.

Client Load Balancing   Collecting and sharing client load and performance metrics between access points to improve overall WLAN operations happens in the control plane.

Mesh Protocols   Routing user data between multiple access points requires some sort of mesh routing protocol. Most WLAN vendors use layer 2 routing methods to move user data between mesh access points. However, some vendors are using layer 3 mesh routing. The 802.11s amendment did define standardized mesh routing mechanisms, but WLAN vendors are currently using proprietary methods and metrics.

Data Plane

The *data plane* is where user data is forwarded. The two devices that usually participate in the data plane are the AP and a WLAN controller. A standalone AP handles all data-forwarding operations locally. In a WLAN controller solution, data is normally forwarded from the centralized controller, but data can also be forwarded at the edge of the network by an AP. As with the management and control planes, each vendor has a unique method and recommendations for handling data forwarding. Data-forwarding models will be discussed in greater detail later in this chapter.

WLAN Architecture

While the acceptance of 802.11 technologies in the enterprise continues to grow, the evolution of WLAN architecture has kept pace. In most cases, the main purpose of 802.11 technologies is to provide a wireless portal into a wired infrastructure network. How an 802.11 wireless portal is integrated into a typical 802.3 Ethernet infrastructure continues to change drastically. WLAN vendors generally offer one of the following three primary WLAN architectures:

- Autonomous WLAN architecture
- Centralized WLAN architecture
- Distributed WLAN architecture

The following sections describe these three architectures in greater detail.

Autonomous WLAN Architecture

For many years, the conventional access point was a standalone WLAN portal device where all three planes of operation existed and operated on the edge of the network architecture. These APs are often referred to as *fat APs* or *standalone APs*. However, the most common industry term for the traditional access point is *autonomous AP*.

All configuration settings exist in the autonomous access point itself, and therefore, the management plane resides individually in each autonomous AP. All encryption and decryption mechanisms and MAC layer mechanisms also operate within the autonomous AP. The data plane also resides in each autonomous AP because all user traffic is forwarded locally by each individual access point. As shown in Figure 11.11, legacy autonomous APs have few shared control plane mechanisms.

FIGURE 11.11 Autonomous WLAN architecture

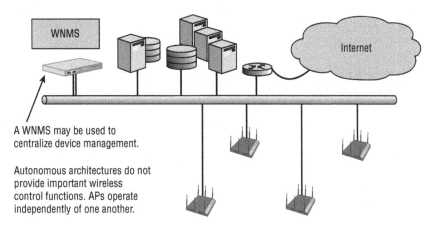

A WNMS may be used to centralize device management.

Autonomous architectures do not provide important wireless control functions. APs operate independently of one another.

An autonomous access point contains at least two physical interfaces: usually a radio frequency (RF) radio and a 10/100/1000 Ethernet port. The majority of the time, these physical interfaces are bridged together by a virtual interface known as a *bridged virtual interface (BVI)*. The BVI is assigned an IP address that is shared by two or more physical interfaces. Access points operate as layer 2 devices; however, they still need a layer 3 address for connectivity to an IP network. The BVI is the management interface of an AP.

An autonomous access point typically encompasses both the 802.11 protocol stack and the 802.3 protocol stack. These APs might support the following features:

- Multiple management interfaces, such as command line, web GUI, and SNMP

- WEP, WPA, and WPA2 security capabilities

- WMM quality-of-service capabilities

- Fixed or detachable antennas

- Filtering options, such as MAC and protocol

- Connectivity modes, such as access, mesh, bridge, or sensor

- Multiple radio and dual-frequency capabilities

- 802.1Q VLAN support

- 802.3af or 802.3at PoE support

Autonomous APs might have some of the following advanced security features:

- Built-in RADIUS and user databases

- VPN client and/or server support

- DHCP server

- Captive web portals

Autonomous APs are deployed at the access layer and typically are powered by a Power-over-Ethernet (PoE)-capable access layer switch. The integration service within an autonomous AP translates the 802.11 traffic into 802.3 traffic. The autonomous AP was the foundation that WLAN architects deployed for many years. However, most enterprise deployments of autonomous APs were replaced by a centralized architecture utilizing a WLAN controller, which is discussed later in this chapter.

Centralized Network Management Systems

One of the challenges for a WLAN administrator using a large WLAN autonomous architecture is management. As an administrator, would you want to configure 300 autonomous APs individually? One major disadvantage of using the traditional autonomous access point is that there is no central point of management. Any intelligent edge WLAN architecture with 25 or more autonomous access points is going to require some sort of *wireless network management system (WNMS)*.

A WNMS moves the management plane out of the autonomous access points. A WNMS provides a central point of management to configure and maintain thousands of autonomous access points. A WNMS can be a hardware appliance or a software solution. WNMS solutions can be vender specific or vender neutral.

As shown previously in Figure 11.11, the whole point of a WNMS server was to provide a central point of management for autonomous access points, which are now considered legacy devices. That definition has changed considerably over the years. Later in this chapter, you will learn about WLAN controllers, which are used as central points of management for controller-based APs. WLAN controllers can effectively replace a WNMS server as a central point of management for access points in small-scale WLAN deployments. However, multiple WLAN controllers are needed in large-scale WLAN enterprise deployments. Currently, most WMNS servers are now used as a central point of management for multiple WLAN controllers in large-scale WLAN enterprises. WNMS servers that are used to manage multiple WLAN controllers from a single vendor may in some cases also be used to manage other vendors' WLAN infrastructure, including standalone access points.

The term WNMS is actually outdated, because many of these centralized management solutions can also be used to manage other types of network devices, including switches, routers, firewalls, and VPN gateways. Therefore, *network management system (NMS)* is now used more often. NMS solutions are usually vendor specific; however, a few exist that can manage devices from a variety of networking vendors.

The main purpose of an NMS is to provide a central point of management and monitoring for network devices. Configuration settings and firmware upgrades can be pushed down to all the network devices. Although centralized management is the main goal, an NMS can have other capabilities as well, such as RF spectrum planning and management of a WLAN. An NMS can also be used to monitor network architecture with alarms and notifications centralized and integrated into a management console. An NMS provides robust monitoring of network infrastructure as well as monitoring of wired and wireless clients connected to the network. As shown in Figure 11.12, NMS solutions usually have extensive diagnostic utilities that can be used for remote troubleshooting.

FIGURE 11.12   NMS diagnostic utilities

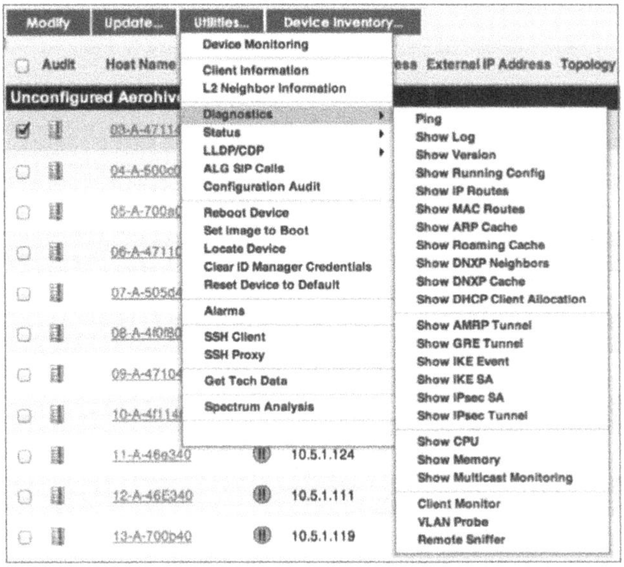

An NMS is a management plane solution; therefore, no control plane or data plane mechanisms exist within an NMS. For example, the only communications between an NMS and an access point are management protocols. Most NMS solutions use the *Simple Network Management Protocol (SNMP)* to manage and monitor the WLAN. Other NMS solutions also use the *Control and Provisioning of Wireless Access Points (CAPWAP)* protocol as strictly a monitoring and management protocol. CAPWAP incorporates *Datagram Transport Layer Security (DTLS)* to provide encryption and data privacy of the monitored management traffic. User traffic is never forwarded by an access point to an NMS; the 802.11 client associations and traffic can still be monitored. Figure 11.13 shows an NMS display of multiple client associations across multiple APs.

NMS solutions can be deployed at a company data center in the form of a hardware appliance or as a virtual appliance that runs on VMware or some other virtualization platform. A network management server that resides in a company's own data center is often referred to as an *on-premises NMS*. NMS solutions are also available in the cloud as a software subscription service. Many WLAN vendors now offer access to their NMS solutions via APIs. An application programming interface (API) is a set of subroutine definitions, protocols, and tools for building application software. Customers and partners can use the WLAN vendor APIs to build their own custom applications to monitor the WLAN. Custom applications can also be built for WLAN device configuration. APIs will be discussed in greater detail later in this chapter.

FIGURE 11.13   NMS client monitoring

Status Health	Hostname	IP	MAC	User Name	OS Type	Channel ⌃	Usage	VLAN
⊗	Coleman-PC	172.16.255.95	74DA3835EAB4		Windows 8/10	11	0 B	1
⊗	aerohives-M...	172.16.255.59	A45E60CB133B		Mac OS X	11	0 B	1
⊗	Blue-Cutter	172.16.255.63	7C5CF8A5DC0F		Windows 8/10	11	537.15 KB	1
⊗	GreenLeaf	172.16.255.62	7C5CF8732E90		Windows 8/10	11	304.66 MB	1
⊗	android-efdf...	172.16.255.88	301966CA471D		Android	149	0 B	1
⊗	Davids-iPhone	172.16.255.90	B844D90E006E		Apple iOS	149	430.96 KB	1
⊗		172.16.255.91	6C29951A2A66		CrOS	149	3.12 MB	1
⊗	iPad	172.16.255.85	9888E392715D		Apple iOS	149	42.97 KB	1
⊗	charkins-mac	172.16.255.84	A4D18CCB9A80		Mac OS X El Capitan	149	0 B	1
⊗	android-bf66...	172.16.255.87	E09971B3E389		Android	149	359.79 KB	1

10 Clients from Friday (July 1, 2016) 08:00 to Friday (July 1, 2016) 10:00

10 | **20** | 50 | 100

Centralized WLAN Architecture

The next progression in the development of WLAN integration is the centralized WLAN architecture. This model uses a central WLAN controller that resides in the core of the network. In the centralized WLAN architecture, autonomous APs have been replaced with *controller-based access points*, also known as *lightweight APs* or *thin APs*. Beginning in 2002, many WLAN vendors decided to move to a WLAN controller model where all three logical planes of operation would reside inside the controller. In a centralized WLAN architecture, the three logical planes exist in a WLAN controller.

Management Plane   Access points are configured and managed from the WLAN controller.

Control Plane   Adaptive RF, load balancing, roaming handoffs, and other mechanisms exist in the WLAN controller.

Data Plane   The WLAN controller exists as a data distribution point for user traffic. Access points tunnel all user traffic to a central controller.

The encryption and decryption capabilities might reside in the centralized WLAN controller or may still be handled by the controller-based APs, depending on the vendor. The distribution system service (DSS) and integration service (IS) both typically function within the WLAN controller. Some time-sensitive operations are still handled by the AP.

WLAN Controller

At the heart of the centralized WLAN architecture model is the *WLAN controller* (see Figure 11.14). WLAN controllers are often referred to as *wireless switches* because they are indeed an Ethernet-managed switch that can process and route data at the Data-Link layer (layer 2) of the OSI model. Many of the WLAN controllers are multilayer switches that can also route traffic at the Network layer (layer 3). However, *wireless switch* has become an outdated term and does not adequately describe the many capabilities of a WLAN controller.

FIGURE 11.14   Centralized WLAN architecture: WLAN controller

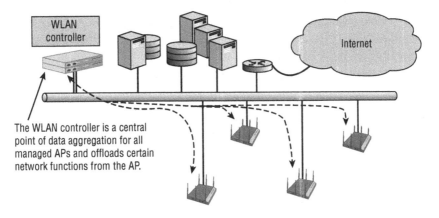

A WLAN controller may offer many of the following features:

AP Management   As mentioned earlier, the majority of the access point functions, such as power, channels, and supported data rates, are configured on the WLAN controller. This allows for centralized management and configuration of APs. Some vendors use proprietary protocols for communications between the WLAN controller and their controller-based APs. These proprietary protocols can transfer configuration settings, update firmware, and maintain keep-alive traffic. A WLAN management protocol has gained acceptance. Many WLAN vendors use the *Control and Provisioning of Wireless Access Points (CAPWAP)* protocol for managing and monitoring access points. CAPWAP can also be used to tunnel user traffic between an AP and a WLAN controller.

WLAN Management   WLAN controllers are capable of supporting multiple WLANs, which are often called *WLAN profiles* or *SSID profiles*. Different groups of 802.11 clients can connect to a different SSID that is unique to each profile. The WLAN profile is a set of configuration parameters that are configured on the WLAN controller. The profile parameters can include the WLAN logical name (SSID), WLAN security settings, VLAN assignment, and quality-of-service (QoS) parameters. WLAN profiles often work together with role-based access control (RBAC) mechanisms. When users connect to a WLAN, users are assigned to specific roles or user profiles.

User Management   WLAN controllers usually provide the ability to control the who, when, and where in terms of using role-based access control (RBAC) mechanisms.

Device Monitoring   WLAN controllers provide visual AP monitoring and client device statistics in terms of connectivity, roaming, uptime, and more.

VLANs   WLAN controllers fully support the creation of VLANs and 802.1Q VLAN tagging. Multiple wireless user VLANs can be created on the WLAN controller so that user traffic can be segmented. VLANs may be assigned statically to WLAN profiles or may be assigned using a RADIUS attribute. User VLANs are usually encapsulated in an IP tunnel.

Layer 2 Security Support   WLAN controllers fully support layer 2 WEP, WPA, and WPA2 encryption. Authentication capabilities include internal databases as well as full integration with RADIUS and LDAP servers.

Layer 3 and 7 VPN Concentrators   Some WLAN controller vendors also offer VPN server capabilities within the controller. The controller can act as a VPN concentrator or endpoint for IPsec or SSL VPN tunnels.

Captive Portal   WLAN controllers have captive portal features that can be used with guest WLANs.

Internal Wireless Intrusion Detection Systems   Some WLAN controllers have integrated WIPS capabilities for security monitoring and rogue AP mitigation.

Firewall Capabilities   Stateful packet inspection is available with an internal firewall in some WLAN controllers.

Automatic Failover and Load Balancing   WLAN controllers usually provide support for Virtual Router Redundancy Protocol (VRRP) for redundancy purposes. Most vendors also offer proprietary capabilities to load-balance wireless clients between multiple controller-based APs.

Adaptive RF Spectrum Management   The majority of WLAN controllers implement some type of *adaptive RF* capability. A WLAN controller is a centralized device that can dynamically change the configuration of the controller-based access points based on accumulated RF information gathered from the access points' radios. In a WLAN controller environment, the access points will monitor their respective channels as well as use off-channel scanning capabilities to monitor other frequencies. Any RF information heard by any of the access points is reported back to the WLAN controller. Based on all the RF monitoring from multiple access points, the WLAN controller will make dynamic changes to the RF settings of the APs. Some access points may be told to change to a different channel, whereas other APs may be told to change their transmit power settings.

Adaptive RF is sometimes referred to as *radio resource management (RRM)* and is considered to be control plane intelligence. All WLAN vendors implement their own proprietary adaptive RF functionality. When implemented, adaptive RF provides automatic cell sizing, automatic monitoring, troubleshooting, and optimization of the RF environment.

Bandwidth Management   Bandwidth pipes can be restricted upstream or downstream.

Layer 3 Roaming Support   Capabilities to allow seamless roaming across layer 3 routed boundaries are fully supported. A more detailed discussion on layer 3 roaming and the Mobile IP standard can be found in Chapter 13, "WLAN Design Concepts."

Power over Ethernet (PoE)   When deployed at the access layer, WLAN controllers can provide direct power to controller-based APs via PoE. However, most controller-based APs are powered by third-party edge switches.

Management Interfaces   Many WLAN controllers offer full support for common management interfaces, such as GUI, CLI, SSH, and so forth.

Split MAC

The majority of WLAN controller vendors implement what is known as a *split MAC architecture*. With this type of WLAN architecture, some of the MAC services are handled by the WLAN controller, and some are handled by the access point. For example, the integration service and distribution system service are handled by the controller. WMM QoS methods are usually handled by the controller. Depending on the vendor, encryption and decryption of 802.11 data frames might be handled by the controller or by the AP.

You have already learned that 802.11 frames are tunneled between the controller-based APs and the WLAN controller. 802.11 data frames are usually tunneled to the controller because the controller's integration service transfers the layer 3–7 MSDU payload of the 802.11 data frames into 802.3 frames that are sent off to network resources. Effectively, the WLAN controller is needed to provide a centralized gateway to network resources for the payload of 802.11 data frames. 802.11 management and control frames do not have an upper-layer payload and therefore are never translated into 802.3 frames. 802.11 management and control frames do not necessarily need to be tunneled to the WLAN controller, because the controller does not have to provide a gateway to network resources for these types of 802.11 frames.

In a split MAC architecture, many of the 802.11 management and control frame exchanges occur only between the client station and the controller-based access point and are not tunneled back to the WLAN controller. For example, beacons, probe responses, and ACKs may be generated by the controller-based AP instead of the controller. It should be noted that most WLAN controller vendors implement split MAC architectures differently. Many WLAN controller solutions use the *Control and Provisioning of Wireless Access Points (CAPWAP)* protocol for monitoring and management. CAPWAP also defines split MAC capabilities. The CAPWAP protocol can be used to tunnel 802.11 traffic between an AP and a WLAN controller.

   Detailed information about the Control and Provisioning of Wireless Access Points (CAPWAP) protocol can be found on IETF's website, in RFC 5415, `https://tools.ietf.org/html/rfc5415`.

Controller Data-Forwarding Models

A key feature of most WLAN controllers is that the integration service (IS) and distribution system services (DSSs) operate within the WLAN controller. In other words, all 802.11 user traffic that is destined for wired-side network resources must first pass through the controller and be translated into 802.3 traffic by the integration service before being sent to the final wired destination. Therefore, controller-based access points send their 802.11 frames to the WLAN controller over an 802.3 wired connection.

The 802.11 frame format is complex and is designed for a wireless medium, not a wired medium. An 802.11 frame cannot travel through an Ethernet 802.3 network by itself. So, how can an 802.11 frame traverse between a controller-based AP and a WLAN controller? The 802.11 traffic is forwarded inside an IP-encapsulated tunnel. Each 802.11 frame is encapsulated entirely within the body of an IP packet. Many WLAN vendors use *Generic Routing Encapsulation (GRE)*, which is a commonly used network tunneling protocol. Although GRE is often used to encapsulate IP packets, GRE can also be used to encapsulate an 802.11 frame inside an IP tunnel. The GRE tunnel creates a virtual point-to-point link between the controller-based AP and the WLAN controller. Although GRE is the most common choice, WLAN vendors might use IPsec or proprietary protocols for IP tunneling. The CAPWAP management protocol can also be used to tunnel user traffic.

As shown in Figure 11.15, the controller-based APs tunnel their 802.11 frames all the way back to the WLAN controller, from the access layer all the way back to the core layer. The distribution system service inside the controller directs the traffic, whereas the integration service translates an 802.11 data MSDU into an 802.3 frame. After 802.11 data frames have been translated into 802.3 frames, they are then sent to their final wired destination.

FIGURE 11.15   Centralized data forwarding

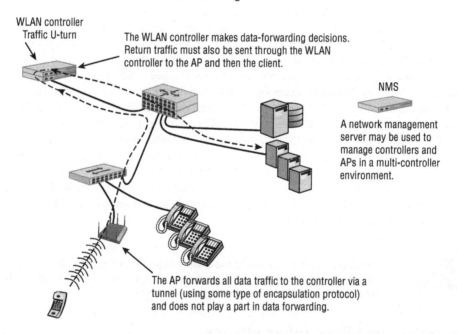

Most WLAN controllers are deployed at the core layer; however, they may also be deployed at either the distribution layer or even the access layer. Exactly where a WLAN controller is deployed depends on the WLAN vendor's solution and the intended wireless integration into the preexisting wired topology. Multiple WLAN controllers that communicate with each other may be deployed at different network layers, providing they can communicate with each other.

There are two types of data-forwarding methods when using WLAN controllers:

Centralized Data Forwarding   Where all data is forwarded from the AP to the WLAN controller for processing. It may be used in many cases, especially when the WLAN controller manages encryption and decryption or applies security and QoS policies.

Distributed Data Forwarding   Where the AP performs data forwarding locally may be used in situations where it is advantageous to perform forwarding at the edge and to avoid a central location in the network for all data, which may require significant processor and memory capacity at the controller.

As shown in Figure 11.15, centralized data forwarding relies on the WLAN controller to forward data. The AP and WLAN controller form an IP encapsulation tunnel, and all user data traffic is passed to the controller for forwarding (or comes from the controller). In essence, the AP plays a passive role in user data handling.

As illustrated in Figure 11.16, with distributed forwarding scenarios, the AP is solely responsible for determining how and where to forward user data traffic. The controller is not an active participant in these processes. This includes the application of QoS or security policies to data. Generally speaking, the device that handles the majority of MAC functions is also likely to handle data forwarding. The decision to use distributed or centralized forwarding is based on a number of factors, such as security, VLANs, and throughput. One major disadvantage of distributed data forwarding is that some control plane mechanisms may be unavailable because they exist only in the WLAN controller. Control plane mechanisms that may be lost include adaptive RF, layer 3 roaming, firewall policy enforcement, and fast secure roaming. However, as the controller architecture has matured, some WLAN vendors have also pushed some of the control plane mechanisms back into the APs at the edge of the network.

FIGURE 11.16   Distributed data forwarding

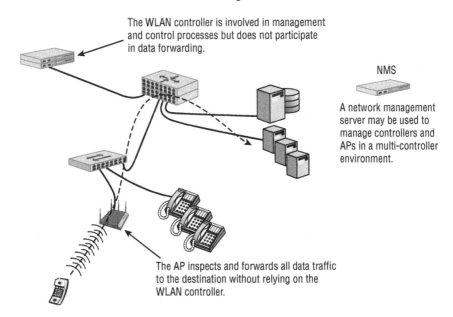

The WLAN controller is involved in management and control processes but does not participate in data forwarding.

NMS

A network management server may be used to manage controllers and APs in a multi-controller environment.

The AP inspects and forwards all data traffic to the destination without relying on the WLAN controller.

As 802.11ac technology and bandwidth become increasingly prevalent in large, enterprise networks, *centralized data forwarding* may become more difficult and expensive due to the traffic loads that can now be generated on the WLAN. Larger controllers with 10 Gbps links will become more commonplace. Additionally, WLAN controller manufacturers are now beginning to embrace *distributed data forwarding* in different ways.

Remote Office WLAN Controller

Although WLAN controllers typically reside on the core of the network, they can also be deployed at the access layer, usually in the form of a remote office WLAN controller. A remote office WLAN controller typically has much less processing power than a core WLAN controller and is also less expensive. The purpose of a remote office WLAN controller is to allow remote and branch offices to be managed from a single location. Remote WLAN controllers typically communicate with a central WLAN controller across a WAN link. Secure VPN tunneling capabilities are usually available between controllers across the WAN connection. Through the VPN tunnel, the central controller will download the network configuration settings to the remote WLAN controller, which will then control and manage the local APs. These remote controllers will allow for only a limited number of controller-based APs. Features typically include Power over Ethernet, internal firewalling, and an integrated router using NAT and DHCP for segmentation.

Distributed WLAN Architecture

A recent trend has been to move away from the centralized WLAN controller architecture toward a distributed architecture. Some WLAN vendors, such as Aerohive Networks, have designed their entire WLAN system around a distributed architecture. Some of the WLAN controller vendors now also offer a distributed WLAN architecture solution, in addition to their controller-based solution. In these systems, cooperative access points are used, and control plane mechanisms are enabled in the system with inter-AP communication via cooperative protocols. A distributed WLAN architecture combines multiple access points with a suite of cooperative protocols, without requiring a WLAN controller. Distributed WLAN architectures are modeled after traditional routing and switching design models, in that the network nodes provide independent distributed intelligence but work together as a system to cooperatively provide control mechanisms.

As shown in Figure 11.17, the protocols enable multiple APs to be organized into groups that share control plane information between the APs to provide functions such as layer 2 roaming, layer 3 roaming, firewall policy enforcement, cooperative RF management, security, and mesh networking. The best way to describe a distributed architecture is to think of it as a group of access points with most of the WLAN controller intelligence and capabilities mentioned earlier in this chapter. The control plane information is shared between the APs using proprietary protocols.

FIGURE 11.17   Distributed WLAN architecture

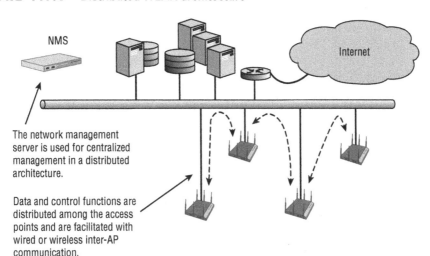

NMS

Internet

The network management server is used for centralized management in a distributed architecture.

Data and control functions are distributed among the access points and are facilitated with wired or wireless inter-AP communication.

In a distributed architecture, each individual access point is responsible for local forwarding of user traffic. As mentioned earlier, since the advent of 802.11n, WLAN controller vendors have begun to offer distributed data-forwarding solutions to handle traffic load. Because a distributed WLAN architecture entirely eliminates a centralized WLAN controller, all user traffic is forwarded locally by each independent AP. In a distributed architecture, the data plane resides in the access points at the edge of the network. No WLAN controller exists; therefore, the data does not need to be tunneled to the core of the network.

Although the control plane and data planes have moved back to the APs in a distributed WLAN architecture, the management plane remains centralized. Configuration and monitoring of all access points in the distributed model is still handled by an NMS server. The NMS server might be an on-premises server or might be offered as a cloud-based service.

Most of the features mentioned in the earlier section about WLAN controllers can also be found in a distributed WLAN architecture even though there is no WLAN controller. For example, a captive web portal that normally resides in a WLAN controller instead resides inside the individual APs. The stateful firewall and RBAC capabilities found in a centralized WLAN controller now exist cooperatively in the APs. Back-end roaming mechanisms and adaptive RF are also cooperative. APs might also function as a RADIUS server with full LDAP integration capabilities. As mentioned earlier, all control plane mechanisms reside in communications between the access points at the edge of the network in a distributed WLAN architecture. The APs implement control plane mechanisms cooperatively using proprietary protocols.

How VLANs are deployed in a WLAN environment depends on the design of the network as well as the type of WLAN architecture that is in place. One very big difference between using a controller-based model versus a noncontroller model is how VLANs are implemented in the network design. In the WLAN controller model, most user traffic is centrally forwarded to the controller from the APs. Because all the user traffic is encapsulated, a controller-based AP typically is connected to an access port on an Ethernet switch that is tied to a single VLAN.

With a WLAN controller architecture, the user VLANs usually reside in the core of the network. The user VLANs are not available at the access layer switch. The controller-based APs are connected to an access port of the edge switch. The user VLANs are still available to the wireless users because all of the user VLANs are encapsulated in an IP tunnel between the controller-based APs at the edge and the WLAN controller in the core.

The noncontroller model, however, requires support for multiple user VLANs at the edge. Each access point is therefore connected to an 802.1Q trunk port on an edge switch that supports VLAN tagging. All of the user VLANs are configured in the access layer switch. The access points are connected to an 802.1Q trunk port of the edge switch. The user VLANS are tagged in the 802.1Q trunk, and all wireless user traffic is forwarded at the edge of the network.

Although the whole point of a cooperative and distributed WLAN model is to avoid centrally forwarding user traffic to the core, the access points may also have IP-tunneling capabilities. Some WLAN customers require that guest VLAN traffic not cross internal networks. In that scenario, a standalone AP might forward only the guest user VLAN traffic in an IP tunnel that terminates at another standalone access point that is deployed in a DMZ. Individual APs can also function as a VPN client or VPN server using IPsec encrypted tunnels across a WAN link.

Another advantage of the distributed WLAN architecture is scalability. As a company grows at one location or multiple locations, more APs will obviously have to be deployed. When a WLAN controller solution is in place, more controllers might also have to be purchased and deployed as the AP count grows. With the controller-less distributed WLAN architecture, only new APs are deployed as the company grows. Many vertical markets such as K–12 education and retail have schools or stores at numerous locations. A distributed WLAN architecture be the better choice as opposed to deploying a WLAN controller at each location.

Hybrid WLAN Architecture

It is important to understand that none of the WLAN architectures described in this chapter are written in stone. Many hybrids of these WLAN architectures exist among the WLAN vendors. As was already mentioned, some of the WLAN controller vendors are pushing some of the control plane intelligence back into the access points. One WLAN controller vendor has a cloud-based controller where much of the control plane intelligence exists in the cloud.

Typically, the data plane is centralized when using WLAN controllers, but distributed data forwarding is also available. Some WLAN vendors have moved the data plane back to the edge of the network, with APs handling the data forwarding of user traffic. With a controller-less distributed WLAN architecture, all data is forwarded locally, but the ability to centralize the data plane is a capability of a distributed WLAN architecture. In general, most WLAN vendors now have the option to either centralize or locally forward the data plane depending on the location of the APs and the traffic routes available.

In a distributed WLAN architecture, the management plane resides in an on-premises or cloud-based network management service. With the WLAN controller model, the management plane normally exists in the WLAN controller. However, the management plane might also be pushed into an NMS that not only manages the controller-based APs but also manages the WLAN controllers.

Specialty WLAN Infrastructure

In the previous sections, we discussed the progression of WLAN network infrastructure devices that are used to integrate an 802.11 wireless network into a wired network architecture. The Wi-Fi marketplace has produced many specialty WLAN devices in addition to APs and WLAN controllers. Many of these devices, such as bridges and mesh networks, have become extremely popular, although they operate outside of the defined 802.11 standards. You will look at these devices in the following sections.

Enterprise WLAN Routers

In addition to the main corporate office, companies often have branch offices in remote locations. A company might have branch offices across a region or an entire country, or they may even be spread globally. The challenge for IT personnel is how to provide a seamless enterprise wired and wireless solution across all locations. A distributed solution using enterprise-grade WLAN routers at each branch office is a common choice.

Keep in mind that WLAN routers are very different from access points. Unlike access points, which use a bridged virtual interface, wireless routers have separate routed interfaces. The radio card exists on one subnet, whereas the WAN Ethernet port exists on a different subnet.

Branch WLAN routers have the ability to connect back to corporate headquarters with VPN tunnels. Employees at the branch offices can access corporate resources across the WAN through the VPN tunnel. Even more important is the fact that the corporate VLANs, SSIDs, and WLAN security can all be extended to the remote branch offices. Employees at a branch office connect to the same SSID that they would connect to at corporate headquarters. The wired and wireless network access policies are therefore seamless across the entire organization. These seamless policies can be extended to the WLAN routers at each branch location.

The enterprise-grade WLAN routers are very similar to the consumer-grade Wi-Fi routers that most of us use at home. However, enterprise WLAN routers are manufactured with better-quality hardware and offer a wider array of features.

The following security features are often supported by enterprise WLAN routers:

- 802.11 layer 2 security for wireless clients
- 802.1X/EAP port security for wired clients

- Network address translation (NAT)
- Port address translation (PAT)
- Port forwarding
- Firewall
- Integrated VPN client
- 3G/4G cellular backhaul

WLAN Mesh Access Points

Almost all WLAN vendors now offer *WLAN mesh access point* capabilities. Wireless mesh APs communicate with each other by using proprietary layer 2 routing protocols and create a self-forming and self-healing wireless infrastructure (a mesh) over which edge devices can communicate, as shown in Figure 11.18. The main purpose of a mesh WLAN is to provide wireless client access in physical areas where an Ethernet cable cannot be connected to an AP. WLAN client traffic can be sent over wireless backhaul links with an eventual destination to mesh portals that are connected to the wired network.

FIGURE 11.18 WLAN mesh network

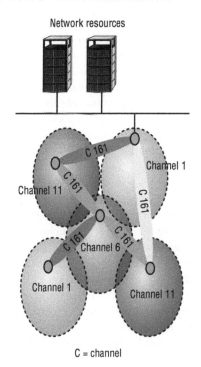

A WLAN mesh network automatically connects access points upon installation and dynamically updates traffic routes as more clients are added. Proprietary layer 2 intelligent routing protocols determine the dynamic routes based on measurement of traffic, signal strength, data rates, hops, and other parameters.

With dual-band WLAN mesh APs, typically the 5 GHz radios are used for the mesh backhaul communications, as shown in Figure 11.18. The mesh backhaul traffic must also be encrypted. In most cases, 802.11 PSK security is used between the mesh radios to provide encryption. The PSK is usually created automatically in most mesh WLAN solutions. A very strong passphrase of 20 characters or more should be used if the WLAN vendor offers the option to manually define mesh backhaul security.

WLAN Bridges

A common specialty deployment of 802.11 technology is the *wireless LAN bridge*. The purpose of bridging is to provide wireless connectivity between two or more wired networks. A bridge generally supports all the same features that an autonomous access point possesses, but the purpose is to connect wired networks, not to provide wireless connectivity to client stations. When facilities are separated from each other and no physical network-capable wiring exists between them, wireless bridges are often employed. Monthly-based fees for Telco circuit costs can be mitigated with the one-time cost of a wireless point-to-point (PtP) bridge. Wireless bridges are also used between communication towers and can sometimes span several miles.

Bridge and backhaul links tend to have very different requirements than do typical AP functions that serve WLAN clients. The first difference is that APs typically serve in the access layer of a network. WLAN bridges operate in the distribution layer and are normally used to connect two or more wired networks together over a wireless link.

Outdoor bridge links are used outdoors to connect the wired networks inside two buildings. An outdoor bridge link is often used as a redundant backup to T1 or fiber connections between buildings. Outdoor wireless bridge links are even more commonly used as replacements to T1 or fiber connections between buildings because of their substantial cost savings.

Wireless bridges support two major configuration settings: *root* and *nonroot*. Bridges work in a parent/child-type relationship, so think of the root bridge as the parent and the nonroot bridge as the child.

One side of the link is usually the root bridge and the other side is the nonroot bridge. The root bridge establishes the channel and beacons for the nonroot bridge to join. The nonroot bridge will then associate with the root bridge in a station-like fashion to establish the link.

A bridge link that connects only two wired networks is known as a *point-to-point (PtP)* bridge. Figure 11.19 shows a PtP connection between two wired networks using two 802.11 bridges and directional antennas. Note that one of the bridges must be configured as the parent root bridge and the other bridge is configured as the child nonroot bridge.

FIGURE 11.19   Point-to-point WLAN bridging

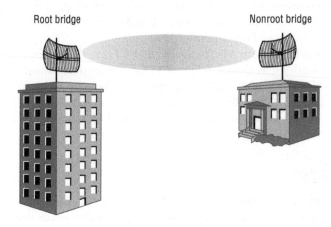

A *point-to-multipoint (PtMP)* bridge link connects multiple wired networks. The root bridge is the central bridge, and multiple nonroot bridges connect back to the root bridge. Figure 11.20 shows a PtMP bridge link between three buildings. Please note that the root bridge is using a high-gain omnidirectional antenna array, whereas the nonroot bridges are all using unidirectional antennas pointing back to the antenna of the root bridge. Also notice that there is only one root bridge in a PtMP connection. There can never be more than one root bridge.

FIGURE 11.20   Point-to-multipoint WLAN bridging

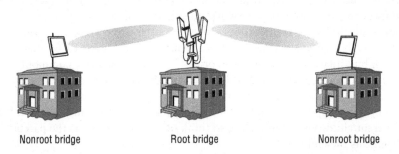

Considerations when deploying outdoor bridge links are numerous, including the Fresnel zone, earth bulge, free space path loss, link budget, and fade margin. There may be other considerations as well, including the IR and EIRP power regulations as defined by the regulatory body of your country.

Point-to-point links in the 2.4 GHz band can be as long as a few miles. A problem that might occur over a long-distance link is an ACK timeout. Because of the half-duplex nature of the medium, every unicast frame must be acknowledged. Therefore, a unicast frame sent across a long-distance PtP link by one bridge must immediately receive an

ACK frame from the opposite bridge, sent back across the same long-distance link. Even though RF travels at the speed of light, the ACK may not be received quickly enough. The original bridge will time out after not receiving the ACK frame for a certain period of microseconds and will assume that a collision has occurred. The original bridge will then retransmit the unicast frame even though the ACK frame is on the way. Retransmitting unicast traffic that does not need to be re-sent can cause throughput degradation of as much as 50 percent. To resolve this problem, most bridges have an ACK timeout setting that can be adjusted to allow a longer period of time for a bridge to receive the ACK frame across the long-distance link.

A common problem with point-to-multipoint bridging is mounting the high-gain omni-directional antenna of the root bridge too high, as pictured in Figure 11.21. The result is that the vertical line of sight with the directional antennas of the nonroot bridges is not adequate. The solution for this problem is to use a high-gain omnidirectional antenna that provides a certain amount of electrical downtilt or to use directional sector antennas aligned to provide omnidirectional coverage.

FIGURE 11.21   Common bridging challenge

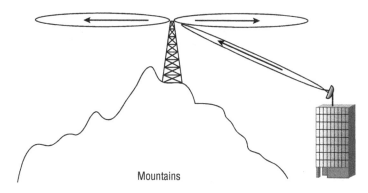

Mountains

Encryption is needed to protect the data privacy of the backhaul communications across bridge links. IPsec VPNs are often used for bridge security, which will be discussed in Chapter 17, "802.11 Network Security Architecture." An 802.1X/EAP solution can also be used for bridge security, with the root bridge assuming the authenticator role and the nonroot bridges assuming the supplicant role. Additionally, PSK authentication is often used for WLAN bridge security and therefore a strong WPA2 passphrase of 20 characters or more is recommended.

WLAN Array

A company called Riverbed offers a unique solution that combines a WLAN controller and multiple access points in a single hardware device known as a *Wi-Fi array*. The CWNP program uses the generic term *WLAN array* to describe this technology.

Their legacy WLAN arrays, as shown in Figure 11.22, supported up to 16 access-point radios using sector antennas and an embedded WLAN controller all residing in one device. The WLAN controller was obviously deployed at the access layer, because the device was mounted on the ceiling. The embedded WLAN controller offered many of the same features and capabilities found in more traditional WLAN controllers.

FIGURE 11.22   WLAN array

Courtesy of Xirrus

One of the key points of a WLAN array is that each AP has a sector antenna providing directional coverage. Each AP therefore provides a sector of coverage. The WLAN array is simply an indoor sectorized array solution that provides 360 degrees of horizontal coverage by combining the directional coverage of all the sector APs. The directional coverage of each AP increases the range much like an outdoor sectorized array. The number of radios that are in a WLAN array often depends on the model and configuration. Riverbed currently offers WLAN arrays with up to four radios.

One advantage of the WLAN array solution is that less physical equipment needs to be deployed; therefore, the number of devices that have to be installed and managed is drastically reduced.

Real-Time Location Systems

NMS solutions, WLAN controllers, and WIDS solutions have some integrated capabilities to track 802.11 clients by using the access points as sensors. However, the tracking capabilities are not necessarily real-time and may be accurate to within only about 25 feet. The tracking capabilities in WLAN controllers and WIDS solutions provide a *near-time* solution and cannot track Wi-Fi RFID tags. Several companies, such as Stanley Healthcare, provide a WLAN *real-time location system (RTLS)*, which can track the location of any 802.11 radio device as well as active Wi-Fi RFID tags with much greater accuracy. The components of an overlay WLAN RTLS solution include the preexisting WLAN infrastructure, preexisting WLAN clients, Wi-Fi RFID tags, and an RTLS server. Additional RTLS WLAN sensors can also be added to supplement the preexisting WLAN APs.

Active RFID tags and/or standard Wi-Fi devices transmit a brief signal at a regular interval, adding status or sensor data if appropriate. Figure 11.23 shows an active RFID tag attached to a hospital IV pump. The signal is received by standard wireless APs (or RTLS sensors), without any infrastructure changes needed, and is sent to a processing engine that resides in the RTLS server at the core of the network. The RTLS server uses signal strength and/or time-of-arrival algorithms to determine location coordinates.

FIGURE 11.23   Active 802.11 RFID tag

As Figure 11.24 shows, a software application interface is then used to see location and status data on a display map of the building's floor plan. The RTLS application can display maps, enable searches, automate alerts, manage assets, and interact with third-party applications.

FIGURE 11.24   RTLS application

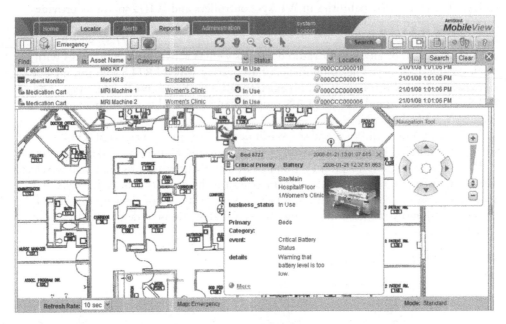

VoWiFi

VoIP communications have been around for many years on wired networks. However, using VoIP on an 802.11 wireless LAN presents many challenges due to the RF environment and QoS considerations. In recent years, the demand for *Voice over Wi-Fi (VoWiFi)* solutions has grown considerably. The WLAN can be used to provide communications for all data applications while at the same time providing for voice communications using the same WLAN infrastructure. The components needed to deploy a VoWiFi solution include the following:

VoWiFi Telephones   A VoWiFi phone is similar to a cell phone, except the radio is an 802.11 radio instead of a cellular radio. VoWiFi phones are 802.11 client stations that communicate through an access point. They fully support WEP, WPA, and WPA2 encryption and WMM quality-of-service capabilities. Figure 11.25 shows Spectralink's 84-Series VoWiFi phones, which has an 802.11a/b/g/n radio and can operate in either the 2.4 GHz or the 5 GHz band. VoWiFi technology can also reside in form factors other than a telephone. As pictured in Figure 11.26, VoWiFi vendor Vocera sells an 802.11 communications badge, a wearable device that weighs less than two ounces. The Vocera badge is a fully functional VoWiFi phone that also uses speech recognition and voiceprint verification software. Currently most VoWiFi solutions use the *Session Initiation Protocol (SIP)* as the signaling protocol for voice communications over an IP network.

FIGURE 11.25   VoWiFi phone (Spectralink 84-Series VoWiFi phone)

Courtesy of Spectralink

FIGURE 11.26   Vocera communications badge

Courtesy of Vocera

802.11 Infrastructure (APs and Controllers)   An existing WLAN infrastructure is used for 802.11 communications between the VoWiFi and access points. Standalone APs and/or WLAN controller solutions can both be used.

PBX   A *private branch exchange (PBX)* is a telephone exchange that serves a particular business or office. PBXs make connections among the internal telephones of a private company and also connect them to the *public switched telephone network (PSTN)* via trunk lines. The PBX provides a dial tone and may provide other features, such as voicemail.

WMM Support   As discussed in earlier chapters, WMM mechanisms are needed to properly support QoS.

Cloud Networking

As mentioned earlier in this chapter, NMS solutions are available in the cloud as a software subscription service. Several WLAN vendors offer a cloud service for management and monitoring. *Cloud computing* and *cloud networking* are catchphrases used to describe the advantages of computer networking functionality when provided under a *Software as a Service (SaaS)* model. The term *the cloud* essentially means a scalable private enterprise network that resides on the Internet. The idea behind cloud networking is that applications and network management, monitoring, functionality, and control are provided as a software service. Amazon is a good example of a company that provides an elastic cloud-based IT infrastructure so that other companies can offer pay-as-you-go subscription pricing for enterprise applications and network services.

The most common cloud networking model is *cloud-enabled networking (CEN)*. With CEN, the management plane resides in the cloud, but data plane mechanisms, such as switching and routing, remain on the local network and usually in hardware. Several WLAN vendors offer cloud-enabled NMS solutions as a subscription service that manages and monitors WLAN infrastructure and clients. Some control plane mechanisms can also be provided with a CEN model. For example, WLAN vendors have begun to also offer subscription-based application services along with their cloud-enabled management solutions. Some examples of these subscriptions services include cloud-enabled guest management, NAC, and MDM solutions.

Distributed computing models billed as "cloud" progressed over time from a basic remote server to a truly cloud-oriented architecture. The innovations in cloud computing were accompanied with a new nomenclature to better describe the roles and relationships. Cloud computing can be defined as a progression of the following three cloud architectures:

Single Server   A single physical server or *virtual machine (VM)* hosts applications for remote access. To meet increasing demands, you can scale up by replacing one machine with a bigger/faster one until you reach a hard scalability limit. To provide high availability (HA), you can add a second standby machine to mirror the primary server so that it can assume operations after a failover. This is the simplest model referred to as a "cloud" solution.

Multiple Servers   When there are more customers than a single server can accommodate, customers are partitioned into multiple server instances. For example, customers A and B are partitioned into instance 1, customer C into instance 2, and so on. This approach scales by adding or removing instances. However, it faces the same scaling and HA limitations as generation 1, and fragmentation challenges can also occur.

Elastic Cloud   *Elastic cloud* is a term used to describe a cloud offering that provides variable service levels based on changing needs. A true elastic cloud is a distributed system where functions are provided by multiple services working collaboratively. As shown in Figure 11.27, services are hosted on pools of servers grouped by clusters in which any member can provide the same set of services as any other. The system scales by adding servers to clusters and removing them as needed. It achieves high performance and availability by distributing loads among cluster members through the use of messaging queues and load balancers. True elastic cloud emphasizes openness and uses APIs extensively.

FIGURE 11.27   Elastic cloud architecture

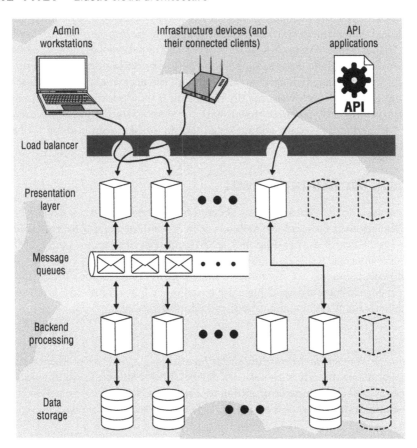

Infrastructure Management

Sound network design dictates placing network devices onto dedicated management VLANs or another out-of-band interface from regular network traffic. Enterprise network gear should allow for this functionality in order to keep infrastructure devices inaccessible to hackers or even to employees who are able to gain access to the network.

As the size of WLANs has grown over the years, so have the challenges of managing them. This includes managing the following:

- Firmware revisions

- Configurations and changes

- Monitoring and incident response

- Managing and filtering of device alerts and alarms

- Performance monitoring

To make the job of managing these devices easier, standard network protocols for device management are typically included in most WLAN hardware and can be integrated with software-based management systems that can span even the largest of networks. The more a network management system (NMS) is incorporated in network designs, the less time is spent performing mundane tasks to support the network. Without exception, every time an NMS is properly implemented, user satisfaction with the network is higher, while the cost of operating the network is lower. Equally important is incorporating the deployment of the NMS into an operational environment with the support staff. Developing processes and procedures around these systems and making them a part of the support staff's daily work are also critical.

Protocols for Management

There are many different types of protocols used for managing network devices. Simple Network Management Protocol (SNMP) has been around for quite some time and has undergone several revisions. In addition to SNMP, most devices can be configured using a command-line interface (CLI) or a graphical user interface (GUI).

The following protocols are common for managing WLANs. Some of these protocols are based on de jure standards and some are based on de facto standards. Either way, they provide the basis for WLAN management and administration.

SNMP

Simple Network Management Protocol (SNMP) is an Application layer protocol (OSI layer 7) used to communicate directly with network devices. SNMP allows for pulling information from devices as well as pushing information to a central SNMP server based on certain, often user-configurable thresholds on network devices. A push from a device might include a message pertaining to an interface reset, a high number of errors, high network or CPU utilization, security alarms, and many other critical factors related to the healthy operations and status of devices.

Components

An SNMP management system contains the following:

- Several (potentially many) nodes, each with an SNMP entity (a.k.a. agent) containing *command responder* and *notification originator* applications, which have access to management instrumentation

- At least one SNMP entity containing command generator and/or notification receiver applications (traditionally called a *manager*)

- A management protocol used to convey management information between the SNMP entities

Structure of Management Information

Management information is structured as a collection of managed objects contained in a database called a *management information base (MIB)*.

The MIB consists of the following definitions: modules, objects, and traps. Module definitions are used when describing information modules. Object definitions are used when describing the managed objects. Trap definitions are notifications used for unsolicited transmissions of MIB information typically to an NMS.

All SNMP-capable devices have a MIB, and in that MIB should reside the configuration and status of the device. However, vendors aren't usually totally complete with their MIBs and SNMP implementations. Often you will find that certain pieces of critical information are not accessible via SNMP, and therefore traps cannot be implemented using that information.

Versions and Differences

SNMP has undergone numerous revisions over the years. This section is not intended to be a complete history of SNMP but rather an overview to guide you in knowing the differences between the various versions. Additionally, this section will help you properly implement different versions into your network designs by understanding the strengths and how to address the weaknesses.

SNMPV1

Version 1 of SNMP hit the scene in 1988. Like many other initial protocol introductions, SNMPv1 did not get it perfect the first time. SNMPv1 was designed to work over a wide range of protocols in use at the time—including IP, UDP, CLNS, AppleTalk, and IPX—but it is most commonly used with UDP.

SNMPv1 used a *community string*, which had to be known by a remote agent. Since SNMPv1 did not implement any encryption, it was subject to packet sniffing to discover the cleartext community string. Therefore, SNMPv1 was heavily criticized as being insecure.

Protocol efficiency was also lacking for this initial protocol introduction. Each MIB object had to be retrieved one by one in an iterative style, which was very inefficient.

SNMPV2

When SNMPv2 was released, several areas of SNMPv1 were addressed, including performance, security, and manager-to-manager communications. Protocol performance became more efficient with the introduction of new functions such as GETBULK, which solved the iterative method of extracting larger amounts of data from MIBs.

Security was improved by the specification of a new party-based security system. Critics accused the party-based system of being too complex, and the system was not widely accepted.

SNMPv2c was later defined in RFCs 1901–1908 and is referred to as the *community-based* version. The community string from SNMPv1 was adopted in SNMPv2c, which essentially dropped any security improvements to the protocol. SNMPv2c does not implement encryption and is subject to packet sniffing of the cleartext community string.

SNMPV3

A great deal of security benefits were added to SNMPv3, including the following:

- Authentication is performed using SHA or MD5.

- Privacy—SNMPv3 uses DES 56-bit encryption based on the CBC-DES (DES-56) standard.

- Access control—Users and groups are used, each with different levels of privileges. Usernames and passwords replace community strings.

Although these features are optional, usually the main driver behind adopting SNMPv3 is to gain these security benefits. It is also optional to have secure authentication but disable encryption.

Even with these features, most network designers still feel it is best to implement the SNMP agents only on secure management interfaces. Specifically, VLAN segmentation and firewall filtering is usually performed on all SNMP traffic to network devices. No sufficiently complex protocol is considered completely secure, and additional safeguards are always highly recommended. If you intend to implement an NMS using SNMP, we highly recommended that you implement SNMPv3. One of the most important security concerns is that most vendor equipment defaults to SNMP being enabled, with default *read* and *write* community strings. This is an *enormous* security threat to the configuration and operation of your network and should be one of the very first lockdown steps to securing network devices.

CLI-Based Management

Command-line interfaces (CLIs) are one of the most common methods used to configure and manage network devices. It seems the age-old debate of GUI versus CLI is still present to this day and is not likely to change any time soon. GUIs do a wonderful job of presenting information, but due to browser incompatibility, JavaScript errors, GUI software bugs, time delays, and more, GUIs still drive many people back to the command line.

CLIs tend to be the raw, unedited configuration of devices and provide the ability to make specific changes quickly to device configurations. Commands issued via a CLI can even be scripted, allowing initial device configuration and even reconfigurations to be performed with a simple copy and paste into a CLI session.

CLIs can be accessed using several methods, which are dependent on the device being used. These commonly include the following:

- Serial and console ports
- Telnet
- SSH1/SSH2

Serial and Console Ports

Serial or console port interfaces can vary from manufacturer to manufacturer and even from model to model. This is extremely frustrating to network engineers. Some of them use a standard DB-9 serial connector interface, whereas others use an RJ-11 or RJ-45 interface. Furthermore, the actual cable might be a proprietary pin-out (namely for the RJ-11 and RJ-45 connectors), a NULL-modem cable, a rollover cable, or a straight-through cable. Baud rates, number of bits, flow control, parity, and other parameters also vary from device to device.

No matter what type of connector or cable you are using to manage your network device, serial or console ports should be locked down and require a user authentication mechanism. Although typically these can be thwarted by a password-recovery routine using instructions that can be readily found on the Internet, a user authentication mechanism will help deter hackers. Typically, a device requires downtime in order to recover the password, and the impact of a service outage may be enough to alert staff of a physical break-in attempt to network devices. It is important to note that most password-recovery routines require direct access to the device via the serial or console port. Securing network equipment in a locked data closet or computer room will help to prevent this type of attack from occurring.

Government regulations such as FIPS 140-2 may require that serial and console ports be secured with a tamper-evident label (TEL) to prevent unauthorized physical access to a WLAN infrastructure device, such as a WLAN controller. TELs cannot be surreptitiously broken, removed, or reapplied without an obvious change in appearance. As shown in Figure 11.28, each TEL has a unique serial number to prevent replacement with similar labels.

FIGURE 11.28   Tamper-evident label

 Applied label

 Removed and reapplied
← Residue

Telnet

Telnet is another protocol that is commonly used, but often it can be used only after a serial port configuration or initial configuration from a factory default state is performed. The IP of the device typically needs to be enabled in order for it to be accessed and managed via the network interface.

Telnet is heavily criticized and usually prohibited from use by enterprise security policies due to its lack of encryption. Telnet is a completely unencrypted protocol, and the payload of each packet can be inspected by packet sniffing. This includes the username and password during the login sequence. We recommend that you disable Telnet after the initial device configuration. Most companies have written policies mandating that Telnet be disabled.

Secure Shell

Secure Shell (SSH) is typically used as the secure alternative to Telnet. SSH implements authentication and encryption using public-key cryptography of all network traffic traversing between the host and user device. The features of Telnet for CLI-based management apply to SSH but include added security benefits. The standard TCP port 22 has been assigned for the SSH protocol. Most WLAN infrastructure devices now support the second version of the SSH protocol, called *SSH2*. As a matter of policy, when WLAN devices are managed via the CLI, an SSH2-capable terminal emulation program should be used. Figure 11.29 shows the configuration screen of the popular freeware program PuTTY, which supports SSH2.

FIGURE 11.29 PuTTY freeware SSH2 client

HTTPS

Hypertext Transfer Protocol Secure (HTTPS) is a combination of the Hypertext Transfer Protocol with the SSL/TLS protocol to provide encryption and secure identification. HTTPS is essentially an SSL session that uses HTTP and is implemented on network devices for management via a graphical user interface (GUI). Not all users prefer CLI-based management methods, and GUIs are commonly used where an NMS is used to manage WLAN infrastructure.

Because HTTP is transmitted in plaintext, it is susceptible to eavesdropping and man-in-the-middle attacks from modifications in transit. Some devices offer both HTTP and HTTPS, but it is important that minimal authentication be performed via HTTPS. If users of devices will be entered into the GUI, not using HTTPS is purely negligent if the device supports it.

Application Programming Interface

For decades, IT systems engineers have been using a *command-line interface (CLI)* or *graphical user interface (GUI)* to interact with physical network devices in order to change the configuration of a device, monitor its operation or troubleshoot an issue. The CLI was traditionally the preferred method for network engineers. The robustness of the CLI was often a key factor when new network equipment was purchased. Provided they were familiar with the CLI syntax, network engineers could quickly configure a network device instead of using the GUI. Network engineers would copy, edit, and paste configuration commands between devices. Additionally, the creation of scripts could automate the configuration and monitoring process of network devices which streamlined the speed and scalability of deployment rollouts.

The evolution of technology has seen a migration of applications to the cloud. These cloud-based applications need to extract data and interact with the corporate network. Legacy network management systems as well as operations support system/business support systems (OSS/BSS) have communicated with network infrastructure using standard management protocols like SNMP. However, these management protocols were not designed to scale in the era of the cloud and were rarely designed for bidirectional communication.

Today, the applications that need to interact with the network are no longer limited to a small subset of users like network engineers and administrators. Applications such as guest management systems, retail analytics, and other analytical engines all require some sort of direct interaction with the network. The communication is often bidirectional and is used for network configuration and extracting network monitoring data. In fact, with the growth of Internet of Things (IoT) devices, some of these applications do not interact with users and instead communicate only with other applications. To enable these applications to interact with one another, an *application programming interface (API)* is needed. An application programming interface (API) is a set of subroutine definitions, protocols, and tools for building application software.

Transport and Data Formats

An API enables applications to interact with other applications in order to exchange data or execute tasks. For example, if you are in a retail store and you would like to use an application on your smartphone to direct you to your favorite shoe brand, the application would need to interact with a *location analytics* system to determine your current location. The data about your location comes from the WLAN, gets stored in the location analytics system, and is requested by the application on your smartphone. For application development, the API replaces the CLI but follows a similar set of rules. The API is evaluated based on its simplicity, performance, and completeness, which means that many of the features a system supports is exposed through the API.

Just as CLI access requires a transport protocol such as SSH to connect the user to the device, APIs require a transport protocol to connect to other applications. The most common transport protocols are HTTP or HTTPS because they are supported across a wide range of devices. Having this support natively avoids the need to install additional libraries or develop new protocols and it also addresses security by supporting encryption through the use of SSL. This enables a secure channel between different applications when exchanging data. A *RESTful API* is an application program interface (API) that uses HTTP requests to GET, PUT, POST and DELETE data.

The data itself can be in various formats but a format called *JavaScript Object Notation (JSON)* is the most popular. Unlike other formats (i.e., XML), the JSON format is self descriptive and is easily readable by humans. The JSON format defines key value pairs where the key describes what the data means and the value is the actual data. Besides being human-readable, JSON is also very effective for transport.

WLAN APIs

WLAN APIs can be categorized into three groups:

Configuration APIs   APIs can be used to change the configuration settings of an AP or other network device. The configuration APIs can be used for something as simple as creating a new set of WLAN user credentials, or more complex settings such as creating a new SSID with VLAN and QoS access policies.

Monitoring APIs   APIs can be used to retrieve network data statistics such as AP status, CPU and memory utilization, traffic counters, and more. The monitoring APIs can also be used to retrieve monitoring data for WLAN client devices such as connection times, roaming events, IP addresses, and application usage.

Notification APIs   Notification APIs, more often called *webhook APIs*, offer a subscription service where an application can subscribe to receive a notification when a specific event happens. Webhook APIs are triggered by an event. For example, when the system detects an AP is no longer responding, it sends a message specifying the device ID, timestamp, and other data to the subscribing application.

Both the configuration and monitoring APIs could be considered synchronous. An application requiring data to initiate a configuration change, calls the API. The system receiving the API call responds with a dataset or simply returns the result of the operation (i.e., configuration change is "success"). However, this type of API call is not conducive for notifications.

For example, to trigger an action when an access point is no longer responding, continuous calls to the monitoring API would be needed to sort through the entire list of returned APs and filter out the APs that are no longer connected. This API approach is called *polling* and is considered a wasteful treatment of system resources. Webhook APIs are instead triggered by an event and polling is not necessary. Using webhook APIs for notifications reduces system load and minimizes the traffic flow between applications.

Common Applications

One of the most common applications that utilize APIs in WLANs is a network management system (NMS). An NMS gathers monitoring data from different WLAN devices and presents the data using dashboards, graphs, and other visualization techniques. An NMS might use the collected data to analyze common behavior, establish baselines, and look for anomalies. A more advanced NMS might use predictive analytics to forecast future events or failures using the collected data. Supported by configuration APIs, an NMS might proactively adapt the configuration of the WLAN to avoid failures, congestion, assign QoS classification for specific client devices and applications.

Another common use of APIs in WLANs is location analytics. WLANs are a rich source of location data, such as physical distribution of devices across a floor during the day, dwell times of customers, or even real-time location tracking of people and assets. The amount of such data generated by a large enterprise WLAN can be overwhelming. Handling and exposing such data requires techniques that will enable the data to be analyzed and stored quickly without overwhelming the NMS. A well-defined and implemented API can help retrieve the location data as well as analyze and store fast growing datasets.

A location API could also be used to aid an application on your smartphone to bring some additional context to the physical space where you currently stand. For example, the location analytics API can trigger an application on a mobile device to display interactive content when a visitor is within near proximity of a museum exhibit. These proximity based solutions are often aided by other RF technologies such as *Bluetooth Low Energy (BLE)* which is discussed in more detail in Chapter 20, "WLAN Deployment and Vertical Markets."

WLAN vendor customers and partners have begun to use WLAN vendor APIs to build their own custom applications to monitor the WLAN. Custom applications can also be built for WLAN device configuration and notifications. Some of the larger WLAN vendors maintain a developer community portal as a repository for API documentation, code examples, and reference applications. With the advent of cloud technology, corporations are leveraging WLAN APIs to meet the growing needs of exposing, analyzing, and storing WLAN data.

Summary

This chapter discussed the various types of radio form factors, their chipsets, and the software interfaces needed for client station configuration. We discussed the three logical planes of telecommunications operations and where they exist within the three most common WLAN architectures. We also showed you the logical progression that WLAN devices have made, starting from autonomous access points, moving to WLAN controllers, and then moving along a path toward a distributed architecture. In addition, we covered specialty WLAN infrastructure devices, which often meet needs that may not be met by more traditional WLAN architecture. We also defined cloud networking and discussed the progression of cloud architecture.

The authors of this book recommend that before you take the CWNA exam, you get some hands-on experience with some WLAN infrastructure devices. We understand that most individuals cannot afford a $10,000 WLAN controller and multiple APs; however, we do recommend that you purchase at least one 802.11ac client adapter and either an access point or a SOHO wireless router. Many enterprise WLAN vendors also offer "free access point" programs to potential customers. Hands-on experience will solidify much of what you have learned in this chapter as well as in many of the other chapters in this book.

Exam Essentials

Know the major radio form factors.   The 802.11 standard does not mandate what type of format can be used by an 802.11 radio. 802.11 radios exist in multiple formats.

Understand the need for client adapters to have an operating system interface and a user interface.   A client adapter requires a special driver to communicate with the operating system and a software client utility for user configuration.

Identify the three major types of client utilities.   The three types of client utilities are enterprise, integrated, and third party.

Define the three logical network planes of operation.   Understand the differences between the management, control, and data plane. Be able to explain where they are used within different WLAN architectures.

Understand types of WLAN architectures.   Know the different types of architectures, including autonomous, centralized, distributed, unified, and hybrid. Understand the various common features and capabilities each architecture has to offer.

Explain the role of WLAN bridges.   Identify the difference between root and nonroot bridges. Be able to explain the differences between point-to-point and point-to-multipoint bridging. Understand bridging problems such as ACK timeout, and study other bridging considerations that are covered in other chapters, such as the Fresnel zone and system operating margin.

Explain WLAN specialty infrastructure.   Be able to explain how RTLS and VoWiFi solutions can all be integrated with a WLAN. Explain other nontraditional WLAN solutions, such as WLAN arrays.

Be familiar with device management features.   Know the various device management methods, features, and protocols available in WLAN devices.

Review Questions

1. Which terms best describe components of a centralized WLAN architecture where the management, control, and data planes all reside in a centralized device? (Choose all that apply.)

   A. WLAN controller

   B. Wireless network management system

   C. Network management system

   D. Distributed AP

   E. Controller-based AP

2. Which logical plane of network operation is typically defined by protocols and intelligence?

   A. User plane

   B. Data plane

   C. Network plane

   D. Control plane

   E. Management plane

3. Which WLAN architectural models typically require support for 802.1Q tagging at the edge on the network when multiple user VLANs are required? (Choose all that apply.)

   A. Autonomous WLAN architecture

   B. Centralized WLAN architecture

   C. Distributed WLAN architecture

   D. None of the above

4. Which type of access points normally use centralized data forwarding?

   A. Autonomous APs

   B. Controller-based APs

   C. Cooperative APs within a distributed WLAN architecture

   D. None of the above

5. Which protocols can be used to tunnel 802.11 user traffic from access points to WLAN controllers or other centralized network servers? (Choose all that apply.)

   A. IPsec

   B. GRE

   C. CAPWAP

   D. DTLS

   E. VRRP

6. Which of these WLAN architectures may require the use of an NMS server to manage and monitor the WLAN?

    A. Autonomous WLAN architecture

    B. Centralized WLAN architecture

    C. Distributed WLAN architecture

    D. All of the above

7. What term best describes a WLAN centralized architecture where the integration service (IS) and distribution system services (DSSs) are handled by a WLAN controller while the generation of certain 802.11 management and control frames is handled by a controller-based AP?

    A. Cooperative control

    B. Distributed data forwarding

    C. Distributed hybrid architecture

    D. Distributed WLAN architecture

    E. Split MAC

8. What are some of the necessary components of a VoWiFi architecture? (Choose all that apply.)

    A. VoWiFi phone

    B. SIP

    C. WMM support

    D. Proxy server

    E. PBX

9. What is the traditional data-forwarding model for 802.11 user traffic when WLAN controllers are deployed?

    A. Distributed data forwarding

    B. Autonomous forwarding

    C. Proxy data forwarding

    D. Centralized data forwarding

    E. All of the above

10. What are some of the security capabilities found in an enterprise WLAN router that is typically deployed in remote branch locations? (Choose all that apply.)

    A. Integrated WIPS server

    B. Integrated VPN server

    C. Integrated NAC server

    D. Integrated firewall

    E. Integrated VPN client

11. Which radio form factors can be used by 802.11 technology?

   A. USB 3.0

   B. Secure Digital

   C. PCMCIA

   D. Mini PCI

   E. ExpressCard

   F. Proprietary

   G. All of the above

12. Which of these protocols can be used to manage WLAN infrastructure devices?

   A. HTTP

   B. SSH

   C. SNMP

   D. Telnet

   E. HTTPS

   F. SNMP

   G. All of the above

13. What are some of the common capabilities of a WLAN controller architecture?

   A. Adaptive RF

   B. AP management

   C. Layer 3 roaming support

   D. Bandwidth throttling

   E. Firewall

   F. All of the above

14. Which management protocols are often used between a network management system (NMS) server and remote access points for the purpose of monitoring a WLAN? (Choose all that apply.)

   A. IPsec

   B. GRE

   C. CAPWAP

   D. DTLS

   E. SNMP

15. What are of some of the common security capabilities often integrated within access points deployed in a distributed WLAN architecture?

   A. Captive web portal

   B. Firewall

   C. Integrated RADIUS

D. WIPS

E. All of the above

16. What RF technology do IoT client devices use for communication?

    A. Wi-Fi

    B. Bluetooth

    C. Zigbee

    D. All of the above

17. What type of 802.11 radio form factor is normally used in mobile devices such as smartphones and tablets?

    A. Integrated single chip

    B. PCMCIA

    C. Express Mini PCI

    D. Mini PCI

    E. Secure Digital

18. Where is redundancy needed if user traffic is being tunneled in a centralized WLAN architecture?

    A. Redundant radios

    B. Redundant controllers

    C. Redundant access switches

    D. Redundant access points

    E. None of the above

19. What are the available form factors for network management system (NMS) solutions? (Choose all that apply.)

    A. Hardware appliance

    B. Virtual appliance

    C. Software subscription service

    D. Integrated access point

20. What planes of operation reside in the access points of a distributed WLAN architecture? (Choose all that apply.)

    A. Radio plane

    B. Data plane

    C. Network plane

    D. Control plane

    E. Management plane

Chapter

12

Power over Ethernet (PoE)

IN THIS CHAPTER, YOU WILL LEARN ABOUT THE FOLLOWING:

✓ **History of PoE**

  ▪ Nonstandard PoE

  ▪ IEEE 802.3af

  ▪ IEEE Std 802.3-2005, Clause 33

  ▪ IEEE 802.3at-2009

  ▪ IEEE Std 802.3-2015, Clause 33

✓ **PoE devices (overview)**

  ▪ Powered device (PD)

  ▪ Power-sourcing equipment (PSE)

  ▪ Endpoint PSE

  ▪ Midspan PSE

  ▪ Power-sourcing equipment pin assignments

✓ **Planning and deploying PoE**

  ▪ Power planning

  ▪ Redundancy

✓ **802.11ac and PoE**

  ▪ IEEE 802.3bt

In this chapter, you will learn about the various ways that an Ethernet cable can be used to provide power to networking devices. *Power over Ethernet (PoE)* is not a Wi-Fi technology, nor is it used specifically for Wi-Fi devices. However, it has become the predominant method for powering enterprise-class access points, thus making it a necessary and important topic when discussing wireless networking.

History of PoE

Before beginning this chapter, we need to explain what PoE is. Originally, computer networking entailed connecting a stationary, electrically powered computer system to a wired network. The computers were anything from desktop PCs to servers and mainframes. As is typical with technology, larger computers gave way to smaller computers, and laptop and portable devices began to appear. Eventually, some of the networking devices became small enough, both physically and electronically, that it became possible and practical not only to use the Ethernet cable to transmit data to the device, but also to send the electricity necessary to power the device.

The concept of providing power from the network dates back to the birth of the telephone, which to this day still receives power from the telephone network. Computer networking devices that are often powered with PoE are desktop Voice over IP (VoIP) phones, cameras, and access points. Ethernet cables consist of four pairs of wires. With 10 Mbps and 100 Mbps Ethernet, two pairs are used for transmitting and receiving data, and the other two pairs are unused. Gigabit Ethernet uses all four pairs of wires to transmit and receive data. As you will see later in this chapter, this is not a problem, since PoE can provide power on the unused wires or on the same wires that are used to transmit and receive data.

When you are providing power to devices via the same Ethernet cable that provides the data, a single low-voltage Ethernet cable is all you need to attach a networked PoE device. The use of PoE devices alleviates the need to run electrical cables and outlets to every location that needs to be connected to the network. Not only does this greatly reduce the cost of installing network devices, it also increases flexibility in terms of where these devices can be installed and mounted. Moving devices is also easier, because all that is required at the new location is a PoE-powered Ethernet cable.

Nonstandard PoE

As with most new technologies, the initial PoE products were proprietary solutions created by individual companies that recognized the need for the technology. The IEEE process to create a PoE standard began in 1999; however, it would take about four years before the standard became a reality. In the meantime, vendor-proprietary PoE continued to proliferate. Proprietary PoE solutions often used different voltages, and mixing proprietary solutions could result in damaged equipment.

IEEE 802.3af

The *IEEE 802.3af* Power over Ethernet committee created the PoE amendment to the 802.3 standard. It was officially referred to as IEEE 802.3 "Amendment: Data Terminal Equipment (DTE) Power via Media Dependent Interface." This amendment to the IEEE 802.3 standard, approved on June 12, 2003, defined how to provide PoE to 10BaseT (Ethernet), 100BaseT (Fast Ethernet), and 1000BaseT (Gigabit Ethernet) devices.

IEEE Std 802.3-2005, Clause 33

In June 2005, the IEEE revised the 802.3 standard, creating IEEE Std 802.3-2005. The 802.3af amendment was one of four amendments that were incorporated into this revised standard. In the 2005 revision of the 802.3 standard, and in the more recent revisions (it was revised in 2008, 2012, and again in 2015), Clause 33 is the section that defines PoE.

IEEE 802.3at-2009

The IEEE 802.3at amendment was ratified in 2009. *802.3at* is also known as PoE+ or *PoE plus*, since it extends the capabilities of PoE as originally defined in the 802.3af amendment. Two of the main objectives of the 802.3at Task Group were to be able to provide more power to powered devices and to maintain backward compatibility with Clause 33 devices. As APs become faster and incorporate newer technologies, such as multiple input, multiple output (MIMO), they require more power to operate. Switches and controllers that incorporate 802.3at technology are able to provide power to legacy APs as well as newer APs that require more power. The IEEE 802.3at amendment is able to provide up to 30 watts of power using two pairs of wires in an Ethernet cable. The 802.3at amendment defines PoE devices as either Type 1 or Type 2. Devices capable of supporting the higher power defined in the 802.3at amendment are defined as Type 2 devices, and devices not capable of supporting the higher power are defined as Type 1 devices.

Typically, when an 802 amendment is created and ratified, the amendment document is essentially a series of additions, deletions, and edits that modify and update the base standard. With 802.3at, the PoE section (Clause 33) of the 802.3-2008 standard was entirely replaced by the 802.3at amendment.

IEEE Std 802.3-2015, Clause 33

In December 2012, the IEEE revised the 802.3 standard again and created IEEE Std 802.3-2012. Just as the 802.3af amendment had been incorporated into the 802.3 standard in 2005, with the release of the 802.3-2012 revised standard, the 802.3at amendment was officially incorporated into this new revision, and the subsequent IEEE Std 802.3-2015 revision. In September 2015, the IEEE revised the 802.3 standard again and created IEEE Std 802.3-2015.

An Overview of PoE Devices

The PoE standard defines two types of PoE devices: powered devices (PDs) and power-sourcing equipment (PSE). These devices communicate with each other and provide the PoE infrastructure.

Powered Device

The *powered device (PD)* either requests or draws power from the power-sourcing equipment. PDs must be capable of accepting up to 57 volts from either the data lines or the unused pairs of the Ethernet cable. The PD must also be able to accept power with either polarity from the power supply in what is known as mode A or mode B, as described in Table 12.1.

TABLE 12.1   PD pinout

Conductor	Mode A	Mode B
1	Positive voltage, negative voltage	
2	Positive voltage, negative voltage	
3	Negative voltage, positive voltage	
4		Positive voltage, negative voltage
5		Positive voltage, negative voltage
6	Negative voltage, positive voltage	
7		Negative voltage, positive voltage
8		Negative voltage, positive voltage

The PD must reply to the power-sourcing equipment with a *detection signature* and notify the power-sourcing equipment whether it is in a state in which it will accept power or will not accept power. The detection signature is also used to indicate that the PD is compliant with the original PoE standard. If the device is determined not to be compliant, power to the device will be withheld. If the device is in a state in which it will accept power, the PD can optionally provide a *classification signature*. This classification signature lets the power-sourcing equipment know how much power the device will need.

Type 2 devices perform a two-event Physical layer classification or Data-Link layer classification, which allows a Type 2 PD to identify whether it is connected to a Type 1 or a Type 2 PSE. If mutual identification cannot be completed, the device can operate only as a Type 1 device.

Table 12.2 lists the current values used to identify the various classification signatures. If none of these current values are measured, the device is considered to be a Class 0 device. If the device is not identified, the PSE does not know how much power the device needs; therefore, it allocates the maximum power. If the device is classified, the PSE has to allocate only the amount of power needed by the PD, thus providing better power management. Proper classification of the devices can lead to a managed reduction in power usage and can also enable you to connect more devices to a single PoE-capable switch.

TABLE 12.2 PD classification signature measured electrical current values

Parameter	Conditions	Minimum	Maximum	Unit
Class 0	14.5 V to 20.5 V	0	4	milliampere (mA)
Class 1	14.5 V to 20.5 V	9	12	mA
Class 2	14.5 V to 20.5 V	17	20	mA
Class 3	14.5 V to 20.5 V	26	30	mA
Class 4	14.5 V to 20.5 V	36	44	mA

In the past, some vendors used proprietary layer 2 discovery protocols to perform classification. Although these techniques are good from the power-management and consumption perspective, they are proprietary and will not work with other manufacturers' products. *Link Layer Discovery Protocol (LLDP)* is a standards-based layer 2 neighbor discovery protocol that can also be used for more detailed power classification. Table 12.3 lists the classes of PoE devices and the range of maximum power that they use. Classes from 0 to 3 are for Type 1 devices that will work with 802.3af power-sourcing equipment.

Class 4 is meant for Type 2 devices that work with 802.3at (PoE+) power-sourcing equipment. The maximum power draw of an 802.3af-compliant device is 12.95 watts, and the maximum power draw of an 802.3at-compliant device is 25.5 watts.

TABLE 12.3 PD power classification and usage

Class	Usage	Range of Maximum Power Used	Class Description
0	Default	0.44 W to 12.95 W	Class unimplemented
1	Optional	0.44 W to 3.84 W	Very low power
2	Optional	3.84 W to 6.49 W	Low power
3	Optional	6.49 W to 12.95 W	Mid power
4	Type 2 devices	12.95 W to 25.5 W	High power

Can 802.3af-Compliant Access Points Use More Power?

Although a PSE port might offer 15.4 watts of power, the maximum draw of an 802.3af-compliant PD is 12.95 W. For example, an access point might use as much as 12.95 W on CAT3 cabling or higher. The power draw is always lower than the original power source because of insertion loss from the cable. In reality, a PD might be able to draw more power if higher grade cabling were used. For example, a 3x3:3 MIMO access point might actually need 14.95 W to operate with full functionality. Most WLAN vendors will recommend the use of CAT5e cabling or better so that 802.11n or 802.11ac access points can draw more than 12.95 W and fully power all the MIMO radios. Conversely, if full power is not available, some WLAN vendor APs offer the ability to manually or automatically downgrade MIMO radio capabilities to reduce power draw. Disabling additional networking interfaces, such as a BLE radio or a secondary Ethernet port, can also be used to reduce power draw.

Power-Sourcing Equipment

The *power-sourcing equipment (PSE)* provides power to the PD. The power supplied is at a nominal 48 volts (44 volts to 57 volts). The PSE searches for powered devices by using a direct current (DC) detection signal. After a PoE-compliant device is identified, the PSE will provide power to that device. If a device does not respond to the detection signature, the PSE will withhold power. This prevents noncompliant PD equipment from becoming damaged.

As you can see in Table 12.4, the amount of power provided by the PSE is greater than what is used by the PD (Table 12.3). This is because the PSE needs to account for the worst-case scenario, in which there may be power loss due to the cables and connectors between the PSE and the PD. The maximum draw of any powered device is 25.5 watts. The PSE can also classify the PD if the PD provided a classification signature. Once connected, the PSE continuously checks the connection status of the PD along with monitoring for other electrical conditions, such as short circuits. When power is no longer required, the PSE will stop providing it. Power-sourcing equipment is divided into two types of equipment: endpoint and midspan.

TABLE 12.4 PSE power

Class	Minimum Power from the PSE
0	15.4 W
1	4.0 W
2	7.0 W
3	15.4 W
4	30.0 W

Endpoint PSE

An *endpoint PSE* provides power and Ethernet data signals from the same device. Endpoint devices are typically PoE-enabled Ethernet switches, such as the 48-port switch shown in Figure 12.1. PoE-enabled switches are used to power access layer devices (such as APs and phones), and therefore the switches are access layer switches, as opposed to distribution or core switches. Some specialty devices, such as WLAN controllers, WLAN branch routers, or wall-plate APs with an extra PoE-enabled port, may also function as endpoint PSE equipment. Most controller-based APs are powered by an access layer switch; however, some of the smaller models or branch WLAN controllers might also be used to power access points.

FIGURE 12.1 48-port Gigabit Ethernet access switch with PoE

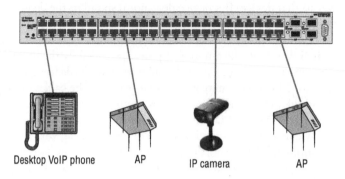

Desktop VoIP phone   AP   IP camera   AP

Endpoint equipment can provide power using two methods referred to as Alternative A and Alternative B:

Alternative A   With *Alternative A*, the PSE places power on the data pair. Figure 12.2 shows how a 10BaseT/100BaseTX endpoint PSE provides power using Alternative A, and Figure 12.3 shows how a 1000BaseT endpoint PSE provides power using Alternative A.

FIGURE 12.2 10BaseT/100BaseTX endpoint PSE, Alternative A

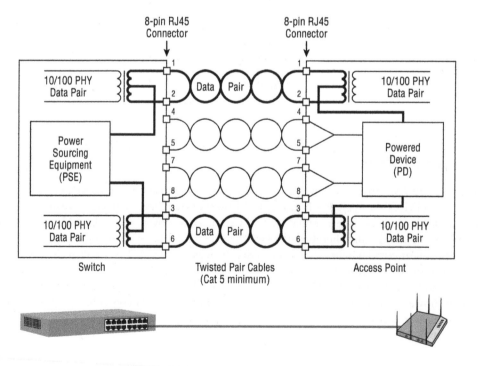

FIGURE 12.3   1000BaseT endpoint PSE, Alternative A

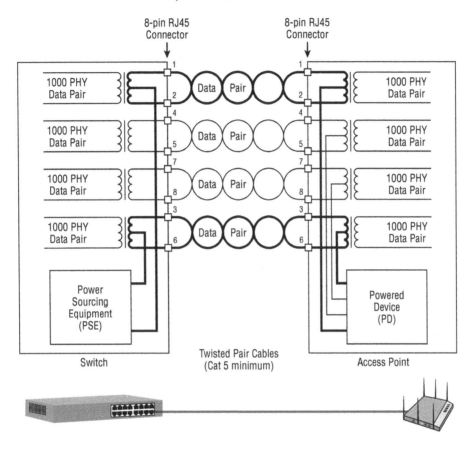

Alternative B   Originally, *Alternative B* was designed to provide power on the spare unused pair of wires in a 10BaseT/100BaseTX cable, as shown in Figure 12.4. A 1000BaseT endpoint PSE can also use Alternative B to provide power to a PD by placing the power on two of the data 1000BaseT data pairs, as depicted in Figure 12.5. Endpoint PSE is compatible with 10BaseT (Ethernet), 100BaseTX (Fast Ethernet), and 1000BaseT (Gigabit Ethernet). When 802.3af was initially ratified, 1000BaseT (Gigabit Ethernet) devices could receive PoE from only endpoint devices. In the next section of this chapter, you will see that that is no longer true. With the ratification of 802.3at, 1000BaseT devices could also be powered using either endpoint PoE or midspan PoE.

FIGURE 12.4   10BaseT/100BaseTX endpoint PSE, Alternative B

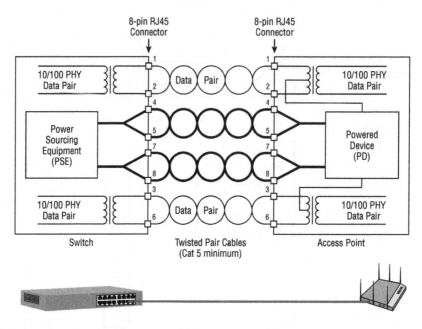

FIGURE 12.5   1000BaseT endpoint PSE, Alternative B

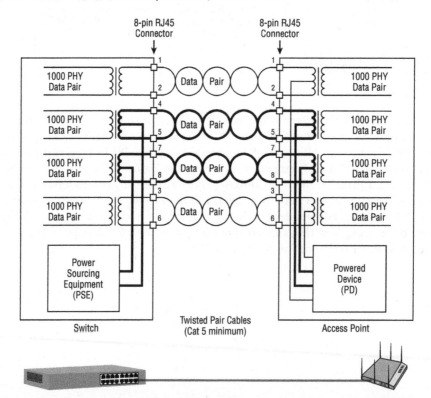

Midspan PSE

A *midspan PSE* acts as a pass-through device, adding power to an Ethernet segment. Midspan equipment enables you to provide PoE to existing networks without having to replace the existing Ethernet switches. A midspan PSE is placed between an Ethernet source (such as an Ethernet switch) and a PD. The midspan PSE acts as an Ethernet repeater while adding power to the Ethernet cable. Originally with 802.3af, midspan devices were only capable of using Alternative B—and only with 10BaseT and 100BaseTX PDs. With the ratification of 802.3at, midspan devices were able to use either Alternative A or Alternative B and could provide support for 1000BaseT devices.

Figure 12.6 shows how a 10BaseT/100BaseTX midspan PSE provides power using Alternative A, and Figure 12.7 shows how a 1000BaseT midspan PSE provides power using Alternative A. Figure 12.8 shows a 10BaseT/100BaseTX midspan PSE providing power using Alternative B, and Figure 12.9 shows how a 1000BaseT midspan PSE provides power using Alternative B.

FIGURE 12.6   10BaseT/100BaseTX midspan PSE, Alternative A

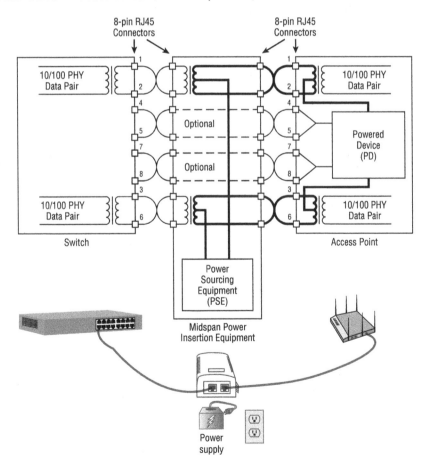

FIGURE 12.7   1000BaseT midspan PSE, Alternative A

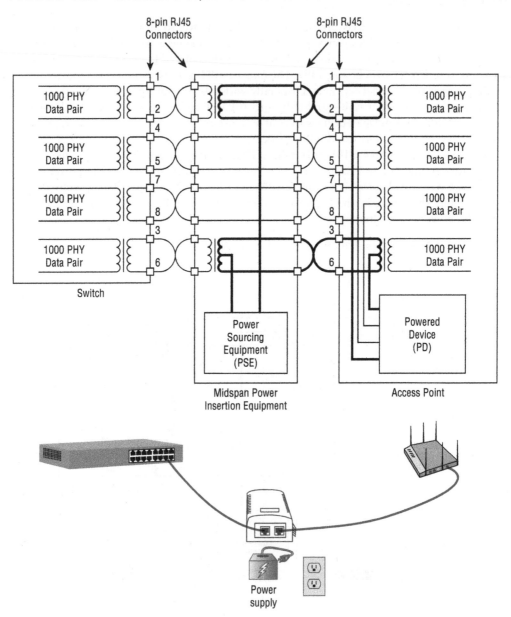

FIGURE 12.8 10BaseT/100BaseTX midspan PSE, Alternative B

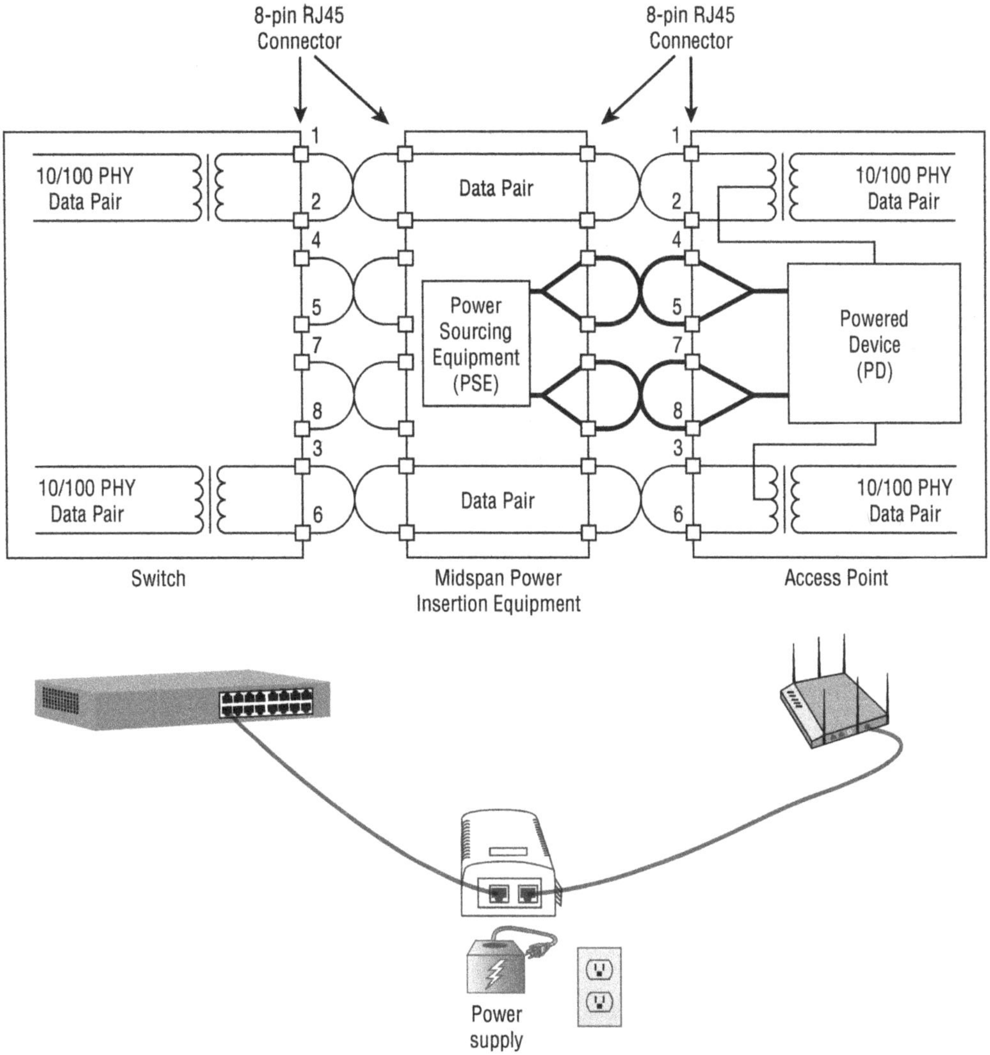

FIGURE 12.9 1000BaseT midspan PSE, Alternative B

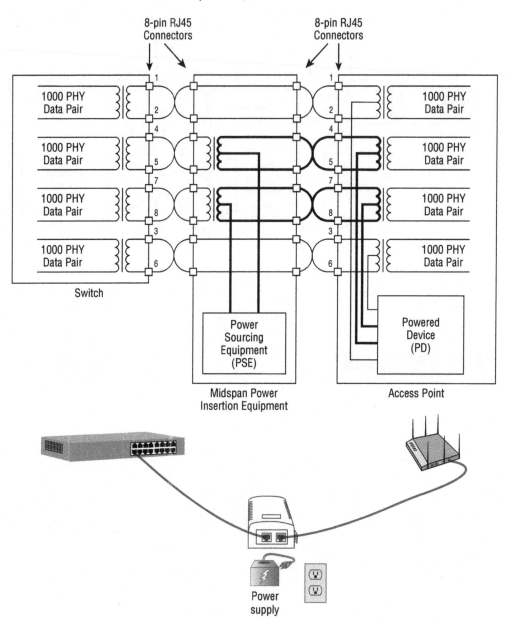

Figure 12.10 shows a single-port midspan device, along with three multiport devices. The midspan PSE is commonly known as a *power injector* (single-port device) or a *PoE hub* (multiport device).

FIGURE 12.10   PowerDsine power injector and PoE hubs

Figure 12.11 shows three typical ways of providing power to a PD. Option 1 illustrates an endpoint PoE-enabled switch with inline power. This switch provides both Ethernet and power to the AP. Option 2 and Option 3 illustrate two methods of providing midspan power. Option 2 shows a multiport midspan PSE commonly referred to as an *inline power patch panel*, and Option 3 shows a single-port midspan PSE commonly referred to as a *single-port power injector*.

FIGURE 12.11   Three PSE solutions

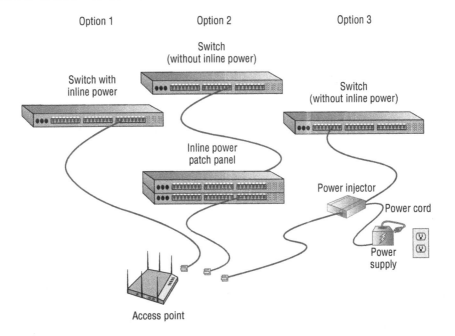

Power-Sourcing Equipment Pin Assignments

The power-sourcing equipment (PSE) must have a *medium dependent interface (MDI)* to carry the current to the powered device (PD). *MDI* is essentially the technical term for the Ethernet cabling connector. Keep in mind that the Ethernet maximum distance limitations of 100 meters (328 feet) still apply when PoE mechanisms are utilized.

There are two valid four-wire pin connections used to provide PoE. In each configuration, the two pairs of conductors carry the same nominal current in both magnitude and polarity. When you power a device using Alternative A, the positive voltage is matched to the transmit pair of the PSE. The input pairs of an Ethernet cable must connect to the output pairs of the device it is connected to. This is known as *medium dependent interface crossover (MDIX or MDI-X)*. Many devices are capable of automatically identifying and providing the crossover connection if needed. If a PSE is configured to automatically configure MDI/MDI-X (also called *Auto MDI-X*, or *automatic crossover*), the port may choose either Alternative A polarity choice, as described in Table 12.5.

TABLE 12.5   PSE pinout alternatives

Conductor	Alternative A (MDI-X)	Alternative A (MDI)	Alternative B (All)
1	Negative voltage	Positive voltage	
2	Negative voltage	Positive voltage	
3	Positive voltage	Negative voltage	
4			Positive voltage
5			Positive voltage
6	Positive voltage	Negative voltage	
7			Negative voltage
8			Negative voltage

Planning and Deploying PoE

In the past, when non-PoE desktop VoIP telephones and APs were connected to the network, each device had to be individually plugged into a power outlet. These outlets were spread around the building or campus, distributing the power needs. PoE consolidates the power source to the wiring closet or data center, requiring that only an Ethernet cable be connected to the PoE-powered device.

Power Planning

Instead of the power being distributed for hundreds or thousands of devices, the power for these devices is now being sourced from either a single or a limited number of locations. At maximum power for a PD, the PSE must be capable of providing 15.4 W or 30 W of power to each PoE device, depending on whether your devices require PoE or PoE+. Assuming that your PDs do not require PoE+, this means that a typical PoE-enabled 24-port Ethernet switch must be able to provide about 370 watts of power to provide PoE to all 24 ports (15.4 watts × 24 ports = 369.6 watts). This does not include the amount of power necessary for the switch to perform its networking duties. A simple way of calculating whether the power supply of the switch is powerful enough is to determine the size of the power supply for the equivalent non-PoE switch and add 15.4 watts for each PoE device that you will be connecting to the switch, or 30 watts for each PoE+ device you will be connecting to the switch.

The maximum power a 110-volt power supply is capable of providing is 3,300 watts (110 volts × 30 amperes). Let's assume that a wiring closet is supplied with a 110 volt, 15-amp circuit (1,650 watts), which is not uncommon. Enterprise-grade PoE-enabled switches often consist of multiple 48-port line cards housed in a chassis. The chassis itself may require 1,000 to 2,000 watts. If the 48-port line cards draw 15.4 watts per port, a total power draw of 740 watts would be required. Depending on the power requirements of the chassis, 3,300 watts would be able to power only the chassis and two to three fully populated 48-port line cards.

Because many devices such as 802.11 APs, video cameras, and desktop VoIP phones may require power, situations often arise where there simply is not enough available wattage to power all the PoE ports. Network engineers have begun to realize the need and importance for a *power budget*. Careful planning is needed to ensure that enough power is available for all the PDs. Powered devices that are capable of classification can greatly assist in conserving energy and subtracting less power from the power budget. A device that needs to draw 3 watts and is not capable of providing a classification signature would be classified as Class 0 by default and subtract 15.4 watts from the power budget. Effectively, 12 watts of power would be wasted. If that same device were capable of providing a classification signature and were classified as a Class 1 device, only 4 watts would be subtracted from the power budget. Classification of PDs will grow in importance as the need for WLAN deployments grows.

Enterprise switch vendors will list the PoE power budget within the switch specification sheet. The PoE power budget listed in a spec sheet is indeed the amount of power that is available to the ports and is not earmarked for other switch functions. When reading the power budget specifications of a switch, be sure to determine how many ports are PoE-capable. For example, Vendor A might have a 24-port gigabit switch with a PoE power budget of 195 watts, but the budget is only available to 8 of the

24 ports. Vendor B might also offer a 24-port gigabit switch with a PoE power budget of 195 watts, but the budget can be available to any of the 24 ports. Vendor C might sell a 24-port switch with a much larger PoE power budget of 408 watts available to all 24 ports. Keep in mind: The larger the power budget, the cost of the PoE-enabled switch rises significantly.

As shown in Figure 12.12, most switches have the ability to designate whether the port is a standard 802.3af port or 802.3at (PoE+) port. An 802.11 access point may need the full 15.4 watts provided by an 802.3af port; however, a VoIP desktop phone might need only 7 watts from the port. The 802.3af port for the phone could be manually configured for only 7 watts and, therefore, save 8.4 watts, which does not need to be subtracted from the overall PoE power budget. PoE ports can often also be configured with a priority level. Higher priority PoE ports take precedence for receiving power in the event that the PoE budget is exceeded. Proper planning of the PoE budget to ensure that the budget is never exceeded is best practice. PoE port priority is also important if there is hardware failure on the switch. PoE switches often require multiple power supplies. If one of the power supplies fails, the switch may not be able to provide power to all of the devices connected to it. PoE port priority allows the network administrator to identify which devices are more critical than others.

FIGURE 12.12   Port-level PoE budgeting

The power budget of a switch or multiple switches should be monitored to make sure that all devices can maintain power. PSE active budget information can usually be seen from the command line of a switch or the GUI interface or monitored by a centralized *network management server (NMS)*. In the example shown in Figure 12.13, a switch has an overall budget of 195 watts. An access point is plugged into port 1 and is currently using 7.4 watts, while another AP plugged into port 2 is currently drawing 3 watts. The total power currently being used is 10.4 watts, which means 184.6 watts can still be used by other devices. In this example, one of the APs has been classified as a Class 0 device, which means it can draw as much as 12.95 watts. If an AP is not very busy, it may need only 3 or 5 watts, but if the AP has many clients connected with heavy traffic, the full 12.95 watts might be needed. Therefore, always plan your power budget based on the maximum draw that devices such as access points might use.

FIGURE 12.13   Power budget monitoring

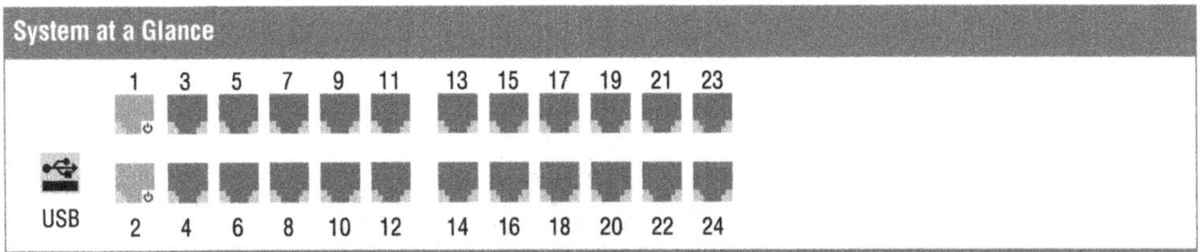

System at a Glance

	Port	Status	Power	Powered Device Type	Powered Device Class
☐	eth1/1	Delivering	7.4 Watts	802.3at	Class 4
☐	eth1/2	Delivering	3.0 Watts	802.3af	Class 0
☐	eth1/3	Searching	0.0 Watts	None	Class not defined
☐	eth1/4	Searching	0.0 Watts	None	Class not defined

Total Power for PoE Devices: 195.0 Watts   **Total Power Used:** 10.4 Watts   **Remaining Power:** 184.6 Watts

---

Why Are My Access Points Randomly Rebooting?

WLAN vendors commonly receive support calls from customers complaining that all of a sudden access points randomly begin to reboot. In many cases, the root cause of random rebooting of APs is that the switch power budget has been eclipsed. Very often, if an AP cannot get the power that it needs, the AP will reboot and try again. Remember that other devices, such as desktop VoIP phones, also use PoE. An extra PoE-powered device might have been plugged into a switch port and the power budget has been exceeded. Proper power budgeting for access points and any other PoE-capable devices is paramount.

Because of the demand for PoE-enabled devices, some switch manufacturers have replaced their 110 volt/30 amp power supplies with 220 volt/20 amp power supplies. The manufacturers are also putting larger power supplies in their switches to handle the additional requirements of PoE. Some PoE switches support power supplies as large as 9,000 watts. As the demand for PoE devices increases, the need to manage and troubleshoot PoE problems will also increase. Test equipment, like that shown in Figure 12.14, can be placed between the PSE and the PD to troubleshoot PoE link issues.

The more PoE devices that you add to the network, the more you concentrate the power requirements in the data center or wiring closet. As your power needs increase, electrical circuits supplying power to the PoE switches might have to be increased. Also, as the power increases, the amount of heat that is generated in the wiring closets increases, often requiring more climate-control equipment. When you are using high-wattage power supplies, we recommend that you also use redundant power supplies.

FIGURE 12.14   Netscout LinkSprinter network tester

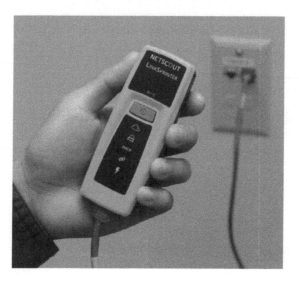

Redundancy

As children, we knew that even when there was an electrical failure, the telephone still worked and provided the ability to call someone. This is a level of service that we have come to expect. As VoIP and VoWiFi telephones replace traditional telephone systems, it is important to still provide this same level of continuous service. To achieve this, you should make sure that all of your PoE PSE equipment is connected to uninterruptible power sources. Additionally, it may be important enough to provide dual Ethernet connections to your PoE PD equipment. Some WLAN vendor APs offer dual Ethernet PoE ports with seamless switchover capabilities for PoE redundancy purposes. The APs might be wired to two separate PoE switches that are mounted in two separate intermediate distribution frame (IDF) cable racks.

Be Careful with PoE

With the increase in popularity of PoE and the requirement to provide power to devices like APs and VoIP telephones, more PoE jacks are being deployed in the office space. One of the nice and necessary features of PoE is that when a device is plugged in, the PSE can determine whether the device is PoE capable and, if so, will provide power to that device. If the device is not PoE capable, then Ethernet without power will be provided to the device.

Depending on the brand and model of your PoE switch, when a PoE-enabled device is unplugged from the switch, it is possible for the port of the switch to maintain its PoE status for a few seconds, even though there is nothing plugged into the switch. If you were to quickly plug another device into the same port, it is possible for the PoE switch to provide power to that device, even if it is not a PoE-capable device. This introduces a risk of damage to the device.

To prevent this risk from occurring, after unplugging any Ethernet device, you should get into the habit of waiting 5 to 10 seconds before plugging another device into that port or jack. This 5- to10-second delay should be long enough for the PoE port to disable itself. Then when another device is plugged into that port, PoE will identify whether the new device is PoE capable.

802.11ac and PoE

In Chapter 10, "MIMO Technology: HT and VHT," you learned that the 802.11ac radios operate only on the 5 GHz band; however, the 2.4 GHz radios still use 802.11n technology. The 802.11ac radios use more complex modulation and can provide for channels up to 160 MHz wide, although 20 and 40 MHz are typically the channel widths that are used. These new 802.11ac mechanisms require more processing resources and therefore require more power. The first generation of 802.11ac 3x3:3 APs typically required more power than 802.3af can provide, and 802.3at power is usually needed for some of the 802.11ac radios to be fully operational. Many of the WLAN vendors devised various techniques so that the first generation of 802.11ac APs can operate using the lower 802.3af power. However, the trade-off is downgraded performance of the first-generation 802.11ac APs that operate in an 802.3af mode. The most common method is to downgrade the MIMO capability of the 802.11ac access points so that 802.3af power can be used. 802.11ac APs with 3x3:3 transmitter capability might use only one or two transmitters on one or both of the radios when using 802.3af PoE, and therefore conserve power. The downside is that not all of the MIMO transmitter capabilities are being used by the APs. Other vendors have chosen to disable processor-intensive 802.11ac functions, such as 80 MHz channel capability and the use of more complex modulation. In other words, the 802.11ac 3x3:3 MIMO radio can still use all three transmitters, but effectively the radio functions as an 802.11n radio when using the lower 802.3af power. As previously mentioned, disabling additional networking interfaces, such as a BLE radio or a secondary Ethernet port, can also be used to reduce power draw. Some WLAN vendors offer 3x3:3 802.11ac access points that are fully functional with 802.3af power.

The second wave of 802.11ac access points uses newer chipsets that enable beamforming capabilities, and require more processing resources and therefore more power. These newer 802.11ac access points often use 4x4:4 MIMO radios, which also require more power than 802.3af power-sourcing equipment can provide.

PoE requirements for access points mainly center on how many radio chains need to be powered in the AP. The 15.4 watts provided by an 802.3af PoE switch port will be plenty of power for a dual frequency 2x2:2 AP. The vast majority of enterprise-grade 3x3:3 APs can also be adequately powered by an 802.3af PoE switch port. However, dual-frequency 4x4:4 APs require the higher power that an 802.3at PoE switch port can provide. In Chapter 19, "802.11ax - High Efficiency (HE)," you will learn that the first generation of 802.11ax APs will be 4x4:4 and even 8x8:8. These APs will obviously require 802.3at (PoE+).

There is no reason you cannot use an available power outlet to provide electrical current to an AP. The downside is that most APs are deployed in areas where a power outlet is not conveniently accessible. The best way to power 802.11ac and 802.11ax APs is to deploy a PoE+ (802.3at) PSE that is capable of providing 30 watts via an Ethernet cable. With the fast growth of wireless, if you are purchasing PoE switches, you should only consider switches that are 802.3at (PoE+) capable.

IEEE 802.3bt

The 802.3bt draft amendment, which will be the third iteration of the PoE standard, is expected to be ratified later in 2018. The new PoE amendment introduces two additional PSE and PD Types 3 and 4. Although new power capabilities are being introduced, 802.3bt devices will be backward compatible with legacy Type 1 and Type 2 devices. In addition to increasing the number of PoE Types, four new power classes will be added, 5 through 8, adding on to the existing classes of 0 through 4.

Four new power levels are also expected to be able to be provided by the power-sourcing equipment. These levels are expected to be 45 W, 60 W, 75 W, and 90 W, as shown in Table 12.6. 802.3bt will also be able to provide power over all four pairs of wires and include support for 10GBase-T devices.

TABLE 12.6   PSE power

Class	Minimum Power from the PSE	Standard
0	15.4 W	802.3af
1	4.0 W	802.3af
2	7.0 W	802.3af
3	15.4 W	802.3af
4	30.0 W	802.3at
5	45.0 W	802.3bt

Class	Minimum Power from the PSE	Standard
6	60.0 W	802.3bt
7	75.0 W	802.3bt
8	90.0 W	802.3bt

Summary

This chapter focused on Power over Ethernet and the equipment and techniques necessary to provide service to PDs. Power over Ethernet can be provided in two general ways: through proprietary PoE or through standards-based PoE (802.3af or 802.3at, integrated into the IEEE Std 802.3 in Clause 33).

Standards-based PoE consists of a few key components:

- Powered device (PD)
- Power-sourcing equipment (PSE)
- Endpoint PSE
- Midspan PSE

These components work together to provide a functioning PoE environment.

The final section of this chapter covered considerations that need to be made when planning and deploying PoE:

- Power planning
- Redundancy

Exam Essentials

Know the history of PoE. Make sure you know the history of PoE, the original 802.3af amendment, the 802.3at amendment, and current references to IEEE Std 802.3, Clause 33.

Be familiar with the various PoE devices and how they interoperate. Make sure you know about the various PoE devices and their roles in providing PoE. Understand how the following devices work: powered device (PD), power-sourcing equipment (PSE), endpoint PSE, and midspan PSE.

Know the different device classes and the classification process. Make sure you know the five device classes and how the classification process works to determine the class of a PD. Know how much current each class of devices uses along with how much power the PSE generates for each class of devices.

Review Questions

1. The IEEE 802.3af and 802.3at amendments have been incorporated into the IEEE Std 802.3-2015 revised standard and are defined in which clause?

    A.   Clause 15

    B.   Clause 17

    C.   Clause 19

    D.   Clause 33

    E.   Clause 43

2. If a classification signature is not provided, the device is considered to be in what class?

    A.   0

    B.   1

    C.   2

    D.   3

    E.   4

3. Which types of PoE devices are defined by the standard? (Choose all that apply.)

    A.   PSE

    B.   PPE

    C.   PD

    D.   PT

4. A powered device (PD) must be capable of accepting up to how many volts from either the data lines or the unused pairs of the Ethernet cable?

    A.   14.5 volts

    B.   20.5 volts

    C.   48 volts

    D.   57 volts

5. To qualify as compliant with the 802.3at amendment (now part of the 802.3 standard), a powered device (PD) must do which of the following? (Choose all that apply.)

    A.   Be able to accept power over the unused data pairs

    B.   Reply to the PSE with a detection signature

    C.   Accept power with either polarity from the PSE

    D.   Reply to the PSE with a classification signature

6. A VoIP telephone is connected to a 24-port PoE midspan PSE. If the telephone does not provide a classification signature, how much power will the PSE provide to the telephone?

 A. 12.95 watts

 B. 4.0 watts

 C. 7.0 watts

 D. 15.4 watts

7. An endpoint PSE that provides power by using Alternative B is capable of providing power to devices using which of the following Ethernet technologies? (Choose all that apply.)

 A. 10BaseT

 B. 100BaseTX

 C. 1000BaseT

 D. 100BaseFX

8. What is the range of maximum power used by a Class 4 PD?

 A. 0.44–12.95 watts

 B. 3.84–6.49 watts

 C. 6.49–12.95 watts

 D. 12.95–25.5 watts

 E. 15–30 watts

9. At maximum power requirements, a 24-port 802.3at-compliant PoE Ethernet switch must be able to provide about how many total watts of power to PoE devices on all ports?

 A. 15.4 watts

 B. 370 watts

 C. 720 watts

 D. 1,000 watts

 E. Not enough information is provided to answer the question.

10. If an 802.3at-compliant AP is equipped with two radios and requires 7.5 watts of power, how much power will the PSE provide to it?

 A. 7.5 watts

 B. 10.1 watts

 C. 15 watts

 D. 15.4 watts

 E. 30.0 watts

11. The PSE provides power within a range of _____ volts, with a nominal value of
_____ volts.

    A. 14.5–20.5, 18

    B. 6.49–12.95, 10.1

    C. 12–19, 15.4

    D. 44–57, 48

12. Tim has installed an Ethernet switch that is compliant with 802.3at. He is having problems
with his APs randomly rebooting. Which of the following could be causing his problems?

    A. Many PoE VoIP telephones are connected to the same Ethernet switch.

    B. Most of the Ethernet cables running from the switch to the APs are 90 meters long.

    C. The Ethernet cables are only Cat 5e.

    D. The switch is capable of 1000BaseT, which is not compatible with VoIP telephones.

13. You are designing an 802.3at-compliant network and are installing a 24-port Ethernet
switch to support 10 Class 1 VoIP phones and 10 Class 0 APs. The switch requires
500 watts to perform its basic switching functions. How much total power will be needed?

    A. 500 watts

    B. 694 watts

    C. 808 watts

    D. 1,000 watts

14. You are designing an 802.3at-capable network and are installing a 24-port Ethernet switch
to support 10 Class 2 cameras and 10 Class 3 APs. The switch requires 1,000 watts to perform its basic switching functions. How much total power will be needed?

    A. 1,080 watts

    B. 1,224 watts

    C. 1,308 watts

    D. 1,500 watts

15. When a PoE network is installed, what is the maximum distance from the PSE to the PD, as
defined in the standard? (Choose all that apply.)

    A. 90 meters

    B. 100 meters

    C. 300 feet

    D. 328 feet

    E. 328 meters

16. What is the maximum power draw of an 802.3at PD?

    A. 12.95 watts

    B. 15 watts

    C. 7.4 watts

    D. 25.5 watts

    E. 30 watts

17. What is the maximum power used by a PD Class 0 device?

    A. 3.84 watts

    B. 6.49 watts

    C. 12.95 watts

    D. 15.4 watts

18. The PSE will apply a voltage of between 14.5 and 20.5 and measure the resulting current to determine the class of the device. Which current range represents Class 2 devices?

    A. 0–4 mA

    B. 5–8 mA

    C. 9–12 mA

    D. 13–16 mA

    E. 17–20 mA

19. A PD must be capable of accepting power with either polarity from the power supply. In mode A, on which conductors/wires does the PD accept power?

    A. 1, 2, 3, 4

    B. 5, 6, 7, 8

    C. 1, 2, 3, 6

    D. 4, 5, 7, 8

20. A Type 2 PSE will perform a two-event Physical layer classification or Data-Link layer classification. If mutual identification cannot be completed, what does the Type 2 device do?

    A. It defaults as a Category 0 device.

    B. It operates as a Type 2 device.

    C. It operates as a Type 1 device.

    D. It provides 15.4 watts of power using Alternative A.

Chapter

13

WLAN Design Concepts

IN THIS CHAPTER, YOU WILL LEARN ABOUT THE FOLLOWING:

✓ **WLAN coverage design**

    ▪ Received signal

    ▪ Signal-to-noise ratio

    ▪ Dynamic rate switching

    ▪ Transmit power

✓ **Roaming design**

    ▪ Primary and secondary coverage

    ▪ Fast secure roaming

    ▪ Layer 3 roaming

✓ **Channel design**

    ▪ Adjacent channel interference

    ▪ 2.4 GHz channel reuse

    ▪ Co-channel interference

    ▪ 5 GHz channel reuse

    ▪ DFS channels

    ▪ 40 MHz channel design

    ▪ Static channels and transmit power vs. adaptive RF

    ▪ Single-channel architecture

✓ **Capacity design**

    ▪ High density

    ▪ Band steering

    ▪ Load balancing

- Airtime consumption
- Voice vs. data

✓ **Dual 5 GHz and software-defined radios**

✓ **Physical environment**

✓ **Antennas**

✓ **Outdoor design**

If you assemble 200 Wi-Fi experts in one room, such as the WLAN Professionals conference (www.wlanpros.com), most likely you will get 200 different opinions as to proper WLAN design for coverage, capacity, and airtime consumption. Experienced WLAN professionals will all agree about the importance of a properly designed WLAN. The bulk of troubleshooting calls can be prevented if a WLAN is well planned and designed prior to deployment. Just as important is a post-deployment validation survey to verify the WLAN design. This chapter discusses the WLAN design concepts that every wireless network administrator should grasp. Although crucial, planning a WLAN for proper coverage is just one aspect of WLAN design. A good WLAN design must also take into consideration user and device capacity to ensure performance needs. Due to the half-duplex nature of the RF medium, a key goal in WLAN design is to reduce airtime consumption. In reality, a cookie-cutter design approach never works, because different vertical markets have different needs. Furthermore, every building has a unique layout and therefore different RF propagation and attenuation properties.

This chapter discusses WLAN coverage, capacity, and integration design aspects from a conceptual basis. However, WLAN professionals do not always agree about WLAN design and each may have their own unique approach. Regardless of the design approach you use, always remember the importance of a validation survey post-installation.

WLAN Coverage Design

When designing a WLAN, probably the first thing that comes to mind will always be the coverage area or zone from which Wi-Fi clients can communicate. The primary coverage goals for any WLAN are to provide high data rate connectivity for connected clients and to provide for seamless roaming. A common mistake is to design a WLAN based solely on an access point's capabilities. The exact opposite should be considered during the design phase. A proper WLAN coverage design should be based on the perspective of the Wi-Fi clients. Therefore, a quality received signal for the client is needed to provide high data rate connectivity.

Received Signal

So what exactly is considered a quality received signal? As shown in Table 13.1, depending on the proximity between an AP and a Wi-Fi client, an 802.11 radio might receive an incoming signal anywhere between –30 dBm and the noise floor. When designing for coverage, the normal recommended best practice is to provide for a –70 dBm or stronger received signal that is well above the noise floor. In other words, a received signal of –70 dBm or higher is considered to be a quality received signal.

TABLE 13.1   Received signal strength

Quality	dBm	mW
Very Strong	–30 dBm	1/1,000th of 1 milliwatt
Very Strong	–40 dBm	1/10,000th of 1 milliwatt
Very Strong	–50 dBm	1/100,000th of 1 milliwatt
Very Strong	–60 dBm	1 millionth of 1 milliwatt
Strong	–70 dBm	1 ten-millionth of 1 milliwatt
Fair	–80 dBm	1 hundred-millionth of 1 milliwatt
Weak	–90 dBm	1 billionth of 1 milliwatt
Very Weak	–95 dBm	Noise floor

Please understand that not all client devices are created equal. For example, the highest data rate possible for a legacy 802.11g client is 54 Mbps, whereas an 802.11n/ac 2×2:2 MIMO radio may be capable of a data rate of 300 Mbps. Furthermore, depending on the chipset vendor, the radios of various Wi-Fi clients have different receive sensitivity thresholds, which are mapped to different data rates. This means that two client radios receiving an RF signal with the same strength may use a different data rate for modulation and demodulation. Despite the variances between devices and sensitivity, there is still a common denominator. A received signal of –70 dBm or higher usually guarantees that a client radio will use one of the highest data rates that the client is capable of.

In Chapter 14, "Site Survey and Validation," you will learn about predictive modeling tools that can assist you when planning for –70 dBm coverage zones for individual APs. Once again, please remember that the actual coverage zone of an AP is from the perspective of a Wi-Fi client, and validating any planned coverage is a necessity. Because RSSI sensitivity varies between WLAN devices, the validation survey is often performed using different types of WLAN clients.

Signal-to-Noise Ratio

Another reason for planning for –70 dBm coverage is because the received signal of –70 dBm is usually well above the noise floor. In Chapter 4, you learned that the *signal-to-noise ratio (SNR)* is an important value because if the background noise is too close to the received signal or the received signal level is too low, data can be corrupted. The SNR is not actually a ratio; it is simply the difference in decibels between the received signal and the background noise (noise floor) measured in dBs, as shown in Figure 13.1. If an 802.11 radio receives a signal of –70 dBm and the noise floor is measured at –95 dBm, the difference between the received signal and the background noise is 25 dB. So, the SNR is 25 dB.

FIGURE 13.1   Signal-to-noise ratio

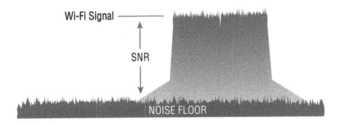

Data transmissions can become corrupted with a very low SNR. If the amplitude of the noise floor is too close to the amplitude of the received signal, data corruption will occur and result in layer 2 retransmissions. An SNR of 25 dB or greater is considered good signal quality, and an SNR of 10 dB or lower is considered poor signal quality. An SNR of below 10 dB will likely result in data corruption and retransmission rates as high as 50 percent. To ensure that frames are not corrupted due to a low SNR, most WLAN vendors recommend a minimum SNR of 20 dB for data WLANs and a minimum SNR of 25 dB for WLANs that require voice-grade communications. In most instances, a received signal of –70 dBm will be 20 dB or higher above the noise floor. In most environments, a –70 dBm signal ensures high rate connectivity, and the 20 dB SNR ensures data integrity. A high SNR also ensures that radios will use modulation and coding schemes (MCSs) that produce higher data rates.

VoWiFi communications are more susceptible to layer 2 retransmissions than other types of application traffic. Therefore, when you are designing for voice-grade WLANs, a –65 dBm or stronger signal is recommended so that the received signal is higher above the noise floor. As shown in Figure 13.2, even if the noise floor were a very high –90 dBm, the SNR of a –65 dBm received signal for a VoWiFi client would still be 25 dB. Always check the recommendations of the manufacturer of the VoWiFi client. One VoWiFi vendor may state that a –67 dBm signal is sufficient, whereas another vendor may suggest an SNR as high as 28 dB. When you are designing for voice, SNR is the most important RF metric. Also keep in mind that as a result of *free space path loss (FSPL),* the effective range for –67 dBm clients will be less distance than clients receiving a –70 dBm signal. Remember that for every 3 dB of loss, the received signal is half strength. For example, a –70 dBm signal is half the power of a −67 dBm signal. A client needs to be in closer proximity to the AP for a −67 dBm received signal.

FIGURE 13.2 VoWiFI vs. high data rate coverage

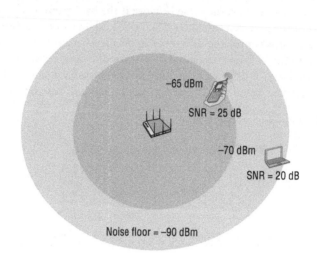

A higher SNR might also be needed to achieve the maximum data rates for 802.11ac clients using 256-QAM modulation. To take advantage of the modulating and coding schemes (MCSs) defined for the use of 256-QAM modulation, an SNR of 29 dB or higher will be needed. In the future, 802.11ax client devices may support 1024-QAM modulation, and an SNR of 35 dB will most likely be needed.

Dynamic Rate Switching

Will a client device be able to communicate with an AP if the signal drops below −70 dBm? The answer is yes, because most client devices can still decode an 802.11 preamble from received signals that are as low as only 4 dB above the noise floor. As mobile client radios move away from an access point, they will shift down to lower-bandwidth capabilities by using a process known as *dynamic rate switching (DRS)*. Data rate transmissions between the access point and the client stations will shift down or up, depending on the quality of the signal between the two radios.

There is a direct correlation between signal quality and distance from the AP. As mobile client stations move farther away from an access point, both the AP and the client will shift down to lower rates that require a less complex modulation and coding scheme (MCS). In the example pictured in Figure 13.3, an 802.11a/g client might connect at 54 Mbps when receiving a −70 dBm signal, but it might shift to transmitting with a lower data rate of 6 Mbps if the signal is much weaker. The transmissions between two radios might be 54 Mbps at 30 feet but 6 Mbps at 90 feet.

FIGURE 13.3   Dynamic rate switching

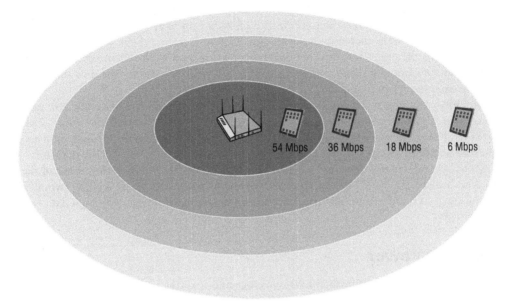

Dynamic rate switching (DRS) is also referred to as *dynamic rate shifting*, *dynamic rate selection*, *adaptive rate selection*, and *automatic rate selection*. All these terms refer to a method of speed fallback on a Wi-Fi radio receiver (Rx) as the incoming signal strength and quality from the transmitting Wi-Fi radio decreases. The objective of DRS is upshifting and downshifting for rate optimization and improved performance. From the client's perspective, the lower data rates will provide larger concentric zones of coverage than the higher data rates.

The thresholds used for dynamic rate switching are proprietary and are defined by 802.11 radio manufacturers. Most vendor radios' DRS capabilities are linked to receive signal strength indicator (RSSI) thresholds, packet error rates, and retransmissions. RSSI metrics are usually based on signal strength and signal quality. In other words, a station might shift up or down between data rates based on received signal strength in dBm or a signal-to-noise ratio (SNR) value. Because vendors implement DRS differently, you may have two different vendor client radios at the same location, while one is communicating with the access point at 300 Mbps, the other is communicating at 270 Mbps. For example, one vendor might shift down from a data rate of 156 Mbps to 52 Mbps at −78 dBm, whereas another vendor might shift between the same two rates at −81 dBm. The data shifting might also be based on SNR. Once again, there is a correlation between signal quality and distance from the AP. Keep in mind that DRS works with all 802.11 Physical layers (PHYs). In other words, a legacy 802.11b radio will shift between the four data rates of 1, 2, 5.5, and 11 Mbps, while an 802.11n/ac radio will shift between a wider range of available data rates.

There often is a big misconception that only client radios use dynamic rate switching. As already mentioned, client radios shift down to lower data rates if there is a weaker received signal from the AP. However, the radio within an access point also uses dynamic rate switching. Based on the incoming received signal strength from a client, an access point radio will shift data rates for the downlink transmission back to the client radio. A weak signal from an incoming client results in a shift to a lower data rate for the downlink transmission from the AP to that client.

Mobility can cause shifts in data rates. DRS provides a method for APs and client radios to continue with lower data rate communications despite a weaker signal and lower SNR. However, one of the main goals of WLAN coverage design is to provide for high data connectivity and limit as much as possible the shifting to lower data rates. Clients that shift down to lower data rates consume more airtime and affect the overall performance of the WLAN. Instead of a client shifting to a much lower data rate, the better scenario would be for the client to roam to another AP with a strong signal and continue with high data rate connectivity.

Transmit Power

A big factor that will affect both WLAN coverage and roaming is the transmit power of the access points. Although most indoor APs may have full transmit power settings as high as 100 mW, they should rarely be deployed at full power. This extends the effective range of the access point; however, designing WLANs strictly for range is an outdated concept. Later in this chapter, we will discuss the higher priorities of WLAN capacity design and reduction of airtime consumption. APs at maximum transmit power will result in oversized coverage and not meet your capacity needs. Access points deployed at full transmit power in an indoor environment will also increase the odds of co-channel interference, which can result in unnecessary medium contention overhead. APs at full power also increase the odds of sticky clients which negatively impact roaming, as discussed further in Chapter 15, "WLAN Troubleshooting." For all these reasons, typical indoor WLAN deployments are designed with the APs set at about one-fourth to one-third maximum transmit power. Higher user density environments may require that the AP transmit power be set at the lowest setting of 1 mW.

Another consideration is the transmit power of the clients. One heavily debated topic is the concept of a balance power link between an AP and a client. In simpler words, the transmit power settings between an AP and a client are the same. Very often WLAN clients transmit at higher power levels than indoor access points. The transmit power of many indoor APs may be 10 mW or less due to high-density design needs. However, most clients, such as smartphones and tablets, may transmit at fixed amplitudes of 15 mW or 20 mW. Because clients often transmit at a higher power than the APs and because clients are mobile, co-channel interference (CCI) is often caused by a power mismatch. Clients and APs that support *transmit power control (TPC)* can usually minimize this issue. A more detailed discussion about TPC can be found in Chapter 15.

Roaming Design

As you have learned throughout this book, *roaming* is the method by which client stations move between RF coverage cells in a seamless manner. Client stations switch communications through different access points. Seamless communications for client stations moving between the coverage zones within an extended service set (ESS) is vital for uninterrupted mobility. One of the most common issues you will need to troubleshoot is a problem with roaming. Roaming problems are usually caused by poor network design.

Client stations, not the access point, decide whether or not the client roams between access points. Some vendors may involve the access point or WLAN controller in the roaming decision, but ultimately the client station initiates the roaming process with a reassociation request frame. The method by which a client station decides to roam is a set of proprietary rules determined by the manufacturer of the 802.11 radio, usually defined by a roaming trigger threshold. Roaming thresholds usually involve signal strength, SNR, and bit-error rate. As the client station communicates on the network, it continues to look for other access points via probing and listening on other channels and will hear received signals from other APs. The most important variable will always be received signal strength: As the received signal from the original AP grows weaker and a station hears a stronger signal from another known access point, the station will initiate the roaming process. However, other metrics, such as SNR, error rates, and retransmissions, may also have a part in what triggers a client to roam. SNR is a metric used by some WLAN clients to trigger roaming events as well as dynamic rate switching.

As shown in Figure 13.4, as the client station moves away from the original access point with which it is associated and the signal drops below a predetermined threshold, the client station will attempt to connect to a new target access point that has a stronger signal. The client sends a frame, called the *reassociation request frame*, to start the roaming procedure. Chapter 9, "802.11 MAC," explains the entire reassociation frame exchange in greater detail.

FIGURE 13.4   Roaming

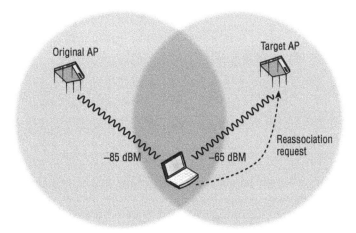

Depending on the type of WLAN client device, roaming trigger thresholds can be very simple or quite complex. For example, a VoWiFi phone may roam to a new AP with a 5 dB stronger signal than the AP that the phone is associated with. Once the phone has roamed to the new AP, it may require a 10 dB better signal to roam back to the original AP. These trigger thresholds are meant to prevent the client from ping-pong associations between two APs. Unfortunately, there is not a lot of published data in regards to roaming trigger thresholds. However, some client vendors do publish useful roaming trigger information via deployment guides or support forums. Following are a few examples:

- Apple macOS devices: `https://support.apple.com/en-us/HT206207`

- Apple iOS devices: `https://support.apple.com/en-us/ht203068`

Because roaming is proprietary, a specific vendor client station may roam sooner than a second vendor client station as they move through various coverage cells. Some vendors like to encourage roaming, whereas others trigger roaming at lower received signal thresholds. In an environment where a WLAN administrator must support multiple vendor radios, different roaming behaviors will most assuredly be seen.

Be aware that some client devices offer the capability to manually adjust their roaming trigger thresholds. As shown in Figure 13.5, the Intel Wi-Fi radios found in many Windows OS laptops have an adjustable roaming aggressiveness setting.

FIGURE 13.5 Roaming aggressiveness

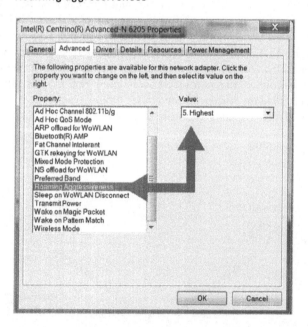

For the time being, a WLAN administrator will always face unique challenges because of the proprietary nature of roaming. However newer clients may utilize additional parameters such as AP neighbor reports or capacity load on an AP to help optimize the roaming process. As discussed in Chapter 2, "IEEE 802.11 Standards and Amendments," the ratified 802.11k amendment defines the use of *radio resource measurement (RRM)* and *neighbor reports* to enhance roaming performance. The ratified 802.11r amendment also defines faster secure handoffs when roaming occurs between cells in a wireless LAN using the strong security defined in a robust security network (RSN). The 802.11v amendment defines mechanisms for clients to learn about AP capacity load which may be factored into client roaming decisions.

Client-Side Support of 802.11k, 802.11r, and 802.11v Mechanisms

Most WLAN infrastructure vendors already support 802.11k and 802.11r technology in their APs and controllers, but many client devices do not. Some aspects of the 802.11r (fast secure roaming), 802.11k (resource management) and 802.11v (wireless network management) amendments are tested by the Wi-Fi Alliance with a certification called Voice-Enterprise. Although the Voice-Enterprise certification is a reality, the majority of legacy clients do not support 802.11k and 802.11r mechanisms. However, support for 802.11k, 802.11r, and 802.11v technology on the client side has grown in recent years for many of the newer 802.11ac clients.

Primary and Secondary Coverage

The best way to ensure that seamless roaming will commence, is through proper design and a thorough site survey. When you are designing an 802.11 WLAN, most vendors recommend a 15–30 percent overlap of –70 dBm coverage cells. For years, WLAN design guides and white papers from various WLAN vendors referenced the 15–30 percent coverage cell overlap, as shown on the left in Figure 13.6. The problem is, how do you calculate and measure cell overlap? Should the cell overlap area be measured by circumference, diameter, or radius? Additionally, WLAN vendor white papers (and even this book) use illustrations to depict the coverage cells as perfectly round and circular. In reality, coverage cells are oddly shaped, like an amoeba or a starburst. How can you measure coverage cell overlap if every coverage cell has a different shape?

Wi-Fi site survey expert Keith Parsons, CWNE #3, has for years preached about the fallacy of measuring access point coverage overlap. Coverage overlap is really duplicate primary and secondary coverage from the perspective of a Wi-Fi client station. A proper site survey should be conducted to make sure that a client always has adequate duplicate coverage from multiple access points. In other words, each Wi-Fi client station (STA) needs to hear at least one access point at a specific RSSI and a backup or secondary access point at a different RSSI. Typically, most vendor RSSI thresholds require a received signal of –70 dBm for the higher data rate communications. Therefore, the client station needs to hear a second AP with a signal of –75 dBm or greater when the signal received from the first AP drops below –70 dBm. The only way to determine whether proper primary/secondary coverage is available for clients is by conducting a coverage analysis site survey. Proper site survey procedures are discussed in detail in Chapter 14.

FIGURE 13.6 Cell overlap

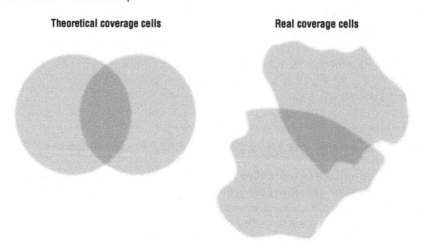

As shown in Figure 13.7, when you are designing for primary and secondary AP coverage, the typical rule is as follows: When associated to an AP, a potential roaming client also hears at least one other AP within a 5 dB range. For example, while a client is connected to an AP with planned coverage of –65 dBm, that same client should always be able to hear at least one more AP at –70 dBm. Some WLAN design professionals prefer matched signal strengths when designing for primary and secondary AP coverage. For example, while a client is associated to an AP with planned coverage of –65 dBm, that same client should also always be able to hear at least one other AP at –65 dBm. This also guarantees redundancy in case of AP failure.

FIGURE 13.7 Primary and secondary coverage

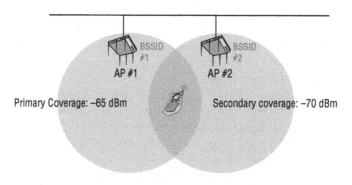

Roaming problems will occur if there is not enough duplicate cell coverage. Too little duplicate coverage will effectively create a roaming dead zone, and connectivity might even temporarily be lost. On the flip side, too much duplicate coverage will also cause roaming problems. What you do not want is a client to be able to hear a strong –70 dBm signal from dozens of APs from any given position. This can also create a situation in which the client device is constantly switching back and forth between two or more APs on different channels. Furthermore, too many APs with strong signals may cause a sticky client problem, where a client station may stay associated with its original AP and not connect to a second access point even though the station is directly underneath the second access point. If a client station can hear dozens of APs on the same channel with very strong signals, degradation in performance will occur due to medium contention overhead.

Fast Secure Roaming

Another roaming design issue of great importance is latency. The 802.11-2016 standard suggests the use of an 802.1X/EAP security solution in the enterprise. The average time involved during the authentication process can be 700 milliseconds or longer. Every time a client station roams to a new access point, reauthentication is required when an 802.1X/EAP security solution has been deployed. The time delay that is a result of the authentication process can cause serious interruptions with time-sensitive applications. VoWiFi requires a roaming handoff of 150 milliseconds or much less when roaming. A *fast secure roaming (FSR)* solution is needed if 802.1X/EAP security and time-sensitive applications are used together in a wireless network. The IEEE defines *fast basic service set transition (FT)* mechanisms as a standard for fast and secure roaming. FT procedures were first defined in the IEEE 802.11r-2008 amendment. The Wi-Fi Alliance has implemented the Voice-Enterprise certification, which defines FT and 802.11r mechanisms. Although 802.11r has been around for more than 10 years, adoption has been slow. However, client-side support for Voice-Enterprise has become much more common.

If 802.1X/EAP security is going to be used for voice-capable clients, a fast secure roaming solution will be needed.

Neither nonstandard fast secure roaming mechanisms, such as opportunistic key caching (OKC), nor standard fast BSS transition (FT) roaming mechanisms are tested on the CWNA exam. Fast secure roaming methods are a heavily tested subject on the Certified Wireless Security Professional (CWSP) exam.

Layer 3 Roaming

One major consideration when designing a WLAN is what happens when client stations roam across layer 3 boundaries. Wi-Fi operates at layer 2, and roaming is essentially a layer 2 process. As shown in Figure 13.8, the client station is roaming between two access points. The roaming is seamless at layer 2, but user VLANs are tied to different subnets on either side of the router. As a result, the client station will lose layer 3 connectivity and must acquire a new IP address. Any connection-oriented applications that are running when the client re-establishes layer 3 connectivity will have to be restarted. For example, a VoIP phone conversation would disconnect in this scenario, and the call would have to be reestablished.

FIGURE 13.8   Layer 3 roaming boundaries

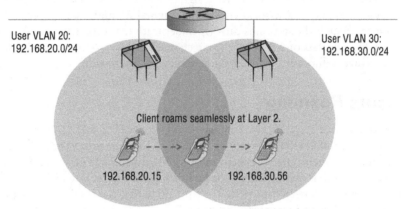

Because 802.11 wireless networks are usually integrated into pre-existing wired topologies, crossing layer 3 boundaries is often a necessity, especially in large deployments. The only way to maintain upper-layer communications when crossing layer 3 subnets is to provide a *layer 3 roaming* solution that is based on the *Mobile IP* standard. Mobile IP is an Internet Engineering Task Force (IETF) standard protocol that allows mobile device users to move from one layer 3 network to another while maintaining their original IP address. Mobile IP is defined in IETF Request for Comments (RFC) 5944. Layer 3 roaming solutions based on Mobile IP use some type of tunneling method and IP header encapsulation to allow packets to traverse between separate layer 3 domains, with the goal of maintaining upper-layer communications. Most WLAN vendors now support some form of layer 3 roaming solution, as shown in Figure 13.9.

FIGURE 13.9   Mobile IP

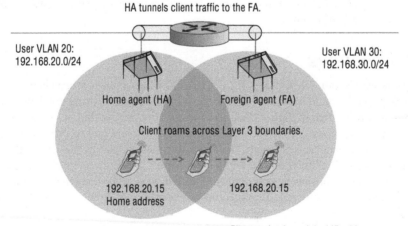

A mobile client receives an IP address, also known as a home address on a home network. The mobile client must register its home address with a device called a *home agent (HA)*. As depicted in Figure 13.9, the client's original associated access point serves as the home agent. The home agent is a single point of contact for a client when it roams across layer 3 boundaries. The HA shares client MAC/IP database information in a table, called a *home agent table (HAT)* with another device, called the *foreign agent (FA)*.

In this example, the foreign agent is another access point that handles all Mobile IP communications with the home agent on behalf of the client. The foreign agent's IP address is known as the *care-of address*. When the client roams across layer 3 boundaries, the client is roaming to a foreign network where the FA resides. The FA uses the HAT tables to locate the HA of the mobile client station. The FA contacts the HA and sets up a Mobile IP tunnel. Any traffic that is sent to the client's home address is intercepted by the HA and sent through the Mobile IP tunnel to the FA. The FA then delivers the tunneled traffic to the client, and the client is able to maintain connectivity using the original home address. In our example, the Mobile IP tunnel is between two APs on opposite sides of a router. If the user VLANs exist at the edge of the network, tunneling of user traffic occurs between access points that assume the roles of HA and FA. The tunneling is often distributed between multiple APs. However, user VLANs may reside back in a DMZ or at the core layer of the network along with a WLAN controller. In a single WLAN controller environment, the layer 3 roaming handoffs exist as control plane mechanisms within the single controller. In a multiple WLAN controller environment, an IP tunnel is created between controllers that are deployed in different routed boundaries with different user VLANs. One controller functions as the home agent, while another controller functions as the foreign agent.

Although maintaining upper-layer connectivity is possible with these layer 3 roaming solutions, increased latency is sometimes an issue. Additionally, layer 3 roaming may not be a requirement for your network. Less complex infrastructure often uses a simpler layer 2 design. Larger enterprise networks often have multiple user and management VLANs linked to multiple subnets; therefore, a layer 3 roaming solution will be required.

Channel Design

Another key component of WLAN design is the selection of the proper channels to be used among multiple APs in the same location. A proper channel pattern or *channel reuse* design is needed to guarantee seamless roaming as well as to prevent two types of interference that are a result of improper channel design. The following sections discuss the fundamentals of WLAN channel design.

Adjacent Channel Interference

Most Wi-Fi vendors use the term *adjacent channel interference (ACI)* to refer to degradation of performance resulting from overlapping frequency space that occurs due to an improper channel reuse design. In the WLAN industry, an adjacent channel is considered to be the next or previous numbered channel. For example, channel 3 is adjacent to channel 2.

As you learned in Chapter 6, "Wireless Networks and Spread Spectrum Technologies," the 802.11-2016 standard requires 25 MHz of separation between the center frequencies of 2.4 GHz channels in order for them to be considered non-overlapping. As pictured in Figure 13.10, if three channels are needed, only channels 1, 6, and 11 can meet these IEEE requirements in the 2.4 GHz ISM band in the United States. Some countries allow the use of all 14 IEEE 802.11–defined channels in the 2.4 GHz ISM band; however, because of the positioning of the center frequencies, no more than 3 channels can be used while avoiding frequency overlap. Even if all 14 channels are available, most WLAN design professionals still choose to use channels 1, 6, and 11 in the 2.4 GHz frequency band.

FIGURE 13.10   2.4 GHz non-overlapping channels

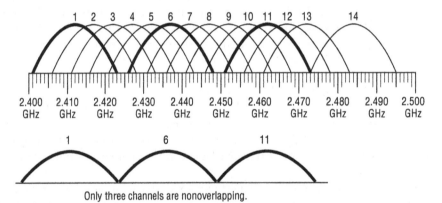

Only three channels are nonoverlapping.

When designing a wireless LAN, you need overlapping coverage cells in order to provide for roaming. However, the overlapping cells should not have overlapping frequencies, and in the United States only channels 1, 6, and 11 should be used in the 2.4 GHz ISM band to get the most available, non-overlapping channels. Overlapping coverage cells with overlapping frequencies cause what is known as adjacent channel interference. If overlapping coverage cells also have frequency overlap from adjacent channels, the transmitted frames will become corrupted, the receivers will not send ACKs, and layer 2 retransmissions will significantly increase.

2.4 GHz Channel Reuse

To avoid adjacent channel interference, a channel reuse design is necessary. Once again, overlapping RF coverage cells are needed for roaming, but overlapping frequencies must be avoided. The only three channels that meet these criteria in the 2.4 GHz ISM band are channels 1, 6, and 11 in the United States. APs in the 2.4 GHz band, therefore, should always be placed in a *channel reuse* pattern similar to the one pictured in Figure 13.11. Any

WLAN channel reuse pattern that uses three or more channels is sometimes referred to as a *multiple-channel architecture (MCA)*.

FIGURE 13.11   2.4 GHz channel reuse pattern

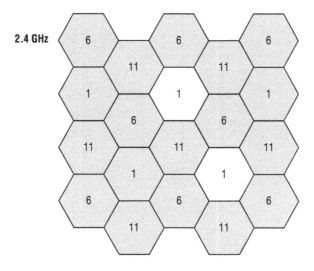

Probably the biggest mistake you can make when deploying a 2.4 GHz channel reuse pattern would be the design shown in Figure 13.12. Notice that all the channels are adjacent. As stated earlier, overlapping coverage cells that also have overlapping frequency space from adjacent cells will result in adjacent channel interference. The result is corrupted data, layer 2 retransmissions, and an extreme degradation in performance. In this scenario, adjacent channel interference is simply RF interference caused by your own APs due to improper channel design. The improper channel design pictured in Figure 13.12 should be avoided at all costs.

It is necessary to always think three-dimensionally when designing a multiple-channel architecture reuse pattern. If access points are deployed on multiple floors in the same building, a reuse pattern will be necessary, such as the one pictured in Figure 13.13. A common mistake is to deploy a cookie-cutter design by performing a site survey on only one floor and then placing the access points on the same channels and same locations on each floor. The access points often need to be staggered to allow for a three-dimensional reuse pattern. Also, the –70 dBm coverage cells of each access point should not extend beyond more than one floor above and below the floor where the access point is mounted. It is inappropriate to always assume that the coverage bleed-over to other floors will provide sufficient signal strength and quality. In some cases, the floors are concrete or steel and allow very little, if any, signal coverage through. As a result, a validation coverage survey is absolutely required. Always remember that RF propagates in all directions. Several commercial predictive RF modeling tools, such as iBwave Design, have the capability to depict RF coverage in a three-dimensional view.

FIGURE 13.12   Improper channel reuse—adjacent channel interference

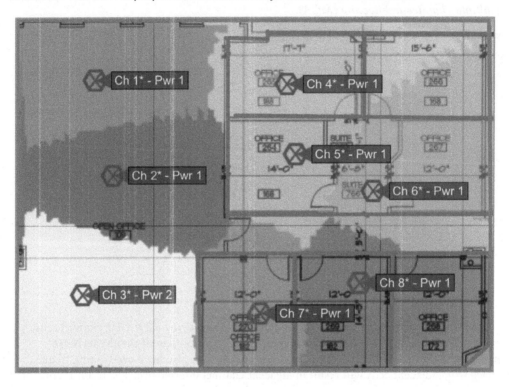

FIGURE 13.13   Three-dimensional channel reuse

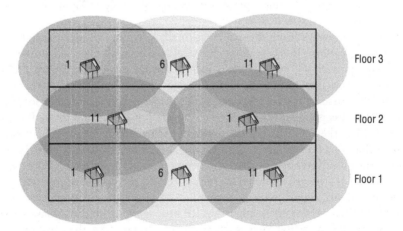

Many more channels are available in the 5 GHz U-NII bands, as shown in Figure 13.14. All these channels are technically considered non-overlapping channels because there is 20 MHz of separation between the center frequencies. In reality, there will be some frequency overlap of the sidebands of neighboring OFDM channels. The good news is that you are not limited to only three channels; many more channels can be used in a 5 GHz channel reuse pattern, which is discussed later in this chapter.

FIGURE 13.14   5 GHz channels

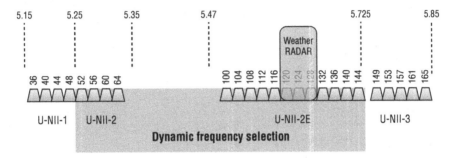

Co-Channel Interference

Another of the most common mistakes many businesses make when first deploying a WLAN is to configure multiple access points all on the same channel. If all the APs are on the same channel, unnecessary medium contention overhead occurs. As you have learned, CSMA/CA dictates half-duplex communications, and only one radio can transmit on the same channel at any given time.

Figure 13.15 depicts multiple nearby APs all configured to transmit on channel 1. If an AP on channel 1 is transmitting, all nearby access points and clients on the same channel within hearing range will defer transmissions. The result is that throughput is adversely affected: Nearby APs and clients have to wait much longer to transmit because they have to take their turn. The unnecessary medium contention overhead that occurs because all the APs are on the same channel is called *co-channel interference (CCI)*. In reality, the 802.11 radios are operating exactly as defined by the CSMA/CA mechanisms, and this behavior should really be called co-channel cooperation. The unnecessary medium contention overhead caused by co-channel interference is a result of APs or clients hearing each other and deferring transmissions.

The good news is that most WLAN designers understand not to configure all the APs on the same channel. Furthermore, the adaptive RF capabilities of many WLAN vendors will also automatically choose channels 1, 6, and 11 for the 2.4 GHz radios in access points. The primary goal of channel reuse patterns is to prevent co-channel interference. A channel reuse plan reduces airtime consumption by isolating frequency domains (channels).

FIGURE 13.15   Improper channel reuse—co-channel interference

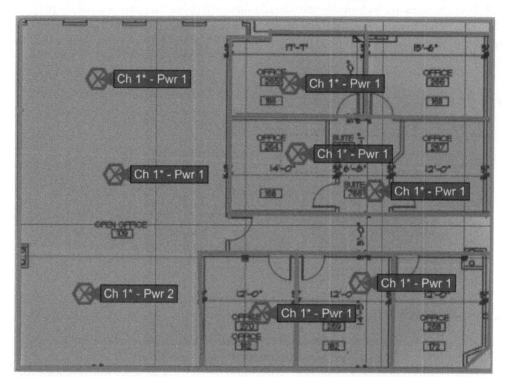

The bad news, however, is that co-channel interference is almost impossible to prevent in the 2.4 GHz band because only three channels are available for a reuse pattern. Does an RF signal just stop at the edge of a coverage cell designed for −70 dBm coverage? The answer is no, the RF signal continues to propagate, and the signal can be heard by other 802.11 radios at a great distance. An 802.11 radio will defer transmissions if it hears the PHY layer preamble transmissions of any other 802.11 radio at a *signal detect (SD)* threshold just four decibels (dB) or greater above the noise floor. Any radio that hears another radio on the same channel will defer, which results in medium contention overhead and delay.

As shown in Figure 13.16, despite a three-channel reuse pattern, APs on the same channel will hear each other and defer. For example, if AP-1 on channel 6 hears the preamble transmission of a nearby AP-2, also transmitting on channel 6, AP-1 will defer and not transmit at the same time. Likewise, all the clients associated to AP-1 must also defer transmission if they hear the preamble transmission of AP-2. All these deferrals create medium contention overhead and consume valuable airtime because you have two basic service sets on the same channel that can hear each other. Co-channel interference (CCI) is also often referred to as an *overlapping basic service set (OBSS)*.

FIGURE 13.16   Co-channel interference—APs

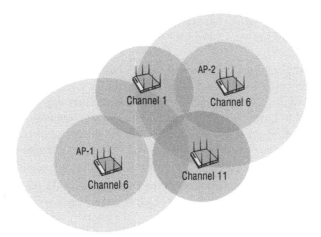

In reality, Wi-Fi clients are the primary cause of OBSS and CCI interference. As shown in Figure 13.17, if a client associated to AP-1 is transmitting on channel 6, it is possible that AP-2 (and any clients associated to AP-2) will hear the PHY preamble of the client and must defer any transmissions. Co-channel interference (CCI) is the top cause of needless airtime consumption that can be minimized with proper WLAN design best practices. What most people do not understand about CCI is the fact that clients are the number one cause of CCI. You should understand that CCI is not static but always changing due to the mobility of client devices.

FIGURE 13.17   Co-channel interference—clients

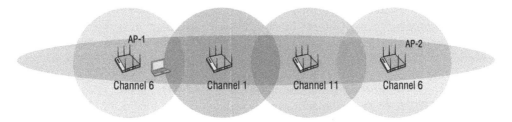

Because of the fact that only three channels are available in the 2.4 GHz band and because of CCI caused by clients, CCI is pretty much inevitable in the 2.4 GHz band. One strategy to reduce CCI in the 2.4 GHz band is to turn off a lot of the 2.4 GHz radios in dual-frequency access points and rely more on the coverage provided by the 5 GHz AP radios to meet density needs. While it is almost impossible to prevent CCI in the 2.4 GHz band, the airtime consumption that is a result of CCI can be minimized and possibly avoided with good 5 GHz WLAN design, which will be discussed later in this chapter.

Please do not confuse adjacent channel interference with co-channel interference. Adjacent channel interference is simply a result of improper channel planning in the 2.4 GHz band and can be avoided by only using channels 1, 6, and 11. Adjacent channel interference is a much more serious problem than co-channel interference because of the corrupted data and layer 2 retries.

In Europe and other regions of the world, more channels are legally available for license-free communications in the 2.4 GHz ISM band. In Europe, a WLAN four-channel reuse pattern of channels 1, 5, 9, and 13 is sometimes deployed. Although there is a small amount of frequency overlap between those four channels, the performance might in some cases be better if the medium contention overhead of co-channel interference can be reduced because there is less bleed-over. The four-channel plan still has the following disadvantages:

- If a nearby business has APs deployed on the traditional 1-6-11 plan, the neighboring business's APs will cause severe adjacent channel interference with your APs deployed with a 1-5-9-13 plan.

- Also, all North American Wi-Fi radios are restricted by firmware and cannot transmit on channel 13. Any visiting customer or employee with a laptop, iPad, or other mobile device that was purchased in North America will not be able to connect to a European access point transmitting on channel 13.

For these reasons, the more traditional 2.4 GHz three-channel plan is usually deployed in Europe and other regions of the world.

5 GHz Channel Reuse

So far we have mainly focused on channel reuse design for the 2.4 GHz band. Channel reuse patterns should also be used in the 5 GHz frequency bands. If all the 5 GHz channels are legally available for transmissions, a total of 25 channels may be available for a channel reuse pattern at 5 GHz. In Europe, fewer channels are available for a 5 GHz reuse pattern because the five U-NII-3 channels are not usually available without a license.

Depending on the region, and other considerations, 8 channels, 12 channels, 17 channels, 22 channels, or other combinations may be used for 5 GHz channel reuse patterns. For example, Figure 13.18 depicts a 5 GHz channel reuse pattern using the channels available in U-NII-1, U-NII-2, and U-NII-3. For the best 5 GHz reuse design, the key is to use as many channels as possible.

FIGURE 13.18   5 GHz channel reuse pattern

Distance to cell with same channel is at least two cells.

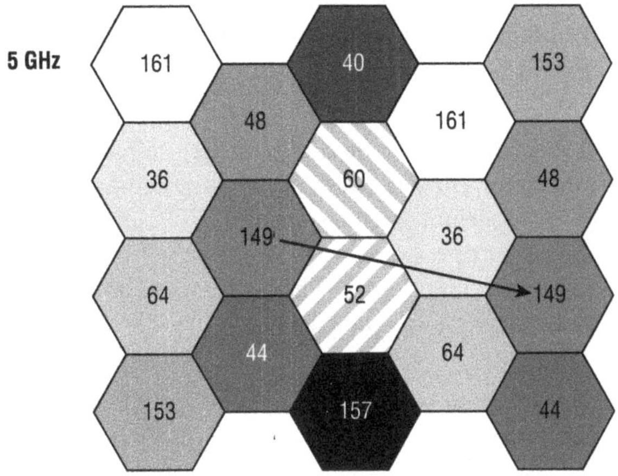

Several factors and best practices should be considered when planning a 5 GHz channel reuse pattern:

- The first factor to consider is which channels are available legally in your country or region. In Europe, a pattern utilizing most of the channels in the U-NII-1, U-NII-2, and U-NII-2E bands is quite common. The U-NII-3 channels are rarely used in the pattern because of regulatory domain restrictions.

- Although by the IEEE's definition, all 5 GHz channels are considered non-overlapping, in reality there is some frequency sideband overlap from adjacent channels. It is a recommended practice that any adjacent coverage cells use frequencies that are at least two channels apart and not use adjacent frequencies. In other words, do not provide coverage with an AP transmitting on channel 36 adjacent to an AP transmitting on channel 40. However, an AP transmitting on channel 36 adjacent to an AP transmitting on channel 48 is acceptable. Following this simple rule will prevent adjacent channel interference from the sideband overlap.

- Figure 13.18 also shows that the second recommended practice for 5 GHz channel reuse design is that there should be at least two cells of coverage space distance between any two access points transmitting on the same channel. Following this rule should minimize co-channel interference from APs. However, this may not necessarily prevent co-channel interference from clients—and remember that client transmissions are the main cause of CCI.

- Whenever possible, use as many channels as possible in 5 GHz to reduce CCI. The more channels that are used, the greater the odds that CCI can be prevented, including co-channel interference that originates from client devices. Figure 13.19 depicts a channel reuse pattern using the 5 GHz channels available in Europe. Notice the spatial distance between the coverage cells on both APs using channel 36. When you also factor in the attenuation from walls, the odds of associated clients to either channel 36 AP hearing each other and deferring transmission, has been most likely eliminated. A channel reuse pattern for the United States could also include the 5 extra channels available in the U-NII-3 band.

- In most cases, you should use the dynamic frequency selection (DFS) channels. The good news is that most current-day client devices are being certified to transmit on the DFS channels, and the inclusion of DFS channels in channel reuse patterns is becoming more commonplace. In Europe, because there are only 4 non-DFS channels, the use of the DFS channels is usually mandatory. The only reason not to use the DFS channels is if the bulk of your client population consists of legacy devices that do not support the DFS channels.

- If a nearby radar transmission is causing your APs and clients to switch to a non-DFS channel, simply eliminate the problematic DFS channel from the 5 GHz channel reuse design.

FIGURE 13.19   Preventing CCI with 5 GHz channel reuse

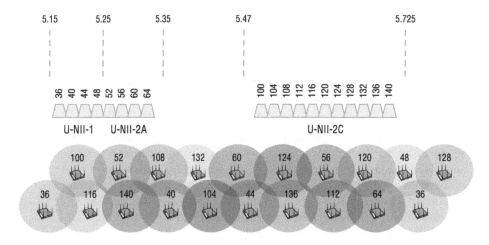

DFS Channels

As depicted earlier in Figure 13.14, all the channels in the U-NII-2 band (5.25–5.35 GHz) and the U-NII-2e band (5.47–5.725 GHz) are known as the *dynamic frequency selection (DFS)* channels. As you previously learned, WLAN radios operating in these 5 GHz bands must support DFS, to protect WLAN communications from interfering with military or weather radar systems. If radar pulses are detected in any of these DFS channels, access points and clients are not allowed to transmit on the same channel. The rules for 802.11 radios to transmit on the DFS channels might vary by region; however, the goal is to avoid inference with radar. Many radar systems are protected by DFS regulations, including radar on boats, weather radar, and military radar. Please note that DFS requirements do not apply just to Wi-Fi radios.

You might have noticed that by default, the DFS channels are not enabled on a WLAN vendor AP. There are really two reasons DFS channels are usually turned off by default. First, the ultimate decision as to whether or not to use the DFS channels in a 5 GHz reuse design is up to the customer or system integrator that is designing and deploying the APs. Second, new APs and client radios must be certified by the FCC and other regulatory agencies. As shown in Figure 13.20, rigorous certification testing is performed to verify that the radios can detect a weather radar pulse as well as abide by the interference avoidance rules. Radios are tested for various levels of required sensitivity for detecting a radar pulse. In the United States, there is typically a 6-month backlog for DFS certification, which means that when WLAN vendors release a new model AP, they may not be able to support the DFS channels for another six months. Once certified, a firmware upgrade of the APs will make the DFS channels available for use. This is why you often simply do not see the DFS channels in use in the enterprise. New APs are originally deployed using only the non-DFS 5 GHz channels, and the DFS channels are never utilized at a later date. However, as you learned earlier in this chapter, incorporating the DFS channels into a 5 GHz channel reuse plan is advantageous because it will help minimize CCI by providing more channels for the channel design.

FIGURE 13.20   DFS Radar Signal Generator

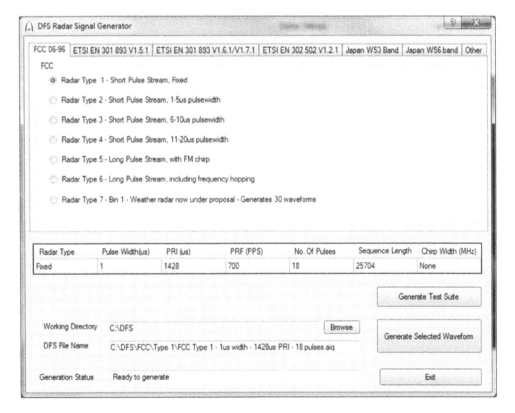

So how exactly does DFS work? Whenever an AP boots up for the first time on a DFS channel, the AP's radio must listen for a period of 60 seconds before being allowed to transmit on the channel. If any radar pulses are detected, the AP cannot use that channel and will have to try a different channel. If no radar is detected during the initial 60-second listening period, the AP can begin transmitting beacon management frames on the channel. In Europe, the rules are even more restrictive for the *Terminal Doppler Weather Radar (TDWR)* channels of 120, 124, and 128. An AP must listen for 10 full minutes before being able to transmit on the TDWR frequency space.

802.11 client radios must also abide by the radar avoidance rules; therefore, they typically will not initially send probe requests on any DFS channel. When client radios scan the DFS channels and hear an AP transmitting a beacon on that DFS channel, the clients assume that the channel is clear of radar and can initiate an authentication and association exchange with the AP.

If APs and clients are already operating on a DFS channel and a radar pulse is detected, the AP and all the associated clients must leave the channel. If radar is detected on the current DFS frequency, the AP will inform all associated client stations to move to another channel using a *channel switch announcement (CSA)* frame. The AP and the clients have 10 seconds to leave the DFS channel. The AP may send multiple CSA frames to ensure that all clients leave. The CSA frame will inform the clients that the AP is moving to a new channel and that they must go to that channel as well. In most cases, the channel is a non-DFS channel and very often is channel 36. Some vendors offer the capability for the WLAN admin to specify a *DFS fallback* channel.

Once an AP and clients switch to a non-DFS channel, they cannot return to the previous DFS channel for at least 30 minutes. One challenge in returning to the original DFS channel is that after the 30-minute waiting period, the AP will once again have to monitor on the DFS channel for 60 seconds before transmitting again. This means there will be at least a 60 second interval when the AP will not be servicing clients. One chipset vendor, Broadcom, offers a solution called *zero-wait DFS* to solve this problem, using the MIMO radio chains of the 5 GHz access point radio. For example, a 3×3:3 AP could be listening on DFS channel 104 with a single MIMO radio chain while still providing access to clients on non-DFS channel 36 with the two remaining radio chains. If channel 104 is clear, the AP can send a new channel switch announcement to all the clients on channel 36, telling them to return to the original channel 104. Even better, the AP could use the single MIMO radio chain to listen to a different DFS channel (for example, channel 64). If the new DFS channel is clear for 60 seconds, the clients can also move. The advantage is that the clients can instead move to channel 64 and not have to wait 30 minutes to return to channel 104.

Historically, the biggest problem with using the DFS channels has been the potential for false-positive detections of radar. In other words, the APs misinterpret a spurious RF transmission as radar and begin changing channels even though they do not need to move. The good news is that most enterprise WLAN vendors have gotten much better at eliminating false-positive detections.

As previously stated, the use of DFS channels is always recommended, unless mission-critical clients do not support them. If real radar exists nearby, simply eliminate the affected DFS channels from the 5 GHz channel plan.

40 MHz Channel Design

802.11n technology introduced the capability of bonding two 20 MHz channels to create a larger 40 MHz channel. As you learned in Chapter 10, "MIMO technology: HT and VHT," channel bonding effectively doubles the frequency bandwidth, meaning double the data rates that can be available to 802.11n/ac radios.

As shown in Figure 13.21, a total of twelve 40 MHz channels are available to be used in a 5 GHz reuse pattern when deploying an enterprise WLAN, depending on the region. The U-NII-3 40 MHz channels are currently not available in Europe.

FIGURE 13.21   40 MHz channels

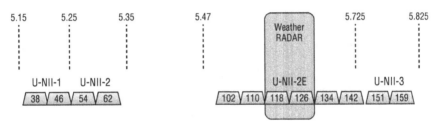

So what are the advantages and disadvantages of using channel bonding and 40 MHz channels? At first glance, you would think that channel bonding should always be enabled because of the higher data rates that are available to 802.11n/ac radios. For example, the highest potential data rate for a 3×3:3 MIMO 802.11n radio transmitting on a 20 MHz channel is 217 Mbps. The highest potential data rate for a 3×3:3 MIMO 802.11n radio transmitting on a 40 MHz channel is 450 Mbps. After looking at these numbers, most administrators assume that channel bonding would be enabled by default. However, many WLAN access point vendors require that channel bonding be manually enabled because there is the potential for channel bonding to negatively impact the performance of the WLAN.

Let us go back to 20 MHz design for one moment. One of the advantages of using 5 GHz instead of 2.4 GHz is that there are many more 5 GHz 20 MHz channels that can be used in a reuse pattern. Only three 20 MHz channels can be used in 2.4 GHz. The problem with using only three 20 MHz channels is that there will always be some amount of co-channel interference even though these channels are non-overlapping. Therefore, a certain amount of medium contention overhead always exists at 2.4 GHz simply because there are not enough channels and frequency space. Medium contention overhead due to APs on the same 20 MHz channel can be almost completely avoided in 5 GHz because there are more channels. A 5 GHz channel reuse plan of eight or more 20 MHz channels will greatly decrease co-channel interference and medium contention overhead.

As depicted in Figure 13.22, consider a 40 MHz reuse pattern using only the non-DFS channels in U-NII-1 and U-NII-3 bands. If only eight 20 MHz channels are being used, then a four channel 40 MHz channel reuse pattern exists. Although the bandwidth is doubled for the 802.11n/ac radios, there will be an increase of medium contention overhead because there are only four 40 MHz channels, and access points and clients on the same 40 MHz channel will likely hear each other. The medium contention overhead may have a negative impact and offset any gains in performance that the extra bandwidth might provide.

FIGURE 13.22   40 MHz channel reuse—4 channels

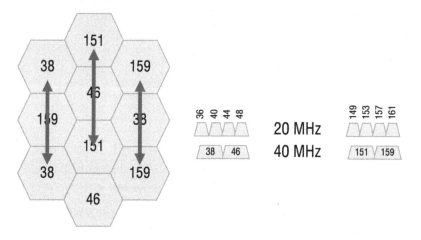

Another problem with channel bonding is that it usually will result in a higher noise floor of about 3 dB. If the noise floor is 3 dB higher, then the SNR is 3 dB lower, which means that the radios may shift down to lower MCS rates and therefore lower modulation data rates. In many cases, this offsets some of the bandwidth gains that the 40 MHz frequency space provides.

So, should you use channel bonding or not? If four or fewer 40 MHz channels are available, you might not want to turn on channel bonding, especially if the 5 GHz radios are transmitting at a higher power level. If the majority of the WLAN clients do not support channel bonding, there is no reason to enable the capability. For example, earlier versions of 802.11n smartphones and tablets did not support bonding. Even if all the 802.11n clients support 40 MHz channel bonding, performance testing would be highly recommended if only four 40 MHz channels are deployed.

However, if the DFS bands are enabled, more 40 MHz channels will be available; therefore, a much better reuse pattern that cuts down on medium contention will be available. The key is that the client radios must support DFS and should support channel bonding. Figure 13.23 depicts a 40 MHz channel reuse pattern of 12 channels, including DFS channels. In this example, CCI is reduced because there are more channels.

FIGURE 13.23 40 MHz channel reuse—12 channels

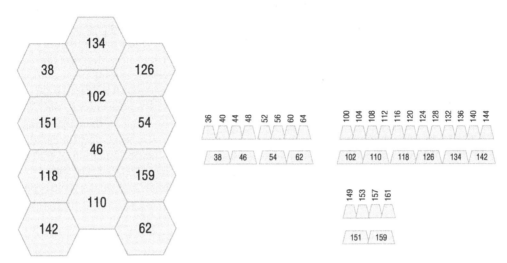

Many WLAN professionals recommend using 20 MHz channels, as opposed to 40 MHz channels, in most 5 GHz WLAN designs. However, 40 MHz channel deployment can work with some careful planning and by abiding by a few general rules:

- Using four or fewer 40 MHz channels in a reuse pattern will not be sufficient. Use 40 MHz channels only if the DFS channels are available. Enabling the DFS channels provides more frequency space and therefore more available 40 MHz channels for the reuse pattern.

- Do not have the AP radios transmit at full power. Transmit power levels of 12 dBm or lower are usually more than sufficient in most indoor environments.

- The walls should be of dense material for attenuation purposes and to cut down on CCI. Cinder block, brick, or concrete walls will attenuate a signal by 10 dB or more. Drywall, however, will attenuate a signal by only about 3 dB.

- If the deployment is a multi-floor environment, consider not using 40 MHz channels unless there is significant attenuation between floors.

As you learned in Chapter 10, 802.11ac introduced the capability of 80 MHz and even 160 MHz channels in the 5 GHz band. Even though 80 MHz and 160 MHz channels are available with 802.11ac radios, they should not be used in enterprise. 80 MHz and 160 MHz channel implementations do not scale in an enterprise WLAN, because there is simply not enough frequency space. Performance levels will drop significantly if 80 MHz channels are deployed on multiple APs. An 80 MHz channel should be used only for a single AP in an isolated area, such as a rural home environment.

Static Channels and Transmit Power vs. Adaptive RF

Probably the most debated topic when it comes to WLAN design is whether to use static channel and power settings for APs versus using adaptive channel and power settings. *Radio resource management (RRM)* is an industry standard term used to describe the automatic and adaptive power and channel configuration of access points. WLAN vendor APs can dynamically change their configuration based on accumulated RF information gathered from the access point radios. Based on the accumulated RF information, the access points adjust their power and channel settings, adaptively changing the RF coverage cells. Radio resource management is also referred to as *adaptive RF*. When implemented, RRM provides automatic cell sizing and automatic monitoring and optimization of the RF environment, which can best be described as a self-organizing wireless LAN. Most RRM protocols also have a *lock-down* capability once the channels and power have been automatically assigned among the APs. Some of the RRM protocols also take into account 802.11k client data gathered from associated clients that support 802.11k.

RRM mechanisms are, for the most part, proprietary, and each WLAN vendor uses their own protocol for adaptive RF capabilities. Each WLAN vendor will likely also have its own marketing name for the technology. RRM is essentially a control plane mechanism. The decision to make adaptive changes in the AP channel and power settings may be distributed between access points or centralized in a WLAN controller or cloud-based management system. Once again, each WLAN vendor has its own adaptive RF protocol; however, the automatic assignment of channel and power settings is based on numerous costs, as defined by the vendor's RRM algorithms. Figure 13.24 depicts one WLAN vendor's RRM cost calculations.

FIGURE 13.24   Adaptive RF cost calculation

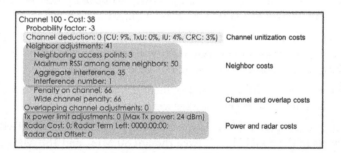

RRM technology has gained wide acceptance because almost all WLAN vendors offer some sort of adaptive RF solution. Many of the vendors' customers have had excellent success with adaptive RF deployments. Very often, sales representatives of the various WLAN vendors claim that WLAN design is no longer necessary because of the dynamic and self-organizing nature of their RRM solution. Although adaptive RF technology has come a long way in recent years, allowing the APs to adapt to the environment, RRM does not by any means replace proper WLAN design. A manual and/or predictive model survey should be the first order of business prior to deployment. Installation and validation of the network design should then be performed. In Chapter 14, you will learn about the importance of a validation survey. Adaptive RF capabilities are used post-deployment to automatically make necessary channel and power changes in a live operational environment.

As previously mentioned, the use of RRM versus a static channel and power design is often a very heated debate. Many old-school WLAN professionals prefer manually setting all channel and power settings of APs, as opposed to relying on an adaptive protocol. Other professionals prefer to use RRM exclusively. So should you use RRM or go with static settings? The answer really depends on the WLAN professional's preference, expertise, and the type of WLAN design. Adaptive RF capabilities are turned on by default on most every WLAN vendor AP. The algorithms for RRM constantly improve year after year. The majority of commercial WLAN customers use RRM because it is easy to deploy. RRM is usually the preferred method in enterprise deployments with thousands of APs. However, careful consideration should be given to using static channel and power settings in complex RF environments. Most WLAN vendors recommend in their own very high-density deployment guides that static power and channels be used, especially when directional antennas are deployed.

We are not going to debate the pros and cons of RRM versus static configuration in this book. Try to think of adaptive RF as one tool in your arsenal for WLAN deployment. You always have the capability to manually set the channel and power settings of every AP. Whichever method is chosen, a well thought out WLAN design and validation survey will always be a requirement.

Single-Channel Architecture

As of this writing, three WLAN vendors—Fortinet, Allied Telesys, and Ubiquiti Networks—offer an alternative configuration for WLAN channel design solution, known as *single-channel architecture (SCA)*. Imagine a WLAN network with multiple access points all transmitting on the same channel and all sharing the same BSSID. A single-channel architecture is exactly what you have just imagined. The client stations see transmissions on only a single channel with one SSID (logical WLAN identifier) and one BSSID (layer 2 identifier). From the perspective of the client station, only one access point exists. In this type of WLAN architecture, all access points in the network can be deployed on one channel in 2.4 GHz or 5 GHz frequency bands. Uplink and downlink transmissions are coordinated by a WLAN controller on a single 802.11 channel in such a manner that the effects of co-channel interference are minimized.

Let us first discuss the single BSSID. Single-channel architecture consists of a WLAN controller and multiple controller-based access points. As shown in Figure 13.25, each AP has its own radio with its own MAC address; however, they all share a *virtual BSSID* that is broadcast from all the access points. Because the multiple access points advertise only one single virtual MAC address (BSSID), client stations believe they are connected to only a single access point, although they may be roaming across multiple physical APs. You have learned that clients make the roaming decisions. In a single-channel architecture (SCA) system, the clients think they are associated to only one AP, so they never initiate a layer 2 roaming exchange. All the roaming handoffs are handled by a central WLAN controller.

FIGURE 13.25 Single-channel architecture

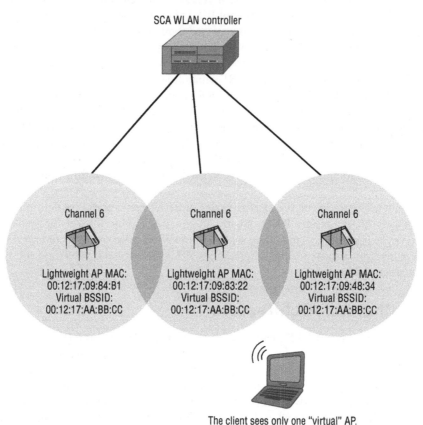

SCA WLAN controller

Channel 6

Channel 6

Channel 6

Lightweight AP MAC:
00:12:17:09:84:B1
Virtual BSSID:
00:12:17:AA:BB:CC

Lightweight AP MAC:
00:12:17:09:83:22
Virtual BSSID:
00:12:17:AA:BB:CC

Lightweight AP MAC:
00:12:17:09:48:34
Virtual BSSID:
00:12:17:AA:BB:CC

The client sees only one "virtual" AP.

As Figure 13.26 shows, the main advantage is that clients experience *zero handoff time*, and the latency issues associated with roaming times are resolved. The *virtual AP* used by SCA solutions is potentially an excellent marriage for VoWiFi phones and 802.1X/EAP solutions. As we discussed earlier, the average time involved during the EAP authentication process can be 700 milliseconds or longer. Every time a client station roams to a new access point, reauthentication is required when an 802.1X/EAP security solution has been deployed. VoWiFi requires a roaming handoff of 150 ms or less. The virtual BSSID eliminates the need for reauthentication while physically roaming within a single-channel architecture. The client does not initiate a reassociation exchange—thus, a zero handoff time.

FIGURE 13.26   Zero handoff time

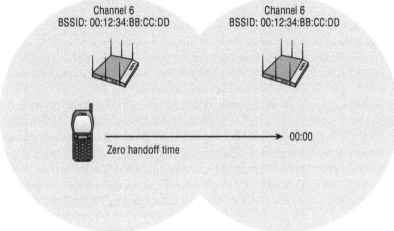

Channel 6

You have learned that client stations make the roaming decision in an MCA environment. However, client stations do not know that they roam in an SCA environment. The clients must still be mobile and transfer layer 2 communications between physical access points. All the client-roaming mechanisms are now handled back on the WLAN controller, and client-side roaming decisions have been eliminated. All station associations are maintained at the SCA WLAN controller, and the SCA controller manages all the APs. The SCA controller assigns to a unique access point the responsibility of handling downlink transmissions for an individual client station. When the controller receives the incoming transmissions of a client, the SCA controller evaluates the RSSI values of the client's transmissions. Based on incoming RSSI measurements, the SCA controller can allocate a specific AP for downlink transmissions. The client believes that it is associated to a single AP. However, the client moves between different physical APs based on RSSI measurements evaluated by the controller.

One big advantage of the single-channel architecture is that adjacent channel interference is no longer an issue. If all the access points are on the same channel, there can be no frequency overlap and, thus, no adjacent channel interference. However, a legitimate question about an SCA WLAN solution is: Why does co-channel interference not occur if all the APs are on the same channel? The answer is that co-channel interference does still occur; however, the WLAN controller attempts to centrally manage it by scheduling transmissions for APs within range of one another. If all the APs are on the same channel in an MCA wireless network, unnecessary medium contention overhead occurs. In a typical MCA environment, each access point has a unique BSSID and a separate channel, and each AP's coverage cell is a single-collision domain. In an SCA wireless environment, the collision domains are managed dynamically by the SCA controller based on RSSI algorithms. The controller ensures that nearby devices on the same channel are not transmitting at the same time. Most of the mechanisms used by SCA vendors are proprietary and beyond the scope of this book.

For many years, the procedures just described were a competitive advantage for the SCA companies selling into verticals where VoWiFi was needed. However, with the wide acceptance of the QoS mechanisms defined by WWM and the fast secure roaming mechanisms defined by Voice-Enterprise, VoWiFi is now deployed extensively within the more traditional MCA architecture. The major SCA companies also offer the capability to disable SCA and use multiple channels, like all the other vendors.

A major disadvantage of the single-channel architecture is the capacity issue, because only one channel is available. In a 2.4 GHz SCA deployment, multiple APs can be co-located by using three channels and three virtual BSSIDs. Co-location design in single-channel architecture is often referred to as *channel layering*. Each layer of multiple APs operating on a single channel and using the same virtual BSSID is known as a *channel blanket* or *channel span*. Although this might sound like a good idea in theory, most customers are not willing to pay for three co-located access points everywhere coverage is needed. Another possible disadvantage with SCA architecture is that the contention domain could be very large. Although AP transmissions are coordinated by an SCA controller to minimize collisions with other APs, some implementations of SCA technology can be highly proprietary, and there is no guarantee that client transmissions can be controlled to perfection.

It is possible, with certain vendors, to combine both methods and use the advantages of SCA in certain locations or network segments, and use the more traditional multiple-channel architecture (MCA) in other areas. For example, SCA might be dedicated for a VoWiFi and high mobility in the 2.4 GHz band, while MCA might be used in the 5 GHz band to take advantage of higher bandwidth and capacity for data-focused clients.

Capacity Design

When a wireless network is designed, two concepts that typically compete with each other are *capacity* and *range*. In the early days of wireless networks, it was common to install an access point with the power set to the maximum level to provide the largest coverage area possible. This was typically acceptable because there were very few wireless devices. Also, access points were very expensive, so companies tried to provide the most coverage while using the fewest access points.

A common question that often gets asked is, "What is the range of the AP?" In theory, an RF signal will travel forever in free space; however, the proper answer to this question is that the "effective" range of an AP really depends on the attenuation environment of the facility. More importantly, the effective range of an AP should really be based on the client's perspective. In other words, range is not just about client connectivity, but also about client performance. Effective range means the client devices can roam efficiently and can communicate with the AP together with many other clients using high data rates.

With the proliferation of wireless devices, network design has changed drastically from the early days. WLANs are rarely designed strictly from a range and coverage perspective anymore. Instead, most WLANs are designed with a primary focus on client capacity needs. This does not mean that coverage design is now ignored. You still have to plan for a –70 dBm or better received signal, high SNR, seamless roaming, and a proper channel reuse pattern. As a matter of fact, how you design for coverage will also impact capacity needs. As previously mentioned, APs configured for full transmit power are no longer ideal. Three APs transmitting at 100 mW may provide –70 dBm coverage for a 10,000 square-foot facility, but what if 1000 or more client devices need Wi-Fi access within that same area. The WLAN bandwidth that three APs could provide simply does not even come close to meeting client capacity requirements.

Adjusting the AP transmit power to limit the effective coverage area is known as *cell sizing* and is one of the most common methods of meeting client capacity needs. Typical indoor WLAN deployments are designed with the APs set at about one-fourth to one-third transmit power. Higher user and client density environments may require that the AP transmit power be set at the lowest setting of 1 mW. In other words, more APs are needed to meet capacity needs, and therefore AP transmit power will need to be lowered. Limiting the transmit power of APs also helps reduce CCI caused by APs, which has a direct impact on performance.

WLANs with high user and client density are becoming a greater concern due to the client population explosion that has occurred. Wi-Fi networks are no longer just about wireless laptop connectivity. Most users now want to connect to the enterprise WLAN with multiple devices, including tablets and smartphones with Wi-Fi radios. Luckily, 802.11n/ac technology has provided greater bandwidth to handle more clients; however, even 802.11n/ac access points can become overwhelmed without proper capacity design.

High Density

The terms *high density (HD)* and *very high density (VHD)* are often used when discussing capacity design and planning for a WLAN. Different WLAN engineers have different opinions as to what constitutes a high-density WLAN; however, due to the abundance of client devices, most WLANs should be considered high density by default. To bring some clarity to the terminology, high-density WLANs can usually be described as three different scenarios:

High Density   Almost all WLANs are high-density (HD) environments due to the proliferation of numerous users with multiple devices. The average person may want to connect to an enterprise WLAN with as many as three or four Wi-Fi devices. Obviously, the density of client devices also depends on the number of users. Most high-density environments consist of multiple areas where roaming is also a top priority. APs are deployed in many different rooms with walls that will often contribute to different levels of attenuation.

Very High Density   Any WLAN environment that has a tremendous amount of people in a single open area is often referred to as a very high-density (VHD) WLAN. Prime examples include auditoriums, gymnasiums, cafeterias, etc. Most VHD environments do not have walls that provide attenuation. All the APs likely hear each other within the open space. Design for a very high-density WLAN is quite complex and different from standard high-density environments with walls. As discussed later in this chapter, a VHD environment usually requires directional antennas to provide sectors of coverage.

Ultra High Density   An ultra high-density WLAN is defined as an environment with tens of thousands of users and devices all within the same space. The best examples of an ultra high-density WLAN are stadiums and sports arenas. Designing for these types of environments requires seasoned WLAN professionals with expertise in stadium Wi-Fi design.

The age-old question that WLAN customers always asked is: How many client devices can connect to an AP radio? The correct answer is, it depends. Nobody likes this answer, but there are simply too many variables to always give the same answer for any WLAN vendor's AP. The default settings of an enterprise WLAN radio might allow as many as 100–250 client connections. Since most enterprise APs are dual-frequency with both a 2.4 GHz and 5 GHz radio, theoretically 200–500 clients could associate to the radios of a single AP. Although more than a hundred devices might be able to connect to an AP radio, these numbers are not realistic for active devices due to the nature of the half-duplex shared medium. The performance needs of this many client devices will not be met and the user experience will be miserable. The perception will be that the Wi-Fi is "slow."

If the access point is using 802.11n/ac radios with 20 MHz channels, a good rule of thumb is that each radio could support 35–50 active devices for average use, such as web browsing and checking email. However, the numbers can vary greatly based on a wide variety of variables. The following are probably the three biggest questions that need to be asked:

What type of applications will be used on the WLAN?   As previously stated, 35–50 active Wi-Fi devices per radio, communicating through a dual-frequency 802.11n/ac access point, with average application use, such as web browsing and email, is realistic. However, bandwidth-intensive applications, such as high-definition video streaming, will have an impact. Different applications require different amounts of TCP throughput, as shown in Table 13.2.

TABLE 13.2   Applications and TCP throughput consumption

Application	Required Throughput
Email/web browsing	500 Kbps to 1 Mbps
Printing	1 Mbps
SD video streaming	1 Mbps to 1.5 Mbps
HD video streaming	2 Mbps to 5 Mbps

How many users and devices are expected?   Three important questions need to be asked with regard to users. First, how many users currently need wireless access and how many Wi-Fi devices will they be using? Second, how many users and devices may need wireless access in the future? These first two questions will help you to begin adequately planning for a good ratio of devices per access point while allowing for future growth. The third question of great significance is, where are the users? Sit down with network management and indicate on the floor plan of the building any areas of high user density. For example, one company might have offices with only 1 or 2 people per room, whereas another company might have 30 or more people in a common area separated by cubicle walls. Other examples of areas with high user density are call centers, classrooms, and lecture halls. You should always plan to conduct a validation survey when the users are present, not during off-hours. A high concentration of human bodies can attenuate the RF signal because of absorption.

What types of client devices are connecting to the WLAN?   Always remember that all client devices are not equal. Many client devices consume more airtime due to lesser MIMO capabilities. For example, an older 802.11n tablet with a 1×1:1 MIMO radio transmitting on a 20 MHz channel can achieve a data rate of 65 Mbps with TCP throughput of 30 Mbps to 40 Mbps. An 802.11n tablet with a 2×2:2 MIMO radio transmitting on a 20 MHz channel might achieve a data rate of 130 Mbps with TCP throughput of 60 Mbps to 70 Mbps. Many laptops also have 3×3:3 MIMO capabilities and thus are capable of higher data rates. The bulk of newer smartphones and tablets are now 2×2:2 MIMO capable. The point is that devices with less MIMO capabilities consume more airtime and therefore affect the aggregate performance of any WLAN. An AP can service more 2×2:2 MIMO clients efficiently as opposed to legacy 1×1:1 MIMO clients, which operate at lower data rates. Enterprise deployments will almost always require some level of backward compatibility to provide access for older 802.11a/b/g radios found in handhelds, VoWiFi phones, or older laptops.

Some of the commercial WLAN predictive modeling tools, such as Ekahau Site Survey, allow you to designate specific high-density areas on a building floor plan. As shown in Figure 13.27, within each area, you can then define the number of devices, the types of devices, as well as what type of application traffic to expect. The algorithms of the modeling software will adjust AP placement, power, and channel settings based on these variables, while still meeting coverage requirements.

Once you have determined the types of devices that are being used and the types of applications, you can then calculate the amount of airtime consumption. For example, an Apple iPad transmitting on a 20 MHz channel can connect at a data rate of 65 Mbps and might achieve a maximum of 30 Mbps of TCP throughput. A 2 Mbps video application running on an iPad will consume 6.67 percent of the airtime of a 20 MHz channel (2 Mbps ÷ 30 Mbps = 6.67%). A 2×2:2 MIMO laptop transmitting on a 20 MHz channel, can connect with a data rate of 130 Mbps might achieve a maximum TCP throughput of close to 70 Mbps. The same 2 Mbps HD video application running on the laptop will consume about 2.86 percent of airtime the (2 Mbps ÷ 70 Mbps = 2.86%).

FIGURE 13.27   Density predictive modeling

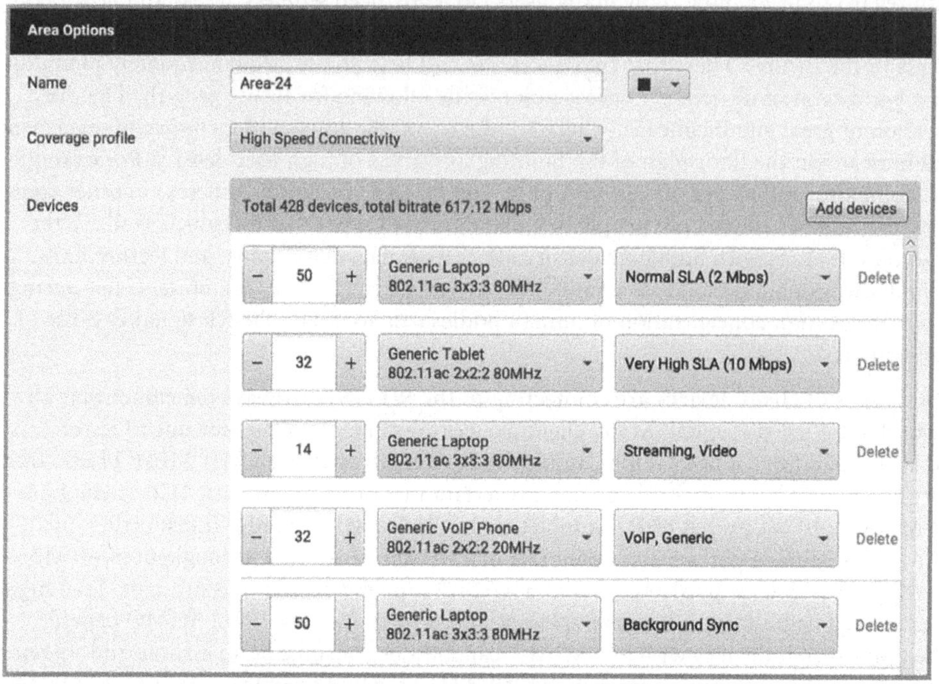

Once you have determined the amount of airtime consumption, you can then calculate the number of active devices that an AP radio can support. Wi-Fi expert Andrew von Nagy, CWNE #84, recommends several good formulas for these calculations. An 802.11 access point is considered to be fully burdened at about 80 percent of airtime utilization. To estimate the number of devices supported on a single AP radio, divide the individual airtime required per device into 80 percent:

$$80 \div \text{single device airtime consumption} = \text{\# devices per AP radio}$$

For example, the 2 Mbps HD video application running on iPads consumes 6.67 percent of airtime per device. Therefore, 80 ÷ 6.67 = 12 iPads that could run the application concurrently on a 20 MHz channel through a single 802.11n/ac AP radio. The AP most likely has a 2.4 GHz and a 5 GHz radio; therefore, 24 iPads could run the same HD video application through a single AP if the devices were balanced between the two frequencies. The same 2 Mbps HD video application running on the laptops consumed 2.86 percent of airtime per device. Therefore, 80 ÷ 2.86 = 28 laptops that could conceivably run the application concurrently on a 20 MHz channel through a single 802.11n/ac AP radio.

To calculate the number of AP radios needed, multiply the number of client devices by the percentage of airtime consumption, and then divide by 80 percent:

$$(\text{\# of devices} \times \text{single device airtime consumption \%}) \div 80\% = \text{number of AP radios}$$

For example, (150 iPads × 6.67%) ÷ 80% = 12.5 AP radios. Therefore, seven dual-band APs could adequately handle 150 iPads concurrently using a very high-bandwidth application. What if you also needed 150 laptops using the same streaming application in the same area with the iPads? Calculate (150 laptops × 2.86%) ÷ 80% = 5.36 AP radios. Therefore, you would probably need three more dual-band access points. A total of 10 dual-band 802.11n APs transmitting with 20 MHz channels would be able to handle concurrent 1 Mbps HD video streams to 150 iPads and 150 laptops.

 Keep in mind that these numbers are estimates and are taking both AP radios into consideration. Extensive testing post-deployment is always a good idea. As shown in Figure 13.28, a free predictive WLAN capacity-planning spreadsheet is available for download at www.revolutionwifi.net.

FIGURE 13.28 Capacity-planning spreadsheet

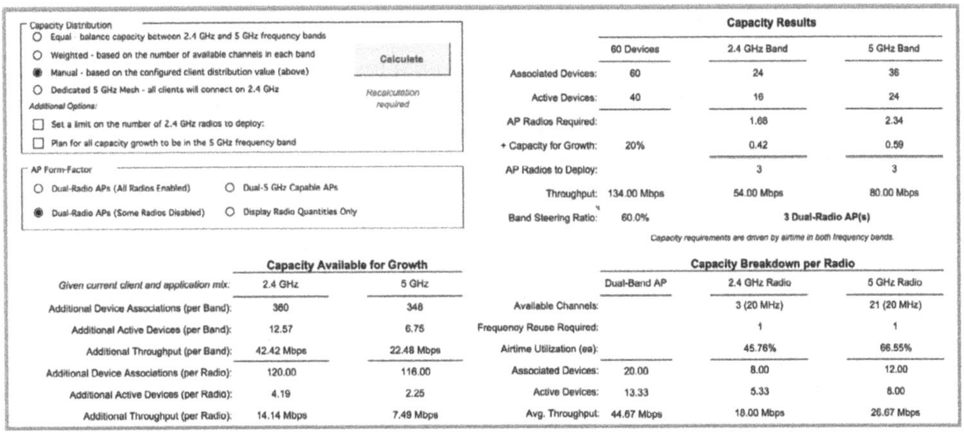

How Many APs per Room?

The location where APs are physically mounted also needs to be taken into consideration based on capacity needs. Some large areas of a building may only have two or three users that require Wi-Fi access. Conversely, other areas, such as an auditorium may have hundreds of users and devices that need Wi-Fi connectivity.

In many verticals, such as K-12 education, due to capacity requirements, it has become commonplace to deploy one AP per room. Please note that one AP per classroom in a K-12 may also be entirely unnecessary. One AP per every two or three classrooms may be sufficient to meet capacity needs. How many APs are needed depends on the capacity requirements as well as customer stipulations. Do you need to deploy one AP in every room? It once again depends on the number of devices, the type of devices, and the application traffic. However, an average of 70 or more Wi-Fi devices per classroom has become prevalent in many education environments. As shown in Figure 13.29, with proper AP placement, low transmit power, and channel reuse, deploying one AP per room using 5 GHz radios is feasible. The 5 GHz radio transmit power is normally 9 dBm (8 mW) or less, and 20 MHz channels are recommended in most cases. The walls must be made of thick material, such as concrete or brick, for attenuation purposes and to help limit CCI.

FIGURE 13.29   One AP per room—5 GHz

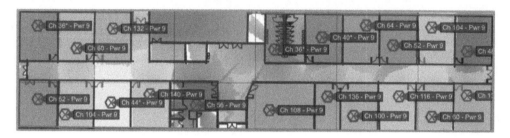

You will notice in Figure 13.29 that there are a total of 20 APs deployed using a 5 GHz channel reuse plan to prevent CCI. Most APs are dual frequency and also have a 2.4 GHz radio. Because there are only three channels and because of the close vicinity of these 20 APs, a majority of the 2.4 GHz AP radios should be disabled to help minimize CCI. In high-density deployments, it has become very common to disable two out of every three or even three out of every four 2.4 GHz radios in dual-frequency APs. In most cases, three or four 2.4 GHz radios will be more than adequate to provide the needed coverage for 2.4 GHz in the same space where twenty 5 GHz radios provide for the bulk of the capacity needs. Most WLAN vendors implement proprietary load balancing, band steering, and other MAC layer mechanisms to further assist capacity needs in a high-density user environment. When properly planning for client capacity, careful consideration needs to be given as to which clients and how many clients connect to the 2.4 GHz band and how many connect to the 5 GHz band. Load-balancing clients between the frequency bands as well as between individual APs should also be taken into consideration.

Band Steering

The unlicensed 5 GHz frequency spectrum offers many advantages over the unlicensed 2.4 GHz frequency spectrum for Wi-Fi communications. The 5 GHz U-NII bands offer a wider range of frequency space and many more channels. A proper 5 GHz channel reuse pattern using multiple channels will greatly decrease medium contention overhead caused by co-channel interference. In the 2.4 GHz band, there is always medium contention overhead due to CCI simply because there are only three channels.

Another major detrimental trait of the 2.4 GHz band is that, in addition to this band being used for WLAN networking, it is heavily used by many other types of devices, including microwave ovens, baby monitors, cordless telephones, and video cameras. With all of these different devices operating in the same frequency range, there is much more RF interference and a much higher noise floor than in the 5 GHz bands.

So, if the use of the 5 GHz bands will provide better throughput and performance, how can we encourage the clients to use this band? For starters, it is the client that decides which AP and which band to connect to, typically based on the strongest signal that it hears for the SSID that it wants to connect to. Most access points have both 2.4 GHz and 5 GHz radios in them, with both of them advertising the same SSIDs. Since the 5 GHz signals naturally attenuate more than the 2.4 GHz signals, it is likely that the client radio will identify the 2.4 GHz radio as having a stronger signal and connect to it by default. In many environments, the client would be capable of making a strong and fast connection with either of the AP's radios but will choose the 2.4 GHz signal because it is the strongest. A technology known as *band steering* has been developed to try to encourage dual-band client radios to connect to a 5 GHz AP radio instead of a 2.4 GHz AP radio.

Band steering is not an IEEE 802.11–developed technology. As of this writing, all implementations of band steering are proprietary. Although band steering implementations are proprietary, most vendors implement this technology using similar techniques by manipulating the MAC sublayer. When a dual-frequency client first starts up, it will transmit probe requests on both the 2.4 and 5 GHz bands, looking for an AP. When a dual-frequency AP hears probe requests on both bands originating from the same client, the AP knows that the client is capable of operating in the 5 GHz band. As depicted in Figure 13.30, the AP will then try to steer the client to the 5 GHz band by responding to the client using only 5 GHz transmissions. Although the client is steered to the 5 GHz AP, there may be reasons for the client to connect to the AP using the 2.4 GHz radio. If the client radio continues to try to connect to the AP using the 2.4 GHz radio, the AP will ultimately allow the connection.

FIGURE 13.30   Band steering to 5 GHz

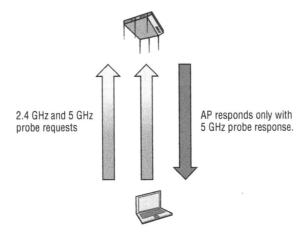

2.4 GHz and 5 GHz
probe requests

AP responds only with
5 GHz probe response.

Although band steering is normally used to encourage clients to connect to 5 GHz access points, clients can also be steered to the 2.4 GHz band. As shown in Figure 13.31, many WLAN vendors can define a percentage of clients to be directed to the 5 GHz band, with the remainder directed to the 2.4 GHz band. In environments where a high density of client devices exists, band steering to both frequencies can be used to balance an almost equal number of clients to both of the radios in the AP. For example, 55 clients connect to the

2.4 GHz radio and 60 clients connect to the 5 GHz radio. Effectively, band steering can be used to load balance clients between the frequencies. Please do not confuse this type of single AP frequency balancing with load balancing clients between multiple access points. Load balancing between multiple access points is described in the next section of this chapter.

FIGURE 13.31 Band steering for frequency balancing

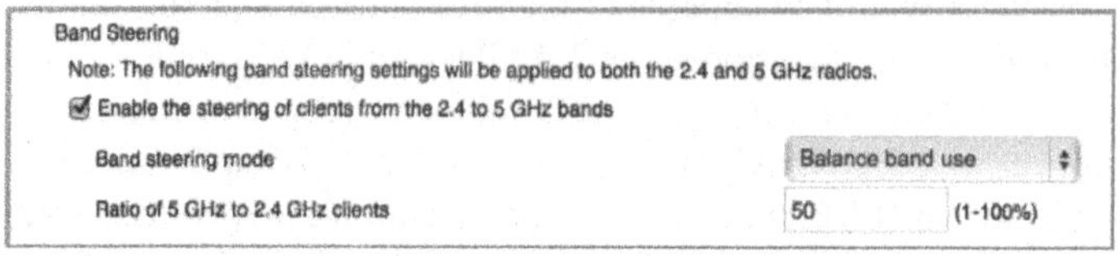

Band Steering

Note: The following band steering settings will be applied to both the 2.4 and 5 GHz radios.

☑ Enable the steering of clients from the 2.4 to 5 GHz bands

| Band steering mode | Balance band use ⬦ |
| Ratio of 5 GHz to 2.4 GHz clients | 50 (1-100%) |

Clients usually have a much better connection and better performance when connected to the 5 GHz band, which is why AP vendors offer band steering capabilities. So should band steering be enabled on a dual-frequency AP? Well, once again, the proper answer is that it depends. Most legacy client devices that have dual frequency capability will prefer the 2.4 GHz band, and band steering to 5 GHz may be necessary. However, many of the newer client devices implement proprietary client-side band selection. For example, macOS and iOS client devices typically prefer to connect to 5 GHz AP radios before they associate with 2.4 GHz AP radios. Some client vendors also offer the capability to configure client-side band preference with software client utilities.

Before enabling band steering on your APs, you might want to monitor what percentage of your client devices associate to 5 GHz radios on their own. If the bulk of the devices still prefer 2.4 GHz, band steering to 5 GHz is probably a good idea. Keep in mind that many legacy clients only have 2.4 GHz radios and cannot connect to 5 GHz radios. Many IoT devices also only have 2.4 GHz radios and do not have 5 GHz capabilities. The best strategy in terms of WLAN design is to provide 2.4 GHz coverage specifically for legacy devices and IoT devices. In the enterprise, the 2.4 GHz band is often considered a "best effort" frequency band, and the 5 GHz channels are reserved for all other clients that require higher performance metrics. Another strategy sometimes used to segment devices between the two frequency bands is SSID segmentation. In other words, mission-critical SSIDs are broadcast only on the 5 GHz band.

Load Balancing

WLAN vendors also use methods to manipulate the MAC sublayer to balance clients between multiple access points. As illustrated in Figure 13.32, load balancing clients between access points ensures that a single AP is not overloaded with too many clients, and that the total client population can be served by numerous APs, with the final result being better performance. When a client wants to connect to an AP, the client will send an association request frame to the AP. If an AP is already overloaded with too many clients, the AP will defer the association response to the client. The hope is that the client will then send another association request to another nearby AP with a lesser client load. Over time, the client associations will be fairly balanced across multiple APs. The client load information will obviously have to be shared among the access points. Load balancing is a control plane mechanism that can exist either in a distributed architecture where all the APs communicate with each other, or within a centralized architecture that utilizes a WLAN controller.

FIGURE 13.32   Load balancing between APs

| 3 clients | 6 clients | 60 clients | 21 clients | 22 clients | 22 clients | 24 clients | 22 clients |

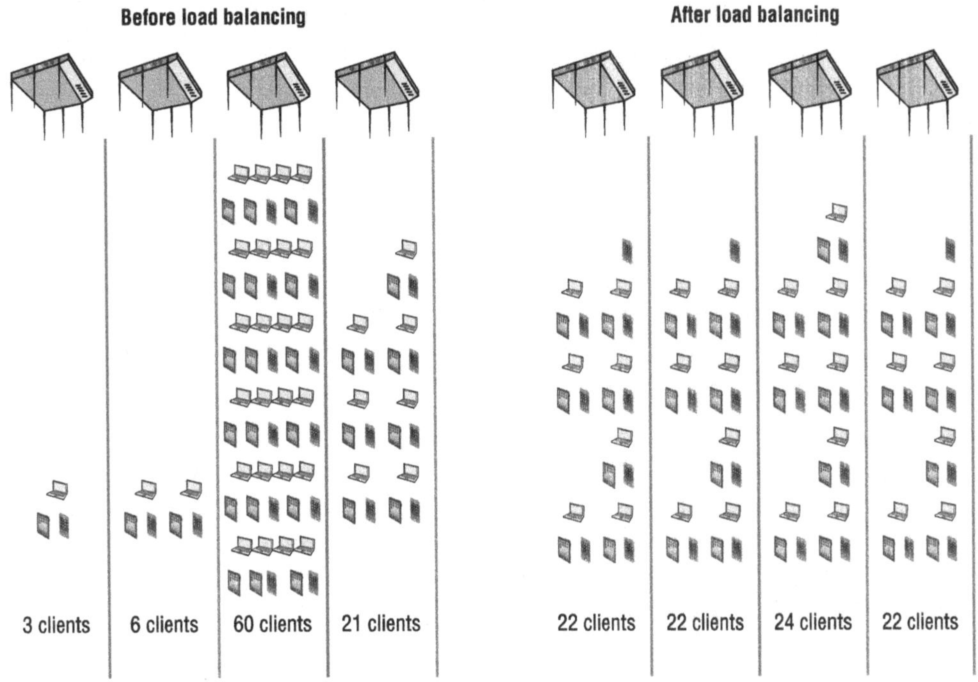

When Should Client Load Balancing Between APs Be Enabled?

Please be warned that enabling a WLAN vendor's load-balancing capabilities should be done only under certain conditions. Load balancing between access points is typically implemented in areas where there is a very high density of clients and roaming is not necessarily the priority—for example, a gymnasium or auditorium with 20 APs deployed in the same open area. In this environment, a client will most likely hear all 20 APs, and load balancing the clients between APs is usually a necessity.

However, in areas where roaming is needed, load balancing is not a good idea, because the mechanisms may cause clients to become sticky and stay associated to the AP too long. If association and reassociation response frames from the APs are being deferred, client mobility will most likely fail. Please understand that load balancing between APs can be detrimental to the roaming process.

Airtime Consumption

Over the years, the cosmic explosion of Wi-Fi client devices combined with enhanced 802.11 technologies has forced us to rethink about how we design WLANs. Designing for client device capacity has now become the norm. WLAN design practices now dictate that you design to minimize *airtime consumption*, which is directly related to capacity design. You have learned throughout this book that Wi-Fi is a half-duplex RF medium and only one radio can transmit on a channel at any given time. Whenever a radio has won a transmission opportunity (TXOP), the radio monopolizes the available airtime until it finishes transmitting. Yes, every radio needs to be able to transmit and deliver data; however, there are some simple WLAN design best practices that can minimize unnecessary airtime consumption.

Co-channel interference (CCI) is the top cause of needless airtime consumption that can be minimized with proper WLAN design best practices, as described earlier in this chapter. Designing for –70 dBm coverage and high SNR also ensures that client devices will transmit 802.11 data frames at high data rates, based on the client radio's capabilities. So what are some other WLAN design best practices that can reduce airtime consumption?

One of the best ways to cut down on airtime consumption is to disable some of the lower data rates on an AP. Figure 13.33 depicts an AP communicating with multiple client stations at 6 Mbps while communicating with one lone client using a 150 Mbps data rate. When 802.11 radios transmit at very low data rates, such as 6 Mbps or even lower, they effectively cause medium contention overhead for higher data rate transmitters due to the long wait time. A radio transmitting a 1,500-byte data frame at 150 Mbps might occupy the medium for 50 microseconds. Another radio transmitting at 6 Mbps may take 1,250 microseconds to deliver the same 1,500 bytes. In other words, the same data payload consumes 2,500 percent more airtime when being delivered at the lower data rate.

FIGURE 13.33   Frame transmission time

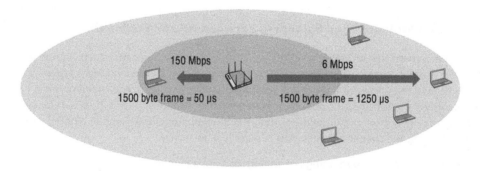

Earlier in this chapter, you learned about dynamic rate switching, which allows AP radios and clients to shift between data rates as a client moves farther away from an AP.

Clients and APs that shift down to lower data rates consume more airtime and affect the overall performance of the WLAN. Instead of a client shifting to a much lower data rate, the better scenario would be for the client to roam to another AP with a strong signal and continue with high data rate connectivity. Proper roaming design with primary and secondary coverage should address this issue.

An even more important reason to disable some of the lower data rates is to cut down on the airtime consumption from 802.11 management frames and control frames. In Chapter 9, "802.11 MAC," you learned that in order for a client station to successfully associate with an AP, the client must be capable of communicating with any of the configured *basic rates* that the AP requires. A basic rate configured on an AP is considered to be a "required" rate for all radios communicating within a BSS. Please understand that an AP will transmit all management frames and many control frames at the lowest configured basic rate. Data frames can be transmitted at much higher supported data rates.

For example, a 5 GHz radio of an AP will transmit all beacon frames and other control and management traffic at 6 Mbps if the radio's basic rate is configured for that rate. This consumes an enormous amount of airtime. Therefore, a common practice is to configure the basic rate of a 5 GHz radio on an AP at either 12 Mbps or, even better, 24 Mbps, as shown in Figure 13.34. Do not configure the AP radio basic rate at 18 Mbps, because some client drivers may not be able to interpret. The airtime consumption of management frames transmitted at 24 Mbps is 400 percent less than if transmitted at 6 Mbps.

FIGURE 13.34   Basic rates—5 GHz

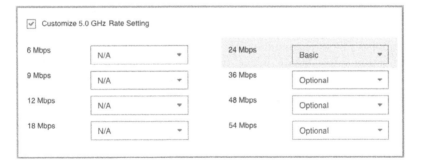

The same can be said for the basic rate configuration of any 2.4 GHz radio of an AP. A basic rate of either 12 Mbps or, even better, 24 Mbps will consume considerably less airtime for 802.11 management traffic. However, legacy 802.11b clients will not be able to connect. This is not necessarily a bad thing. 802.11b technology is more than 18 years old, and ideally all 802.11b clients have been replaced or eliminated a long time ago. In the real world, however, this is not always the case. If connectivity for 802.11b dinosaur radios is needed, the 2.4 GHz basic rate should be 11 Mbps, as shown in Figure 13.35.

FIGURE 13.35 Basic rates - 2.4 GHz

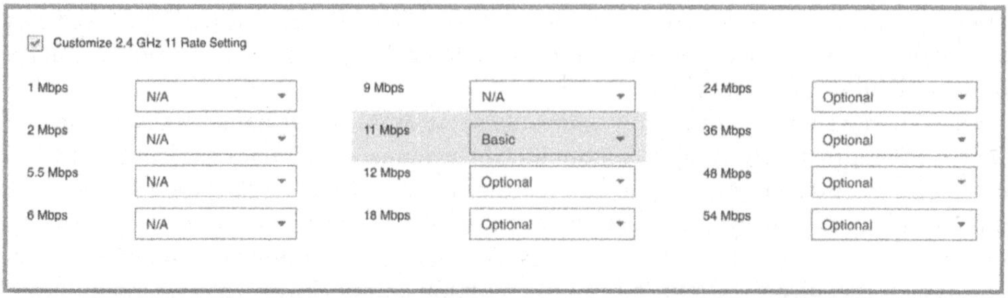

Disabling lower data rates and designating a higher basic rate will reduce airtime consumption. As described in Chapter 15, "WLAN Troubleshooting," a side benefit of this WLAN design practice is that sticky client roaming issues and hidden node problems will decrease.

Another WLAN design basic practice is to reduce the number of transmitted SSIDs per AP. Most WLAN vendors offer the capability of transmitting as many as 16 SSIDs per radio. The problem is that when an AP has been configured for multiple SSIDs, the AP will transmit beacon frames for each SSID. When client stations transmit probe request frames, the AP will also reply with multiple probe response frames. The airtime consumption overhead consequences of transmitting multiple beacons and probe responses are significant. In the early days of WLAN design, multiple SSIDs were needed and matched to unique user VLANs and IP subnets to segment traffic. As you will learn in Chapter 17, "802.11 Network Security Architecture," SSIDs can be consolidated to help eliminate overhead. Users can be associated to a single SSID and assigned to different VLANs, or assigned other access policies by leveraging RADIUS attributes.

Standard best practices dictate that no more than 3–4 SSIDs be broadcast. As shown in Figure 13.36, SSID overhead calculators can assist you in determining the airtime consumed by excessive beacon transmissions. You can download a free SSID overhead calculator at http://bit.ly/SSIDcalc.

The authors of this book recently attended a Certified Wireless Design Professional (CWDP) training class offered by Divergent Dynamics (www.divdyn.com). One of the exercises during the class required students to come up with as many unique ways as possible to limit airtime consumption. The WLAN design best practices discussed throughout this chapter were obviously correct answers for the exercise. However, depending on the individual WLAN deployment, the WLAN vendor, and even the WLAN professional, many other creative methods might be used to reduce airtime consumption. WLAN vendor configuration tweaks, such as probe suppression, broadcast traffic suppression, IPv6 suppression, and client isolation, may be appropriate in certain WLAN environments. Always remember that airtime is a precious commodity. Any method that can be used to reduce airtime consumption while still providing needed WLAN performance and mobility is highly encouraged.

FIGURE 13.36   SSID overhead calculator

Wi-Fi SSID Overhead Calculator

64.49%

It is likely that you are in a highly congested area
and will need to coordinate the WLAN

Data Rate	802.11b 1 Mbps	Select
Frame Size	380.0 Bytes	− +
Interval	102.4 ms	− +
CCI	2 APs	− +
SSIDs	10 SSIDs	− +

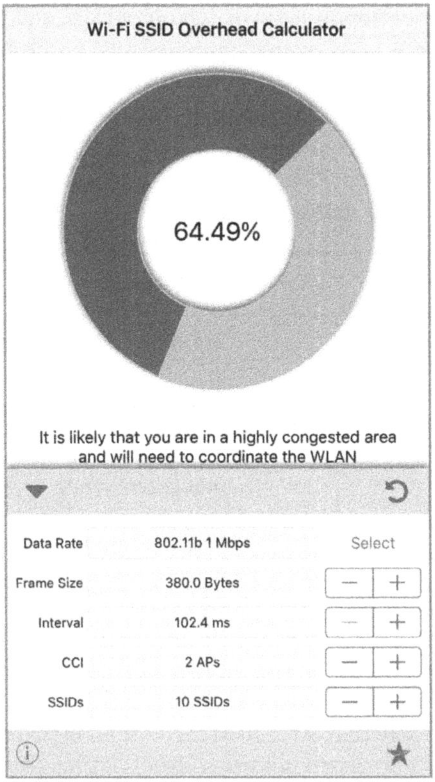

Voice vs. Data

As you have already learned, most data applications in a Wi-Fi network can handle a layer
2 retransmission rate of up to 10 percent without any noticeable degradation in performance. However, time-sensitive applications, such as VoIP, require that higher-layer IP
packet loss be no greater than 2 percent. Therefore, Voice over Wi-Fi (VoWiFi) networks
need to limit layer 2 retransmissions to 5 percent or less to guarantee the timely and consistent delivery of VoIP packets. When layer 2 retransmissions exceed 5 percent, latency
problems may develop and jitter problems will most likely surface. VoWiFi communications
are more susceptible to layer 2 retransmissions; therefore, when designing for voice-grade
WLANs, a −65 dBm or stronger signal is recommended so that the received signal is high
above the noise floor.

Many WLANs are initially designed to provide coverage only for data applications and not
for voice. Even in a poorly designed WLAN, enterprise data applications might still function,

although not optimally. Many companies that decide to add a VoWiFi solution to their WLAN at a later date quickly discover that the WLAN was not optimized for voice communications. The VoWiFi phones may have choppy audio or echo problems, and voice calls may even disconnect. Adding voice to the WLAN often exposes existing problems: Because data applications can withstand a much higher layer 2 retransmission rate, problems that existed within the WLAN may have gone unnoticed. As shown in Table 13.3, IP voice traffic is more susceptible to late or inconsistent packet delivery due to layer 2 retransmissions.

TABLE 13.3 IP voice and IP data comparison

IP Voice	IP Data
Small, uniform-size packets	Variable-size packets
Even, predictable delivery	Bursty delivery
Highly affected by late or inconsistent packet delivery	Minimally affected by late or inconsistent packet delivery
"Better never than late"	"Better late than never"

Optimizing the WLAN to support voice traffic will optimize the network for all wireless clients, including the clients running data applications other than voice. A proper WLAN design and validation site survey will reduce layer 2 retransmissions and provide an environment with the seamless coverage that is required for VoWiFi networks. All the possible causes of layer 2 retransmissions are discussed in further detail in Chapter 15. Because voice is so susceptible to the negative impact of layer 2 retransmissions, it is highly recommended to offer VoWiFi communication only on the 5 GHz frequency band and to avoid 2.4 GHz altogether.

Although voice traffic will usually be mixed with other application data traffic that traverses through the same access point, one standard practice is to segment the voice traffic on a separate SSID. The voice SSID may adjust several QoS settings and power-save settings to optimize for voice traffic. Always consult with the VoWiFi client vendor for recommended DTIM, U-APSD, and other per-SSID settings. WMM Admission Control settings may also be configured to designate the number of permitted active VoWiFi calls.

 Real World Scenario

How Many Simultaneous VoWiFi Calls Can an Access Point Support?

Several factors come into play, including available bandwidth, average use, and vendor specifics. When deploying over 5 GHz, WLAN vendor Cisco recommends a maximum of 27 simultaneous bidirectional voice calls when connected at 24 Mbps or higher. Because of medium contention, that number drops to a recommended maximum of 20 calls when connected at 12 Mbps. Different vendor-specific access point characteristics can also affect the number of concurrent calls, and extensive testing is recommended. Probability models also exist for predicting VoWiFi traffic. Not every Wi-Fi phone user will be making a call at the same time. Probabilistic traffic formulas use a telecommunications unit of measurement known as an *erlang*. An erlang is equal to one hour of telephone traffic in one hour of time. Some online VoWiFi erlang traffic calculators can be found at www.erlang.com.

Dual 5 GHz and Software-Defined Radios

Earlier in this chapter, you learned that in many high-density WLAN designs, disabling multiple 2.4 GHz radios in dual-frequency APs is often necessary to limit CCI in the 2.4 GHz band. One AP may be deployed per room to provide for adequate 5 GHz coverage and to meet capacity needs. However, 60–75 percent of the 2.4 GHz radios might be disabled, as shown in Figure 13.37. In reality, you really just want to disable the transmit capabilities of many of the 2.4 GHz radios. Enterprise APs usually have a radio sensor mode that effectively converts the radio to a listen-only state. This is useful for triangulation purposes when detecting rogue APs as well as any client location services that a WLAN vendor may provide.

FIGURE 13.37   2.4 GHz coverage vs. 5 GHz coverage

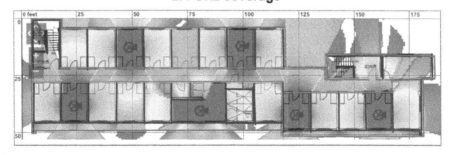

As previously mentioned, enterprise WLAN vendors use adaptive RF protocols to automatically assign channel and power settings for AP radios. Based on RF conditions, some of these protocols also have the ability to automatically disable a 2.4 GHz radio, as shown in Figure 13.38. As previously stated, changing the 2.4 GHz radio to sensor mode is preferred.

FIGURE 13.38   Disabling 2.4 GHz radio transmit capability

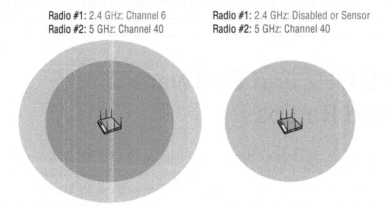

Radio #1: 2.4 GHz: Channel 6
Radio #2: 5 GHz: Channel 40

Radio #1: 2.4 GHz: Disabled or Sensor
Radio #2: 5 GHz: Channel 40

A newer trend with WLAN vendors is to offer a *software defined-radio (SDR)* along with a fixed 5 GHz radio within a dual-frequency AP. The radio that has SDR functionality can operate as either a 2.4 GHz or a 5 GHz radio. This means a dual-radio AP can either offer 2.4 GHz and 5 GHz coverage or offer coverage on two different 5 GHz channels, as shown in Figure 13.39. So what exactly are the advantages and WLAN implications of providing *dual 5 GHz* coverage from the same AP? Consider the high-density design we have discussed where 60–75 percent of the 2.4 GHz radios are disabled or converted into a sensor. An alternative would be to convert the bulk of those 2.4 GHz radios into 5 GHz radios. The SDR radios could be configured, either manually or automatically, to operate on 5 GHz channels. For example, suppose that 20 APs are deployed. The fixed 5 GHz radios in 20 of the APs would all be enabled. Additionally, 5 of the software-selectable radios would transmit on 2.4 GHz channels, while the remaining 15 SDRs would provide coverage on 5 GHz channels. The whole point behind dual 5 GHz coverage is to provide more capacity. A 2.4 GHz radio that has been disabled cannot service clients; however, an SDR operating as either a 2.4 GHz or 5 GHz radio can service clients and provide more airtime.

Although great for capacity needs, dual 5 GHz WLAN design is not without challenges. The main challenge is ensuring that the two radios within each AP do not interfere with each other when both AP radios are transmitting on 5 GHz channels. Enterprise WLAN vendors all use different hardware and software capabilities so that dual 5 GHz radios can operate within the same physical AP. Some vendors use expensive band-pass filters to prevent interference within the AP. Other vendors leverage smart antenna technology, with multiple antenna elements creating the required separation of the RF signals between the two, concurrently operating 5 GHz radios. One constant is that there needs to be some frequency separation between the two transmitting 5 GHz radios in the same AP. Vendor recommendations may require anywhere from 60 to 100 MHz of frequency separation between these two 5 GHz radios.

FIGURE 13.39   Dual 5 GHz coverage

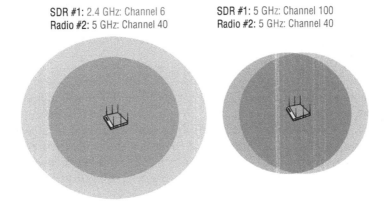

Regardless of whether the radios are configured manually or automatically, careful channel planning is needed with a dual 5 GHz design. Despite what some marketing literature might tell you, only use 20 MHz channels in a dual 5 GHz design. The use of 40 MHz channels in a dual 5 GHz design is usually not recommended. Because you will need as many 20 MHz channels as possible, enabling the DFS channels will almost always be required. Because of the needed channel separation within each AP, a channel pairing strategy is needed, as shown in Table 13.4. If possible, pairing non-DFS channels with DFS channels is a good strategy, just in case some clients do not support DFS.

TABLE 13.4   Dual 5 GHz AP channel pairings

AP	Channel Pairing	AP	Channel Pairing
AP #1	36/100	AP #5	149/116
AP #2	40/104	AP #6	153/132
AP #3	44/108	AP #7	157/136
AP #4	48/112	AP #8	161/140

Another issue to look out for when multiple dual 5 GHz APs are deployed is a greater potential for CCI. As previously mentioned, one of the advantages of 5 GHz is that CCI can usually be eliminated by a channel reuse pattern if all the channels are used. However, since up to twice the number of AP radios are using 5 GHz channels, the odds are greater that APs and clients on the same channel might hear each other. You would not want to have an AP with a 36/100 channel pairing within physical vicinity of another AP with a 100/48 pairing. The radios on channel 100 would hear each other and defer transmissions. As you can see, dual 5 GHz channel planning can be complex, and any RRM adaptive RF protocol also needs to factor in all these variables.

As client and user capacity needs continue to grow at a rocket pace, expect dual 5 GHz technology to get better and dual 5 GHz designs to become more commonplace.

Physical Environment

Always keep in mind that the physical environment of every building and floor is different. In the early days of Wi-Fi, coverage was the only concern, and the greater the effective range the better. The attenuation caused by walls usually had a negative impact if range was your primary goal. Now that capacity is of equal concern, wall attenuation is actually preferred. As shown in Figure 13.40, different materials cause different levels of attenuation when a Wi-Fi signal passes through the walls. For example, drywall attenuates a signal by about 3 dB, whereas concrete attenuates a signal by 12 dB.

FIGURE 13.40   Wall attenuation

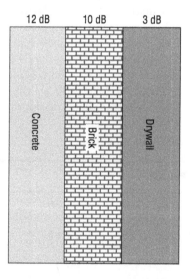

Why is wall attenuation good? In the high-density designs that are now commonplace, with APs placed in every room or every other room, wall attenuation helps reduce CCI for both 2.4 GHz and 5 GHz channels. You can use the attenuation properties of the walls to isolate contention domains and maximize channel reuse patterns. If all the interior walls are drywall, the one AP per room design discussed earlier will not be possible, even for the 5 GHz band, due to CCI. However, concrete or cinder block walls will attenuate a signal three times or more than drywall, making it easier to reduce CCI.

Where you mount an AP indoors is also very important. Most APs have internal omnidirectional antennas that provide 360 degrees of horizontal coverage but only a finite amount of vertical coverage. For this reason, these APs should never be mounted higher than 25 feet (8 meters) from the ground. If the ceilings are higher than 25 feet (for example, in an atrium or a warehouse), directional antennas will most likely be needed. Always check the recommendations of the WLAN vendor for AP mounting.

An important aspect of AP mounting is the "pretty factor." Many businesses prefer that all wireless hardware remain completely out of sight. Aesthetics is extremely important in retail stores, hospitals, and in hospitality industries (restaurants and hotels). However, WLAN customers often sacrifice good WLAN design in favor of aesthetics. For example, hospitals and hotels may insist that APs not be mounted in patient rooms or guest rooms. Have you ever wondered why the Wi-Fi at many hotels is so bad? The Wi-Fi is often bad because all the APs are mounted in a straight line throughout the hallway of the hotel, as opposed to inside the rooms. As shown in Figure 13.41, mounting all of the APs in the hallway is a common mistake because it causes a CCI nightmare. APs and clients on the same channel are guaranteed to hear each other despite great physical distance between the devices. Mounting APs in the hallway also usually will not provide adequate coverage in the rooms, especially hotels that have multi-room suites. Always mount APs inside rooms, where the bulk of the client devices reside, and use the attenuation aspects of the walls to your advantage.

FIGURE 13.41   Hallways are bad!

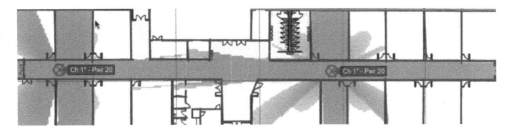

Antennas

WLAN design determines the proper placement of access points and power settings. The location of all the wiring closets should be noted, and care should be taken to ensure that the placement of any access point is within a 100-meter (328-foot) cable run back to the wiring closet because of copper Ethernet cabling distance limitations. Be sure to account for vertical cabling distances as well as horizontal runs.

Another often-overlooked component in WLAN design is the use of directional antennas. Many WLAN deployments use only the manufacturer's default low-gain omnidirectional antenna, which typically has about 2 to 5 dBi of gain. Buildings come in many shapes and sizes and often have long corridors or hallways, where the coverage of an indoor directional antenna may be much more advantageous. It is common for patch antennas to be connected to access points to provide directional coverage within a building. Because Omnidirectional antennas often have difficulty providing effective RF coverage in areas with shelving. MIMO patch antennas, such as the one shown in Figure 13.42, can be used effectively in libraries, warehouses, and retail stores with long aisles of shelves.

FIGURE 13.42   MIMO patch antenna

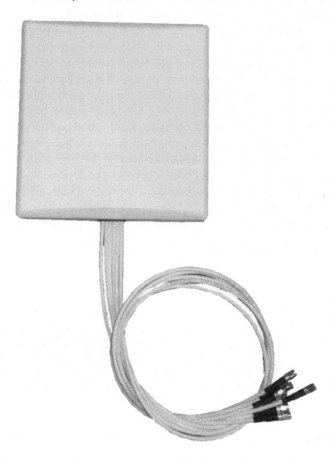

Figure 13.43 depicts the use of directional antennas in a warehouse with long corridors and metal racks that line the corridors. Patch antennas are mounted on the walls and pointed through the aisles and metal racks to provide coverage. Note that the patch antennas are staggered on the opposite sides of the building.

Coverage, not capacity, is usually the main concern in a warehouse environment. The client devices are usually handheld barcode scanners or other wireless data-collection devices that are used for inventory management. VoWiFi is also common in many warehouse WLAN deployments. Because most warehouses have very high ceilings, coverage is primarily provided with directional antennas mounted on the walls and pointing down the aisles. However, because many aisles are very long, directional antennas are often also mounted from the ceiling. As shown in Figure 13.44, the ceiling-mounted directional antennas are mounted in the center of the aisles to provide coverage in combination with the directional antennas mounted on the walls.

FIGURE 13.43   Warehouse WLAN—directional antennas

FIGURE 13.44   Warehouse WLAN—ceiling and wall mount coverage

Another common use case for deploying MIMO patch antennas indoors is in very high-density (VHD) environments. An example might be a school gymnasium or a meeting hall packed with people using multiple Wi-Fi radios. In a VHD scenario, an omnidirectional antenna is not the best solution for coverage, because multiple APs are deployed in the same open area. MIMO patch antennas are often mounted from the wall or ceiling to provide tight "sectors" of coverage. An even more effective method would be to mount the patch antennas below the floor or under seating or bleachers. The MIMO patch antennas point upward, and the attenuation properties of the people are also taken into account. Designing in auditoriums, gymnasiums, and other VHD environments can be quite difficult. Most WLAN vendors have specific recommendations in their VHD deployment guides. Whenever directional antennas are used, most WLAN vendors will recommend using static channel and power settings.

WLAN professionals with years of experience usually use a variety of antennas, both directional and omnidirectional, depending on the building and the WLAN requirements. Another example of a different kind of antenna is Ventev's directional Junction Box antenna, as shown in Figure 13.45. This antenna has a 75-degree horizontal and vertical beamwidth and is useful for the deployment of multiple APs, providing concentrated coverage cells in small areas. The use of directional antennas reduces CCI, especially when a 40 MHz channel reuse pattern is deployed.

FIGURE 13.45   Directional Junction Box antenna

Courtesy of Ventev

This directional antenna can articulate on one axis from 0 to 45 degrees, making it ideal for lecture halls, conference centers, large boardrooms, and any other high-density environments with close proximity designs. The Junction Box antenna has a 6-foot antenna cable lead, which alleviates installation headaches by allowing for the access point to be placed away from the mounting location of the antenna. The physical dimensions of the Ventev Junction Box antenna are large enough that some APs can also fit inside of the junction box along with the antenna elements. In regards to aesthetics, the Junction Box antenna's cover plate visually disappears into any interior design scheme. The antenna can be mounted into hard ceiling or in drop ceiling tile. The articulation options of this antenna allow it to be mounted parallel or perpendicular to the floor, and horizontally or vertically mounted on a wall.

Other WLAN environments that might also require specialty antennas include stadiums and arenas. As mentioned earlier, ultra high-density design requires WLAN professionals with expertise with stadium Wi-Fi. Several WLAN vendors manufacture specialty stadium APs that are meant to be mounted under the stadium seating. MIMO patch antennas are also designed to be mounted on posts and railings.

Cellular LTE coverage for stadiums and arenas is often planned by the same integrator that is creating the Wi-Fi design. Some predictive modeling solutions, such as iBwave Design, offer the capabilities to design for both technologies. As shown in Figure 13.46, premiere predictive modeling software offers both two- and three-dimensional views of RF coverage.

FIGURE 13.46   Stadium coverage - 3D view

Courtesy of iBwave

Outdoor Design

A whole separate chapter could be written about outdoor Wi-Fi design and deployment. Outdoor Wi-Fi is often used for coverage in areas such as parking lots, parks, marinas, etc. Outdoor coverage is often planned for the surrounding areas of buildings that also offer indoor coverage. User density is usually not as big of a concern when designing for outdoor WLAN coverage. Outdoor WLAN deployments often require the use of a mesh WLAN because Ethernet cabling may not be readily available in the outdoor coverage area. A mesh point AP needs to be able to connect via a backhaul link provided by a mesh portal that is typically mounted to a building with wired access.

The other use case for outdoor Wi-Fi deployment is for point-to-point wireless bridge links between buildings. WLAN bridging requires many calculations, such as Fresnel zone, free space path loss, link budget, and fade margin. More information about the challenges and aspects for both outdoor mesh and WLAN bridge links is discussed in multiple chapters in this book. When deploying a Wi-Fi mesh network outdoors or perhaps an outdoor bridge link, a WLAN administrator must take into account the adverse affect of weather conditions. The following weather conditions must be considered:

Lightning   Direct and indirect lightning strikes can damage WLAN equipment. Lightning arrestors should be used for protection against transient currents. Solutions such as lightning rods or copper/fiber transceivers may offer protection against lightning strikes.

Wind   Because of the long distances and narrow beamwidths, highly directional antennas are susceptible to movement or shifting caused by wind. Even slight movement of a highly directional antenna can cause the RF beam to be aimed away from the receiving antenna, interrupting the communications. In high-wind environments, a grid antenna will typically remain more stable than a parabolic dish. Other mounting options may be necessary to stabilize the antenna from movement.

Water   Conditions such as rain, snow, and fog present two unique challenges. First, all outdoor equipment must be protected from damage caused by exposure to water. Water damage is often a serious problem with cabling and connectors. Connectors should be protected with drip loops and coax seals to prevent water seepage and damage. Cables and connectors should be checked on a regular basis for damage. A radome (weatherproof protective cover) should be used to protect antennas from water damage or snow buildup.

Outdoor bridges, access points, and mesh APs should be protected from the weather elements by using appropriate National Electrical Manufacturers Association (NEMA) enclosure units. Precipitation can also cause an RF signal to attenuate. A torrential downpour can attenuate a signal as much as 0.08 dB per mile (0.05 dB per kilometer) in both the 2.4 GHz and 5 GHz frequency ranges. Over long-distance bridge links, a system operating margin (SOM) of 20 dB is usually recommended to compensate for attenuation due to rain, fog, or snow.

UV/Sun   UV rays and ambient heat from rooftops can damage cables over time if proper cable types are not used.

Summary

This chapter discussed WLAN coverage, capacity, and integration design aspects from a conceptual basis. WLAN professionals do not always agree about WLAN design and may have their own unique approach. The recommendations in this chapter are based on years of experience from many highly skilled WLAN design specialists. However, there are always different WLAN design strategies that may be successful. We discussed proper WLAN design for both high data rate connectivity and voice-grade Wi-Fi. We discussed the importance of a –70 dBm or stronger received signal and an SNR of 25 dB or higher. Always remember that proper coverage design is based on the perspective of WLAN clients. We discussed the concept of dynamic rate switching and the many aspects of roaming design.

We also discussed both adjacent channel interference and co-channel interference. Much of this chapter discussed channel reuse design strategies for both the 2.4 and 5 GHz bands. 20 MHz channels are usually preferred; however, a 40 MHz channel reuse design can work, with the recommended caveats discussed in this chapter. Designing for coverage is only part of the equation. Just as important is proper capacity planning. Almost all WLANs have high-density user requirements; therefore, planning on ways to reduce airtime consumption is imperative. We also discussed the DFS channels of the 5 GHz band and how they are usually needed to meet high-density capacity needs. Dual 5 GHz design is becoming a common WLAN design option for capacity, and very often directional antennas are needed to provide sectors of coverage to reduce CCI.

Always remember that the main goal of a WLAN is to provide mobility and wireless access to network resources. For this reason, integration planning is also a major component of proper WLAN design. Integration aspects not discussed in this chapter include VLAN design, WLAN security, access control, client onboarding, guest access, and PoE. All of these integration aspects are discussed at length in other chapters of the book.

Exam Essentials

Define dynamic rate switching.   Understand the process of stations shifting between data rates based on RSSI and SNR. Understand that AP radios also shift between data rates.

Explain the various aspects of roaming.   Understand that clients make the roaming decision. Know that client RSSI, SNR, and other metrics are used for roaming trigger thresholds. Understand the importance of primary and secondary coverage. Describe latency issues that can occur with roaming. Understand why crossing layer 3 boundaries can cause problems and the solutions that exist.

Define the differences between adjacent channel interference and co-channel interference.   Understand the negative effects of both adjacent channel interference (ACI) and co-channel interference (CCI). Explain why channel reuse patterns minimize the problems.

Understand the importance of channel reuse.   Explain why channel reuse patterns are necessary to minimize CCI and eliminate ACI. Understand that CCI is usually impossible to prevent in the 2.4 GHz band and that clients are the main cause of CCI.

Explain strategies to reduce airtime consumption.   Explain the benefits of disabling lower data rates and defining a higher basic rate, such as 12 Mbps or 24 Mbps.

Explain when to use directional antennas.   Understand the importance of using directional antennas for high-density environments. Explain how MIMO patch antennas can help reduce contention domains by providing indoor sector coverage.

Review Questions

1. How long do all Wi-Fi radios have to exit a dynamic frequency selection (DFS) channel when a radar pulse is detected?

    A. 10 seconds

    B. 30 seconds

    C. 60 seconds

    D. 30 minutes

    E. 60 minutes

2. What WLAN client connectivity may be negatively impacted by client load balancing between access points?

    A. Capacity

    B. Range

    C. Roaming

    D. Throughput

    E. Security

3. What are some potential problems when enabling 40 MHz channels in the 5 GHz band? (Choose all that apply.)

    A. Adjacent channel interference

    B. Co-channel interference

    C. Higher SNR

    D. Lower SNR

    E. Decreased range

4. What are some recommendations when using 40 MHz channels in the 5 GHz band? (Choose all that apply.)

    A. Enable the DFS channels

    B. Lower AP transmit power

    C. Verify wall attenuation

    D. All of the above

5. How many client devices can successfully connect and communicate to an AP?

    A. 35

    B. 50

    C. 100

    D. 250

    E. It depends.

6. What are some questions that should be considered during WLAN client capacity planning?

    A. How many users and devices need access currently?

    B. How many users and devices will need access in the future?

    C. Where are the users and devices located?

    D. What are the MIMO capabilities of the client devices?

    E. What type of applications will be used for the WLAN?

    F. All of the above

7. Which is the preferred channel and power configuration method for enterprise WLAN APs? (Choose all that apply.)

    A. Adaptive RF in standard WLAN environments

    B. Static channel and power settings in standard WLAN environments

    C. Adaptive RF in complex RF environments using idirectional antennas

    D. Static channel and power settings in complex RF environments using directional antennas

8. What is the biggest problem when enabling 5 GHz dynamic frequency selection (DFS) channels?

    A. Channel switch announcements

    B. False positives

    C. Co-channel interference

    D. Adjacent channel interference

    E. The 60-second wait time

9. To reduce airtime consumption and provide for better capacity in the 5 GHz band, which data rates are recommended to be selected as basic rates? (Choose all that apply.)

    A. 6 Mbps

    B. 9 Mbps

    C. 12 Mbps

    D. 18 Mbps

    E. 24 Mbps

10. How long must an AP listen on a DFS channel before transmitting?

    A. 10 seconds

    B. 30 seconds

    C. 60 seconds

    D. 30 minutes

    E. 60 minutes

11. What is the main cause of co-channel interference?

    A. Microwave interference on the same channel within hearing range

    B. APs on the same channel within hearing range

    C. APs on a different channel within hearing range

    D. Clients on the same channel within hearing range

    E. Clients on a different channel within hearing range

12. What percentage of cell overlap is needed to provide for seamless roaming?

    A. 10 percent

    B. 15 percent

    C. 20 percent

    D. 25 percent

    E. This is a trick question.

13. How many channels should be used in a 5 GHz channel reuse pattern of 20 MHz channels?

    A. 3

    B. 4

    C. 8

    D. 12

    E. As many as available

14. What is the recommended received signal and SNR when providing for VoWiFi client communications?

    A. −70 dBm and 20 dB

    B. −70 dBm and 25 dB

    C. −70 dBm and 15 dB

    D. −65 dBm and 15 dB

    E. −65 dBm and 25 dB

15. What type of interference is a result of APs on the same channel or clients on the same channel hearing each other despite being members of different basic service sets?

    A. Intersymbol interference

    B. Adjacent channel interference

    C. All-band interference

    D. Narrowband interference

    E. Co-channel interference

16. What type of interference is caused by overlapping coverage cells with overlapping frequencies?

    A. Intersymbol interference

    B. Adjacent channel interference

C. All-band interference

D. Narrowband interference

E. Co-channel interference

17. How many channels should be used in a channel reuse plan for the 2.4 GHz frequency band?

A. 3

B. 4

C. 6

D. 11

E. 13

18. Based on RSSI metrics, concentric zones of variable data rate coverage exist around an access point due to the upshifting and downshifting of client stations between data rates. What is the correct name of this process, according to the IEEE 802.11-2016 standard?

A. Dynamic rate shifting

B. Dynamic rate switching

C. Automatic rate selection

D. Adaptive rate selection

E. All of the above

19. Given: Wi-Fi clients can roam seamlessly at layer 2 if all the APs are configured with the same SSID and same security settings. However, if clients cross layer 3 boundaries, a layer 3 roaming solution will be needed. Which device functions as the home agent if a Mobile IP solution has been implemented in an enterprise WLAN environment where no WLAN controller is deployed?

A. Wireless network management server (WNMS)

B. Access layer switch

C. Layer 3 switch

D. Access point on the original subnet

E. Access point on the new subnet

20. What is a telecommunications unit of measurement of traffic equal to one hour of telephone traffic in one hour of time?

A. Ohm

B. dBm

C. Erlang

D. Call hour

E. Voltage standing wave ratio

Chapter

14

Site Survey and Validation

IN THIS CHAPTER, YOU WILL LEARN ABOUT THE FOLLOWING:

✓ **WLAN site survey and design interview**

- Customer briefing
- Business requirements
- Capacity and coverage requirements
- Existing wireless network
- Upgrading an existing WLAN
- Infrastructure connectivity
- Security expectations
- Guest access
- Aesthetics
- Outdoor surveys

✓ **Vertical market considerations**

- Government
- Education
- Healthcare
- Retail
- Warehouses and manufacturing
- Multi-tenant buildings

✓ **Legacy AP-on-a stick survey**

- Spectrum analysis
- Coverage analysis

✓ **Hybrid survey**

- Initial site visit
- Predictive design

✓ **Validation survey**

- Capacity and throughput
- Roaming
- Delay and jitter
- Connectivity
- Aesthetics

✓ **Site survey tools**

- Indoor site survey tools
- Outdoor site survey tools

✓ **Documents and reports**

- Forms and customer documentation
- Deliverables
- Additional reports

When most people are asked to define a wireless site survey, the usual response is that a site survey is for determining RF coverage. In the early days of wireless networking, when there were far fewer wireless client stations connecting to the wireless network, that definition was absolutely correct. However, not only do present-day wireless networks need to provide the coverage that was sought when early site surveys were performed, but they often also need to provide higher throughput for denser deployments of stations. To achieve these goals, the site survey must encompass much more than just determining coverage; including looking for potential sources of interference as well as the proper placement, installation, and configuration of 802.11 hardware and related components.

Depending on whom you speak to, the definition of a WLAN site survey might be different. Site survey professionals often have their own unique technical approach for executing a site survey. Some prefer an AP-on-a-stick method, whereas others prefer a hybrid method that involves predictive modeling. Whichever method is used, it should be understood that the site survey process is fully interwoven with planned WLAN design. Proper WLAN design cannot be achieved unless an extensive interview is conducted to predetermine WLAN requirements and expectations.

One component that all site survey professionals will agree upon is the importance of a validation survey. Coverage verification, capacity performance, and roaming testing are key components of a proper validation survey. Depending on the purpose of the wireless network, different tools can be used to assist with the site survey.

In this chapter, you will learn about the site survey components and process. This chapter explains the client interview, necessary documentation, and considerations that may need to be addressed for specific vertical markets. Much preparation must take place before the WLAN site survey is conducted. The needs of the WLAN must be predetermined and the proper questions must be asked. The chapter will then proceed to explain the procedures and tools to perform the necessary tasks, explaining both the legacy AP-on-a-stick method along with the now more common hybrid method. Lastly, you will learn about the deliverable client documentation and reports.

WLAN Site Survey and Design Interview

Is a site survey even needed? The answer to that question is a resounding *yes*. If an owner of a small retail flower shop desires a wireless network, the site survey that is conducted may be as simple as placing a small office, home office (SOHO) Wi-Fi router in the middle of the shop, turning the transmit power to a lower setting, and making sure you have connectivity. Performing a site survey in a medium or large business entails much more physical work and time. Before the actual survey is conducted, a proper *site survey and design interview* should occur to both educate the customer and properly determine their needs.

Whether you are performing an AP-on-a-stick survey or a hybrid survey, it is imperative to perform an initial interview with the stakeholders to determine the objectives, requirements, and goals of the WLAN. Different vertical markets have different requirements, and WLAN concerns will likely be unique on a customer-by-customer basis.

The interview process may consist of a single meeting (in the case of a small customer) to many in-depth meetings (in the case of a large or complex customer). During this process you need to learn about their existing network and WLAN environment, their security requirements, their application needs, their current and planned mobile devices, and their objectives and requirements. This is one of the key components of designing the network: working with the client to identify the requirements for the network. It is critical to identify and document these requirements, as these will be the key deliverables for the network, and what you will use to verify and validate the network upon delivery of the network and during the validation survey.

In addition to collecting information about the customer and their network, you also need to have the customer provide you with electronic scaled floor plans of all the buildings and possible outdoor space that will require RF coverage. Notes need to be made regarding any specific or unique requests or requirements on any of the floor plans, including aesthetic or mounting restrictions. You will also need to schedule time and access (special security clearance, approval, or attire may be needed) so that you can either perform an AP-on-a-stick survey or to collect RF data that will be used to perform a hybrid survey with a predictive design.

Asking the correct questions during a site survey and design interview not only ensures that the proper tools are used during the survey, but it also makes the survey more productive. Most important, the end result of a thorough interview and thorough survey will be a properly designed WLAN that meets all the intended mobility, coverage, and capacity needs. The following sections cover the questions that should be carefully considered during the site survey interview.

Customer Briefing

Even though 802.11 technologies have been around since 1997, much misunderstanding and misinformation about wireless networking still exists. Because many businesses and individuals are familiar with Ethernet networks, a "just plug it in and turn it on" mentality is prevalent. If a wireless network is being planned for your company or for a prospective client, it is highly recommended that you sit down with management, give them an overview of 802.11 wireless networking, and talk with them about how and why site surveys are conducted. You do not need to explain the inner workings of MIMO or CSMA/CA; however, a conversation about the advantages of Wi-Fi, as well as the limitations of a WLAN, is a good idea.

There is a good chance that the company already has a WLAN and the customer briefing is about an upgrade to the existing WLAN. A brief explanation about the advantages of mobility would be an excellent start for a customer that is looking to deploy Wi-Fi for the very first time.

A discussion about the bandwidth and throughput capabilities and expectations of 802.11 technology is also very important. Enterprise users are accustomed to 1 Gbps full-duplex speeds on the wired network. Because of vendor hype and marketing, people often believe that a Wi-Fi network will provide them with similar or even better bandwidth and throughput. Management will need to be educated that because of many factors, the aggregate throughput of a WLAN is 50 percent or less of the advertised data rate. It should also be explained that the medium is a half-duplex shared medium and not full-duplex. The average customer usually has many misconceptions regarding WLAN bandwidth versus actual throughput.

Another appropriate discussion is why a site survey and WLAN design is needed. A very brief explanation on how RF signals propagate and attenuate will provide management with a better understanding of why an RF site survey is needed to ensure the proper coverage and enhance performance. A discussion and comparison of a 2.4 GHz vs. a 5 GHz WLAN might also be necessary. If management is properly briefed on the basics of Wi-Fi as well as the importance of a site survey, the forthcoming technical questions will be answered in a more suitable fashion.

Business Requirements

The first question that should be posed is, "What is the purpose of the WLAN?" If you have a complete understanding of the intended use of a wireless network, the result will be a better-designed WLAN. For example, a VoWiFi network has different requirements than a heavily used data network. If the purpose of the WLAN is only to provide users a gateway to the Internet, security and integration recommendations will be different. A warehouse environment with 200 handheld scanners is very different from an office environment. A hospital's wireless network will have different business requirements than an airport's wireless network. Here are some of the business requirement questions that should be asked:

What applications will be used over the WLAN?   This question could have both capacity and quality-of-service (QoS) implications. A wireless network for graphic designers moving huge graphics files across the WLAN would obviously need more bandwidth than a wireless network for nothing but wireless bar code scanners. If time-sensitive applications such as voice or video are required, wired QoS design may need to be addressed for proper integration with 802.11e/WMM wireless QoS capabilities.

Who will be using the WLAN?   Different types of users have different capacity and performance needs. Users may also need to be separated for organizational purposes. Groups of users might be segmented into separate SSIDs and VLANs or even segmented by different frequencies. This is also an important consideration for security roles.

What types of devices will be connecting to the WLAN?   Are the bulk of the devices laptops or handheld mobile devices? What are the MIMO capabilities of the devices? Will employees be allowed to connect their personal devices to the network? Does the company have a *bring your own device (BYOD)* strategy, and is a *mobile device management (MDM)* solution needed? Handheld wireless barcode scanners may also be segmented into separate VLANs or by frequency. VoWiFi phones are always put in a separate VLAN than data users with laptops. What types of legacy devices are currently deployed? If the WLAN infrastructure is being upgraded, this is a good time to also upgrade the client population. Will wireless IoT devices be deployed? Many Wi-Fi IoT devices can only transmit on the 2.4 GHz band. The capabilities of the devices may force decisions in security, frequency, technology, and overall WLAN design.

Are there any aesthetic or mounting restrictions?   Aesthetics is extremely important in many verticals, including retail, hotels, and hospitals. There may be restrictions as to where an AP can be physically mounted, and the restrictions may not be conducive to good WLAN design. For example, hospitals often prefer that APs not be mounted in guest and patient rooms but instead be mounted in hallways. As you learned in Chapter 13, "WLAN Design Concepts" mounting APs in a hallway is not a good idea in most cases. Compromises often must be made regarding where APs can be installed.

We discuss the varying business requirements of different vertical markets later in this chapter. Defining the purpose of the WLAN in advance will lead to a more productive site survey and is imperative to the eventual design of the WLAN.

Capacity and Coverage Requirements

After the purpose of the WLAN has been clearly defined, the next step is to begin asking all the necessary questions for planning and designing the wireless network. Although the final design of a WLAN is completed after the site survey is conducted, some preliminary design based on the *capacity* and *coverage* needs of the customer is recommended. You will need to sit down with a copy of the building's floor plan and ask the customer where they want RF coverage. The answer will almost always be everywhere. If a VoWiFi deployment is planned, that answer is probably legitimate, as VoWiFi phones will need mobility and connectivity throughout the building. Furthermore, because of the proliferation of handheld mobile devices , broad coverage is usually a necessity.

However, the need for blanket coverage might not be necessary. Do laptop data users need access in a storage area? Do they need connectivity in the outdoor courtyard? Do handheld bar code scanners used in a warehouse area need access in the front office? The answer to these questions may vary depending on the earlier questions that were asked regarding the purpose of the WLAN. If you can determine that certain areas of the facility do not require coverage, you will save the customer money and yourself time when conducting the physical survey.

Considering the Proliferation of Wireless Devices

The Wi-Fi client population explosion usually dictates that RF coverage be wide-ranging throughout most buildings and locations. Initially, Wi-Fi networks were mostly used to provide access to laptop users. In recent years, an unbridled growth in mobile devices with Wi-Fi radios has occurred.

Wi-Fi radios are now a common component in most smartphones, tablets, scanners, and many other mobile devices. Although mobile devices initially were intended for personal use, most employees now use them in the corporate workplace as well. Employees now have expectations of being able to connect to a corporate WLAN with multiple personal mobile devices. Because of the proliferation of personal mobile devices, a bring your own device (BYOD) policy is needed to define how employees' personal devices may access the corporate WLAN. A mobile device management (MDM) solution might also be needed for onboarding both personal mobile devices and company-issued devices (CIDs) onto the WLAN. Chapter 18, "Bring Your Own Device and Guest Access," discusses BYOD strategies and MDM solutions in great detail.

The Wi-Fi industry is currently on the cusp of an expansion of IoT devices that will likely dwarf what has occurred in the past. Companies are embedding Wi-Fi, Bluetooth and Zigbee radios in just about any type of electronic or electrical device, such as HVAC, refrigerators, lighting, and just about any sensor or monitoring device they currently have or can think of. Data collection and remote device management will continue to drive growth of the IoT market. Although IoT devices will have limited user interaction, they can provide for either continuous or periodic data collection that can be used for analytic purposes.

Depending on the layout and materials used inside the building, some preplanning might be necessary regarding the type of antennas to use in certain areas of the facility. A high-density area may require semidirectional patch antennas for sectorized coverage, as opposed to using omnidirectional antennas. When the survey is performed, this will be confirmed or adjusted accordingly.

The most often neglected aspect prior to the site survey is determining capacity needs of the WLAN. As you learned in Chapter 13, "WLAN Design Concepts," you must not consider coverage alone; you must also plan for client capacity. The need for high-density WLAN design is now commonplace. In order for the wireless end user to experience acceptable performance, a ratio of average number of users per access point must be established. The answer to the capacity question depends on a host of variables, including answers to earlier questions about the purpose of the WLAN. Capacity will not be as big of a concern in a warehouse environment using mostly handheld data scanners. However, if the WLAN has average-to-heavy data requirements, capacity will absolutely be a concern. The following are among the many factors to consider when planning for capacity:

- Data applications
- Number of users and devices
- Client device capabilities
- Peak on/off usage
- Backward compatibility for legacy devices

Carefully planning coverage and capacity during the WLAN design phase will help you determine AP placement and power settings, types of antennas, and coverage zones. The physical site survey will still have to be conducted to validate and further determine coverage and capacity requirements.

Existing Wireless Network

Quite often the reason you are conducting a WLAN site survey is that you have been called in as a consultant to fix an existing deployment. Professional site survey companies are often hired to troubleshoot existing WLANs, which often requires conducting a second site survey or discovering that one was never conducted to begin with.

As more corporations and individuals become educated in 802.11 technologies, the percentage will obviously drop. Sadly, many untrained integrators or customers just install the access points wherever they can mount them and leave the default power and channel settings on every AP. A diagnostic survey will be conducted either because of performance problems or difficulty roaming. Performance problems are often caused by RF interference, low SNR, adjacent cell interference, or co-channel interference. Roaming problems may also be related to interference or caused by a lack of adequate coverage and/or by a lack of primary/secondary cell coverage for roaming. Here are some of the questions that should be asked prior to the reparative site survey:

What are the current problems with the existing WLAN?   Ask the customer to clarify the problems. Are they related to throughput? Are there frequent disconnections? Is there any difficulty roaming? In what part of the building do the problems occur most often? Are the problems happening with one WLAN device or multiple devices? How often do the problems occur, and have any steps been taken to duplicate the issues?

Are there any known sources of RF interference?   More than likely the customer will have no idea, but it does not hurt to ask. Are there any microwave ovens? Do people use cordless phones or headsets? Does anyone use Bluetooth for keyboards or mice? After asking these interference questions, you should always perform a spectrum analysis, which is the *only* way to determine whether there is any RF interference in the area that may inhibit future transmissions.

Are there any known coverage dead zones?   This is related to the roaming questions, and areas probably exist where proper coverage is not being provided. Remember, this could be too little or too much coverage. Both create roaming and connectivity problems.

Does a prior site survey and WLAN design data exist?   It is possible that an original site survey was not even conducted. If old site survey and WLAN design documentation exists, it may be helpful when troubleshooting existing problems. It is important to note that unless quantifiable data was collected that shows dBm strengths, the survey report should be viewed with extreme caution. Additionally, changes may have been made to the network design since the original survey was performed.

What equipment is currently installed?   Ask what type of equipment is being used, such as 802.11ac (5 GHz) or 802.11b/g/n (2.4 GHz), and which vendor has been used. Is the customer looking to upgrade to an 802.11n/ac or 802.11ax network? Again, the customer might have no idea, and it will be your job to determine what has been installed and why it is not working properly. Also check the configurations of the devices, including service set identifiers (SSIDs), WEP or WPA keys, channels, power levels, and firmware versions. Often issues can be as simple as all the access points are transmitting on the same channel or there is a buffer issue that can be resolved with the latest firmware update.

Depending on the level of troubleshooting that is required on the existing wireless network, a second site survey consisting of coverage and spectrum analysis will often be necessary. After the new site survey has been conducted, adjustments to the existing WLAN equipment may be adequate. However, the worst-case scenario would involve a complete redesign of the WLAN. Keep in mind that whenever a second site survey is necessary, all the same questions that are asked as part of a survey for a new installation (Greenfield survey) should also be asked prior to the second site survey. If wireless usage requirements have changed, a redesign might be the best course of action. For more information about Wi-Fi problems and diagnostics, please read Chapter 15, "WLAN Troubleshooting."

Upgrading an Existing WLAN

Upgrading an existing wireless network is a task that needs to be evaluated carefully. A company may consider installing new APs in the same location of the existing APs. This is usually not recommended. The RF coverage and patterns of the new APs will be different from the existing equipment. It is important to state this to the customer and make them realize the importance of performing a new site survey and WLAN redesign.

Different companies use different strategies when performing WLAN upgrades. Some companies may use a salt-and-pepper design approach, where APs in only certain areas of a building are upgraded first. For example, one area of a building might have a higher density of users and clients, so the AP upgrades would be focused on those areas first. Another approach would be to upgrade one entire building with new APs, to test the new 802.11 technology in a live enterprise environment. If the technology is solid, then upgrades to all other buildings and locations could proceed.

Typically, WLAN upgrade cycles in the enterprise are every 4–5 years. As mentioned several times in this book, WLAN customers usually upgrade their APs and WLAN infrastructure, but often fail to upgrade the client population. To take full advantage of new 802.11 technology, upgrading the client population is essential. Legacy clients can have a negative impact on overall WLAN performance.

Infrastructure Connectivity

You have already learned that the usual purposes of a WLAN are to provide client mobility and to provide access via an AP into a pre-existing wired network infrastructure. Part of the interview process includes asking the correct questions so that the WLAN will integrate properly into the existing wired architecture. Asking for a copy of the wired network physical and logical topology maps is highly recommended.

For security reasons, the customer may not want to disclose the wired topology, and you may need to sign a nondisclosure agreement. It is a good idea to request that an agreement be signed to protect you legally as the integrator. Be sure that someone in your organization with the authority to sign finalizes the agreement.

Understanding the existing topology will also be of help when planning WLAN segmentation and security proposals and recommendations. With or without a topology map, the following topics are important to ensure the desired infrastructure connectivity:

Roaming   Is roaming required? In most cases, the answer will be *yes*, because mobility is a key advantage of wireless networking. Any devices that run connection-oriented applications will need seamless roaming. Seamless roaming is mandatory if handheld devices and/or VoWiFi phones are deployed. Most smartphone and tablet users expect mobility. Providing for secure seamless roaming is pretty much always a requirement.

It should also be understood that there might be certain areas where the WLAN was designed with roaming as a very low priority, such as areas with a high density of users. For example, a gymnasium filled with 800 people might have APs on the ceiling with MIMO patch antennas to provide for unidirectional sectorized coverage. This is a WLAN design with high density as the priority, as opposed to mobility and roaming.

Another important roaming consideration is whether users will need to roam across layer 3 boundaries. A Mobile IP solution or a proprietary layer 3 roaming solution will be needed if client stations need to roam across subnets. Special consideration has to be given to roaming with VoWiFi devices because of the issues that can arise from network latency.

Wiring Closets   Where are the wiring closets located? Will the locations that are being considered for AP installations be within a 100-meter (328-foot) cable drop from the wiring closets?

Antenna Structure   If an outdoor network or point-to-point bridging application is requested, some additional structure might have to be built to mount the antennas. Asking for building diagrams of the roof to locate structural beams and existing roof penetrations is a good idea. You may also need permission to install equipment on the roof or mount it to the building. Depending on the weight of the installation, you may also need to consult a structural engineer.

Switches   Will the access points be connected by category 6 (CAT6) cabling to unmanaged switches or managed switches? CAT5e or higher grade cabling is needed for 802.3at PoE. An unmanaged switch will support only a single VLAN. A managed switch will be needed if multiple VLANs are required. Are there enough switch ports? What is the power budget of the switch? Who will be responsible for configuring the VLANs? What port speeds are supported? Modern switch ports support 1,000 Mbps; however, older switches may still only have 100 Mbps ports. 100 Mbps is not enough bandwidth for uplink traffic from an AP. Do the switches support 2.5 Gbps for switch ports for future-proofing purposes?

PoE How will the access points be powered? Because APs are typically mounted to the ceiling, Power over Ethernet (PoE) will likely be required to remotely power the access points. The customer may not yet have a PoE solution in place, so further investment will be needed. If the customer already has a PoE solution installed, it must be determined whether the PoE solution is compliant with 802.3af or 802.3at (PoE Plus). Also, is the solution an endpoint or midspan solution? If the customer is migrating toward an 802.11ax deployment of access points, more power will be needed, and 802.3at power sourcing equipment will be necessary.

Regardless of what the customer has, it is important to make sure that it is compatible with the system you are proposing to install. If PoE injectors need to be installed, you will need to make sure there are sufficient power outlets. Who will be responsible for installing those? If you are installing dual frequency 4×4:4 access points, they most likely will require an 802.3at PoE Plus solution to properly power all the MIMO radios.

Segmentation How will the WLAN and/or users of the WLAN be segmented from the wired network? Will the entire wireless network be on a separate IP subnet tied to unique VLANs? Will VLANs be used, and is a guest VLAN necessary? Will firewalls be used for segmentation? What type of user access policies will be applied to different groups of users or devices? Or will the wireless network be a natural extension of the wired network and follow the same wiring, VLAN numbering, and design schemes as the wired infrastructure? In the existing network design, are the VLANs at the core of the network or do the VLANs reside at the edge of the network? All these questions are also directly related to security expectations.

Naming Convention Does the customer already have a naming convention for cabling and network infrastructure equipment, and will one need to be created for the WLAN? Many enterprise APs now have the ability to advertise the AP hostname in the beacon frame, which makes conducting a site survey much easier.

User Management Considerations regarding RBAC, bandwidth throttling, and load balancing should be discussed. Do they have an existing RADIUS server, or does one need to be installed? What type of LDAP user database is being used? Where will usernames and passwords be stored? Will usernames and passwords be used for authentication, or will they be using client certificates? Will guest user access be provided?

Device Management Will employees be allowed to access the WLAN with their own personal devices? How will personal and company-issued mobile devices be managed? Do they want to provide different levels of access based on device type—for example, smartphone, tablet, personal laptop, or corporate laptop? A BYOD strategy may be needed as well as an MDM solution. How will IoT devices be secured and managed?

Infrastructure Management How will the WLAN remote access points be managed? Is a central cloud management solution a requirement? Is an on-premises management solution required? Will devices be managed using SSH2, SNMP, or HTTP/HTTPS? Do they have standard credentials that they would like to use to access these management interfaces?

IPv6 Considerations Does the existing wired network support or require IPv6 connectivity? Will the enterprise WLAN infrastructure and WLAN clients support IPv6? Will IPv6 be a requirement for the network at a later date?

A comprehensive interview that provides detailed feedback about infrastructure connectivity requirements will result in a more thorough site survey and a well-designed wireless network. Seventy-five percent of the work for a good wireless network is in the pre-engineering. It creates the road map for all the other pieces.

Security Expectations

Network management personnel should absolutely be interviewed about security expectations. All data privacy and encryption needs should be discussed. All AAA requirements must be documented. It should be determined whether the customer plans to implement a wireless intrusion detection or prevention system (WIDS or WIPS) for protection against rogue APs and the many other types of wireless attacks. Older devices may not support fast secure roaming mechanisms, and 802.1X/EAP might not be an option for those devices.

A comprehensive interview regarding security expectations will provide the necessary information to make competent security recommendations after the site survey has been conducted and prior to deployment. Industry-specific regulations, such as the Health Insurance Portability and Accountability Act (HIPAA), Gramm-Leach-Bliley, and Payment Card Industry (PCI), may have to be taken into account when making security recommendations. U.S. government installations may have to abide by the strict Federal Information Processing Standards (FIPS) 140-2 regulations, and all security solutions may need to be FIPS-compliant. In Europe, General Data Protection Regulation (GDPR) privacy requirements will have to be addressed.

All of these answers should also assist in determining whether the necessary hardware and software exists to perform these functions. If not, it will be your job to consider the requirements and recommendations that may be necessary.

Guest Access

Although the primary purpose of enterprise WLANs has always been to provide employees with wireless mobility, WLAN access for company guests can be just as important. In today's world, business customers have come to expect guest WLAN access. Free guest access is often considered a value-added service. Because of the widespread acceptance of Wi-Fi in business environments, most companies offer some sort of wireless guest access to the Internet. Guest users access the WLAN via the same access points; however, they usually connect via a unique guest SSID.

The primary purpose of a guest WLAN is to provide a wireless gateway to the Internet for company visitors and/or customers. Generally, guest users do not need access to company network resources. Therefore, the most important security aspect of a guest WLAN is to protect the company network infrastructure from the guest users.

During the interview, the various types of WLAN guest access and security solutions will need to be discussed. At a minimum, there should be a separate guest SSID, a unique guest VLAN, and a guest firewall policy. Additionally, there may be the need for a captive web portal with a guest WLAN. Encrypted guest access is also becoming more common. Many other WLAN guest access options are available, including guest self-registration and employee sponsorship. For a more detailed discussion about Wi-Fi guest access, please refer to Chapter 18.

Aesthetics

An important aspect of the installation of wireless equipment is the "pretty factor." Many businesses prefer that all wireless hardware remain completely out of sight. Aesthetics is extremely important in retail environments and in the hospitality industry (restaurants and hotels). Any business that deals with the public will often require that the Wi-Fi hardware be hidden or at least secured. WLAN vendors continue to design more aesthetic-looking access points and antennas. Most indoor APs use internal antennas for aesthetic purposes. Some vendors have even camouflaged access points to resemble smoke detectors. Indoor ceiling-mount enclosures can also be used to streamline the appearance of the AP and securely mount it. Most enclosure units can be locked to help prevent the theft of expensive Wi-Fi hardware.

Outdoor Surveys

Some of the focus of this book and the CWNA exam is on outdoor site surveys for establishing bridge links. Calculations necessary for outdoor bridging surveys are numerous, including the Fresnel zone, earth bulge, free space path loss, link budget, and fade margin. However, outdoor site surveys for the purpose of providing general outdoor wireless access for users are becoming more commonplace. As the popularity of wireless mesh networking continues to grow, outdoor wireless access has become more commonplace. Outdoor site survey kits using outdoor mesh APs will be needed.

Weather conditions—such as lightning, snow and ice, heat, and wind—must also be contemplated. The most important consideration is the apparatus that the antennas will be mounted to. Unless the hardware is designed for outdoor use, the outdoor equipment must ultimately be protected from the weather elements by using NEMA-rated enclosure units. (*NEMA* stands for National Electrical Manufacturers Association.) NEMA weatherproof enclosures are available with a wide range of options, including heating, cooling, and PoE interfaces.

Safety is also a big concern for outdoor deployments. Consideration should be given to hiring professional installers. Certified tower climbing courses and tower safety and rescue training courses are available.

> Information about RF health and safety classes can be found at
> www.sitesafe.com. Tower climbing can be dangerous work.
> Information about tower climbing and safety training can be found
> at www.comtrainusa.com.

All RF power regulations, as defined by the regulatory body of your country, will need to be considered. If towers are to be used, you may have to contact several government agencies. Local and state municipalities may have construction regulations, and a permit is almost always required. In the United States, if any tower exceeds a height of 200 feet above ground level (AGL) or is within a certain proximity to an airport, both the FCC and Federal Aviation Administration (FAA) must be contacted. If a roof mount is to be installed that is greater than 20 feet above the highest roof level, the FCC and FAA may have to be consulted as well. Other countries have similar types of height restrictions. Contact the proper RF regulatory authority and aviation authority to find out the details.

Vertical Market Considerations

No two site surveys will ever be exactly alike. Every business has its own needs, issues, and considerations when conducting a survey. Some businesses may require an outdoor site survey instead of an indoor survey. A *vertical market* is a particular industry or group of businesses that develops and markets similar products or services. A detailed discussion about vertical markets can be found in Chapter 20, "WLAN Deployment and Vertical Markets." The following sections outline the distinctive subjects that must be examined when a WLAN is being considered in specialized vertical markets.

Government

The key concern during government wireless site surveys is security. When security expectations are addressed during the interview process, careful consideration should be given to all aspects of planned security. Many U.S. government agencies, including the military, require that all wireless solutions be FIPS 140-2 compliant. Other government agencies may require that the wireless network be completely shielded or shut off during certain times of the day. Be sure to check export restrictions before traveling to other countries with certain equipment. The United States forbids the export of AES encryption technology to some countries. Other countries have their own regulations and customs requirements.

Obtaining the proper security credentials will most likely be a requirement before conducting the government survey. An identification badge or pass often is required. In some government facilities, an escort is needed in certain sensitive areas.

Education

As with government facilities, obtaining the proper security credentials in an education environment usually is necessary. Properly securing access points in lockable enclosure units is also necessary to prevent theft or tampering. Because of the high concentration of students, user density should be accounted for during capacity and coverage planning. K–12 schools across the United States are implementing 1:1 tablet deployments, where every student in every classroom has access to a tablet. Because of these 1:1 programs, it is not uncommon to deploy an access point in every classroom to meet the device density needs. This may or may not be the correct solution. A proper site survey will help you determine the best deployment scenario. More information about the education vertical market can be found at `www.apple.com/education` or `edu.google.com`.

In campus environments, wireless access is required in most buildings, and very often bridging solutions are needed between buildings across the campus. Some older educational facilities were constructed in such a manner as to serve as disaster shelters. That means that propagation in these areas is limited. Most school buildings use dense wall materials, such as cinderblock or brick, to attenuate the sound between classrooms. These materials also heavily attenuate RF signals.

Healthcare

One of the biggest concerns in a healthcare environment is sources of interference from the vast array of biomedical equipment that exists on site. Many biomedical devices operate in the ISM bands. For example, cauterizing devices in operating rooms have been known to cause problems with wireless networks. There is also a concern with 802.11 radios possibly interfering with biomedical equipment.

A meeting with the department that maintains and services all biomedical equipment will be necessary. Some hospitals have a person responsible for tracking and monitoring all RF devices in the facility.

A thorough spectrum analysis survey using a spectrum analyzer is extremely important and necessary. We recommend that you conduct several sweeps of these areas and compare them to ensure the greatest probability of capturing all the possible RF interferers. The dense deployment environment of a healthcare facility will require 5 GHz because you will need more channel options to prevent co-channel interference. Hospitals are usually large in scale, and a physical site survey may take many weeks; a predictive site survey can save a lot of time. Long hallways, multiple floors, fire safety doors, reflective materials, concrete construction, lead-lined X-ray rooms, and wire mesh safety glass are some of the physical conditions that you will encounter.

The applications used in the medical environment should all be considered during the interview and the survey. Numerous healthcare applications exist for handheld iOS and Android devices. Tablets and smartphones are being used by doctors and nurses to access these mobile applications. Mobile devices are also used to transfer large files, such as X-ray graphics. Medical carts use radios to transfer patient data back to the nursing stations. VoWiFi phone deployments are commonplace in hospitals because of the communication mobility that they provide to nurses. Wi-Fi real-time location systems (RTLSs) using active 802.11 RFID tags are commonplace in hospitals for asset management tracking. Because of the presence of medical patients, proper security credentials and/or an escort will often be necessary. Many applications are connection-oriented, and drops in connectivity can be detrimental to the operation of these applications.

Retail

A retail environment often has many potential sources of 2.4 GHz interference. Demonstration models of cordless phones, baby monitors, and other ISM band devices can cause problems. The inventory storage racks and bins and the inventory itself are all potential sources of signal attenuation. Heavy user density should also be considered. If possible, a retail site survey should be done in the height of the shopping season, as opposed to late January, when the malls are empty.

Wireless applications that are used in retail stores include handheld scanners used for data collection and inventory control. Retailers may require Wi-Fi analytic solutions to monitor and track customer movement and behavior. Point-of-sale devices, such as cash registers, may also have Wi-Fi radios.

Warehouses and Manufacturing

Some of the earliest deployments of 802.11 technology were in warehouses for the purpose of inventory control and data collection. A 2.4 GHz WLAN may still be deployed because many handheld devices still exist that use legacy 802.11b/g radios. Coverage, not capacity, is usually the main objective when designing a wireless network in a warehouse. Warehouses are filled with metal racks and all sorts of inventory that can cause reflections or attenuation. The use of directional antennas in a warehouse environment may be a requirement. High ceilings often cause mounting problems as well as coverage issues. Indoor chain-link fences that are often used to secure certain areas can scatter and block an RF signal. Seamless roaming is also mandatory because the handheld devices will be mobile. Forklifts that can move swiftly through the warehouse often have computing devices with Wi-Fi radios. Handheld WLAN barcode scanners are often being replaced with smartphones or tablets that use barcode-scanning applications.

A manufacturing environment is often similar to a warehouse environment in terms of interference and coverage design. However, a manufacturing plant presents many unique site survey challenges, including safety and the presence of employee unions. Heavy machinery and robotics may present safety concerns to the surveyor, and special care should be taken so as not to mount access points where they might be damaged by other machines. Many manufacturing plants also work with hazardous chemicals and materials. Proper protection gear may need to be worn, and ruggedized access points or enclosures may have to be installed. Technology manufacturing plants often have clean rooms, and the surveyor will have to wear a clean suit and follow clean-room procedures, if they are even allowed in the room.

Many manufacturing plants are union shops with union employees. A meeting with the plant's union representative may be necessary to make sure that no union policies will be violated by the site survey team.

Multi-Tenant Buildings

By far the biggest issue when conducting a survey in a multi-tenant building is the presence of other WLAN equipment used by nearby businesses. Office building environments are extremely cluttered with 802.11b/g/n wireless networks that operate at 2.4 GHz. Almost assuredly, all the other tenants' WLANs will be powered to full strength, and some equipment will be on nonstandard channels, such as 2 and 8, which will likely interfere with your WLAN equipment. It is also likely that 40 MHz or 80 MHz channels are being improperly used by other tenants. Careful channel and power planning in both frequency bands will be needed due to the bad Wi-Fi implementation of other tenants. Discussion with other tenants may be necessary.

Legacy AP-on-a-Stick Survey

The definition of a WLAN site survey has changed over the years. For many years, the legacy method of *AP-on-a-stick* was the sole method used for performing site surveys. The AP-on-a-stick method entails temporarily mounting an AP and performing a walk-through site survey to determine proper coverage zone. Then, you move the AP to the next location and determine the next coverage zone, and repeat this process throughout the entire building. While the method is still effective, it is often very time-consuming and expensive.

In the following sections, we cover the spectrum analysis requirement of the AP-on-a-stick survey as well as the coverage analysis requirement. During the coverage analysis process, a determination will be made for the proper placement of access points, the transmission power of the access point radios, and the proper use of antennas.

Spectrum Analysis

Before conducting the coverage analysis survey, locating sources of potential interference is a must. Some companies and consultants still ignore *spectrum analysis* because of the cost associated with purchasing the necessary spectrum analyzer hardware; however, with the prices of PC-based analyzers decreasing over recent years, spectrum analysis has become more of the norm with site surveys.

Spectrum analyzers are frequency domain measurement devices that can measure the amplitude and frequency space of electromagnetic signals. Dedicated bench-top spectrum analyzer hardware can cost tens of thousands of dollars (in U.S. dollars), making them cost-prohibitive for many businesses. The good news is that several companies have solutions, both hardware- and software-based, that are designed specifically for 802.11 site survey spectrum analysis and are drastically less expensive. To conduct a proper 802.11 spectrum analysis survey, the *spectrum analyzer* needs to be capable of scanning both the 2.4 GHz ISM band and the 5 GHz U-NII bands. Several companies sell software-based solutions that work with USB adapters. As shown in Figure 14.1, MetaGeek (www.metageek.net) offers a cost-effective, USB-based spectrum analyzer that is capable of monitoring both the 2.4 GHz and 5 GHz spectrums.

FIGURE 14.1   Wi-Spy DBx 2.4 GHz and 5 GHz USB spectrum analyzer

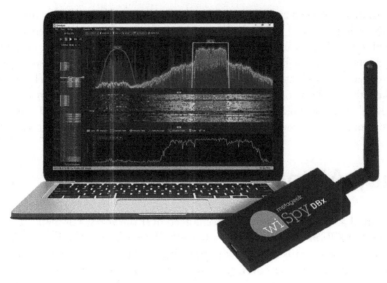

Courtesy of MetaGeek

As shown in Figure 14.2, Ekahau (www.ekahau.com) sells an all-in-one Wi-Fi site survey diagnostics and measurement hardware device that has an integrated spectrum analyzer as well as multiple 802.11 radios.

FIGURE 14.2   Ekahau Sidekick

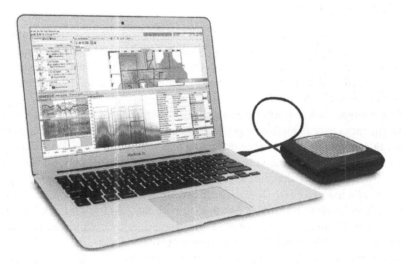

Courtesy of Ekahau

So, why is spectrum analysis even necessary? If the background noise level exceeds −85 dBm in either the 2.4 GHz ISM band or the 5 GHz U-NII bands, the performance of the wireless network can be severely degraded. A noisy environment can cause the data in 802.11 transmissions to become corrupted. Consider the following:

- If the data is corrupted, the cyclic redundancy check (CRC) will fail and the receiving 802.11 radio will not send an ACK frame to the transmitting 802.11 radio.

- If an ACK frame is not received by the original transmitting radio, the unicast frame will not be acknowledged and will have to be retransmitted.

- If an interfering device, such as a microwave oven, causes retransmissions above 10 percent, the performance or throughput of the wireless LAN will suffer significantly.

Most data applications in a Wi-Fi network can handle a layer 2 retransmission rate of up to 10 percent without any noticeable degradation in performance. However, time-sensitive applications such as VoIP require that higher-layer IP packet loss be no greater than 1 percent. Therefore, Voice over Wi-Fi (VoWiFi) networks need to limit retransmissions at layer 2 to about 10 percent or less to guarantee the timely delivery of VoIP packets.

Interfering devices might also prevent an 802.11 radio from transmitting. If another RF source is transmitting with strong amplitude, an 802.11 radio can sense the energy during the clear channel assessment (CCA) and defer transmission. If the source of the interference is a constant signal, an 802.11 radio will continuously defer transmissions until the medium is clear. In other words, a strong source of RF interference could actually prevent 802.11 client stations and access point radios from transmitting at all.

It is a recommended practice to conduct spectrum analysis of all 802.11 frequency ranges. The 2.4 to 2.5 GHz ISM band is an extremely crowded frequency space. The following are potential sources of RF interference in the 2.4 GHz ISM band:

- Microwave ovens
- 2.4 GHz cordless phones, DSSS, and FHSS
- Fluorescent bulbs
- 2.4 GHz video cameras
- Elevator motors
- Cauterizing devices
- Plasma cutters
- Bluetooth radios

A common everyday source of RF interference that should be documented during the site survey interview is microwave ovens. Microwave ovens typically operate at 800–1,000 watts. Although microwave ovens are shielded, they can become leaky over time. Commercial-grade microwave ovens will be shielded better than a discount microwave oven that you can buy at many retail outlets. A received signal of −40 dBm is about 1/10,000 of a milliwatt (mW) and is considered a strong signal for 802.11 communications. If a 1,000 watt microwave oven is as little as 0.0000001 percent leaky, the oven will interfere with the 802.11 radio. Figure 14.3 shows a spectrum view of a microwave oven. Note that this microwave operates near channel 11 in the 2.4 GHz ISM band. Some microwave ovens can congest the entire frequency band.

FIGURE 14.3 Spectrum analyzer view of a microwave oven

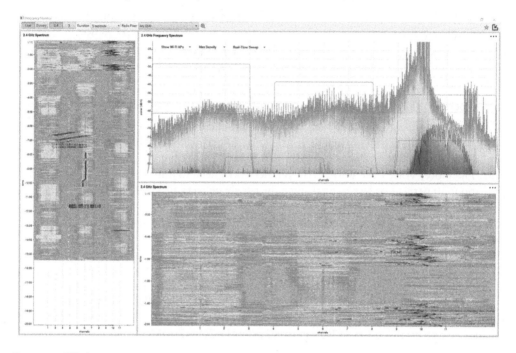

Courtesy of Ekahau

You should also check whether the call centers, receptionist, or other employees use Bluetooth mice, keyboards, or headsets. These can also cause interference.

With the introduction of 802.11ac, and because of the extreme crowding of the 2.4 GHz ISM band, most enterprise deployments have switched to 802.11n/ac equipment, which operates in the 5 GHz U-NII bands. Switching to a 5 GHz WLAN is a wise choice in the enterprise because the 5 GHz U-NII bands are currently not very crowded, and there are more choices for channel reuse patterns. Not nearly as many interfering devices exist. Although there is much less interference present at 5 GHz as compared to 2.4 GHz, this is beginning to change. Just as everyone moved from 900 MHz to 2.4 GHz to avoid interference, the band-jumping effect may also catch up with 5 GHz.

It is also important to note that the evolution of Wi-Fi technology is transitioning away from the 2.4 GHz spectrum. Very high throughput (VHT) technology defined by the 802.11ac amendment operates only in the 5 GHz U-NII bands. Since most enterprises deploy dual-frequency access points that have multiple radios (effectively installing both a 2.4 GHz and a 5 GHz network simultaneously), the 2.4 GHz radio will continue to support 802.11b/g/n communications, while the 5 GHz radio will support 802.11a/n/ac communications. These dual-frequency APs are important to provide backward compatibility to older 2.4 GHz–only devices. Newer 802.11n/ac–capable devices benefit from connecting to the less congested 5 GHz U-NII bands, while compatibility is still provided for 2.4 GHz–only devices. Current potential sources of interference in 5 GHz U-NII bands include the following:

- 5 GHz cordless phones
- Radar
- Perimeter sensors
- Digital satellite
- Nearby 5 GHz WLANs
- Outdoor wireless 5 GHz bridges
- Unlicensed LTE

The 802.11-2016 standard defines *dynamic frequency selection (DFS)* and *transmit power control (TPC)* mechanisms to satisfy regulatory requirements for operation in the 5 GHz band to avoid interference with 5 GHz radar systems. As you learned in earlier chapters, 802.11h–compliant radios are required to detect radar at 5 GHz and not transmit, to avoid interfering with the radar systems. Using a 5 GHz spectrum analyzer during a site survey may help determine in advance whether radar transmissions exist in the area where the WLAN deployment is planned. After locating the sources of interference, the best and simplest solution is to eliminate them entirely. If a microwave oven is causing problems, consider purchasing a more expensive commercial-grade oven that is less likely to be a nuisance. Other devices, such as 2.4 GHz cordless phones, should be removed and a policy that bans them should be strictly enforced. 5.8 GHz cordless phones operate in the 5.8 GHz ISM band, which overlaps with the upper U-NII band (5.725 GHz to 5.850 GHz). Indoor use of 5.8 GHz phones will cause interference with 5 GHz radios transmitting in the upper U-NII band.

In the past, VoWiFi phones operated only in the very crowded 2.4 GHz ISM band. 5 GHz VoWiFi phones are now widely available and are the better choice for VoIP transmissions. If your WLAN is being used for either data or voice, or for both, a proper and thorough spectrum analysis is mandatory in an enterprise environment.

Coverage Analysis

After you conduct a spectrum analysis, your next step is the all-important determination of proper 802.11 RF coverage inside your facility. During the site survey and design interview, capacity and coverage requirements are discussed and determined before the actual site survey is performed. In certain areas of your facility, more APs may be required because of a high density of users or heavy application bandwidth requirements.

After all the capacity and coverage needs have been determined, RF measurements must be taken to guarantee that these needs are met and to determine the proper placement and configuration of the access points and antennas. Proper *coverage analysis* must be performed using some type of *received signal strength* measurement tool or planning tool. This tool could be something as simple as the received signal strength meter in your Wi-Fi radio's client utility, or it could be a more expensive and complex site survey software package. All of these measurement tools are discussed in more detail later in this chapter.

So how do you conduct proper coverage analysis? That question is often debated by industry professionals. Many site survey professionals have their own techniques; however, we will try to describe a basic procedure for coverage analysis using the AP-on-a-stick method.

One mistake that many people make during the site survey is leaving the access point radio at the default full-power setting. A good starting point for an indoor access point is 25 mW transmit power. After the site survey is performed, the power can be increased, if needed, to meet unexpected coverage needs, or it can be decreased to meet capacity needs. Most APs are dual frequency, with both 2.4 GHz and 5 GHz radios. Measurements are usually taken using the 5 GHz radio because the effective range of 5 GHz is normally smaller than 2.4 GHz. Therefore, using the lowest common denominator of the 5 GHz range is preferred.

When you are designing for coverage during a site survey, the normal recommended best practice is to provide for a –70 dBm or stronger received signal, which is well above the noise floor. When you are designing for WLANs with VoWiFi clients, a –65 dBm or stronger signal, which is even higher above the noise, is recommended.

The hardest part of physically performing a coverage analysis site survey is often finding where to place the first access point and determining the boundaries of the first RF cell. The following procedure explains how this can be achieved (and is further illustrated in Figure 14.4):

FIGURE 14.4   Starting coverage cell

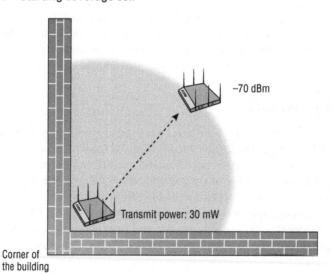

1. Place an access point with a power setting of 25 mW (or the power level that you determine is ideal for your environment) in the corner of the building.

2. Walk diagonally away from the access point toward the center of the building until the received signal drops to –70 dBm, or the signal strength that you are planning for. During this process, you should be transmitting data from the client to the AP, ensuring not only signal strength but actual transmission capabilities. This is the location where you place your first access point. (–70 dBm will be used as the desired signal level throughout the rest of this example. If you are using a different desired signal level, use it instead.)

3. Temporarily mount the access point in the first location and begin walking throughout the facility to find the –70 dBm endpoints, also known as *cell boundaries* or *cell edges*.

4. Depending on the shape and size of the first coverage cell, you may want to change the power settings and/or move the initial access point.

After the first coverage cell and boundaries have been determined, the next question is where to place the next access point. The placement of the next access point is performed by using a technique that is similar to the one you used to place the first access point.

Think of the cell boundary of the first access point, where the signal is –70 dBm, as the initial starting point, similar to the way you used the corner of the building as your initial starting point, and do the following:

1. From the first access point, walk parallel to the edge of the building and place a temporary access point at the location where the received signal is –70 dBm, as pictured in Figure 14.5.

FIGURE 14.5   Second AP location

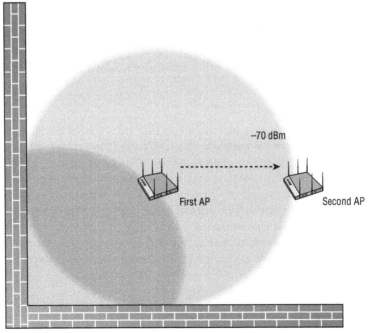

2. Now walk away from this access point, parallel to the edge of the building, until the received signal drops to −70 dBm.

3. Move to that location and temporarily mount the access point. The AP mounted at this location will provide for the second coverage cell.

4. Begin walking throughout the facility to find the −70 dBm endpoints, or cell boundaries.

5. Again, depending on the shape and size of the first coverage cell, you may want to change the power settings and/or move this access point.

It is important to avoid excessive overlap, because it can cause frequent roaming and performance degradation. The shape and size of the building and the attenuation caused by the various materials of walls and obstacles will require you to change the distances between access points to ensure proper cell overlap. After finding the proper placement of the second access point and all of its cell boundaries, repeat the procedure over again. The rest of a manual site survey like this one is basically repeating this procedure over and over again, effectively daisy-chaining throughout the building until all coverage needs are determined.

WLAN design guides and white papers from various WLAN vendors often reference 15 percent to 30 percent coverage cell overlap for roaming purposes. However, there is no way to measure coverage cell overlap. Coverage overlap is really duplicate coverage from the perspective of a Wi-Fi client station. A proper site survey should be conducted to ensure that a client always has proper primary and secondary coverage from multiple access points. In other words, each Wi-Fi client station needs to hear at least one access point at a specific received signal strength indicator (RSSI) and a backup or secondary access point at the same RSSI. Typically, vendor RSSI thresholds require a received signal of greater than −70 dBm for the higher data rate communications. Therefore, a client station needs to see at least two access points at the desired signal level so that the client can roam if necessary.

The following cell edge measurements are taken during the site survey:

▪ Received signal strength (dBm), also known as received signal level (RSL)

▪ Noise level (dBm)

▪ Signal-to-noise ratio, or SNR (dB)

The received signal strength measurements that are recorded during a site survey typically depend on the intended use of the WLAN. If the intent of the WLAN is primarily to provide low-density data service versus capacity, a lower received signal of −73 dBm might be used as the boundary for overlapping cells. When throughput and capacity are a higher priority, using a received signal of −70 dBm or higher is recommended. When you are designing for WLANs with VoWiFi clients, a −65 dBm or stronger signal, which is even higher above the noise, is recommended. The SNR is an important value because if the background noise is too close to the received signal, data can be corrupted and retransmissions will increase. The SNR is simply the difference in decibels between the received signal and the background noise, as shown in Figure 14.6. Many vendors recommend a minimum SNR of 20 dB for data networks and a minimum of 25 dB for voice networks.

FIGURE 14.6   Signal-to-noise ratio

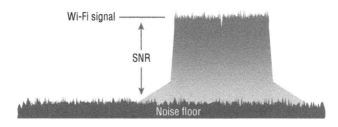

Manual coverage analysis involves the techniques described earlier to find the cell boundaries. There are two major types of manual coverage analysis surveys:

Passive   During a *passive manual survey*, the radio collects RF measurements, including received signal strength (dBm), noise level (dBm), and signal-to-noise ratio (dB). Although the client adapter is not associated to the access point during the survey, information is received from radio signals that exist at layer 1 and layer 2.

Active   During an *active manual survey*, the radio is associated to the access point and has layer 2 connectivity, allowing for low-level frame transmissions. If layer 3 connectivity is also established, low-level data traffic, such as Internet Control Message Protocol (ICMP) pings, are sent in 802.11 data frame transmissions. Layer 1 RF measurements can also be recorded during the active survey. However, upper-layer information, such as packet loss and layer 2 retransmission percentages, can be measured because the client card is associated to a single access point.

Some vendors recommend conducting both passive and active manual site surveys. The information from both manual surveys can then be compared, contrasted, and/or merged into one final coverage analysis report. What measurement software tools can be used to collect the data required for both passive and active manual surveys? There are numerous free and commercial WLAN discovery tools for a variety of operating systems and devices.

Some handheld devices, such as VoWiFi phones and Wi-Fi barcode scanners, may have site survey capabilities built into their internal software. A common mistake that surveyors make is to hold the VoWiFi phone in a horizontal position when measuring RF signals during a manual site survey. The internal antenna of the VoWiFi phone is typically polarized vertically, so holding the phone in a horizontal position can result in misleading signal measurements. We suggest holding the phone as it will be used, not in a way that creates the best signal readings.

Commercial RF site survey applications, like the one shown in Figure 14.7, have gained wide acceptance and generally provide better results.

FIGURE 14.7 Commercial coverage analysis site survey software

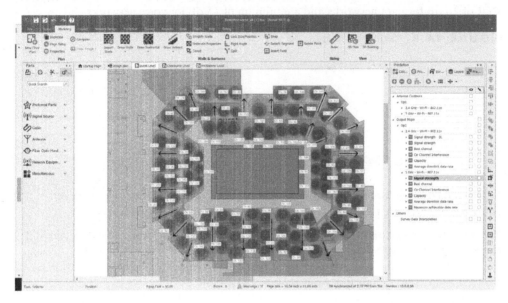

Courtesy of iBwave

These commercial packages allow the site survey engineer to import a graphic of the building's floor plans into the application. A variety of graphic formats are usually supported, and the floor plan typically must be to scale. The commercial application works with an 802.11 client radio and takes measurements in either a passive manual mode or an active manual mode. The site survey engineer walks through the building, capturing the RF information while also recording the location on the graphic of the floor plan that is displayed in the software. The information collected during both active and passive modes can then be merged, and a visual representation of the RF footprints or coverage cells can be displayed over the graphic floor plan. It should be noted that these commercial site survey applications usually also offer dual-functionality for creating predictive WLAN models, which will be discussed later in this chapter.

Hybrid Survey

Although the physical AP-on-a-stick survey method is still used, the majority of WLAN design and survey professionals use a hybrid survey process. The hybrid method follows many of the same steps and principles as the AP-on-a-stick method, which makes sense since the ultimate goal is the same: a well-designed and functional WLAN. In this section, you will learn about the components and steps used during the hybrid survey process, and you will learn the similarities and differences between the two methods.

The main premise of a hybrid survey is to use RF predictive analysis software to model the RF coverage within the building or area where Wi-Fi is needed. The predictive analysis software uses the RF power of the AP along with the antenna radiation pattern, then using free space path loss along with signal attenuation properties of the walls; it predicts the coverage area of each AP within the building. The more accurately you enter the information into the analysis software, the more accurate the predictive model will be. Therefore, an initial site visit is highly recommended.

Initial Site Visit

After conducting the initial interviews and meetings, you will need to visit the site to learn about the RF environment that the APs and clients will be operating in. Prior to visiting the site, make sure that you have multiple copies of all the necessary floor plans and/or digital copies on a laptop or tablet. The purpose of your visit is to take notes and document the environment. The data that you collect from the site visit will be used to generate the predictive model.

You also need to make sure that you will have access to the facility and all rooms and closets. You may need a security escort, or at least have ready access to someone with the keys to provide you with the necessary physical access. This may not be possible in some environments; however, it is important for the customer to realize that the more access and information you have, the better the design will be.

Another concern with the site visit is whether you can perform it on your own, or if you will need a second person to assist you. If the site is a public location or facility, you will definitely need two people, since there will likely be times when you will have to leave your test equipment to walk into another room or to another floor to take RF measurements. If the customer is providing you with an escort, make sure that they realize that the site visit will involve a great deal of walking and the escort will need to be in reasonable shape to keep up with you during the process.

Spectrum Analysis

The spectrum analysis process for a hybrid site survey is the same as described previously in this chapter in the AP-on-a-stick survey section. The goal of the spectrum analysis is to identify any devices that are causing RF interference in the Wi-Fi bands or that can cause interference. The information from the spectrum analysis will be used to identify how to prevent interference when the WLAN is installed, either by working around any interfering devices or removing any interfering devices. A thorough spectrum analysis needs to be performed throughout the entire facility.

Attenuation Spot Checks

As opposed to an AP-on-a-stick survey that is used to determine coverage zones on the initial site visit, the hybrid method first requires attenuation spot checks. Prior to performing the predictive design, you need to walk the property and take notes and RF measurements. Most buildings have consistent construction throughout. Hallway walls, bathroom walls, and stairwells are each typically constructed the same throughout the building. During the walkthrough of the property, you will be documenting the wall types, along with any differences that exist. You will also need to take RF measurements of the property to identify how much attenuation is caused by each of the different wall types. This information will be used in the predictive design.

Attenuation spot checks are essentially periodic measurements of the signal loss that occurs as the signal travels through a wall. Performing a measurement is easy. Temporarily mount an AP in a room in a location where the AP might normally be deployed. Configure the AP with medium power, transmitting on any 5 GHz channel with a 20 MHz channel width. You will use a handheld RF measurement tool, or even a handheld smartphone or a laptop, to measure the received signal from the AP in dBm value.

As shown in Figure 14.8, the first signal evaluation is a *free space path loss (FSPL)* measurement, and there should be no obstructions between your measuring tool and the AP. The measurement device should be about 15 feet or 5 meters from the AP, and 1 meter from the wall. For the second measurement, stand in a path on the exact opposite side of the wall with the measurement device about 3 feet (1 meter) from the wall. The difference between these two measurements is the attenuation, or signal loss, that is caused by the wall. For example, the first measurement in Figure 14.8 is –60 dBm and the second measurement is –72 dBm. Therefore, the attenuation of the wall is 12 dB.

FIGURE 14.8 Measuring wall loss

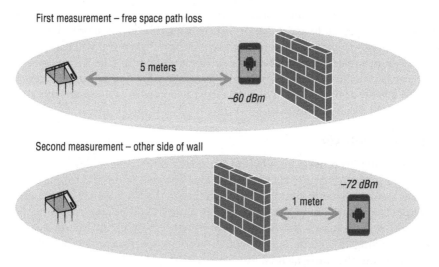

First measurement – free space path loss

5 meters

–60 dBm

Second measurement – other side of wall

–72 dBm

1 meter

You should take these measurements periodically to make sure that the signal loss is consistent throughout the facility for the same wall type. If you suspect that a wall is constructed differently from the other walls, you should take a measurement to confirm the signal loss. These values should be documented on your floor plan so that you can use them later when you are creating the predictive design. Accurate attenuation measurements are imperative to improve the accuracy of the predictive design.

Building and Infrastructure

During the initial site visit, you should also be observing the ceiling and infrastructure to identify how much effort will be needed to mount the APs to the ceiling, along with running the cables. Additionally, there are numerous other questions to consider that may need to be addressed. Does the location have drop ceiling tiles, smooth plaster ceilings, or open exposed beams? In addition to mounting the APs, how easy or difficult will it be to run the Ethernet cable to the APs? In most cases, this will also be a significant concern. Will the APs need to be hidden? Will you be limited in any way as to where you can mount them? Is there any other equipment that you will have to avoid or work around? You may not have answers to these questions, but you need to observe and document any concerns or questions you have. Make sure to take plenty of pictures, and make sure that the pictures are well documented so that you know what you are looking at later. Taking pictures at the initial site visit will assist during the predictive design phase as well as during the validation survey.

Predictive Design

Multiple enterprise WLAN design and survey solutions are available to create a predictive RF design. The following is a list of some of the well-known software products that can be used to perform this type of design:

- Ekahau Site Survey
- iBwave Wi-Fi Suite
- AirMagnet Survey
- TamoGraph Site Survey

The predictive design process begins by adding the floor plan to the design software. As part of this process, it is important that the floor plan scale and measurements be accurate. If the building has multiple floors, each floor plan must also be properly aligned with the floors above and below. RF information must also be entered, specifying how much RF loss occurs between each of the floors.

Blueprints and floor plans often use vector graphic formats (.dwg, .dwf) and can contain layer information, including the type of building materials that are used. Predictive analysis software often supports both vector and raster graphics (.bmp, .jpg, .tif) and allows for the import of building floor plans. The software creates forecast models using the predictive algorithms and the attenuation information. The modeling forecast can include the following:

- Channel reuse patterns
- Coverage cell boundaries
- Access point placement
- Access point power settings
- Number of access points
- CCI

After the floor plans are entered, all of the walls need to be entered, and RF attenuation values must be specified for every wall. The WLAN design engineer will indicate in the software what materials are used in the floor plan. The predictive application already has attenuation values for various materials, such as drywall, concrete, and glass, programmed into the software. Custom attenuation values can also be assigned to walls. If the floor plan was imported as a *computer-aided design (CAD)* drawing, it may be possible to select all the walls of a specific type and universally assign the attenuation value to all walls of that type. If not, you will need to select or define wall types, assign attenuation parameters, and manually draw the walls by tracing over the floor plan.

After all the walls are drawn and have RF attenuation values assigned to them, you can now place APs on the floor plans. It is important to know the make and model of the APs that you intend to use, as the antenna pattern varies from one AP to another. It is also important to select a power level that you are planning to use for the APs. AP and antenna settings can be modified throughout the design process, allowing you to evaluate different scenarios and options.

As shown in Figure 14.9, predictive modeling software can assist you in planning for AP placement and desired coverage, and can visualize potential co-channel interference. Depending on the solution, you may have a two-dimensional or three-dimensional view of the predicted coverage. All predictive modeling software can also automatically suggest channel reuse plans for both the 2.4 and 5 GHz frequency bands. Most enterprise predictive modeling software also offers the capability to plan for device and user capacity. You can designate high-density areas on a floor plan and enter in the number of WLAN client devices, the types of devices, as well as anticipated application usage.

FIGURE 14.9   Predictive model

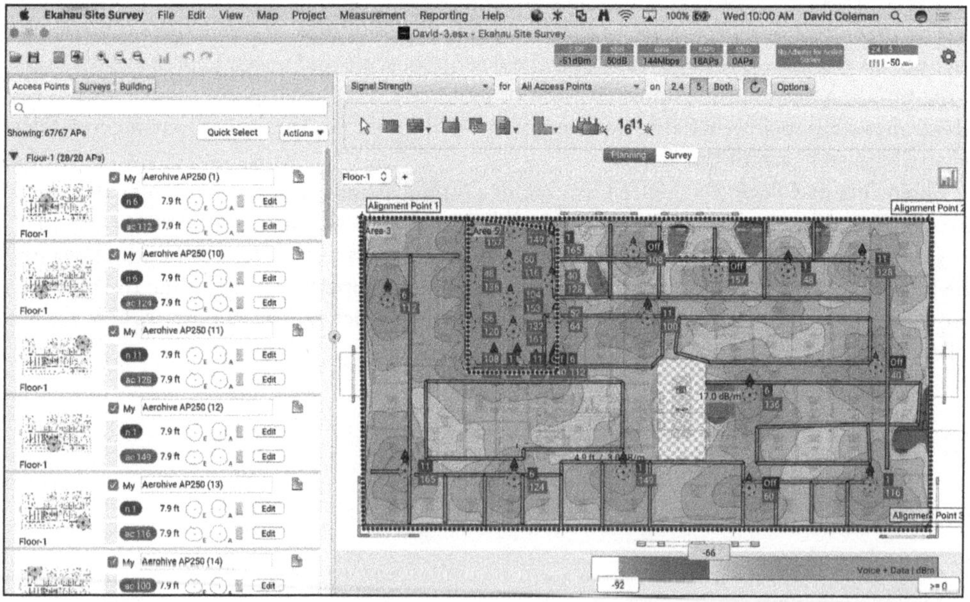

Courtesy of Ekahau

After all your data entry and manipulation, you will ultimately have the predictive design that you believe will achieve your coverage and capacity requirements. The design software will be able to generate a list of APs and antennas (if you planned for any external antennas). The predictive design, along with the bill of materials (BoM), can be used to help order the equipment and install the equipment.

Validation Survey

One part of the WLAN design and survey process that is often overlooked is the final *validation survey.* After a wireless network has been installed, and before it has been placed into service, it is important to audit or validate the installation. This validation allows you to verify that the RF coverage and other WLAN design objectives have been met or exceeded. As part of the validation survey, you can compare the actual values with the expected values from your network design plans—whether the design was performed using the AP-on-a-stick method or with a predictive model. Hopefully, these numbers meet or exceed your expectations. If they do not, you will need to further analyze why, and then determine whether the actual coverage, roaming, and performance is acceptable or whether you will need to modify your installation. If you hired a company to design and install your network for you, this WLAN validation is important to ensure that they delivered what was promised and expected.

Unfortunately, a wireless network does not always behave as expected. Over time or even overnight, the performance of the network could degrade. This degradation could be caused by a change in how the network is used, hardware or software problems, failure with access points or the WLAN controller, or a change in the physical operating environment of the network. Any or all of these can affect the RF coverage. In this situation, a wireless network validation should be able to help you with identifying the cause of your problem.

A wireless network validation is typically performed by systematically walking through the building or coverage area of the wireless network and taking RF and network measurements. These measurements are then documented on the floor plan or map. This information should help you to identify where and why your problem exists.

There are many products on the market that can help you perform a wireless network validation. Many of the same site survey modeling tools can also be used to perform wireless network validation. The site survey software will allow you to take network measurements and, using those measurements, provide you with visual RF heat maps of your environment. This often involves an extensive and tedious process of walking through your building while carrying a laptop. In lieu of walking around with a laptop, or as an additional resource, you may use a professional handheld validation survey tool, like the one shown in Figure 14.10.

FIGURE 14.10 NetScout AirCheck G2

Courtesy of NetScout

Handheld validation survey devices are typically ruggedized to protect them from accidents and misuse. The validation tools can identify both access points and client devices. They can provide extensive information about the access points, SSIDs, RF signals, security, and network traffic, along with many other pieces of information. A good handheld tool can provide you with an extensive array of information and allow you to understand what a client device sees on the network, hopefully helping you to understand if the network is operating correctly and, if not, why.

Because RSSI sensitivity varies between WLAN devices, the validation survey is often performed using different types of WLAN clients. For example, validation RF measurements might be logged using both a professional handheld survey tool as well as a less expensive smartphone with RF measurement software.

During the design phase, you should have documented all of the areas that required Wi-Fi coverage. Additionally, you should have documented the minimum level of coverage that was needed—specifically, the required dBm received signal level and SNR. In most cases, a received signal of at least –70 dBm will be required as well as a 20 dB or higher SNR. The validation survey requires you to walk throughout the facility and validate that these RF signal requirements have been met.

Keep in mind that the validation survey is not just about verifying AP placement and coverage. You also need to validate capacity, roaming, and other performance metrics.

Capacity and Throughput

Addressing WLAN capacity is done early in the design phase of the network. Capacity is about providing enough APs for the number of client devices, distributing a large number of devices across many APs. During the validation survey, you need to make sure that the

APs implemented will satisfy the customer's needs. Therefore, you also need to ensure that the network design is validated to meet client capacity and requirements.

Throughput testing is important not only to ensure that you have a signal, but that the signal is strong enough to provide a certain level of throughput. In addition to making sure that the specific 802.11 data rates are achieved, it is important to validate that the wired infrastructure performs as designed and is capable of handling the amount of data that is passed on to it from the APs and wireless clients.

Throughput testing tools are used to evaluate bandwidth and performance throughout a network. Throughput testers normally work on a client/server model to measure data streams between the two ends and in both directions. When testing downlink WLAN throughput, the 802.11 client should be configured as the server. When testing uplink WLAN throughput, the 802.11 client should be configured as the client communicating with a server behind the AP. iPerf is an open-source utility that is commonly used to generate TCP or UDP data streams to test throughput. Many WLAN vendors offer iPerf as a CLI test utility from within the OS of access points. As shown in Figure 14.11, TamoSoft (www.tamos.com) offers a freeware GUI-based throughput tester that is available for Windows, macOS, iOS, and Android clients.

FIGURE 14.11   TamoSoft throughput tester

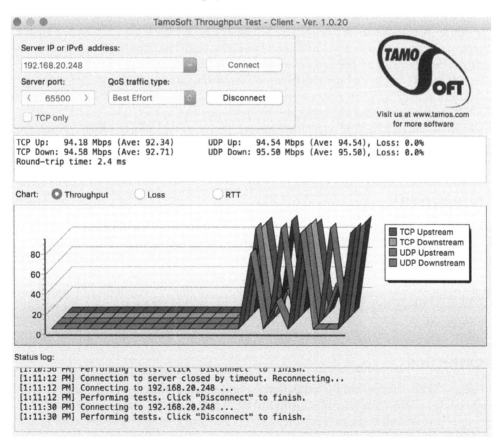

Roaming

Unless you are installing a single AP network, seamless roaming will be a requirement of the WLAN. Depending on the design of the network, a Mobile IP layer 3 roaming solution may be required, allowing seamless roaming between layer 3 networks while maintaining an IP address. Although this is technically not a component of the RF design, it is potentially part of the network design, and something that may need to be tested and validated. Another aspect of roaming that may need to be tested and validated is client roaming while using an 802.1X/EAP authenticated and encrypted client connection. This will likely include validation of fast secure roaming mechanisms. The handheld validation tools mentioned earlier often also have built-in roaming test capabilities. Roaming performance might also be evaluated using the management platform of the APs.

Delay and Jitter

Network delay, or *latency*, is the time it takes to deliver a packet from a device to its final destination. For many applications, such as web browsing and email, delay is a minor, typically unnoticed phenomenon. For applications such as VoIP or streaming video, however, any retransmission of packets or delay can be very noticeable and annoying. Depending on the network design and requirement, you may need to test and validate that the infrastructure can support the necessary delivery requirements.

Jitter is a variation of latency. Jitter measures how much the latency of each packet varies from the average. If all the packets travel at the same speed through a network, jitter is zero. If the WLAN has a high rate of layer 2 retransmissions, jitter is a common result, resulting in choppy audio or video. Most WLAN applications can tolerate up to 10 percent of retransmissions without any noticeable performance degradation. However, time-sensitive applications such as VoIP need much lower retry rates, typically less than 5 percent and preferably closer to 2 percent. A good end-to-end network design should help achieve these numbers; however, testing is needed to verify that they are achieved.

Connectivity

Another critical part of the WLAN is its connectivity to the core corporate network, and the connectivity of the WLAN client devices to the corporate network. The wired infrastructure needs to be able to handle the load placed on it by the WLAN. The infrastructure also needs to be able to support any necessary segmentation, routing, and PoE requirements, along with any other behind the scenes functions that are required. Any features that you are demanding of the infrastructure in support of the WLAN need to be tested and validated.

Aesthetics

Whether it is actually part of the validation survey or part of the entire installation process, aesthetics is an important part of a successful WLAN installation. Some environments, such as historical buildings, may require that APs and antennas not be visible. This can be accomplished by actually mounting the devices above the ceiling, behind walls, or below the floor. This may also be done by camouflaging the APs to look like other parts of the building, such as incorporating them into molding or lighting. Whatever the installation method, make sure that it is clean and professional looking. This should be monitored and corrected during the installation process; however, you should make sure to also pay attention to it during the validation survey, to make sure that nothing was missed.

Site Survey Tools

Anyone who is serious about deploying wireless networks will put together a site survey toolbox with a multitude of products that can aid the site survey process. The main tool will be some sort of signal measurement software utility that interfaces with your wireless client card and is used for signal analysis. Prepackaged site survey kits can be found for sale on the Internet, but many site survey professionals prefer to put together their own kit. Indoor and outdoor site surveys are very different. The following sections discuss the various tools that are used in both types of surveys.

Indoor Site Survey Tools

As stated earlier, you will need a spectrum analyzer to locate potential sources of interference. Your main weapon in your coverage analysis arsenal will be a received signal strength measurement tool. If you are performing a simple site survey, this tool could be something as basic as the received signal strength meter in your wireless card's client utility. For most site surveys, however, it is recommended that you use a more expensive and complex site survey software package. There are many other tools, though, that can assist you when you are conducting the physical site survey. Here are some of the tools that you might use for an indoor site survey:

Spectrum Analyzer   This analyzer is needed for frequency spectrum analysis.

Blueprints   Blueprints or floor plans of the facility are needed to map coverage and mark RF measurements. CAD software may be needed to view and edit digital copies of the blueprints.

Signal Strength Measurement Software   You will need this software for RF coverage analysis.

WLAN Client or Handheld Validation Tool   You will need a WLAN laptop, tablet, or smartphone with the signal measurement software. Optionally, you could use a more expensive handheld validation survey tool.

Access Point   At least one AP is needed, preferably more. Access points can be used as standalone devices during the site survey or initial site visit. Controller-based APs require a controller, but some can be configured to operate without the use of a controller for this purpose.

WLAN Controller   Most WLAN controller vendors manufacture small controllers that are designed for use in branch and remote offices. When you are performing a site survey and a controller is required, a small remote office WLAN controller that weighs 2 pounds will be easier and cheaper to work with than a core WLAN controller that weighs 30 pounds.

Battery Pack   A battery pack is a necessity because the site survey engineer does not want to have to run electrical extension cords to power the access point while it is temporarily mounted for the site survey. Not only does the battery pack provide power to the access point, it also provides a safer environment because you do not have to run a loose power cord across the floor, and it makes it easier and quicker to move the access point to a new location.

Binoculars   It may seem strange to have binoculars for an indoor sight survey, but they can be very useful in tall warehouses and convention centers. They can also be handy for looking at things in the plenum space above the ceiling.

Flashlight   A powerful, directional flashlight can come in handy in a dark corner or in a ceiling.

Walkie-Talkies or Cell Phones   When performing a site survey in an office environment, it is often necessary to be as quiet and unobtrusive as possible. Walkie-talkies or cell phones are typically preferred over yelling across the room. You must also remember that RF is three-dimensional, and it is common for one person to be on one floor with the access point while the other person is on another floor checking the received signal.

Antennas   A variety of both indoor omnidirectional and indoor directional antennas historically has been common in indoor Wi-Fi site survey kits. Although external antennas are still used, their use is not as common as in the past. Most enterprise AP vendors integrate antennas directly into their APs, with the placement and antenna radiation pattern designed for the AP to be mounted on the ceiling. If the internal antennas do not meet your design needs, most enterprise AP vendors also have AP models that support external antennas.

Temporary Mounting Gear   During the site survey, you will be temporarily mounting the access point—often high up, just below the ceiling. Some sort of solution is needed to temporarily mount the AP. Bungee cords and plastic ties are often used, as well as good old-fashioned duct tape. Tripods can also be used to temporarily mount and move APs during a site survey. The mast or tripod can be moved within the building, bypassing the need to temporarily mount the access point to a wall or ceiling. Figure 14.12 shows a professional site survey kit from HiveRadar, www.hiveradar.com, with a mast that can extend up to 9 feet tall. A search on the Internet will show you many professionally produced site survey kits for sale, along with examples and even directions on how to build your own custom site survey cart or tripod. Some of the features you should consider include the transportability or "shipability" of the kit, how high can the AP be placed, whether the unit is self-contained (especially power), and the ease with which you can move or roll the unit around the location of the site survey.

Digital Camera   A digital camera should be used to record the exact location of the access point placement. Recording this information visually will assist whoever does the final installation at a later date. Setting the date/time on the pictures may also come in handy when viewing the pictures later. With the incredible optical zoom capabilities available on moderately priced consumer cameras, a digital camera can also be used in lieu of binoculars.

FIGURE 14.12   WLAN mobile site survey tripod

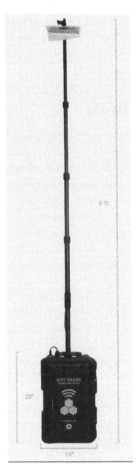

Courtesy of HiveRadar

Measuring Wheel or Laser Measuring Meter   A tool is needed to make sure the access point will in fact be close enough for a 100-meter cable to be run back to the wiring closet. Keep in mind that a 100-meter cable run includes running the CAT5E or CAT6 cabling through the plenum. A measuring wheel or a laser distance measuring tool could be used to help easily measure the distance back to the wiring closet or for documenting the distance from walls for mounting an AP in a large room.

Markers   Colored tape or sticky dots can be used to leave markers where you want to mount the access points. Leave a small piece of colored electrical tape at the location where the access point was temporarily mounted during the site survey. This will assist whoever does the final AP installation at a later date.

Ladder or Forklift   Ladders and/or forklifts may be needed to temporarily mount the access point to the ceiling.

When conducting a site survey, you should use the same 802.11 access point hardware that you plan on deploying. Keep in mind that every vendor is different and implements RSSI differently. It is not advisable to conduct a coverage analysis survey using one vendor's access point and then deploy a completely different vendor's hardware. Many established site survey companies have put together vendor site survey kits so that they can offer their customers several options.

Outdoor Site Survey Tools

Outdoor site surveys for the purpose of providing general outdoor wireless access for users are becoming more commonplace. Outdoor site surveys are conducted using outdoor mesh access points, which are used to provide access for client stations in an outdoor environment. These outdoor Wi-Fi surveys will use most of the same tools as an indoor site survey but may also use a global positioning system (GPS) device to record latitude and longitude coordinates. Although outdoor 802.11 deployments can be used to provide access, very often a discussion of outdoor site surveys is about wireless bridging or wireless backhaul for surveillance cameras or electronic monitoring equipment. Wi-Fi bridging exists at the distribution layer and is used to provide a wireless link between two or more wired networks.

An entirely different set of tools is needed for an outdoor WLAN bridging site survey, and many more calculations are required to guarantee the stability of the bridge link. In earlier chapters, you learned that the calculations necessary when deploying outdoor bridge links are numerous, including the Fresnel zone, earth bulge, free space path loss, link budget, and fade margin. Other considerations may include the intentional radiator (IR) and equivalent isotropically radiated power (EIRP) limits, as defined by the regulatory body of your country. Weather conditions are another major consideration in any outdoor site survey, and proper protection against lightning and wind will need to be deployed. An outdoor wireless bridging site survey usually requires the cooperative skills of two individuals. The following list includes some of the tools that you might use for an outdoor bridging site survey:

Topographic Map   Instead of a building floor plan, a topographic map that outlines elevations and positions may be needed.

Link Analysis Software   Point-to-point link analysis software can be used with topographic maps to generate a bridge link profile and also perform many of the necessary calculations, such as Fresnel zone and EIRP. The bridge link analysis software is a predictive modeling tool.

Calculators   Software calculators and spreadsheets can be used to provide necessary calculations for link budget, Fresnel zone, free space path loss, and fade margin. Other calculators can provide information about cable attenuation and voltage standing wave ratio (VSWR). In Exercise 14.1, you will use a calculator to determine cable attenuation.

Maximum Tree Growth Data   Trees are a potential source of obstruction of the Fresnel zone, and unless a tree is fully mature, it will likely grow taller. A chainsaw is not always the answer, and planning antenna height based on potential tree growth might be necessary. The regional or local agricultural government agency should be able to provide you with the necessary information regarding the local foliage and what type of growth you can expect.

Binoculars   Visual line of sight can be established with the aid of binoculars. However, please remember that determining RF line of sight means calculating and ensuring Fresnel zone clearance.

Walkie-Talkies or Cell Phones   802.11 bridge links may span up to (or at times exceed) a mile. Two site survey engineers working as a team will need some type of device for communicating during the survey.

Signal Generator and Wattmeter   A signal generator can be used together with a wattmeter to test cabling, connectors, and accessories for signal loss and VSWR. This testing gear can be used for testing cabling and connectors before deployment. The testing gear can also be used after deployment to check that water and other environmental conditions have not damaged the cabling and connectors.

Variable-Loss Attenuator   A variable-loss attenuator has a dial that enables you to adjust the amount of energy that is absorbed. It can be used during an outdoor site survey to simulate different cable lengths or cable losses.

Inclinometer   This device is used to determine the height of obstructions. Doing so is crucial when you need to ensure that a link path is clear of obstructions.

GPS   Recording the latitude and longitude of the transmit sites and any obstructions or points of interest along the path is important for planning. A GPS can easily provide this information.

Digital Camera   You will want to take pictures of outdoor mounting locations, cable paths, grounding locations, obstructions, and so on. You will likely need a camera with a good optical zoom lens. If you have a strong enough zoom lens on your camera, you may also be able to use it to identify and document the visual line of sight of your link.

Spectrum Analyzer   This device should be used to test ambient RF levels at transmit sites.

High-Power Spotlight or Sunlight Reflector   In the case of a wireless bridge, you will need to make sure you are surveying in the right direction. As the path gets farther away, the ability to identify a specific rooftop or tower becomes harder and harder. To aid in this task, a high-power (3 million candle or greater) spotlight or a sunlight reflector may be used. Because light travels so well, it can be used to narrow in on the actual remote site and ensure that the survey is conducted in the correct direction.

Antennas and access points are not typically used during the bridging site survey. Bridging hardware is rarely installed during the survey because most times a mast or some other type of structure has to be built. If all the bridging measurements and calculations are accurate, the bridge link will most likely work. An outdoor site survey for a mesh network will require mesh APs and antennas.

EXERCISE 14.1

Calculating Cable Loss

To perform this exercise, you need to go to the Times Microwave website (www.timesmicrowave.com). On the website, look for the link to the free online calculator.

1. In the Product text box, choose a grade of cable called LMR-1700-DB.

2. In the Frequency (MHz) text box, enter **2500**, and in the Run Length (Feet) text box, enter **200** feet.

3. Click the Calculate button.

Note the amount of dB loss per 100 feet for this specific cable.

4. Under the Product text box, choose a lower grade of cable called LMR-400.

5. In the Frequency (MHz) text box, enter **2500**, and in the Run Length (Feet) text box, enter **200** feet.

6. Click the Calculate button.

Note that this grade of cabling is rated at a much higher dB loss per 100 feet.

Documents and Reports

During the WLAN design interview (and prior to the site survey), proper documentation about the facility and network must be obtained. Additionally, site survey checklists should be created and adhered to during the physical survey. After the physical survey is performed, you will deliver a professional and comprehensive final report to the customer. Additional reports and customer recommendations may also be included with the final report. This report should provide detailed instructions on how to install and configure the proposed network so that anyone could read the report and understand your intent.

Forms and Customer Documentation

Before the site survey interview, you must obtain the following critical documentation from the customer:

Blueprints   You need a floor plan layout in order to discuss coverage and capacity needs with network administration personnel. As discussed earlier in this chapter, while reviewing floor plan layouts, keep in mind that capacity and coverage requirements will be preplanned. Photocopies of the floor plan will also need to be created and used to record the RF measurements that are taken during the physical site survey, as well as to record the locations of hardware placement. Some software survey tools allow you to import floor plans, and the software will record the survey results on the floor plan for you. Figure 14.13 shows an example of a typical floor plan. These are highly recommended and make the final report much easier to compile.

FIGURE 14.13   Typical floor plan

What if the customer does not have a set of blueprints? Blueprints can be located via a variety of sources. The original architect of the building will probably still have a copy of the blueprints. Many public and private buildings' floor plans might also be located at a public government resource, such as city hall or the fire department. Businesses are usually required to post a fire escape plan. Many site surveys have been conducted using a simple fire escape plan that has been drawn to scale, if blueprints cannot be located. In a worst-case scenario, you may have to use some graph paper and map out the floor plan manually. Predictive analysis tools require detailed information about building materials that may be found in blueprints. Blueprints may already be in a vector graphic format (.dwg and .dwf) for importing into a predictive analysis application, or they may have to be scanned.

Topographic Map   If an outdoor site survey is planned, a topographic map, also called a *contour map*, may be needed. Contour maps display terrain information, such as elevations, forest cover, and the locations of streams and other bodies of water. Figure 14.14 depicts a typical topographic map. A topographic map will be a necessity when performing bridging calculations, such as Fresnel zone clearance.

FIGURE 14.14 Topographic map

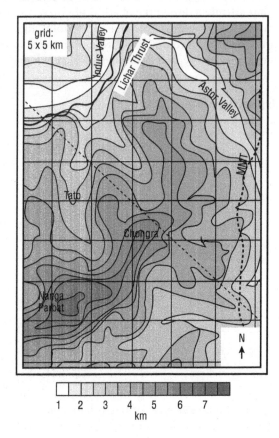

Network Topology Map Understanding the layout of the customer's current wired network infrastructure will speed up the site survey process and allow for better planning of the WLAN during the design phase. A computer network topology map will provide necessary information, such as the location of the wiring closets and layer 3 boundaries. The WLAN topology will be integrated as seamlessly as possible into the wired infrastructure. VLANs will normally be used for segmentation for both the wired and wireless networks.

Acquiring a network topology map from the customer is a highly recommended practice that will result in a well-designed and properly integrated WLAN. Some organizations may not wish to reveal their wired network topology for security reasons. It may be necessary to obtain security clearance and/or sign nondisclosure agreements to gain access to these documents.

Security Credentials You might need proper security authorization to access facilities when conducting the site survey. Hospitals, government facilities, and many businesses require badges, passes, and maybe even an escort for entrance into certain areas. A meeting with security personnel and/or the facilities manager in advance of the survey will be necessary in order to meet all physical security requirements. You do not want to show up at the customer site and be asked to return at another time because somebody forgot to schedule a security escort. Regardless of the security requirements, it is always a good idea to have the network administrator alert everyone that you will be in the area.

As a site survey and WLAN design professional, you will have created your own documentation or necessary checklists that will be used during the interview as well as during the actual physical survey. There are several types of survey and design checklists:

Interview Checklist   A detailed checklist containing all the questions to be asked during the site survey interview should be created in advance. The many detailed interview questions discussed earlier in this chapter will all be outlined in the interview checklist.

Installation Checklist   Many site survey professionals prefer to record all installation details on the floor plan documents. An installation checklist detailing hardware placement and mounting for each individual access point is also an option. Information about AP location, antenna type, antenna orientation, mounting devices, and power sources may be logged.

Equipment Checklist   For organizational purposes, a checklist of all the hardware and software tools used during the survey might also be a good idea. All the necessary tools needed for both indoor and outdoor site surveys are covered in this chapter.

Deliverables

After the interview process has been completed and the survey has been conducted, a final report must be delivered to the customer. Information gathered during the site survey will be organized and formatted into a professional technical report for the customer's review. Compiled information contained in the *deliverables* will include the following:

Purpose Statement   The final report should begin with a WLAN purpose statement that stipulates the customer requirements and business justification for the WLAN.

Spectrum Analysis   Be sure to identify potential sources of interference.

RF Coverage Analysis   Define RF cell boundaries.

Hardware Placement and Configuration   Recommend AP placement, antenna orientation, channel reuse pattern, power settings, and any other AP-specific information, such as installation techniques and cable routing.

Capacity and Performance Analysis   Include results from application throughput testing, which is sometimes an optional analysis report included with the final survey report.

A detailed site survey report may be hundreds of pages, depending on the size of the facility. Site survey reports often include pictures that were taken with a digital camera during the survey. Pictures can be used to record AP placement as well as to identify problems, such as interfering RF devices, or potential installation problems, such as a solid ceiling or concrete walls. Professional site survey software applications exist that generate professional-quality reports using preformatted templates.

Additional Reports

Along with the site survey report, other recommendations will be made to the customer so that appropriate equipment and security are deployed. Usually, the individuals and/or company that performed the site survey are also hired to install the wireless network. However, the customer might use the information from the site survey report to conduct their own deployment. Regardless of who handles the installation, other recommendations and reports will be provided along with the site survey report:

Vendor Recommendations   Many enterprise wireless vendors exist in the marketplace. It is a highly recommended practice to conduct the site survey using equipment from the same vendor that will supply the equipment that will later be deployed on site. Although the IEEE has set standards in place to ensure interoperability, every Wi-Fi vendor's equipment operates in some sort of proprietary fashion. You have already learned that many aspects of roaming are proprietary. The mere fact that every vendor's radios use proprietary RSSI thresholds is reason enough to stick with the same vendor during surveying and installation. Many site survey professionals have different vendor kits for the survey work. It is not unheard of for a survey company to conduct two surveys with equipment from two different vendors and present the customer with two separate options. However, the interview process will usually determine in advance the vendor recommendations that will be made to the customer.

Implementation Diagrams   Based on information collected during the site survey, a final design diagram will be presented to the customer. The implementation diagram is basically a wireless topology map that illustrates where the access points will be installed and how the wireless network will be integrated into the existing wired infrastructure. AP placement, VLANs, and layer 3 boundaries will all be clearly defined.

Bill of Materials   Along with the implementation diagrams will be a detailed *bill of materials (BOM)* that itemizes every hardware and software component necessary for the final installation of the wireless network. The model number and quantity of each piece of equipment will be necessary. This includes access points, bridges, wireless switches, antennas, cabling, connectors, and lightning arrestors.

Project Schedule and Costs   A detailed deployment schedule should be drafted that outlines all timelines, equipment costs, and labor costs. Particular attention should be paid to the schedule dependencies, such as delivery times and licensing, if applicable.

Security Solution Recommendations   As mentioned earlier in this chapter, security expectations should be discussed during the site survey interview. Based on these discussions, the surveying company will make comprehensive wireless security recommendations. All aspects of authentication, authorization, accounting, encryption, and segmentation should be included in the security recommendations documentation.

Wireless Policy Recommendations   An addendum to the security recommendations might be corporate wireless policy recommendations. You might need to assist the customer in drafting a wireless network security policy if they do not already have one.

Training Recommendations One of the most overlooked areas when deploying new solutions is proper training. It is highly recommended that wireless administration and security training sessions be scheduled with the customer's network personnel. Additionally, condensed training sessions should be scheduled with all end users.

Summary

In this chapter, you have learned about all the preparations and questions that must be asked prior to conducting a wireless site survey. The site survey and WLAN design interview is an important process that is necessary to both educate the customer and determine the customer's wireless needs. Defining the business purpose of the wireless network leads to a more productive survey. Capacity and coverage planning, as well as planning for infrastructure connectivity, is all part of the site survey and WLAN design interview. Before the interview, you should obtain critical documentation, such as blueprints or topographical maps, from the customer. Interview and installation checklists are used during the site survey interview and during the actual physical survey. Different survey considerations are required for different vertical markets.

You also learned about the mandatory and optional aspects of a wireless site survey. We discussed the importance of locating potential sources of interference by using a spectrum analyzer, and we defined the steps necessary to conduct both a manual and passive coverage analysis site survey, and for performing a predictive survey.

This chapter also provided a discourse of the tools necessary for either an indoor or outdoor site survey. After deployment, a validation site survey is required and will ensure that your planned design is operating as expected.

After the site survey is completed, you will deliver to the customer a final site survey report, as well as additional reports and recommendations.

Exam Essentials

Define the site survey interview. Be able to explain the importance of the interview process prior to the wireless site survey. Understand that the interview is for educating the customer and clearly defining all their wireless needs.

Identify the questions necessary to determine capacity and coverage needs. Understand the importance of proper capacity and coverage planning. Define all the numerous considerations when planning for RF coverage, bandwidth, and throughput.

Define infrastructure connectivity issues. Understand all the necessary questions that must be asked in order to guarantee proper integration of the WLAN into the existing wired infrastructure.

Identify sources of WLAN interference.   Describe all of the various devices that are potential sources of interference in both the 2.4 GHz ISM and the 5 GHz U-NII bands.

Understand the different site survey methodologies.   Understand the difference between the AP-on-a-stick method and the hybrid methods, which rely on predictive modeling.

Explain RF measurements.   Be able to explain the procedure used while conducting coverage analysis and the different types of RF measurements recorded, including received signal strength and signal-to-noise ratio.

Identify all site survey tools.   Understand the difference between an outdoor and indoor site survey, and identify all the necessary tools.

Explain the two types of coverage analysis.   Describe the differences between manual and predictive site surveys, and explain self-organizing WLAN technology.

Understand the importance of performing a wireless network validation.   Explain the importance of a wireless network validation to verify a newly installed network or to help troubleshoot a network that is not operating as expected.

Identify site survey documentation and forms.   Identify all the documentation that must be created and assembled prior to the site survey. Be familiar with all the information and documentation that is needed in the final deliverables.

Explain considerations for vertical markets.   Understand the business requirements of different vertical markets and how these requirements will alter the site survey and final deployment.

Review Questions

1. Which of the following statements best describe security considerations during a wireless site survey? (Choose all that apply.)

    A. Questions will be asked to define the customer's security expectations.

    B. Wireless security recommendations will be made after the survey.

    C. Recommendations about wireless security policies may also be made.

    D. During the survey, both mutual authentication and encryption should be implemented.

2. ACME Hospital uses a connection-oriented telemetry monitoring system in the cardiac care unit. Management wants the application available over a WLAN. Uptime is very important because of the critical nature of the monitoring system. What should the site survey and design engineer be looking for that might cause a loss of communication over the WLAN? (Choose all that apply.)

    A. Medical equipment interference

    B. Safety glass containing metal mesh wire

    C. Patients

    D. Bedpans

    E. Elevator shafts

3. Which of the following tools might be used in an outdoor site survey of an area designed to provide outdoor coverage? (Choose all that apply.)

    A. Spectrum analyzer

    B. Outdoor blueprints or topography map

    C. Mesh routers

    D. GPS

    E. Oscilloscope

4. Name a unique consideration when deploying a wireless network in a hotel or other hospitality business. (Choose the best answer.)

    A. Equipment theft

    B. Aesthetics

    C. Segmentation

    D. Roaming

    E. User management

5. Which documents might be needed prior to performing an indoor site survey for a new wireless LAN? (Choose all that apply.)

    A. Blueprints

    B. Network topography map

    C. Network topology map

    D. Coverage map

    E. Frequency map

6. After conducting a simple site survey in the office building where your company is located on the fifth floor, you have discovered that other businesses are also operating access points on nearby floors on channels 2 and 8. What is the best recommendation you will make to management about deploying a new WLAN for your company?

    A. Install a 2.4 GHz access point on channel 6 and use the highest available transmit power setting to overpower the WLANs of the other businesses.

    B. Speak with the other businesses. Suggest that they use channels 1 and 6 at lower power settings. Install a 2.4 GHz access point using channel 9.

    C. Speak with the other businesses. Suggest that they use channels 1 and 11 at lower power settings. Install a 2.4 GHz access point using channel 6.

    D. Recommend installing a 5 GHz access point.

    E. Install a wireless intrusion prevention system (WIPS). Classify the other businesses' access points as interfering and implement deauthentication countermeasures.

7. The Harkins Corporation has hired you to make recommendations about a future wireless deployment that will require more than 300 access points to meet all coverage requirements. What is the most cost-efficient and practical recommendation in regard to providing electrical power to the access points?

    A. Recommend that the customer replace older edge switches with new switches that have inline PoE.

    B. Recommend that the customer replace the core switch with a new core switch that has inline PoE.

    C. Recommend that the customer use single-port power injectors.

    D. Recommend that the customer hire an electrician to install new electrical outlets.

8. During the interview process, which topics will be discussed so that the WLAN will integrate properly into the existing wired architecture?

    A. PoE

    B. Segmentation

    C. User management

    D. AP management

    E. All of the above

9. The Jackson County Regional Hospital has hired you for a wireless site survey. Prior to conducting the site survey, you should consult with employees from which departments? (Choose all that apply.)

   A. Network management

   B. Biomedical department

   C. Hospital security

   D. Custodial department

   E. Marketing department

10. What additional documentation is usually provided along with the final site survey deliverable? (Choose all that apply.)

   A. Bill of materials

   B. Implementation diagrams

   C. Network topology map

   D. Project schedule and costs

   E. Access point user manuals

11. Which type of coverage analysis requires a radio card to be associated to an access point?

   A. Associated

   B. Passive

   C. Predictive

   D. Assisted

   E. Active

12. Which of the following tools can be used in an indoor site survey? (Choose all that apply.)

   A. Measuring wheel

   B. GPS

   C. Ladder

   D. Battery pack

   E. Microwave oven

13. Name potential sources of interference in the 5 GHz U-NII bands. (Choose all that apply.)

   A. Microwave ovens

   B. Cordless phones

   C. FM radios

   D. Radar

   E. Unlicensed LTE transmissions

14. Which of these measurements are taken during a passive manual site survey? (Choose all that apply.)

A. SNR

B. dBi

C. dBm signal strength

D. dBd

15. Name potential sources of interference that might be found during a 2.4 GHz site survey. (Choose all that apply.)

A. Toaster ovens

B. Nearby 802.11 FHSS access points

C. Plasma cutters

D. Bluetooth headsets

E. 2.4 GHz video cameras

16. Which access point settings should be recorded during an AP-on-a-stick site survey? (Choose all that apply.)

A. Power settings

B. Encryption settings

C. Authentication settings

D. Channel settings

E. IP addresses

17. Which type of site survey uses modeling algorithms and attenuation values to create visual models of RF coverage?

A. Associated

B. Passive

C. Predictive

D. Assisted

E. Active

18. The ACME Corporation has hired you to design a wireless network that will have data clients, VoWiFi phones that do not support 802.11r, and access for guest users. The company wants the strongest security solution possible for the data clients and phones. Which design best fits the customer's requirements?

A. Create one wireless VLAN. Segment the data clients, VoWiFi phones, and guest users from the wired network. Use 802.1X/EAP authentication and CCMP/AES encryption for a wireless security solution.

B. Create three separate VLANs. Segment the data clients, VoWiFi phones, and guest users into three distinct VLANs. Use 802.1X/EAP authentication and TKIP encryption for security in the data VLAN. Use WPA2-Personal in the voice VLAN. The guest VLAN will have no security, other than possibly a captive portal.

   C. Create three separate VLANs. Segment the data clients, VoWiFi phones, and guest users into three distinct VLANs. Use 802.1X/EAP authentication with CCMP/AES encryption for security in the data VLAN. Use WPA2-Personal in the voice VLAN. The guest VLAN traffic will require a captive web portal and a guest firewall policy for security.

   D. Create two separate VLANs. The data and voice clients will share one VLAN, while the guest users will reside in another. Use 802.1X/EAP authentication and CCMP/AES encryption for security in the data/voice VLAN. The guest VLAN will have no security, other than possibly a captive portal.

19. Which type of WLAN site survey is the most important survey that should always be performed regardless of the vertical deployment?

   A. AP-on-a-stick survey

   B. Hybrid survey

   C. Predictive model survey

   D. Validation survey

20. Name the necessary calculations for an outdoor bridging survey under 5 miles. (Choose all that apply.)

   A. Link budget

   B. Free space path loss

   C. Fresnel zone

   D. Fade margin adjustment

   E. Height of the antenna beamwidth

Chapter
15

WLAN
Troubleshooting

IN THIS CHAPTER, YOU WILL LEARN ABOUT THE FOLLOWING:

✓ **Five tenets of WLAN troubleshooting**

- Troubleshooting best practices
- Troubleshoot the OSI model
- Most Wi-Fi problems are client issues
- Proper WLAN design reduces problems
- WLAN always gets the blame

✓ **Layer 1 troubleshooting**

- WLAN design
- Transmit power
- RF interference
- Drivers
- PoE
- Firmware bugs

✓ **Layer 2 troubleshooting**

- Layer 2 retransmissions
- RF interference
- Low SNR
- Adjacent channel interference
- Hidden node
- Mismatched power
- Multipath

✓ **Security troubleshooting**

- PSK troubleshooting
- 802.1X/EAP troubleshooting
- VPN troubleshooting

✓ **Roaming troubleshooting**

✓ **Channel utilization**

✓ **Layers 3–7 troubleshooting**

✓ **WLAN troubleshooting tools**

- WLAN discovery applications
- Spectrum analyzers
- Protocol analyzers
- Throughput test tools
- Standard IP network test utilities
- Secure Shell

Throughout this book, you have learned about the building blocks of WLAN fundamentals. As with any type of communications network, however, problems with WLAN networks arise that might require attention from an administrator. Client connectivity issues often arise that might be the result of improper implementation of WLAN security. In this chapter, you will learn about troubleshooting best practices and how to focus on layer 1 and layer 2 when investigating Wi-Fi problems. Being able to investigate roaming problems as well as monitor channel utilization is key to maintaining good WLAN health. You will also learn WLAN troubleshooting strategies from a security perspective. Although this chapter is not meant to be a step-by-step diagnostic guide, you will learn about many common WLAN problems and suggested resolutions. We will also discuss many of the freeware and commercial WLAN troubleshooting tools that are available.

Five Tenets of WLAN Troubleshooting

Before we discuss specific WLAN troubleshooting strategies, you should understand the following five basic tenets for troubleshooting any type of WLAN problem:

- Implement troubleshooting best practices.
- Troubleshoot the OSI model.
- Most problems are client side.
- Proper WLAN design/planning is important.
- The WLAN will always get the blame.

We will now review these WLAN troubleshooting doctrines in greater detail.

Troubleshooting Best Practices

The fundamentals of troubleshooting best practices are to ask questions and collect information. When troubleshooting any type of computer network, you must ask the correct questions to collect information that is relevant to the problem. It is easy to get sidetracked when troubleshooting, so asking the proper questions will help an IT administrator focus on the pertinent data with a goal of isolating the root cause of the problem. For example, WLAN security problems often result in WLAN client connectivity issues; asking the appropriate questions will point you in the right direction toward solving the problem. The following basic questions are among those that need to be asked:

- When is the problem happening?

  At what time did the problem occur? Does this problem happen during a very specific time period? This information can be easily determined by looking at the log files of APs, WLAN controllers, and applicable servers, such as RADIUS. Best practices mandate that all *Network Time Protocol (NTP)* and time zone settings be correctly configured on all network hardware.

- Where is the problem happening?

  Is the problem widespread, or does it exist only in one physical area? Is the problem occurring on a single floor or in the entire building? Does the problem affect just one access point or a group of access points? Determining the location of the problem will help you gather better information toward solving the problem.

- Does the problem affect one client or numerous clients?

  If the problem is affecting only a single client, you might have a simple driver issue or an incorrectly configured supplicant. If the issue is affecting numerous clients, then the problem is obviously of greater concern. Most connectivity problems are client side, whether they are detrimental to a single client or multiple clients.

- Does the problem reoccur or did it just happen once?

  Troubleshooting a problem that happens only one time or only a few times can be difficult. Collecting data is much easier with recurring problems. You may have to enable debug commands on APs or WLAN controllers to hopefully capture the problem again in a log file.

- Did you make any changes recently?

  This is a question that the support personnel of WLAN vendors always ask their customers. And the answer is almost always no, despite the fact that changes to the network indeed take place. Best practices dictate that any network configuration changes be planned and scheduled. WLAN infrastructure security audit logs will always leave a paper trail of which administrator made which changes at any specific time.

Once you have asked numerous questions, you can begin the process of solving the problem. Troubleshooting best practices include the following:

1. Identify the issue.

   Because the WLAN always seems to get the blame, it is even more important to correctly identify the problem. Determine that a problem actually exists. Asking questions and collecting information will help you identify the true issue.

2. Re-create the problem.

   Having the ability to duplicate the problem either on-site or in a remote lab enables you to collect more information to diagnose the problem. If you cannot re-create a problem, you may need to ask more questions.

3. Locate and isolate the cause.

   The whole point of asking the pointed questions and gathering data is so that you can isolate the root cause of the problem. Troubleshooting up the OSI model will also help you identify the culprit. Identify if the problem is at the access layer, distribution layer or core of your network design.

4. Solve the problem.

   Formulate and implement a plan to solve the problem. This may require network changes, firmware updates, and so forth.

5. Test to verify the problem is solved.

   Always be sure to test in different areas during different times and with multiple devices. Extensive testing will ensure that the problem is indeed resolved.

6. Document the problem and the solution.

   Troubleshooting best practices dictate that you document all problems, diagnostics, and resolutions. A reference help desk database will assist you in solving problems in a timely fashion should any problem reoccur.

7. Provide feedback.

   As a professional courtesy, always be sure to follow up with the individual(s) who first alerted you to the problem.

Troubleshoot the OSI Model

The diagnostic approach that is used to troubleshoot wired 802.3 networks should also be applied when troubleshooting a wireless local area network (WLAN). A bottom-up approach to analyzing the OSI reference model layers also applies to wireless networking. Remember that 802.11 technology is similar to 802.3 in that it operates at the first two layers of the OSI model. For that reason, a WLAN administrator should always try to first determine whether problems exist at layer 1 and layer 2. If the first two layers of the OSI model have been eliminated as the cause of the problem, the problem is not a Wi-Fi problem and the higher layers of the OSI model should be investigated.

As with most networking technologies, most problems usually exist at the Physical layer. Simple layer 1 problems, such as nonpowered access points or client radio driver problems, are often the root cause of connectivity or performance issues. Disruption of RF signal propagation and RF interference will affect both the performance and coverage of your WLAN. Inadequate WLAN coverage, capacity, and performance are often layer 1 problems that are a result of poor WLAN design. Client driver issues and misconfigured supplicants are also common layer 1 problems.

After eliminating layer 1 as the source of the problem, a WLAN administrator should try to determine whether the problem exists at the Data-Link layer. Basic 802.11 communications, such as discovery, authentication, association, and roaming, all occur at the MAC sublayer of layer 2. As shown in Figure 15.1, WLAN security mechanisms also operate at layer 2. Modern 802.11 radios use CCMP encryption that provides data privacy for layers 3–7. The chosen encryption method must match on both the AP and client radios. For example, if an AP has disabled backward compatibility for TKIP encryption, a legacy client that supports only TKIP will not be able to connect. Remember that only CCMP encryption can be used for 802.11n (HT) and 802.11ac (VHT) data rates. An access point might be configured to transmit an SSID that supports both TKIP and CCMP encryption. In this situation, a common support call may be that the legacy TKIP clients seem slow because of the lack of support for higher data rates. The simple solution is to replace the legacy clients with modern-day clients that support CCMP.

FIGURE 15.1 OSI model

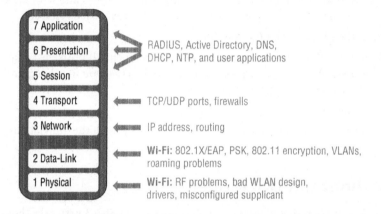

In Chapter 17, "802.11 Network Security Architecture" you will learn that there is a symbiotic relationship between the creation of dynamic encryption keys and authentication. A *pairwise master key (PMK)* is used to seed the 4-Way Handshake that generates the unique dynamic encryption keys employed by any two 802.11 radios. The PMK is generated as a byproduct of either PSK or 802.1X/EAP authentication. Therefore, if authentication fails, no encryption keys are generated. We will discuss troubleshooting both 802.11 authentication methods later in this chapter.

As stated earlier, if the first two layers of the OSI model have been eliminated, the problem is not a Wi-Fi problem, and therefore the problem exists within layers 3–7. It is likely the problem is either a TCP/IP networking issue or an application issue. As shown in Figure 15.1, TCP/IP problems should be investigated at layers 3–4, whereas most application issues exist between layers 5 and 7.

Most Wi-Fi Problems Are Client Issues

As previously mentioned, whenever you troubleshoot a WLAN, you should start at the Physical layer. Additionally, 70 percent of the time the problem will reside on the WLAN client. If there are any client connectively problems, WLAN Troubleshooting 101 dictates that you disable and re-enable the WLAN network adapter. The driver for the WLAN network interface card (NIC) is the interface between the 802.11 radio and the operating system (OS) of the client device. For whatever reason, the WLAN driver and the OS of the device may not be communicating properly. A simple disable/re-enable of the WLAN NIC will reset the driver. Always eliminate this potential problem before investigating anything else. Additionally, first-generation radio drivers and firmware are notorious for possible bugs. Always make sure the WLAN client population has the latest available drivers installed. Another change that is quick and easy to make is to reconfigure the client configuration profile. Most client supplicants allow the user to define a WLAN configuration profile or connection parameters. Sometimes troubleshooting a problem is as easy as deleting the old profile and configuring a new profile.

As mentioned earlier, client-side security issues usually evolve from improperly configured supplicant settings. This could be something as simple as a mistyped WPA2-Personal passphrase or as complex as 802.1X/EAP digital certificate problems.

Real World Scenario

Is There a Master Database of Wi-Fi Client Capabilities?

The short answer is that there is no official IEEE database of 802.11 client devices and their capabilities. There are, however, a few resources, including the Wi-Fi Alliance, which maintains a *Wi-Fi CERTIFIED Product Finder* database at www.wi-fi.org/product-finder. Although most WLAN infrastructure vendors submit their access points for certification, it should be understood that many manufacturers of WLAN client devices do not go through the certification process. As shown in Figure 15.2, Wi-Fi expert Mike Albano (CWNE #150) maintains a free public listing of WLAN client capabilities at clients.mikealbano.com. Mike has put together a good database of many of the modern popular WLAN client devices. You can also download 802.11 frame captures of the client devices as well as submit WLAN client information. Often, a laptop or mobile device manufacturer will list the radio model in the specification sheet for the laptop or mobile device. However, some manufacturers may not list detailed radio specifications and capabilities. Another method of identifying the Wi-Fi radio in your device is from the FCC ID. In the United States, all Wi-Fi radios must be certified by the Federal Communications Commission (FCC) government agency. The FCC maintains a searchable equipment authorization database at www.fcc.gov/fccid. You can enter the FCC ID of your device into the database search engine and find documentation and pictures submitted by the manufacturer to the FCC. The FCC database is very useful in helping identity Wi-Fi radio models and specifications if the information is not available on the manufacturer's website.

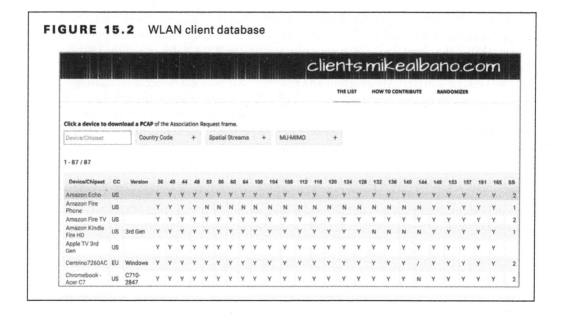

FIGURE 15.2   WLAN client database

Device/Chipset	CC	Version	36	40	44	48	52	56	60	64	100	104	108	112	116	120	124	128	132	136	140	144	148	153	157	161	165	SS
Amazon Echo	US		Y	Y	Y	Y	Y	Y	Y	Y	Y	Y	Y	Y	Y	Y	Y	Y	Y	Y	Y	Y	Y	Y	Y	Y	Y	2
Amazon Fire Phone	US		Y	Y	Y	Y	N	N	N	N	N	N	N	N	N	N	N	N	N	N	N	N	Y	Y	Y	Y	Y	1
Amazon Fire TV	US		Y	Y	Y	Y	Y	Y	Y	Y	Y	Y	Y	Y	Y	Y	Y	Y	Y	Y	Y	Y	Y	Y	Y	Y	Y	2
Amazon Kindle Fire HD	US	3rd Gen	Y	Y	Y	Y	Y	Y	Y	Y	Y	Y	Y	Y	Y	Y	Y	Y	N	N	N	N	Y	Y	Y	Y	Y	1
Apple TV 3rd Gen	US		Y	Y	Y	Y	Y	Y	Y	Y	Y	Y	Y	Y	Y	Y	Y	Y	Y	Y	Y	Y	Y	Y	Y	Y	Y	
Centrino7260AC	EU	Windows	Y	Y	Y	Y	Y	Y	Y	Y	Y	Y	Y	Y	Y	Y	Y	Y	Y	Y	Y	/	Y	Y	Y	Y	Y	2
Chromebook - Acer C7	US	C710-2847	Y	Y	Y	Y	Y	Y	Y	Y	Y	Y	Y	Y	Y	Y	Y	Y	Y	Y	N	Y	Y	Y	Y	Y	Y	2

Proper WLAN Design Reduces Problems

Two of the most important real-world chapters in this book are Chapter 13, "WLAN Design Concepts," and Chapter 14, "Site Survey and Validation." These chapters are important because a huge percentage of WLAN support phone calls are a symptom of a lack of WLAN design. Proper capacity and coverage planning, spectrum analysis, and a validation site survey will eliminate the majority of WLAN support tickets in regard to performance. Proper WLAN design will minimize problems such as co-channel interference (CCI) in advance. Additionally, many WLAN security holes can be eliminated in advance with proper WLAN security planning. If 802.1X/EAP is deployed, one of the biggest challenges is how to provision the root CA certificates for mobile devices, such as smart phones and tablets. A well-thought-out security strategy for employee WLAN devices, BYOD devices, and guest WLAN access is essential. Proper WLAN planning and design in advance will reduce time spent troubleshooting WLAN problems at a later juncture.

WLAN Always Gets the Blame

Despite all your best WLAN troubleshooting practices and best efforts, you should resign yourself to the fact the WLAN will always get the blame. Experienced WLAN administrators know the WLAN will be blamed for problems that have nothing to do with the WLAN. This is another reason that troubleshooting up the OSI stack is important. If the problem is not a layer 1 or a layer 2 issue, then Wi-Fi is not the culprit. However, put yourself in the shoes of the end user who is connected to the WLAN. 802.11 technology exists at the access layer. The whole point of an AP is to provide a wireless portal to a preexisting network infrastructure. Your employees and guests who connect to the WLAN expect seamless wireless mobility; they have no concept of problems that exist at layers 3–7. A WLAN end user does not know that the DHCP server is out of leases. A WLAN end user is not aware the Internet service provider (ISP) is experiencing difficulty and the WAN link is down. The WLAN end user just knows that they cannot access www.facebook.com through the WLAN, so they blame the Wi-Fi network.

Layer 1 Troubleshooting

As previously discussed, most networking problems usually exist at the Physical layer. In this section, we will delve deeper into layer 1 problems often caused by bad WLAN design or RF interference. We will also discuss layer 1 issues related to radio drivers, firmware bugs, and Power over Ethernet (PoE).

WLAN Design

You learned earlier that the bulk of WLAN problems can be avoided with good WLAN design prior to deployment. Probably the two most common layer 1 problems that arise due to poor design are coverage holes and co-channel interference. WLAN coverage holes are usually a result of a lack of a site survey validation. Confirmation of –70 dBm coverage for high data rate connectivity and –65 dBm coverage for voice-quality WLANs is imperative. Always remember that the AP radio's receive sensitivity is usually much stronger than the client device's receive sensitivity. Measurements to validate received signal strength should be from the perspective of the client devices. Because of wide variances in client radio RSSI, the least sensitive client device is often used for validation of proper signal strength. A lower than desired received signal will result in radios shifting to lower data rates, which consume more airtime and negatively impact performance. Coverage dead zones often arise post-deployment when furniture and even walls are moved.

Bad WLAN coverage often is a result of improper placement of APs as well as improper antenna orientation. Always check the technical specifications of your WLAN vendor, although most indoor access points with low-gain omnidirectional antennas should be mounted on the ceiling no higher than 3 meters from the floor. When using external omnidirectional antennas, it is important that they be positioned vertically. A common mistake is to position them horizontally. You would be shocked at how many APs are improperly installed, often mounted in the plenum with the antennas facing upward. You can view many pictures of improper AP placement at https://badfi.com/bad-fi, which is a fun blog written by Eddie Forero, CWNE #160.

Co-channel interference (CCI) is the top cause of needless airtime consumption that can be minimized with proper WLAN design best practices. Carrier Sense Multiple Access with Collision Avoidance (CSMA/CA) dictates half-duplex communications, and only one radio can transmit on the same channel at any given time. An 802.11 radio will defer transmissions if it hears the PHY preamble transmissions of any other 802.11 radio at an SNR of just 4 dB or greater. Unnecessary medium contention overhead that occurs when too many APs and clients hear each other on the same channel is called *co-channel interference (CCI)*. In reality, the 802.11 radios are operating exactly as defined by the CSMA/CA mechanisms, and this behavior should really be called co-channel cooperation. However, the unnecessary medium contention overhead caused by co-channel interference is usually a result of improper channel reuse design. While it is almost impossible to prevent CCI in the 2.4 GHz band, the airtime consumption that is a result of CCI can be minimized—and possibly avoided—with good 5 GHz WLAN design best practices. Making good use of proper 5 GHz channel reuse patterns and enabling the *dynamic frequency selection (DFS)* channels will reduce CCI. Lower transmit power will also reduce CCI.

Transmit Power

Another common layer 1 problem often made when deploying access points is to have the APs transmit at full power. Although most indoor APs may have full transmit power settings as high as 100 mW, they should rarely be deployed at full power. Effectively, this extends the effective range of the access point; however, designing WLANs strictly for coverage is an outdated concept. WLAN capacity and reducing airtime consumption are higher priorities. Access points at maximum transmit power will result in oversized coverage but not meet your capacity needs. Access points at full power will also increase the odds of co-channel interference due to bleed-over transmissions. The following is a quick summary of all the problems caused by APs transmitting at maximum power:

- Capacity needs not met
- Increase in CCI and airtime consumption due to unnecessary medium contention overhead
- Increase in hidden node issues
- Increase in sticky clients and roaming problems

For all these reasons, typical indoor WLAN deployments are designed with the APs set at about one-fourth to one-third full transmit power. Environments with a high user density may require that the AP transmit power be set at the lowest setting of 1 mW.

RF Interference

RF interference from non-802.11 transmitters is by far the most common external cause of WLAN problems that exist at layer 1. The *energy detect (ED)* thresholds for non-802.11 transmissions are much higher than the *signal detect (SD)* thresholds for detecting 802.11 radio transmissions. However, various types of RF interference can still greatly affect the performance of an 802.11 WLAN. Interfering devices may exceed the energy detect threshold and prevent an 802.11 radio from transmitting, thereby causing a denial of service. If another RF source is transmitting with strong amplitude, 802.11 radios can sense the RF energy during the *clear channel assessment (CCA)* and defer transmission entirely. The other typical result of RF interference is that 802.11 frame transmissions become corrupted. If frames are corrupted due to RF interference, excessive retransmissions will occur and therefore throughput will be reduced significantly. There are several different types of RF interference, as described in the following sections.

Narrowband Interference

A narrowband RF signal occupies a smaller and finite frequency space and will not cause a denial of service (DoS) for an entire band, such as the 2.4 GHz ISM band. A narrowband signal is usually very high amplitude and will absolutely disrupt communications in the frequency space in which it is being transmitted. Narrowband signals can disrupt one or several 802.11 channels.

Narrowband RF interference can also result in corrupted frames and layer 2 retransmissions. The only way to eliminate narrowband interference is to locate the source of the interfering device with a spectrum analyzer and remove the interfering device. To work around interference, use a spectrum analyzer to determine the affected channels and then design the channel reuse plan around the interfering narrowband signal. Figure 15.3 shows a spectrum analyzer capture of a narrowband signal close to channel 11 in the 2.4 GHz ISM band.

FIGURE 15.3   Narrowband RF interference

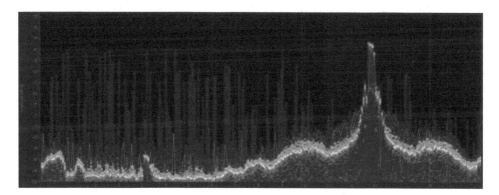

Wideband Interference

A source of interference is typically considered wideband if the transmitting signal has the capability to disrupt the communications of an entire frequency band. Wideband jammers exist that can create a complete DoS for the 2.4 GHz ISM band. The only way to eliminate wideband interference is to locate the source of the interfering device with a spectrum analyzer and remove the interfering device. Figure 15.4 shows a spectrum analyzer capture of a wideband signal in the 2.4 GHz ISM band with average amplitude of –70 dBm, well above the defined energy detect thresholds of all 802.11 radios.

FIGURE 15.4 Wideband RF interference

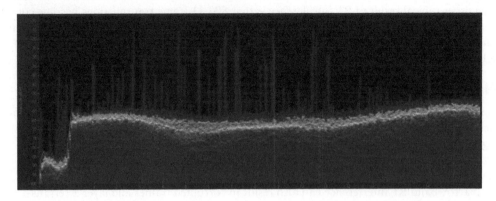

All-Band Interference

The term *all-band interference* is typically associated with frequency-hopping spread spectrum (FHSS) communications that usually disrupt the 802.11 communications at 2.4 GHz. As you learned in earlier chapters, FHSS constantly hops across an entire band, intermittently transmitting on very small subcarriers of frequency space. A legacy 802.11 FHSS radio, for example, transmits on hops that are 1 MHz wide in the 2.4 GHz band. 802.11b radios transmit in a stationary 22 MHz of frequency space, and 802.11g/n radios transmit on fixed channels of 20 MHz of spectrum. While hopping and dwelling, an FHSS device will transmit in sections of the frequency space occupied by an 802.11b/g/n channel. Although an FHSS device will not typically cause a denial of service, the frame transmissions from the 802.11b/g/n devices can be corrupted from the all-band transmissions of a legacy 802.11 FHSS interfering radio.

 **Real World Scenario**

Which Devices Cause RF Interference?

Numerous devices—including cordless phones, microwave ovens, and video cameras—can cause RF interference and degrade the performance of an 802.11 WLAN. The 2.4 GHz ISM band is extremely crowded, with many known interfering devices. Interfering devices also transmit in the 5 GHz U-NII bands, but the 2.4 GHz frequency space is much more crowded. The tool that is necessary to locate sources of RF interference is a spectrum analyzer. Luckily, most enterprise WLANs require the use of the 5 GHz band, which is less crowded and has much more frequency space. However, 5 GHz RF interferers also exist, so spectrum analysis monitoring is needed.

Bluetooth (BT) is a short-distance RF technology used in WPANs. Bluetooth uses FHSS and hops across the 2.4 GHz ISM band at 1,600 hops per second. Older Bluetooth devices were known to cause severe all-band interference. Newer Bluetooth devices utilize adaptive mechanisms to avoid interfering with 802.11 WLANs. Bluetooth adaptive frequency hopping is most effective at avoiding interference with a single AP transmitting on one 2.4 GHz channel. If multiple 2.4 GHz APs are transmitting on channels 1, 6, and 11 in the same physical area, it is impossible for the Bluetooth transmitters to avoid interfering with the WLAN. Digital Enhanced Cordless Telecommunications (DECT) cordless telephones also use frequency-hopping transmissions. A now-defunct WLAN technology known as HomeRF also used FHSS; therefore, HomeRF devices can potentially cause all-band interference. Other frequency-hopping devices that you may run across include various types of medical telemetry units. Although all the FHSS interferers mentioned so far transmit in the 2.4 GHz ISM band, 5 GHz frequency-hopping transmitters that can cause interference also exist.

Frequency-hopping transmitters do not usually result in as much data corruption as fixed-channel transmitters; however, the existence of a high number of frequency-hopping transmitters in a finite space can result in a high amount of 802.11 data corruption and is especially devastating to VoWiFi communications. The only way to eliminate all-band interference is to locate the interfering device with a spectrum analyzer and remove the interfering device. Figure 15.5 shows a spectrum analyzer capture of a frequency-hopping transmission in the 2.4 GHz ISM band. After locating the sources of interference, the best and simplest solution is to eliminate them entirely.

FIGURE 15.5   All-band RF interference

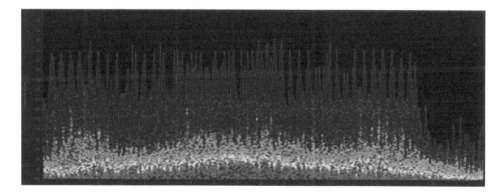

Drivers

As mentioned previously, first generation drivers for client devices often cause connectivity and roaming problems. Always check with the client device manufacturer to make sure you are using the most up-to-date drivers.

Another thing to keep in mind is backward compatibility between newer access points and older client devices. Although the 802.11 amendments make provisions for backward compatibility, the opposite is often true in the real world. Legacy client drivers do not know how to handle the new fields in 802.11 information elements found in beacon and other management frames transmitted by an AP. When new technology is enabled on an AP, legacy clients often can no longer connect. For example, roaming and connectivity problems may be a direct result of lack of support for 802.11k/r/v mechanisms on the client. Figure 15.6 depicts the two information elements that are seen in management frames sent by an AP with 802.11r enabled. The *mobility domain information element (MDIE)* and the *fast BSS transition information element (FTIE)* are fields of information necessary for APs and clients that support Voice-Enterprise roaming capabilities. The drivers of legacy clients that do not support 802.11r may ignore these information fields, and everything will be fine. But the legacy client drivers may also be disrupted by the 802.11r information elements and client connectivity problems may occur.

FIGURE 15.6   Fast BSS transition information element

```
▼ Tag: Mobility Domain
    Tag Number: Mobility Domain (54)
    Tag length: 3
    Mobility Domain Identifier: 0x3b4d
    FT Capability and Policy: 0x00
    .... ...0 = Fast BSS Transition over DS: 0x0
    .... ..0. = Resource Request Protocol Capability: 0x0
▼ Tag: Fast BSS Transition
    Tag Number: Fast BSS Transition (55)
    Tag length: 101
    MIC Control: 0x0000
    0000 0000 .... .... = Element Count: 0
    MIC: 00000000000000000000000000000000
    ANonce: 0000000000000000000000000000000000000000000000000000...
    SNonce: 0000000000000000000000000000000000000000000000000000...
    Subelement ID: PMK-R1 key holder identifier (R1KH-ID) (1)
    Length: 6
    PMK-R1 key holder identifier (R1KH-ID): 08ea4476b568
    Subelement ID: PMK-R0 key holder identifier (R0KH-ID) (3)
    Length: 9
    PMK-R0 key holder identifier (R0KH-ID): AH-76b540
```

Most businesses and corporations can eliminate many of the client connectivity and performance problems by simply upgrading company-owned client devices before updating the WLAN infrastructure. Sadly, the opposite is often more common, with companies spending many hundreds of thousands of dollars on technology upgrades with new access points while still deploying legacy clients.

PoE

In Chapter 12, "Power over Ethernet (PoE)," we discussed the importance of proper PoE power budget planning. WLAN vendors commonly receive support calls from customers complaining that all of a sudden access points randomly begin to reboot. In most cases, the root cause of random rebooting of APs is that the switch power budget has been eclipsed. Very often, if an AP cannot get the power that it needs, the AP will reboot and try again. The power budget of a switch or multiple switches should be monitored to ensure that all devices can maintain power. Active power budget information can usually be seen from the command line of a switch or the GUI interface, or monitored by a centralized network management server (NMS).

Although proper power budgeting can prevent this problem during the design phase, remember that other devices, such as desktop VoIP phones, also use PoE. An extra PoE-powered device might have been plugged into a switch port and the power budget exceeded. Proper power budgeting and monitoring for access points and any other PoE-capable devices is paramount. Power budget problems will grow with the introduction of more 4x4:4 access points, which will require more than the 15.4 watts defined by 802.3af. As we move toward the next generation of 802.11ax access points, the use of 802.3at power will be a necessary requirement.

PoE provided by switches can also be your friend when trying to troubleshoot an AP that for whatever reason may be unreachable from a remote location. For example, an AP may be inoperable due to some sort of processor overhead and can no longer be monitored from an NMS or reachable via SSH. A simple forced reboot of the AP may restore communications. One of the oldest tricks used by network administrators is to *power cycle* the PSE port of the access switch that is providing power to the unresponsive AP. Enabling and disabling the power will force a reboot of the AP that may be locked up.

Firmware Bugs

As mentioned previously, older client firmware and drivers often causes connectivity issues with newer model APs. Conversely, updating access points with new firmware may also result in unexpected WLAN connectivity and more often performance problems. As with any type of networking device, an upgrade of the AP operating system is often needed when WLAN vendors introduce new features and capabilities. Prior to the release of the newer AP code, WLAN vendors perform *regression testing*, which verifies that previously developed features and capabilities still operate and perform in the same manner. Despite the regression testing, new performance bugs may arise once the newer firmware is deployed in the field in enterprise environments.

A suggested deployment best practice would be to upgrade APs in a staging area for testing prior to wide-scale deployment. Another strategy would be to update the APs in one building with active clients and see if any new problems arise. Once the firmware has been validated to be stable, a full upgrade of all the company APs can occur. Larger enterprise companies often have clearly defined change management processes for making any type of network change, including firmware updates. Whenever possible, proper testing should also take place before the introduction of new client devices to the company WLAN.

When troubleshooting possible bugs, it will be necessary to engage with the support personnel of your WLAN vendor. You will most likely be requested to supply tech data logs and possibly packet captures. Many WLAN vendors may also offer an older golden version of firmware that has been thoroughly field tested in enterprise environments and is often considered the best bet in terms of quality assurance. Please understand that older versions of AP or WLAN controller firmware may not offer you the new features that you desire.

Another advantage of updating APs to newer firmware is that previously discovered bugs may be fixed in the new release.

If a new bug is discovered once your APs are updated, the WLAN vendor's support team may recommend that you roll back your APs to the previous version of code until the bug can be addressed. When updating any access points, WLAN controllers, or other networking devices, always read the release notes to verify new features, fixed bugs, and known issues.

Layer 2 Troubleshooting

Wi-Fi radios communicate via 802.11 frame exchanges at the MAC sublayer of the Data-Link layer. Therefore, the next logical layer to troubleshoot in the OSI model is layer 2. This section will cover the many causes of layer 2 retransmissions and the significant adverse effects they cause. When troubleshooting a WLAN, you should always monitor layer 2 retransmission metrics.

Layer 2 Retransmissions

The mortal enemy of WLAN performance is layer 2 retransmissions that occur at the MAC sublayer. As shown in Figure 15.7, all unicast 802.11 frames must be acknowledged. In the trailer of each frame is the *cyclic redundancy check (CRC)*. The receiver 802.11 radio uses the CRC of the frame to confirm the data integrity of the payload of the incoming frame. If the CRC passes, the frame has not been corrupted during transit. The receiver 802.11 radio will then send an 802.11 acknowledgement (ACK) frame back to the original transmitter. Layer 2 ACK frames are used as a method of delivery verification.

FIGURE 15.7 Layer 2 ACK

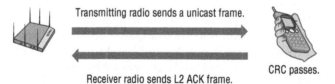

Transmitting radio sends a unicast frame.

CRC passes.

Receiver radio sends L2 ACK frame.

If a collision occurs or any portion of a unicast frame is corrupted, the CRC will fail and the receiving 802.11 radio will not return an ACK frame to the transmitting 802.11 radio. As shown in Figure 15.8, if an ACK frame is not received by the original transmitting radio, the unicast frame is not acknowledged and will have to be retransmitted. Additionally, aggregate frames are acknowledged with a Block ACK; if one of the aggregated frames is corrupted, it will also have to be retransmitted. Any frames that must be retransmitted create extra MAC layer overhead and consume extra airtime in the half-duplex medium.

FIGURE 15.8   Layer 2 retransmission

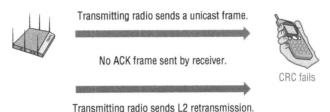

Transmitting radio sends a unicast frame.

No ACK frame sent by receiver.

CRC fails

Transmitting radio sends L2 retransmission.

Excessive layer 2 retransmissions adversely affect the WLAN in two ways. First, layer 2 retransmissions increase airtime consumption overhead and therefore decrease throughput. Although other factors may affect throughput, abundant layer 2 retransmissions are usually the culprit.

Second, if application data has to be retransmitted at layer 2, the delivery of application traffic becomes delayed or inconsistent. Applications such as VoIP depend on the timely and consistent delivery of the IP packet. Excessive layer 2 retransmissions usually result in latency and jitter problems for time-sensitive applications, such as voice and video. When discussing VoIP, people are often confused about the difference between latency and jitter.

Latency   *Latency* is the time it takes to deliver a packet from the source device to the destination device. Ideally, latency should not exceed 50 milliseconds for a VoIP packet. A delay in the delivery (increased latency) of a VoIP packet due to layer 2 retransmissions can result in echo problems.

Jitter   *Jitter* is a variation of latency. Jitter measures how much the latency of each packet varies from the average. If all packets travel at exactly the same speed through the network, jitter will be zero. A high variance in the latency (jitter) is a common result of 802.11 layer 2 retransmissions. Jitter will result in choppy audio communications, and constant retransmissions will result in reduced battery life for VoWiFi phones. Although client devices use jitter buffers to compensate for varying delay, they are usually effective only on delay variations of less than 100 milliseconds. Jitter buffers will not compensate for a high percentage of layer 2 retransmission. Jitter variation of less than 5 milliseconds is the ideal goal for VoWiFi.

Most data applications in a Wi-Fi network can handle a layer 2 retransmission rate of up to 10 percent without any noticeable degradation in performance. However, time-sensitive applications such as VoIP require that higher-layer IP packet loss be no greater than 2 percent. Therefore, Voice over Wi-Fi (VoWiFi) networks need to limit layer 2 retransmissions to 5 percent or less to ensure the timely and consistent delivery of VoIP packets. VoWiFi communication usually is restricted to 5 GHz because maintaining a 5 percent layer 2 retry rate in the over-crowded 2.4 GHz band is rarely possible.

How can you measure layer 2 retransmissions? Any good 802.11 protocol analyzer can track layer 2 retry statistics for the entire WLAN. 802.11 protocol analyzers can also track retry statistics for each individual WLAN access point and client station. As shown in Figure 15.9, layer 2 retry statistics can also usually be centrally monitored using APs across an entire WLAN enterprise from a WLAN controller or from a network management server (NMS). Because the retry statistics are from the perspective of the AP radios, transmit (TX) statistics indicate a measure of downlink retransmissions from the AP radio, whereas receive (RX) statistics indicate a measure of uplink client retransmissions. Even pristine RF environments will always have some layer 2 retransmissions. The goal should be 10 percent or less and 5 percent or less for voice-grade WLANS. Exceeding a 20 percent retry rate will almost always impact performance.

FIGURE 15.9   Layer 2 retransmission statistics

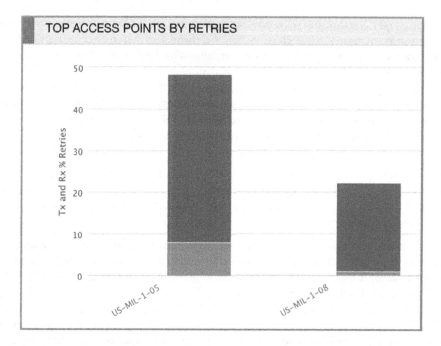

Unfortunately, layer 2 retransmissions are a result of many possible problems. Multipath, RF interference, and low signal-to-noise ratio (SNR) are problems that exist at layer 1 yet result in layer 2 retransmissions. Other causes of layer 2 retransmissions include hidden nodes, mismatched power settings, and adjacent channel interference, which are all usually a symptom of improper WLAN design.

RF Interference

RF interference from a non-802.11 transmitter is the number one cause of layer 2 retransmissions. If frames are corrupted due to RF interference, excessive retransmissions will occur and therefore throughput will be reduced significantly. When layer 2 retransmissions reach excessively high levels intermittently or at different times of the day, the culprit is most likely some sort of interfering device, such as a microwave oven. A good WLAN spectrum analyzer will use RF signature files to help you identify the source of RF inference. To stop the layer 2 retransmissions, locate the interfering device with a spectrum analyzer, and remove the interfering device.

Low SNR

Probably the number two most common cause of layer 2 retransmissions is a low SNR. The *signal-to-noise ratio (SNR)* is an important value because if the background noise is too close to the received signal or the received signal level is too low, data can be corrupted and layer 2 retransmissions will increase. The SNR is not actually a ratio. It is simply the difference in decibels between the received signal and the background noise (noise floor), as shown in Figure 15.10. If an 802.11 radio receives a signal of –70 dBm and the noise floor is measured at –95 dBm, the difference between the received signal and the background noise is 25 dB. The SNR is therefore 25 dB.

FIGURE 15.10   Signal-to-noise ratio

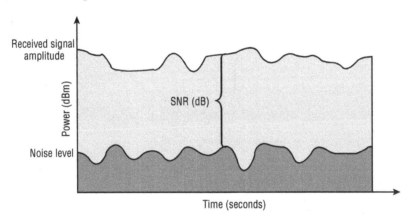

Data transmissions can become corrupted with a very low SNR. If the amplitude of the noise floor is too close to the amplitude of the received signal, data corruption will occur and result in layer 2 retransmissions. An SNR of 25 dB or greater is considered good signal quality, and an SNR of 10 dB or lower is considered poor signal quality. To ensure that frames are not corrupted, many vendors recommend a minimum SNR of 20 dB for data WLANs and a minimum SNR of 25 dB for voice WLANs. SNR of lower than 20 dB will result in AP and client radios shifting to a lower *modulation and coding scheme (MCS)* and lower data rates. The lower data rates consume more airtime and degrade performance. Additionally, an SNR of 10 dB or lower will almost guarantee a higher layer 2 retry rate due to data corruption and thus poor performance.

As you learned in Chapter 13, when designing for coverage, the normal recommended best practice is to provide for a –70 dBm or stronger received signal that is normally well above the noise floor. This will ensure a high SNR. When designing for WLANs with VoWiFi clients, a −65 dBm or stronger signal that is even higher above the noise floor is recommended. Figure 15.11 shows a noise floor of −95 dBm. When a client station receives a −70 dBm signal from an access point, the SNR is 25 dB and; therefore, no data corruption results. However, another client receives a weaker −88 dBm signal and a very low SNR of 7 dB. Because the received signal is so close to the noise floor, data corruption will occur and result in layer 2 retransmissions.

FIGURE 15.11   High and low signal-to-noise ratio

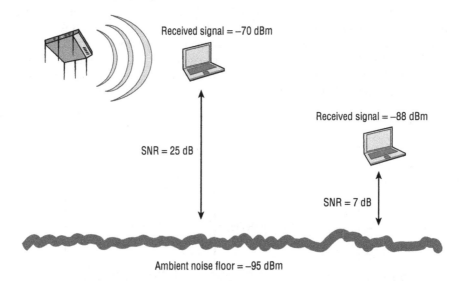

Adjacent Channel Interference

Most Wi-Fi vendors use the term *adjacent channel interference* to refer to the degradation of performance resulting from overlapping frequency space that occurs due to an improper channel reuse design. In the WLAN industry, an adjacent channel is considered to be the next or previous numbered channel. For example, channel 3 is adjacent to channel 2. Figure 15.12 depicts overlapping coverage cells that also have overlapping frequency space that will result in corrupted data and layer 2 retransmissions. Channels 1 and 4, channels 4 and 7, and channels 7 and 11 all have overlapping frequency space in the 2.4 GHz band. Adjacent channel interference can cause both deferred 802.11 transmissions as well as corrupted data, which results in layer 2 retries. The performance issues that result from adjacent cell interference normally occur due to a poorly planned 2.4 GHz WLAN. Using a 2.4 GHz channel reuse pattern of channels 1, 6, and 11 is a standard WLAN design practice that will prevent adjacent cell interference.

FIGURE 15.12   Adjacent cell interference

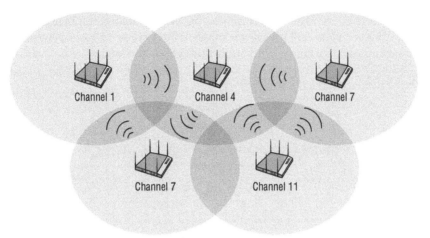

Hidden Node

In Chapter 8, "802.11 Medium Access," you learned about physical carrier sense and the clear channel assessment (CCA). CCA involves listening for 802.11 RF transmissions at the Physical layer; the medium must be clear before a station can transmit. The problem with physical carrier sense is that all stations may not be able to hear each other. Remember that the medium is half-duplex and, at any given time, only one radio can be transmitting. What would happen, however, if one client station that was about to transmit performed a CCA but did not hear another station that was already transmitting? If the station that was about to transmit did not detect any RF energy during its CCA, it would transmit. The problem is that you then have two stations transmitting at the same time. The end result is a collision, and the frames will become corrupted and have to be retransmitted.

The *hidden node* problem occurs when one client station's transmissions are heard by the access point but are not heard by any or all of the other client stations in the basic service set (BSS). The clients would not hear each other and therefore could transmit at the same time. Although the access point would hear both transmissions, because two client radios are transmitting at the same time on the same frequency, the incoming client transmissions would be corrupted.

Figure 15.13 shows the coverage area of an access point. Note that a thick block wall resides between one client station and all the other client stations that are associated to the access point. The RF transmissions of the lone station on the other side of the wall cannot be heard by all of the other 802.11 client stations, even though all the stations can hear the AP. That unheard station is the hidden node. What keeps occurring is that every time the hidden node transmits, another station is also transmitting, and a collision occurs.

The hidden node continues to have collisions with the transmissions from all the other stations that cannot hear it during the clear channel assessment. The collisions continue on a regular basis and so do the layer 2 retransmissions, with the final result being a decrease in throughput. A hidden node can drive retransmission rates above 15 to 20 percent or even higher. Retransmissions, of course, will affect throughput and latency.

FIGURE 15.13 Hidden node—obstruction

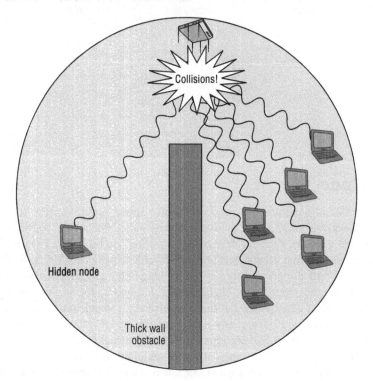

The hidden node problem may exist for several reasons—for example, poor WLAN design or obstructions, such as a newly constructed wall or a newly installed bookcase. A user moving behind some sort of obstruction can cause a hidden node problem. Smartphones and other mobile Wi-Fi devices often become hidden nodes because users take their mobile devices into quiet corners or areas where the RF signal of the phone cannot be heard by other client stations. Users with wireless desktops often place their devices underneath a metal desk and effectively transform the desktop radio into an unheard hidden node.

The hidden node problem can also occur when two client stations are at opposite ends of an RF coverage cell and cannot hear each other, as shown in Figure 15.14. This often happens when the effective coverage cells are too large as a result of the access point's radio transmitting at an excessive power level.

FIGURE 15.14 Hidden node—large coverage cell

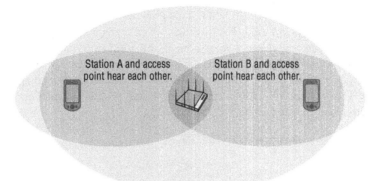

Station A and access point hear each other.

Station B and access point hear each other.

Station A and Station B cannot hear each other.

Another cause of the hidden node problem is distributed antenna systems. Some manufacturers design distributed systems, which are basically made up of a long coaxial cable with multiple antenna elements. Each antenna in the distributed system has its own coverage area. Many companies purchase a *distributed antenna system (DAS)* for cost-saving purposes. Distributed antenna systems and leaky cable systems are specialty solutions that are sometimes deployed because they can also provide coverage for cellular phone frequencies. The hidden node problem, as shown in Figure 15.15, will almost always occur if only a single access point is connected to the DAS. If a DAS solution is deployed, multiple APs will still be needed.

FIGURE 15.15 Hidden node—distributed antenna system

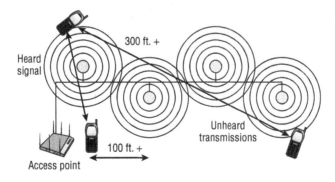

300 ft. +

Heard signal

Unheard transmissions

100 ft. +

Access point

So how do you troubleshoot a hidden node problem? If your end users complain of a degradation of throughput, one possible cause is a hidden node. A protocol analyzer is a useful tool for determining hidden node issues. If the protocol analyzer indicates a higher retransmission rate for the MAC address of one station when compared to the other client stations, chances are a hidden node has been found. Some protocol analyzers even have hidden node alarms based on retransmission thresholds.

Another way is to use request-to-send/clear-to-send (RTS/CTS) to diagnose the problem. If a client device can be configured for RTS/CTS, try lowering the RTS/CTS threshold on a suspected hidden node to about 500 bytes. This level may need to be adjusted, depending on the type of traffic being used. For instance, let us say you have deployed a terminal emulation application in a warehouse environment and a hidden node problem exists. In this case, the RTS/CTS threshold should be set for a much lower size, such as 50 bytes. Use a protocol analyzer to determine the appropriate size. As you learned in Chapter 9, "802.11 MAC," RTS/CTS is a method in which client stations can reserve the medium. Figure 15.16 shows a hidden node initiating an RTS/CTS exchange.

FIGURE 15.16   Hidden node and RTS/CTS

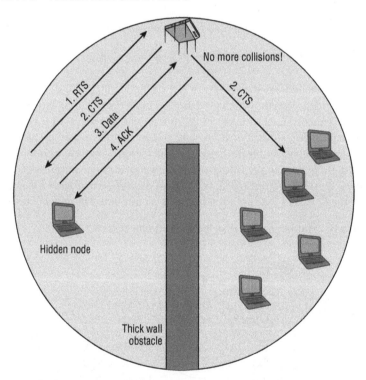

The stations on the other side of the obstacle may not hear the RTS frame from the hidden node, but they will hear the CTS frame sent by the access point. The stations that hear the CTS frame will reset their NAV for the period of time necessary for the hidden node to transmit the data frame and receive its ACK frame. Implementing RTS/CTS on a hidden node will reserve the medium and force all other stations to pause; thus, the collisions and retransmissions will decrease.

Collisions and retransmissions as a result of a hidden node will cause throughput to decrease. RTS/CTS usually decreases throughput as well. However, if RTS/CTS is implemented on a suspected hidden node, throughput will probably *increase* due to the stoppage of the collisions and retransmissions. If you implement RTS/CTS on a suspected hidden node and throughput increases, you have confirmed the existence of a hidden node.

Many legacy 802.11 client devices had the ability to adjust RTS/CTS thresholds. In reality, most current client devices cannot be manually configured for RTS/CTS. Therefore, RTS/CTS as a diagnostic tool from the client side usually is not an option. It should be noted that because hidden node problems occur often, WLAN radios may automatically use RTS/CTS to alleviate hidden node problems. Automatic use of RTS/CTS will more likely occur from AP radios as opposed to client-side radios.

RTS/CTS thresholds can always be manually adjusted on access points. A common use of manually adjusted RTS/CTS is *point-to-multipoint (PtMP)* bridging. The nonroot bridges in a PtMP scenario will not be able to hear each other because they may be miles apart. RTS/CTS should be implemented on nonroot PtMP bridges to eliminate collisions caused by hidden node bridges that cannot hear each other.

The following methods can be used to fix a hidden node problem:

Use RTS/CTS.   Use either a protocol analyzer or RTS/CTS to diagnose the hidden node problem. RTS/CTS can also be used as an automatic or manual fix to the hidden node problem.

Increase power to all stations.   Most client stations have a fixed transmission power output. However, if power output is adjustable on the client side, increasing the transmission power of client stations will increase the transmission range of each station. If the transmission range of all stations is increased, the likelihood of the stations hearing each other also increases. This is usually a bad idea and not a recommended fix, because increasing client power can increase co-channel interference.

Remove the obstacles.   If it is determined that some sort of obstacle is preventing client stations from hearing each other, simply removing the obstacle will solve the problem. Obviously, you cannot remove a wall, but if a metal desk or file cabinet is the obstacle, it can be moved to resolve the problem.

Move the hidden node station.   If one or two stations are in an area where they become unheard, simply moving them within transmission range of the other stations will solve the problem.

Add another access point.   If moving the hidden nodes is not an option, adding another access point in the hidden area to provide coverage will rectify the problem. The best fix for a continuous hidden node problem is to add another AP.

Mismatched Power

Another potential cause of layer 2 retransmissions is mismatched transmit power settings between an access point and a client radio. Communications can break down if a client station's transmit power level is less than the transmit power level of the access point. As a client moves to the outer edges of the coverage cell, the client can "hear" the AP; however, the AP cannot "hear" the client. The good news is that this problem does not occur much in high-density indoor environments. In recent years, there have been significant improvements to access point hardware. Improved receive sensitivity of AP radios has essentially fixed many issues with client and AP mismatched power settings in indoor environments. Problems that occur because of mismatched power settings are more likely to occur outdoors.

As shown in Figure 15.17, if an outdoor access point has a transmit power of 100 mW and a client has a transmit power of 20 mW, the client will hear a unicast frame from the AP because the received signal is within the client station's receive sensitivity capabilities. However, when the client sends an ACK frame back to the AP, the amplitude of the client's transmitted signal has dropped well below the receive sensitivity threshold of the AP's radio. The ACK frame is not "heard" by the AP, which then must retransmit the unicast frame. All the client's transmissions are effectively seen as noise by the AP, and layer 2 retransmissions are the result.

FIGURE 15.17   Mismatched AP and client power

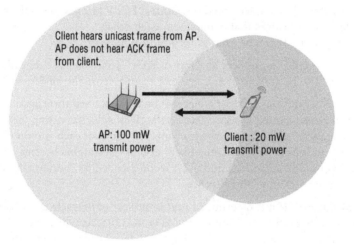

AP/client power problems usually occur because APs are often deployed at full power in order to increase range. Increasing the power of an access point is the wrong way to increase range. If you want to increase the range for the clients, the best solution is to increase the antenna gain of the access point. Most people do not understand the simple concept of *antenna reciprocity*, which means that antennas amplify received signals just as they amplify transmitted signals. A high-gain antenna on an access point will amplify the AP's transmitted signal and extend the range at which the client is capable of hearing the signal. The AP's high-gain antenna will also amplify the received signal from a distant client station.

One way to test whether the mismatched AP/client power problem exists is to listen with a protocol analyzer. An AP/client power problem exists if the frame transmissions of the client station are corrupted when you listen near the access point but are not corrupted when you listen near the client station.

How do you prevent layer 2 retries that are caused by mismatched power settings between the AP and clients? The best solution is to ensure that all of the client transmit power settings match the access point's transmit power. However, significant improvements in AP receive sensitivity have essentially fixed many issues with client and AP mismatch power settings. With that in mind, configuring an access point to transmit at full power is usually not a good idea and may cause this problem as well as many other problems mentioned earlier in this chapter.

In this section, we have focused on mismatched power settings being a symptom of too much transmit power from the access point. In reality, the much bigger problem of mismatched power is usually the fact that clients transmit at higher power than access points deployed indoors. The transmit power of many indoor APs may be 10 mW or less due to high-density design needs. However, most clients, such as smartphones and tablets, may transmit at fixed amplitude of 15 mW or 20 mW. Because clients often transmit at a higher power than the APs and because clients are mobile, the result will be co-channel interference (CCI), as shown in Figure 15.18. As previously mentioned, CCI results in medium contention overhead, which consumes valuable airtime. What most people do not understand about CCI is the fact that clients are the number one cause of CCI. You should understand that CCI is not static and is always changing due to the mobility of client devices.

FIGURE 15.18 Client-based co-channel interference

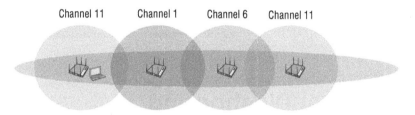

An access point with 802.11k capabilities enabled can inform associated clients to use *transmit power control (TPC)* capabilities to change their transmit amplitude dynamically to match the AP's power. Clients that support TPC will adjust their power to match the AP transmit power, as shown in Figure 15.19. Implementing TPC settings on an AP will greatly reduce co-channel inference caused by clients. It should be noted, however, that legacy clients do not support TPC, and some legacy clients may have connectivity issues if TPC is enabled on the AP.

FIGURE 15.19 Transmit power control

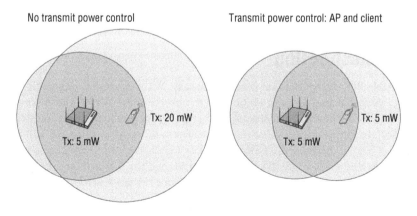

Multipath

As discussed in Chapter 3, "Radio Frequency Fundamentals," *multipath* can cause *intersymbol interference (ISI)*, which causes data corruption. Because of the difference in time between the primary signal and the reflected signals, known as the *delay spread*, the receiver can have problems demodulating the RF signal's information. The delay spread time differential results in corrupted data. If the data is corrupted because of multipath, layer 2 retransmissions will result.

Multipath can be a serious problem when working with legacy 802.11a/b/g equipment. The use of directional antennas will often reduce the number of reflections, and antenna diversity can also be used to compensate for the negative effects of multipath. Multipath is an RF phenomenon that for many years caused destructive effects when older 802.11a/b/g technology was deployed. However, because most WLAN deployments have upgraded to 802.11n or 802.11ac technology, multipath is no longer our enemy. Multipath has a constructive effect with 802.11n/ac transmissions that utilize *multiple-input, multiple-output (MIMO)* antennas and *maximum ratio combining (MRC)* signal-processing techniques.

Security Troubleshooting

802.11 security defines layer 2 authentication methods and layer 2 dynamic encryption. Therefore, WLAN security problems will usually occur at layer 2 and result in WLAN client connection failures. Many WLAN vendors offer layer 2 diagnostic tools to troubleshoot client device authentication and association. These diagnostic tools may be accessible directly from an AP, a WLAN controller, or a cloud-based network management system (NMS). Better diagnostic tools may even offer suggested remediation for detected problems. Security and AAA log files from the WLAN hardware and the RADIUS server are also a great place to start when troubleshooting either PSK or 802.1X/EAP authentication problems. Log files may also be gathered from individual WLAN supplicants.

PSK Troubleshooting

Troubleshooting PSK authentication is relatively easy. WLAN vendor diagnostic tools, log files, or a protocol analyzer can all be used to observe the 4-Way Handshake process between a WLAN client and an access point. Let us first take a look at a successful PSK authentication. In Figure 15.20, you can see the client associate with the AP and then PSK authentication begins. Because the PSK credentials matched on both the access point and the client, a pairwise master key (PMK) is created to seed the 4-Way Handshake. The 4-Way Handshake process is used to create the dynamically generated unicast encryption key that is unique to the AP radio and the client radio.

FIGURE 15.20   Successful PSK authentication

Device Name	Device BSSID	Event Type	Description
12-A-3BD500	08EA443BD514	Basic	Rx assoc req (rssi 40dB)
12-A-3BD500	08EA443BD514	Basic	Tx assoc resp <accept> (status 0, pwr 3dBm)
12-A-3BD500	08EA443BD514	Info	WPA-PSK auth is starting (at if=wifi0.1)
12-A-3BD500	08EA443BD514	Info	Sending 1/4 msg of 4-Way Handshake (at if=wifi0.1)
12-A-3BD500	08EA443BD514	Info	Sending 1/4 msg of 4-Way Handshake (at if=wifi0.1)
12-A-3BD500	08EA443BD514	Info	Received 2/4 msg of 4-Way Handshake (at if=wifi0.1)
12-A-3BD500	08EA443BD514	Info	Sending 3/4 msg of 4-Way Handshake (at if=wifi0.1)
12-A-3BD500	08EA443BD514	Info	Received 4/4 msg of 4-Way Handshake (at if=wifi0.1)
12-A-3BD500	08EA443BD514	Info	PTK is set (at if=wifi0.1)
12-A-3BD500	08EA443BD514	Basic	Authentication is successfully finished (at if=wifi0.1)
12-A-3BD500	08EA443BD514	Info	station sent out DHCP DISCOVER message
12-A-3BD500	08EA443BD514	Info	DHCP server sent out DHCP OFFER message to station
12-A-3BD500	08EA443BD514	Info	DHCP server sent out DHCP OFFER message to station
12-A-3BD500	08EA443BD514	Info	station sent out DHCP REQUEST message
12-A-3BD500	08EA443BD514	Info	DHCP server sent out DHCP ACKNOWLEDGE message to station
12-A-3BD500	08EA443BD514	Basic	DHCP session completed for station
12-A-3BD500	08EA443BD514	Basic	IP 10.5.1.162 assigned for station

Figure 15.20 shows that the 4-Way Handshake process was successful and that the unicast *pairwise transient key (PTK)* is installed on the AP and the client. The layer 2 negotiations are now complete, and it is time for the client to move on to higher layers. So, of course, the next step is that the client obtains an IP address via DHCP. If the client does not get an IP address, there is a networking issue and therefore the problem is not a Wi-Fi issue.

Perhaps a Wi-Fi administrator receives a phone call from an end user who cannot get connected using WPA2-Personal. The majority of problems are at the Physical layer; therefore, Wi-Fi Troubleshooting 101 dictates that the end user first enable and disable the Wi-Fi network card. This should ensure the Wi-Fi NIC drivers are communicating properly with the operating system. If the connectivity problem persists, the problem exists at layer 2. You can then use diagnostic tools, log files, or a protocol analyzer to observe the failed PSK authentication of the WLAN client.

In Figure 15.21, you can see the client associate and then start PSK authentication. However, the 4-Way Handshake process fails. Notice that only two frames of the 4-Handshake complete.

FIGURE 15.21   Unsuccessful PSK authentication

2016-02-22 16:06:48	05-A-764fc0	08EA44764FD4	Info	WPA-PSK auth is starting (at if=wifi0.1)
2016-02-22 16:06:48	05-A-764fc0	08EA44764FD4	Info	Sending 1/4 msg of 4-Way Handshake (at if=wifi0.1)
2016-02-22 16:06:49	05-A-764fc0	08EA44764FD4	Info	Received 2/4 msg of 4-Way Handshake (at if=wifi0.1)
2016-02-22 16:06:52	05-A-764fc0	08EA44764FD4	Info	Sending 1/4 msg of 4-Way Handshake (at if=wifi0.1)
2016-02-22 16:06:52	05-A-764fc0	08EA44764FD4	Info	Received 2/4 msg of 4-Way Handshake (at if=wifi0.1)

The problem is almost always a mismatch of the PSK credentials. If the PSK credentials do not match, a *pairwise master key (PMK)* seed is not properly created, and therefore the 4-Way Handshake fails entirely. The final pairwise transient key (PTK) is never created. A symbiotic relationship exists between authentication and the creation of dynamic encryption keys. If PSK authentication fails, so does the 4-Way Handshake that is used to create the dynamic encryption keys. There is no attempt by the client to get an IP address because the layer 2 process did not complete.

When PSK security is configured, an 8–63 character case-sensitive passphrase is entered by the user or administrator. This passphrase is then used to create the PSK. The passphrase could possibly be improperly configured on the access point; however, the majority of the time, the problem is simple: The end user is incorrectly typing in the passphrase. The administrator should make a polite request to the end user to retype the passphrase slowly and carefully, which is a well-known cure for what is known as *fat-fingering.*

Another possible cause of the failure of PSK authentication could be a mismatch of the chosen encryption methods. An access point might be configured to support only WPA2 (CCMP-AES), which a legacy WPA (TKIP) client does not support. A similar failure of the 4-Way Handshake would occur.

802.1X/EAP Troubleshooting

PSK authentication (also known as WPA2-Personal) is simple to troubleshoot because the authentication method was designed to be uncomplicated. However, troubleshooting the more complex 802.1X/EAP authentication (also known as WPA2-Enterprise) is a bigger challenge because multiple points of failure exist.

802.1X is a port-based access control standard that defines the mechanisms necessary to authenticate and authorize devices to network resources. The 802.1X authorization framework consists of three main components, each with a specific role. These three 802.1X components work together to ensure that only properly validated users and devices are authorized to access network resources. The three 802.1X components are known as the supplicant, authenticator, and authentication server. The supplicant is the user or device that is requesting access to network resources. The authentication server's job is to validate the supplicant's credentials. The authenticator is a gateway device that sits in the middle between the supplicant and authentication server, controlling or regulating the supplicant's access to the network.

802.1X/EAP Troubleshooting Zones

In the example shown in Figure 15.22, the supplicant is a Wi-Fi client, an AP is the authenticator, and an external RADIUS server functions as the authentication server. The RADIUS server can maintain an internal user database or query an external database, such as an LDAP database. Extensible Authentication Protocol (EAP) is used within the 802.1X/EAP framework to validate users at layer 2. The supplicant will use an EAP protocol to communicate with the authentication server at layer 2. The Wi-Fi client will not be allowed to communicate at the upper layers of 3–7 until the RADIUS server has validated the supplicant's identity at layer 2.

FIGURE 15.22 802.1X/EAP

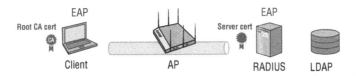

The AP blocks all of the supplicant's higher-layer communications until the supplicant is validated. When the supplicant is validated, higher layer communications are allowed through a virtual "controlled port" on the AP (the authenticator). Layer 2 EAP authentication traffic is encapsulated in RADIUS packets between the authenticator and the authentication server. The authenticator and the authentication server also validate each other with a *shared secret*.

Better versions of EAP, such as EAP-PEAP and EAP-TTLS, use *tunneled authentication* to protect the supplicant credentials from offline dictionary attacks. Certificates are used within the EAP process to create an encrypted SSL/TLS tunnel and to ensure a secure authentication exchange. As illustrated in Figure 15.23, a server certificate resides on the RADIUS server, and the root CA public certificate must be installed on the supplicant. As mentioned earlier, there are many potential points of failure in an 802.1X/EAP process. However, as depicted in Figure 15.23, there are effectively two troubleshooting zones within the 802.1X/EAP framework where failures will occur. Troubleshooting zone 1 consists of the backend communications between the authenticator, the authentication server, and the LDAP database. Troubleshooting zone 2 resides solely on the supplicant device that is requesting access.

FIGURE 15.23 802.1X/EAP troubleshooting zones

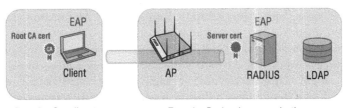

Zone 1: Backend Communication Problems

Zone 1 should always be investigated first. If an AP and a RADIUS server cannot communicate with each other, the entire authentication process will fail. If the RADIUS server and the LDAP database cannot communicate, the entire authentication process will also fail.

Figure 15.24 shows a capture of a supplicant (Wi-Fi client) trying to contact a RADIUS server. The authenticator forwards the request to the RADIUS server, but the RADIUS server never responds. The AP (authenticator) then sends a deauthentication frame to the Wi-Fi client because the process failed. This is an indication that there is a backend communication problem in the first troubleshooting zone.

FIGURE 15.24 The RADIUS server does not respond.

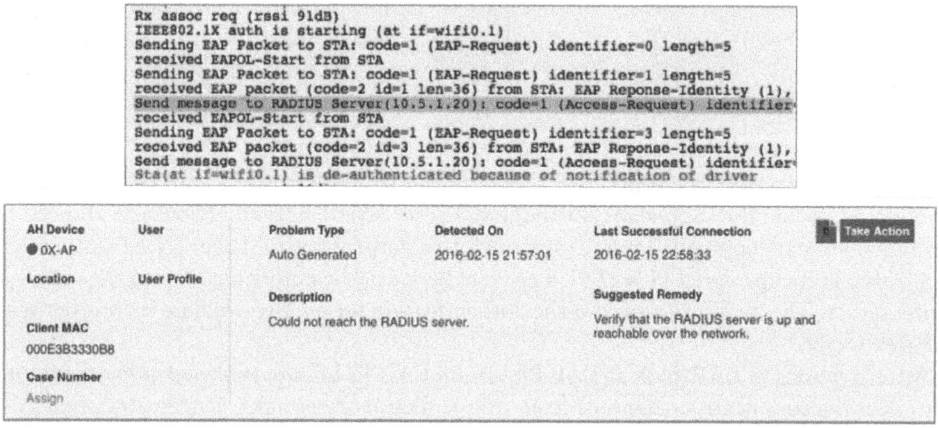

```
Rx assoc req (rssi 91dB)
IEEE802.1X auth is starting (at if=wifi0.1)
Sending EAP Packet to STA: code=1 (EAP-Request) identifier=0 length=5
received EAPOL-Start from STA
Sending EAP Packet to STA: code=1 (EAP-Request) identifier=1 length=5
received EAP packet (code=2 id=1 len=36) from STA: EAP Reponse-Identity (1),
Send message to RADIUS Server(10.5.1.20): code=1 (Access-Request) identifier
received EAPOL-Start from STA
Sending EAP Packet to STA: code=1 (EAP-Request) identifier=3 length=5
received EAP packet (code=2 id=3 len=36) from STA: EAP Reponse-Identity (1),
Send message to RADIUS Server(10.5.1.20): code=1 (Access-Request) identifier
Sta(at if=wifi0.1) is de-authenticated because of notification of driver
```

AH Device	User	Problem Type	Detected On	Last Successful Connection	Take Action
● 0X-AP		Auto Generated	2016-02-15 21:57:01	2016-02-15 22:58:33	
Location	User Profile				
		Description		**Suggested Remedy**	
Client MAC		Could not reach the RADIUS server.		Verify that the RADIUS server is up and reachable over the network.	
000E3B3330B8					
Case Number					
Assign					

As shown in Figure 15.25, if the RADIUS server never responds to the supplicant, there are four possible points of failure in the first troubleshooting zone:

- Shared secret mismatch
- Incorrect IP settings on the AP or the RADIUS server
- Authentication port mismatch
- LDAP query failure

The first three possible points of failure are between the authenticator and the RADIUS server. The authenticator and the authentication server validate each other with a *shared secret*. The most common failure in RADIUS communications is that the shared secret has been typed in wrong on either the RADIUS server or the AP functioning as the authenticator.

The second most common failure in RADIUS communications is simply misconfigured IP networking settings. The AP must know the correct IP address of the RADIUS server. Likewise, the RADIUS server must be configured with the IP addresses of any APs or WLAN controllers functioning as authenticators. Incorrect IP settings will result in miscommunications.

FIGURE 15.25 Points of failure—802.1X/EAP troubleshooting zone 1

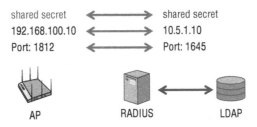

shared secret ←——→ shared secret
192.168.100.10 ←——→ 10.5.1.10
Port: 1812 ←——→ Port: 1645

AP   RADIUS   LDAP

• Shared secret mismatch
• Incorrect IP settings on AP or RADIUS server
• Authentication port mismatch (default is 1812)
• LDAP communications error

The third point of failure between an authenticator and an authentication server is a mismatch of RADIUS authentication ports. UDP ports 1812 and 1813 are defined as the industry standard ports used for RADIUS authentication and accounting. However, some older RADIUS servers may be using UDP ports 1645 and 1646. UDP ports 1645 and 1646 are rarely used anymore but do occasionally show up on older RADIUS servers. Although not a common point of failure, if the authentication ports do not match between a RADIUS server and the AP, the authentication process will fail.

The final point of failure on the backside is a failure of the LDAP query between a RADIUS server and the LDAP database. A standard domain account can be used for LDAP queries; however, if the account has expired or if there is a networking issue between the RADIUS server and the LDAP server, the entire 802.1X/EAP authentication process will fail.

 Real World Scenario

Which Tools Can Be Used to Troubleshoot 802.1X/EAP Backend Communications?

The good news is that multiple troubleshooting resources are available to troubleshoot zone 1. Several WLAN vendors offer built-in diagnostic tools to test the communications between an authenticator and a RADIUS server as well as LDAP communications. Depending on the WLAN vendor and architecture, the authenticator may be either an access point or a WLAN controller. As shown in Figure 15.26, a standard domain account and password can be used to test the RADIUS and EAP communications.

Several software utilities are also available to test backend 802.1X/EAP communications. EAPTest is a commercial test utility available for macOS. More information can be found at www.ermitacode.com/eaptest.html. RADLogin is a free test utility for the Windows and Linux platforms. More information can be found at www.iea-software.com/products/radlogin4.cfm. RADIUS server and LDAP database logs are also great resources for troubleshooting 802.1X/EAP backend communication problems. In worst-case scenarios, a wired protocol analyzer may be needed to capture RADIUS packets. Many of these test tools can also be used to troubleshoot issues with RADIUS attributes, which can be leveraged during 802.1X/EAP authentication for role-based access control (RBAC).

FIGURE 15.26   802.1X/EAP backend diagnostic tool

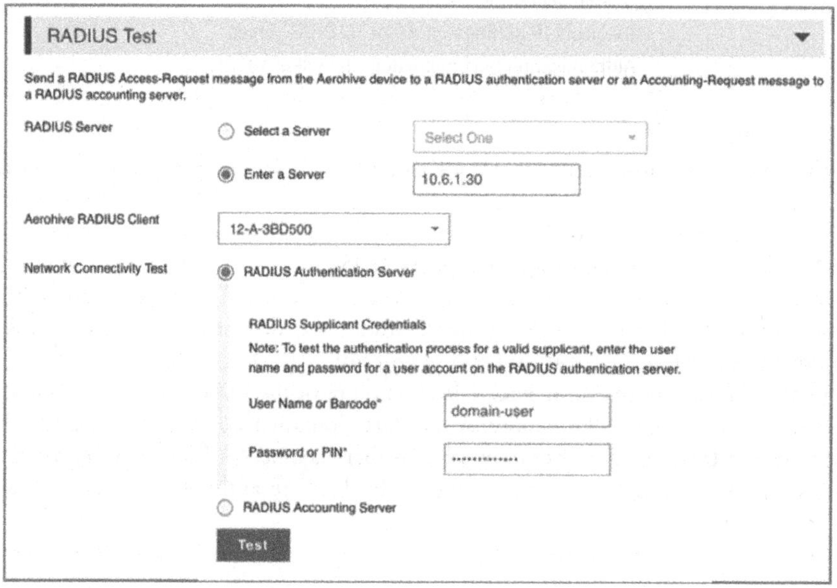

Zone 2: Supplicant Certificate Problems

If all backend communications between the authenticator and the RADIUS server are functioning properly, then the 802.1X/EAP troubleshooting focus should now be redirected to zone 2. In simpler words, the culprit is the client (supplicant). Problems with the supplicant usually revolve around either certificate issues or client credential issues. Take a look at Figure 15.27. Note that the RADIUS server is responding and therefore verifying that the backend communications are good. Also notice an SSL tunnel negotiation starts and finishes successfully. This 802.1X/EAP diagnostic log confirms that the certificate exchange

was successful and that an SSL/TLS tunnel was successfully created to protect the supplicant credentials.

FIGURE 15.27   Successful SSL/TLS tunnel creation

```
Send message to RADIUS Server(10.5.1.129): code=1 (Access-Request) identifier=
RADIUS: SSL negotiation, send server certificate and other message
Receive message from RADIUS Server: code=11 (Access-Challenge) identifier=109
Sending EAP Packet to STA: code=1 (EAP-Request) identifier=3 length=280
received EAP packet (code=2 id=3 len=208) from STA: EAP Reponse-PEAP (25)
Send message to RADIUS Server(10.5.1.129): code=1 (Access-Request) identifier=
RADIUS: SSL connection established
Receive message from RADIUS Server: code=11 (Access-Challenge) identifier=110
Sending EAP Packet to STA: code=1 (EAP-Request) identifier=4 length=65
received EAP packet (code=2 id=4 len=6) from STA: EAP Reponse-PEAP (25)
Send message to RADIUS Server(10.5.1.129): code=1 (Access-Request) identifier=
RADIUS: SSL negotiation is finished successfully
Receive message from RADIUS Server: code=11 (Access-Challenge) identifier=111
Sending EAP Packet to STA: code=1 (EAP-Request) identifier=5 length=43
received EAP packet (code=2 id=5 len=59) from STA: EAP Reponse-PEAP (25)
Send message to RADIUS Server(10.5.1.129): code=1 (Access-Request) identifier=
RADIUS: PEAP inner tunneled conversion
```

Figure 15.28 displays an 802.1X/EAP diagnostic log where you can see the SSL negotiation begin and the server certificate sent from the RADIUS server to the supplicant.
However, the SSL/TLS tunnel is never created, and EAP authentication fails. If the SSL/
TLS tunnel cannot be established, this is an indication that there is some sort of certificate
problem.

FIGURE 15.28   Unsuccessful SSL/TLS tunnel creation

```
Rx assoc req (rssi 95dB)
IEEE802.1X auth is starting (at if=wifi0.1)
Sending EAP Packet to STA: code=1 (EAP-Request) identifier=0 length=5
received EAP packet (code=2 id=0 len=16) from STA: EAP Reponse-Identity (1),
Send message to RADIUS Server(10.5.1.129): code=1 (Access-Request) identifier
RADIUS: EAP start with type peap
Receive message from RADIUS Server: code=11 (Access-Challenge) identifier=50
Sending EAP Packet to STA: code=1 (EAP-Request) identifier=1 length=6
received EAP packet (code=2 id=1 len=105) from STA: EAP Reponse-PEAP (25)
Send message to RADIUS Server(10.5.1.129): code=1 (Access-Request) identifier
RADIUS: SSL negotiation, receive client hello message
Receive message from RADIUS Server: code=11 (Access-Challenge) identifier=51
Sending EAP Packet to STA: code=1 (EAP-Request) identifier=2 length=1024
received EAP packet (code=2 id=2 len=6) from STA: EAP Reponse-PEAP (25)
Send message to RADIUS Server(10.5.1.129): code=1 (Access-Request) identifier
RADIUS: SSL negotiation, send server certificate and other message
Receive message from RADIUS Server: code=11 (Access-Challenge) identifier=52
Sending EAP Packet to STA: code=1 (EAP-Request) identifier=3 length=280
received EAP packet (code=2 id=3 len=6) from STA: EAP Reponse-PEAP (25)
Send message to RADIUS Server(10.5.1.129): code=1 (Access-Request) identifier
RADIUS: SSL negotiation, send server certificate and other message
Receive message from RADIUS Server: code=11 (Access-Challenge) identifier=53
Sending EAP Packet to STA: code=1 (EAP-Request) identifier=4 length=6
Sta(at if=wifi0.1) is de-authenticated because of notification of driver
```

You can usually verify that there is a certificate problem by editing the supplicant client
software settings and temporarily disabling the validation of the server certificate, as shown
in Figure 15.29. If EAP authentication is successful after you temporarily disable the validation of the server certificate, then you have confirmed there is a problem with the implementation of the certificates within the 802.1X/EAP framework. Please note that this is not
a fix but an easy way to verify that some sort of certificate issue exists.

FIGURE 15.29 Server certificate validation

A whole range of certificate problems could be causing the SSL/TLS tunnel not to be successfully created. The most common certificate issues are as follows:

- The root CA certificate is installed in the incorrect certificate store.

- The incorrect root certificate is chosen.

- The server certificate has expired.

- The root CA certificate has expired.

- The supplicant clock settings are incorrect.

The root CA certificate needs to be installed in the Trusted Root Certificate Authorities store of the supplicant device. A common mistake is to install the root CA certificate in the default location, which is typically the personal store of a Windows machine. Another common mistake is to select the incorrect root CA certificate with the supplicant configuration. The SSL/TLS tunnel will fail because the incorrect root CA certificate will not be able to validate the server certificate. Digital certificates are also time-based, and a common problem is that the server certificate has expired. Although not as common, the root CA certificate can also have expired. The clock settings on the supplicant might be incorrect and predate the creation of either certificate.

Because of all the possible points of failure involving certificates, troubleshooting 802.1X/EAP certificate problems in zone 2 can be difficult. Additionally, there are more potential problems with certificates. The server certificate configuration may be incorrect on the RADIUS server. In other words, the certificate problem exists back in troubleshooting zone 1. What if EAP-TLS is the deployed authentication protocol? EAP-TLS requires the provisioning of client-side certificates in addition to server certificates. Client certificates add an additional layer of possible certificate troubleshooting on the supplicant as well as within the private PKI infrastructure that has been deployed.

There is one final complication that might result in the failure of tunneled authentication. The chosen layer 2 EAP protocols must match on both the supplicant and the authentication server. For example, the authentication will fail if PEAPv0 (EAP-MSCHAPv2) is selected on the supplicant while PEAPv1 (EAP-GTC) is configured on the RADIUS server. Although the SSL/TLS tunnel might still be created, the inner tunnel authentication protocol does not match and authentication will fail. Although it is possible for multiple flavors of EAP to operate simultaneously over the same 802.1X framework, the EAP protocols must match on both the supplicant and the authentication server.

Zone 2: Supplicant Credential Problems

If you can verify that you do not have any certificate issues and the SSL/TLS tunnel is indeed established, the supplicant problems are credential failures. Figure 15.30 displays an 802.1X/EAP diagnostic log where the RADIUS server is rejecting the supplicant credentials. The following supplicant credential problems are possible:

- Expired password or user account

- Wrong password

- User account does not exist in LDAP.

- Machine account has not been joined to the Windows domain.

FIGURE 15.30   RADIUS server rejects supplicant credentials

```
RADIUS: SSL connection established
Receive message from RADIUS Server: code=11 (Access-Challenge) identifier=127 length=123
Sending EAP Packet to STA: code=1 (EAP-Request) identifier=5 length=65
received EAP packet (code=2 id=5 len=6) from STA: EAP Reponse-PEAP (25)
Send message to RADIUS Server(10.5.1.129): code=1 (Access-Request) identifier=128 length=176
RADIUS: SSL negotiation is finished successfully
Receive message from RADIUS Server: code=11 (Access-Challenge) identifier=128 length=101
Sending EAP Packet to STA: code=1 (EAP-Request) identifier=6 length=43
received EAP packet (code=2 id=6 len=43) from STA: EAP Reponse-PEAP (25)
Send message to RADIUS Server(10.5.1.129): code=1 (Access-Request) identifier=129 length=213
RADIUS: PEAP inner tunneled conversion
Receive message from RADIUS Server: code=11 (Access-Challenge) identifier=129 length=117
Sending EAP Packet to STA: code=1 (EAP-Request) identifier=7 length=59
received EAP packet (code=2 id=7 len=91) from STA: EAP Reponse-PEAP (25)
Send message to RADIUS Server(10.5.1.129): code=1 (Access-Request) identifier=130 length=261
RADIUS: PEAP Tunneled authentication was rejected. NTLM_auth failed for Logon failure (0xc00(
Receive message from RADIUS Server: code=11 (Access-Challenge) identifier=130 length=101
Sending EAP Packet to STA: code=1 (EAP-Request) identifier=8 length=43
received EAP packet (code=2 id=8 len=43) from STA: EAP Reponse-PEAP (25)
Send message to RADIUS Server(10.5.1.129): code=1 (Access-Request) identifier=131 length=213
RADIUS: rejected user 'user' through the NAS at 10.5.1.129.
Authentication is terminated (at if=wifi0.1) because it is rejected by RADIUS server
Sending EAP Packet to STA: code=4 (EAP-Failure) identifier=8 length=4
Sta(at if=wifi0.1) is de-authenticated because of notification of driver
```

If the user credentials do not exist in the LDAP database or the credentials have expired, authentication will fail. Unless single sign-on capabilities have been implemented on the supplicant, there is always the possibility that the domain user password can be incorrectly typed by the end user.

Another common error is that the Wi-Fi supplicant has been improperly configured for machine authentication, and the RADIUS server has been configured only for user authentication. In Figure 15.31, you see a diagnostic log that clearly shows the machine credentials being sent to the RADIUS server and not the user credentials. The RADIUS server was expecting a user account and therefore rejected the machine credentials because no machine accounts had been set up for validation. In the case of Windows, the machine credentials are based on a *System Identifier (SID)* value that is stored on a Windows domain computer after being joined to a Windows domain with Active Directory.

FIGURE 15.31   Machine authentication failure

```
Send message to RADIUS Server(10.5.1.129): code=1 (Access-Request) identifier=151 length=203,
RADIUS: SSL negotiation, send server certificate and other message
Receive message from RADIUS Server: code=11 (Access-Challenge) identifier=151 length=340
Sending EAP Packet to STA: code=1 (EAP-Request) identifier=4 length=280
received EAP packet (code=2 id=4 len=17) from STA: EAP Reponse-PEAP (25)
Send message to RADIUS Server(10.5.1.129): code=1 (Access-Request) identifier=152 length=214,
RADIUS:
RADIUS: rejected user 'host/TRAINING-PC16.ah-lab.local' through the NAS at 10.5.1.129.
Authentication is terminated (at if=wifi0.1) because it is rejected by RADIUS server
Sending EAP Packet to STA: code=4 (EAP-Failure) identifier=4 length=4
Sta(at if=wifi0.1) is de-authenticated because of notification of driver
```

Of course, a WLAN administrator can always verify that all is well with an 802.1X/EAP client session. Always remember that a byproduct of the EAP process is the generation of the pairwise master key (PMK) that seeds the 4-Way Handshake exchange. Figure 15.32 shows the EAP process completing; the pairwise master key (PMK) is sent to the AP from the RADIUS server. The 4-Way Handshake process then begins to dynamically generate the pairwise transient key (PTK) that is unique between the radios of the AP and the client device. When the 4-Way Handshake completes, the encryption keys are installed, and the layer 2 connection is completed. The virtual controlled port on the authenticator opens up for this Wi-Fi client. The supplicant can now proceed to higher layers and get an IP address. If the client does not get an IP address, there is a networking issue, and therefore the problem is not a Wi-Fi issue.

FIGURE 15.32   4-Way Handshake

```
Receive message from RADIUS Server: code=2 (Access-Accept) identifier=125
PMK is got from RADIUS server (at if=wifi0.1)
Sending EAP Packet to STA: code=3 (EAP-Success) identifier=5 length=4
Sending 1/4 msg of 4-Way Handshake (at if=wifi0.1)
Received 2/4 msg of 4-Way Handshake (at if=wifi0.1)
Sending 3/4 msg of 4-Way Handshake (at if=wifi0.1)
Received 4/4 msg of 4-Way Handshake (at if=wifi0.1)
PTK is set (at if=wifi0.1)
Authentication is successfully finished (at if=wifi0.1)
IP 10.5.10.100 assigned for station
station sent out DHCP REQUEST message
DHCP server sent out DHCP ACKNOWLEDGE message to station
DHCP session completed for station
```

One final consideration when troubleshooting 802.1X/EAP is RADIUS attributes. RADIUS attributes can be leveraged during 802.1X/EAP authentication for role-based access control, providing custom settings for different groups of users or devices. For example, different groups of users may be assigned to different VLANs even though they are connected to the same 802.1X/EAP SSID. If the RADIUS attribute configuration does not match on the authenticator and the RADIUS server, users might be assigned to default role or VLAN assignments. In worst-case scenarios, a RADIUS attribute mismatch might result in authentication failure.

VPN Troubleshooting

VPNs are rarely used anymore as the primary method of security for WLANs. Occasionally, a VPN may be used to provide data privacy across a point-to-point 802.11 wireless bridge link. IPsec VPNs are still commonly used to connect remote branch offices with corporate offices across WAN links. Although a site-to-site VPN link is not necessarily a WLAN security solution, the wireless user traffic that originated at the remote location may be required to traverse through a VPN tunnel. Most WLAN vendors also offer VPN capabilities within their solution portfolio. For example, a WLAN may offer a VPN solution where user traffic is tunneled from a remote AP or a WLAN branch router to a VPN server gateway. Third-party VPN overlay solutions are often also used.

The creation of an IPsec tunnel involves two phases, called *Internet Key Exchange (IKE)* phases:

IKE Phase 1   The two VPN endpoints authenticate one another and negotiate keying material. The result is an encrypted tunnel used by Phase 2 for negotiating the *Encapsulating Security Payload (ESP)* security associations.

IKE Phase 2   The two VPN endpoints use the secure tunnel created in Phase 1 to negotiate ESP *security associations (SAs)*. The ESP SAs are used to encrypt user traffic that traverses between the endpoints.

The good news is that any quality VPN solution offers diagnostic tools and commands to troubleshoot both IKE phases. The following are some of the common problems that can occur if IKE Phase 1 fails:

- Certificate problems
- Incorrect networking settings
- Incorrect NAT settings on the external firewall

Figure 15.33 shows the results of an IKE Phase 1 diagnostic command executed on a VPN server. IPsec uses digital certificates during Phase 1. If IKE Phase 1 fails due to a certificate problem, ensure that you have the correct certificates installed properly on the VPN endpoints. Also remember that certificates are time based. Very often, a certificate problem during IKE Phase 1 is simply an incorrect clock setting on either VPN endpoint.

FIGURE 15.33   IPsec Phase 1—certificate failure

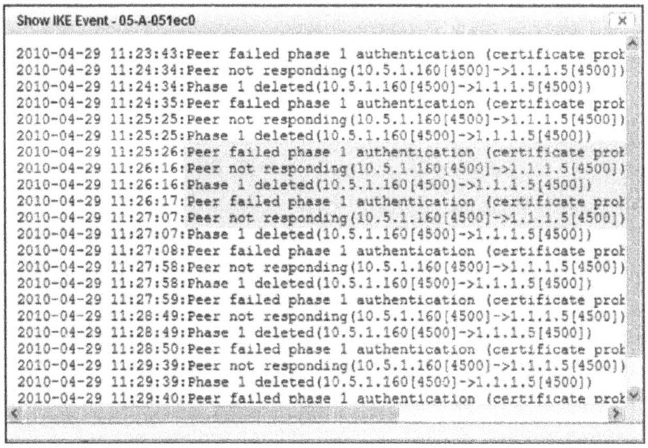

In Figure 15.34 you see the results of an IKE Phase 1 diagnostic command executed on a VPN server that indicates a possible networking error due to incorrect configuration. IPsec uses private IP addresses for tunnel communications and also uses external IP addresses, which are normally the public IP address of a firewall. If an IKE Phase 1 failure occurs as shown in Figure 15.34, check the internal and external IP settings on the VPN devices. If an external firewall is being used, also check the *Network Address Translation (NAT)* settings. Another common networking problem that causes VPNs to fail is that needed firewall ports are blocked. Ensure that the following ports are open on any firewall that the VPN tunnel may traverse:

- UDP 500 (IPsec)
- UDP 4500 (NAT Transversal)

FIGURE 15.34 IPsec Phase 1—networking failure

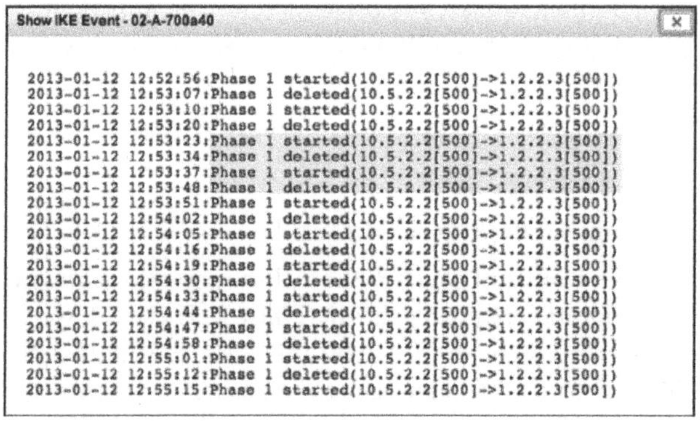

If you can confirm that IKE Phase 1 is successful yet the VPN is still failing, then IKE Phase 2 is the likely culprit. The following are some of the common problems that can occur if IKE Phase 2 fails:

- Mismatched transform sets between the client and server (encryption algorithm, hash algorithm, and so forth)
- Mixing different vendor solutions

Figure 15.35 shows the successful results of an IKE Phase 2 diagnostic command executed on a VPN server. If this command had indicated a failure, be sure to check both encryption and hash settings on the VPN endpoints. Check other IPsec settings, such as *tunnel mode*. You will need to verify that all settings match on both ends. IKE Phase 2 problems often occur when different VPN vendors are used on opposite sides of the intended VPN tunnel. Although IPsec is a standards-based suite of protocols, mixing different VPN vendor solutions often results in more troubleshooting.

FIGURE 15.35   IPsec Phase 2—Success

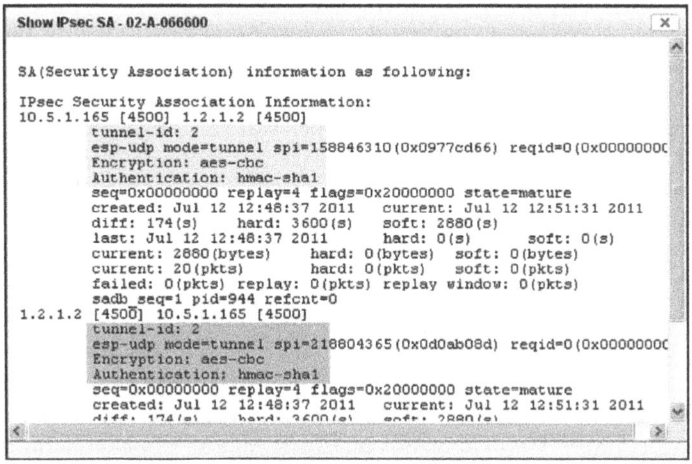

```
Show IPsec SA - 02-A-066600                                              ☒

SA(Security Association) information as following:

IPsec Security Association Information:
10.5.1.165 [4500] 1.2.1.2 [4500]
        tunnel-id: 2
        esp-udp mode=tunnel spi=158846310(0x0977cd66) reqid=0(0x0000000C
        Encryption: aes-cbc
        Authentication: hmac-sha1
        seq=0x00000000 replay=4 flags=0x20000000 state=mature
        created: Jul 12 12:48:37 2011   current: Jul 12 12:51:31 2011
        diff: 174(s)     hard: 3600(s)    soft: 2880(s)
        last: Jul 12 12:48:37 2011       hard: 0(s)        soft: 0(s)
        current: 2880(bytes)     hard: 0(bytes)    soft: 0(bytes)
        current: 20(pkts)        hard: 0(pkts)     soft: 0(pkts)
        failed: 0(pkts) replay: 0(pkts) replay window: 0(pkts)
        sadb_seq=1 pid=944 refcnt=0
1.2.1.2 [4500] 10.5.1.165 [4500]
        tunnel-id: 2
        esp-udp mode=tunnel spi=218804365(0x0d0ab08d) reqid=0(0x0000000C
        Encryption: aes-cbc
        Authentication: hmac-sha1
        seq=0x00000000 replay=4 flags=0x20000000 state=mature
        created: Jul 12 12:48:37 2011   current: Jul 12 12:51:31 2011
        diff: 174(s)     hard: 3600(s)    soft: 2880(s)
```

Roaming Troubleshooting

Mobility is the whole point behind wireless network access. 802.11 clients need the ability to seamlessly roam between access points without any interruption of service or degradation of performance. As shown in Figure 15.36, seamless roaming has become even more important in recent years because of the proliferation of handheld personal Wi-Fi devices, such as smartphones and tablets.

FIGURE 15.36   Seamless roaming

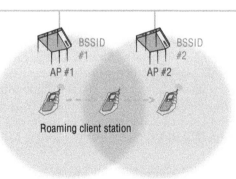

Roaming client station

The most common roaming problems are a result of either bad client drivers or bad WLAN design. The very common *sticky client problem* is when client stations stay connected to their original AP and do not roam to a new AP of closer vicinity and stronger signal. The sticky client problem is often a result of APs in close physical vicinity with transmit power levels that are too high. The sticky client problem and other roaming performance issues can usually be avoided with proper WLAN design and site surveys. Good roaming design entails defining proper primary coverage and secondary coverage from the client perspective, as discussed in Chapter 13.

Client stations, not the access point, decide whether or not to roam between access points. Some vendors may involve the access point or WLAN controller in the roaming decision, but ultimately, the client station initiates the roaming process with a reassociation request frame. The method by which a client station decides to roam depends on unique thresholds determined by the manufacturer of the 802.11 client radio. Roaming thresholds are usually defined by RSSI and SNR; however, other variables, such as error rates and retransmissions, may also have a part in the roaming decision. Client stations that support 802.11k may obtain neighbor reports from 802.11k-compliant APs, which provide the client stations with additional input so that they can make better roaming decisions. Support for 802.11k is becoming increasingly important in today's complex RF environments.

Roaming problems will occur if there is not enough duplicate secondary coverage. No secondary coverage will effectively create a roaming dead zone, and connectivity might even temporarily be lost. On the flip side, too much secondary coverage will also cause roaming problems. For example, a client station may stay associated with its original AP and not connect to a second access point even though the station is directly underneath the second access point. As previously mentioned, this is commonly referred to as the sticky client problem. Too many potential APs heard by a client may also result in a situation in which the client device is constantly switching back and forth between two or more APs on different channels. If a client station can also hear dozens of APs on the same channel with very strong signals, degradation in performance will occur due to medium contention overhead.

Roaming performance also has a direct relationship to WLAN security. Every time a client station roams, new encryption keys must be generated between the AP and the client station radios via the 4-Way Handshake. When using 802.1X/EAP security, roaming can be especially troublesome for VoWiFi and other time-sensitive applications. Due to the multiple frame exchanges between the authentication server and the supplicant, an 802.1X/EAP authentication can take 700 milliseconds (ms) or longer for the client to authenticate. VoWiFi requires a handoff of 150 ms or less to avoid a degradation of the quality of the call, or even worse, a loss of connection. Therefore, faster, secure roaming handoffs are required.

Changes in the WLAN environment can also cause roaming headaches. RF interference will always affect the performance of a wireless network and can make roaming problematic as well. Very often new construction in a building will affect the coverage of a WLAN and create new dead zones. If the physical environment where the WLAN is deployed changes, the coverage design might have to change as well. It is always a good idea to periodically conduct a validation survey to monitor changes in coverage patterns.

Troubleshooting roaming by using a protocol analyzer is tricky because the reassociation roaming exchanges occur on multiple channels. For example, in order to troubleshoot a client roaming between channels 1, 6, and 11, you would need three separate protocol analyzers on three separate laptops that would produce three separate frame captures. As shown in Figure 15.37, three USB radios can be configured to capture frames on channels 1, 6, and 11 simultaneously. All three radios are connected to a USB hub and save the frame captures of all three channels into a single time-stamped capture file. Several WLAN analyzer vendors offer multichannel monitoring capabilities for both the 2.4 GHz and 5 GHz frequency band. The roaming history of a WLAN client can also be collected from AP log files and visualized in WLAN network management solutions.

FIGURE 15.37 Multichannel monitoring and analysis

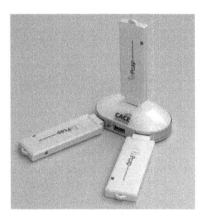

Sybex Publishing's *CWSP - Certified Wireless Security Professional Study Guide: Exam CWSP-205, 2nd Edition (2016)* devotes an entire chapter to fast secure roaming, such as *opportunistic key caching (OKC)* and *fast BSS transition (FT)*. Both OKC and FT produce roaming handoffs of closer to 50 ms even when 802.1X/EAP is the chosen security solution. Both OKC and FT use key distribution mechanisms so that roaming clients do not have to reauthenticate every time they roam. OKC is now considered a legacy method of fast secure roaming. The FT roaming mechanisms defined in both 802.11r and Voice-Enterprise are considered the standard. Many WLAN enterprise vendor APs are now certified for Voice-Enterprise by the Wi-Fi Alliance. Please note that any client devices that were manufactured before 2012 simply will not support 802.11k/r/v operations. Although the bulk of legacy client devices do not support Voice-Enterprise capabilities, client-side support is growing and more commonplace.

Most security-related roaming problems are based on the fact that many clients simply do not support either OKC or fast BSS transition (FT). Client-side support for any device that will be using voice applications and 802.1X/EAP is critical. Proper planning and verification of client-side and AP support for OKC or FT will be necessary. Figure 15.38 shows the results of a diagnostic command that displays the roaming cache of an access point. This type of diagnostic command can verify if PMKs are being forwarded between access points. In this situation, FT is enabled on the AP and supported on the client radio. You can verify the MAC address of the supplicant and the authenticator as well as the PMKR0 and the PMKR0 holder. Always remember that the supplicant must also support FT; otherwise, the suppliant will reauthenticate every time the client roams.

FIGURE 15.38   Roaming cache

```
sh roam cache mac  b844:d90e:006e
Supplicant Address(SPA): b844:d90e:006e
PMK(1st 2 bytes): n/a
PMKID(1st 2 bytes): n/a
Session time: -1 seconds
(-1 means infinite)
PMK Time left in cache: 3581
PMK age: 1040
Roaming cache update interval: 60
last time logout: 1221 seconds ago
Authenticator Address: MAC=9c5d:122e:c124, IP=172.16.255.93
Roaming entry is got from neighbor AP: 9c5d:122e:c124
PMK is got(Flag): Locally
Station IP address: 172.16.255.90 (from DHCP)
Station hostname: Davids-iPhone
Station default gateway: 172.16.255.1
Station DNS server: 172.16.255.1
Station DHCP lease time: 85349 seconds
Hops: 0
WPA key mgmt: 64
R0KH: 9c5d:1263:6464
R0KH IP: 172.16.255.94
PMKR0 Name: 19D2*
```

As mentioned previously, enabling Voice-Enterprise mechanisms on an access point may actually create connectivity problems for legacy clients. When FT is configured on an access point, the AP will broadcast management frames with new information elements. For example, the *mobility domain information element (MDIE)* will be in all beacon and probe response frames. Unfortunately, the drivers of some older legacy client radios may not be able to process the new information in these management frames. The result is that legacy clients may have connectivity problems when an AP is configured for FT. Always test the legacy client population when configuring APs for fast BSS transition. If connectivity problems arise, consider using a separate SSID solely for fast BSS transition devices. Remember, however, that every SSID consumes airtime due to the layer 2 management overhead. Additionally, as more devices begin to support Voice-Enterprise, upgrade your client devices.

Because 802.11 wireless networks are usually integrated into preexisting wired topologies, crossing layer 3 boundaries is often a necessity, especially in large enterprise deployments. The only way to maintain upper-layer communications when crossing layer 3 subnets is to provide a *layer 3 roaming* solution. When clients roam to a new subnet, a Generic Routing Encapsulation (GRE) tunnel must be created to the original subnet so that the WLAN client can maintain its original IP address. As shown in Figure 15.39, the major WLAN vendors offer diagnostic tools and commands to verify that layer 3 roaming tunnels are being successfully created.

FIGURE 15.39   GRE tunnel

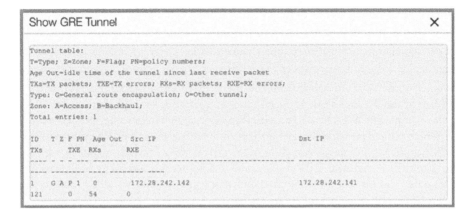

Channel Utilization

An important statistic when troubleshooting performance on a WLAN is *channel utilization,* as shown in Figure 15.40. Remember that RF is a shared medium and that 802.11 radios must take turns transmitting on any Wi-Fi channel. If the channel is oversaturated with 802.11 transmissions, performance will be negatively impacted. When they are not transmitting, both AP and client 802.11 radios listen to a channel every 9 microseconds for both 802.11 transmission activity as well as non-802.11 transmissions.

FIGURE 15.40   Channel utilization

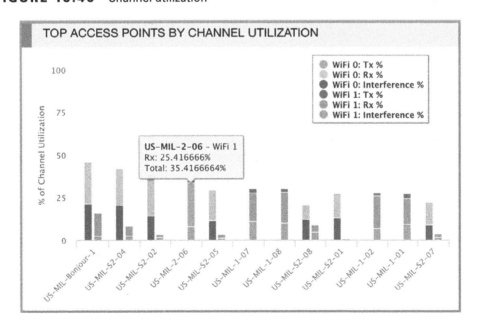

Some good channel utilization thresholds to live by include the following:

- 80 percent channel utilization impacts all 802.11 data transmissions.
- 50 percent channel utilization impacts video traffic.
- 20 percent channel utilization impacts voice traffic.

Monitoring and troubleshooting channel utilization is important because of perceived performance of the Wi-Fi network from end users. A common support call is that a user complains that the Wi-Fi is slow. If the channel utilization is over 80 percent, the Wi-Fi is in fact slow. Improper WLAN design, with improper channel planning, very often leads to CCI, which will cause high channel utilization. Oversaturation of clients and high-bandwidth applications can consume too much airtime on a channel, which is why proper capacity planning is important. As you learned in Chapter 13, too many broadcast SSIDs, low basic data rates on APs, and any abundance of legacy clients are all airtime consumption culprits that will affect channel utilization.

The QBSS information element found in 802.11 beacon and probe response frames sent by APs is a good indicator of channel utilization from the perspective of the AP radio. The information found in the QBSS information element, as shown in Figure 15.41, is often used by WLAN vendor monitoring solutions and other applications to visualize channel utilization in the form of graphs or charts. Large enterprise customers mostly rely on the monitoring/troubleshooting capabilities and perspective of the radio within an access point. RF statistics gathered from incoming client transmission can also be centrally monitored. The information gathered is from the perspective of the AP radio.

FIGURE 15.41   QBSS information element

▼ 🏷 QBSS Load	Stations: 6, Channel Utilization: 48%
Element ID:	11
Length:	5 bytes
Station Count:	6
Channel Utilization:	124 (48%)
Available Admission Capacity:	0

In reality, the best view of the RF network will always be from the client's perspective, which is why WLAN design and survey validation is so important. One approach that some enterprise WLAN vendors have taken is to have sensor APs act in place of client devices, whereby they log into other APs as client devices and then perform health checks. Keep in mind that clients have different receive sensitivities, and radios in APs are usually much more sensitive. Centralized monitoring and diagnostics using an AP radio is usually a great start, but additional information may need to be collected from the client's perspective when troubleshooting client issues.

Slow performance and bandwidth bottlenecks can indeed be a result of bad Wi-Fi design, and poor channel utilization is a good indicator of a Wi-Fi performance problem. However, the reason the Wi-Fi seems slow to the end user most often has nothing to do with the WLAN or channel utilization. Bandwidth bottlenecks are very often on the wired network due to poor wired network design. The number one bandwidth bottleneck is usually the WAN uplink at any remote site. But remember that the Wi-Fi will always be blamed first, despite the inadequate WAN bandwidth.

Layers 3–7 Troubleshooting

Although this chapter has focused on troubleshooting the WLAN at layers 1 and 2 of the OSI model, upper-layer troubleshooting may still be necessary. WLANs very often get blamed for causing problems that actually exist in the wired network at higher layers. If an employee cannot connect to the corporate WLAN, the employee will blame the WLAN even though the actual problem is somewhere else on the corporate network. If it can be determined that the problem is not a layer 1 or layer 2 problem, then the problem is usually a networking issue or a problem with an application.

The good news is that many WLAN vendors offer upper-layer troubleshooting tools that are available in network management servers and WLAN controllers or from the command line of APs. A common support call is that a user complains that they have a Wi-Fi connection but cannot connect to the network. If you have already determined that the problem is not a Wi-Fi problem, move up the OSI stack to layer 3 to check for IP connectivity.

Consider the diagram of a school WLAN shown in Figure 15.42. An AP is deployed in a school and is transmitting three SSIDS, one each for teachers, students, and guests. The teacher SSID is mapped to VLAN 2; the student SSID is mapped to VLAN 5; and the guest SSID is mapped to VLAN 8. The management interface of the AP is mapped to VLAN 1. All four VLANs are tagged across an 802.1Q trunk between the AP and the access switch. All four VLANs are mapped to respective subnets, and all IP addresses are supplied from defined scopes on the network DHCP server.

FIGURE 15.42   School WLAN diagram

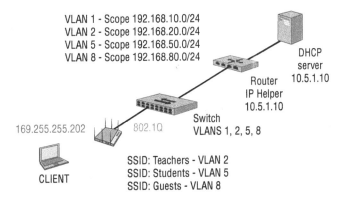

As previously mentioned, a common support call is that a user complains they have a Wi-Fi connection but cannot connect to the network. In this scenario, a student should be getting an IP address on the 192.168.50.0/24 network. A quick check determines that the student is connected to the proper SSID; however, the student receives an *automatic private IP address (APIPA)* in the 169.254.0.0–169.254.255.255 range. This would be your first indication that the problem is most likely a wired-side network problem.

WLAN vendors might offer a diagnostic tool that can be used to report back if the VLANs are operational on the wired network as well as the subnet of each VLAN. As shown in Figure 15.43, an administrator can select an AP to perform a probe across a designated range of VLANs. Please note that VLAN 5 (the student VLAN) failed.

FIGURE 15.43   VLAN probe

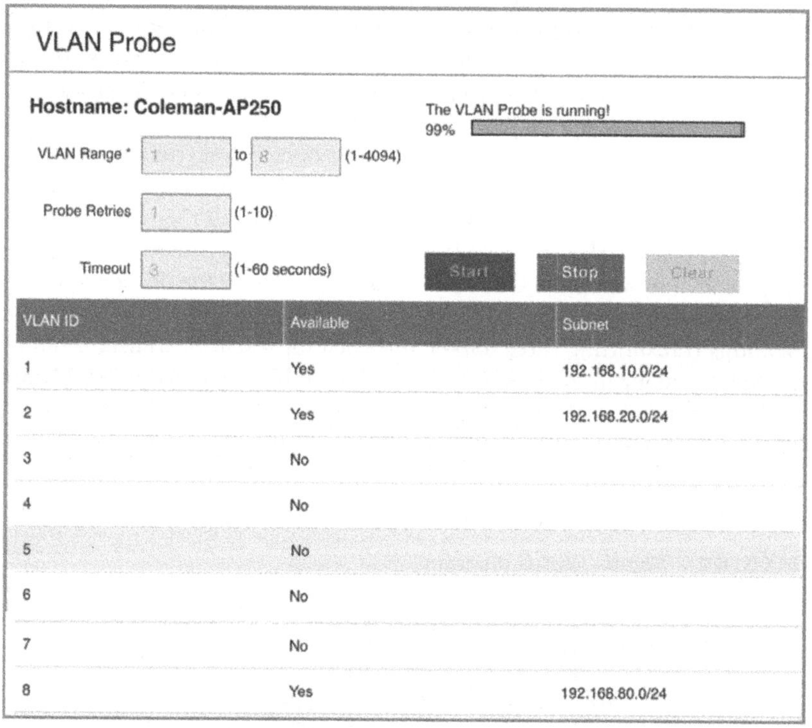

The diagnostic tool leverages the ability of the management interface of any access point to send out DHCP requests, as shown in Figure 15.44. Once the probe starts, the management interface of the AP will send out multiple DHCP requests across all the designated VLANs. Each DHCP request is sent up the 802.1Q trunk and onto the wired network. Once the DHCP request finally reaches the DHCP server, a lease offer is sent back to the AP. The management interface of the AP does not need another IP address, therefore a NAK is sent back to the DHCP server. If the DHCP lease offer reaches the AP, then there is not an issue with the wired network. However, if the DHCP lease offer does not reach the AP, then there is absolutely a wired-side problem, and the diagnostic probe will show a negative result.

FIGURE 15.44   DHCP probe

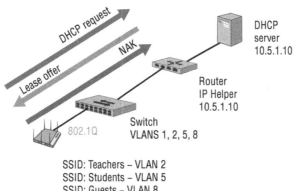

As shown in Figure 15.45, two common points of failure are the upstream router and the DHCP server. DHCP requests use a broadcast address and therefore an IP Helper (DHCP-Relay) address needs to be configured on the upstream router to convert the DHCP request into a unicast packet. If the router does not have the correct IP Helper address, then the DHCP request never makes it to the DHCP server. The DHCP server is more likely to be a point of failure. The DHCP server may have crashed, the scopes might not be configured correctly, or the server could simply be out of leases.

FIGURE 15.45   Backend DHCP failures

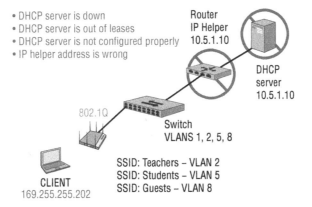

Although these two points of failure are certainly possible, the most likely culprit is the access switch, as shown in Figure 15.46. Almost 90 percent of the time, the problem is an improperly configured access switch. The VLANs might not be configured on the switch, the VLANs might not be tagged on the 802.1Q trunk port, or the port might have been misconfigured as an access port.

FIGURE 15.46 Misconfigured switch

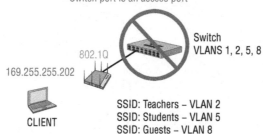

- VLANs not configured on the access switch
- VLANs not tagged on the 802.1Q port
- Switch port is an access port

Switch
VLANS 1, 2, 5, 8

802.1Q

169.255.255.202

CLIENT

SSID: Teachers – VLAN 2
SSID: Students – VLAN 5
SSID: Guests – VLAN 8

Even if the WLAN clients are successfully receiving IP addresses, there could still be layer 3 network issues. The ping and traceroute/tracert commands are your next step to diagnosing your network. The ping command and other network query commands are available from every client operating system (OS) as well as the OS that runs on APs, switches, and routers.

Once you have determined that there is not a layer 3 networking problem, you can begin to investigate layers 4–7. Scott Adams created a funny Dilbert comic strip in 2013 that blames the firewall for all network problems: http://dilbert.com/strip/2013-04-07. This cartoon actually mirrors real life because an incorrectly configured firewall policy can be blocking TCP or UDP ports. In addition to stateful firewall capability, WLAN vendors have begun to build Application-layer firewalls capable of *deep packet inspection (DPI)* into access points or WLAN controllers. DPI provides visibility into applications being used over the WLAN, and Application-layer firewalls can block specific applications or groups of applications. Wherever the firewall may be deployed in your network, firewall log files may need to be reviewed if a higher layer problem is suspected.

Always remember that an access point is a wireless portal to the complete network infrastructure. If the Wi-Fi network is not the problem, troubleshooting layers 3–7 will be necessary.

WLAN Troubleshooting Tools

Although WLAN vendors provide significant diagnostic capabilities from their network management systems, every WLAN professional usually carries a wide array of toys in their personal WLAN troubleshooting toolkit. This section will describe some of the tools that are available.

WLAN Discovery Applications

To start troubleshooting WLANs, you will need an 802.11 client NIC and a WLAN discovery application, such as WiFi Explorer (shown in Figure 15.47). WLAN discovery applications are a quick and easy way to give you a broad overview of an existing WLAN. WLAN discovery tools find existing Wi-Fi networks by sending out null probe request frames and listening for the 802.11 probe response frames and beacon frames sent by APs. Although a WLAN discovery tool will not give you the deep analysis that a protocol analyzer may provide, a lot of useful information can be gathered. For example, a WLAN discovery tool could immediately tell you that 80 MHz channels have been enabled on an AP, and performance is negatively impacted. A good WLAN discovery tool can give you a quick view of the number of transmitting APs and their channels, channel sizes, and security capabilities. Other available information includes signal strength, SNR, channel utilization statistics, and much more.

FIGURE 15.47   WLAN discovery tool

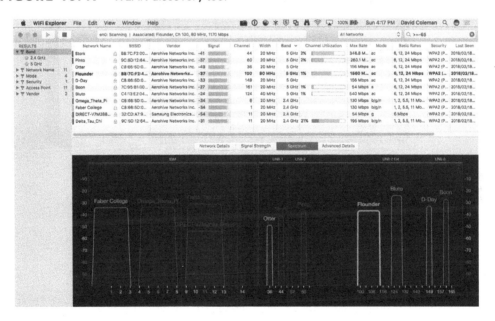

Numerous freeware and commercial discovery tools exist, including inSSIDer Office for Windows, Acrylic Wi-Fi Home or Professional for Windows, WiFi Explorer for macOS, and WiFi Analyzer for Android. You can download inSSIDer Office from www.metageek .net, Acrylic Wi-Fi Home or Professional from www.acrylicwifi.com, WiFi Explorer from https://www.adriangranados.com, and WiFi Analyzer from http://bit.ly/ WiFIAnalyze.

Spectrum Analyzers

Spectrum analyzers are frequency domain measurement devices that can measure the amplitude and frequency space of electromagnetic signals. Figure 15.48 depicts a PC-based spectrum analyzer that uses a USB-based adapter that is capable of monitoring both the 2.4 GHz and 5 GHz spectrums. MetaGeek's (www.metageek.net) Wi-Spy spectrum analyzer was used to identify the sources of RF interference shown earlier in the chapter in Figures 15.3–15.5. A spectrum analyzer is a layer 1 diagnostic tool that is most often used to identify sources of RF interference that originate from non-802.11 transmitters.

FIGURE 15.48   Wi-Spy DBx 2.4 GHz and 5 GHz PC-based spectrum analyzer

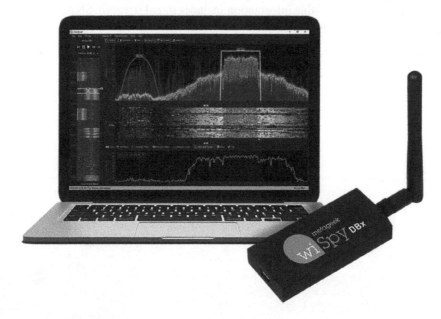

Protocol Analyzers

Protocol analyzers provide network visibility into exactly what traffic is traversing a network. Protocol analyzers capture and store network packets, providing you with a protocol decode for each packet captured, which is a readable display showing the individual fields and values for each packet. The power of a protocol analyzer is that it allows you to see conversations between various networking devices at many layers of the OSI model. Protocol analysis is sometimes the only way to troubleshoot a difficult problem. Many commercial WLAN protocol analyzers are available, such as TamoSoft's CommView for WiFi (www.tamos.com), Savvius's Omnipeek (www.savvius.com), as well as the popular freeware protocol analyzer Wireshark (www.wireshark.org).

Wired protocol analyzers are often called packet analyzers because they are used to troubleshoot IP packets that traverse wired networks. Remember, if the problem is not a layer 1 problem or a layer 2 problem, the problem is not a Wi-Fi problem. Packet analysis of wired traffic is often necessary to troubleshoot problems that occur at layers 3–7.

WLAN protocol analysis is mostly used to look at layer 2: 802.11 frame exchanges between APs and client devices. Wi-Fi radios communicate via 802.11 frame exchanges at the MAC sublayer. Unlike many wired network standards, such as IEEE 802.3, which uses a single data frame type, the IEEE 802.11 standard defines three major frame types: management, control, and data. These frame types are further subdivided into multiple subtypes, as you learned in Chapter 9. When using a WLAN protocol analyzer to view 802.11 frame conversations, you will normally not yet be looking at layers 3–7. Hopefully, all your 802.11 data traffic is encrypted.

A WLAN protocol analyzer and some WLAN discovery tools can also provide insight to some layer 1 and RF statistical information. *Radiotap* headers provide additional link-layer information that is added to each 802.11 frame when they are captured. The drivers of an 802.11 radio supply additional information via the Radiotap header. Please understand that the Radiotap header is not part of the 802.11 frame format. However, the ability to see additional information, such as signal strength associated to each 802.11 frame heard by the WLAN protocol analyzer radio, is quite useful. Wi-Fi expert Adrian Granados provides a more detailed explanation about the Radiotap header in his blog (`https://www.adriangranados.com/blog/link-layer-header-types`).

We always tell people that one of the best things we did early in our Wi-Fi careers was to teach ourselves 802.11 frame analysis. Eighteen years later, 802.11 protocol analysis skills are still important and invaluable. Modern WLAN protocol analyzer tools are more robust, but the 802.11 frame exchanges are constantly becoming more complex. As new 802.11 technologies, such as 802.11ax, become reality, the information inside 802.11 frame exchanges creates a Wi-Fi mosaic that is tough to interpret. No matter how much you might think you know about 802.11 frame analysis, you can always learn more.

Although there are many commercial WLAN protocol analyzers, Wireshark (`www.wireshark.org`) is open source and the tool of choice for many WLAN professionals. The authors of this book highly recommend two Wireshark video training courses created by Jerome Henry (CWNE #45) and James Garringer (CWNE #179). If you are a Wireshark novice, we highly recommend the *Wireshark Fundamentals LiveLessons* video training course (`http://bit.ly/WShark1`), which offers nearly five hours of instruction on using Wireshark to troubleshoot Ethernet and Wi-Fi networks and the protocols they transport. You will learn Wireshark capture basics, customization, filters, command-line options, and much more.

If you are interested in a deep-dive of 802.11 analysis, then we highly recommend the *Wireshark for Wireless LANs LiveLessons* video training course (`http://bit.ly/WShark2`), which offers more than eight hours of expert instruction on troubleshooting Wi-Fi networks using Wireshark. Nine lessons and sub-lessons will take you through learning the 802.11 MAC header, dissecting captured frames, advanced tools, and common WLAN problems that can be solved with proper analysis.

When using a protocol analyzer, utilizing traffic-filtering capabilities to focus on the networking conversations that you are trying to troubleshoot is essential. A reference guide for the most common 802.11 filters used in Wireshark is available for download at http://bit.ly/WFilter, courtesy of François Vergès (CWNE #180). Commercial protocol analyzers will have more advanced traffic filtering capabilities. Any good protocol analyzer will also have the ability to visualize traffic conversations as well as the ability to possibly provide intelligent diagnostics and suggested steps for remediation. Figure 15.49 is a screen capture from MetaGeek's EyePA WLAN protocol analyzer diagnosing an unacceptable percentage of layer 2 retransmissions on an AP, along with suggestion steps to investigate the cause of the problem.

FIGURE 15.49 EyePA analysis and remediation

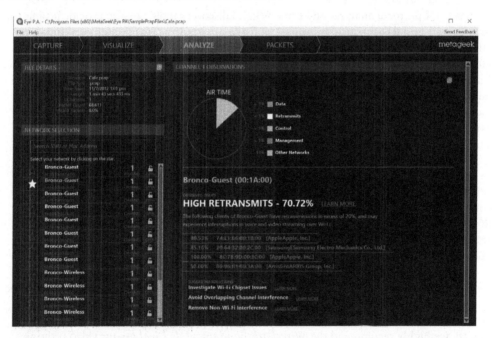

Identifying the correct location to place a network analyzer is an essential step in performing successful wireless network analysis. Incorrect placement of the WLAN protocol analyzer can lead to false conclusions. For example, if you are capturing traffic too far away from the source and destination, you might see a lot of corrupted frames; however, the intended recipient may not be experiencing any frame anomalies. An access point acts as the central point in an 802.11 wireless network, and all traffic must flow through the access point. Enterprise WLAN vendors offer direct packet capture from access points. In this scenario, if the analyzer reports a corrupted frame, it is more than likely that the AP also saw the frame as corrupted.

Throughput Test Tools

Throughput test tools are used to evaluate bandwidth and performance throughout a network. Throughput testers normally work on a client/server model to measure data streams between two ends or in both directions. When you are testing downlink WLAN throughput, the 802.11 client should be configured as the server. When you are testing uplink WLAN throughput, the 802.11 client should be configured as the client communicating with a server behind the AP. *iPerf* is an open-source command-line utility that is commonly used to generate TCP or UDP data streams to test throughput. Many WLAN vendors offer iPerf as a CLI test utility from within the OS of access points or WLAN controllers. As shown in Figure 15.50, TamoSoft (www.tamos.com) offers a freeware GUI-based throughput tester that is available for Windows, macOS, iOS, and Android clients.

FIGURE 15.50   TamoSoft throughput tester

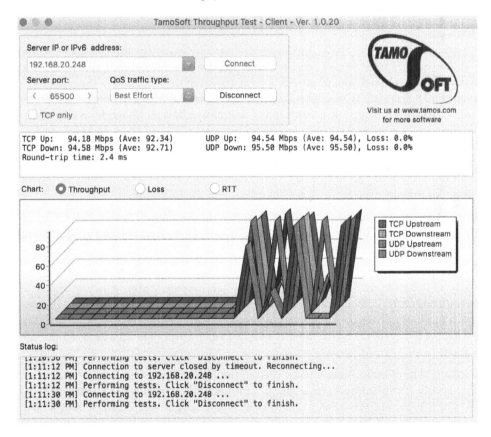

When performing throughput testing of the wireless link, always remember that you are not testing 802.11 data rates. Depending on the WLAN network conditions, the aggregate WLAN throughput is usually 50 percent of the advertised 802.11 data rate due to normal medium contention overhead. 802.11 data rates are not TCP throughput. The medium contention protocol of CSMA/CA consumes much of the available bandwidth. In laboratory conditions, the TCP throughput in an 802.11n/ac environment is 60–70 percent of the data rate between one AP and one client. The aggregate throughput numbers are considerably less in real-world environments with active participation of multiple WLAN clients communicating through an AP.

Using client/server throughput test tools on the wired-side of the network is very often necessary. Remember, the reason the Wi-Fi seems slow to the end user most often has nothing to do with the WLAN or channel utilization. Bandwidth bottlenecks are very often on the wired network due to poor wired network design. Once again, the number one bandwidth bottleneck is usually the WAN uplink at any remote site.

Standard IP Network Test Commands

Always remember that you have standard network troubleshooting tools available within the various operating systems. Everyone knows that you always start with ping, the most commonly used network tool to provide a basic connectivity test between the requesting host and a destination host. The ping command uses the *Internet Control Message Protocol (ICMP)* to send an echo packet to a destination host and listen for a response from the host. Use ping to test IP connectivity between a WLAN client and a local network server. Use ping to see if the WLAN client can reach the default gateway address. Ping the public Google DNS server at 8.8.8.8 to see if the WLAN client can access the Internet via the WAN.

Other commonly used network commands include the following:

Arp   The arp command is used to display the *Address Resolution Protocol (ARP)* cache, which is a mapping of IP addresses to MAC addresses. Every time a device's TCP/IP stack uses ARP to determine the MAC address for an IP address, it records the mapping in the ARP cache to speed up future ARP lookups. Viewing an ARP cache on access points is often helpful when troubleshooting.

Tracert/Traceroute   The tracert or traceroute command is available in most operating systems to determine detailed information about the path to a destination host, including the route an IP packet takes, the number of hops, and the response time between the various hops.

Nslookup   The nslookup command is used to troubleshoot problems with *Dynamic Name System (DNS)* address resolution. DNS is used to resolve domain names to IP addresses. Use the nslookup command to look up a specific IP address associated with a domain name. Many WLAN captive web portals used for WLAN guest access rely on DNS redirection. If a WLAN captive web portal suddenly stops working, you should probably suspect that there is a DNS issue.

Netstat   The `netstat` command displays network statistics for active TCP sessions for both incoming and outgoing ports, Ethernet statistics, IPv4 and IPv6 statistics, and more. The `netstat` command is often useful when troubleshooting suspected application problems and firewall issues.

When you are troubleshooting from a WLAN client perspective, these commands will be easily available from the command line of devices running Windows, macOS, or Linux. Many freeware applications are also available using both iOS and Android, so you can access these troubleshooting capabilities for smartphone and tablet mobile devices.

Are There Any CLI Commands to Troubleshoot WLAN Client Radios?

The simple answer is yes, depending on the operating system of the WLAN client device. The netsh (network shell) command can be used to configure and troubleshoot both wired and wireless network adapters on a Windows computer. The `netsh wlan show` commands will expose detailed information about the Wi-Fi radio being used by the Windows computer. For example, `netsh wlan show networks` will display all the visible Wi-Fi networks that the client radio can see. A comparable command-line utility to configure and troubleshoot 802.11 network adapters on macOS computers is the `airport` command-line tool. Take the time to familiarize yourself with both the `netsh wlan` commands for Windows and the `airport` commands for macOS. As with any CLI command, execute the `?` command to view full options.

Secure Shell

When connecting to network hardware such as an access point or a switch, an SSH or serial client is required. *Secure Shell (SSH)* is used as the secure alternative to Telnet. SSH implements authentication and encryption using public-key cryptography of all network traffic traversing between the host and user device. The standard TCP port 22 has been assigned for the SSH protocol. Most WLAN infrastructure devices now support the second version of the SSH protocol, called *SSH2*. As a matter of policy, when WLAN devices are managed via the CLI, an SSH2-capable terminal emulation program should be used. Figure 15.51 shows the configuration screen of the popular freeware program PuTTY, which supports SSH2 and terminal emulation. PuTTy is often the freeware program of choice when having to climb a ladder and connect to the console port of an access point. Additionally, some operating systems, such as macOS, support SSH natively from the command line, or using a program such as iTerm2.

FIGURE 15.51   PuTTy—freeware SSH and serial client

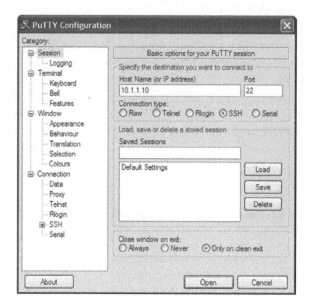

Summary

Troubleshooting WLANs can be very challenging. Much of WLAN troubleshooting revolves around performance or connectivity issues that are a result of improper WLAN design. However, because of the ever-changing RF environment, problems such as roaming, hidden nodes, and interference are bound to surface. Never forget that Wi-Fi operates at layer 1 and layer 2 of the OSI model, and understand that the WLAN will always receive the blame no matter where the problem exists. Always remember to use troubleshooting best practices, analyze the problems at the different layers of the OSI model, and utilize all diagnostic tools that might be available.

Exam Essentials

Understand troubleshooting basics.   Recognize the importance of asking the correct questions and gathering the proper information to determine the root cause of the problem.

Explain where in the OSI model various WLAN problems occur.   Remember that troubleshooting up the OSI model is a recommended strategy. WLAN security issues almost always reside at layers 1 and 2. Remember that most WLAN connectivity problems also exist on the client devices as opposed to the WLAN infrastructure.

Explain how to troubleshoot PSK authentication.   Understand that the usual causes of failed PSK authentication are client driver issues and mismatched passphrase credentials. The 4-Way Handshake will fail if PSK authentication fails.

Define the multiple points of failure of 802.1X/EAP authentication.   Explain all the potential backend communications points of failure and possible supplicant failures. Understand how to analyze the 802.1X/EAP process to pinpoint the exact point of failure.

Explain potential WLAN security problems with roaming.   Understand that both the WLAN infrastructure and the WLAN clients must support fast secure roaming mechanisms such as OKC or Voice-Enterprise.

Identify the causes of layer 1 WLAN problems.   Understand that most networking problems usually exist at the Physical layer. Explain all layer 1 problems often caused by bad WLAN design, RF interference, radio drivers, firmware issues, or PoE problems.

Understand that layer 2 retransmissions are evil.   Recognize the many causes of layer 2 retransmissions and the significant adverse affects when layer 2 retransmissions exceed the level of 10 percent.

Review Questions

1. At which layer of the OSI model do most networking problems occur?

   A. Physical

   B. Data-Link

   C. Network

   D. Transport

   E. Session

   F. Presentation

   G. Application

2. What can cause PSK authentication to fail? (Choose all that apply.)

   A. Passphrase mismatch

   B. Expired root CA certificate

   C. WLAN client driver problem

   D. Expired LDAP user account

   E. Encryption mismatch

3. When the Wi-Fi network is the actual source of a connectivity, security, or performance problem, which WLAN device is usually where the problem resides?

   A. WLAN controller

   B. Access point

   C. WLAN client

   D. Wireless network management server

4. What are some problems that can occur when an indoor access point is transmitting at full power? (Choose all that apply.)

   A. Hidden node

   B. Co-channel interference

   C. Sticky clients

   D. Intersymbol interference

   E. Band hopping

5. A single user is complaining that her VoWiFi phone has choppy audio. The WLAN administrator notices that the user's MAC address has a retry rate of 25 percent when observed with a protocol analyzer. However, all the other users have a retry rate of about 5 percent when also observed with the protocol analyzer. What is the most likely cause of this problem?

   A. Near/far

   B. Multipath

   C. Co-channel interference

   D. Hidden node

   E. Low SNR

6. Andrew Garcia, the WLAN administrator, is trying to explain to his boss that the WLAN is not the reason that Andrew's boss cannot post on Facebook. Andrew has determined that the problem does not exist at layer 1 or layer 2 of the OSI model. What should Andrew say to his boss? (Choose the best answer.)

   A. Wi-Fi operates only at layer 1 and layer 2 of the OSI model. The WLAN is not the problem.

   B. The problem is most likely a networking problem or an application problem.

   C. Do not worry, boss; I will fix it.

   D. Why are you looking at Facebook during business hours?

7. What are some of the negative effects of layer 2 retransmissions? (Choose all that apply.)

   A. Decreased range

   B. Excessive MAC sublayer overhead

   C. Decreased latency

   D. Increased latency

   E. Jitter

8. You have been tasked with troubleshooting a client connectivity problem at your company's headquarters. All the APs and employee iPads are configured for PSK authentication. An employee notices that he cannot connect his iPad to the AP in the reception area of the main building but can connect to other APs. View the following graphic and describe the cause of the problem.

```
BASIC   Rx assoc req (rssi 93dB)
INFO    WPA-PSK auth is starting (at if=wifi0.1)
INFO    Sending 1/4 msg of 4-Way Handshake (at if=wifi0.1)
INFO    Received 2/4 msg of 4-Way Handshake (at if=wifi0.1)
INFO    Sending 1/4 msg of 4-Way Handshake (at if=wifi0.1)
INFO    Received 2/4 msg of 4-Way Handshake (at if=wifi0.1)
BASIC   Sta(at if=wifi0.1) is de-authenticated because of notification of driver
```

   A. The WLAN client driver is not communicating properly with the device's OS.

   B. The APs are configured for CCMP encryption only. The client only supports TKIP.

   C. The client has been configured with the wrong WPA2-Personal passphrase.

   D. The AP in the reception area has been configured with the wrong WPA2-Personal passphrase.

9. You have been tasked with configuring a secure WLAN for 400 APs at the corporate offices. All the APs and employee Windows laptops have been configured for 802.1X/EAP using EAP-MSCHAPv2. The domain user accounts are failing authentication with every attempt. After viewing the graphic shown here, determine the possible causes of the problem. (Choose all that apply.)

```
Rx assoc req (rssi 95dB)
IEEE802.1X auth is starting (at if=wifi0.1)
Sending EAP Packet to STA: code=1 (EAP-Request) identifier=0 length=5
received EAP packet (code=2 id=0 len=16) from STA: EAP Reponse-Identity (1),
Send message to RADIUS Server(10.5.1.129): code=1 (Access-Request) identifier
RADIUS: EAP start with type peap
Receive message from RADIUS Server: code=11 (Access-Challenge) identifier=50
Sending EAP Packet to STA: code=1 (EAP-Request) identifier=1 length=6
received EAP packet (code=2 id=1 len=105) from STA: EAP Reponse-PEAP (25)
Send message to RADIUS Server(10.5.1.129): code=1 (Access-Request) identifier
RADIUS: SSL negotiation, receive client hello message
Receive message from RADIUS Server: code=11 (Access-Challenge) identifier=51
Sending EAP Packet to STA: code=1 (EAP-Request) identifier=2 length=1024
received EAP packet (code=2 id=2 len=6) from STA: EAP Reponse-PEAP (25)
Send message to RADIUS Server(10.5.1.129): code=1 (Access-Request) identifier
RADIUS: SSL negotiation, send server certificate and other message
Receive message from RADIUS Server: code=11 (Access-Challenge) identifier=52
Sending EAP Packet to STA: code=1 (EAP-Request) identifier=3 length=280
received EAP packet (code=2 id=3 len=6) from STA: EAP Reponse-PEAP (25)
Send message to RADIUS Server(10.5.1.129): code=1 (Access-Request) identifier
RADIUS: SSL negotiation, send server certificate and other message
Receive message from RADIUS Server: code=11 (Access-Challenge) identifier=53
Sending EAP Packet to STA: code=1 (EAP-Request) identifier=4 length=6
Sta(at if=wifi0.1) is de-authenticated because of notification of driver
```

   A. The networking settings on the AP are incorrect.

   B. The Windows OS laptops' supplicant has been configured for machine authentication.

   C. The supplicant clock settings are incorrect.

   D. An authentication port mismatch exists between the AP and the RADIUS server.

   E. The networking settings on the RADIUS server are incorrect.

   F. The incorrect root certificate is selected in the supplicant.

10. The network administrator of the WonderPuppy Coffee Company calls the support hotline for his WLAN vendor and informs the support personnel that the WLAN is broken. The support personnel ask the customer a series of questions so that they can isolate and identify the cause of a potential problem. What are some common Troubleshooting 101 questions? (Choose all that apply.)

   A. When is the problem happening?

   B. What is your favorite color?

   C. What is your quest?

   D. Does the problem reoccur or did it just happen once?

   E. Did you make any changes recently?

11. The corporate IT administrators, Hunter, Rion, and Liam, are huddled together to try to solve an issue with the newly deployed VoWiFi phones. The chosen security solution is PEAPv0 (EAP-MSCHAPv2) for the voice SSID that also has Voice-Enterprise enabled on the access points. The VoWiFi phones are authenticating flawlessly and voice calls are stable when the employees use the devices from their desk. However, there seem to be gaps in the audio and sometimes disconnects when the employees are talking on the VoWiFi phones

and move to other areas of the building. What are the possible causes of the interruption of service for the voice calls while the employees are mobile? (Choose all that apply.)

A. VoWiFi phones should be configured only for PSK authentication when roaming is a requirement.

B. VoWiFi phones are reauthenticating every time they roam to a new AP.

C. VoWiFi phones do not use opportunistic key caching.

D. VoWiFi phones do not support fast BSS transition.

12. Which of the following can cause roaming problems? (Choose all that apply.)

A. Not enough secondary coverage

B. Too much secondary coverage

C. Free space path loss

D. CSMA/CA

E. Hidden node

13. Adrian White has an Ethernet switch that is compliant with 802.3at. He is having problems with his APs randomly rebooting. Which of the following could be causing his problems?

A. Multiple desktop PoE VoIP telephones are connected to the same Ethernet switch.

B. Most of the Ethernet cables running from the switch to the APs are 90 meters long.

C. The Ethernet cables are only Cat 5e.

D. The switch is capable of 1000BaseT, which is not compatible with the AP.

14. Which Windows CLI command can display a WLAN client's authentication method, encryption method, channel, signal strength, and data rate?

A. `netsh wlan show drivers`

B. `airport -S`

C. `netsh wlan show interfaces`

D. `airport -I`

E. `nslookup`

F. `traceroute`

15. The network administrator of the Holy Grail Corporation calls the support hotline for his WLAN vendor and informs the support personnel that the WLAN is no longer working. The support personnel asks the customer a series of questions so that they can isolate and identify the cause of a potential problem. What are some common Troubleshooting 101 questions? (Choose all that apply.)

A. When is the problem happening?

B. Where is the problem happening?

C. Does the problem affect one client or numerous clients?

D. What is the airspeed velocity of an unladen swallow?

16. WLAN administrator Marko Tisler is troubleshooting an IPsec VPN problem between a remote WLAN branch router and a VPN gateway server at corporate headquarters. Marko cannot get the VPN tunnel to establish and notices that there is a certificate error during the IKE Phase 1 exchange. What are the possible causes of this problem? (Choose all that apply.)

   A. The VPN server at corporate headquarters is using AES-256 encryption, and the remote WLAN branch router is using AES-192 encryption.

   B. The VPN server at corporate headquarters is using SHA-1 hash for data integrity, and the remote WLAN branch router is using MD5 for data integrity.

   C. The root CA certificate installed on the VPN remote WLAN branch router was not used to sign the server certificate on the corporate VPN server.

   D. The clock settings of the corporate VPN server predate the creation of the server certificate.

   E. The public/private IP address settings are misconfigured on the remote WLAN branch router.

17. You have been tasked with configuring a secure WLAN for 600 APs at the corporate offices. All the APs and employee Windows laptops have been configured for EAP-MSCHAPv2. Connectivity is failing for one of the employee laptops. After viewing the graphic shown here, determine the possible causes of the problem. (Choose all that apply.)

```
Receive message from RADIUS Server: code=2 (Access-Accept) identifier=125
PMK is got from RADIUS server (at if=wifi0.1)
(63)Sending 1/4 msg of 4-Way Handshake (at if=wifi0.1)
(64)Received 2/4 msg of 4-Way Handshake (at if=wifi0.1)
(65)Sending 3/4 msg of 4-Way Handshake (at if=wifi0.1)
(66)Received 4/4 msg of 4-Way Handshake (at if=wifi0.1)
(67)PTK is set (at if=wifi0.1)
(68)Authentication is successfully finished (at if=wifi0.1)
(69)station sent out DHCP REQUEST message
(70)station sent out DHCP REQUEST message
(71)station sent out DHCP REQUEST message
```

   A. The VLAN on the access layer switch is incorrectly configured.

   B. The machine accounts were not joined to the domain.

   C. The server certificate has expired.

   D. The supplicant has been configured only for user authentication.

   E. The root certificate has expired.

   F. The DHCP server has run out of leases.

18. What can be done to fix the hidden node problem? (Choose all that apply.)

   A. Increase the power on the access point.

   B. Move the hidden node station.

   C. Decrease power on all client stations.

   D. Remove the obstacle.

   E. Decrease power on the hidden node station.

   F. Add another AP.

19. Layer 2 retransmissions occur when frames become corrupted. What are some of the causes of layer 2 retransmissions? (Choose all that apply.)

   A. High SNR

   B. Low SNR

   C. Co-channel interference

   D. RF interference

   E. Adjacent channel interference

20. When you are troubleshooting client connectivity problems with a client using 802.1X/EAP security, what is the first action you should take to investigate a potential layer 1 problem?

   A. Reboot the WLAN client.

   B. Verify the root CA certificate.

   C. Verify the EAP protocol.

   D. Disable and re-enable the client radio network interface.

   E. Verify the server certificate.

Chapter

16

Wireless Attacks, Intrusion Monitoring, and Policy

IN THIS CHAPTER, YOU WILL LEARN ABOUT THE FOLLOWING:

✓ **Wireless attacks**

- Rogue wireless devices
- Peer-to-peer attacks
- Eavesdropping
- Encryption cracking
- KRACK attack
- Authentication attacks
- MAC spoofing
- Management interface exploits
- Wireless hijacking
- Denial-of-service (DoS) attacks
- Vendor-specific attacks
- Social engineering

✓ **Intrusion monitoring**

- Wireless intrusion prevention system (WIPS)
- Rogue detection and mitigation
- Spectrum analyzers

✓ **Wireless security policies**

- General security policies
- Functional security policies
- Legislative compliance
- 802.11 wireless policy recommendations

In this chapter, we cover the wide variety of attacks that can be launched against 802.11 wireless networks. Some of these attacks can be mitigated by using the strong encryption and mutual authentication solutions that we discuss in Chapter 17, "802.11 Network Security Architecture." Other attacks, however, cannot be prevented and can only be detected. Therefore, we also discuss the wireless intrusion detection systems that can be implemented to expose both layer 1 and layer 2 attacks. The most important component for a secure wireless network is a properly planned and implemented corporate security policy. This chapter also discusses some of the fundamental components of a wireless security policy that are needed to cement a foundation of Wi-Fi security.

Wireless Attacks

As you have learned throughout this book, the main function of an 802.11 WLAN is to provide a portal into a wired network infrastructure. The portal must be protected with strong authentication methods so that only legitimate users and devices with the proper credentials will be authorized to have access to network resources. If the portal were not properly protected, unauthorized users could gain access to these resources. The potential risks of exposing these resources are endless. An intruder could gain access to financial databases, corporate trade secrets, or personal health information. Network resources could be damaged.

What would be the financial cost to an organization if an intruder used the wireless network as a portal to disrupt or shut down a VoIP server or email server? If the Wi-Fi portal were not protected, any individual wishing to cause harm could upload data such as viruses, Trojan horse applications, keystroke loggers, or remote control applications. Spammers have already figured out that they can use open wireless gateways to the Internet to commence spamming activities. Other illegal activities, such as software theft and remote hacking, may also occur through an unsecured gateway.

While an intruder can use the wireless network to attack wired resources, equally at risk are all of the wireless network resources. Any information that passes through the air can be captured and possibly compromised. If not properly secured, the management interfaces of Wi-Fi equipment can be accessed. Many wireless users are fully exposed for peer-to-peer attacks. Finally, the possibility of denial-of-service (DoS) attacks against a wireless network always exists. With the proper tools, any individual with ill intent can temporarily disable a Wi-Fi network, thus denying legitimate users access to the network resources.

In the following sections, you will learn about many of the potential attacks that can be launched against 802.11 wireless networks.

Rogue Wireless Devices

The big buzz-phrase in Wi-Fi security has always been the *rogue access point*: a potential open and unsecured gateway straight into the wired infrastructure that the company wants to protect. In Chapter 17, you will learn about 802.1X/EAP authentication solutions that can be put in place to prevent unauthorized access. But what prevents an individual from installing his own wireless portal onto the network backbone? A rogue access point is any unauthorized Wi-Fi device that is not under the management of the proper network administrators. The greatest WLAN security threat is any type of unauthorized rogue Wi-Fi device that is connected to the wired network infrastructure, as depicted in Figure 16.1. The skull and crossbones icon is a common symbol used to represent rogue APs, as well as pirates. Any consumer-grade Wi-Fi access point or router can be plugged into a live data port. The rogue device will just as easily act as a portal into the wired network infrastructure. Because the rogue device will likely be configured with no authorization and authentication security in place, any intruder could use this open portal to gain access to network resources.

FIGURE 16.1 Rogue access point

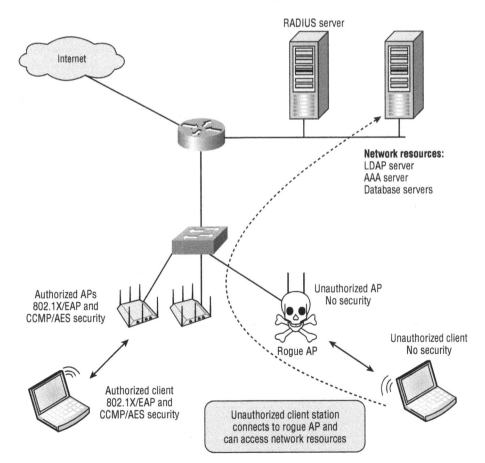

The individuals most responsible for installing rogue access points are typically not hackers; they are employees not realizing the consequences of their actions. Wi-Fi networking has become ingrained in our society, and the average employee has become accustomed to the convenience and mobility that Wi-Fi offers. As a result, it is not uncommon for an employee to install their own wireless devices in the workplace because the employee believes installing their own wireless device is easier or more reliable than using the corporate WLAN. The problem is, although these self-installed access points might provide the wireless access that the employees desire, they are often unsecured. Only a single open portal is needed to expose network resources, and many large companies have discovered literally dozens of rogue access points that have been installed by employees.

Ad hoc wireless connections also have the potential of providing rogue access into the corporate network. Very often an employee will have a laptop or desktop computer plugged into the wired network via an Ethernet network card. On that same computer, the employee has a Wi-Fi radio and has set up an ad hoc Wi-Fi connection with another employee. This connection may be set up on purpose or may be accidental and occur as an unwitting result of the manufacturer's default configurations. As shown in Figure 16.2, the Ethernet connection and the Wi-Fi network interface controller (NIC) can be bridged together; an intruder might access the ad hoc wireless network and then potentially route their way to the Ethernet connection and get onto the wired network.

FIGURE 16.2   Bridged ad hoc WLAN

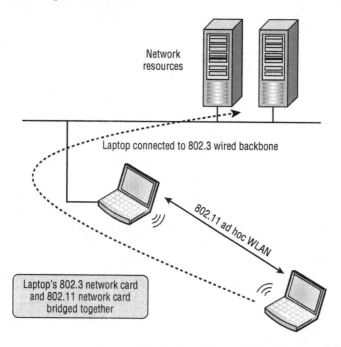

Network resources

Laptop connected to 802.3 wired backbone

802.11 ad hoc WLAN

Laptop's 802.3 network card and 802.11 network card bridged together

Many government agencies and corporations ban the use of ad hoc networks on laptops for this very reason. The ability to configure an ad hoc network can and should be disabled on most enterprise client devices. On some computers, it is possible to limit the use of multiple NICs simultaneously. This is a great feature that can prevent bridged networks from occurring while allowing flexibility for the user. When the user plugs an Ethernet cable into the laptop, the wireless adapter is automatically disabled, eliminating the risk of an intentional or unintentional bridged network.

Another common rogue type is the wireless printer. Many printers now have 802.11 radios with a default configuration of ad hoc mode. Attackers can connect to these printers using the printer manufacturer's administrative tools, downloadable from their website. Using these tools, attackers can upload their own firmware to your printer, allowing them to bridge the wired and wireless connections of your printer to gain access to your wired network, without the use of an access point. Many 802.11 wireless camera security systems can be breached in a similar manner.

As stated earlier, most rogue APs are installed by employees not realizing the consequences of their actions, and any malicious intruder can use these open portals to gain access. Furthermore, besides physical security, there is nothing to prevent an intruder from also connecting their own rogue access point via an Ethernet cable into any live data port provided in a wall plate. Later in this chapter, we discuss intrusion prevention systems that can both detect and disable rogue access points as well as ad hoc clients.

If an 802.1X/EAP solution is deployed for the wireless network, it can also be used to secure the network ports on the wired network. The best way of preventing rogue access is wired-side port control. 802.1X/EAP can also be used to authenticate and authorize access through wired ports on an access layer switch. A rogue device cannot act as a wireless portal to network resources if the rogue device is plugged into a managed port that is blocking upper-layer traffic. When 802.1X/ EAP is used for port control on an access layer switch, desktop clients function as the supplicants requesting access. Some WLAN vendor APs can also function as supplicants and cannot forward user traffic unless the approved AP is authenticated. Therefore, a wired 802.1X/EAP solution is an excellent method for preventing rogue access. Some WLAN vendors have also begun to support MACsec for wired-side port control. The IEEE 802.1AE Media Access Control Security standard, often referred to as MACsec, specifies a set of protocols to meet the security requirements for protecting data traversing Ethernet LANs. In that case, any new device, including APs, would need to be authenticated to the network prior to being given access. This is a good way to not only utilize existing resources but also provide better security for your wired network by protecting against rogue APs.

Many businesses do not use an 802.1X/EAP solution for wired-side port control. Therefore, a WLAN monitoring solution known as a wireless intrusion detection system (WIDS) is usually recommended to detect potential rogue devices. Most WIDS vendors prefer to call their products a wireless intrusion prevention system (WIPS). The reason that they refer to their products as prevention systems is that they are also capable of mitigating attacks from rogue APs and rogue clients.

Peer-to-Peer Attacks

A commonly overlooked risk is the *peer-to-peer attack*. As you learned in earlier chapters, an 802.11 client station can be configured in either infrastructure mode or ad hoc mode. When configured in ad hoc mode, the wireless network is officially known by the 802.11 standard as an *independent basic service set (IBSS)*, and all communications are peer-to-peer without the need for an access point. Because an IBSS is by nature a peer-to-peer connection, any user who can connect wirelessly with another user can potentially gain access to any resource available on either computer. A common use of ad hoc networks is to share files on the fly. When shared access is provided, files and other assets can accidentally be exposed. A personal firewall is often used to mitigate peer-to peer attacks. Some client devices can also disable this feature so that the device will connect to only certain networks and will not associate to a peer-to-peer without approval.

Users associated to the same access point are potentially just as vulnerable to peer-to-peer attacks as IBSS users. Properly securing your wireless network often involves protecting authorized users from each other, because hacking at companies is often performed internally by employees. Any users associated to the same AP that are members of the same basic service set (BSS) and are in the same VLAN are susceptible to peer-to-peer attacks because they reside in the same layer 2 and layer 3 domains. In most WLAN deployments, Wi-Fi clients communicate only with devices on the wired network, such as email or web servers, and peer-to-peer communications are not needed. Therefore, most enterprise AP vendors provide some proprietary method of preventing users from inadvertently sharing files with other users or bridging traffic between the devices. When connections are required to other wireless peers, the traffic is routed through a layer 3 switch or other network device before passing to the desired destination station.

Client isolation is a feature that can often be enabled on WLAN access points or controllers to block wireless clients from communicating with other wireless clients on the same wireless VLAN. Client isolation, or the various other terms used to describe this feature, usually means that packets arriving at the AP's wireless interface are not forwarded back out of the wireless interface to other clients. This isolates each user on the wireless network to ensure that a wireless station cannot be used to gain layer 3 or higher access to another wireless station. The client isolation feature is usually a configurable setting per SSID linked to a unique VLAN. As shown in Figure 16.3, with client isolation enabled, client devices cannot communicate directly with other client devices on the wireless network.

Although client isolation is the most commonly used term, some vendors use the term *peer-to-peer blocking* or *public secure packet forwarding (PSPF)*. Not all vendors implement client isolation in the same fashion. Some WLAN vendors can only implement client isolation on an SSID/VLAN pair on a single access point, whereas others can enforce the peer-blocking capabilities across multiple APs.

FIGURE 16.3   Client isolation

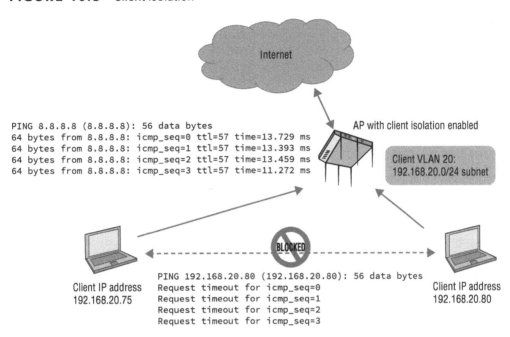

```
PING 8.8.8.8 (8.8.8.8): 56 data bytes
64 bytes from 8.8.8.8: icmp_seq=0 ttl=57 time=13.729 ms
64 bytes from 8.8.8.8: icmp_seq=1 ttl=57 time=13.393 ms
64 bytes from 8.8.8.8: icmp_seq=2 ttl=57 time=13.459 ms
64 bytes from 8.8.8.8: icmp_seq=3 ttl=57 time=11.272 ms
```

AP with client isolation enabled

Client VLAN 20:
192.168.20.0/24 subnet

BLOCKED

Client IP address
192.168.20.75

```
PING 192.168.20.80 (192.168.20.80): 56 data bytes
Request timeout for icmp_seq=0
Request timeout for icmp_seq=1
Request timeout for icmp_seq=2
Request timeout for icmp_seq=3
```

Client IP address
192.168.20.80

Some applications require peer-to-peer connectivity. Many Voice over Wi-Fi (VoWiFi) phones offer push-to-talk capabilities that use multicasting. VoWiFi phones are typically segmented in a separate wireless VLAN from the rest of wireless data clients. Client isolation should not be enabled in the VoWiFi VLAN if push-to-talk multicasting is required, because it can prevent these devices from functioning properly.

Eavesdropping

As we have mentioned throughout this book, 802.11 wireless networks operate in license-free frequency bands, and all data transmissions travel in the open air. Access to wireless transmissions is available to anyone within listening range, and therefore strong encryption is mandatory. Wireless communications can be monitored via two eavesdropping methods: casual eavesdropping and malicious eavesdropping.

Casual eavesdropping, sometimes referred to as *WLAN discovery*, is accomplished by simply exploiting the 802.11 frame exchange methods that are clearly defined by the 802.11-2016 standard. Software utilities known as *WLAN discovery tools* exist for the purpose of finding open WLAN networks. As we discussed in Chapter 9, "802.11 MAC," in order for an 802.11 client station to be able to connect to an access point, it must first discover the access point. A station discovers an access point by either listening for an AP (passive scanning) or searching for an AP (active scanning). In *passive scanning*, the client station listens for 802.11 beacon management frames, which are continuously sent by the access points.

A casual eavesdropper can simply use any 802.11 client radio to listen for 802.11 beacon management frames and to discover layer 2 information about the WLAN. Some of the information found in beacon frames includes the service set identifier (SSID), MAC addresses, supported data rates, and other basic service set (BSS) capabilities. All of this layer 2 information is in cleartext and can be seen by any 802.11 radio.

In addition to scanning passively for APs, client stations can scan actively for them. In *active scanning*, the client station transmits management frames known as *probe requests*. The access point then answers back with a *probe response frame*, which basically contains all the same layer 2 information found in a beacon frame. A probe request without the SSID information is known as a *null probe request*. If a directed probe request is sent, all APs that support that specific SSID and hear the request should reply by sending a probe response. If a null probe request is heard, all APs, regardless of their SSID, should reply with a probe response.

Many wireless client software utilities instruct the radio to transmit probe requests with null SSID fields when actively scanning for APs. Additionally there are numerous freeware and commercial WLAN discovery tool applications. WLAN discovery tools send out null probe requests across all license-free 802.11 channels with the hope of receiving probe response frames containing wireless network information, such as SSID, channel, encryption, and so on. Some WLAN discovery tools may also use passive scanning methods. Shown in Figure 16.4, a very popular Windows-based WLAN discovery tool is inSSIDer Office, which is available from www.metageek.net.

FIGURE 16.4 MetaGeek inSSIDer Office

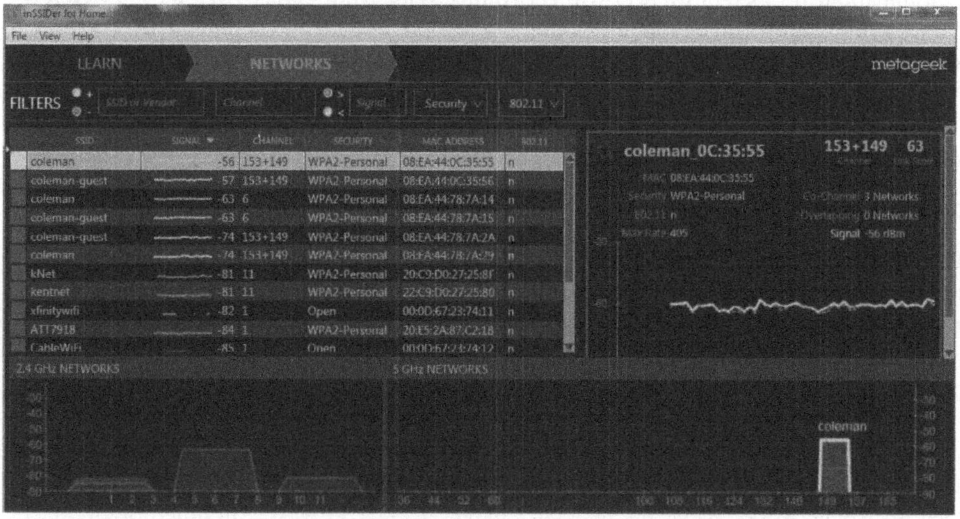

Casual eavesdroppers can discover 802.11 networks by using software tools that send null probe requests. Casual eavesdropping is typically considered harmless and is often referred to as *wardriving*. Wardriving is strictly the act of looking for wireless networks, usually while in a moving vehicle. The term wardriving was derived from wardialing from the 1983 film *WarGames*. Wardialing was a technique employed by hackers using computer modems to scan thousands of telephone numbers automatically to search for other computers with which they could connect.

Wardiving is now considered an outdated term and concept. In the very early days of Wi-Fi, wardriving was a hobby and sport for techno-geeks and hackers looking to find WLANs. Wardriving competitions were often held at hacker conventions to see who could find the most WLANs.

While the sport of wardriving has faded, millions of people still use WLAN discovery tools to find available Wi-Fi networks. A more current term would be "WLAN discovery." In the early days of Wi-Fi, the original WLAN discovery software tool was a freeware program called NetStumbler. Although still available as a free download, NetStumbler has not been updated in many years. However, many newer WLAN discovery tools exist that operate on a variety of operating systems. Figure 16.5 depicts the Android-based WLAN discovery tool WiFi Analyzer. Numerous WLAN discovery tools are available for Android mobile devices, but currently there are not many tools for iOS mobile devices.

FIGURE 16.5   WiFi Analyzer WLAN discovery tool

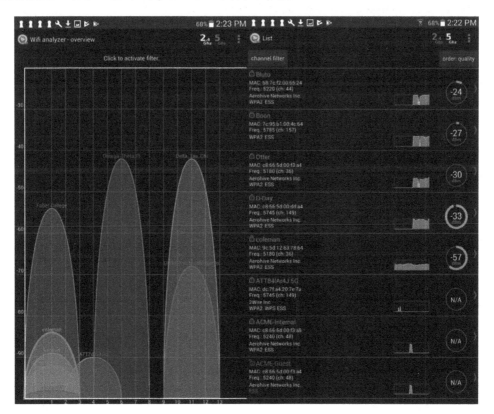

By design, the very nature of 802.11 passive and active scanning is to provide the identifying network information that is accessible to anyone with an 802.11 radio. Because this is an inherent and necessary function of 802.11, wardriving is not a crime. The legality of using someone else's wireless network without permission is often unclear, but be warned that people have been arrested and prosecuted as a result of these actions. An alarming decision about the use of networks owned by others was reached in March 2011. The Hague Court ruled that they were no longer going to prosecute as a criminal offense the unauthorized use of networks belonging to others if the person using the network without permission only used it to access the Internet, even if the access was gained by extraordinary means. The ruling does leave the opportunity for civil action. Every nation has its own laws covering such actions.

WARNING We do not encourage or support the efforts of using wireless networks that you are not authorized to use. We recommend that you connect only to wireless networks that you are authorized to access.

What Tools Are Needed for WLAN Discovery?

To start finding WLANs, you will need an 802.11 client NIC and a WLAN discovery application. Numerous freeware and commercial discovery tools exist, including inSSIDer Office for Windows, Acrylic Wi-Fi for Windows, WiFi Explorer for the macOS, and WiFi Analyzer for Android. You can download inSSIDer Office from www.metageek.net, Acrylic Wi-Fi Home or Professional from www.acrylicwifi.com, WiFi Explorer or WiFi Explorer Pro from www.adriangranados.com, and WiFi Analyzer from bit.ly/WiFIAnalyze.

Global positioning system (GPS) devices, in conjunction with WLAN discovery tools, can be used to pinpoint longitude and latitude coordinates of the signal from APs that are discovered. WLAN discovery capture files with GPS coordinates can be uploaded to large dynamic mapping databases on the Internet. The Wireless Geographic Logging Engine (WIGLE) maintains a searchable database of more than 362 million Wi-Fi networks. Go to www.wigle.net and type in your address to see whether any wireless access points have already been discovered in your neighborhood.

While casual eavesdropping is considered harmless, *malicious eavesdropping*, the unauthorized use of 802.11 protocol analyzers to capture wireless communications, is typically considered illegal. Most countries have some type of wiretapping law that makes it a crime to listen in on someone else's phone conversation. Additionally, most countries have laws making it illegal to listen in on any type of electromagnetic communications, including 802.11 wireless transmissions.

An 802.11 protocol analyzer application allows wireless network administrators to capture 802.11 traffic for the purpose of analyzing and troubleshooting their own wireless networks. A protocol analyzer is a passive device that operates in RF monitoring mode to capture any 802.11 frame transmissions within range. Because protocol analyzers capture 802.11 frames passively, a wireless intrusion prevention system (WIPS) cannot detect malicious eavesdropping. Commercial WLAN protocol analyzers are available, such as Savvius Omnipeek, as well as the popular freeware protocol analyzer Wireshark (www.wireshark.org).

A WLAN protocol analyzer is meant to be used as a diagnostic tool. However, an attacker can use a WLAN protocol analyzer as a malicious listening device for unauthorized monitoring of 802.11 frame exchanges. Although all layer 2 information is always available, all layer 3–7 information can be exposed if WPA2 encryption is not in place. Any cleartext communications, such as email, FTP, and Telnet passwords, can be captured if no encryption is provided. Furthermore, any unencrypted 802.11 frame transmissions can be reassembled at the upper layers of the OSI model. Email messages can be reassembled and, therefore, read by an eavesdropper. Web pages and instant messages can also be reassembled. VoIP packets can be reassembled and saved as a WAV sound file. Malicious eavesdropping of this nature is highly illegal.

Because of the passive and undetectable nature of this attack, encryption must always be implemented to provide data privacy. Encryption is the best protection against unauthorized monitoring of the WLAN. WPA2 encryption provides data privacy for all layer 3–7 information.

The most common targets of malicious eavesdropping attacks are public access hotspots. Public hotspots rarely offer security and usually transfer data without encryption, making hotspot users prime targets. As a result, it is imperative that a VPN security solution be implemented for all mobile users who connect outside of your company's network.

Encryption Cracking

Wired Equivalent Privacy (WEP) is a legacy 802.11 encryption method that was compromised many years ago. WEP-cracking tools are freely available on the Internet and can crack WEP encryption in less than 5 minutes. There are several methods used to crack WEP encryption. However, an attacker usually needs only to capture several hundred thousand encrypted packets with a protocol analyzer and then run the captured data through a WEP-cracking software program, as shown in Figure 16.6. The software utility will usually then be able to derive the secret 40-bit or 104-bit key in a matter of seconds. After the secret key has been revealed, the attacker can decrypt any and all encrypted traffic. In other words, an attacker can then eavesdrop on the WEP-encrypted network. Because the attacker can decrypt the traffic, they can reassemble the data and read it as if there were no encryption whatsoever.

FIGURE 16.6   WEP-cracking utility

```
* Got   286716! unique IVs | fudge factor = 2
* Elapsed time [00:00:03] | tried 1 keys at 20 k/m

KB    depth    votes
 0    0/  1    DA(  60) 70(  23) 55(  15) A2(   5) CD(   5) 3E(   4)
 1    0/  2    BD(  57) 2A(  32) 29(  22) 1D(  13) F9(  13) 9F(  12)
 2    0/  1    8C(  51) 67(  23) 48(  15) DD(  15) D6(  13) FA(  12)
 3    0/  3    1D(  30) A5(  17) 07(  15) 7B(  12) 4B(  10) 63(  10)
 4    0/  1    43(  66) B1(  15) D2(   6) 1A(   5) 20(   5) 21(   5)
 5    0/  5    92(  27) 23(  25) 02(  18) 2F(  17) C1(  16) 36(  12)
 6    0/  1    C6(  51) 54(  17) 50(  15) 66(  15) 01(  13) 4A(  13)
 7    0/  2    84(  29) C0(  17) EE(  13) 80(  12) 49(  11) F6(  11)
 8    0/  1    81(1808) 09( 119) 99( 116) 32(  75) 49(  75) 9D(  65)
 9    0/  1    C4(1947) E1( 125) FC( 123) BD( 105) 8C(  98) 2F(  85)
10    0/  1    8A( 580) 41( 120) 18(  93) ED(  85) B0(  65) 97(  60)
11    0/  1    08(  97) FF(  29) 5D(  20) 1E(  17) 18(  15) 5E(  15)
12    0/  1    1B( 145) DD(  21) 46(  20) 1C(  15) 76(  15) 07(  13)

         KEY FOUND! [ DABD8C1D4392C68481C48A081B ]
```

KRACK Attack

In October of 2017, Belgian researchers Mathy Vanhoef and Frank Piessens of the University of Leuven, published details of the key reinstallation attack (KRACK). This replay attack targets the 4-Way Handshake used to establish dynamic encryption keys in the WPA2 protocol. The KRACK vulnerability received a lot of press because of the potential of compromised encryption keys for many existing Wi-Fi devices. It is beyond the scope of this book to explain how the attack works, however, information can be found at www.krackattacks.com. The good news is that the KRACK vulnerability can be very easily fixed via firmware patches. All the major WLAN vendors responded quickly in 2017 and released firmware updates. The bigger concern is updated firmware for client devices. Although the vulnerability has been patched in all the major client device operating systems, many legacy clients might not have patched firmware updates available.

Authentication Attacks

As you have already learned, authorization to network resources can be achieved by either an 802.1X/EAP authentication solution or the use of PSK authentication. The 802.11-2016 standard does not define which type of EAP authentication method to use, and all flavors of EAP are not created equal. Some types of EAP authentication are more secure than others. *Lightweight Extensible Authentication Protocol (LEAP)*, once one of the most commonly deployed 802.1X/EAP solutions, is susceptible to offline dictionary attacks. The hashed password response during the LEAP authentication process is easily crackable.

An attacker merely has to capture a frame exchange when a LEAP user authenticates and then run the capture file through an offline dictionary attack tool, as shown in Figure 16.7. The password can be derived in a matter of seconds. The username is also seen in cleartext during the LEAP authentication process. After the attacker gets the username and password, they are free to impersonate the user by authenticating onto the WLAN and then accessing any network resources that are available to that user. Stronger EAP authentication protocols that use tunneled authentication are not susceptible to offline dictionary attacks.

FIGURE 16.7   Offline dictionary attack

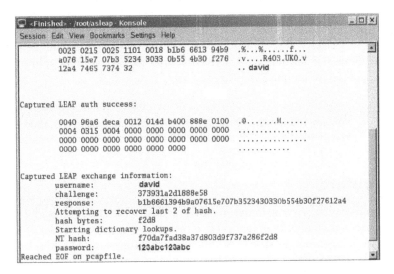

The biggest risk with any authentication attack is that all network resources become vulnerable if the authentication credentials are compromised. The risks of authentication attacks are similar to rogue access points. If an authorized WLAN portal can be compromised and the authentication credentials can be obtained, network resources are exposed. Because of these severe risks, the corporate WLAN infrastructure must be secured properly with an 802.1X/EAP solution that uses a RADIUS server and the tunneled authentication EAP protocols discussed in Chapter 17.

Because most home users do not have a RADIUS server, they typically use the weaker WPA/WPA2-Personal authentication methods. WPA/WPA2-Personal, using a passphrase (sometimes referred to as a preshared key or PSK), is a weak authentication method that is vulnerable to an offline *brute-force dictionary attack*. Shared keys or passphrases are also easily obtained through social engineering techniques. *Social engineering* is the act of manipulating people into performing actions or divulging confidential information. An attacker who obtains the passphrase can associate with the WPA/WPA2 access point and access network resources. To help mitigate offline brute-force dictionary attacks, both the IEEE and the Wi-Fi Alliance recommend very strong passphrases of 20 characters or more whenever a WPA/WPA2-Personal solution is deployed. The length of a WPA/WPA2 passphrase can range from 8 to 63 characters. The biggest risk with any authentication attack is that all network resources could become vulnerable if the authentication credentials were compromised.

Even worse is that after obtaining the passphrase, the hacker can decrypt the dynamically generated TKIP/ARC4 or CCMP/AES encryption key. The passphrase is used to derive the *pairwise master key (PMK)*, which is used with the *4-Way Handshake* to create the final dynamic encryption keys. If a hacker has the passphrase and captures the 4-Way Handshake, they can re-create the dynamic encryption keys and decrypt traffic. WPA/WPA2-Personal is not considered a strong security solution for the enterprise, because if the passphrase is compromised, the attacker can not only access network resources, they can also decrypt traffic. Because of these risks, a static PSK authentication solution should not be used in the enterprise. In situations where there is no AAA server or the client devices do not support 802.1X/EAP, a proprietary PSK authentication solution implementing unique PSKs is recommended. Several enterprise WLAN vendors offer proprietary PSK solutions that provide the capability of unique PSKs for each user.

You will learn more about the 4-Way Handshake and dynamic encryption key generation in Chapter 17, "802.11 Network Security Architecture."

MAC Spoofing

All 802.11 radios have a physical address known as a *MAC address*. This address is a 12-digit hexadecimal number that appears in cleartext in the layer 2 header of 802.11 frames. Wi-Fi vendors often provide MAC filtering capabilities on their APs. Usually, MAC filters are configured to apply restrictions that will allow traffic only from specific client stations to pass through. These restrictions are based on their unique MAC addresses. All other client stations whose MAC addresses are not on the allowed list will not be able to pass traffic through the virtual port of the access point and onto the distribution system medium. MAC filtering is often used as a security mechanism for legacy client devices, such as mobile handheld scanners, that do not support the stronger authentication and encryption techniques.

Unfortunately, MAC addresses can be *spoofed*, or impersonated, and any amateur hacker can easily bypass any MAC filter by spoofing an allowed client station's address.

Because of spoofing and because of all of the administrative work involved with setting up MAC filters, MAC filtering is not considered a reliable means of security for wireless enterprise networks and should be implemented only if stronger security is not available, or used together with some stronger form of security as part of a multifactor security plan.

Management Interface Exploits

One of the main goals of attackers is to gain access to administrative accounts or root privilege. Once they gain that access, they can run several attacks against networks and individual devices. On wired networks these attacks are launched against firewalls, servers, and infrastructure devices. In wireless attacks, these are first launched against access points or WLAN controllers and subsequently against the same targets as in wired attacks.

Wireless infrastructure hardware such as access points and WLAN controllers can be managed by administrators via a variety of interfaces, much like managing wired infrastructure hardware. Devices can typically be accessed via a web interface, a command-line interface, a serial port, a console connection, and/or Simple Network Management Protocol (SNMP). It is imperative that these interfaces be protected. Interfaces that are not used should be disabled. Strong passwords should be used, and encrypted login capabilities using SSH2 (Secure Shell) or Hypertext Transfer Protocol Secure (HTTPS) should always be used.

Lists of all the default settings of every major manufacturer's access points exist on the Internet and are often used for security exploits by hackers. It is not uncommon for attackers to use security holes left in management interfaces to reconfigure APs. Legitimate users and administrators can find themselves locked out of their own Wi-Fi equipment. After gaining access via a management interface, an attacker might even be able to initiate a firmware upgrade of the wireless hardware and, while the upgrade is being performed, power off the equipment. This attack could likely render the hardware useless, requiring it to be returned to the manufacturer for repair.

Policy often dictates that all WLAN infrastructure devices be configured from only the wired side of the network. If an administrator attempts to configure a WLAN device while connected wirelessly, the administrator could lose connectivity due to configuration changes being made. Some WLAN vendors offer secure wireless console connectivity capabilities for troubleshooting and configuration.

Wireless Hijacking

An attack that often generates a lot of press is *wireless hijacking*, also known as the *evil twin attack*. The attacker configures access point software on a laptop, effectively turning a Wi-Fi client radio into an access point. Some small Wi-Fi USB devices also have the ability to operate as an AP. The access point software on the attacker's laptop is configured with the same SSID that is used by a public-access hotspot. The attacker's access point is now functioning as an evil twin AP with the same SSID, but it is transmitting on a different channel. The attacker then sends spoofed disassociation or deauthentication frames, forcing users associated with the hotspot access point to roam to the evil twin access point. At this point, the attacker has effectively hijacked wireless clients at layer 2 from the original AP. Although deauthentication frames are usually used as one way to start a hijacking attack, an RF jammer can also be used to force any clients to roam to an evil twin AP.

The evil twin will typically be configured with a Dynamic Host Configuration Protocol (DHCP) server available to issue IP addresses to the clients. At this point, the attacker will have hijacked the users at layer 3 and will now have a private wireless network and be free to perform peer-to-peer attacks on any of the hijacked clients. The user's computer could, during the process of connecting to the evil twin, fall victim to a DHCP attack, an attack that exploits the DHCP process to dump root kits or other malware onto the victim's computer in addition to giving them an IP address as expected.

The attacker may also be using a second wireless NIC with their laptop to execute what is known as a *man-in-the-middle attack*, as shown in Figure 16.8. The second WLAN radio is associated to the original access point as a client. In operating systems, networking interfaces can be bridged together to provide routing. The attacker has bridged together their second wireless NIC with the Wi-Fi radio that is being used as the evil twin access point. After the attacker hijacks the users from the original AP, the traffic is then routed from the evil twin AP through the second Wi-Fi radio, right back to the original AP from which the users have just been hijacked. The result is that the users remain hijacked; however, they still have a route back through the gateway to their original network, so they never know they have been hijacked. The attacker can therefore sit in the middle and execute peer-to-peer attacks indefinitely while remaining completely unnoticed.

FIGURE 16.8   Wireless hijacking/man-in-the-middle attack

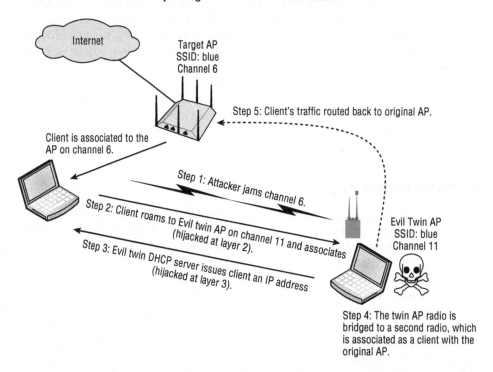

These attacks can take another form in what is known as a *Wi-Fi phishing attack*. The attacker may also have web server software and captive portal software. After the users have been hijacked to the evil twin access point, they will be redirected to a login web page that looks exactly like the hotspot's login page. Then the attacker's fake login page may request a credit card number from the hijacked user. Phishing attacks are common on the Internet and are now appearing at your local hotspot.

The only way to prevent a hijacking, man-in-the-middle, or Wi-Fi phishing attack is to use a mutual authentication solution. Mutual authentication solutions not only validate the user connecting to the network, they also validate the network to which the user is connecting. 802.1X/EAP authentication solutions require that mutual authentication credentials be exchanged before a user can be authorized. A user cannot get an IP address unless authorized; therefore, users cannot be hijacked.

Denial-of-Service Attacks

The attack on wireless networks that seems to receive the least amount of attention is the *denial-of-service (DoS)* attack. With the proper tools, any individual with ill intent can temporarily disable a Wi-Fi network by preventing legitimate users from accessing network resources. The good news is that monitoring systems exist that can detect and identify DoS attacks immediately. The bad news is that usually nothing can be done to prevent DoS attacks other than locating and removing the source of the attack.

DoS attacks can occur at either layer 1 or layer 2 of the OSI model. Layer 1 attacks are known as *RF jamming attacks*. The two most common types of RF jamming attacks are intentional jamming and unintentional jamming:

Intentional Jamming   Intentional jamming attacks occur when an attacker uses some type of signal generator to cause interference in the unlicensed frequency space. Both narrowband and wideband jammers exist that will interfere with 802.11 transmissions, either causing all data to become corrupted or causing the 802.11 radios to continuously defer when performing a *clear channel assessment (CCA)*.

Unintentional Jamming   Whereas an intentional jamming attack is malicious, unintentional jamming is more common. Unintentional interference from microwave ovens, cordless phones, and other devices can also cause denial of service. Although unintentional jamming is not necessarily an attack, it can cause as much harm as an intentional jamming attack.

The best tool to detect any type of layer 1 interference, whether intentional or unintentional, is a spectrum analyzer. A good example of a standalone spectrum analyzer is the Wi-Spy USB spectrum analyzer, which is available from www.metageek.net.

The more common types of denial-of-service attacks that originate from hackers are layer 2 DoS attacks. A wide variety of layer 2 DoS attacks exist that are a result of manipulating 802.11 frames. The most common involves spoofing disassociation or deauthentication frames. The attacker can edit the 802.11 header and spoof the MAC address of an access point or a client in either the transmitter address (TA) field or the receiver address (RA) field. The attacker then retransmits the spoofed deauthentication frame repeatedly. The station that receives the spoofed deauthentication frame thinks the spoofed frame is coming from a legitimate station and disconnects at layer 2.

Many more types of layer 2 DoS attacks exist, including association floods, authentication floods, PS-Poll floods, and virtual carrier attacks. Luckily, any good wireless intrusion detection system will be able to alert an administrator immediately to a layer 2 DoS attack. The 802.11w-2009 amendment defined *management frame protection (MFP)* mechanisms for the prevention of spoofing certain types of 802.11 management frames. These 802.11w frames are referred to as *robust management frames*. Robust management frames can be protected by the management frame protection service and include disassociation, deauthentication, and robust action frames. Action frames are used to request a station to take action on behalf of another station, and not all action frames are robust.

It should be noted that the 802.11w amendment did not put an end to all layer 2 DoS attacks. Numerous layer 2 DoS attacks cannot be prevented. Furthermore, 802.11w MFP mechanisms are not widely supported on the client side. However, enterprise WLAN vendors do implement 802.11w mechanisms on access points; therefore, some of the more common layer 2 DoS attacks can be prevented if the clients support 802.11w.

A spectrum analyzer is your best tool to detect a layer 1 DoS attack, and a protocol analyzer or wireless IDS is your best tool to detect a layer 2 DoS attack. The best way to prevent any type of denial-of-service attack is physical security. The authors of this book recommend guard dogs and barbed wire fencing. If that is not an option, then several vendor solutions provide intrusion detection at layers 1 and 2.

Where Can You Learn More about WLAN Security Risk Assessment?

This chapter covers the basics of Wi-Fi security attacks and intrusion monitoring. Although numerous books have been written about wireless hacking, a good starting point is *CWSP Certified Wireless Security Professional Official Study Guide: Exam CWSP-205* (Sybex, 2016). Many WLAN security auditing tools are also available for Wi-Fi penetration testing.

One of the more popular Wi-Fi penetration testing tools is the Wi-Fi Pineapple. The Wi-Fi Pineapple is a WLAN auditing tool from Hak5 that uses custom hardware and software with a web interface. More information about the Wi-Fi Pineapple can be found at www.wifipineapple.com.

The majority of effective wireless auditing tools run on Linux platforms, many of which can be accessed from a bootable CD, such as Kali Linux. Kali Linux, which has 600+ tools, is a Debian-based distribution with a collection of security and forensics tools. You can download Kali Linux at www.kali.org.

Vendor-Specific Attacks

Hackers often find holes in the firmware code used by specific WLAN access point and WLAN controller vendors. New WLAN vulnerabilities and attacks are discovered on a regular basis, including vendor proprietary attacks. Many of these vendor-specific exploits are in the form of buffer overflow attacks. When these vendor-specific attacks become known, the WLAN vendor usually makes a firmware fix available in a timely manner. Once the exploits are discovered, the affected WLAN vendor will make safeguard recommendations on how to avoid the exploit. In most cases the WLAN vendor will quickly release a patch that can fix the problem. These attacks can be best avoided by staying informed through your WLAN vendor's support services and maintaining your WLAN infrastructure with recent and fully supported firmware. As with most network infrastructure hardware, the firmware running on a WLAN controller or AP will also have a lifecycle. Before an upgrade, an administrator should always examine the impact of the latest version of firmware. However, security updates should always be taken seriously.

Social Engineering

Hackers do not compromise most wired or wireless networks with the use of hacking software or tools. The majority of breaches in computer security occur due to social engineering attacks. *Social engineering* is a technique used to manipulate people into divulging confidential information, such as computer passwords. The best defense against social engineering attacks is strictly enforced policies to prevent confidential information from being shared.

Any information that is static is extremely susceptible to social engineering attacks. WEP encryption uses a static key, and WPA/WPA2-Personal requires the use of a static PSK or passphrase. You should avoid both of these security methods because of their static nature.

Intrusion Monitoring

When people think of wireless networking, they tend to think only in terms of access and not in terms of attacks or intrusions. However, it has become increasingly necessary to monitor constantly for many types of WLAN attacks because of the potential damage they can cause. Businesses of all sizes deploy 802.11 wireless networks for mobility and access. Many of these networks are running a *wireless intrusion detection system (WIDS)* to monitor for attacks. Wireless intrusion monitoring has evolved since its creation in 2006. Today most systems have methods to prevent and mitigate several of the better known wireless attacks. Therefore, most WLAN vendors instead call their solutions a *wireless intrusion prevention system (WIPS)*. In this book, we will use the terminology of wireless intrusion prevention system (WIPS).

Distributed monitoring solutions may also have prevention capabilities, including mitigating rogue APs and clients. The use of distributed WLAN monitoring and rogue prevention reduces the time and expense required to maintain a healthy and secure wireless network.

Wireless Intrusion Prevention System

In today's world, a wireless intrusion prevention system (WIPS) might be necessary even if there is no authorized 802.11 Wi-Fi network on site. Wireless can be an intrusive technology, and if wired data ports at a business are not controlled, any individual (including an employee) can install a rogue access point. Because of this risk, many companies—such as banks, other financial institutions, and hospitals—choose to install a WIPS before deploying a Wi-Fi network for employee access. After an 802.11 network is installed for access, it has become almost mandatory to also have a WIPS because of the other numerous attacks against Wi-Fi, such as DoS, hijacking, and so on. The typical WIPS is a client-server model that consists of the following two primary components:

WIPS Server A WIPS server is a software server or hardware server appliance acting as a central point of monitoring security and performance data collection. The server uses signature analysis, behavior analysis, protocol analysis, and RF spectrum analysis to detect potential threats. Signature analysis looks for patterns associated with common WLAN attacks. Behavior analysis looks for 802.11 anomalies. Protocol analysis dissects the MAC layer information from 802.11 frames. Protocol analysis may also look at the layer 3–7 information of 802.11 data frames that are not encrypted. Spectrum analysis monitors RF statistics, such as signal strength and signal-to-noise ratio (SNR). Performance analysis can be used to gauge WLAN health statistics, such as capacity and coverage.

Sensors Hardware- or software-based sensors may be placed strategically to listen to and capture all 802.11 communications. Sensors are the eyes and ears of a WIPS monitoring solution. Sensors use 802.11 radios to collect information used in securing and analyzing WLAN traffic. Figure 16.9 depicts the client-server model used by most wireless intrusion prevention systems.

FIGURE 16.9 Wireless intrusion prevention system (WIPS)

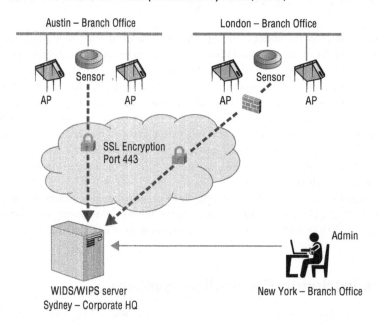

Sensors are basically radio devices that are in a constant listening mode as passive devices. The sensor devices are usually hardware based and resemble an access point. The sensors have some intelligence but must communicate with the centralized WIPS server.

The centralized server can collect data from literally thousands of sensors from many remote locations and thus meet the scalability needs of large corporations.

Standalone sensors do not provide access to WLAN clients, because they are configured in a listen-only mode. The sensors constantly scan all the channels in the 2.4 GHz ISM band, as well as all the channels in the 5 GHz U-NII bands. On rare occasions, the sensors can also be configured to listen on only one channel or a select group of channels. Access points can also be used as part-time sensors. An AP can use off-channel scanning methods to monitor other channels while still spending the majority of time on the AP's home channel to provide client access. Some WLAN vendor's APs offer a third radio that functions full time as a sensor in listen-only mode.

Some solutions might additionally require a software-based management console that is used to communicate back to a WIPS server from a desktop station. The management console is the software interface used for administration and configuration of the server and sensors. The management console can also be used for 24/7 monitoring of 802.11 wireless networks. However, the majority of the WIPS solutions do not require an additional console, and all security monitoring is viewed directly from the server. As shown in Figure 16.10, a WLAN administrator can monitor all the potential WLAN security threats from the graphical user interface (GUI) of the WIPS server.

FIGURE 16.10   WIPS monitoring

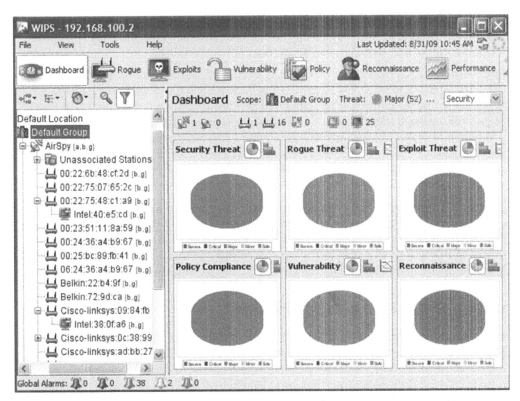

WIPS are best at monitoring layer 2 attacks, such as MAC spoofing, disassociation attacks, and deauthentication attacks. Most WIPS have alarms for as many as 100 potential security risks. An important part of deploying a WIPS is setting the policies and alarms. False positives are often a problem with intrusion detection systems, but they can be less of a problem if proper policies and thresholds are defined. Policies can be created to define the severity of various alerts as well as provide for alarm notifications. For example, an alert for broadcasting the SSID might not be considered severe and might even be disabled. However, a policy might be configured that classifies a deauthentication spoofing attack as severe, and an email message or SMS test message might be sent automatically to the network administrator.

Although most of the scrutiny performed by a WIPS is for security purposes, many WIPS also have performance-monitoring capabilities. For example, performance alerts might be in the form of excessive bandwidth utilization or excessive reassociation and roaming of VoWiFi phones.

The components of a WLAN security monitoring solution are usually deployed within one of the following two major WIPS architectures:

Overlay The most secure model is an overlay WIPS, which is deployed on top of the existing wireless network. This model uses an independent vendor's WIPS and can be deployed to monitor any existing or planned WLAN. The overlay systems typically have more extensive features, but they are usually more expensive. The overlay solutions consist of a WIPS server and sensors that are not part of the WLAN solution that provides access to clients. Dedicated overlay systems are not as common as they used to be; many of the WIPS capabilities have been integrated into most enterprise WLAN products.

Integrated Most WLAN vendors have fully integrated WIPS capabilities. A centralized WLAN controller or a centralized *network management server (NMS)* functions as the WIPS server. Access points can be configured in a full-time sensor-only mode or can act as part-time sensors when not transmitting as access points. The APs use off-channel scanning procedures for dynamic RF spectrum-management purposes. The APs are also effectively part-time sensors for the integrated WIPS server when listening off channel. A recommended practice would be to also deploy some APs as full-time sensors. The integrated solution is a less expensive solution but may not have all the capabilities of an overlay WIPS.

Of the two WIPS architectures, the integrated WIPS is by far the most widely deployed. The overlay WIPS is usually cost prohibitive for most WLAN customers. The more robust overlay WIPS solutions are usually deployed in defense, finance, and big-box retail vertical markets, where the budget for an overlay solution may be available.

Rogue Detection and Mitigation

As already mentioned, a rogue access point is any unauthorized Wi-Fi device that is not under the management of the proper network administrators. The most worrisome type of unauthorized rogue Wi-Fi device is one that is connected to the wired network infrastructure. WLAN vendors use a variety of wireless and wired detection methods to determine whether a rogue access point is plugged into the wired infrastructure. Some rogue detection and classification methods are published, whereas many remain proprietary and trade secrets. Any 802.11 device that is not already authorized will automatically be classified as an unauthorized device. However, rogue classification is a little more complex. A WIPS characterizes access points and client radios in four or more classifications. Although various WIPS vendors use different terminology, some examples of classifications include the following:

Authorized Device   This classification refers to any client station or access point that is an authorized member of the company's wireless network. A network administrator can manually label each radio as an authorized device after detection from the WIPS or can import a list of all the company's WLAN radio MAC addresses into the system. Devices may also be authorized in bulk from a comma-delimited file. Integrated solutions automatically classify any APs as authorized devices. An integrated solution will also automatically classify client stations as authorized if the client stations are properly authenticated.

Unauthorized or Unknown Device   The unauthorized device classification is assigned automatically to any new 802.11 radios that have been detected but not classified as rogues. Unknown devices are considered to be unauthorized and are usually investigated further to determine whether they are a neighbor's device or a potential future threat. Unauthorized devices may later be manually classified as a known neighbor device.

Neighbor Device   This classification refers to any client station or access point that is detected by the WIPS and whose identity is known. This type of device initially is detected as an unauthorized or unknown device. The neighbor device label is then typically assigned manually by an administrator. Devices manually classified as known are most often 802.11 access points or client radio devices of neighboring businesses that are not considered a threat.

Rogue Device   The rogue classification refers to any client station or access point that is considered an interfering device and a potential threat. Most WIPS solutions define rogue access points as devices that are actually plugged into the wired network backbone and are not known or managed by the organization. Most of the WIPS vendors use a variety of methods to determine whether a rogue access point is actually plugged into the wired infrastructure.

Most WIPS vendors use different terminology when classifying devices. For example, some wireless intrusion prevention systems classify all unauthorized devices as rogue devices, whereas other WIPS solutions assign the rogue classification only to APs or WLAN devices that have been detected with a connection to the wired network. After a client station or AP has been classified as a rogue device, the WIPS can effectively mitigate an attack. WIPS vendors have several ways of accomplishing this. One of the most common methods is to use spoofed deauthentication frames. As shown in Figure 16.11, the WIPS will have the sensors go active and begin transmitting deauthentication frames that spoof the MAC addresses of the rogue APs and rogue clients. The WIPS uses a known layer 2 denial-of-service attack as a countermeasure. The effect is that communications between the rogue AP and clients are rendered useless. This countermeasure can be used to disable rogue APs, individual client stations, and rogue ad hoc networks.

FIGURE 16.11   Wireless rogue containment

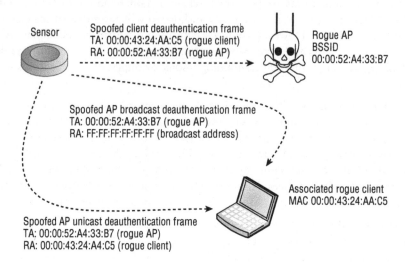

Many WIPSs also use a wired-side termination process to effectively mitigate rogue devices. The wired-side termination method of rogue mitigation uses the Simple Network Management Protocol (SNMP) for *port suppression*. Many WIPSs can determine that the rogue AP is connected to the wired infrastructure and may be able to use SNMP to disable the managed switch port that is connected to the rogue AP. If the switch port were closed, the attacker could access network resources that are behind the rogue AP.

WIPS vendors have other, often unpublished, proprietary methods of disabling rogue APs and client stations. Currently, the main purpose of a WIPS is to contain and disable rogue devices. In the future, other wireless attacks might be mitigated as well.

 Real World Scenario

Will a WIPS Protect Against All Known Rogue Devices?

The simple answer is no. Although wireless intrusion prevention systems are outstanding products that can mitigate many rogue attacks, some rogue devices will go undetected. The radios inside the WIPS sensors typically monitor the 2.4 GHz ISM band and the 5 GHz U-NII frequencies. Channels in the 4.9 GHz range, which is reserved for public safety in the United States and is a valid channel band in Japan, are often also monitored for potential rogue devices. However, legacy wireless networking equipment exists that transmits in the 900 MHz ISM band as well as other frequencies. These devices will not be detected. The only tool that will 100 percent detect a 900 MHz access point is a spectrum analyzer that also operates at 900 MHz. Rogue APs that do not transmit in either the 2.4 GHz or 5 GHZ frequency bands will not be detected by a standard 802.11 WIPS solution.

Not all WIPSs have spectrum analysis capabilities, although distributed spectrum analysis is becoming more common. Even if a WIPS has spectrum analysis capabilities, it can only perform spectrum analysis within a range of supported frequencies—typically the same frequencies that it monitors as a WIPS sensor. The WIPS should also monitor all the available 2.4 and 5 GHz channels, not just the ones permitted in your resident country.

Spectrum Analyzer

A *spectrum analyzer* is a frequency domain tool that can detect any RF signal in the frequency range being scanned. A spectrum analyzer that monitors the 2.4 GHz ISM band will be able to detect both intentional jamming and unintentional jamming devices. Some spectrum analyzers can look at the RF signature of the interfering signal and classify the device. For example, the spectrum analyzer might identify the signal as a microwave oven, a Bluetooth transmitter, or an 802.11 FHSS radio. Two forms of spectrum analysis systems are available: mobile and distributed. Most spectrum analyzers are standalone mobile solutions. Many enterprise WLAN vendors are providing distributed spectrum analysis by using the RF listening capabilities of the access points that function as sensors. A *distributed spectrum analysis system (DSAS)* is effectively a layer 1 wireless intrusion detection system that can detect and classify RF interference. The DSAS has the ability to categorize interference types based on frequency signatures. This can be useful to help classify and locate interfering devices. Most DSAS solutions use access points for the distributed spectrum analysis. Some vendor APs use an integrated spectrum analyzer chipset that operates independently from the 802.11 radio. Other vendors use the 802.11 radio in the access point to accomplish a lower grade of spectrum analysis.

Wireless Security Policies

Securing a wireless network and monitoring for threats are absolute necessities, but both are worthless unless proper security policies are in place. What good is an 802.1X/EAP solution if the end users share their passwords? Why purchase an intrusion detection system without having a policy to deal with rogue APs?

More and more businesses have started to amend their network usage policies to include a wireless policy section. If you have not done so already, you should absolutely add a WLAN section to your corporate security policy. Two good resources for learning about best practices and computer security policies are the SANS Institute and the National Institute of Standards and Technology (NIST).

General Security Policies

When establishing a wireless security policy, you must first define a *general policy*. A general wireless security policy establishes why a wireless security policy is needed for an organization. Even if a company has no plans for deploying a wireless network, there should be at a minimum a policy for how to deal with rogue wireless devices. A general wireless security policy defines the following items:

Statement of Authority   The statement of authority defines who put the wireless policy in place and the executive management that backs the policy.

Applicable Audience   The applicable audience is the audience to whom the policy applies, such as employees, visitors, and contractors.

Violation Reporting Procedures   Violation reporting procedures define how the wireless security policy will be enforced, including what actions should be taken and who is in charge of enforcement.

Risk Assessment and Threat Analysis   The risk assessment and threat analysis defines the potential wireless security risks and threats and the financial impact on the company if a successful attack occurs.

Security Auditing   Internal auditing procedures, as well as the need for independent outside audits, should also be defined.

Functional Security Policies

A *functional policy* is also needed to define the technical aspects of wireless security. The functional security policy establishes how to secure the wireless network in terms of what solutions and actions are needed. A functional wireless security policy defines the following items:

Policy Essentials   Basic security procedures, such as password policies, training, and proper usage of the wireless network, are policy essentials and should be defined.

Baseline Practices   Baseline practices define minimum wireless security practices, such as configuration checklists, staging and testing procedures, and so on.

Design and Implementation   The actual authentication, encryption, and segmentation solutions that are to be put in place are defined.

Monitoring and Response   All wireless intrusion detection procedures and the appropriate response to alarms are defined.

Legislative Compliance

Most countries have mandated regulations on how to protect and secure data communications within all government agencies. In the United States, NIST maintains the Federal Information Processing Standards (FIPS). Of special interest to wireless security is the FIPS 140-2 standard, which defines security requirements for cryptography modules. The U.S. government requires the use of validated cryptographic modules for all unclassified communications. Other countries also recognize the FIPS 140-2 standard or have similar regulations.

In the United States, other legislation exists for protecting information and communications in certain industries, including the following:

HIPAA   The Health Insurance Portability and Accountability Act (HIPAA) establishes national standards for electronic healthcare transactions and national standards for providers, health insurance plans, and employers. The goal is to protect patient information and maintain privacy.

Sarbanes-Oxley   The Sarbanes-Oxley Act of 2002 defines stringent controls on corporate accounting and auditing procedures, with a goal of corporate responsibility and enhanced financial disclosure.

GLBA   The Gramm-Leach-Bliley Act (GLBA) requires banks and financial institutions to notify customers of policies and practices disclosing customer information. The goal is to protect personal information, such as credit card numbers, Social Security numbers, names, addresses, and so forth.

All the FIPS publications can be found online at csrc.nist.gov/publications/PubsFIPS.html. Learn more about HIPAA at https://www.hhs.gov/hipaa/index.html. You can find general information about Sarbanes-Oxley at www.sarbanes-oxley-101.com and about GLBA at www.ftc.gov.

In 2015, the European Union (EU) adopted legislation designed to protect the privacy rights of EU citizens in regards to the collection and processing of their personal data. Enforcement of the *General Data Protection Regulation (GDPR)* began May 25, 2018. GDPR is a complex law with 99 articles. Its principle goal is to ensure that organizations with access to the personal data of EU citizens provide protection from privacy intrusion and data breaches. GDPR strengthens the security and protection of personal data by giving EU citizens a higher degree of control over of their personal data, and how it is used in the digital economy. Independent of their location, organizations that collect, store, or process the personal data of EU residents must abide by GDPR. More information about GDPR can be found at www.eugdpr.org.

PCI Compliance

As more of us rely on credit cards as our primary method of payment, more of us risk losing our card numbers to attackers and identity thieves through unsecure processing and/or storage of our cardholder information. The Payment Card Industry (PCI) realizes that in order to sustain continued business growth, measures must be taken to protect customer data and card numbers. The PCI Security Standards Council (SSC) has implemented regulations for organizations processing and storing cardholder information. This is commonly referred to as the *PCI Data Security Standard*. The current documented revision of the standard is PCI-DSSS ver 3.2; however, the standard is updated about every three years. Within this standard are components governing the use of wireless devices. More information about the PCI standard can be found at www.pcisecuritystandards.org.

802.11 Wireless Policy Recommendations

Although a detailed and thorough policy document should be created, we highly recommend the following six wireless security policies:

BYOD Policy   Employees like to bring their personal Wi-Fi devices, such as tablets and smartphones, to the workplace. Employees usually expect to be able to use their personal Wi-Fi devices on the secure corporate WLAN. Each employer needs to define a *bring your own device (BYOD)* policy that clearly states how personal devices will be onboarded onto the secure corporate WLAN. The BYOD policy should also state how the personal devices can be used while connected to the company WLAN and which corporate network resources will be accessible. BYOD is discussed in greater detail in Chapter 18, "Bring Your Own Device (BYOD) and Guest Access."

Guest Access Policy   Guess WLAN users should be restricted from accessing most company network resources. Guest users should be restricted from accessing company network resources with strong firewall enforcement and network segmentation practices. Bandwidth and QoS restrictions might also be implemented. Guest WLANs are discussed in greater detail in Chapter 18.

Remote-Access WLAN Policy   End users take their laptops and handheld devices off site and away from company grounds. Most users likely use wireless networks at home and at wireless hotspots to access the Internet. By design, many of these remote wireless networks have absolutely no security in place, and it is imperative that a remote access WLAN policy be strictly enforced. This policy should include the required use of an IPsec or SSL VPN solution to provide device authentication, user authentication, and strong encryption of all wireless data traffic. Hotspots are prime targets for malicious eavesdropping attacks. Personal firewalls should also be installed on all remote computers to prevent peer-to-peer attacks. Personal firewalls will not prevent hijacking attacks or peer-to-peer attacks, but they will prevent attackers from accessing your most critical information. Endpoint WLAN policy-enforcement software solutions exist that force end users to use VPN and firewall security when accessing any wireless network other than the corporate WLAN. The remote access policy is mandatory because the most likely and vulnerable location for an attack to occur is at a public-access hotspot.

Rogue AP Policy   No end users should ever be permitted to install their own wireless devices on the corporate network. This includes APs, wireless routers, wireless hardware USB clients, and other WLAN NICs. Any users installing their own wireless equipment could open unsecured portals into the main infrastructure network. This policy should be strictly enforced. End users should not be permitted to set up ad hoc or peer-to-peer networks. Peer-to-peer networks are susceptible to peer attacks and can serve as unsecured portals to the infrastructure network if the computer's Ethernet port is also in use.

Wireless LAN Proper Use Policy   A thorough policy should outline the proper use and implementation of the main corporate wireless network. This policy should include proper installation procedures, proper security implementations, and allowed application use on the wireless LAN.

WIPS Policy   Policies should be written defining how to properly respond to alerts generated by the wireless intrusion prevention system. An example would be how to deal with the discovery of rogue APs and all the necessary actions that should take place.

These six policies are simple but are a good starting point in writing a wireless security policy document.

Summary

In this chapter, we discussed all the potential wireless attacks and threats. The rogue access point has always been the biggest concern in terms of wireless threats, followed immediately by social engineering. We discussed many other serious threats—such as peer-to-peer attacks and eavesdropping—that can have just as serious consequences. We also discussed denial-of-service (DoS) attacks, which cannot be mitigated, only monitored. We covered the various solutions that are available for intrusion monitoring. Most intrusion detection solutions use a distributed client-server model, and some offer rogue prevention capabilities. Finally, we discussed the need for sound wireless security policies, which act as a foundation for the wireless security solutions you implement.

Exam Essentials

Understand the risk of the rogue access point.   Be able to explain why the rogue AP provides a portal into network resources. Understand that employees are often the source of rogue APs.

Define peer-to-peer attacks.   Understand that peer-to-peer attacks can happen via an access point or through an ad hoc network. Explain how to defend against this type of attack.

Know the risks of eavesdropping. Explain the difference between casual and malicious eavesdropping. Explain why encryption is needed for protection.

Define authentication and hijacking attacks. Explain the risks behind these types of attacks. Understand that a strong 802.1X/EAP solution is needed to mitigate them.

Explain wireless denial-of-service attacks. Know the difference between layer 1 and layer 2 DoS attacks. Explain why these attacks cannot be mitigated, only monitored.

Understand the types of wireless intrusion solutions. Explain the purpose of a WIPS. Understand that most solutions are distributed client-server models. Know the various components of an intrusion monitoring solution as well as the various models. Understand which attacks can be monitored and which can be prevented.

Understand the need for a wireless security policy. Explain the difference between general and functional policies.

Review Questions

1. Which of these attacks are considered denial-of-service attacks? (Choose all that apply.)
   A. Man-in-the-middle
   B. Jamming
   C. Deauthentication spoofing
   D. MAC spoofing
   E. Peer-to-peer

2. Which of these attacks would be considered malicious eavesdropping? (Choose all that apply.)
   A. NetStumbler
   B. Peer-to-peer
   C. Protocol analyzer capture
   D. Packet reconstruction
   E. PS-Poll floods

3. Which of these attacks will not be detected by a wireless intrusion prevention system (WIPS)?
   A. Deauthentication spoofing
   B. MAC spoofing
   C. Rogue access point
   D. Eavesdropping with a protocol analyzer
   E. Association flood

4. Which of these attacks can be mitigated with a mutual authentication solution? (Choose all that apply.)
   A. Malicious eavesdropping
   B. Deauthentication
   C. Man-in-the-middle
   D. Wireless hijacking
   E. Authentication flood

5. What type of security can be used to stop attackers from seeing the MAC addresses used by your legitimate 802.11 WLAN devices?
   A. MAC filtering
   B. CCMP/AES encryption
   C. MAC spoofing
   D. Rogue mitigation
   E. Rogue detection
   F. None of the above

6. When you are designing a wireless policy document, what two major areas of policy should be addressed?

    A.   General policy

    B.   Functional policy

    C.   Rogue AP policy

    D.   Authentication policy

    E.   Physical security

7. What can happen when an intruder compromises the PSK or passphrase used during WPA/WPA2-Personal authentication? (Choose all that apply.)

    A.   Decryption

    B.   ASLEAP attack

    C.   Spoofing

    D.   Encryption cracking

    E.   Access to network resources

8. Which of these attacks are considered layer 2 DoS attacks? (Choose all that apply.)

    A.   Deauthentication spoofing

    B.   Jamming

    C.   Virtual carrier attacks

    D.   PS-Poll floods

    E.   Authentication floods

9. Which of these can cause unintentional RF jamming attacks against an 802.11 wireless network? (Choose all that apply.)

    A.   Microwave oven

    B.   Signal generator

    C.   2.4 GHz cordless phones

    D.   900 MHz cordless phones

    E.   Deauthentication transmitter

10. Rogue WLAN devices are commonly installed by whom? (Choose all that apply.)

    A.   Attackers

    B.   Wardrivers

    C.   Contractors

    D.   Visitors

    E.   Employees

11. Which two solutions help mitigate peer-to-peer attacks from other clients associated to the same 802.11 access point?

    A. Personal firewall

    B. WPA2 encryption

    C. Client isolation

    D. MAC filter

12. What type of solution can be used to perform countermeasures against a rogue access point?

    A. CCMP

    B. PEAP

    C. WIPS

    D. TKIP

    E. WINS

13. A WIPS uses which four labels to classify an 802.11 device?

    A. Authorized

    B. Neighbor

    C. Enabled

    D. Disabled

    E. Rogue

    F. Unauthorized/unknown

14. Scott is an administrator at the Williams Lumber Company, and his WIPS has detected a rogue access point. What actions should he take after the WIPS detects the rogue AP? (Choose the best two answers.)

    A. Enable the layer 2 rogue containment feature that his WIPS provides.

    B. Unplug the rogue AP from the electrical outlet upon discovery.

    C. Call the police.

    D. Call his mother.

    E. Unplug the rogue AP from the data port upon discovery.

15. Which of these attacks are wireless users susceptible to at a public-access hotspot? (Choose all that apply.)

    A. Wi-Fi phishing

    B. Happy AP attack

    C. Peer-to-peer attack

    D. Malicious eavesdropping

    E. 802.11 sky monkey attack

    F. Man-in-the-middle attack

    G. Wireless hijacking

16. Which two components should be mandatory in every remote access wireless security policy?

   A. Encrypted VPN

   B. 802.1X/EAP

   C. Personal firewall

   D. Captive portal

   E. Wireless stun gun

17. MAC filters are typically considered to be a weak security implementation because of what type of attack?

   A. Spamming

   B. Spoofing

   C. Phishing

   D. Cracking

   E. Eavesdropping

18. Which WIPS architecture is the most commonly deployed?

   A. Integrated

   B. Overlay

   C. Access

   D. Core

   E. Cloud

19. Which of these encryption technologies have been cracked? (Choose all that apply.)

   A. 64-bit WEP

   B. 3DES

   C. CCMP/AES

   D. 128-bit WEP

20. What is another name for a wireless hijacking attack?

   A. Wi-Fi phishing

   B. Man-in-the-middle

   C. Fake AP

   D. Evil twin

   E. AirSpy

Chapter 17

802.11 Network Security Architecture

IN THIS CHAPTER, YOU WILL LEARN ABOUT THE FOLLOWING:

✓ **802.11 security basics**

- Data privacy and integrity
- Authentication, authorization, and accounting (AAA)
- Segmentation
- Monitoring and policy

✓ **Legacy 802.11 security**

- Legacy authentication
- Static WEP encryption
- MAC filters
- SSID cloaking

✓ **Robust security**

- Robust security network (RSN)
- Authentication and authorization
- PSK authentication
- Proprietary PSK authentication
- Simultaneous Authentication of Equals (SAE)
- 802.1X/EAP framework
- EAP types
- Dynamic encryption-key generation
- 4-Way Handshake
- WLAN encryption
- TKIP encryption
- CCMP encryption
- GCMP encryption

✓ **Traffic segmentation**

- VLANs
- RBAC

✓ **WPA3**

✓ **VPN wireless security**

- VPN 101
- Layer 3 VPNs
- SSL VPNs
- VPN deployment

In this chapter, you will learn about one of the most often discussed topics relating to 802.11 wireless networks: security. In this chapter, we discuss legacy 802.11 security solutions as well as more robust solutions that are now defined by the 802.11-2016 standard. WLAN security had a bad reputation in its early years—and deservedly so. The legacy security mechanisms originally defined by the IEEE did not provide the adequate authentication and data privacy that are needed in a mobility environment. Although there is no such thing as 100 percent security, properly installed and managed solutions do exist that can fortify and protect your wireless network.

As you learned in Chapter 16, "Wireless Attacks, Intrusion Monitoring, and Policy," numerous wireless security risks exist. Many of the attacks against an 802.11 network can be defended with the proper implementation of the security architectures discussed in this chapter. However, many attacks cannot be mitigated and can merely be monitored and hopefully responded to.

Although less than 10 percent of the CWNA exam covers 802.11 security, the CWNP program offers another certification, Certified Wireless Security Professional (CWSP), which focuses on just the topic of wireless security. The CWSP certification exam requires a more in-depth understanding of 802.11 security. However, this chapter will give you a foundation of wireless security that should help you pass the security portions of the CWNA exam as well as give you a head start in the knowledge you will need to implement proper wireless security.

802.11 Security Basics

When you are securing a wireless 802.11 network, the following five major components are typically required:

- Data privacy and integrity
- Authentication, authorization, and accounting (AAA)
- Segmentation
- Monitoring
- Policy

Because data is transmitted freely and openly in the air, proper protection is needed to ensure data privacy, so strong encryption is needed. The function of most wireless networks is to provide a portal into some other network infrastructure, such as an 802.3 Ethernet backbone. The wireless portal must be protected, and therefore an authentication solution is needed to ensure that only authorized devices and users can pass through the portal via a wireless access point (AP). After users have been authorized to pass through the wireless portal, users and client devices require further identity-based access restrictions to access network resources. 802.11 wireless networks can be further protected with continuous monitoring by a wireless intrusion prevention system. All these security components should also be cemented together with policy enforcement.

For wired or wireless networks, never take network security lightly. Unfortunately, WLAN security still has a bad reputation with some people because of the weak legacy 802.11 security mechanisms that were originally deployed. However, in 2004, the 802.11i amendment was ratified and defined strong encryption and better authentication methods. The 802.11i amendment is part of the 802.11-2016 standard and fully defines a robust security network (RSN), which is discussed later in this chapter. If proper encryption and authentication solutions are deployed, a wireless network can be just as secure, if not more secure, than the wired segments of a network. When properly implemented, the five components of 802.11 security discussed in this chapter will lay a solid foundation for protecting your WLAN.

Data Privacy and Integrity

802.11 wireless networks operate in license-free frequency bands, and all data transmissions travel in the open air. Protecting data privacy in a wired network is much easier because physical access to the wired medium is more restricted, whereas access to wireless transmissions is available to anyone in listening range. Therefore, using cipher encryption technologies to obscure information is mandatory to provide proper data privacy. A *cipher* is an algorithm used to perform encryption. The term "cryptology" is derived from Greek and translates to "hidden word." The goal of cryptology is to take a piece of information, often referred to as *plaintext*, and, using a process or algorithm, also referred to as a key or cipher, transform the plaintext into encrypted text, also known as *ciphertext*. The science of concealing the plaintext and then revealing it is known as *cryptography*. In the computer and networking industries, the process of converting plaintext into ciphertext is commonly referred to as *encryption*, and the process of converting ciphertext back to plaintext is commonly referred to as *decryption*.

The two most common ciphers used to protect data are the *ARC4 algorithm* and the *Advanced Encryption Standard (AES)* algorithm. Some ciphers encrypt data in a continuous stream, whereas others encrypt data in groupings known as *blocks*.

ARC4 Cipher   The ARC4 algorithm is a streaming cipher used to protect 802.11 wireless data in two legacy encryption methods, known as WEP and TKIP, both of which are discussed later in this chapter. ARC4 is short for Alleged RC4. RC4 was created in 1987 by Ron Rivest of RSA Security. It is known as either "Rivest Cipher 4" or "Ron's Code 4." RC4 was initially a trade secret; however, in 1994 a description of it was leaked onto the Internet. Comparison testing confirmed that the leaked code was genuine. RSA has never officially released the algorithm, and the name "RC4" is trademarked—hence, the reference to it as ARCFOUR or ARC4.

Advanced Encryption Standard Cipher   The AES algorithm, originally named the Rijndael algorithm, is a block cipher that offers much stronger protection than the ARC4 streaming cipher. AES is used to encrypt 802.11 wireless data by using an encryption method known as *Counter Mode with Cipher Block Chaining Message Authentication Code Protocol (CCMP)*, which will also be discussed later in this chapter. The AES algorithm encrypts data in fixed data blocks with choices in encryption key strength of 128, 192, or 256 bits. The AES cipher is the mandated algorithm of the U.S. government for protecting both sensitive and classified information. The AES cipher is also used for encryption in many other networking technologies, such as IPsec VPNs.

In Chapter 9, "802.11 MAC," you learned about the three major types of 802.11 wireless frames. Inside the body of a management frame is layer 2 information necessary for the basic operation of the BSS, and historically, 802.11 management frames have not been encrypted. However, the need to protect some critical network functions, such as authentication and association, was introduced with the 802.11w amendment, which provides protection for certain types of management frames. Control frames have no body and also are not encrypted. The information that needs to be protected is the upper-layer information inside the body of 802.11 data frames. If data encryption is enabled, the *MAC service data unit (MSDU)* inside the body of any 802.11 data frame is protected by layer 2 encryption. Most of the encryption methods discussed in this chapter use layer 2 encryption, which is used to protect the layer 3–7 information found inside the body of an 802.11 data frame. In Exercise 17.1, you will use an 802.11 protocol analyzer to view the MSDU payload of an 802.11 data frame.

EXERCISE 17.1

Using Unencrypted and Encrypted Data Frames

1. To perform this exercise, you need to first download the CWNA_CHAPTER17.PCAP file from the book's web page at www.wiley.com/go/cwnasg.

2. After the file is downloaded, you will need packet analysis software to open the file. If you do not already have a packet analyzer installed on your computer, you can download Wireshark from www.wireshark.org.

3. Using the packet analyzer, open the CWNA_CHAPTER17.PCAP file. Most packet analyzers display a list of capture frames in the upper section of the screen, with each frame numbered sequentially in the first column.

4. Scroll down the list of frames and click frame #8, which is an unencrypted simple data frame. Look at the frame body and notice the upper-layer information, such as IP addresses and TCP ports.

5. Click frame #136, which is an encrypted simple data frame. Look at the frame body and notice that WEP encryption is being used and that the upper-layer information cannot be seen.

WEP, TKIP, and CCMP all use a data integrity check to ensure that the data has not been maliciously altered. WEP uses an integrity check value (ICV), and TKIP uses a message integrity check (MIC). CCMP also uses a message integrity check (MIC) that is much stronger than the data integrity methods used in TKIP or WEP.

Authentication, Authorization, and Accounting

Authentication, authorization, and accounting (AAA) is a key computer security concept that defines the protection of network resources.

Authentication   Authentication is the verification of identity and credentials. Users or devices must identify themselves and present credentials, such as usernames and passwords or digital certificates. More secure authentication systems use multifactor authentication, which requires at least two sets of different types of credentials to be presented.

Authorization   Authorization determines if the device or user is authorized to have access to network resources. This can include identifying whether you can have access based upon the type of device you are using (laptop, tablet, or smartphone), time of day restrictions, or location. Before authorization can be determined, proper authentication must occur.

Accounting   Accounting is tracking the use of network resources by users and devices. It is an important aspect of network security, used to keep a historical trail of who used what resource, when, and where. A record is kept of user identity, which resource was accessed, and at what time. Keeping an accounting trail is often a requirement of many industry regulations, such as the payment card industry (PCI).

Remember that the usual purpose of an 802.11 wireless network is to act as a portal into an 802.3 wired network. Therefore, it is necessary to protect that portal with strong authentication methods so that only legitimate users with the proper credentials will be authorized onto network resources.

Segmentation

Although it is of the utmost importance to secure an enterprise wireless network by utilizing both strong encryption and an AAA solution, an equally important aspect of wireless security is segmentation. Segmentation is the chosen method of separating user traffic within a network. Prior to the introduction of stronger authentication and encryption techniques, wireless was viewed as an untrusted network segment. Therefore, before the ratification of the 802.11i security amendment, the entire wireless segment of a network was commonly treated as an untrusted segment, and the wired 802.3 network was considered the trusted segment.

Now that better security solutions exist, properly secured WLANs are more seamlessly and securely integrated into the wired infrastructure. It is still important to separate users and devices into proper groups, much like what is done on any traditional network. Once authorized onto network resources, users and devices can be further restricted as to which resources can be accessed and where they can go. Segmentation can be achieved through a variety of means, including firewalls, routers, VPNs, VLANs, and encapsulation or tunneling techniques, such as Generic Routing Encapsulation (GRE). The most common wireless segmentation strategy used in 802.11 enterprise WLANs is segmentation using virtual LANs (VLANs). Segmentation is also intertwined with role-based access control (RBAC), which is discussed later in this chapter.

Monitoring and Policy

After you have designed and installed your wireless network, it is important to monitor it. In addition to monitoring it to make sure that it is performing up to your expectations and those of your users, it may be necessary to constantly monitor it for attacks and intrusions. Similar to a business placing a video camera on the outside of its building to monitor the traffic going in and out of a locked door, it is important for the wireless network administrator to monitor the wireless traffic of a secured network. To monitor potentially malicious wireless activity on your network, you should install a wireless intrusion prevention system (WIPS). WLAN security monitoring can be an integrated solution or an overlay solution. Network management solutions that include WLAN security monitoring can be cloud-based or run on a private data center server. As discussed in Chapter 16, a WIPS can also mitigate attacks from rogue access points and rogue clients by performing attacks against the rogue devices, effectively disabling their ability to communicate with your network.

Legacy 802.11 Security

The original 802.11 standard defined little in terms of security. The authentication methods first outlined in 1997 basically provided an open door into the network infrastructure. The encryption method defined in the original 802.11 standard has long been cracked and is considered inadequate for data privacy. In the following sections, you will learn about the legacy authentication and encryption methods that were the only defined standards for 802.11 wireless security from 1997 until 2004. Later in this chapter, you will learn about the more robust security that was defined in the 802.11i security amendment, which is now part of the current 802.11-2016 standard.

Legacy Authentication

You already learned about legacy authentication in Chapter 9. The original 802.11 standard specified two methods of authentication: *Open System authentication* and *Shared Key authentication*. When discussing authentication, we often think of validating the identity of a user when they are connecting or logging into a network. 802.11 authentication is very different from this. These legacy authentication methods were not so much an authentication of user identity but more of an authentication of capability. Think of these authentication methods as verification between the two devices that they are both valid 802.11 devices.

Open System authentication provides authentication without performing any type of user verification. It is essentially a two-way exchange between the client radio and the access point:

1. The client sends an authentication request.

2. The access point then sends an authentication response.

Because Open System authentication does not require the use of any credentials, every client gets authenticated and therefore authorized onto network resources after they have been associated. Static WEP encryption is optional with Open System authentication and may be used to encrypt the data frames after Open System authentication and association occur.

As you learned in Chapter 9, Shared Key authentication used Wired Equivalent Privacy (WEP) to authenticate client stations and required that a static WEP key be configured on both the station and the access point. In addition to WEP being mandatory, authentication would not work if the static WEP keys did not match. The authentication process was similar to Open System authentication but included a challenge and response between the radio cards. Shared Key authentication was a four-way authentication frame handshake:

1. The client station sent an authentication request to the access point.

2. The access point sent a cleartext challenge to the client station in an authentication response.

3. The client station encrypted the cleartext challenge and sent it back to the access point in the body of another authentication request frame.

4. The access point decrypted the station's response and compared it to the challenge text:

   ▪ If they matched, the access point would respond by sending a fourth and final authentication frame to the station confirming the success.

   ▪ If they did not match, the access point would respond negatively. If the access point could not decrypt the challenge, it would also respond negatively.

If Shared Key authentication were successful, the same static WEP key that was used during the Shared Key authentication process would also be used to encrypt the 802.11 data frames.

Open System vs. Shared Key

Although Shared Key authentication might seem to be a more secure solution than Open System authentication, in reality Shared Key could be the bigger security risk. During the Shared Key authentication process, anyone who captures the cleartext challenge phrase and then captures the encrypted challenge phrase in the response frame could potentially derive the static WEP key. If the static WEP key were compromised, a whole new can of worms would be opened, because now all the data frames could be decrypted and the attacker could gain direct access to the network. Neither of the legacy authentication methods is considered strong enough for enterprise security. Shared Key authentication has been deprecated and is no longer recommended. More secure 802.1X/EAP authentication methods are discussed later in this chapter.

Static WEP Encryption

Wired Equivalent Privacy (WEP) is a layer 2 encryption method that uses the ARC4 streaming cipher. The original 802.11 standard initially defined only 64-bit WEP as a supported encryption method. Shortly thereafter, 128-bit WEP was also defined as a supported encryption process. The three main goals of WEP encryption are as follows:

Confidentiality   The primary goal of confidentiality is to provide data privacy by encrypting the data before transmission.

Access Control   WEP also provides access control, which is basically a crude form of authorization. Client stations that do not have the same matching static WEP key as an access point are refused access to network resources.

Data Integrity   A data integrity checksum known as the *integrity check value (ICV)* is computed on data before encryption and used to prevent data from being modified.

Although 128-bit WEP was feasible, initially the U.S. government allowed the export of only 64-bit technology. After the U.S. government loosened export restrictions on key size, WLAN radio manufacturers began to produce equipment that supported 128-bit WEP encryption. The 802.11-2016 standard refers to the 64-bit version as *WEP-40* and the 128-bit version as *WEP-104*. As shown in Figure 17.1, 64-bit WEP uses a secret 40-bit static key, which is combined with a 24-bit number selected by the radio's device drivers. This 24-bit number, known as the *initialization vector (IV)*, is sent in cleartext and a new IV is created for every frame. Although the IV is said to be new on every frame, there are only 16,777,216 different IV combinations; therefore, you are forced to reuse the IV values. The effective key strength of combining the IV with the 40-bit static key is 64-bit encryption. 128-bit WEP encryption uses a 104-bit secret static key, which is also combined with a 24-bit IV.

FIGURE 17.1   Static WEP encryption key and initialization vector

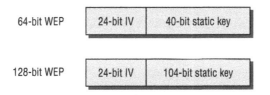

A static WEP key can usually be entered either as hexadecimal (hex) characters (0–9 and A–F) or as ASCII characters. The static key must match on both the access point and the client device. A 40-bit static key consists of 10 hex characters or 5 ASCII characters, whereas a 104-bit static key consists of 26 hex characters or 13 ASCII characters. Not all client stations or access points support both hex and ASCII. Many clients and access points support the use of up to four separate static WEP keys from which a user can choose one as the default transmission key (Figure 17.2 shows an example). The *transmission key* is the static key that is used to encrypt data by the transmitting radio. A client or access point may use one key to encrypt outbound traffic and a different key to decrypt received traffic. However, each of the four keys must match exactly on both sides of a link for encryption/decryption to work properly.

FIGURE 17.2 Transmission key

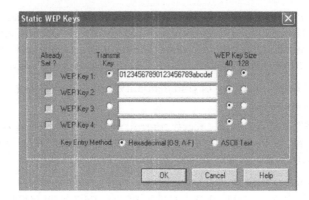

How does WEP work?

1. WEP runs a cyclic redundancy check (CRC) on the plaintext data that is to be encrypted and then appends the integrity check value (ICV) to the end of the plaintext data.

2. A 24-bit cleartext initialization vector (IV) is then generated and combined with the static secret key.

3. WEP then uses both the static key and the IV as seeding material through a pseudorandom algorithm that generates random bits of data known as a *keystream*.

   These pseudorandom bits are equal in length to the plaintext data that is to be encrypted.

4. The pseudorandom bits in the keystream are then combined with the plaintext data bits by using a Boolean XOR process.

   The end result is the WEP ciphertext, which is the encrypted data.

5. The encrypted data is then prefixed with the cleartext IV.

   Figure 17.3 illustrates this process.

FIGURE 17.3 WEP encryption process

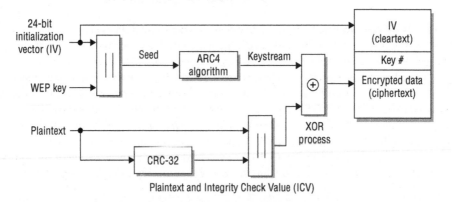

Unfortunately, WEP has quite a few weaknesses, including the following four main attacks:

IV Collisions Attack   Because the 24-bit initialization vector is in cleartext and is different in every frame, all 16 million IVs will eventually repeat themselves in a busy WEP encrypted network. Because of the limited size of the IV space, IV collisions occur, and an attacker can recover the secret key much easier when IV collisions occur in wireless networks.

Weak Key Attack   Because of the ARC4 key-scheduling algorithm, weak IV keys are generated. An attacker can recover the secret key much easier by recovering the known weak IV keys.

Reinjection Attack   Hacker tools exist that implement a packet reinjection attack to accelerate the collection of weak IVs on a network with little traffic.

Bit-Flipping Attack   The ICV data integrity check is considered weak. WEP encrypted packets can be tampered with.

WEP cracking tools have been available for many years. These cracking tools may use a combination of the first three mentioned attacks and can crack WEP in less than 5 minutes. After an attacker has compromised the static WEP key, any data frame can be decrypted with the newly discovered key. Later in this chapter, we discuss TKIP, which is an enhancement of WEP. CCMP encryption uses the AES algorithm and is an even stronger encryption method. As defined by the original 802.11 standard, WEP encryption is considered optional. WEP has been cracked and has been an unacceptable encryption method for the enterprise for more than 15 years. If legacy devices that support only WEP encryption are still being deployed, these devices should be replaced immediately.

Dynamic WEP Encryption

Prior to 2004, many vendors implemented solutions that generated dynamic WEP encryption keys as a result of 802.1X/EAP authentication. Dynamic WEP was never standardized but was used by vendors until TKIP and CCMP became available to the marketplace.

Dynamic WEP was a short-lived encryption-key management solution that was often implemented prior to the release of WPA-certified WLAN products. The generation and distribution of dynamic WEP keys as a by-product of the EAP authentication process had many benefits and was preferable to the use of static WEP keys. Static keys were no longer used and did not have to be entered manually. Also, every user had a separate and independent key. If a user's dynamic WEP key were compromised, only that one user's traffic could be decrypted. However, a dynamic WEP key could still be cracked, and if compromised, it could indeed be used to decrypt data frames. Dynamic WEP still had risks.

Please understand that a dynamic WEP key is not the same as TKIP or CCMP encryption keys which are also generated dynamically. Later in this chapter, you will learn about TKIP/ARC4 or CCMP/AES encryption keys which are dynamically generated by a process called the 4-Way Handshake.

MAC Filters

Every network card has a physical address known as a *MAC address*. This address is a 12-digit hexadecimal number. Every 802.11 radio has a unique MAC address. Most vendors provide MAC filtering capabilities on their access points. MAC filters can be configured to either allow or deny traffic from specific client MAC addresses to associate and connect to an AP.

The 802.11-2016 standard does not define MAC filtering, and any implementation of MAC filtering is vendor specific. Most vendors use MAC filters to deny client associations to an AP. Other vendors use MAC firewall filters to apply restrictions that will allow traffic only from specific client stations to pass through based on their unique MAC addresses. Any other client stations whose MAC addresses are not on the allowed list will not be able to pass traffic through the virtual port of the access point and on to the distribution system medium. It should be noted that MAC addresses can be *spoofed*, or impersonated, and any amateur hacker can easily bypass any MAC filter by spoofing an allowed client MAC address. Because of spoofing and because of all the administrative work involved with setting up MAC filters, MAC filtering is not considered a reliable means of security for wireless enterprise networks. MAC filters can be used as a security measure to protect legacy radios that do not support stronger security. For example, older handheld bar code scanners may use 802.11 radios that support only static WEP. Best practices dictate an extra layer of security by segmenting the handheld devices in a separate VLAN with a MAC filter based on the manufacturer's OUI address (the first three octets of the MAC address, which are manufacturer specific).

SSID Cloaking

Remember in *Star Trek* when the Romulans cloaked their spaceship, but somehow Captain Kirk always found the ship anyway? Well, there is a way to "cloak" your service set identifier (SSID). Access points typically have a setting called *closed network*, *hidden SSID*, or *stealth mode*. By enabling this feature, you can hide, or cloak, your wireless network name.

When you implement a closed network, the SSID field in the beacon frame is null (empty), and therefore passive scanning will not reveal the SSID to client stations that are listening to beacons. The SSID, which is also often called the ESSID, is the logical identifier of a WLAN. The idea behind cloaking the SSID is that any client station that does not know the SSID of the WLAN will not be able to discover the WLAN and therefore will not associate.

Many wireless client software utilities transmit probe requests with null SSID fields when actively scanning for access points. Additionally, there are many popular WLAN discovery applications, such as inSSIDer, WiFi Explorer, and WiFi Analyzer, that can be used by individuals to discover wireless networks. Most of these discovery applications also send out null probe requests actively scanning for access points. When you implement a closed network, the access point responds to null probe requests with probe responses; however, as in the beacon frame, the SSID field is null, and therefore the SSID is hidden to client stations that are using active scanning. Implementing a closed network varies between WLAN vendors; some vendor access points may simply ignore null probe requests when a closed network has been configured. Effectively, your wireless network is temporarily invisible, or cloaked. Note that an access point in a closed network will respond to any configured client station that transmits directed probe requests with the properly configured SSID. This ensures that legitimate end users will be able to authenticate and associate to the AP. However, any client stations that are not configured with the correct SSID will not be able to authenticate or associate.

Although implementing a closed network may hide your SSID from some of these WLAN discovery tools, anyone with a layer 2 wireless protocol analyzer can capture the frames transmitted by any legitimate end user and discover the SSID, which is transmitted in cleartext. In other words, a hidden SSID can be found usually in seconds with the proper tools. Many wireless professionals will argue that hiding the SSID is a waste of time, whereas others view a closed network as just another layer of security.

Although you can hide your SSID to cloak the identity of your wireless network from novice hackers (often referred to as *script kiddies*) and nonhackers, it should be clearly understood that SSID cloaking is by no means an end-all wireless security solution. The 802.11 standard does not define SSID cloaking, so all implementations of a closed network are vendor specific. As a result, incompatibility can potentially cause connectivity problems. Some wireless clients will not connect to a hidden SSID, even when the SSID is manually entered in the client software. Therefore, be sure to know the capabilities of your devices before implementing a closed network. Cloaking the SSID can also become an administrative and support issue. Requiring end users to configure the SSID in the radio software interface often results in more calls to the help desk because of misconfigured SSIDs. We highly recommend that you never cloak your SSID and instead broadcast your SSID for anyone and everyone to see.

Robust Security

In 2004, the 802.11i security amendment was ratified and is now part of the 802.11-2016 standard. The 802.11-2016 standard defines an enterprise authentication method as well as a method of authentication for home use. The current standard defines the use of an 802.1X/EAP authentication and also the use of a preshared key (PSK) or a passphrase. 802.1X/EAP is a strong authentication method most often deployed in the enterprise. The less complex PSK authentication is normally used in small office, home office (SOHO) environments but can be deployed in the enterprise as well. The 802.11-2016 standard also requires the use of strong, dynamic encryption-key generation methods. CCMP/AES encryption is the default encryption method, and TKIP/ARC4 is an optional encryption method.

Prior to the ratification of the 802.11i amendment, the Wi-Fi Alliance introduced the *Wi-Fi Protected Access (WPA)* certification as a snapshot of the not-yet-released 802.11i amendment, supporting only TKIP/ARC4 dynamic encryption-key generation. 802.1X/EAP authentication was intended for the enterprise, and passphrase authentication was suggested in a SOHO environment.

After 802.11i was ratified, the Wi-Fi Alliance introduced the WPA2 certification. *WPA2 is a more complete implementation* of the 802.11i amendment and supports both CCMP/AES and TKIP/RC4 dynamic encryption-key generation. 802.1X/EAP authentication is more complex and meant for the enterprise, whereas passphrase authentication is simpler and meant for a SOHO environment. Any 802.11 radios manufactured after 2005 are most likely WPA2 compliant. If a radio is WPA compliant, it most likely only supports TKIP/ARC4 encryption. If the radio is WPA2 compliant, it supports the stronger CCMP/AES dynamic encryption. Table 17.1 offers a valuable comparison of the various security standards and certifications.

TABLE 17.1   Security standards and certifications comparison

802.11 Standard	Wi-Fi Alliance Certification	Authentication Method	Encryption Method	Cipher	Key Generation
802.11 legacy	None	Open System or Shared Key	WEP	ARC4	Static
	WPA-Personal	WPA Passphrase (also known as WPA PSK and WPA Preshared Key)	TKIP	ARC4	Dynamic
	WPA-Enterprise	802.1X/EAP	TKIP	ARC4	Dynamic
802.11-2016 (RSN)	WPA2-Personal	WPA2 Passphrase (also known as WPA2 PSK and WPA2 Preshared Key)	CCMP (mandatory)	AES (mandatory)	Dynamic
			TKIP (optional)	ARC4 (optional)	
	WPA2-Enterprise	802.1X/EAP	CCMP (mandatory)	AES (mandatory)	Dynamic
			TKIP (optional)	ARC4 (optional)	

Robust Security Network

The 802.11-2016 standard defines what are known as *robust security network (RSN)* and *robust security network associations (RSNAs)*. Two stations (STAs) must authenticate and associate with each other, as well as create dynamic encryption keys through a process known as the 4-Way Handshake. This association between two stations is referred to as an RSNA. In other words, any two radios must share dynamic encryption keys that are unique between those two radios. CCMP/AES encryption is the mandated encryption method, and TKIP/ARC4 is an optional encryption method.

A robust security network (RSN) is a network that allows for the creation of only robust security network associations (RSNAs). An RSN can be identified by a field found in 802.11 management frames. This field is known as the *RSN information element (IE)*. An information element is an optional field of variable length that can be found in 802.11 management frames. The RSN information element field is always found in four different 802.11 management frames: beacon management frames, probe response frames, association request frames, and reassociation request frames. The RSN information element can also be found in reassociation response frames if 802.11r capabilities are enabled on an AP and roaming client. This field will identify the cipher suite and authentication capabilities of each station. The 802.11-2016 standard allows for the creation of pre-robust security network associations (pre-RSNAs) as well as RSNAs. In other words, legacy security measures can be supported in the same basic service set (BSS) along with RSN-security-defined mechanisms. A *transition security network (TSN)* supports RSN-defined security, as well as legacy security such as WEP, within the same BSS, although most vendors do not support a TSN.

Authentication and Authorization

As you learned earlier in this chapter, authentication is the verification of user or device identity and credentials. Users and devices must identify themselves and present credentials, such as passwords or digital certificates. Authorization involves whether a user or device is granted access to network resources and services. Before authorization to network resources can be granted, proper authentication must occur.

The following sections detail more advanced authentication and authorization defenses. You will also learn that dynamic encryption capabilities are also possible as a by-product of these stronger authentication solutions.

PSK Authentication

The 802.11-2016 standard defines authentication and key management (AKM) services. AKM services require both authentication processes and the generation and management of encryption keys. An *authentication and key management protocol (AKMP)* can be either a preshared (PSK) or an EAP protocol used during 802.1X/EAP authentication. 802.1X/EAP requires a RADIUS server and advanced skills to configure and support it. The average home or small business Wi-Fi user has no knowledge of 802.1X/EAP and does not have a RADIUS server in their living room. PSK authentication is meant to be used in SOHO environments because the stronger enterprise 802.1X/EAP authentication solutions are not available. Furthermore, many consumer-grade Wi-Fi devices, such as printers, do not support 802.1X/EAP and support only PSK authentication. Therefore, the security used in SOHO environments is *PSK authentication*. WPA/WPA2-Personal utilizes PSK authentication. On the other hand, WPA/WPA2-Enterprise refers to the 802.1X/EAP authentication solution.

Most SOHO wireless networks are secured with WPA/WPA2-Personal mechanisms. Prior to the IEEE ratification of the 802.11i amendment, the Wi-Fi Alliance introduced the Wi-Fi Protected Access (WPA) certification as a snapshot of the not-yet-released 802.11i amendment, but it supported only TKIP/ARC4 dynamic encryption-key generation. 802.1X/EAP authentication was suggested for the enterprise, while a passphrase authentication method called WPA-Personal was the suggested security mechanism in a SOHO environment.

The intended goal of WPA-Personal was to move away from static encryption keys to dynamically generated keys using a simple passphrase as a seed. A preshared key (PSK) used in a robust security network is 256 bits in length, or 64 characters when expressed in hex. A PSK is a static key that is configured on the access point and on all the clients. The same static PSK is used by all members of the basic service set (BSS). The problem is that the average home user is not comfortable with entering a 64-character hexadecimal PSK on both a SOHO Wi-Fi router and a laptop client utility. Even if home users did enter a 64-character PSK on both ends, they probably would not be able to remember the PSK and would have to write it down. Most home users are, however, very comfortable configuring short ASCII passwords or passphrases. WPA/WPA2-Personal allows an end user to enter a simple ASCII character string, dubbed a passphrase, any length from 8 to 63 characters in size. Behind the scenes, a passphrase-to-PSK-mapping function takes care of the rest, converting the passphrase to a 256-bit PSK. As shown in Figure 17.4, the user enters a static passphrase, an 8- to 63-character string, into the client software utility on the end-user device and also on the access point. The passphrase must match on each device.

FIGURE 17.4   Client configured with static passphrase

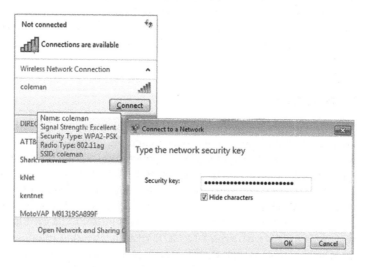

As previously mentioned, a *passphrase-PSK mapping* formula is defined by the 802.11-2016 standard to allow end users to use a simple ASCII passphrase, which is then converted to the 256-bit PSK. Here is a quick review of the formula to convert the passphrase to a PSK, followed by an explanation of the process:

PSK = PBKDF2(PassPhrase, ssid, ssidLength, 4096, 256)

A simple passphrase is combined with the SSID and hashed 4,096 times to produce a 256-bit (64-character) PSK. Table 17.2 illustrates some examples of how the formula uses both the passphrase and SSID inputs to generate the PSK.

TABLE 17.2 Passphrase-PSK mapping

Passphrase (8–63 Characters)	SSID	256-Bit/64-Character PSK
Carolina	cwna	7516b6d5169ca633ece6aa43e0ca9d5c0afa 08268ab9fde47c38a627546b71c5
certification	cwna	51da37d0c6ebba86123a13fb1ab0a1755a22fc 9791e53fab7208a5fceb6038a2
seahawks	cwna	20829812270679e481067e149dbe90ab59b 5179700c6359ba534b240acf410c3

The whole point of the passphrase-PSK mapping formula is to simplify configuration for the average home end user. Most people can remember an 8-character passphrase as opposed to a 256-bit PSK. Later in this chapter, you will learn that there is a symbiotic relationship between both PSK and 802.1X/EAP authentication with the generation of dynamic encryption keys via a 4-Way Handshake process. The 256-bit PSK is also used as the pairwise master key (PMK). The PMK is the seeding material for the 4-Way Handshake that is used to generate dynamic encryption keys. Therefore, the PSK in WPA/WPA2-Personal mode is quite literally the same as the PMK.

In June 2004, the IEEE 802.11 TGi working group formally ratified 802.11i, which added support for CCMP/AES encryption. The Wi-Fi Alliance revised the previous WPA specification to WPA2 and incorporated the CCMP/AES cipher. Therefore, the only practical difference between WPA and WPA2 has to do with the encryption cipher. WPA-Personal and WPA2-Personal both use the PSK authentication method; however, WPA-Personal specifies TKIP/ARC4 encryption, whereas WPA2-Personal specifies CCMP/AES. TKIP encryption has slowly been phased out over the years and is not supported for any of the 802.11n and 802.11ac data rates. In other words, older 802.11/a/b/g radios that support only WPA-Personal and TKIP might still be deployed. Any 802.11 radios manufactured after 2006 will be certified for WPA2-Personal and use CCMP/AES encryption. If PSK authentication is the chosen security method, WPA2-Personal with CCMP/AES encryption should always be used.

The Wi-Fi Alliance name for PSK authentication is WPA-Personal or WPA2-Personal. However, WLAN vendors have many marketing names for PSK authentication, including WPA/WPA2-Passphrase, WPA/WPA2-PSK, and WPA/WPA2-Preshared Key.

Proprietary PSK Authentication

Keep in mind that the simple PSK authentication method defined by WPA/WPA2-Personal can be a weak authentication method that is vulnerable to brute-force offline dictionary attacks. Because the passphrase is static, PSK authentication is also susceptible to social engineering attacks.

Although passphrases and PSK authentication are intended for use in a SOHO environment, in reality WPA/WPA2-Personal is often still used in the enterprise. For example, even though fast secure roaming (FSR) mechanisms have been possible for a while, some older VoWiFi phones and other handheld devices still do not yet support 802.1X/EAP. As a result, the strongest level of security used with these devices is PSK authentication. Cost issues may also drive a small business to use the simpler WPA/WPA2-Personal solution, as opposed to installing, configuring, and supporting a RADIUS server for 802.1X/EAP.

In Chapter 16, you learned that PSK authentication is vulnerable to brute-force offline dictionary attacks. However, the bigger problem with using PSK authentication in the enterprise is social engineering. Because the WPA/WPA2 passphrase is static, PSK authentication is highly susceptible to social engineering attacks. The PSK is the same on all WLAN devices. If an end user accidentally gives the PSK to a hacker, WLAN security is compromised. If an employee leaves the company, to maintain a secure environment, all the devices have to be reconfigured with a new 256-bit PSK. Because the passphrase or PSK is shared by everyone, a strict policy should be mandated stating that only the WLAN security administrator is aware of the passphrase or PSK. That, of course, creates another administrative problem because of the work involved in manually configuring each device.

Several enterprise WLAN vendors have come up with a creative solution to using WPA/WPA2-Personal that solves some of the biggest problems of using a single passphrase for WLAN access. Each computing device will have its own unique PSK for the WLAN. Therefore, the MAC address of each STA will be mapped to a unique WPA/WPA2-Personal passphrase. A database of unique PSKs mapped to usernames or client stations must be stored on all access points or on a centralized management server. The individual client stations are then assigned individual PSKs, which are created either dynamically or manually. As shown in Figure 17.5, the multiple per-user/per-device PSKs can be tied to a single SSID. The PSKs that are generated can also have an expiration date. Unique time-based PSKs can be used in a WLAN environment as a replacement for more traditional username/password credentials. Unlike static PSKs, per-user/per-device PSKs can provide unique identity credentials.

FIGURE 17.5 Proprietary PSK

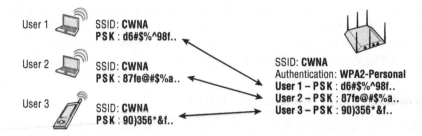

Currently, five enterprise WLAN vendors offer proprietary PSK solutions, which provide the capability of unique PSKs for each user or each device. These vendors are Aerohive Networks, Cisco, Fortinet, Ruckus Wireless, and Xirrus. Proprietary PSK solutions provide a way to implement unique identity credentials without the burden of deploying a more complex 802.1X/EAP solution. Social engineering and brute-force dictionary attacks are still possible, but they are harder to accomplish if strong and lengthy passphrase credentials are implemented. If a unique PSK is compromised, an administrator only has to revoke the single PSK credential and no longer has to reconfigure all access points and end-user devices.

Some WLAN client devices have limited support for 802.1X/EAP. In such situations, proprietary PSK solutions may be of benefit for those classes of devices and a vast improvement over standard static single PSK authentication. A proprietary PSK solution provides unique user or device credentials that standard PSK cannot provide. Additionally, proprietary PSK solutions with unique credentials do not require anywhere near the complex configuration needed for 802.1X/EAP.

Proprietary implementations of PSK authentication is not meant to be a replacement for 802.1X/EAP. However, multiple use cases for per-user and per-device PSK credentials have gained popularity in the enterprise. Some of these use cases include the following:

Legacy Devices   A per-user and per-device PSK solution can be used to supplement 802.1X/EAP security to provide legacy devices with unique PSK credentials.

Personal Devices   Proprietary implementations of PSK authentication can be used for bring your own device (BYOD) security to provide unique PSK credentials for employee personal devices.

Guest Access   Providing unique PSK credentials for access to the guest WLAN provides unique identity for guest users as well as the value-added service of encrypted guest access.

IoT Devices   Machines and sensors equipped with 802.11 radios often do not support 802.1X/EAP. Per-user and per-device PSK can provide Internet of Thing (IoT) devices with unique PSK credentials.

Simultaneous Authentication of Equals

The 802.11s amendment was ratified in 2011 with the intention of standardizing mesh networking of 802.11 WLANs. The 802.11s amendment is now part of the 802.11-2016 standard. The amendment defined a *Hybrid Wireless Mesh Protocol (HWMP)* that 802.11 mesh portals and mesh points could use to dynamically determine the best path selection for traffic flow through a meshed WLAN. HWMP and other mechanisms defined by the 802.11s-2011 amendment have not been embraced by WLAN vendors because of competitive reasons. The majority of WLAN vendors do not want their APs to accept mesh communications from a competitor's APs. As a result, the major WLAN vendors offer proprietary mesh solutions using their own mesh protocols and metrics.

The 802.11s-2011 amendment also defined RSN security methods that could be used by mesh portals and mesh points. An *Authenticated Mesh Peering Exchange (AMPE)* is used to securely create and exchange pairwise master keys (PMKs). 802.1X/EAP could be one method used to derive a PMK in a mesh environment. This method is not ideal for mesh points, because the RADIUS server resides on the wired network. Therefore, the 802.11s-2011 amendment proposed a new peer-to-peer authentication method called *Simultaneous Authentication of Equals (SAE)*. SAE is based on a *Dragonfly key exchange*. Dragonfly is a patent-free and royalty-free technology that uses a *zero-knowledge proof* key exchange, which means a user or device must prove knowledge of a password without having to reveal a password.

Although SAE has yet to be implemented for 802.11 mesh networks, the Wi-Fi Alliance views SAE as a future replacement for PSK authentication. As you have already learned, PSK authentication is susceptible to brute-force dictionary attacks and is very insecure when weak passphrases are used. With the current implementation of PSK authentication, brute-force dictionary attacks can be circumvented by using very strong passphrases between 20 and 63 characters. Dictionary attacks against strong passphrases could take years before they are successful. However, dictionary attacks are feasible and more easily achieved via the combined resources of distributed cloud computing. The bigger worry is that most users usually create only an 8-character passphrase, which can usually be compromised in several hours or even minutes. The ultimate goal of SAE is to prevent dictionary attacks altogether. As of this writing, the Wi-Fi Alliance has a proposed program of interoperability certification for SAE. Some of the proposals from the Wi-Fi Alliance regarding an SAE certification program include the following:

- WEP and TKIP must not be used for SAE.

- For transition purposes, WPA2-Personal and SAE must be supported simultaneously within the same BSS.

- For transition purposes, WPA2-Personal devices and SAE devices should use the same passphrase.

Think of SAE as a more secure PSK authentication method. The goal is to provide the same user experience by still using a passphrase. However, the SAE protocol exchange protects the passphrase from brute-force dictionary attacks. The passphrase is never sent between 802.11 stations during the SAE exchange.

The SAE process consists of a commitment message exchange and a confirmation message exchange. The commitment exchange is used to force each radio to commit to a single guess of the passphrase. The confirmation exchange is used to prove that the password guess was correct. SAE authentication frames are used to perform these exchanges. The passphrase is used in SAE to deterministically compute a secret element in the negotiated group, called a password element, which is then used in the authentication and key exchange protocol.

SAE has been proposed by the Wi-Fi Alliance as an eventual replacement for PSK authentication as part of the recently announced WPA3 security certification.

802.1X/EAP Framework

The IEEE *802.1X* standard is not specifically a wireless standard and is often mistakenly referred to as 802.11x. The 802.1X standard is a *port-based access control* standard. 802.1X-2001 was originally developed for 802.3 Ethernet networks. Later, 802.1X-2004 provided additional support for 802.11 wireless networks and Fiber Distributed Data Interface (FDDI) networks. The current version of the port-based access control standard, 802.1X-2010, defined further enhancements. 802.1X provides an authorization framework that allows or disallows traffic to pass through a port and thereby access network resources. An 802.1X framework may be implemented in either a wireless or wired environment.

The 802.1X authorization framework consists of three main components, each with a specific role. These three 802.1X components work together to ensure that only properly validated users and devices are authorized to access network resources. A layer 2 authentication protocol called Extensible Authentication Protocol (EAP) is used within the 802.1X framework to validate users at layer 2. The three major components of an 802.1X framework are as follows:

Supplicant   A host with software that requests authentication and access to network resources is known as a *supplicant*. Each supplicant has unique authentication credentials that are verified by the authentication server. In a WLAN, the supplicant is often the laptop or wireless handheld device trying to access the network.

Authenticator   An *authenticator* device blocks traffic or allows traffic to pass through its port entity. Authentication traffic is normally allowed to pass through the authenticator, while all other traffic is blocked until the identity of the supplicant has been verified. The authenticator maintains two virtual ports: an *uncontrolled port* and a *controlled port*. The uncontrolled port allows EAP authentication traffic to pass through, whereas the controlled port blocks all other traffic until the supplicant has been authenticated. In a WLAN, the authenticator is usually either an AP or a WLAN controller.

Authentication Server   An *authentication server (AS)* validates the credentials of the supplicant that is requesting access and notifies the authenticator that the supplicant has been authorized. The authentication server maintains a native database or may proxy query with an external database, such as an LDAP database, to authenticate the supplicant credentials. A RADIUS server normally functions as the authentication server. Within an 802.3 Ethernet network, the supplicant would be a desktop host, the authenticator would be a managed switch, and the authentication server would typically be a Remote Authentication Dial-In User Service (RADIUS) server. In an 802.11 wireless environment, the supplicant would be a client station requesting access to network resources. As shown in Figure 17.6, a standalone access point would be the authenticator, blocking access via virtual ports, and the authentication server is typically an external RADIUS server. Figure 17.6 also shows that when an 802.1X security solution is used with a WLAN controller solution, the WLAN controller is typically the authenticator—and not the controller-based access points. In either case, directory services are often provided by a Lightweight Directory Access Protocol (LDAP) database that the RADIUS server communicates with directly.

Active Directory would be an example of an LDAP database that is queried by a RADIUS server. Note that some WLAN vendors offer solutions where either a standalone AP or a WLAN controller can dual-function as a RADIUS server and perform direct LDAP queries, thus eliminating the need for an external RADIUS server.

FIGURE 17.6   802.1X comparison—standalone vs. controller-based access points

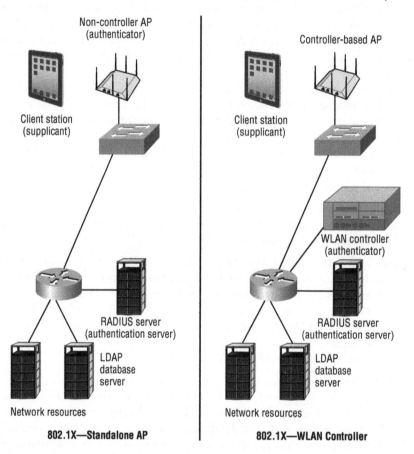

Although the supplicant, authenticator, and authentication server work together to provide the framework for 802.1X port-based access control, an authentication protocol is needed to perform the authentication process. *Extensible Authentication Protocol (EAP)* is used to provide user authentication. EAP is a flexible layer 2 authentication protocol used by the supplicant and the authentication server to communicate. The authenticator allows the EAP traffic to pass through its virtual uncontrolled port. After the authentication server has verified the credentials of the supplicant, the server sends a message to the authenticator that the supplicant has been authenticated; the authenticator is then authorized to open the virtual controlled port and allow all other traffic to pass through. Figure 17.7 depicts the generic 802.1X/EAP frame exchanges.

FIGURE 17.7   802.1X/EAP frame exchange

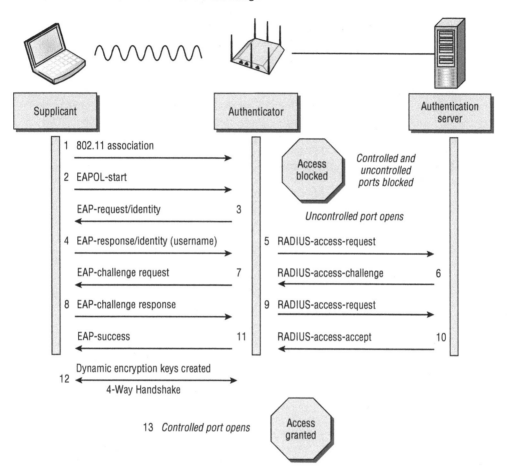

The 802.1X framework, together with EAP, provides the necessary means of validating user and device identity as well as authorizing client stations' access to network resources.

EAP Types

As noted earlier, *EAP* stands for *Extensible Authentication Protocol.* The key word in EAP is *extensible.* EAP is a layer 2 protocol that is very flexible, and many different flavors of EAP exist. Some, such as Cisco's Lightweight Extensible Authentication Protocol (LEAP), are proprietary, whereas others, such as Protected Extensible Authentication Protocol (PEAP), are considered standards-based. Some provide for only one-way authentication; others provide two-way authentication. Mutual authentication not only requires that the authentication server validate the client credentials, but the supplicant must also authenticate the validity of the authentication server. By validating the authentication server, the supplicant can ensure that the username and password are not inadvertently given to a rogue authentication server. Most types of EAP that require mutual authentication use a server-side digital certificate to validate the authentication server. A server-side certificate is installed on the RADIUS server, while the *certificate authority (CA)* root certificate resides on the supplicant. During the EAP exchange, the supplicant's root certificate is used to verify the server-side certificate. As depicted in Figure 17.8, the certificate exchange also creates an encrypted *Secure Sockets Layer (SSL)/Transport Layer Security (TSL)* tunnel in which the supplicant's username/password credentials or client certificate can be exchanged. Many of the secure forms of EAP use *tunneled authentication*. The SSL/TLS tunnel is used to encrypt and protect the user credentials during the EAP exchange.

FIGURE 17.8   Tunneled authentication

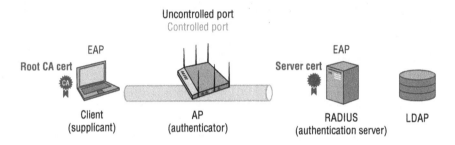

Table 17.2 provides a comparison chart of many of the various types of EAP. It is beyond the scope of this book to discuss in detail all the authentication mechanisms and differences between the various flavors of EAP. The CWSP exam will test you heavily on the operations of the various types of EAP authentication. The CWNA exam will not test you on the specific EAP functions.

TABLE 17.3   EAP comparison table

	EAP-MD5	EAP-LEAP	EAP-TLS	EAP-TTLS	PEAPv0 (EAP-MSCHAPv2)	PEAPv0 (EAP-TLS)	PEAPv1 (EAP-GTC)	EAP-FAST
Security Solution	RFC-3748	Cisco proprietary	RFC-5216	RFC-5281	IETF draft	IETF draft	IETF draft	RFC-4851
Digital Certificates—Client	No	No	Yes	Optional	No	Yes	Optional	No
Digital Certificates—Server	No	No	Yes	Yes	Yes	Yes	Yes	No

	EAP-MD5	EAP-LEAP	EAP-TLS	EAP-TTLS	PEAPv0 (EAP-MSCHAPv2)	PEAPv0 (EAP-TLS)	PEAPv1 (EAP-GTC)	EAP-FAST
Client Password Authentication	Yes	Yes	N/A	Yes	Yes	No	Yes	Yes
PACs—Client	No	No	No	No	No	No	No	Yes
PACs—Server	No	No	No	No	No	No	No	Yes
Credential Security	Weak	Weak (depends on password strength)	Strong	Strong	Strong	Strong	Strong	Strong (if Phase 0 is secure)
Encryption Key Management	No	Yes	Yes	Yes	Yes	Yes	Yes	Yes
Mutual Authentication	No	Debatable	Yes	Yes	Yes	Yes	Yes	Yes
Tunneled Authentication	No	No	Optional	Yes	Yes	Yes	Yes	Yes
Wi-Fi Alliance Supported	No	No	Yes	Yes	Yes	No	Yes	Yes

Dynamic Encryption-Key Generation

Although the 802.1X framework does not require encryption, it highly suggests the use of encryption to provide data privacy. You have already learned that the purpose of 802.1X/EAP is authentication and authorization. If a supplicant is properly authenticated by an authentication server using a layer 2 EAP protocol, the supplicant is allowed through the controlled port of the authenticator, and communication at the upper layers of 3–7 can begin for the supplicant. 802.1X/EAP protects network resources so that only validated supplicants are authorized for access.

However, as shown in Figure 17.9, an outstanding by-product of 802.1X/EAP can be the generation and distribution of dynamic encryption keys. Later in this chapter, you will see that dynamic encryption keys can also be generated as a by-product of PSK authentication. EAP protocols that utilize mutual authentication provide "seeding material" that can be used to generate encryption keys dynamically. A symbiotic relationship exists between 802.1X/EAP or PSK authentication and the creation of dynamic encryption keys. Mutual authentication is required to generate unique dynamic encryption keys.

FIGURE 17.9   802.1X/EAP and dynamic keys

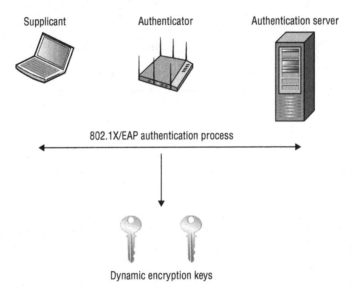

Legacy WLAN security involved the use of static WEP keys. The use of static keys is typically an administrative nightmare, and when the same static key is shared among multiple users, the static key is easy to compromise via social engineering. The first advantage of using dynamic keys rather than static keys is that they cannot be compromised by social engineering attacks because the users have no knowledge of the keys. The second advantage of dynamic keys is that every user has a different and unique key. If a single user's encryption key were compromised, none of the other users would be at risk, because every user has a unique key. The dynamically generated keys are not shared between the users.

4-Way Handshake

As you just learned, there is a symbiotic relationship between the creation of dynamic encryption keys and authentication. A pairwise master key (PMK) is used to seed the 4-Way Handshake that generates the unique dynamic encryption keys employed by any two 802.11 radios. The PMK is generated as a by-product of either PSK or 802.1X/EAP authentication.

As we explained earlier, the 802.11-2016 standard defines what is known as a robust security network (RSN) and robust security network associations (RSNAs). Two stations (STAs) must establish a procedure to authenticate and associate with each other as well as create dynamic encryption keys through a process known as the *4-Way Handshake*, which is depicted in Figure 17.10.

FIGURE 17.10   4-Way Handshake

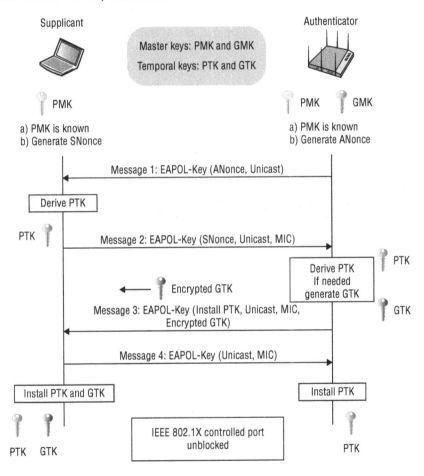

RSNAs utilize a dynamic encryption-key management method that involves the creation of five separate keys. It is beyond the scope of this book to fully explain this entire process, but a brief explanation is appropriate. Part of the RSNA process involves the creation of two master keys, known as the *group master key (GMK)* and the *pairwise master key (PMK)*. As previously discussed, there is a symbiotic relationship between the creation of dynamic encryption keys and authentication. The PMK is created as a result of the 802.1X/EAP authentication. A PMK can also be created from PSK authentication instead of the

802.1X/EAP authentication. These master keys are the seeding material used to create the final dynamic keys that are used for encryption and decryption. The final encryption keys are known as the *pairwise transient key (PTK)* and the *group temporal key (GTK)*. The PTK is used to encrypt/decrypt unicast traffic, and the GTK is used to encrypt/decrypt broadcast and multicast traffic.

These temporal keys are created during a four-way EAP frame exchange known as the 4-Way Handshake. The 4-Way Handshake will always be the final four frames exchanged during either an 802.1X/EAP authentication or a PSK authentication. Whenever TKIP/ARC4 or CCMP/AES dynamic keys are created, the 4-Way Handshake must occur. Also, every time a client radio roams from one AP to another, a new 4-Way Handshake must occur so that new unique dynamic keys can be generated. An example of the 4-Way Handshake can be seen in Exercise 17.2.

> The CWNA exam currently does not test on the mechanics of the dynamic encryption-key creation process, which was originally defined by the 802.11i amendment. The process is heavily tested in the CWSP exam.

EXERCISE 17.2

802.1X/EAP and the 4-Way Handshake Process

1. To perform this exercise, you need to first download the CWNA_CHAPTER17.PCAP file from the book's web page at www.wiley.com/go/cwnasg.

2. After downloading the file, you will need packet analysis software to open the file. If you do not already have a packet analyzer installed on your computer, you can download Wireshark from www.wireshark.org.

3. Using the packet analyzer, open the CWNA_CHAPTER17.PCAP file. Most packet analyzers display a list of capture frames in the upper section of the screen, with each frame numbered sequentially in the first column.

4. Scroll down the list of frames and observe the EAP frame exchange from frame #209 to frame #246.

5. Scroll down the list of frames and observe the 4-Way Handshake from frame #247 to frame #254.

WLAN Encryption

The 802.11-2016 standard defines four encryption methods that operate at layer 2 of the OSI model: WEP, TKIP, CCMP, and GCMP. The information that is being protected by these layer 2 encryption methods is data found in the upper layers of 3–7. Layer 2 encryption methods are used to provide data privacy for 802.11 data frames. The technical name for an 802.11 data frame is a *MAC protocol data unit (MPDU)*. The 802.11 data frame, as shown in Figure 17.11, contains a layer 2 MAC header, a frame body, and a trailer, which is a 32-bit CRC known as the *frame check sequence (FCS)*. The layer 2 header contains MAC addresses and the duration value. Encapsulated inside the frame body of an 802.11 data frame is an upper-layer payload called the *MAC service data unit (MSDU)*. The MSDU contains data from the Logical Link Control (LLC) and layers 3–7. A simple definition of the MSDU is that it is the data payload that contains an IP packet plus some LLC data. When encryption is enabled, the MSDU payload within an 802.11 data frame is encrypted.

FIGURE 17.11   802.11 data frame

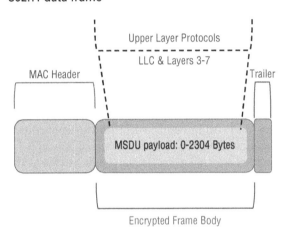

WEP, TKIP, CCMP, GCMP, and other proprietary layer 2 encryption methods are used to encrypt the MSDU payload of an 802.11 data frame. Therefore, the information that is being protected is the upper layers of 3–7, which is more commonly known as the *IP packet*.

It should be noted that many types of 802.11 frames are either never encrypted or typically not encrypted. 802.11 management frames only carry a layer 2 payload in their frame body, so encryption is not necessary for data security purposes. Some management frames are protected if 802.11w *management frame protection (MFP)* mechanisms are enabled. 802.11w provides a level of cryptographic protection for some 802.11 management frames. The vast majority of WLAN clients do not support 802.11w and therefore the capability is rarely enabled on access points.

802.11 control frames have only a header and a trailer; therefore, encryption is not necessary. Some 802.11 data frames, such as the null function frame, actually do not have an MSDU payload. Non-data-carrying data frames have a specific function, but they do not require encryption. Only 802.11 data frames with an MSDU payload can be encrypted with WEP, TKIP, or CCMP. As a matter of corporate policy, 802.11 data frames should always be encrypted for data privacy and security purposes.

WEP, TKIP, CCMP, and GCMP are encryption methods that all use symmetric algorithms. WEP and TKIP use the ARC4 cipher, whereas CCMP and GCMP use the AES cipher. The current 802.11-2016 standard defines WEP as a legacy encryption method for pre-RSNA security. TKIP, CCMP, and GCMP are considered to be robust security network (RSN) encryption protocols. TKIP is being deprecated, however, and GCMP has not yet been used in the enterprise WLAN marketplace.

TKIP Encryption

The optional encryption method defined for a robust security network is *Temporal Key Integrity Protocol (TKIP)*. This method uses the ARC4 cipher, just as WEP encryption does. As a matter of fact, TKIP is an enhancement of WEP encryption that addresses many of the known weaknesses of WEP. The problem with WEP was not the ARC4 cipher but how the encryption key was created. TKIP was developed to rectify the problems that were inherent in WEP.

TKIP starts with a 128-bit temporal key, which is combined with a 48-bit initialization vector (IV) and source and destination MAC addresses in a complicated process known as *per-packet key mixing.* This key-mixing process mitigates the known IV collision and weak key attacks used against WEP. TKIP also uses a sequencing method to mitigate the reinjection attacks used against WEP. Additionally, TKIP uses a stronger data integrity check, known as the *message integrity check (MIC),* to mitigate known bit-flipping attacks against WEP. The MIC is sometimes referred to by the nickname *Michael.* All TKIP encryption keys are dynamically generated as a final result of the 4-Way Handshake.

WEP encryption adds an extra 8 bytes of overhead to the body of an 802.11 data frame. When TKIP is implemented, because of the extra overhead from the extended IV and the MIC, a total of 20 bytes of overhead is added to the body of an 802.11 data frame. Because TKIP uses the ARC4 cipher and is simply WEP that has been enhanced, most vendors released a WPA firmware upgrade that gave legacy WEP-only radios the capability of using TKIP encryption. The 802.11-2016 standard does not permit the use of WEP encryption or TKIP encryption for the *high throughput (HT)* and *very high throughput (VHT)* data rates. The Wi-Fi Alliance will only certify 802.11n radios that use CCMP encryption for the higher data rates. For backward compatibility, newer radios will still support TKIP and WEP for the slower data rates defined for legacy 802.11/a/b/g radios. Although both WEP and TKIP are defined in the IEEE 802.11-2016 standard, they have been deprecated due to security risks and are not supported for 802.11n and 802.11ac data rates. WEP and TKIP still remain as optional security protocols to provide support for older legacy devices.

 Real World Scenario

Can WEP and TKIP Still Be Used?

The 802.11n and higher amendments do not permit the use of WEP encryption or TKIP encryption for high throughput (HT) or very high throughput (VHT) data rates. HT data rates were introduced with 802.11n, and VHT data rates were introduced with 802.11ac. The Wi-Fi Alliance will certify 802.11n and 802.11ac radios to use CCMP encryption only for the higher data rates. For backward compatibility, newer radios can still support TKIP and WEP for the slower data rates defined by legacy 802.11/a/b/g. WEP and TKIP still remain optional security protocols to provide support for older legacy devices. As you have learned, these encryption protocols are outdated and have security risks. Legacy 802.11/a/b/g client devices should be replaced to take advantage of the higher 802.11n/ac data rates and use CCMP/AES encryption.

CCMP Encryption

The default encryption method defined under the 802.11i amendment is known as *Counter Mode with Cipher Block Chaining Message Authentication Code Protocol (CCMP)*. This method uses the Advanced Encryption Standard (AES) algorithm (Rijndael algorithm). CCMP/AES uses a 128-bit encryption-key size and encrypts in 128-bit fixed-length blocks. An 8-byte message integrity check (MIC) is used, which is considered much stronger than the one used in TKIP. Also, because of the strength of the AES cipher, per-packet key mixing is unnecessary.

WEP and TKIP both use ARC4, which is a streaming cipher. CCMP uses AES, which is a block cipher. Unlike stream ciphers, which operate on one bit at a time, a block cipher takes a fixed-length block of plaintext and generates a block of ciphertext of the same length. A block cipher is a symmetric key cipher operating on fixed-length groups of bits, called blocks. For example, a block cipher will use a 128-bit block of input of plaintext, and the resulting output would be a 128-bit block of ciphertext.

CCMP/AES encryption is designated by the Wi-Fi Alliance for WPA2 certification and has been the prevailing WLAN encryption method for more than 10 years. All CCMP encryption keys are dynamically generated as a final result of the 4-Way Handshake.

GCMP Encryption

The 802.11ad-2012 amendment standardized the use of *Galois/Counter Mode Protocol (GCMP)*, which uses AES cryptography. The extremely high data rates defined by 802.11ad require GCMP because it is more efficient than CCMP. GCMP is also considered an optional encryption method for 802.11ac radios. CCMP uses a 128-bit AES key, whereas GCMP can use either a 128-bit or 256-bit AES key.

GCMP is based on the GCM of the AES encryption algorithm. GCM protects the integrity of both the 802.11 data frame body and selected portions of the 802.11 header. GCMP calculations can be run in parallel and are computationally less intensive than the cryptographic operations of CCMP. GCM is significantly more efficient and faster than CCM. Like CCM, GCM uses the same AES encryption algorithm, although it is applied differently. GCMP needs only a single AES operation per block, immediately reducing the encryption process by half. Additionally, GCM does not link or chain the blocks together.

Since each block is not dependent on the previous block, they are independent of each other and can be processed simultaneously using parallel circuits.

As of this writing, none of the chipset vendors have implemented GCMP in the first several generations of 802.11ac radios. GCMP is not backward compatible with existing Wi-Fi equipment, and thus hardware upgrades for both access points and client radios will be required. It remains to be seen if GCMP will become the eventual replacement for CCMP.

Traffic Segmentation

As discussed earlier in this chapter, segmentation is a key part of a network design. Once authorized onto network resources, user traffic can be further restricted as to which resources may be accessed and where user traffic is destined. Segmentation can be achieved through a variety of means, including firewalls, routers, VPNs, and VLANs. A common wireless segmentation strategy used in 802.11 enterprise WLANs is layer 3 segmentation, which employs VLANs mapped to different subnets. Segmentation is also often intertwined with role-based access control (RBAC).

VLANs

Virtual local area networks (VLANs) are used to create separate broadcast domains in a layer 2 network and are often used to restrict access to network resources without regard to physical topology of the network. VLANs are a layer 2 concept and are used extensively in switched 802.3 networks for both security and segmentation purposes. VLANs are mapped to unique layer 3 subnets, although it is possible to match a VLAN to multiple subnets. VLANs are used to support multiple layer 3 networks on the same layer 2 switch.

As you learned in Chapter 11, "WLAN Architecture," where VLANs are used in a WLAN environment depends on the design of the network as well as the type of WLAN architecture that is in place. One very big difference between using a controller model versus a noncontroller model is how VLANs are implemented in the network design. In the WLAN controller model, most user traffic is centrally forwarded to the controller from the APs. The user VLANs are still available to the wireless users because all the user VLANs are encapsulated in an IP tunnel between the controller-based APs at the edge and the WLAN controller in the core.

The non-controller model, however, requires support for multiple user VLANs at the edge. Each access point is therefore connected to an 802.1Q trunk port on an edge switch that supports VLAN tagging. All the user VLANs are configured in the access layer switch. The access points are connected to an 802.1Q trunk port of the edge switch. The user VLANS are tagged in the 802.1Q trunk, and all wireless user traffic is forwarded at the edge of the network.

In a WLAN environment, individual SSIDs can be mapped to individual VLANs, and users can be segmented by the SSID/VLAN pair, all while communicating through a single access point. Each SSID can also be configured with separate security settings. Most enterprise access points have the ability to broadcast multiple SSIDs, and each SSID can be mapped to a unique VLAN. A common strategy is to create guest, voice, and employee SSID/VLAN pairs, as shown in Figure 17.12. Management access to the WLAN controllers or APs should also be isolated on a separate VLAN.

FIGURE 17.12   Wireless VLANs

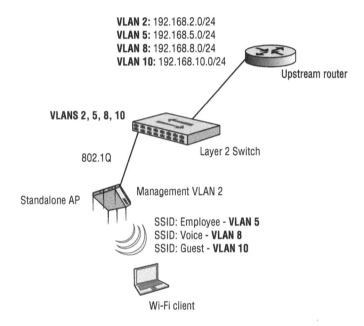

Guest SSID/VLAN   The SSID mapped to the guest VLAN often is an open SSID, although all guest users should be restricted via a firewall policy. Guest users are denied access to local network resources and routed off to an Internet gateway.

Voice SSID/VLAN   The voice SSID might be using a security solution, such as a WPA2 Passphrase, and the VoWiFi client traffic is typically routed to a VoIP server or private branch exchange (PBX).

Employee SSID/VLAN   The employee SSID uses a stronger security solution, such as WPA2-Enterprise, and access control lists (ACLs) or firewall policies allow the employees to access full network resources once authenticated.

Most WLAN vendors allow a radio to broadcast as many as 16 SSIDs. However, broadcasting 16 SSIDs is a bad practice because of the layer 2 overhead created by the

802.11 management and control frames for each SSID. The broadcast of 16 SSIDs will result in degraded performance. The best practice is to never broadcast more than three or four SSIDs.

What if you want your employees segmented into multiple VLANs? Can a single employee SSID be mapped to multiple VLANs? RADIUS attributes can be leveraged for VLAN assignment when using 802.1X/EAP authentication on the employee SSID. As you have already learned, when a RADIUS server provides a successful response to an authentication request, the Access-Accept response can contain a series of *attribute-value pairs (AVPs)*. One of the most popular uses of RADIUS AVPs is assigning users to VLANs, based on the identity of the user authenticating. Instead of segmenting users to different SSIDs that are each mapped to a unique user VLAN, all the users can be associated to a single SSID and assigned to different VLANs. RADIUS servers can be configured with different access policies for different groups of users. The RADIUS access policies are usually mapped to different LDAP groups.

RBAC

Using RADIUS attributes for user VLAN assignment has been a network design strategy for many years. However, RADIUS attributes can be further leveraged to assign different groups of users to all kinds of different user traffic settings, including VLANs, firewall policies, bandwidth policies, and much more.

Role-based access control (RBAC) is an approach to restricting system access to authorized users. The majority of enterprise WLAN vendor solutions have RBAC capabilities. The three main components of an RBAC approach are users, roles, and permissions. Separate roles can be created, such as the sales role or the marketing role. User traffic permissions can be defined as layer 2 permissions (MAC filters), VLANs, layer 3 permissions (access control lists), layers 4–7 permissions (stateful firewall rules), and bandwidth permissions. All these permissions can also be time-based. The user traffic permissions are mapped to the roles. Some WLAN vendors use the term "roles," whereas other vendors use the term "user profiles."

When a user authenticates using 802.1X/EAP, RADIUS attribute value pairs (AVPs) can be used to assign users to a specific role automatically. All users can associate to the same SSID but be assigned to unique roles. This method is often used to assign users from certain *Active Directory (AD)* groups into predefined roles created on a WLAN controller or access point. Each role has unique access restrictions. Once users are assigned to roles, they inherit the user traffic permissions of whatever roles they have been assigned.

Figure 17.13 depicts a RADIUS server with three unique access policies mapped to three different Active Directory groups. For example, user-2 belongs to the marketing AD group. Based on the RADIUS access policy for that AD group, when user-2 authenticates, the RADIUS server will send the AP a RADIUS packet with an attribute that contains a value relevant to Role-B, which has been configured on the AP. The user-2 WLAN client will then be assigned to VLAN 20, firewall-policy-B, and a bandwidth policy of 4 Mbps.

FIGURE 17.13   RADIUS attributes for role assignment

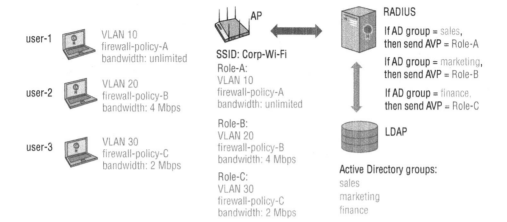

WPA3

On January 8th, 2018, the Wi-Fi Alliance announced *Wi-Fi Protected Access 3*, also known as *WPA3*, which defines enhancements to the existing WPA2 security capabilities for 802.11 radios. New capabilities for both personal and enterprise networks have been announced:

SAE   WPA3 defines stronger authentication with Simultaneous Authentication of Equals (SAE). SAE uses a zero-knowledge proof key exchange, which means a user or device must prove knowledge of a password without having to reveal a password. As you learned earlier in this chapter, SAE is viewed as an upgrade from exiting PSK authentication.

MFP   WPA3 defines mandatory support for management frame protection (MFP) as a safeguard against many of the more common layer 2 denial-of-service (DoS) attacks.

CNSA   WPA3 defines an optional 192-bit cryptographic security suite aligned with the *Commercial National Security Algorithm (CNSA)* suite. The goal will be to further protect Wi-Fi networks with highly sensitive security requirements, such as government, defense, and industrial.

Several other capabilities were also initially proposed to be part of the WPA3 security certification, however, they have been removed and now exist as separate Wi-Fi Alliance certifications. These two new certifications include:

Easy Connect   The Easy Connect certification defines simplified onboarding security with a new *Device Provisioning Protocol (DPP)*. The goal is to simplify the process of configuring security of Wi-Fi devices that have limited or no display interface. DPP could be used for wearable devices, such as watches, and may play a role in improving the future security of IoT devices. WPA3-certified devices would allow users to tap their smartphone against a sensor device or IoT device, or scan a QR code, and then provision the device onto the network.

Enhanced Open   The Enhanced Open certification defines improved data privacy in open networks using a protocol called *Opportunistic Wireless Encryption (OWE)*. The goal is to improve security at open Wi-Fi hotspots. OWE will define methods to provide each Wi-Fi hotspot user with an individual encryption key without any required authentication. Both clients and APs will have to support OWE for successful functionality.

The Wi-Fi Alliance could begin certification of WPA3-capable devices as early as Q4 of 2018. However, broad adoption in the marketplace of the capabilities defined by WPA3 is expected to take a very long time for the client population to catch up. WPA2 will continue to be deployed for the foreseeable future. Please note that you will not be tested about any of the WPA3 security capabilities on the current CWNA exam.

VPN Wireless Security

Although the 802.11-2016 standard clearly defines layer 2 security solutions, upper-layer *virtual private network (VPN)* solutions can also be deployed with WLANs. VPNs are typically not recommended to provide wireless security in the enterprise due to the overhead and because faster, more secure layer 2 solutions are now available. Although not usually a recommended practice, VPNs were often used for WLAN security because the VPN solution was already in place inside the wired infrastructure—especially at the time when WEP was cracked and proven to be unsecure. VPNs do have their place in Wi-Fi security and should definitely be used for remote access. They are also sometimes used in wireless bridging environments. The two major types of VPN topologies are router-to-router and client-server based.

The use of VPN technology is mandatory for remote access. Your employees will take their laptops off site and will most likely use public access Wi-Fi hotspots. Most public access WLANs provide no security, so a VPN solution is needed. The VPN user will need to bring the security to the hotspot in order to provide a secure, encrypted connection. It is imperative that users implement a VPN solution coupled with a personal firewall whenever accessing any public access Wi-Fi networks.

VPN 101

Before discussing the ways that VPNs are used in a WLAN, it is important to review what a VPN is, what it does, how it works, and the components that are configured to construct one. By now you know that a VPN is a virtual private network. But what does that really mean? As shown in Figure 17.14, a VPN is essentially a private network that is created or extended across a public network. In order for a VPN to work, two computers or devices communicate to establish what is known as the *VPN tunnel*. Typically, a VPN client initiates the connection by trying to communicate with the VPN server.

FIGURE 17.14   VPN components

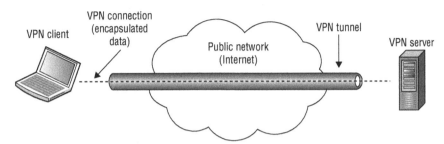

The VPN client can be a computer, router, WLAN controller, or even an AP, which you will learn about later in this chapter. When the client and the server are able to communicate with each other, the client will attempt to authenticate with the server by sending its credentials. The server will take the client's credentials and validate them. If the client's credentials are valid, the server and client will create the VPN tunnel between them. Any data that is sent from the VPN client to the VPN server is encapsulated in the VPN tunnel. The client and the server also agree on if and how the data will be encrypted. Prior to the data being encapsulated in the tunnel, the data is encrypted to ensure that it cannot be compromised as it travels through the tunnel. Since the underlying premise of a VPN is that the data is traveling across an insecure public network, security is one of the primary reasons for implementing a VPN.

When the client and server build the tunnel, it is their responsibility to route the data across the public network between the two devices. They take the data from the local network, encrypt and encapsulate it, and then send it to the other device, where it is decapsulated and decrypted and then placed on the local network of the other device.

Layer 3 VPNs

VPNs have several major characteristics. They provide encryption, encapsulation, authentication, and data integrity. VPNs use *secure tunneling*, which is the process of encapsulating one IP packet within another IP packet. The first packet is encapsulated inside the second, or outer, packet. The original destination and source IP address of the first packet is encrypted along with the data payload of the first packet. VPN tunneling, therefore, protects your original private layer 3 addresses and also protects the data payload of the original packet. Layer 3 VPNs use layer 3 encryption; therefore, the payload that is being encrypted is the layer 4–7 information. The IP addresses of the second or outer packet are seen in cleartext and are used for communications between the tunnel endpoints. The destination and source IP addresses of the second or outer packet will point to the public IP address of the VPN server and VPN client.

The most commonly used layer 3 VPN technology is *Internet Protocol Security (IPsec)*. IPsec VPNs use stronger encryption methods and more secure methods of authentication and are the most commonly deployed VPN solution. IPsec supports multiple ciphers, including DES, 3DES, and AES. Device authentication is achieved by using either a server-side certificate or a preshared key. IPsec VPNs require client software to be installed on the remote devices that connect to a VPN server. Most IPsec VPNS are NAT-transversal, but any firewalls at a remote site require (at a minimum) that UDP ports 4500 and 500 be open. A full explanation of IPsec technology is beyond the scope of this book, but IPsec is usually the choice for VPN technology in the enterprise.

SSL VPNs

VPN technologies do exist that operate at other layers of the OSI model, including SSL tunneling. Unlike an IPsec VPN, an SSL VPN does not necessarily require the installation and configuration of client software on the end user's computer. A user can connect to a *Secure Sockets Layer (SSL)* VPN server via a web browser. The traffic between the web browser and the SSL VPN server is encrypted with the SSL protocol or Transport Layer Security (TLS). TLS and SSL encrypt data connections above the Transport layer, using asymmetric cryptography for privacy and a keyed message authentication code for message reliability.

Although most IPsec VPN solutions are NAT-transversal, SSL VPNs are often chosen because of issues with NAT or restrictive firewall policies at remote locations.

VPN Deployment

VPNs are most often used for client-based security when connected to public access WLANs and hotspots that do not provide security. Because most hotspots do not provide layer 2 security, it is imperative that end users provide their own security. VPN technology can provide the necessary level of security for remote access when end users connect to public access WLANs. Since no encryption is used at public access WLANs, a VPN solution is usually needed to provide for data privacy, as shown in Figure 17.15.

FIGURE 17.15   VPN established from a public hotspot

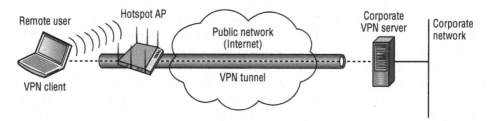

Another common use of VPN technology is to provide site-to-site connectively between a remote office and a corporate office. Most WLAN vendors now offer VPN client-server capabilities in either their APs or WLAN controllers. As shown in Figure 17.16, a branch office WLAN controller with VPN capabilities can tunnel WLAN client traffic and bridged wired-side traffic back to the corporate network. Other WLAN vendors can also tunnel user traffic from a remote AP to a VPN server gateway.

FIGURE 17.16   Site-to-site VPN

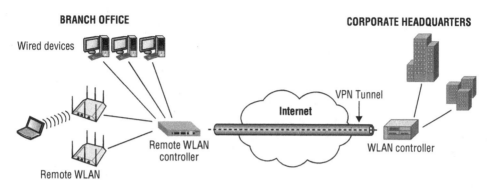

Another use of VPNs is to provide security for 802.11 WLAN bridges links. In addition to being used to provide client access, 802.11 technology is used to create bridged networks between two or more locations. When WLAN bridges are deployed for wireless backhaul communications, VPN technology can be used to provide the necessary level of data privacy. Depending on the bridging equipment used, VPN capabilities may be integrated into the bridges, or you may need to use other devices or software to provide the VPN. Figure 17.17 shows an example of a point-to-point wireless bridge network using dedicated VPN devices. A site-to-site VPN tunnel is used to provide encryption of the 802.11 communications between the two WLAN bridges.

FIGURE 17.17   WLAN bridging and VPN security

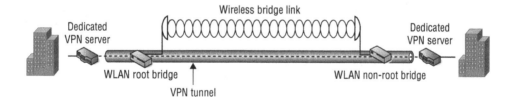

Summary

In this chapter, you learned that five major facets are needed for wireless security. A strong encryption solution is needed to protect the data frames. A secure authentication solution is needed to ensure that only legitimate users are authorized to use network resources. A segmentation solution is necessary to further restrict users as to which resources they can access and where they can go. 802.11 wireless networks can be further protected with continuous monitoring and enforcement through a WLAN security policy.

We discussed legacy 802.11 authentication and encryption solutions and why they are weak. We covered the stronger 802.1X/EAP authentication solutions and the benefits of dynamic encryption-key generation, as well as what is defined by the 802.11-2016 standard and the related WPA/WPA2 certifications. The 802.11-2016 standard defines a layer 2 robust security network using either 802.1X/EAP or PSK authentication. CCMP/AES encryption is designated by the Wi-Fi Alliance for WPA2 certification, and has been the prevailing WLAN encryption method for more than 10 years. Although not yet a testable CWNA exam subject, the WPA3 security and the Simultaneous Authentication of Equals (SAE) protocol are coming soon to the market.

Finally, we discussed how VPN technology is used in WLAN environments. It is important to understand the capabilities and limitations of the devices that will be deployed within your 802.11 wireless networks. Ideally, devices will be segmented into separate VLANs and access policies by using 802.1X/EAP authentication and CCMP/AES encryption. VoIP phones, mobile scanners, tablets, handheld devices, and so on are often not equipped with the ability to handle more advanced security capabilities. Proper security designs must take into account all these components to ensure the most dynamic and secure network.

Exam Essentials

Define the concept of AAA.   Be able to explain the differences between authentication, authorization, and accounting and why each is needed for a WLAN network.

Explain why data privacy is needed.   Be able to discuss why data frames must be protected with encryption. Know the differences between the various encryption ciphers.

Understand legacy 802.11 security.   Identify and understand Open System authentication and Shared Key authentication. Understand how WEP encryption works and know its weaknesses.

Explain the 802.1X/EAP framework.   Be able to explain all the components of an 802.1X solution and the EAP authentication protocol. Understand that dynamic encryption key generation is a by-product of mutual authentication.

Define the requirements of a robust security network.   Understand what the 802.11-2016 standard specifically defines for robust security and be able to contrast what is defined by both the WPA and WPA2 certifications.

Understand TKIP/ARC4 and CCMP/AES.   Be able to explain the basics of both dynamic encryption types and why they are the end result of an RSN solution.

Explain why segmentation is needed.   Understand how VLANs and role-based access policies are used to further restrict network resources.

Explain VPN WLAN security.   Define the basics of VPN technology and when it might be used in a WLAN environment.

Review Questions

1. Which WLAN security mechanism requires that each WLAN user have unique authentication credentials?

   A. WPA-Personal

   B. 802.1X/EAP

   C. Open System

   D. WPA2-Personal

   E. WPA-PSK

2. Which wireless security standards and certifications call for the use of CCMP/AES encryption? (Choose all that apply.)

   A. WPA

   B. 802.11-2016

   C. 802.1X

   D. WPA2

   E. 802.11 legacy

3. 128-bit WEP encryption uses a user-provided static key of what size?

   A. 104 bytes

   B. 64 bits

   C. 124 bits

   D. 128 bits

   E. 104 bits

4. Which three main components constitute an 802.1X authorization framework? (Choose all that apply.)

   A. Supplicant

   B. Authorizer

   C. Authentication server

   D. Intentional radiator

   E. Authenticator

5. Which of these security methods is a replacement for PSK authentication as defined by WPA3?

   A. Per-user/per-device PSK

   B. Wi-Fi Protected Setup (WPS)

   C. Simultaneous Authentication of Equals (SAE)

   D. EAP-PSK

   E. WPA2 Personal

6. The ACME Company is using WPA2-Personal to secure handheld barcode scanners that are not capable of 802.1X/EAP authentication. Because an employee was recently fired, all the barcode scanners and APs had to be reconfigured with a new static 64-bit PSK. What type of WLAN security solution may have avoided this administrative headache?

   A. MAC filter

   B. Hidden SSID

   C. Changing the default settings

   D. Proprietary PSK

7. Which of the following encryption methods use symmetric ciphers? (Choose all that apply.)

   A. WEP

   B. TKIP

   C. Public-key cryptography

   D. CCMP

8. The IEEE 802.11-2016 standard states which of the following regarding 802.11ac data rates and encryption? (Choose all that apply.)

   A. WEP and TKIP must not be used.

   B. CCMP and GCMP can be used.

   C. WEP cannot be used; however, TKIP can be used if also using 802.1X.

   D. Any encryption method defined by the standard can be used.

9. When 802.1X/EAP security is deployed, RADIUS attributes can also be leveraged for role-based assignment of which type of user access permissions? (Choose all that apply.)

   A. Stateful firewall rules

   B. Time

   C. VLANS

   D. ACLs

   E. Bandwidth

10. How are IPsec VPNs used to provide security in combination with 802.11 WLANs?

   A. Client-based security on public access WLANs

   B. Point-to-point wireless bridge links

   C. Connectivity across WAN links

   D. All of the above

11. When enabled, WLAN encryption provides data privacy for which portion of an 802.11 data frame?

   A. MPDU

   B. MSDU

   C. PPDU

   D. PSDU

12. Which of the following methods of authentication must occur along with the 4-Way Handshake in order to generate dynamic TKIP/ARC4 or CCMP/AES encryption keys? (Choose all that apply.)

    A. Shared Key authentication and 4-Way Handshake

    B. 802.1X/EAP authentication and 4-Way Handshake

    C. Static WEP and 4-Way Handshake

    D. PSK authentication and 4-Way Handshake

13. For an 802.1X/EAP solution to work properly, which two components must both support the same type of EAP? (Choose all that apply.)

    A. Supplicant

    B. Authorizer

    C. Authenticator

    D. Authentication server

14. When you are using an 802.11 wireless controller solution, which device would usually function as the authenticator?

    A. Access point

    B. LDAP server

    C. WLAN controller

    D. RADIUS server

15. Which of these use cases for a per-user/per-device implementation of PSK authentication is not recommended?

    A. Unique credentials for BYOD devices

    B. Unique credentials for IoT devices

    C. Unique credentials for guest WLAN access

    D. Unique credentials for legacy enterprise devices without 802.1X/EAP support

    E. Unique credentials for enterprise devices with 802.1X/EAP support

16. What does 802.1X/EAP provide when implemented for WLAN security? (Choose all that apply.)

    A. Access to network resources

    B. Verification of access point credentials

    C. Dynamic authentication

    D. Dynamic encryption-key generation

    E. Verification of user credentials

17. CCMP encryption uses which AES key size?

    A. 192 bits

    B. 64 bits

    C. 256 bits

    D. 128 bits

18. Identify the security solutions that are defined by WPA2. (Choose all that apply.)

   A. 802.1X/EAP authentication

   B. Dynamic WEP encryption

   C. Optional CCMP/AES encryption

   D. PSK authentication

   E. DES encryption

19. Which encryption methods do the IEEE 802.11-2016 standard mandate for robust security network associations, and which method is optional?

   A. WEP, AES

   B. IPsec, AES

   C. MPPE, TKIP

   D. TKIP, WEP

   E. CCMP, TKIP

20. Which layer 2 protocol is used for authentication in an 802.1X framework?

   A. RSN

   B. SAE

   C. EAP

   D. PAP

   E. CHAP

Chapter

18

Bring Your Own Device (BYOD) and Guest Access

IN THIS CHAPTER, YOU WILL LEARN ABOUT THE FOLLOWING:

✓ **Mobile device management**

- Company-issued devices vs. personal devices
- MDM architecture
- MDM enrollment
- MDM profiles
- MDM agent software
- Over-the-air management
- Application management

✓ **Self-service device onboarding for employees**

- Dual-SSID onboarding
- Single-SSID onboarding
- MDM vs. self-service onboarding

✓ **Guest WLAN access**

- Guest SSID
- Guest VLAN
- Guest firewall policy
- Captive web portals
- Client isolation, rate limiting, and web content filtering
- Guest management
- Guest self-registration

- Employee sponsorship
- Social login
- Encrypted guest access

✓ **Hotspot 2.0 and Passpoint**

- Access Network Query Protocol
- Hotspot 2.0 architecture
- 802.1X/EAP and Hotspot 2.0
- Online sign-up
- Roaming agreements

✓ **Network access control (NAC)**

- Posture
- OS Fingerprinting
- AAA
- RADIUS Change of Authorization
- Single Sign-On
- SAML
- OAuth

For many years, the primary purpose of enterprise WLANs was to provide wireless access for company-owned laptop computers used by employees. Some vertical markets, such as healthcare, retail, and manufacturing, also required WLAN access for company-owned mobile devices, such as VoWiFi phones and wireless barcode scanners. Over the last 10 years, however, there has been a massive population explosion of Wi-Fi–enabled personal mobile devices. Wi-Fi radios are now the primary communications component in smartphones, tablets, PCs, and many other mobile devices.

Although mobile devices initially were intended for personal use, organizations found ways of deploying corporate mobile devices with custom software to improve productivity or functionality. Employees also increasingly want to use their personal mobile devices in the workplace. Employees expect to be able to connect to a corporate WLAN with multiple personal mobile devices. The catchphrase *bring your own device (BYOD)* refers to the policy of permitting employees to bring personally owned mobile devices, such as smartphones, tablets, and laptops, to their workplace. A BYOD policy dictates which corporate resources can or cannot be accessed when employees connect to the company WLAN with their personal devices. BYOD policy usually also defines how employee devices are allowed to connect to the WLAN.

The main focus of this chapter is to explain how security is used to control and monitor BYOD access to a WLAN. *Mobile device management (MDM)* solutions can be used to remotely manage and control company-owned as well as personal mobile Wi-Fi devices. MDM solutions use server software or cloud services to configure client settings, along with client applications, and to control and monitor what users can do. Additionally there is a growing trend to use self-service BYOD solutions where employees can securely provision their personal WLAN devices. *Network access control (NAC)* integrates different security technologies, such as AAA, RADIUS, client health checks, guest services, and client self-registration and enrollment. Using these technologies, NAC can control and monitor client access on the network. NAC can be used to provide authentication and access control of MDM-managed devices, corporate Wi-Fi devices, or BYOD and guest devices.

This chapter also covers the many components of WLAN guest access and how it has evolved over the years. Guest access technology includes support for visitor devices along with employee BYOD devices.

Mobile Device Management

Consumerization of IT is a phrase used to describe a shift in information technology (IT) that begins in the consumer market and moves into business and government facilities. It has become commonplace for employees to introduce consumer market devices into the workplace after already embracing this new technology at home. In the early days of Wi-Fi, most businesses did not provide wireless network access to the corporate network. Due to the limited wireless security options available at that time, along with a general mistrust of the unknown, it was common for companies to avoid implementing WLANs. However, because employees enjoyed the flexibility of Wi-Fi at home, they began to bring small office/ home office (SOHO) wireless routers into the office and install them, despite the objections of the IT department. Eventually, businesses and government agencies realized that they needed to deploy WLANs to take advantage of the technology as well as manage the technology.

Personal mobile Wi-Fi devices, such as smartphones and tablets, have been around for quite a few years. The Apple iPhone was first introduced in June 2007, and the first iPad debuted in April 2010. HTC introduced the first Android smartphone in October 2008. These devices were originally meant for personal use, but in a very short time, employees wanted to also use their personal devices on company WLANs. Additionally, software developers began to create enterprise mobile business applications for smartphones and tablets. Businesses began to purchase and deploy tablets and smartphones to take advantage of these mobile enterprise applications. Tablets and smartphones provided the true mobility that employees and businesses desired, and within a few years, the number of mobile devices connecting to corporate WLANs surpassed the number of laptop connections. This trend is continuing, with many, if not most, devices shipping with Wi-Fi as the primary network adapter. Many laptop computers now ship without an Ethernet adapter because the laptop Wi-Fi radio is used for network access.

Because of the proliferation of personal mobile devices, a BYOD policy is needed to define how employees' personal devices may access the corporate WLAN. A mobile device management (MDM) solution might be needed for onboarding personal mobile devices as well as company-issued devices to the WLAN. Corporate IT departments can deploy MDM servers to manage, secure, and monitor the mobile devices. An MDM solution can manage devices across multiple mobile operating systems and across multiple mobile service providers. Most MDM solutions are used to manage iOS and Android mobile devices. However, mobile devices that use other operating systems, such as BlackBerry OS and Windows Phone, can also be managed by MDM solutions. Although the main focus of an MDM solution is the management of smartphones and tablets, some MDM solutions can also be used to onboard personal macOS and Chrome OS laptops. A few of the devices that can be managed by an MDM solution are shown in Figure 18.1.

FIGURE 18.1 Personal mobile devices with Wi-Fi radios

Some WLAN infrastructure vendors have developed small-scale MDM solutions that are specific to their WLAN controller and/or access point solution. However, the bigger MDM companies sell overlay solutions that can be used with any WLAN vendor's solution.

Following are some of the major vendors selling overlay MDM solutions:

VMware AirWatch—www.air-watch.com

Citrix—www.citrix.com

IBM—www.maaS360.com

JAMF Software—www.jamfsoftware.com

MobileIron—www.mobileiron.com

Company-Issued Devices vs. Personal Devices

An MDM solution can be used to manage both company-issued devices and personal devices. However, the management of company-issued devices and personal devices is quite different. A company mobile device was purchased by the company with the intent of enhancing employee performance. A tablet or smartphone might be issued to an individual employee or shared by employees on different shifts. Commercial business applications, and often industry-specific applications, are deployed on these devices. Many companies even develop in-house applications unique to their own business needs. Company mobile devices often are deployed to replace older hardware. For example, inventory control software running on a tablet might replace legacy handheld barcode scanners. A software Voice over Internet Protocol (VoIP) application running on a smartphone might be used to replace WLAN VoWiFi handsets. The IT department will usually choose one model of mobile device that runs the same operating system.

The management strategy for company mobile devices usually entails more in-depth security because often the company devices have corporate documents and information stored on them. When company devices are provisioned with an MDM solution, many configuration settings are enabled, such as virtual private network (VPN) client access, email account settings, Wi-Fi profile settings, passwords, and encryption settings. The ability for employees to remove MDM profiles from a company-owned device is disabled, and the

MDM administrator can remotely wipe company mobile devices if they are lost or stolen. The MDM solution is also used for hardware and software inventory control. Because these devices are not personal devices, the IT department can also dictate which applications can or cannot be installed on tablets and/or smartphones.

The concept of BYOD emerged because personally owned mobile devices are difficult to control and manage, while allowing access to the enterprise network. Access and control may be managed using an MDM or network access control (NAC) solution, but BYOD needs are different from corporate needs. Employees, visitors, vendors, contractors, and consultants bring a wide range of personal devices—different makes and models loaded with a variety of operating systems and applications—to the workplace. Therefore, a different management strategy is needed for BYOD. Every company should have its own unique BYOD containment strategy while still allowing access to the corporate WLAN. For example, when the personal devices are provisioned with an MDM solution, the camera may be disabled so that pictures cannot be taken within the building. As shown in Figure 18.2, many restrictions can be enforced on a company or personal device after it has been enrolled in the MDM solution.

FIGURE 18.2 Device restrictions

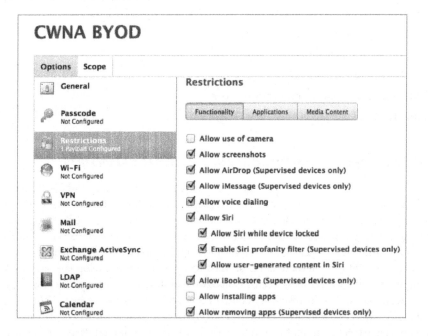

As an alternative to MDM solutions, NAC solutions can be used to authenticate personal devices along with controlling access to the enterprise network without having to necessarily install software on the client device. NAC solutions do not provide control of the individual device but can provide extensive control over the level of access the personal device has on the network. NAC solutions can dictate minimum security requirements, such as anti-virus protection, before a device is permitted access to a network. A more detailed discussion about NAC can be found later in this chapter.

MDM Architecture

The basic architecture of any MDM solution consists of the following four main components:

Mobile Device   The mobile Wi-Fi device requires access to the corporate WLAN. The mobile device can be either a company-owned or an employee-owned device. Depending on the MDM vendor, multiple operating systems may be supported, including iOS, Android, Chrome OS, and macOS, among others. The mobile devices are not allowed onto the corporate network until an enrollment process has been completed and an MDM profile has been installed.

AP/WLAN Controller   All Wi-Fi communications are between the mobile devices and the access point to which they connected. If the devices have not been enrolled via the MDM server, the AP or WLAN controller quarantines the mobile devices within a restricted area of the network known as a *walled garden*. Mobile devices that have been taken through the enrollment process are allowed outside of the walled garden.

MDM Server   The MDM server is responsible for enrolling client devices. The MDM server provisions the mobile devices with MDM profiles that define client device restrictions as well as configuration settings. Certificates can be provisioned from the MDM server. MDM servers can also be configured for either enrollment whitelisting or blacklisting. Whitelisting policies restrict enrollment to a list of specific devices and operating systems. Blacklisting policies allow all devices and operating systems to enroll except for those specifically prohibited by the blacklist. Although the initial role of an MDM server is to provision and onboard mobile devices to the WLAN, the server is also used for client device monitoring. Device inventory control and configuration are key components of any MDM solution. The MDM server usually is available as either a cloud-based service or as an on-premises server that is deployed in the company data center. On-premises MDM servers can be in the form of a hardware appliance or can run as software in a virtualized server environment.

Push Notification Servers   The MDM server communicates with push notification servers, such as *Apple Push Notification service (APNs)* and *Google Cloud Messaging (GCM)*, for over-the-air management of mobile Wi-Fi devices. Over-the-air management is discussed in greater detail later in this chapter.

An MDM architecture deployment includes other key components. MDM servers can be configured to query *Lightweight Directory Access Protocol (LDAP)* databases, such as Active Directory. Typically, a corporate firewall also will be in place. Proper outbound ports need to be open to allow for communications between all the various components of the MDM architecture. For example, Transmission Control Protocol (TCP) port 443 needs to be open for encrypted SSL communications between the AP and the MDM server as well as SSL communications between the mobile device and the MDM server. TCP port 5223 needs to be open so that mobile devices can communicate with APNs. TCP ports 2195 and 2196 are needed for traffic between the MDM server and APNs. TCP ports 443, 5223, 5229, and 5330 are required for communication between mobile devices and GCM. Communications between the MDM server and GCM require TCP port 443 to be open.

MDM Enrollment

When MDM architecture is in place, mobile devices must go through an enrollment process in order to access network resources. The enrollment process can be used to onboard both company-issued devices and personal devices. Figure 18.3 illustrates the initial three steps of the MDM enrollment process.

FIGURE 18.3 MDM enrollment—initial steps

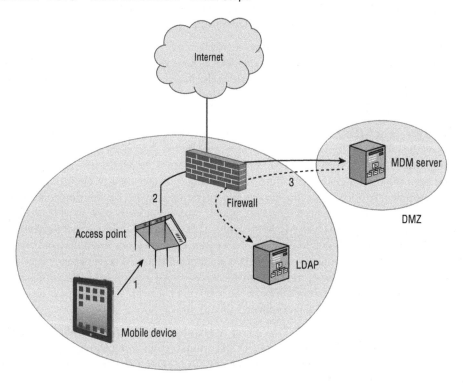

Step 1: The mobile device connects with the access point. The mobile device must first establish an association with an AP. The Wi-Fi security could be open, but usually the company-issued or personal devices are trying to establish a connection with a secure corporate SSID that is using 802.1X/EAP or preshared key (PSK) security. At this point, the AP holds the mobile client device inside a *walled garden*. Within a network deployment, a *walled garden* is a closed environment that restricts access to web content and network resources while still allowing access to some resources. A walled garden is a closed platform of network services provided for devices and/or users. While inside the walled garden designated by the AP, the only services that the mobile device can access are Dynamic Host Configuration Protocol (DHCP), Domain Name System (DNS), push notification services, and the MDM server. To escape from the walled garden, the mobile device must find the proper exit point, much like in a real walled garden. The designated exit point for a mobile device is the MDM enrollment process.

Step 2: The AP checks whether the device is enrolled.   The next step is to determine if the mobile device has been enrolled. Depending on the WLAN vendor, the AP or a WLAN controller queries the MDM server to determine the enrollment status of the mobile device. If the MDM is provided as a cloud-based service, the enrollment query crosses a WAN link. An on-premises MDM server typically will be deployed in a demilitarized zone (DMZ). If the mobile device is already enrolled, the MDM server will send a message to the AP to release the device from the walled garden. Unenrolled devices will remain quarantined inside the walled garden.

Step 3: The MDM server queries LDAP.   Although an open enrollment process can be deployed, administrators often require authentication. The MDM server queries an existing LDAP database, such as Active Directory. The LDAP server responds to the query, and then the MDM enrollment can proceed.

Step 4: The device is redirected to the MDM server.   Although the unenrolled device has access to DNS services, the quarantined device cannot access any web service other than the MDM server. When the user opens a browser on the mobile device, it is redirected to the captive web portal for the MDM server, as shown in Figure 18.4. The enrollment process can then proceed. For legal and privacy reasons, captive web portals contain a legal disclaimer agreement that gives the MDM administrator the ability to restrict settings and remotely change the capabilities of the mobile device. The legal disclaimer is particularly important for a BYOD situation where employees are onboarding their own personal devices. If users do not agree to the legal disclaimer, they cannot proceed with the enrollment process and will not be released from the walled garden.

FIGURE 18.4   MDM server—Enrollment captive web portal—step 4

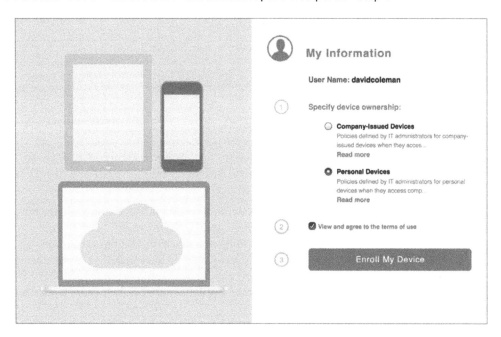

Step 5: The device installs the certificate and MDM profile.   Once enrollment begins, a secure *over-the-air provisioning* process for installing the MDM profile is needed. Over-the-air provisioning differs between device operating systems, but using trusted certificates and SSL encryption is the norm. For this example, we will describe how iOS devices are provisioned. For iOS devices, the *Simple Certificate Enrollment Protocol (SCEP)* uses certificates and Secure Sockets Layer (SSL) encryption to protect the MDM profiles. The user of the mobile device accepts an initial profile that is installed on the device. After installation of the initial profile, device-specific identity information can be sent to the MDM server. The MDM server then sends an SCEP payload that instructs the mobile device about how to download a trusted certificate from the MDM certificate authority (CA) or a third-party CA. Once the certificate is installed on the mobile device, the encrypted MDM profile with the device configuration and restrictions payload is sent securely to the mobile device and installed. Figure 18.5 depicts the installation of MDM profiles using SCEP on an iOS device.

FIGURE 18.5   Certificate and MDM profile installation—step 5

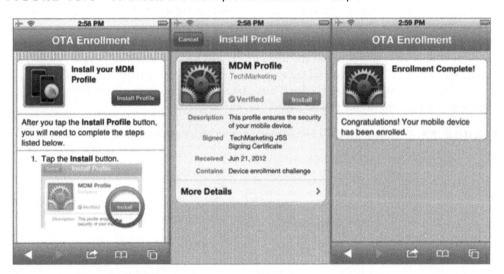

Step 6: The MDM server releases the mobile device.   As shown in Figure 18.6, once the device has completed the MDM enrollment, the MDM server sends a message to the AP or WLAN controller to release the mobile device from the walled garden.

Step 7: The mobile device exits the walled garden.   The mobile device now abides by the restrictions and configuration settings defined by the MDM profile. For example, the use of the mobile device's camera may no longer be allowed. Configuration settings, such as email or VPN settings, also may have been provisioned. The mobile device is now free to exit the walled garden and access the Internet and corporate network resources, as illustrated in Figure 18.6. Access to available network resources is dictated by the type of device or the identity of the user. For example, company-owned devices may have access to all network servers, whereas personal devices may have access only to specific servers, such as the email server. Once released from the walled garden, personal devices might be placed in a VLAN with only access to the Internet, whereas company-owned devices may be placed in a less restrictive VLAN.

FIGURE 18.6 Mobile device released from the walled garden

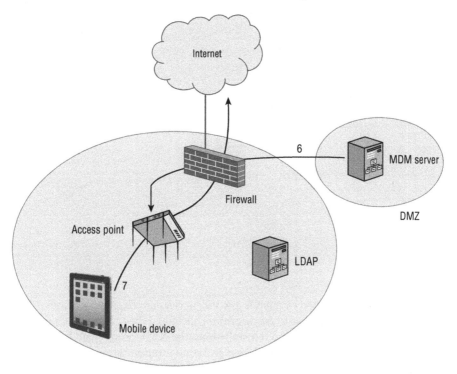

MDM Profiles

You have already learned that MDM profiles are used for mobile device restrictions. The MDM profiles can also be used to globally configure various components of a mobile device. MDM profiles are essentially configuration settings for a mobile device. As the example in Figure 18.7 shows, MDM profiles can include device restrictions, email settings, VPN settings, LDAP directory service settings, and Wi-Fi settings. MDM profiles can also include *webclips*, which are browser shortcuts that point to specific URLs. A webclip icon is automatically installed on the desktop screen of the mobile device. For example, a company-issued device could be provisioned with a webclip link to the company's internal intranet.

FIGURE 18.7   MDM profile settings

The configuration profiles used by macOS and iOS devices are *Extensible Markup Language (XML)* files. Apple has several tools to create profiles, including the Apple Configurator and the iPhone Configuration Utility. For manual installations, the XML profiles can be delivered via email or through a website. Manual installation and configuration is fine for a single device, but what about in an enterprise, where thousands of devices might need to be configured? In the enterprise, a method is needed to automate the delivery of configuration profiles, which is where an MDM solution comes into play. MDM configuration profiles are created on the MDM server and installed onto the mobile devices during the enrollment process.

As mentioned, one aspect of an MDM profile is that the Wi-Fi settings can be provisioned. Company-owned devices can be locked down with a specific Wi-Fi profile that designates the corporate SSID and proper security settings. An MDM profile can also be used to deploy Wi-Fi settings to an employee's personal device. If 802.1X/EAP is deployed, a root CA certificate must be installed on the supplicant mobile device. An MDM solution is an effective way to securely provision root CA certificates on mobile devices. Client certificates can also be provisioned if EAP-TLS is the chosen 802.1X security protocol. Some companies use an MDM solution solely for the purpose of onboarding certificates to WLAN client devices because of the wide variance of operating systems.

MDM profiles can be removed from the device locally or can be removed remotely through the Internet via the MDM server.

Can Employees Remove the MDM Profiles from the Mobile Device?

Once a mobile device has gone through an enrollment process, the MDM configuration profiles and related certificates are installed on the mobile device. The following figure shows the settings screen of an iPad with installed MDM profiles.

Can an employee remove the MDM profiles? The answer to this question is a matter of company policy. Company-owned mobile devices usually have the MDM profiles locked and they cannot be removed. This prevents the employee from making unauthorized changes to the device. If the mobile device is stolen and sensitive information resides on the device, the MDM administrator can remotely wipe the mobile device if it is connected to the Internet. The BYOD policy of personal devices is usually less restrictive. When an employee enrolls their personal device through the corporate MDM solution, typically the employee retains the ability to remove the MDM profiles because they own the device. If the employee removes the MDM profiles, the device is no longer managed by the corporate MDM solution. The next time the employee tries to connect to the company's WLAN with the mobile device, the employee will have to once again go through the MDM enrollment process.

MDM Agent Software

The operating systems of some mobile devices require *MDM agent* application software. For example, Android devices require an MDM agent application like the one shown in Figure 18.8. The Android OS is an open-source operating system that can be customized by the various mobile device manufacturers. Although this provides much more flexibility, managing and administering Android devices in the enterprise can be challenging due to the sheer number of hardware manufacturers. An MDM agent application can report unique information about the Android device back to an MDM server, which can later be used in MDM restriction and configuration policies. An MDM agent must support multiple Android device manufacturers.

FIGURE 18.8   MDM agent application

An employee downloads the MDM agent from a public website or company website and installs it on their Android device. The MDM agent contacts the MDM server over the WLAN and is typically required to authenticate to the server. The MDM agent must give the MDM server permission to make changes to the device and function as the administrator of the device. Once this secure relationship has been established, the MDM agent software enforces the device restriction and configuration changes. MDM administration on an Android device is handled by the agent application on the device. Changes can, however, be sent to the MDM agent application from the MDM server via the Google Cloud Messaging (GCM) service.

Although iOS devices do not require MDM agent software, some MDM solutions do offer iOS MDM agents. The MDM agent on the iOS device could potentially send information back to the MDM server that is not defined by the Apple APIs.

Over-the-Air Management

Once a device has been provisioned and enrolled with an MDM server, a permanent management relationship exists between the MDM server and the mobile device. As shown in Figure 18.9, the MDM server can monitor such device information as its name, serial number, capacity, battery life, and the applications that are installed on the device. Information that cannot be seen includes SMS messages, personal emails, calendars, and browser history.

FIGURE 18.9   Device information

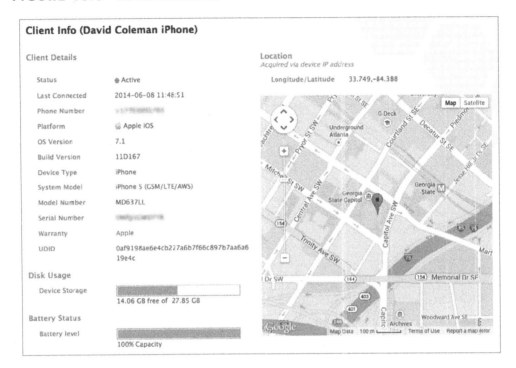

The mobile device can still be managed remotely, even if the mobile device is no longer connected to the corporate WLAN. The MDM server can still manage the device as long as the device is connected to the Internet from any location. The communication between the MDM server and the mobile devices requires push notifications from a third-party service. Both Google and Apple have APIs that allow applications to send push notifications to mobile devices. iOS applications communicate with the Apple Push Notification service (APNs) servers, and Android applications communicate with the Google Cloud Messaging (GCM) servers.

As shown in Figure 18.10, the first step is for the MDM administrator to make changes to the MDM configuration profile on the MDM server. The MDM server then contacts push notification servers. A previously established secure connection already exists between the push notification servers and the mobile device. The push notification service then sends a message to the mobile device telling the device to contact the MDM server over the Internet. Once the mobile device contacts the MDM server, the MDM server sends the configuration changes and/or messages to the mobile device.

FIGURE 18.10   Over-the-air management

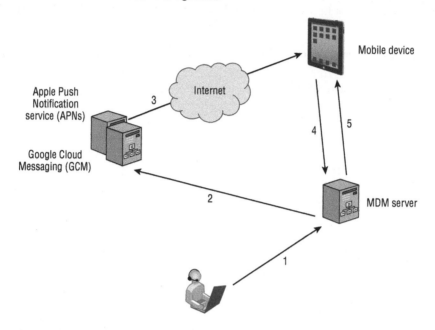

What kind of remote actions can an MDM administrator accomplish over the Internet?

- Make changes to the configuration
- Make changes to the device restrictions
- Deliver a message to the device
- Lock the device
- Wipe the device
- Make application management changes

Stop, Thief!

A stolen company-owned device can be remotely wiped. MDM vendors implement different types of remote wipes.

Enterprise Wipe   Wipes all corporate data from the selected device and removes the device from the MDM. All the enterprise data contained on the device is removed, including MDM profiles, policies, and internal applications. The device will return to the state it was in prior to the enrollment with the MDM.

Device Wipe   Wipes all data from the device, including all data, email, profiles, and MDM capabilities, and returns the device to factory default settings.

Application Management

Enterprise MDM solutions also offer various levels of management of the applications that run on mobile devices. Once an MDM profile is installed, all the applications installed on the device can be viewed from the MDM server, as shown in Figure 18.11. The MDM server can manage applications by whitelisting and/or blacklisting specific applications that can be used on the mobile devices. Managing applications on company-owned devices is commonplace; however, application management on employees' personal devices is not as prevalent.

FIGURE 18.11 Mobile device applications

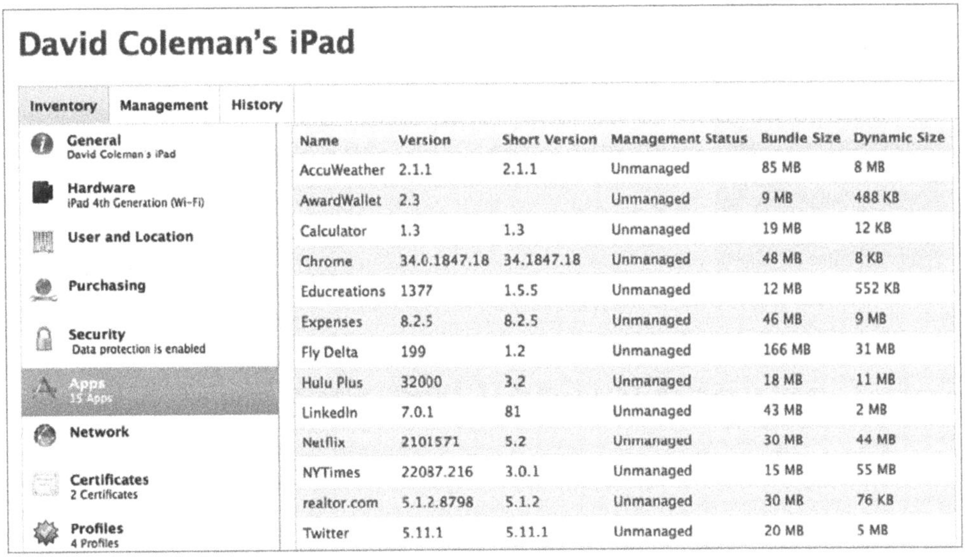

David Coleman's iPad

Inventory	Management	History

	Name	Version	Short Version	Management Status	Bundle Size	Dynamic Size
General David Coleman s iPad	AccuWeather	2.1.1	2.1.1	Unmanaged	85 MB	8 MB
Hardware iPad 4th Generation (Wi-Fi)	AwardWallet	2.3		Unmanaged	9 MB	488 KB
User and Location	Calculator	1.3	1.3	Unmanaged	19 MB	12 KB
	Chrome	34.0.1847.18	34.1847.18	Unmanaged	48 MB	8 KB
Purchasing	Educreations	1377	1.5.5	Unmanaged	12 MB	552 KB
	Expenses	8.2.5	8.2.5	Unmanaged	46 MB	9 MB
Security Data protection is enabled	Fly Delta	199	1.2	Unmanaged	166 MB	31 MB
Apps 15 Apps	Hulu Plus	32000	3.2	Unmanaged	18 MB	11 MB
	LinkedIn	7.0.1	81	Unmanaged	43 MB	2 MB
Network	Netflix	2101571	5.2	Unmanaged	30 MB	44 MB
	NYTimes	22037.216	3.0.1	Unmanaged	15 MB	55 MB
Certificates 2 Certificates	realtor.com	5.1.2.8798	5.1.2	Unmanaged	30 MB	76 KB
Profiles 4 Profiles	Twitter	5.11.1	5.11.1	Unmanaged	20 MB	5 MB

MDM solutions integrate with public application stores, such as iTunes and Google Play, in order to allow access to public applications. The MDM server communicates with the push notification server, which then places an application icon on the mobile device. The mobile device user can then install the application. The Apple Volume Purchase Program (VPP) provides a way for businesses and educational institutions to purchase apps in bulk and distribute them across their organization. Applications can be purchased and pushed silently to the remote devices. An MDM server can also be configured to deliver custom in-house applications that might be unique to the company.

As shown in Figure 18.12, eBooks can also be managed and distributed to mobile devices via an MDM platform. We suggest that your company make a bulk purchase of the *CWNA Study Guide eBook*.

FIGURE 18.12 MDM distribution of the *CWNA Study Guide eBook*

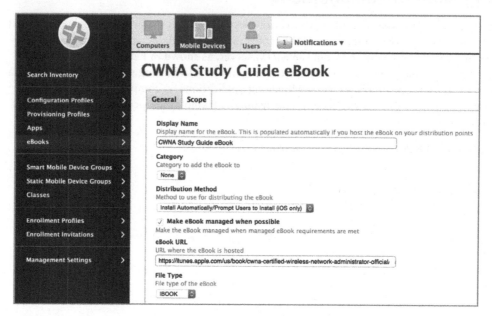

Self-Service Device Onboarding for Employees

As you have learned, MDM solutions can be used to manage and provision both company-owned devices and employee-owned WLAN devices. The typical configuration for BYOD management is a *self-service device onboarding* solution, as opposed to a robust enterprise MDM. The main purpose of a self-service device onboarding solution is to provide an inexpensive and simple way to provision employee personal WLAN devices onto a secure corporate SSID. A self-service device onboarding solution is not meant to offer all the monitoring and restriction aspects of a full-blown MDM. Instead, a device onboarding solution provides a self-service method for an employee to configure the BYOD supplicant and install security credentials, such as an 802.1X/EAP root CA certificate.

Consider this scenario: Jeff logs in to the enterprise network as an employee from his corporate computer using 802.1X/EAP. His username is verified in the LDAP database using RADIUS. In this scenario, Jeff is trusted as the user because his username and password are valid. His corporate laptop is trusted because his machine can be validated. However, Jeff using his corporate laptop is different from Jeff using his smartphone, Jeff using his personal laptop, or Jeff using his tablet.

In the world of authentication and encryption, 802.1X/EAP is typically the required method to provide secure access to corporate networks and data. However, configuration of the client supplicant is typically not a task that can be easily performed by a nontechnical user. Additionally, the root CA certificate needs to be securely transferred to the client device and installed. This can be a problem for corporations and employee BYOD users, since the corporation will not provide access without a properly configured device. As you can imagine, requiring a trained IT help desk person to configure employee personal devices is not practical. The solution is a process known as *onboarding*.

A properly configured and secured 802.1X/EAP network requires that a root CA certificate be installed on the supplicant. Installing the root certificate onto Windows laptops can be easily automated using a *Group Policy Object (GPO)* if the Windows laptop is part of the Active Directory (AD) domain. However, a GPO cannot be used for macOS, iOS, or Android mobile devices, or for personal Windows BYOD devices that are not joined to the AD domain. Manually installing certificates on mobile devices and employee-owned devices is an administrative nightmare.

The onboarding solution is most often used to install the root CA certificates on mobile devices to be used with an 802.1X/EAP-enabled SSID. Client certificates can also be provisioned with an onboarding solution. Some of the Wi-Fi vendors that offer dynamic PSK solutions also offer onboarding solutions that can provision mobile devices with Wi-Fi client profiles configured with unique individual PSKs.

Self-service onboarding solutions for personal employee devices can come in many different forms and are often WLAN vendor-specific. Third-party self-service onboarding solutions, such as SecureW2 (www.securew2.com), are also available. Onboarding solutions typically use an application that uses similar over-the-air provisioning aspects of MDM solutions to securely install certificates and Wi-Fi client profiles onto mobile devices. Self-service onboarding solutions may also use custom applications built using a WLAN vendor's *application programming interface (API)*. Regardless of the solution, the device onboarding normally requires an initial WLAN connection to complete the self-service process.

Dual-SSID Onboarding

Dual-SSID onboarding is performed using an open SSID and an 802.1X/EAP secure corporate SSID. The employee initially connects to the open SSID and will be prompted with a captive portal page. Depending on how onboarding is implemented, the employee may log in directly through the captive portal using their corporate username and password, or the employee may click a link that would take them to an onboard login screen, where they would enter their corporate username and password. The captive web portal authentication validates the employee's username and password via RADIUS and LDAP. The captive web portal authentication is protected via HTTPS. After the employee logs in to the network, an onboarding application is typically downloaded to the mobile device. The onboarding application is then executed to securely download the 802.1X/EAP root certificate and/or other security credentials as well as provision the supplicant on the mobile device. Custom onboarding applications are often distributed via the open SSID. An onboarding solution for different types of users might also be available as a web-based application. Figure 18.13 shows a custom onboarding web application used by the Calgary Board of Education to provision different security credentials to guests, students, and staff.

FIGURE 18.13   BYOD onboarding application

Courtesy of the Calgary Board of Education

After the employee device is provisioned using the onboarding application, it is ready to connect to the secure network. Because the secure network is a separate SSID, the employee would have to manually disconnect from the open SSID and reconnect to the secure SSID.

Single-SSID Onboarding

Single-SSID onboarding uses a single SSID that is capable of authenticating 802.1X/EAP-PEAP clients and 802.1X/EAP-TLS clients. The client initially logs in to the SSID using an 802.1X/EAP-PEAP connection, using their corporate username and password. After the device is logged in to the network, the employee would bring up a captive portal page, requiring the user to log in again, this time to validate that they are allowed to perform the onboard process. As with the dual-SSID process, an onboarding program is downloaded to the device and executed, and then the application downloads the server certificate via SSL and provisions the supplicant on the device.

After the employee device is configured, the RADIUS server will initiate a Change of Authorization (CoA) to the employee device, disconnecting the device from the network. The device will immediately reconnect to the same SSID, using either 802.1X/EAP-PEAP or 802.1X/EAP-TLS, depending on the wireless profile that was installed on the employee device. This time, the client will also validate the server certificate.

MDM vs. Self-Service Onboarding

MDM solutions are often the preferred choice for large corporations. An enterprise MDM gives a corporation the ability to manage and monitor company WLAN devices as well as provide a provisioning solution for employee personal WLAN devices. However, an MDM solution is not always the best choice for a BYOD solution. Enterprise MDM deployments are often cost-prohibitive for medium-sized and small businesses. As previously mentioned, employees often do not like to use MDM due to privacy issues.

Self-service device onboarding solutions are typically much cheaper and simpler to deploy as an employee BYOD solution. Self-service onboarding solutions are used primarily to provision employee WLAN devices and are not used to enforce device restrictions or for over-the-air management. The privacy concerns are no longer an issue for the employee personal devices.

Depending on a company's security requirements, MDM, self-service onboarding, or a combination of the two solutions can be chosen for a BYOD solution.

Guest WLAN Access

Although the primary purpose for enterprise WLANs has always been to provide employees wireless mobility, WLAN access for company guests can be just as important. Customers, consultants, vendors, and contractors often need access to the Internet to accomplish job-related duties. When they are more productive, employees will also be more productive. Guest access can also be a value-added service and often breeds customer loyalty. In today's world, business customers have come to expect guest WLAN access. Free guest access is often considered a value-added service. There is a chance that your customers will move toward your competitors if you do not provide guest WLAN access. Retail, restaurants, and hotel chains are all prime examples of environments where wireless Internet access is often expected by customers.

The primary purpose of a guest WLAN is to provide a wireless gateway to the Internet for company visitors and/or customers. Generally, guest users do not need access to company network resources. Therefore, the most important security aspect of a guest WLAN is to protect the company network infrastructure from the guest users. In the early days of Wi-Fi, guest networks were not very common, because of fears that the guest users might access corporate resources. Guest access was often provided on a separate infrastructure. Another common strategy was to send all guest traffic to a separate gateway that was different from the Internet gateway for company employees. For example, a T1 or T3 line might have been used for the corporate gateway, whereas all guest traffic was segmented on a separate DSL phone line.

WLAN guest access has grown in popularity over the years, and the various types of WLAN guest solutions have evolved to meet the need. In the following sections, we will discuss the security aspects of guest WLANs. At a minimum, there should be a separate guest SSID, a unique guest VLAN, and a guest firewall policy. Additionally, we will discuss the use of captive web portals with a guest WLAN. We will also discuss the many guest access options that are available, including guest self-registration.

Guest SSID

In the past, a common SSID strategy was to segment different types of users—even employees—on separate SSIDs; each SSID was mapped to an independent VLAN. For example, a hospital might have unique SSID/VLAN pairs for doctors, nurses, technicians, and administrators. That strategy is rarely recommended now because of the layer 2 overhead created by having multiple SSIDs. Today, the more common method is to place all employees on the same SSID and leverage Remote Authentication Dial-In User Service (RADIUS) attributes to assign different groups of users to different VLANS. What has not changed over time is the recommendation that all guest user traffic be segmented onto a separate SSID. The guest SSID will always have different security parameters from the employee SSID, so the necessity of a separate guest SSID continues. For example, employee SSIDs commonly are protected with 802.1X/EAP security, whereas guest SSIDs are most often an open network that uses a captive web portal for authentication. Although encryption is not usually provided for guest users, some WLAN vendors have begun to offer encrypted guest access and provide data privacy using dynamic PSK credentials. Encrypted guest access can also be provided with 802.1X/EAP with Hotspot 2.0, which is discussed later.

Like all SSIDs, a guest SSID should never be hidden and should have a simple name, such as CWNA-Guest. In most cases, the guest SSID is prominently displayed on a sign in the lobby or entrance of the company offices.

Guest VLAN

Guest user traffic should be segmented into a unique VLAN tied to an IP subnet that does not mix with the employee VLANs. Segmenting your guest users into a unique VLAN is a security and management best practice. The main debate about the guest VLAN is whether it should be supported at the edge of the network. As shown in Figure 18.14, a frequent design scenario is that the guest VLAN does not exist at the edge of the network and instead is isolated in what is known as a *demilitarized zone (DMZ)*. As depicted in Figure 18.14, the guest VLAN (VLAN 10) does not exist at the access layer; therefore, all guest traffic must be tunneled from the AP back to the DMZ, where the guest VLAN does exist. An IP tunnel, commonly using the Generic Routing Encapsulation (GRE) protocol, transports the guest traffic from the edge of the network back to the isolated DMZ. Depending on the WLAN vendor solution, the tunnel destination in the DMZ can be either a WLAN controller, GRE server appliance, or a router.

FIGURE 18.14 GRE tunneling guest traffic to a DMZ

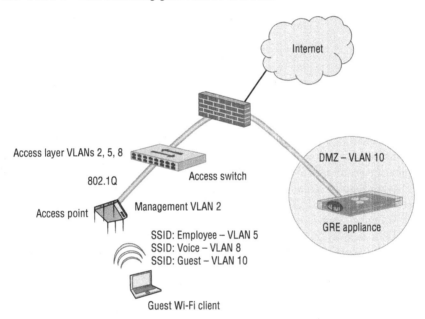

Although isolating the guest VLAN in a DMZ has been a common practice for many years, it is no longer necessary if guest firewall policies are being enforced at the edge of the network. Various WLAN vendors are now building enterprise-class firewalls into access points. If the guest firewall policy can be enforced at the edge of the network, the guest VLAN can also reside at the access layer and no tunneling is needed.

Guest Firewall Policy

The most important security component of a guest WLAN is the firewall policy. The guest WLAN firewall policy prevents guest user traffic from getting near the company network infrastructure and resources. Figure 18.15 shows a very simple guest firewall policy that allows DHCP and DNS but restricts access to private networks 10.0.0.0/8, 172.16.0.0/12, and 192.168.0.0/16. Guest users are not allowed on these private networks, because corporate network servers and resources often reside on that private IP space. The guest firewall policy should simply route all guest traffic straight to an Internet gateway and away from the corporate network infrastructure.

FIGURE 18.15   Guest firewall policy

Source IP	Destination IP	Service	Action
Any	Any	DHCP-Server	PERMIT
Any	Any	DNS	PERMIT
Any	10.0.0.0/255.0.0.0	Any	DENY
Any	172.16.0.0/255.240.0.0	Any	DENY
Any	192.168.0.0/255.255.0.0	Any	DENY
Any	Any	Any	PERMIT

Firewall ports that should be permitted include the DHCP server (UDP port 67), DNS (UDP port 53), HTTP (TCP port 80), and HTTPS (TCP port 443). This allows the guest user's wireless device to receive an IP address, perform DNS queries, and browse the web. Many companies require their employees to use a secure VPN connection when connected to an SSID other than the company SSID. Therefore, it is recommended that IPsec IKE (UDP port 500) and IPsec NAT-T (UDP port 4500) also be permitted.

The firewall policy shown in Figure 18.15 represents the minimum protection needed for a guest WLAN. The guest firewall policy can be much more restrictive. Depending on company policy, many more ports can be blocked. One practice is to force the guest users to use webmail and block SMTP and other email ports so that users cannot "spam through" the guest WLAN. However, now that most mail services use SSL, this practice is not as common. It is up to the security policy of the company to determine which ports need to be blocked on the guest VLAN. If the policy forbids the use of SSH on the guest WLAN, then TCP port 22 will need to be blocked. In addition to blocking UDP and TCP ports, several WLAN vendors now have the ability to block applications. In addition to stateful firewall capability, WLAN vendors have begun to build application-layer firewalls capable of *deep packet inspection (dpi)* into access points or WLAN controllers. An application-layer firewall can block specific applications or groups of applications. For example, some popular video streaming applications can be blocked on the guest SSID, as shown in Figure 18.16. The company security policy will also determine which applications should be blocked or rate-limited on a guest WLAN.

FIGURE 18.16   Application firewall policy

Source IP	Destination IP	Service	Action
Any	Any	YOUTUBE	DENY
Any	Any	NETFLIX VIDEO STREAM	DENY
Any	Any	FACETIME	DENY
Any	Any	GOOGLE VIDEO	DENY
Any	Any	INSTAGRAM VIDEO	DENY
Any	Any	Any	PERMIT

Captive Web Portals

Often, guest users must log in through a captive web portal page before they are given access to the Internet. One of the most important aspects of the captive web portal page is the legal disclaimer. A good legal disclaimer informs guest users about acceptable behavior when using the guest WLAN. Businesses are more likely to be legally protected if something bad, such as being infected by a computer virus, should happen to a guest user's WLAN device while connected through the portal. A *captive portal* solution effectively turns a web browser into an authentication service. To authenticate, the user must first connect to the WLAN and launch a web browser. After the browser is launched and the user attempts to go to a website, no matter what web page the user attempts to browse to, the user is redirected to a different URL, which displays a captive portal login page. Captive portals can redirect unauthenticated users to a login page using an IP redirect, DNS redirection, or redirection by HTTP. As shown in Figure 18.17, many captive web portals are triggered by DNS redirection. The guest user attempts to browse to a web page, but the DNS query redirects the browser to the IP address of the captive web portal.

FIGURE 18.17   Captive web portal—DNS redirection

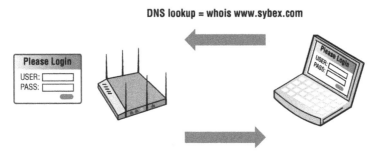

Captive portals are available as a standalone server solution or as a cloud-based service. Additionally, most WLAN vendors offer integrated captive portal solutions. The captive portal may exist within a WLAN controller, or it may be deployed at the edge within an access point. WLAN vendors that support captive portals provide the ability to customize the captive portal page. You can typically personalize the page by adding graphics, such as a company logo, inserting an acceptable use policy, or configuring the login requirements. Depending on the chosen security of the guest WLAN, different types of captive web portal login pages can be used. A user authentication login page requires the AP or WLAN controller to query a RADIUS server with the guest user's name and password. If the guest user does not already have an account, the login page may provide a link, allowing the user to create a guest account, as shown in Figure 18.18. The guest registration page allows the user to enter the necessary information for them to self-register, as shown in Figure 18.19. The guest user may also be connected to a captive portal web page requiring them to simply acknowledge a user policy acceptance agreement, as shown in Figure 18.20.

FIGURE 18.18   Captive web portal—guest login

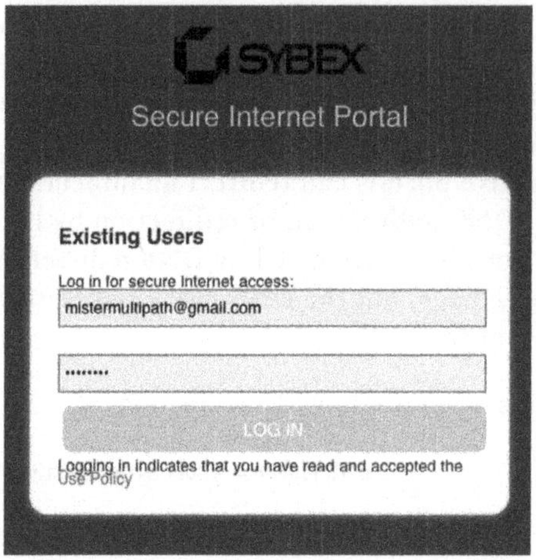

FIGURE 18.19   Captive web portal—guest self-registration

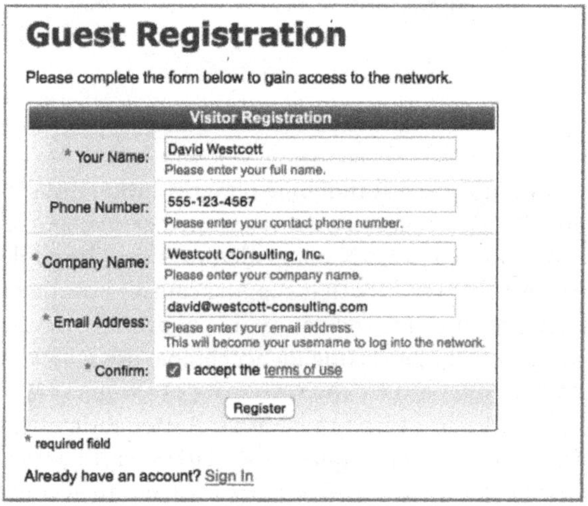

FIGURE 18.20   Captive web portal—policy acceptance

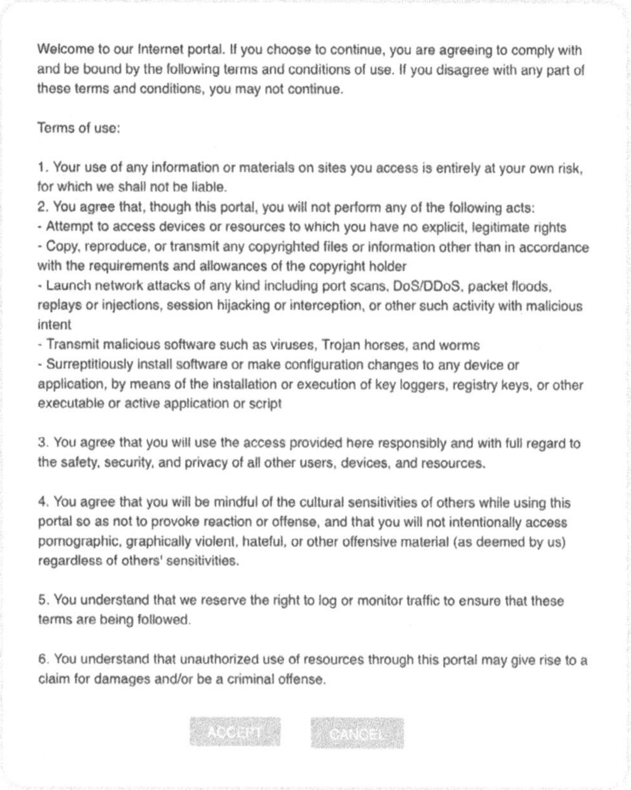

Welcome to our Internet portal. If you choose to continue, you are agreeing to comply with and be bound by the following terms and conditions of use. If you disagree with any part of these terms and conditions, you may not continue.

Terms of use:

1. Your use of any information or materials on sites you access is entirely at your own risk, for which we shall not be liable.
2. You agree that, though this portal, you will not perform any of the following acts:
- Attempt to access devices or resources to which you have no explicit, legitimate rights
- Copy, reproduce, or transmit any copyrighted files or information other than in accordance with the requirements and allowances of the copyright holder
- Launch network attacks of any kind including port scans, DoS/DDoS, packet floods, replays or injections, session hijacking or interception, or other such activity with malicious intent
- Transmit malicious software such as viruses, Trojan horses, and worms
- Surreptitiously install software or make configuration changes to any device or application, by means of the installation or execution of key loggers, registry keys, or other executable or active application or script

3. You agree that you will use the access provided here responsibly and with full regard to the safety, security, and privacy of all other users, devices, and resources.

4. You agree that you will be mindful of the cultural sensitivities of others while using this portal so as not to provoke reaction or offense, and that you will not intentionally access pornographic, graphically violent, hateful, or other offensive material (as deemed by us) regardless of others' sensitivities.

5. You understand that we reserve the right to log or monitor traffic to ensure that these terms are being followed.

6. You understand that unauthorized use of resources through this portal may give rise to a claim for damages and/or be a criminal offense.

ACCEPT   CANCEL

RADIUS servers are often used with captive portal authentication to validate guest user credentials for a guest SSID. A captive web portal solution will query a RADIUS server with a username and password, using a weak authentication protocol such as MS-CHAPv2. As opposed to using a preexisting user database, such as Active Directory, the guest credentials are usually created during the guest registration process and often stored in the native database of the RADIUS server. Captive web portal authentication is also often used together with BYOD solutions for validating employee credentials. The employee database would most likely be Active Directory, which, in turn, would be queried by the RADIUS server.

Keep in mind that a captive web portal requires user interaction, and sometimes the guest user experience can be adversely affected. Captive web portals often fail after browser updates or mobile device operating system updates. DNS problems also cause captive web portal failures. Furthermore, the design of many captive web portals is not always user-friendly. At some point in time, just about everyone has had a bad experience with a captive web portal. Captive web portals should be of simple design, easy to understand, and thoroughly tested to provide the best quality of guest user experience.

Client Isolation, Rate Limiting, and Web Content Filtering

When guest users are connected to the guest SSID, they are all in the same VLAN and the same IP subnet. Because they reside in the same VLAN, the guests can perform peer-to-peer attacks against each other. Client isolation is a feature that can be enabled on WLAN access points or controllers to block wireless clients from communicating directly with other wireless clients on the same wireless VLAN. *Client isolation* (or the various other terms used to describe this feature) usually means that packets arriving at the AP's wireless interface are not allowed to be forwarded back out of the wireless interface to other clients. This isolates each user on the wireless network to ensure that a wireless station cannot be used to gain layer 3 or higher access to another wireless station. The client isolation feature is usually a configurable setting per SSID linked to a unique VLAN. Client isolation is highly recommended on guest WLANs to prevent peer-to-peer attacks.

Enterprise WLAN vendors also offer the capability to throttle the bandwidth of user traffic. *Bandwidth throttling*, also known as *rate limiting*, can be used to curb traffic at either the SSID level or the user level. Rate limiting is often utilized on guest WLANs. It can ensure that the majority of the bandwidth is preserved for employees. Rate limiting the guest user traffic to 1024 Kbps is a common practice. However, because guest access is usually an expected value-added service, rate-limiting on guest SSIDs may not be a good strategy. Some businesses that attempt to monetize WLAN guest access often provide two levels of guest access. The free level of guest access is rate-limited, whereas the paid guest access has no bandwidth restrictions.

Enterprise companies often deploy *web content filter* solutions to restrict the type of websites that their employees can view while at the workplace. A web content filtering solution blocks employees from viewing websites based on content categories. Each category contains websites or web pages that have been assigned based on their primary web content. For example, the company might use a web content filter to block employees from viewing any websites that pertain to gambling or violence. Content filtering is most often used to block what employees can view on the Internet, but web content filtering can also be used to block certain types of websites from guest users. All guest traffic might be routed through the company's web content filter.

Guest Management

As Wi-Fi has evolved, so have WLAN guest management solutions. Most guest WLANs require guest users to authenticate with credentials via a captive web portal. Therefore, a database of user credentials must be created. Unlike user accounts in a preexisting Active Directory database, guest user accounts are normally created on the fly and in a separate guest user database. Guest user information is usually collected when the guests arrive at company offices. Someone has to be in charge of managing the database and creating the guest user accounts. IT administrators are typically too busy to manage a guest database; therefore, the individual who manages the database is often a receptionist or the person who greets guests at the front door. This individual requires an administrative account to the guest management solution, which might be a RADIUS server or some type of other guest database server. The guest management administrators have the access rights to create guest user accounts in the guest database and issue the guest credentials, which are usually usernames and passwords.

A guest management server can be cloud-based or reside as an on-premises server in the company data center. Although most guest management systems are built around a RADIUS server, the guest management solution offers features in addition to providing RADIUS services. Modern WLAN guest management solutions offer robust report-generation capabilities for auditing and compliance requirements. As shown in Figure 18.21, a guest management solution can also be used as a 24/7 full-time monitoring solution. An IT administrator usually configures the guest management solution initially; however, a company receptionist will have limited access rights to provision guest users. Guest management solutions can also be integrated with LDAP for employee sponsorships and usually have some method for guest users to self-register. Most often, guest management solutions are used for wireless guests, but they might also be used to authenticate guests connected to wired ports.

FIGURE 18.21   Guest management and monitoring

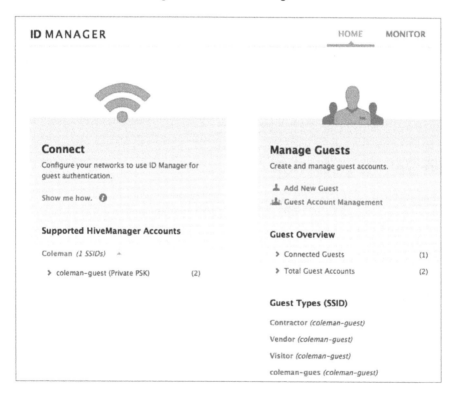

As you can see in Figure 18.22, there can be multiple ways to deliver the guest credentials to the guest user. The credentials can be delivered via an electronic wallet, SMS text message, an email message, or a printed receipt. The SMS, email, and receipt can also be customized with company information. The guest registration login pages can all be customized with company logos and information.

FIGURE 18.22 Guest credential delivery methods

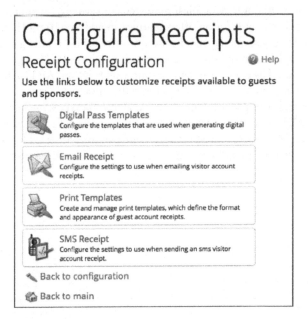

Guest Self-Registration

Guest management solutions have traditionally relied on a company receptionist or lobby ambassador to register the guest users. A good guest management solution allows the receptionist to register a single guest user or groups of users. Over the past few years, there has also been a greater push for guest users to create their own account, what is commonly referred to as *self-registration*. When guest users are redirected to the captive web portal, if they do not already have a guest account, a link on the login web page redirects them to a self-registration page. Simple self-registration pages allow guests to fill out a form, and their guest account is created and displayed or printed for them. More advanced self-registration pages require guests to enter an email or SMS address, which is then used by the registration system to send users their login credentials.

As shown in Figure 18.23, some guest management solutions now offer kiosk applications, where the self-registration login page runs on a tablet that functions as the kiosk. Self-registration via a kiosk is quite useful when the kiosk is deployed in the main lobby or at the entrance to the company. An advantage of self-registration kiosks is that the receptionist does not have to assist the users and can concentrate on other duties.

FIGURE 18.23   Kiosk mode

Employee Sponsorship

Guest users can also be required to enter the email address of an employee, who in turn must approve and sponsor the guest. The sponsor typically receives an email with a link that allows them to easily accept or reject the guest's request. Once the user is registered or sponsored, they can log in using their newly created credentials. A guest management solution with *employee sponsorship* capabilities can be integrated with an LDAP database, such as Active Directory.

As you already learned, a receptionist can register guest users, or a company may choose to use a registration kiosk so that guests can self-register. For larger or distributed organizations, a central registration kiosk does not scale well. Self-registration with employee sponsorship is becoming popular for many organizations.

When guest users initially connect to the guest network, they are redirected to a captive portal page. The captive portal page prompts them to log in if they already have an account, or it allows them to click a link that allows them to create their own guest account. As shown in Figure 18.24, the guest must enter the email address of the employee who is sponsoring them. Typically, the guest has a business meeting with the employee who is providing sponsorship.

FIGURE 18.24   Employee sponsorship registration

When the registration form is completed and submitted, the sponsor receives an email notifying them that the guest would like network access. As shown in Figure 18.25, the email typically contains a link that the sponsor must click to approve network access. Once the link is clicked, the guest account is approved and the guest receives confirmation, either by email or SMS, and they will then be allowed to log in to the network. If the sponsor does not click the link, the guest account is never created and the guest is denied access to the network.

FIGURE 18.25   Employee sponsorship confirmation email

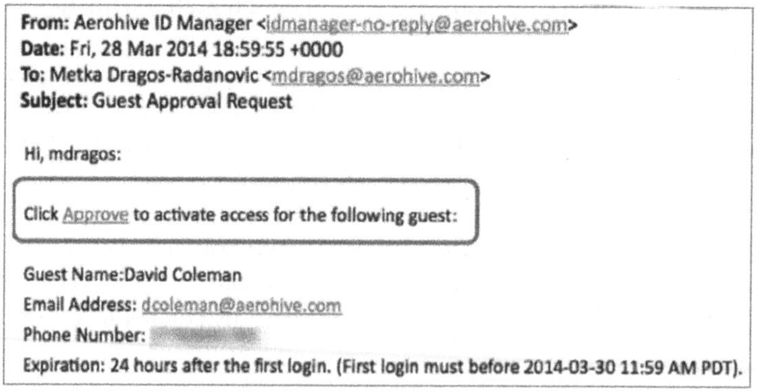

Employee sponsorship ensures that only authorized guest users are allowed onto the guest WLAN and that the company employees are actively involved in the guest user authorization process.

Social Login

A new trend in guest networks in retail and service industries is *social login*. Social login is a method of using existing login credentials from a social networking service (such as Twitter, Facebook, or LinkedIn) to register on a third-party website. Social login allows a user to forgo the process of creating new registration credentials for the third-party website. Social login is often enabled using the *OAuth* protocol. OAuth (Open Standard for Authorization) is a secure authorization protocol that allows access tokens to be issued to third-party clients by an authorization server. As shown in Figure 18.26, the OAuth 2.0 authorization framework enables a third-party application to obtain limited access to an HTTP service and can be used for social login for Wi-Fi guest networks.

FIGURE 18.26   OAuth 2.0 application

As shown in Figure 18.27, social login can be tied to an open guest SSID. Guest users are redirected to a captive web portal page, where they can then log in to the guest WLAN using their existing social media login credentials. Retail and service businesses like the idea of social login because it allows them to obtain meaningful marketing information about guest users from the social networking services. Businesses can then build a database of the type of customers who are using the guest Wi-Fi while shopping at the business. It should be noted that there are serious privacy concerns with social login, and the login captive web portal always has a legal disclaimer stating that customer information might be gathered if the customer agrees to use the social login registration to the guest WLAN.

FIGURE 18.27 Social login

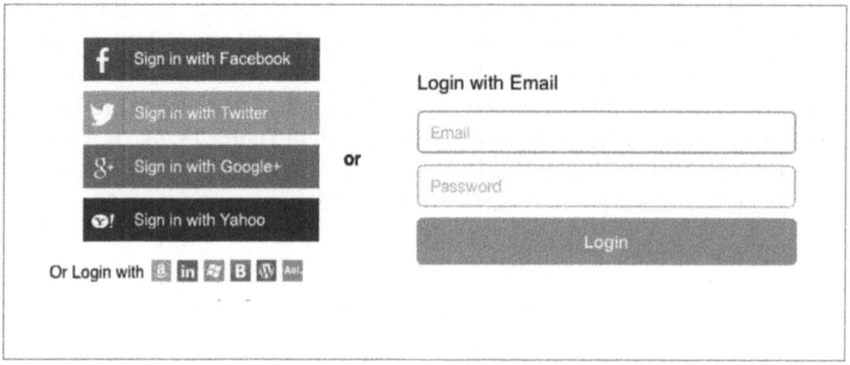

Encrypted Guest Access

Most guest networks are open networks that do not use encryption; thus, there is no data privacy for guest users. In Chapter 16, you learned about the numerous wireless attacks that make unsecured Wi-Fi users vulnerable. Because most guest WLANs do not use encryption, guest users are low-hanging fruit and often targets of skilled hackers or attackers. For that reason, many corporations require their employees to use an IPsec VPN solution when connected to any kind of public or open guest SSID. Because the guest SSID does not provide data protection, guest users must bring their own security in the form of a VPN connection that provides encryption and data privacy.

The problem is that many consumers and guest users are not savvy enough to know how to use a VPN solution when connected to an open guest WLAN. As a result, there is a recent trend to provide encryption and better authentication security for WLAN guest users. Protecting the company network infrastructure from attacks from a guest user still remains a top security priority. However, if a company can also provide encryption on the guest SSID, the protection provided to guest users is a value-added service.

One simple way to provide encryption on a guest SSID is to use a static PSK. Although encryption is provided when using a static PSK, this is not ideal because of brute-force dictionary attacks and social engineering attacks. Some WLAN vendors offer cloud-based solutions to distribute secure guest credentials in the form of unique per-user PSKs. A guest management solution that utilizes unique PSKs as credentials also provides data privacy for guest users with WPA2 encryption.

Hotspot 2.0 and Passpoint

Another growing trend with public access networks is the use of 802.1X/EAP with *Hotspot 2.0* technology. Hotspot 2.0 is a Wi-Fi Alliance technical specification that is supported by the Passpoint certification program. With Hotspot 2.0, the WLAN client device is equipped by a cellular carrier service provider with one or more credentials, such as a SIM card, username/password pair, or X.509 certificate. Much of the Hotspot 2.0 technical specification is based on mechanisms originally defined by the IEEE 802.11u-2011 amendment. The two main goals of the Hotspot 2.0 technical specification are as follows:

- To make public/commercial Wi-Fi networks as secure and easy to use as enterprise/home Wi-Fi networks
- To offload 3G/4G cellular network traffic onto Wi-Fi networks

Passpoint is the brand for the certification program operated by Wi-Fi Alliance. The Passpoint certification is based on the Wi-Fi Alliance Hotspot 2.0 specification. Devices that pass this certification testing can be referred to as "Passpoint devices."

Access Network Query Protocol

Passpoint devices can query the WLAN prior to connecting in order to discover the cellular service providers supported by the network. Passpoint devices use the *Access Network Query Protocol (ANQP)*, a query and response protocol defined by the 802.11u. As shown in Figure 18.28, an ANQP server is needed to respond to the ANQP queries from the Passpoint clients. The ANQP server can be a standalone server or embedded in the access points.

FIGURE 18.28   ANQP

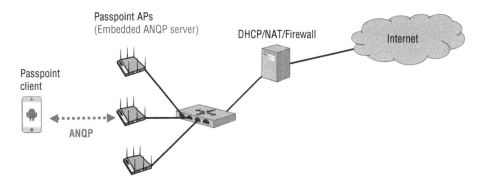

ANQP is used by Passpoint mobile devices to discover a range of information, including the following:

- Venue name
- Required authentication types, such as EAP-AKA or EAP-TLS

- Information about 3GPP cellular networks available through the access point
- Roaming consortium (for hotspots with roaming agreements with other service providers)
- Network Address Identifier (NAI) home
- NAI realms accessible through the AP
- WAN metrics
- Much more

The ANQP information in the Beacon frame will not usually be sufficient for the Passpoint client to connect to the Passpoint SSID. Therefore, the Passpoint client uses *Generic Advertisement Service (GAS)* query frames to gather most of the needed information from the ANQP server, as shown in Figure 18.29.

FIGURE 18.29   GAS queries

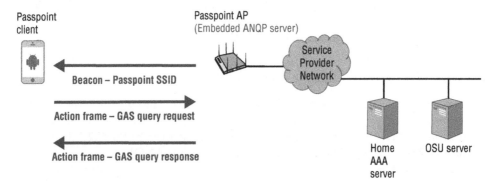

Hotspot 2.0 Architecture

The complexity of the Hotspot 2.0 technical specification is the integration with all the backend servers of the cellular service provider network, as shown in Figure 18.30. The service provider components include the following:

- Online Sign Up (OSU) server is used to create a new account for a new user/device during registration. HTTPS is used between a mobile device and OSU server to register and provision a new Hotspot subscriber.

- Subscription Remediation (Sub Rem) server is used to remediate the subscription information for a user after registration. Example: password expiration or delinquent payment.

- Policy server is used to provision the mobile device with network policy after registration.

- EAP and RADIUS are used to verify Passpoint client credentials to the Home operator's AAA server. Passpoint clients use an 802.1X/EAP protocol, and the EAP frames are forwarded to the Home AAA server in RADIUS packets.

- Home Location Register (HLR) server is the local registration server for the 3G/4G cellular network.

- Home AAA server communicates with the Home HLR server for devices with SIM or USIM cards.
- Mobile Application Part (MAP) is the telephone signaling system protocol for Home Location Register (HLR).
- Certificate Authority (CA) issues certificates to the AAA server, OSU server, Sub Rem server, and Policy server.
- If needed, the CA is also used to provision Passpoint clients with certificates.

FIGURE 18.30   Hotspot 2.0 WLAN

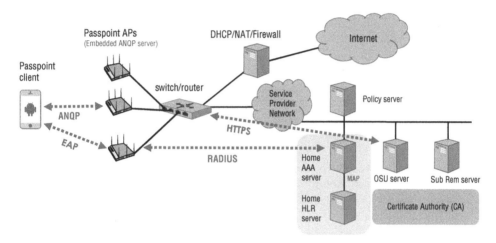

802.1X/EAP and Hotspot 2.0

Because Hotspot 2.0 requires 802.1X/EAP security, all devices have a unique authentication identity, and traffic is encrypted on the Passpoint SSIDs. Table 18.1 lists the supported credentials and EAP methods for a Hotspot 2.0 WLAN.

TABLE 18.1   HotSpot 2.0—supported client credentials and EAP protocols

Credential Type	EAP Method
Client certificate	EAP-TLS
SIM card	EAP-SIM
USIM card	EAP-AKA
Username/password	EAP-TTLS

EAP-Subscriber Identity Module (EAP-SIM) was primarily developed for the mobile phone industry and more specifically for second-generation (2G) mobile networks. Many of us who have mobile phones are familiar with the concept of a *Subscriber Identity Module (SIM)* card. A SIM card is an embedded identification and storage device very similar to a smart card. SIM cards are smaller and fit into small mobile devices like cellular or mobile phones with a 1:1 relationship to a device at any given time. The Global System for Mobile Communications (GSM) is a second-generation mobile network standard. EAP-SIM, outlined in the IETF RFC 4186, specifies an EAP mechanism that is based on 2G mobile network GSM authentication and key agreement primitives. For mobile phone carriers, this is a valuable piece of information that can be utilized for authentication. EAP-SIM does not offer mutual authentication, and key lengths are much shorter than the mechanisms used in third-generation (3G) mobile networks.

EAP-Authentication and Key Agreement (EAP-AKA) is an EAP type primarily developed for the mobile phone industry and more specifically for 3G mobile networks. EAP-AKA, outlined in the IETF RFC 4187, defines the use of the authentication and key agreement mechanisms already being used by the two types of 3G mobile networks. The 3G mobile networks include the Universal Mobile Telecommunications System (UTMS) and CDMA2000. AKA typically runs in a SIM module. The SIM module may also be referred to as a *Universal Subscriber Identity Module (USIM)* or *Removable User Identity Module (R-UIM)*. AKA is based on challenge-response mechanisms and symmetric cryptography, and runs in the USIM or R-UIM module. Encryption key lengths can be substantially longer, and mutual authentication has now been included. EAP-AKA can also be used in 4G mobile networks that typically use *Long Term Evolution (LTE)* cellular technology.

Non-SIM Passpoint devices, such as laptops, will use either EAP-TLS or EAP-TTLS security. *EAP Transport Layer Security (EAP-TLS)* is defined in IETF RFC 5216 and is a widely used security protocol. EAP-TLS is considered to be one of the most secure EAP methods used in WLANs today. EAP-TLS requires the use of client-side certificates in addition to a server certificate. The client-side certificate is used as the credential for the client device. *EAP-Tunneled Transport Layer Security (EAP-TTLS)* requires only the use of server-side certificates and is defined in IETF RFC 5281. With EAP-TTLS, username and password credentials are securely authenticated within an SSL tunnel. As with most methods of EAP, the secure provisioning of root certificates and possibly client certificates is required. The secure provisioning of the certificates occurs during the *online sign-up (OSU)* process in any Hotspot 2.0 WLAN.

Online Sign-Up

There are two options for online sign-up (OSU) that a Passpoint capable client can use to initially register and then connect to a secure Passpoint SSID. As shown in Figure 18.31, both methods require two SSIDs—one SSID for the initial sign-up and a second secure Passpoint SSID.

FIGURE 18.31   Online sign-up

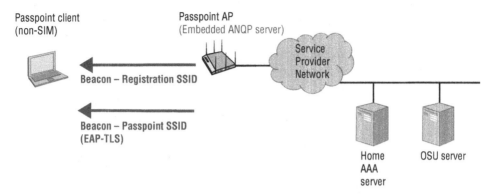

The first option allows hotspot operators to use a legacy open SSID that does not use any encryption. The Passpoint client will first connect to the open SSID to register with the service provider. The client will then be redirected to a captive web portal that is secured via HTTPS. From the captive web portal, the client will select a service provider and continue the registration process. The registration between the client device and the OSU server is protected by HTTPS. During the registration process, EAP-TLS certificates and client credentials provided by the service provider can be provisioned and installed on the client device. Upon completion of registration, the client is disconnected from the sign-up SSID and then must connect to the secure Passpoint SSID using 802.1X/EAP.

The second option for online sign-up also uses two SSIDs; however, the initial registration SSID is known as an *OSU Server-Only Authenticated L2 Encryption Network (OSEN)* SSID. Just like the first option, the registration between the client device and the OSU server is protected by HTTPS. The main difference is that an OSEN registration SSID is meant to protect other mobile device communications not related to the client registration. An OSEN SSID uses *Anonymous EAP-TLS*, which authenticates only the service provider network, not the client. The intent is to ensure that the service provider network is legitimate. During the registration process, certificates and client credentials provided by the service provider can be provisioned and installed on the client device. Upon completion of registration, the client is disconnected from the OSEN SSID and then must connect to the secure Passpoint SSID using 802.1X/EAP. If OSEN SSIDs are used for Passpoint clients to register, a third open SSID will still most likely exist for WLAN connectively for legacy non-Passpoint clients.

Roaming Agreements

The cellular carrier service providers may also have roaming agreements with other service providers. Do not confuse this with 802.11 roaming mechanisms that have been discussed previously in this book. A service provider roaming agreement is simply a business relationship between different cellular service providers. A customer of one hotspot provider may be able to also connect to a hotspot WLAN service of a different cellular provider with no additional charges. Applicable roaming agreement information is provided to Passpoint clients via the ANQP protocol. Passpoint enables inter-carrier roaming, with discovery, authentication, and accounting. Hotspot 2.0 and Passpoint technology was initially intended for cellular carrier networks; however, there are some possibilities for ANQP being used for guest access WLANs in private enterprise networks.

The goal of Hotspot 2.0 is to provide secure authentication and encryption of public access WLANs. Although open networks are still the norm today, growing interest in security and automated connectivity in public access networks may motivate wider adoption and use of Hotspot 2.0 technology. Keep in mind that clients and access points must be Passpoint-certified and able to use the Access Network Query Protocol (ANQP). Legacy WLAN devices do not support ANQP and cannot take advantage of Hotspot 2.0 technology, although recent versions of most operating systems support the earliest versions of Passpoint. As you hopefully have also surmised, the backend integration with the service provider networks is extremely complex. Because of this complexity, the adoption of the technology has been very slow, with many carrier providers still not using the technology. The advent of forthcoming 5G cellular technology may also have future deployment implications.

Network Access Control

Network access control (NAC) evaluates the capability or state of a computer to determine the potential risk of the computer on the network and to determine the level of access to allow. NAC has changed over the years from an environment that primarily assessed the virus and spyware health risk to an environment where checks and fingerprinting are performed on a computer, extensively identifying its capabilities and configuration. These checks are integrated with 802.1X/EAP and RADIUS to authenticate and authorize network access for the user and the computer.

Posture

NAC began as a response to computer viruses, worms, and malware that appeared in the early 2000s. The early NAC products date back to around 2003 and provided what is known as *posture assessment*. Posture is a process that applies a set of rules to check the health and configuration of a computer and determine whether it should be allowed access to the network. NAC products do not perform the health checks themselves but rather validate that the policy is adhered to. A key task of posture assessment is to verify that security software (antivirus, antispyware, and a firewall) is installed, up-to-date, and operational. Figure 18.32 shows an example of some of the antivirus settings that can be checked. Essentially, posture assessment "checks the checkers." In addition to checking security software status, posture assessment can check the state of the operating system. Posture policy can be configured to make sure that specific patches or updates are installed, verify that certain processes are running or not running, or even check to determine whether or not specific hardware (such as USB ports) is active.

FIGURE 18.32   Antivirus posture settings

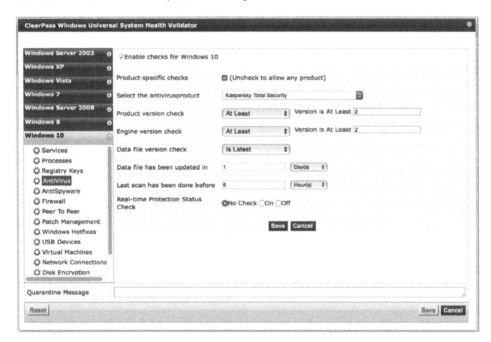

A posture check is performed by a *persistent agent* (software that is permanently installed on the computer) or by a *dissolvable agent* (software that is temporarily installed). If a company deploys posture software, a persistent agent will likely be installed on all of the corporate laptops to make sure that they are healthy. The company may also want to check guest computers that are trying to connect to the network; however, the guest is not likely to allow your company to install software on their computer. When the guest connects to the captive portal, a posture assessment process can temporarily run and check the guest computer for compliance.

After the posture check is performed, if a computer is considered unhealthy, the ideal scenario would be for the posture agent to automatically fix or remediate the problem so that the computer can pass the check and gain network access. Since the persistent agent is installed on the corporate computer and typically has permissions to make changes, automatic remediation can be performed. Computers that are running dissolvable agents typically cannot be automatically updated. The guest user must resolve the problems before network access will be allowed.

OS Fingerprinting

The operating system of WLAN client devices can be determined by a variety of fingerprinting methods, including DHCP and HTTP snooping. After a client successfully establishes a layer 2 connection, the next action is to send a DHCP request to obtain an IP address. As part of that request, the client device includes DHCP option information and requests a list of DHCP parameters or options from the DHCP server. These options may include subnet mask, domain name, default gateway, and the like. When a client sends DHCP discover and request messages, each type of client requests different parameters under the DHCP option 55 portion of the request. The parameters within DHCP option 55 create a fingerprint that can be used to identify the operating system of the client.

For example, iOS devices include a common set of parameters when performing a DHCP request, thus making it possible to identify that the device is most likely an iOS device. DHCP fingerprinting is not perfect, and it is often not possible to discern the difference between similar devices, such as an iPod, an iPhone, or an iPad. Depending on the NAC vendor, the DHCP fingerprint is referenced as an ASCII list of parameter request options, such as 1, 3, 6, 15, 119, 252. Or it might be displayed as a hexadecimal string, such as 370103060F77FC. In the string, the first two hex digits are equal to ASCII 55 (option 55), and each of the following two digits pairs are the hex values of each option.

You can find an extensive list of DHCP fingerprints at www.fingerbank.org. Although the parameter request list is not guaranteed to be unique, it can typically be used along with other fingerprinting techniques to identify devices.

Another OS detection method is *HTTP fingerprinting*. The user-agent header within an HTTP packet identifies the client operating system. During captive portal authentication, NAC solutions are able to inspect HTTP/HTTPS frames while handling the client requests. This fingerprinted information is combined with the information obtained through other methods to paint a better picture of the client device.

Other ways of obtaining client information are active methods such as SNMP and TCP scanning.

AAA

Earlier in the book, we mentioned authentication, authorization, and accounting (AAA). AAA is a key component of NAC. Authentication obviously is used to identify the user who is connecting to the network. We often refer to this as identifying "who you are." Although "who you are" is a very important piece of the process for allowing access to the network, an equally important component of the connection is authorization. We often refer to this as identifying "what you are." Authorization is used to analyze information such as the following:

- User type (admin, help desk, staff)
- Location, connection type (wireless, wired, VPN)
- Time of day

- Device type (smartphone, tablet, computer)
- Operating system
- Posture (system health or status)

When configuring AAA for authentication, one of the tasks is to define or specify the database that will be used to verify the user's identity. Historically, we have referred to this as the user database; however, the user's identity could be verified by something other than a user account and password, such as a MAC address or a certificate. If you are not sure of the identity type that is being used, or if you want to maintain a more neutral stance, the term *identity store* is a good one to use.

By utilizing both authentication and authorization, a NAC can distinguish between Jeff using his smartphone and Jeff using his personal laptop. From this information, the NAC can control what Jeff can do with each device on the network.

To use an analogy to explain authorization, say that George is a member of a country club. As he drives onto the property, the guard at the entrance checks his identification card and verifies his membership. He has been authenticated. After parking his car, he decides to go to the restaurant since it is 6:30 p.m. and he is hungry. When he arrives at the restaurant, he is told that he is not allowed in the restaurant. Confused, George questions why since he has already verified that he is a member of the club at the entrance. The hostess then explains to him that after 6 p.m., the restaurant has a policy that all male guests must be wearing slacks and a sport or suit jacket. Unfortunately, George was wearing shorts and did not have a jacket; therefore, he was not authorized to eat at the restaurant. The hostess was polite and did tell George that he was authorized to go to the lounge and have a meal there, since the requirements for the lounge were not as strict.

As you can see from the analogy, authentication is about who you are, whereas authorization is about other parameters, such as what, where, when, and how. Also, unlike authentication, where you are or are not authenticated, authorization varies depending on the parameters and the situation.

RADIUS Change of Authorization

Prior to RADIUS *Change of Authorization (CoA)*, if a client were authenticated and assigned a set of permissions on the network, the client authorization would not change until the client logged out and logged back in. This only allowed the authorization decision to be made during the initial connection of the client.

RADIUS accounting (the final *A* in *AAA*) is used to monitor the user connection. In the early days of AAA, it typically tracked client connection activity: logging in and logging off events, which in some environments may be all you want or need to track. Enhancements to accounting allow the AAA server to also provide interim accounting. Interim accounting can track resource activity such as time and bytes used for the connection. If the user exceeds or violates the allowed limits of resources, RADIUS CoA can be used to dynamically change the permissions that the user has on the network.

To use an analogy to explain RADIUS CoA, let us say that Jack is going to a club with friends to enjoy some cocktails and dancing. When they arrive, a bouncer at the door admits them into the club but tells them that they are not allowed to become drunk or cause trouble in the club. While telling them this, the bouncer checks to make sure that they are not already drunk or causing trouble. Unfortunately, the bouncer must stand at the door and monitor the guests only as they enter the club; the bouncer cannot monitor the guests once they are inside the club. After a few nights of experiencing some problems in the club, the manager decides to hire additional bouncers, who walk around the club and monitor the guests who are already in the club. Anyone who is found to be drunk or causing problems is either restricted within the club (maybe they are no longer allowed to purchase alcoholic drinks), or the guest may be removed from the club. Once the guest is outside the club, the bouncer at the door can reevaluate the status of the guest, possibly denying reentry into the club, allowing the guest back in the club, or allowing the guest to reenter but with a different set of permissions.

RADIUS CoA was originally defined by RFC 3576 and later updated in RFC 5176. Before you begin to worry, no, you do not need to know this for the CWNA exam. We are mentioning it because many of the AAA servers, NAC servers, and enterprise wireless equipment reference "RADIUS RFC 3576" on configuration menus without referring to CoA. Therefore, from a practical perspective, you should be aware that if you see RFC 3576 on any configuration menu, that is the section where RADIUS CoA is configured.

Single Sign-On

In the early days of networking, users had to log in to the file or print server in order to get access to the network resources. The user accounts were managed and stored on each server. Initially, this was rarely a problem since networks were smaller, but as the number of internal servers and server types increased, logging in to multiple servers became a hassle. To simplify the process, companies began to implement *single sign-on (SSO)* within the organization, allowing users to access many if not all of the internal resources using a single network login. Not only did this simplify the login process for the user, it also simplified network management by consolidating user accounts into one central user database.

Within the organization, single sign-on worked well for many years, until corporate resources began migrating to Internet- and cloud-based servers and services. User logins now had to extend outside the corporate network, and many cloud-servers were actually services provided by other companies, such as CRM systems, office applications, knowledge bases, and file-sharing servers. Authentication and authorization across organizational boundaries introduces more complexity and a greater security risk.

Two technologies, *Security Assertion Markup Language (SAML)* and OAuth, can be used to provide the access security needed to expand outside the organization's network. The following sections briefly explain the components of these technologies and how they work.

SAML

SAML provides a secure method of exchanging user security information between your organization and an external service provider, such as third-party cloud-based *customer relationship management (CRM)* platform. When a user attempts to connect to the CRM platform, instead of requiring the user to log in, a trust relationship between your authentication server and the CRM server will validate the user's identity and provide access to the application or service. This allows the users to log in once to the enterprise network and then seamlessly and securely access external services and resources without having to revalidate their identities.

The SAML specification defines three roles that participate in the SSO process: the identity provider (IdP), which is the asserting party; the service provider (SP), which is the relying party; and the user. This section will briefly explain two scenarios in which SAML can be used to provide SSO.

The first scenario is the service provider–initiated login, as illustrated in Figure 18.33. Here, the user attempts to access a resource on the CRM server (SP). In this scenario, if the user has not been authenticated, the user is redirected to the enterprise authentication server (IdP) using a SAML request. After successful authentication, the user is then redirected to the CRM server using a SAML assertion, at which time the user will have access to the requested resources.

FIGURE 18.33 Service provider–initiated login

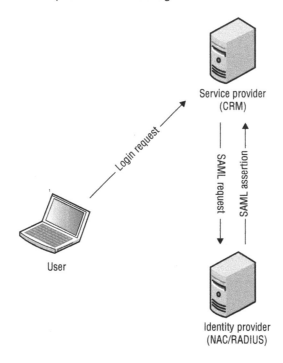

The second scenario is the identity provider–initiated login, as illustrated in Figure 18.34. Here, the user logs in to the enterprise authentication server (IdP) first. Once logged in, the user is redirected to the CRM server and a SAML assertion verifies the user's access.

FIGURE 18.34   Identity provider–initiated login

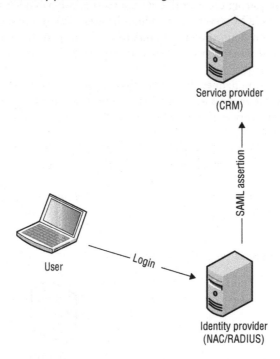

There are many ways of configuring SAML. The key concept is that it provides access to resources outside of the enterprise network, using the user's enterprise credentials and without requiring the user to log in multiple times.

OAuth

OAuth is different from SAML; it is an authorization standard and not an authentication standard. With OAuth, a user logs in to the authenticating application. Once logged in, the user (resource owner) can authorize a third-party application to access specific user information or resources, providing an authorization flow for web and desktop applications, along with mobile devices. NACs can use OAuth to communicate with external resources and systems.

Summary

In this chapter, we discussed BYOD policies and the MDM solutions that are needed to manage company-issued mobile devices as well as employee-owned mobile devices. We examined the differences between company-issued and personal devices and the MDM policy considerations for both. We discussed the various components of an MDM architecture and how the components interact. We explained the MDM enrollment process and over-the-air provisioning. We reviewed the types of mobile devices that use MDM profiles and those that use MDM agent software. We also discussed over-the-air management and application management when using an MDM solution for mobile devices.

We discussed self-service device onboarding solutions for employees. Self-service onboarding for employee devices is the fastest growing trend for BYOD provisioning.

We reviewed guest WLAN access and the key security components needed to protect corporate network infrastructure from guest users. We examined the various methods of guest management, including employee sponsorship, self-registration, and social login. We also discussed the Hotspot 2.0 technical specification and the Wi-Fi Alliance Passpoint certification. Finally, we discussed how NAC can be used to provide access control by monitoring posture and by fingerprinting the client device prior to it connecting to the network. AAA services can authenticate the user connecting to the network and can authorize the device onto the network. RADIUS CoA can be used to modify the authorization of a user if a new set of permissions needs to be assigned.

Although MDM, self-service device onboarding, WLAN guest management, and NAC are separate components of a WLAN, we chose to write about all four together in this chapter because several WLAN vendors package these security solutions together as one application suite. MDM, device onboarding, WLAN guest management, and NAC can be deployed as separate components or in unison to provide mobile device security management, guest user security, and network access security.

Exam Essentials

Define the differences between company-issued devices and personal mobile devices.   Be able to explain the MDM policy concerns for both company-issued and personal mobile devices.

Describe the four main components of an MDM architecture.   Define the roles of a mobile device, an MDM server, an AP, and push notification servers. Explain how they interact.

Explain how MDM profiles and MDM agents are used within an MDM solution.
Describe how MDM profiles can be used for restrictions and mobile device configurations. Describe the role of MDM agents and which mobile devices require MDM agent software.

Discuss MDM over-the-air management and MDM application management.   Be able to explain how push notification servers are used to manage mobile devices across the Internet. Explain how an MDM solution can manage mobile device applications.

Explain self-service device management.   Be able to discuss both dual- and single-SSID methods used to provision employee devices. Explain the advantages and differences between self-service device management versus MDM.

Define the four main security objectives of a guest WLAN. Discuss the importance of guest SSIDs, guest VLANs, guest firewall policies, and captive web portals.

Explain the many components and methods of WLAN guest management. Be able to explain self-registration, employee sponsorship, social login, and other ingredients of guest management.

Understand the Hotspot 2.0 technical specification. Describe how Passpoint-certified client devices can securely connect to public access networks via 802.1X/EAP .

Explain NAC and how it is used to control access to the network. Describe how posture, RADIUS attributes, and DHCP fingerprinting are used along with AAA to authenticate and authorize a user and device onto the network. Describe how RADIUS CoA can be used to modify the authorization of the user.

Review Questions

1. In a guest firewall policy, what are some of the ports that are recommended to be permitted? (Choose all that apply.)

   A. TCP 22

   B. UDP 53

   C. TCP 443

   D. TCP 110

   E. UDP 4500

2. In a guest firewall policy, which IP networks should be restricted? (Choose all that apply.)

   A. 172.16.0.0/12

   B. 20.0.0.0/8

   C. 192.16.0.0/16

   D. 172.10.0.0/24

   E. 10.0.0.0/8

3. What are some of the components within an MDM architecture? (Choose all that apply.)

   A. AP

   B. RADIUS

   C. BYOD

   D. APNs

   E. GCM

4. What are some of the methods that can be used to provision a root certificate onto Wi-Fi clients that function as 802.1X/EAP supplicants? (Choose all that apply.)

   A. GPO

   B. RADIUS

   C. MDM

   D. APNs

   E. GCM

5. Which protocol is used by Passpoint-certified WLAN client devices to discover the cellular service providers supported by a Hotspot 2.0 WLAN?

   A. DNS

   B. SCEP

   C. OAuth

   D. ANQP

   E. IGMP

6. What type of information can be seen on a mobile device that is monitored by an MDM server? (Choose all that apply.)

   A. SMS messages

   B. Battery life

   C. Web browsing history

   D. Installed applications

   E. Device capacity

7. Which methods of EAP can be used for authentication for a Passpoint client to connect to a secure Passpoint SSID? (Choose all that apply.)

   A. EAP-PEAP

   B. EAP-TLS

   C. Anonymous EAP-TLS

   D. EAP-AKA

   E. EAP-LEAP

8. What are some of the methods that can be used by a captive web portal to redirect a user to the captive portal login page? (Choose all that apply.)

   A. HTTP redirection

   B. IP redirection

   C. UDP redirection

   D. TCP redirection

   E. DNS redirection

9. During the MDM enrollment process, what resources can a mobile client reach while quarantined inside a walled garden? (Choose all that apply.)

   A. SMTP server

   B. DHCP server

   C. DNS server

   D. MDM server

   E. Exchange server

10. Which protocol is used by iOS and macOS devices for over-the-air provisioning of MDM profiles using certificates and SSL encryption?

    A. OAuth

    B. GRE

    C. SCEP

    D. XML

    E. HTTPS

11. What mechanism can be used if the guest VLAN is not supported at the edge of the network and resides only in a DMZ?

    A. GRE

    B. VPN

    C. STP

    D. RTSP

    E. IGMP

12. Which type of guest management solution needs to integrate with LDAP?

    A. Social login

    B. Kiosk mode

    C. Receptionist registration

    D. Self-registration

    E. Employee sponsorship

13. An employee has enrolled a personal device with an MDM server over the corporate WLAN. The employee removes the MDM profile while at home. What will happen with the employee's personal device the next time the employee tries to connect to the company SSID?

    A. The MDM server will reprovision the MDM profile over the air.

    B. The push notification service will reprovision the MDM profile over the air.

    C. The device will be quarantined in the walled garden and will have to re-enroll.

    D. The device will be free to access all resources because the certificate is still on the mobile device.

14. Which phrase best describes a policy of permitting employees to connect personally owned mobile devices, such as smartphones, tablets, and laptops, to their workplace network?

    A. MDM

    B. NAC

    C. DMZ

    D. BYOD

15. Which method of guest management can a company use to gather valuable personal information about guest users?

    A. Social login

    B. Kiosk mode

    C. Receptionist registration

    D. Self-registration

    E. Employee sponsorship

16. What kind of remote actions can an MDM administrator send to the mobile device over the Internet?

   A. Configuration changes

   B. Restrictions changes

   C. Locking the device

   D. Wiping the device

   E. Application changes

   F. All of the above

17. What are some extra restrictions that can be placed on a guest user other than those defined by the guest firewall policy? (Choose all that apply.)

   A. Encryption

   B. Web content filtering

   C. DHCP snooping

   D. Rate limiting

   E. Client isolation

18. Within a WLAN infrastructure, where can the guest captive web portal operate? (Choose the best answer.)

   A. AP

   B. WLAN controller

   C. Third-party server

   D. Cloud-based service

   E. All of the above

19. When an MDM solution is deployed, after a mobile device connects to an access point, where does the mobile device remain until the MDM enrollment process is complete?

   A. DMZ

   B. Walled garden

   C. Quarantine VLAN

   D. IT sandbox

   E. None of the above

20. To calculate the capability Jeff should have on the network, which of the following can the NAC server use to initially identify and set his permission? (Choose all that apply.)

   A. Posture

   B. DHCP fingerprinting

   C. RADIUS attributes

   D. RADIUS CoA

   E. MDM profiles

Chapter 19

802.11ax: High Efficiency (HE)

IN THIS CHAPTER, YOU WILL LEARN ABOUT THE FOLLOWING:

✓ **HE overview**

✓ **Multi-user (MU)**

✓ **OFDMA**

- Subcarriers
- Resource units
- Trigger frames
- Downlink OFDMA
- Uplink OFDMA

✓ **MU-MIMO**

✓ **BSS color**

- Adaptive CCA
- Dual NAV timers

✓ **Target wake time**

✓ **Additional 802.11ax PHY and MAC capabilities**

- 1024-QAM
- Long symbol time and guard intervals
- 802.11ax PPDU formats
- 20 MHz-only
- Multi-TID AMPDU

✓ **802.11ax design considerations**

With each draft amendment, the IEEE proposes enhancements to 802.11 technologies. In 2009, the ratification of the 802.11n amendment was a momentous change in the way Wi-Fi radios functioned. 802.11n made the change from single-input, single-output (SISO) radios to multiple-input, multiple-output (MIMO) radios. Prior to 802.11n technology, the RF phenomena of multipath was destructive; however, multipath is actually useful for 802.11n/ac MIMO radios. When 802.11n technology entered the marketplace, WLAN engineers had to relearn how 802.11 technology worked and rethink WLAN design. In 2013, 802.11ac introduced further enhancements using MIMO radios. The end result of 802.11n/ac has been higher data rates but not necessarily better efficiency. While both 802.11n and 802.11ac did improve performance, the technologies were both more complex, and therefore new WLAN design and troubleshooting challenges arose.

The IEEE currently has a draft 802.11ax amendment that defines fundamental changes to how Wi-Fi radios communicate. The 802.11ax draft amendment references high efficiency (HE) technology that could be the most significant change to Wi-Fi since the debut of 802.11n MIMO radios in 2009. As of this writing the IEEE had released draft version 3.0 of the 802.11ax amendment, although the final ratification is not expected until late 2019. The Wi-Fi Alliance will also have a matching 802.11ax certification with a similar timeline. Although 802.11ax is not yet ratified, some major chipset manufacturers, such as Broadcom and Qualcomm, have already adopted the technology. Several major enterprise WLAN vendors are shipping access points with 802.11ax radios prior to ratification of the amendment. 802.11ax clients are also expected in the marketplace in late 2018. This chapter is a high-level overview of the efficiency enhancements of 802.11ax technology.

HE Overview

802.11ax is an IEEE draft amendment that defines modifications to the 802.11 Physical layer (PHY) and the Medium Access Control (MAC) sublayer for high efficiency (HE) operations in frequency bands between 1 GHz and 6 GHz. Much like *very high throughput (VHT)* is the technical term for 802.11ac, *high efficiency (HE)* is the technical term for 802.11ax. Historically, previous 802.11 amendments defined technologies that gave us higher data rates and wider channels but did not address efficiency. An often-used analogy is that faster cars and bigger highways have been built, but traffic jams still exist. Despite the higher data rates and 40/80/160 MHz channels used by 802.11n/ac radios, multiple factors contribute to the 802.11 traffic congestion, which do not provide for an efficient use of the medium.

You have previously learned that 802.11 data rates are not TCP throughput. Always remember that RF is a half-duplex medium and that the 802.11 medium contention protocol of CSMA/CA consumes much of the available bandwidth. In laboratory conditions, the TCP throughput in an 802.11n/ac environment can achieve 60 percent of the data rate between one AP and one client. The aggregate throughput numbers are considerably less in real-world environments with active participation of multiple WLAN clients communicating through an AP. As more clients contend for the medium, the medium contention overhead increases significantly and efficiency drops. Therefore, the aggregate WLAN throughput is usually at best 50 percent of the advertised 802.11 data rate.

What else contributes to 802.11 traffic congestion? Because legacy clients often still participate in enterprise WLANs, RTS/CTS protection mechanisms are needed, which contributes to the inefficiency. As shown in Figure 19.1, about 60 percent of all WLAN traffic is 802.11 control frames, and 15 percent is 802.11 management frames. Control and management frames consume 75 percent of the usable airtime, and only 25 percent of WLAN traffic is used for 802.11 data frames. Additionally, layer 2 retransmissions as a result of either RF interference or a poorly designed WLAN can also contribute to 802.11 inefficiency.

FIGURE 19.1   802.11 traffic

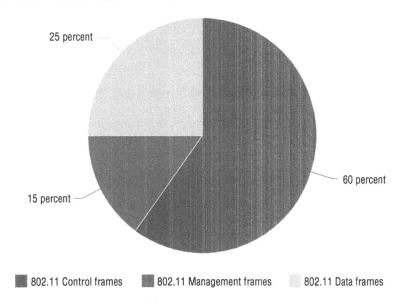

High data rates are useful for a large data payload; however, the bulk of 802.11 data frames (75–80 percent) are small and under 256 bytes, as shown in Figure 19.2. Each small frame requires a PHY header, a MAC header, and a trailer. The result is excessive PHY/MAC overhead as well as medium contention overhead for each small frame. Small frames can be aggregated to reduce the overhead; however, in most cases, the small frames are not aggregated, because they must be delivered sequentially due to the higher-layer application protocols. For example, VoIP packets cannot be aggregated, because they must arrive sequentially.

FIGURE 19.2   802.11 data frame size

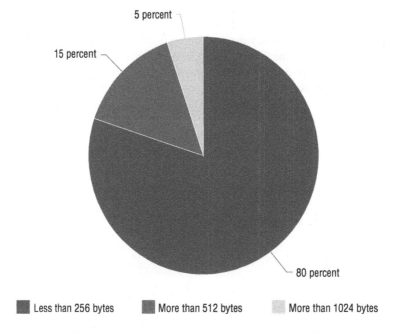

Despite the higher data rates and wide channels that can be used by 802.11n/ac radios, the result is 802.11 traffic congestion. We all know that automobile traffic congestion can result in drivers becoming frustrated and thereby engage in road rage. 802.11ax technology is all about better 802.11 *traffic management* and hopefully eliminating Wi-Fi radio rage. Higher data rates and wider channels are not the goals of 802.11ax. The goal is better and more efficient 802.11 traffic management. Most of the 802.11ax enhancements are at the Physical layer and involve a new multi-user version of OFDM technology, as opposed to the single-user OFDM technology already used by 802.11a/g/n/ac radios. Another significant 802.11ax change is that an 802.11ax access point can actually supervise both downlink and uplink transmissions to multiple client radios while the AP has control of the medium. 802.11ax radios will be backward compatible with 802.11a/b/g/n/ac radios. Table 19.1 shows a high-level comparison of 802.11n, 802.11ac, and 802.11ax capabilities. Please note that unlike 802.11ac radios, which can transmit only on the 5 GHz frequency band, 802.11ax radios can operate on both the 2.4 GHz and 5 GHz frequency bands.

TABLE 19.1   Comparison of 802.11n, 802.11ac, and 802.11ax

	802.11n	**802.11ac**	**802.11ax**
Frequency bands	2.4 GHz and 5 GHz	5 GHz only	2.4 GHz and 5 GHz
Channel size (MHz)	20, 40	20, 40, 80, 80 + 80, and 160	20, 40, 80, 80 + 80, and 160
Frequency multiplexing	ODFM	OFDM	OFDM and OFDMA
Subcarrier spacing (KHz)	312.5	312.5	78.125
OFDM symbol time (μs)	3.2	3.2	12.8
Guard interval (μs)	0.4 or 0.8	0.4 or 0.8	0.8, 1.6, or 3.2
Total symbol time (μs)	4.0	3.6 or 4.0	13.6, 14.4, or 16.0
Modulation	BPSK, QPSK, 16-QAM, 64-QAM	BPSK, QPSK, 16-QAM, 64-QAM	BPSK, QPSK, 16-QAM, 64-QAM, 1024-QAM
MU-MIMO	N/A	Downlink	Downlink and uplink

As you can see in Table 19.1, 802.11ax does support 40 MHz, 80 MHz, and 160 MHz channels. For the bulk of the discussion of 802.11ax operations, however, we will focus on 20 MHz channels. As a matter of fact, the key benefits of 802.11ax will be a result of the partitioning of a 20 MHz channel into smaller sub-channels using a multi-user version of OFDM called orthogonal frequency-division multiple access (OFDMA).

Multi-User

The term *multi-user (MU)* simply means that transmissions between an AP and multiple clients can occur at the same time, depending on the supported technology. However, the MU terminology can be *very* confusing when discussing 802.11ax. MU capabilities exist for both OFDMA and MU-MIMO. Please understand the key differences, as explained further in this chapter.

802.11ax defines the use of both multi-user technologies, OFDMA and MU-MIMO. But please do not *confuse* OFDMA with MU-MIMO. OFDMA allows for multiple-user access by subdividing a channel. MU-MIMO allows for multiple-user access by using different spatial streams. If we reference the car and road analogy discussed earlier, OFDMA uses a single road subdivided into multiple lanes for use by different cars at the same time, whereas MU-MIMO uses different single lane roads to arrive at the same destination.

When discussing 802.11ax, there will most likely be a lot of confusion because many people may already be somewhat familiar with MU-MIMO technology introduced with 802.11ac. What most people are not familiar with is the multi-user technology of OFDMA. You will learn that most of the efficiency benefits of 802.11ax are a result of multi-user OFDMA. The 802.11ax amendment does allow for the combined use of multi-user OFDMA and MU-MIMO at the same time, but this is not expected to be widely implemented.

OFDMA

Orthogonal frequency-division multiple access (OFDMA) is a multi-user version of the OFDM digital-modulation technology. 802.11a/g/n/ac radios currently use OFDM for single-user transmissions on an 802.11 frequency. OFDMA subdivides a channel into smaller frequency allocations, called *resource units (RUs)*. By subdividing the channel, parallel transmissions of small frames to multiple users can happen simultaneously. Think of OFDMA as a technology that partitions a channel into smaller *sub-channels* so that simultaneous multiple-user transmissions can occur. For example, a traditional 20 MHz channel might be partitioned into as many as nine smaller sub-channels. Using OFDMA, an 802.11ax AP could simultaneously transmit small frames to nine 802.11ax clients. OFDMA is a much more efficient use of the medium for smaller frames. The simultaneous transmission cuts down on excessive overhead at the MAC sublayer as well as medium contention overhead. The goal of OFDMA is better use of the available frequency space. OFDMA technology has been time-tested with other RF communications. For example, OFDMA is used for downlink LTE cellular communications.

For backward compatibility, 802.11ax radios will still support OFDM. Keep in mind that 802.11 management and control frames will still be transmitted at a basic data rate using OFDM technology that 802.11a/g/n/ac radios can understand. Therefore, the transmission of management and control frames will be transmitted across all the subcarriers of an entire primary 20 MHz channel. OFDMA is only for 802.11 data frame exchanges between 802.11ax APs and 802.11ax clients.

Subcarriers

OFDM divides a channel into subcarriers through a mathematical function known as an *inverse fast Fourier transform (IFFT)*. The spacing of the subcarriers is orthogonal, so they will not interfere with one another despite the lack of guard bands between them. This creates signal nulls in the adjacent subcarrier frequencies, thus preventing *inter-carrier interference (ICI)*.

So what are some of the key differences between OFDM and OFDMA? As shown in Figure 19.3, a 20 MHz 802.11n/ac channel consists of 64 subcarriers. Fifty-two of the subcarriers are used to carry modulated data; four of the subcarriers function as pilot carriers; and eight of the subcarriers serve as guard bands. OFDM subcarriers are sometimes also referred to as OFDM *tones*. In this chapter, we will use both terms interchangeably. Each OFDM subcarrier is 312.5 KHz.

FIGURE 19.3   802.11n/ac 20 MHz channel—OFDM subcarriers

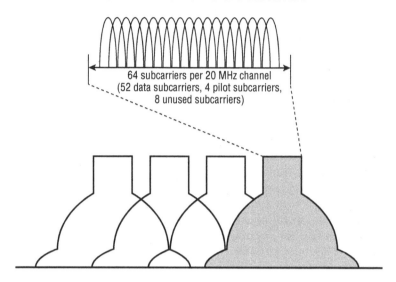

You will learn later in this chapter that 802.11ax introduces a longer OFDM symbol time of 12.8 μs, which is four times longer than the legacy symbol time of 3.2 μs. As a result of the longer symbol time, the subcarrier size and spacing decreases from 312.5 KHz to 78.125 KHz, as shown in Figure 19.4. The narrow subcarrier spacing allows better equalization and therefore enhanced channel robustness. Because of the 78.125 KHz spacing, an OFDMA 20 MHz channel consists of a total of 256 subcarriers (tones).

FIGURE 19.4   Subcarrier spacing

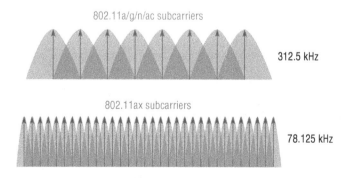

Just like with OFDM, there are three types of subcarriers for OFDMA, as follows:

Data Subcarriers   These subcarriers will use the same modulation and coding schemes (MCSs) as 802.11ac and two new MCSs with the addition of 1024-QAM.

Pilot Subcarriers   The pilot subcarriers do not carry modulated data; however, they are used for synchronization purposes between the receiver and transmitter.

Unused Subcarriers   The remaining unused subcarriers are mainly used as guard carriers or null subcarriers against interference from adjacent channels or sub-channels.

The data and pilot subcarriers within each resource unit are both adjacent and contiguous with the OFDMA channel. These tones are grouped into smaller sub-channels, known as resource units (RUs).

Resource Units

To further illustrate the difference between OFDM and OFDMA, please reference both Figures 19.5 and 19.7. When an 802.11n/ac AP transmits downlink to 802.11n/ac clients on an OFDM channel, the entire frequency space of the channel is used for each independent downlink transmission. In the example shown in Figure 19.5, the AP transmits to six clients independently over time. All 64 subcarriers are used when an OFDM radio transmits on a 20 MHz channel. In other words, the entire 20 MHz channel is needed for the downlink communication between the AP and a single OFDM client. The same holds true for any uplink transmission from a single 802.11n/ac client to the 802.11n/ac AP. The entire 20 MHz OFDM channel is needed for the client transmission to the AP.

FIGURE 19.5   OFDM transmissions over time

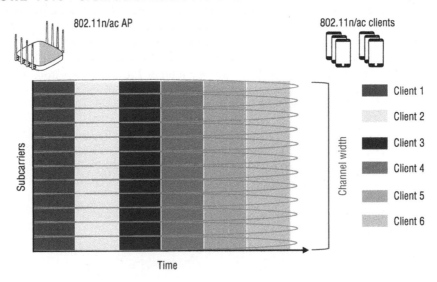

As previously stated, an OFDMA channel consists of a total of 256 subcarriers (tones). These tones are grouped into smaller sub-channels, known as *resource units (RUs)*. As shown in Figure 19.6, when subdividing a 20 MHz channel, an 802.11ax access point designates 26, 52, 106, and 242 subcarrier resource units (RUs), which equates roughly to 2 MHz, 4 MHz, 8 MHz, and 20 MHz channels, respectively. The 802.11ax AP dictates how many RUs are used within a 20 MHz channel, and different combinations can be used. The AP may allocate the whole channel to only one client at a time, or it may partition the channel to serve multiple clients simultaneously. For example, an 802.11ax AP could simultaneously communicate with one 802.11ax client using 8 MHz of frequency space while communicating with three other 802.11ax clients using 4 MHz sub-channels. These simultaneous communications can be either downlink or uplink.

FIGURE 19.6   OFDMA resource units

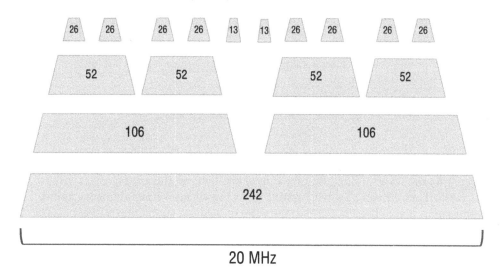

In the example shown in Figure 19.7, the 802.11ax AP first simultaneously transmits downlink to 802.11ax clients 1 and 2. The 20 MHz OFDMA channel is effectively partitioned into two sub-channels. Remember that an ODFMA 20 MHz channel has a total of 256 subcarriers; however, the AP simultaneously transmitted to clients 1 and 2 using two different 106-tone resource units. In the second transmission, the AP simultaneously transmits downlink to clients 3, 4, 5, and 6. In this case, the ODFMA channel had to be partitioned into four separate 52-tone sub-channels. In the third transmission, the AP uses a single 242-tone resource unit to transmit downlink to a single client (5). Using a single 242-tone resource unit is effectively using the entire 20 MHz channel. In the fourth transmission, the AP simultaneously transmits downlink to clients 4 and 6 using two 106-tone resource units. In the fifth transmission, the AP once again transmits only downlink to a single client, with a single 242-tone RU utilizing the entire 20 MHZ channel. In the sixth transmission, the AP simultaneously transmits downlink to clients 3, 4, and 6. In this instance, the 20 MHz channel is partitioned into three sub-channels; two 52-tone RUs are used for clients 3 and 4, and a 106-tone RU is used for client 6.

FIGURE 19.7   OFDMA transmissions over time

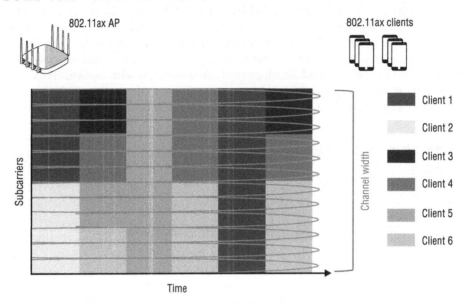

As shown in Figure 19.7, the 802.11ax AP partitions the 20 MHz OFDMA channel on a continuous basis for downlink transmissions. Later in this chapter, you will learn that an 802.11ax AP can also synchronize 802.11ax clients for simultaneous uplink transmissions. It should be noted that the rules of medium contention still apply. The AP still has to compete against legacy 802.11 stations for a *transmission opportunity (TXOP)*. Once the AP has a TXOP, the AP is then in control of up to nine 802.11ax client stations for either downlink or uplink transmissions. The number of RUs used can vary on a per TXOP basis.

OFDMA combines different user data within the 20 MHz channel. The AP assigns RUs to associated clients on a per-TXOP basis to maximize the download and upload efficiency. Transmit power can be adjusted for the resource units for both downlink and uplink to improve the *signal-to-interference-plus-noise ratio (SINR)*.

Can resource units be used for 40 MHz or 80 MHz channels? The answer is yes; 40 MHz, 80 MHz, and even 160 MHz channels can also be partitioned into various combinations of RUs, as shown in Table 19.2. If an 80 MHz channel were subdivided using strictly 26-tone RUs, theoretically 37 802.11ax clients could communicate simultaneously using their OFDMA capabilities. However, we anticipate that most real-world 802.11ax deployments will use 20 MHz channels, with a maximum of 9 clients participating in multi-user OFDMA transmissions per TXOP. Remember, the whole point of OFDMA is to make use of smaller sub-channels.

TABLE 19.2   Resource units and wide channels

Resource Unit (RU)	20 MHz Channel	40 MHz Channel	80 MHz Channel	160 MHz Channel	80 + 80 MHz Channel
996 (2x) tones	n/a	n/a	n/a	1 client	1 client
996 tones	n/a	n/a	1 client	2 clients	2 clients
484 tones	n/a	1 client	2 clients	4 clients	4 clients
242 tones	1 client	2 clients	4 clients	8 clients	8 clients
106 tones	2 clients	4 clients	8 clients	16 clients	16 clients
52 tones	4 clients	8 clients	16 clients	32 clients	32 clients
26 tones	9 clients	18 clients	37 clients	74 clients	74 clients

Trigger Frames

When referencing downlink and uplink OFDMA transmissions, the acronyms of DL-OFDMA and UL-OFDMA are often used. In the sections to follow, you will learn that a series of frame exchanges are used for both DL-OFDMA and UL-OFDMA. In both cases, *trigger frames* are needed to bring about the necessary frame exchanges for multi-user communications. For example, a trigger frame is used to allocate resource units (RUs) to 802.11ax clients. Multiple types of 802.11 control frames can function as trigger frames, as shown in Table 19.3.

TABLE 19.3   Trigger frames

Trigger Type Subfield Value	Trigger Frame Variant
0	Basic
1	Beamforming report poll (BRP)
2	MU-BAR
3	MU-RTS
4	Buffer status report poll (BSRP)

TABLE 19.3   Trigger frames *(continued)*

Trigger Type Subfield Value	Trigger Frame Variant
5	GCR MU-BAR
6	Bandwidth query report poll (BQRP)
7	NDP feedback report poll (NFRP)
8–15	Reserved

As previously mentioned, trigger frames contain information about RU allocation. RU allocation information is communicated to clients at both the PHY and MAC layers. RU allocation information can be found in the *HE-SIG-B* field of the PHY header of an 802.11 trigger frame. A more detailed discussion about the PHY header will follow later in this chapter. Additionally, RU allocation information is delivered in the *user information* field in the body of a trigger frame. Figure 19.8 displays a table of how RU allocation information is communicated at the MAC layer. The table highlights all the possible RUs within a 20 MHz channel and the subcarrier range for each RU. Each specific RU is defined by a unique combination of 7 bits within the user information field of the trigger frame, known as the RU allocation bits. In the example in Figure 19.9, the trigger frame allocates specific RUs to three client stations for simultaneous uplink transmission within a 20 MHZ OFDMA channel. Clients STA-1 and STA-2 are each assigned to a 52-tone RU, whereas client STA-3 is assigned to a 106-tone RU.

For UL-OFDMA, the trigger frame sent by the AP is also used to tell the clients how many spatial streams and which modulation and coding scheme (MCS) to use when transmitting uplink on their assigned RUs. This information can be found in the SS Allocation and UL MCS subfields of the user information field within the body of a trigger frame.

Trigger frames can also be used by an AP to tell clients to adjust their power settings for synchronized uplink transmissions. Within a trigger frame, the UL Target RSSI subfield indicates, in a dBm value, the expected receive power at the AP across all antennas for the assigned resource unit (RU) transmissions from the uplink 802.11ax clients. The UL Target RSSI subfield uses values of 0 to 90 which are directly mapped to –110 dBm to –20 dBm. A value of 127 indicates to the client station to transmit at its maximum power for the assigned MCS. Based on this information provided by the trigger frame, the transmit power could be adjusted by the uplink clients. Please note that a client station might be unable to satisfy the target RSSI due to its hardware or regulatory limitations.

FIGURE 19.8 RU index and subcarrier range for 20 MHz channel

26 tone RU	RU-1	RU-2	RU-3	RU-4	RU-5	RU-6	RU-7	RU-8	RU-9
Subcarrier range	-121:-96	-95:-70	-68:-43	-42:-17	-16:-4, 4:16	17:42	43:68	70:95	96:121
RU allocation bits	0000000	0000001	0000010	0000011	0000100	0000101	0000110	0000111	0001000

52 tone RU	RU-1		RU-2			RU-3		RU-4	
Subcarrier range	-121:-70		-68:-17			17:68		70:121	
RU allocation bits	0100101		0100110			0100111		0101000	

106 tone RU	RU-1					RU-2			
Subcarrier range	-122:-17					17:122			
RU allocation bits	0110101					0110110			

242 tone RU	RU-1								
Subcarrier range	-122:-2, 2:122								
RU allocation bits	0111101								

FIGURE 19.9 Trigger frame RU allocation

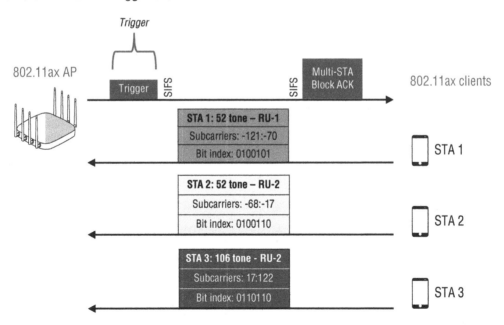

Trigger

802.11ax AP

Trigger — SIFS — Multi-STA Block ACK — SIFS

802.11ax clients

STA 1: 52 tone – RU-1
Subcarriers: -121:-70
Bit index: 0100101
STA 1

STA 2: 52 tone – RU-2
Subcarriers: -68:-17
Bit index: 0100110
STA 2

STA 3: 106 tone - RU-2
Subcarriers: 17:122
Bit index: 0110110
STA 3

Downlink OFDMA

Let us first take a look at how multi-user DL-OFDMA communications can occur between an 802.11ax AP and 802.11ax clients. Please remember that OFDMA is only for 802.11 data frame exchanges between 802.11ax APs and 802.11ax clients. An 802.11ax AP will first need to contend for the medium and win a TXOP for the entire DL-OFDMA frame exchange. As shown in Figure 19.10, once an 802.11ax AP has won a TXOP, the AP will first send a multi-user request-to-send (MU-RTS) frame. The MU-RTS frame is transmitted using OFDM (not OFDMA) across the entire 20 MHz channel so that legacy clients can also understand the MU-RTS. The duration value of the MU-RTS frame is needed to reserve the medium and reset the NAV timers of all legacy clients for the remainder of the DL-OFDMA frame exchange. The legacy clients must remain idle while the multi-user OFDMA data frames are transmitted between the 802.11ax AP and 802.11ax clients. The MU-RTS frame is also an extended trigger frame from the AP used to synchronize uplink clear-to-send (CTS) client responses for 802.11ax clients. The AP uses the MU-RTS as a trigger frame to allocate resource units (RUs). The 802.11ax clients will send CTS responses in parallel using their assigned RUs.

FIGURE 19.10 Downlink OFDMA

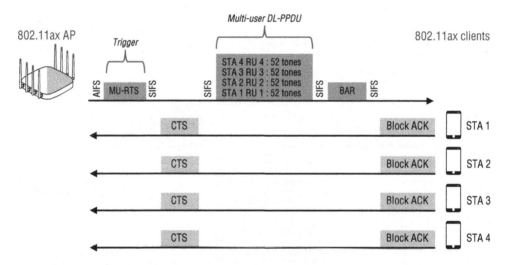

After the parallel CTS response from the clients, the AP will begin multi-user DL-PPDU transmissions from the AP to the OFDMA-capable clients. Keep in mind that the AP determined how to partition the 20 MHz channel into multiple RUs. Once the 802.11ax clients receive their data via their assigned RUs, they will need to send a Block ACK to the AP. The AP will send a Block ACK request (BAR) frame followed by the clients replying with Block ACKs in parallel. Optionally, an automatic Block ACK can be sent by the clients in parallel.

Once the frame exchange is over, the AP or clients that win the next TXOP will then be able to transmit on the medium. For example, if an 802.11n/ac client wins the next TXOP, the 802.11n/ac will send a data frame uplink to the AP using OFDM. But how does uplink transmission work for OFDMA clients? In the next section, you will learn that the AP must once again win a TXOP to coordinate synchronized UL-OFDMA communication.

Uplink OFDMA

In the original 802.11 standard, the IEEE proposed an operational mode called *Point Coordination Function (PCF)*, which defined operations where the AP could control the medium for uplink client transmission. With PCF mode, the AP could poll clients for uplink transmissions during a contention-free period of time when the AP controlled the medium. However, PCF never caught on and was never implemented in the real world. 802.11ax now introduces mechanisms where the AP can once again control the medium for uplink transmissions using UL-OFDMA. You should understand that UL-OFDMA has nothing to do with PCF; the methods are very different. You should also understand that the 802.11ax AP must first contend for the medium and win a TXOP. Once the 802.11ax AP wins a TXOP, it can then coordinate uplink transmissions from 802.11ax clients that support UL-OFDMA.

UL-OFDMA is more complex than DL-OFDMA and may require the use of as many as three trigger frames. Each trigger frame is used to solicit a specific type of response from the clients. UL-OFDMA also requires the use of *buffer status report (BSR)* frames from the clients. Clients use BSR frames to inform the AP about the client's buffered data and about the QoS category of data. The information contained in BSR frames assists the AP in allocating RUs for synchronized uplink transmissions. The AP will use the information gathered from the clients to build uplink window times, client RU allocation, and client power settings for each RU. Buffer status reports (BSRs) can be unsolicited or solicited. If solicited, the AP will poll the clients for BSRs.

Let us now take a look at how multi-user UL-OFDMA communications can occur between an 802.11ax AP and 802.11ax clients. An 802.11ax AP will first need to contend for the medium and win a TXOP for the entire UL-OFDMA frame exchange. As shown in Figure 19.11, once an 802.11ax AP has won a TXOP, the AP will first send a buffer status report poll (BSRP) frame to solicit information from the 802.11ax clients about their need to send uplink data. The clients will then respond with buffer status reports (BSRs). The AP will use this information to decide how to best allocate RUs to the clients for uplink transmissions.

FIGURE 19.11   Uplink OFDMA

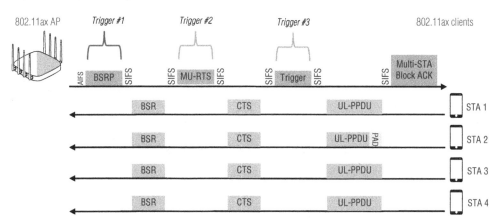

The AP will first send a multi-user request-to-send (MU-RTS) frame, which functions as a second type of trigger frame. The MU-RTS frame is transmitted using OFDM (not OFDMA) across the entire 20 MHz channel so that legacy clients can also understand the MU-RTS. The duration value of the MU-RTS frame is needed to reserve the medium and reset the NAV timers of all legacy clients for the remainder of the UL-OFDMA frame exchange. The AP uses the MU-RTS as a trigger frame to allocate resource units (RUs). The 802.11ax clients will send CTS responses in parallel using their assigned RUs.

A third *basic trigger frame* is needed to tell the 802.11ax clients to begin uplink transmission of their data with their assigned RUs. The basic trigger frame also dictates the length of the uplink window. The uplink client devices must all start and stop at the same time. The basic trigger frame also contains power control information so that individual clients can increase or decrease their transmit power. This will help equalize the received power to the AP from all uplink clients and improve reception. Once the uplink data is received from the clients, the AP will send a single *multi-user Block ACK* to the clients. The AP also has the option of sending separate Block ACKs to each individual client.

To summarize the UL-OFDMA frame exchange just discussed, the following three trigger frames were used:

- **Trigger 1:** BSRP to solicit buffer status reports from the clients

- **Trigger 2:** MU-RTS to allocate RUs and set every client's NAV

- **Trigger 3:** Basic trigger to signal the clients to begin their parallel uplink transmissions

In addition to the scheduled UL-OFDMA, 802.11ax also provides for an optional *UL-OFDMA random access (UORA)* method. A random-access method is favorable in conditions where the AP is unaware of traffic buffered on the clients. The AP sends a random-access trigger frame to allocate RUs for random access. Clients that want to transmit will use an *OFDMA back-off (OBO)* procedure. Initially, a client chooses a random value, with each trigger frame the client decrements the value by the number of RUs specified in the trigger frame until it reaches zero. The client will then randomly select an RU and then transmit.

Can an 802.11ax client station suspend participation for synchronized uplink OFDMA and contend for the medium for an independent uplink transmission? The 802.11ax draft proposes an *operating mode indication (OMI)* procedure for this purpose. A client station can signal to the AP the maximum number of spatial streams and the maximum channel bandwidth that the client can support for either uplink or downlink transmission. Additionally, a client can switch between single-user or multi-user UL-OFDMA operations. An 802.11ax client uses the OM Control subfield in 802.11 data and management frames to indicate a change of either transmission or receiver mode of operation. For example, a client can switch between single-user or multi-user UL-OFDMA operations. A client can suspend and resume responses to the trigger frames sent by an AP during the UL-OFDMA process.

MU-MIMO

As you learned in Chapter 10, "MIMO Technology: HT and VHT" downlink MU-MIMO capabilities were introduced in the second generation of 802.11ac access points; however, widespread use of MU-MIMO technology is rare. Although MU-MIMO sounds great on paper, real-world implementation is not practical for the following reasons:

- Very few MU-MIMO-capable 802.11ac clients exist in the current marketplace, and the technology is rarely used in the enterprise.

- MU-MIMO requires spatial diversity; therefore, physical distance between the clients is necessary. Most modern-day enterprise deployments of Wi-Fi involve a high density of users, which is not conducive for MU-MIMO conditions.

- Because MU-MIMO requires spatial diversity, a sizable distance between the clients and the AP is necessary. Most modern-day enterprise deployments of Wi-Fi involve a high density of users, which is not conducive for MU-MIMO conditions.

- MU-MIMO requires transmit beamforming (TxBF), which requires sounding frames. The sounding frames add excessive overhead, especially when the bulk of data frames are small. The overhead from the sounding frames usually negates any performance gained from an 802.11ac AP transmitting downlink simultaneously to multiple 802.11ac clients. To address this issue, there are some MU-MIMO enhancements proposed in the 802.11ax draft amendment including grouping sounding frames, data frames, and other frames among multiple users to reduce overhead.

Although MU-MIMO has received fanfare for its technology and capabilities, WLANs rarely benefit from downlink MU-MIMO. That aside, the 802.11ax draft amendment proposes the possible use of uplink MU-MIMO. Trigger frames are used to signal 802.11ax clients to participate in uplink MU-MIMO communications. It should be noted that if MU-MIMO is used for either downlink or uplink, the minimum RU size is 106 subcarriers or greater.

In theory, MU-MIMO would be a favorable option in very low client density but high-bandwidth environments. In pristine conditions, MU-MIMO can improve efficiency when large packets and high-bandwidth applications are used. Table 19.4 shows a quick comparison of potential benefits from MU-MIMO when compared to multi-user OFDMA.

TABLE 19.4   Multi-user OFDMA vs. multi-user MIMO comparison

MU-OFDMA	MU-MIMO
Increased efficiency	Increased capacity
Reduced latency	Higher data rates per user
Best for low-bandwidth applications	Best for high-bandwidth applications
Best with small packets	Best with large packets

As stated earlier, the 802.11ax amendment does allow for the combined use of MU-OFDMA and MU-MIMO at the same time, but this is not expected to be widely implemented. The 802.11ax draft amendment currently specifies that uplink MU-MIMO is optional. Support for uplink MU-MIMO in 802.11ax clients will not be in the first generation of 802.11ax client chipsets.

One key difference between 802.11ac MU-MIMO and 802.11ax MU-MIMO is how many MU-MIMO clients communicate to an AP at the same time. 802.11ac is limited to a MU-MIMO group of only four clients, whereas eight clients can participate in an 802.11ax MU-MIMO group. Because of this new capability, expect some WLAN vendors to manufacture APs with 8x8:8 radios.

BSS Color

Carrier Sense Multiple Access with Collision Avoidance (CSMA/CA) dictates half-duplex communications, and only one radio can transmit on the same channel at any given time. An 802.11 radio will defer transmissions if it hears the PHY preamble transmissions of any other 802.11 radio at an SNR of just 4 dB greater. Unnecessary medium contention overhead that occurs when too many APs and clients hear each other on the same channel is called an *overlapping basic service set (OBSS)*. The more commonly used terminology for OBSS is *co-channel interference (CCI)*. For this discussion, we will use the term overlapping basic service set (OBSS). If AP-1 on channel 36 hears the preamble transmission of a nearby AP-2 also transmitting on channel 36, the original AP-1 will defer and cannot transmit at the same time. Likewise all the associated clients of AP-1 cannot transmit at the same time as AP-2; if any nearby clients on channel 36 hear the preamble transmission of the nearby AP-2, the clients must also defer. All of the deferral creates medium contention overhead and consumes valuable airtime because you have two basic service sets on the same channel that can hear each other—thus, the term OBSS. What most people do not understand is that clients are the number one cause of OBSS interference. As shown in Figure 19.12, if client-A, which is associated to AP-1, is transmitting on channel 36, then it is possible that AP-2 and any of AP-2's clients might hear the PHY preamble of client-A, and then all radios must defer. You should understand that OBSS interference is not static and is always changing due to the mobility of client devices.

FIGURE 19.12 OBSS interference caused by a client

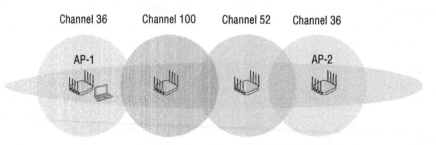

Channel 36   Channel 100   Channel 52   Channel 36

AP-1   AP-2

Adaptive CCA

In Chapter 13, "WLAN Design Concepts," you learned that the main reason for channel reuse patterns is to minimize airtime consumption and the degradation of performance due to OBSS. The 802.11ax amendment defines a method that has the potential to possibly increase the channel reuse by a factor of eight. *BSS color*, also known as *BSS coloring*, is a method for addressing medium contention overhead due to an overlapping basic service set (OBSS). BSS color was first defined in the 802.11ah-2016 amendment and is now also defined in the 802.11ax draft amendment. BSS color is a numerical identifier of the BSS. 802.11ax radios are able to differentiate between BSSs using BSS color identifier when other radios transmit on the same channel.

BSS color information is communicated at both the PHY layer and the MAC sublayer. In the preamble of an 802.11ax PHY header, the SIG-A field contains a 6-bit BSS color field. This field can identify as many as 63 BSSs. At the MAC sublayer, BSS color information is seen in 802.11 management frames. The HE operation information element contains a subfield for BSS color information. Six bits can be used to identify as many as 63 different colors (numerical values) and represent 63 different BSSs. The goal is for 802.11ax radios to differentiate between BSSs using a BSS color identifier when other radios transmit on the same channel.

Channel access is dependent on the color detected. If the color is the same, this is considered to be an *intra-BSS* frame transmission. In other words, the transmitting radio belongs to the same BSS as the receiver; therefore, the listening radio will defer. However, if the detected frame has a different BSS color from its own, then the STA considers that frame as an *inter-BSS* frame from an overlapping BSS. Using a procedure called *spatial reuse operation*, 802.11ax radios will be able to apply adaptive clear channel assessment (CCA) thresholds for detected OBSS frame transmissions. The goal of BSS color and spatial reuse is to ignore transmissions from an OBSS and therefore be able to transmit at the same time. In the example shown in Figure 19.13, any radio on channel 36 that detected the black color would defer because that would be considered an intra-BSS. However, deferral may not be necessary if an AP detected a green or purple color from nearby OBSS transmissions also on channel 36.

Keep in mind that this figure is a visual illustration and that the color information is actually a numerical value. An 802.11ax AP has the ability to change its BSS color if it detects a BSS color collision. An AP might decide to change its BSS color if it hears a frame from an OBSS AP or OBSS client with the same color. Associated clients might also autonomously report a BSS collision to their AP if the clients detect frames from an OBSS station of the same color.

In Chapter 8, "802.11 Medium Access" you learned that 802.11 radios use a *clear channel assessment (CCA)* to appraise the RF medium. If the RF medium is busy, an 802.11 radio will not transmit and instead defer for a period of time, called a *slot time*. The CCA involves listening for RF transmissions at the Physical layer. 802.11 radios use two separate CCA thresholds when listening to the RF medium. As shown in Figure 19.14, the *signal detect (SD)* threshold is used to identify the incoming 802.11 preamble transmission from another transmitting 802.11 radio. The preamble is a component of the Physical layer header of 802.11 frame transmissions. The signal detect (SD) threshold is statistically around 4 dB signal-to-noise ratio (SNR) for most 802.11 radios to detect and decode an 802.11 preamble. In other words, an 802.11 radio can usually decode any incoming 802.11 preamble transmissions at a received signal at about 4 dB above the noise floor. The *energy detect (ED)* threshold is used to detect any other type of RF transmissions during the clear channel assessment (CCA). As shown in Figure 19.14, the ED threshold is 20 dB higher than the signal detect threshold. Think of the signal detect threshold as a method of detecting and deferring for 802.11 radio transmissions. Think of the energy detect threshold as a method of detecting and deferring for any signals from non-802.11 transmitters. Both thresholds are used together during the CCA to determine whether the medium is busy and therefore must defer transmissions.

FIGURE 19.13 BSS color

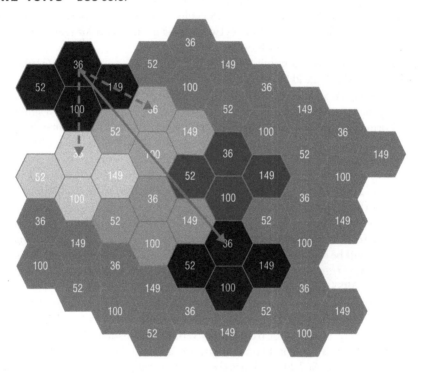

FIGURE 19.14 Clear channel assessment—signal detect and energy detect thresholds

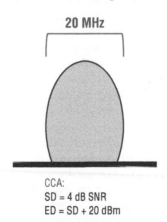

OBSS interference is a result of radios deferring based on the signal detect (SD) threshold being so low. Statistically, most radios can decode an 802.11 preamble if the received signal is only 4 DB above the noise floor. As a result of this very low signal detect threshold, APs and clients on the same channel hear each other and will defer, despite being separated by significant physical distance.

Based on the detected color bits in the PHY header, 802.11ax radios could implement an adaptive CCA implementation that could raise the signal detect threshold for inter-BSS frames while maintaining a lower threshold for intra-BSS traffic. If the signal detect threshold is raised higher for incoming OBSS frames, a radio might not need to defer, despite being on the same channel. An adaptive signal detect threshold as a result of spatial reuse operation potentially decreases the channel contention problem, which is a symptom of the existing very low signal detect (SD) threshold, which statistically is only about 4 dB above the noise floor. The adaptive signal detect threshold could be adjusted on a per-color and per-frame basis for inter-BSS traffic. In the example in Figure 19.15, an SD threshold of −96 dBm might be used for the reception of intra-BSS traffic, while an adaptive SD threshold of −96 dBm to −83 dBm might be used for inter-BSS traffic.

FIGURE 19.15 Spatial reuse operation – Adaptive CCA

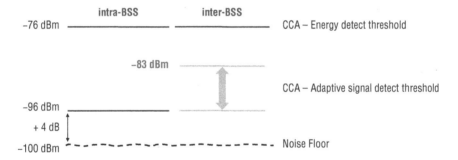

Dual NAV timers

In Chapter 8, "802.11 Medium Access," you learned that the CCA is a physical carrier sense mechanism which works together with MAC layer virtual carrier sense. Virtual carrier sense uses a timer mechanism known as the *network allocation vector (NAV)*. The adaptive CCA capabilities of spatial reuse operation will also work together with virtual carrier sense, with a new requirement for two NAV timers. The 802.11ax draft amendment defines the use of two *network allocation vector (NAV)* timers for 802.11ax client stations: an *intra-BSS NAV timer* and a *basic NAV timer*. The intra-BSS NAV timer of an 802.11ax station can only be reset by the Duration/ID value from frame transmissions of stations that belong to the same BSS. The basic NAV timer of an 802.11ax station is reset by the Duration/ID value from frame transmissions of stations that belong to a different BSS (inter-BSS). If either NAV has a non-zero value, the medium is considered to be busy.

Maintaining two NAVs is beneficial in dense deployment scenarios in which an 802.11ax station requires protection from frames transmitted by other stations within its BSS (intra-BSS), and to avoid interference from frames transmitted by stations in a neighboring BSS, (inter-BSS). As shown in Figure 19.16, AP-1 sends an RTS frame with a duration value of 200 μs to protect the remaining frame exchange with client STA #1. Because STA #1 is associated to AP-1, the client resets its intra-BSS NAV for 200 μs. However, during the frame exchange, the initial client that belongs to BSS #1 may also hear an RTS frame from a nearby client station that belongs to a different BSS. The duration value of 125 μs in the RTS frame sent by client STA #2 will reset the basic NAV timer of STA #1. The intra-BSS NAV timer will expire first; however, the basic NAV timer of STA #1 will continue to decrement, and STA #1 cannot transmit until both timers have expired. This ensures that the frame exchange in nearby BSS #2 is protected. Also please remember that any static or adaptive CCA of any client station will also have to be clear. When and how these adaptive CCAs and dual NAV timers are triggered remains to be seen.

FIGURE 19.16   Dual NAV timers

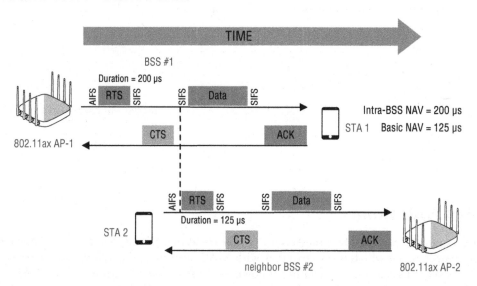

Please understand that BSS color is based on the detection of a color in the PHY header of an 802.11ax frame transmission. This means that any legacy 802.11/a/b/g/n clients will not be able to interpret the color, because they use a different PHY header format. Legacy clients will not be able to differentiate between intra-BSS and inter-BSS.

In theory, BSS color could provide the capability to take advantage of 80 MHz channels in the enterprise. Once again, however, this would be practical only if legacy clients were not in the same physical location. It remains to be seen how effective BSS color will be in real-world enterprise deployments.

Target Wake Time

Target wake time (TWT) is a power-saving mechanism originally defined in the 802.11ah-2016 amendment. A TWT is a negotiated agreement, based on expected traffic activity between clients and the AP, to specify a scheduled target wake-up time for the clients in power-save (PS) mode. The negotiated TWTs allow an AP to manage client activity by scheduling client stations to operate at different times in order to minimize contention between the clients. A TWT reduces the required amount of time that a client station in power-save (PS) mode needs to be awake. This allows the client to "sleep" longer and reduce energy consumption. As opposed to legacy client power-saving mechanisms such as delivery traffic indication map (DTIM), which require sleeping client devices to wake up in microsecond intervals, target wake time (TWT) could theoretically allow client devices to sleep for hours.. TWT is an ideal power-saving method for IoT devices that need to conserve battery life.

TWT setup frames are used between the client and the AP to negotiate a scheduled target wake time. For each client, there can be as many as eight separate negotiated scheduled wake-up agreements for different types of application traffic. 802.11ax has also extended TWT functionality to include a non-negotiated TWT capability. An AP can create wake-up schedules and deliver TWT values to the 802.11ax clients via a *broadcast TWT* procedure.

As previously mentioned, TWT was original defined in the 802.11ah amendment, which defined the use of Wi-Fi in frequencies below 1 GHz. A likely use for 802.11ah is sensor networks, along with backhaul for sensor networks, and extended range Wi-Fi. The extended range and the TWT power-saving capabilities are ideal for *Internet of Things (IoT)* devices. Despite the goal of moving IoT devices to a lower frequency band, most IoT devices with a Wi-Fi radio still currently transmit in the 2.4 GHz frequency band. Because 802.11ax now also defines TWT, these same extended power-saving capabilities could be available to IoT devices with 802.11ax radios that transmit in the 2.4 GHz band.

It remains to be seen if IoT device companies will manufacture IoT devices with 802.11ax radios; however, the potential to conserve battery life is attractive if the devices are TWT-capable. Additionally, less airtime is consumed if 802.11ax IoT devices can sleep for extended periods of time. As previously mentioned, an 802.11ax sensor device could sleep for hours while still remaining associated to the AP. Association tables on an AP could therefore be quite large if numerous IoT devices are in the vicinity on an AP. Any IoT device that sleeps for hours will most likely need a static IP address because their sleep interval might actually exceed their DHCP lease interval. Static IP addresses would be needed to avoid the DHCP lease renewal exchange when the IoT devices wake up.

In addition to TWT, the 802.11ax draft amendment proposes multiple power saving mechanisms, some of which can be quite complex. *Intra-PPDU power save* is a power save mechanism for any HE client STA to enter a doze state until the end of a received PPDU which is identified as an Intra-BSS frame. For this to happen, the BSS color of the transmitted PPDU must be the same color of the BSS to which the STA is associated. The destination MAC address of the PPDU must also not be the STA. In other words, a client could doze while intra-BSS transmissions are occurring between the AP and other associated clients that belong to the BSS.

Earlier in this chapter, you learned about an optional UL-OFDMA random access (UORA) method that 802.11ax clients can use for uplink OFDMA communications. The 802.11ax draft amendment also proposes a *power save with UORA* for client stations that need power-save capabilities and while using the UORA procedure described earlier. Effectively this allows for TWT power-save operations during the UORA process. Finally, an *opportunistic power save* mechanism also allows 802.11ax client STAs to opportunistically enter a doze state for a defined period. The opportunistic power save mechanism has two modes: unscheduled and scheduled.

Additional 802.11ax PHY and MAC Capabilities

Multi-user OFDMA, BSS coloring, MU-MIMO enhancements and new power-saving mechanisms are all core components of 802.11ax. However, the 802.11ax draft amendment defines numerous other PHY and MAC layer capabilities that will result in better and more efficient Wi-Fi communication.

1024-QAM

Although the goal of 802.11ax is better efficiency, as opposed to higher data rates, there is one exception. First-generation 802.11ax radios will support *1024-QAM* modulation, which also means some new *modulation and code schemes (MCSs)* that define some higher data rates. 256-QAM modulates 8 bits per symbol, whereas 1024-QAM modulates 10 bits per symbol, meaning there is a potential 20 percent increase in throughput. As a result, 802.11ax introduces two new modulation and coding schemes, MSC-10 and MSC-11, which will most likely be optional. 1024-QAM can be used only with 242-tone resource units or larger. This means that at least a full 20 MHz channel will be needed for 1024-QAM.

Much like 256-QAM, very high SNR thresholds (35 dB) will be needed in order for 802.11ax radios to use 1024-QAM modulation. Pristine RF environments with a low noise floor and close proximity between an 802.11ax AP and 802.11ax client will most likely be needed. Figure 19.17 shows a comparison of constellation charts between 256-QAM and 1024-QAM modulation. As you can see, 1024-QAM has many more constellation points. *Error vector magnitude (EVM)* is a measure used to quantify the performance of a radio receiver or transmitter in regard to modulation accuracy. With QAM modulation, EVM is a measure of how far a received signal is from a constellation point. 802.11ax radios that use 1024-QAM modulation will need strong EVM and receive sensitivity capabilities.

FIGURE 19.17 256-QAM and 1024-QAM

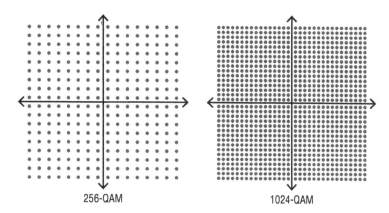

256-QAM   1024-QAM

Long Symbol Time and Guard Intervals

For digital signals, data is modulated onto the carrier signal in bits or collections of bits called *symbols*. 802.11ax introduces a longer *OFDM symbol time* of 12.8 µs, which is four times the legacy symbol time of 3.2 µs. The increase in the number of subcarriers (tones) also increases the OFDM symbol duration. Subcarrier spacing is equal to the reciprocal of the symbol time. The symbol time used is four times longer, as 802.11ax uses subcarrier spacing of 78.125 KHz, which is one quarter the size of legacy 802.11n/ac subcarrier spacing.

802.11a/g defined the use of a 0.8 µs (800 nanosecond) *guard interval (GI)*, while 802.11n/ac also added the option for a 0.4 µs (400 nanosecond) short guard interval, which was intended for use in indoor environments. When the legacy symbol time of 3.2 µs, which is used for the modulated data, is combined with the standard 0.8 µs guard interval, the total symbol duration is 4.0 µs. When the legacy data symbol time of 3.2 µs is combined with the 0.4 µs short guard interval, the total symbol duration is 3.6 µs.

802.11ax defines three different guard intervals that can be used together with the 12.8 µs symbol time that is used for the modulated data:

0.8 µs Guard Interval   This guard interval will likely be used for most indoor environments. When combined with the time of 12.8 µs that is used for the data, the total symbol time for indoor communications will be 13.6 µs.

1.6 µs Guard Interval   This guard interval is intended for outdoor communications. When combined with the time of 12.8 µs that is used for the data, the total symbol time will be 14.4 µs. The guard interval may be needed in high multipath indoor environments to ensure the stability of uplink MU-OFDMA or uplink MU-MIMO communication.

3.2 µs Guard Interval   This guard interval is also intended for outdoor communications. When combined with the time of 12.8 µs that is used for the data, the total symbol time will be 16.0 µs. The longer symbol time and longer guard intervals will provide for more robust outdoor communications.

802.11ax PPDU Formats

The 802.11ax draft amendment defines four *PLCP protocol data unit (PPDU)* formats. In simpler words, there are four new PHY headers and preambles for high efficiency (HE) radio transmission. As shown in Figure 19.18, legacy training fields are prepended to the HE PHY header to allow for backward compatibility with the legacy 802.11/a/b/g/n/ac radios, which use a different PHY format. The four 802.11ax PPDUs are as follows:

HE SU   The high efficiency single-user PPDU is used when there is a single-user transmission.

HE MU   The high efficiency multi-user PPDU is used for transmission to one or more users. Please note that this PPDU format contains the HE-SIG-B field, which is needed for both MU-MIMO or MU-OFDMA and resource unit allocation. The HE MU PPDU is not used as a response to a trigger, which means this PPDU format is used for trigger frames or downlink transmissions.

HE ER SU   The high efficiency extended-range single-user PPDU format is intended for a single user; however, the HE-SIG-A and the HE training fields are boosted by 3 dB. The HE-SIG-A field is twice as long as the other HE PPDUs. This PPDU format is meant to enhance outdoor communications and range.

HE TB   The high efficiency trigger-based PPDU is for a transmission that is a response to a trigger frame. In other words, this PPDU format is used for uplink communications.

FIGURE 19.18   HE PPDU formats

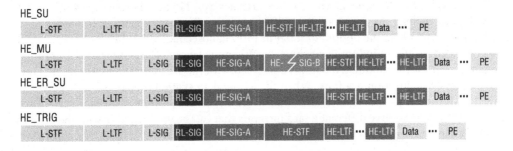

The HE-SIG-A field in all four of these PHY headers, contains information necessary to interpret the HE PPDUs. Various bits in this PHY header field are used as indicators of this information. For example, the SIG-A field can indicate whether the transmission is uplink or downlink. Modulation and coding scheme (MCS) information, BSS color information, guard interval size, and more are all contained in this PHY header field.

As previously mentioned, RU allocation information is communicated to clients at both the PHY and MAC layers. RU allocation information can be found in the *HE-SIG-B* field of the PHY header of an HE MU PPDU. The HE-SIG-B field is used to communicate RU assignments to clients. As shown in Figure 19.19, the HE-SIG-B field consists of two sub-fields: the *common field* and *user specific field*. A subfield of the common field is used to indicate how a channel is partitioned into various RUs. For example, a 20 MHz channel might be subdivided into one 106-tone RU and five 26-tone RUs. The user-specific field comprises *multiple* user fields that are used to communicate which users are assigned to each individual RU.

FIGURE 19.19   HE-SIG-B

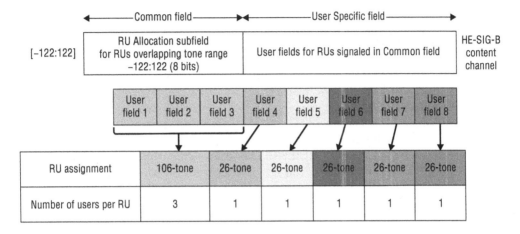

20 MHz-Only

The 802.11ax draft amendment also proposes a *20 MHz-only* mode of operation for 802.11ax clients. Client stations will be able to inform an AP that they are operating as 20 MHz-only clients. As shown in Figure 19.20, a 20 MHz-only client can still operate within a 40 MHz or 80 MHz channel. However, with one rare operational exception, the 20 MHz-only clients must communicate via RUs of the primary channel.

FIGURE 19.20   20 MHz-only 802.11ax client

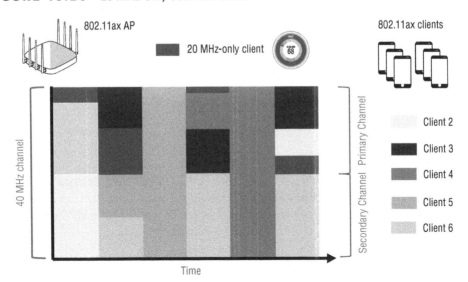

What this effectively means is that the clients can support only certain tone mappings for OFDMA resource units. If the AP is transmitting on a standard 20 MHz channel, the 20 MHz-only client will obviously be able to support tone mappings of a 26-tone RU, a 52-tone RU, a 106-tone RU, and a 242-tone RU within a 20 MHz channel. If the AP is transmitting on a 40 MHz channel, the 20 MHz-only client will be able to support only the 40 MHz tone mappings of a 26-tone RU, a 52-tone RU, or a 106-tone RU. Very specific 26-tone RU, 52-tone RU, or 106-tone mappings are also supported for a 20 MHz-only client, if the AP is transmitting on an 80 MHz channel or a 160 MHz channel. For any channels larger than 20 MHz, a 242-tone RU is optional for 20 MHz-only clients.

The whole purpose behind these rules is to ensure that a 20 MHz-only client is only assigned the proper OFDMA tone mappings and RU allocations that the client can support even if larger channels are being used. The 20 MHz-only operational mode is ideal for IoT clients that could take advantage of the 802.11ax power-saving capabilities but not necessarily need the full capabilities that 802.11ax has to offer. This will allow client manufacturers to design less complex chipsets at a lower cost which is ideal for IoT devices.

Multi-TID AMPDU

In Chapter 10, you learned that frame aggregation is used with 802.11n/ac radios. As you learned in earlier chapters, an 802.11 MPDU is an entire 802.11 frame, including the MAC header, body, and trailer. As shown in Figure 19.21, multiple MPDUs can be aggregated into a single PPDU transmission. A-MPDU comprises multiple MPDUs and is prepended with a PHY header.

FIGURE 19.21   Aggregate MPDU

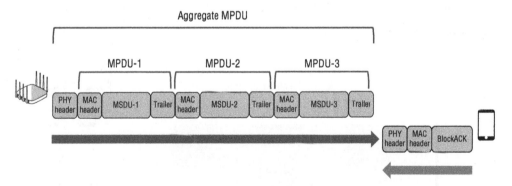

Prior to 802.11ax, all the individual MPDUs must have been of the same 802.11e QoS access category when A-MPDU frame aggregation is used. Voice MPDUs cannot be mixed with Best Effort or Video MPDUs within the same aggregated frame. 80211ax introduces *multi-traffic identifier aggregated MAC protocol data unit (multi-TID AMPDU)*, which allows the aggregation of frames from multiple traffic identifiers (TIDs), from the same or different QoS access categories. The ability to mix MPDUs of different QoS traffic classes allows 802.11ax radios to aggregate more efficiently, reducing overhead and thus increasing throughput and therefore overall network efficiency.

802.11ax Design Considerations

So will there be any performance benefit for legacy 802.11n/ac clients when 802.11ax APs are deployed? 802.11ax APs will not improve the performance or range of any legacy Wi-Fi clients (802.11/a/b/g/n/ac). 802.11ax clients will be needed to take full advantage of 802.11ax high efficiency capabilities, such as multi-user OFDMA. While there will be no PHY improvements with legacy clients, there will be performance improvements as a result of newer hardware capabilities of the new 802.11ax APs, such as stronger CPUs, better memory handling, and other normal hardware advancements. However, as we see more 802.11ax clients mixed into the client population, the efficiency improvements gained by 802.11ax client devices will free valuable airtime for those older clients, therefore improving the overall efficiency of the WLAN. Enterprise WLAN administrators should be highly encouraged to upgrade their client population to 802.11ax clients to take advantage of all that 802.11ax has to offer. It has been estimated that by 2019, 40 percent of all Wi-Fi client shipments will be 802.11ax clients.

When deploying 802.11ax APs, should 20 MHz, 40 MHz, or 80 MHz channels be used? In theory, BSS color could provide the capability to take better advantage of 80 MHz channels. However, this is assuming no legacy devices co-exist in the same environment. In reality, designing for 20 MHz channels will still be the best practice. Remember, the whole point of OFDMA is to make use of smaller sub-channels. If deploying, 40 MHz channels in the enterprise, design best practices for 40 MHz channels will most likely remain the same. 40 MHz channels should be used only if the dynamic frequency selection (DFS) channels are available.

What about recommended configured basic rates for 802.11ax APs? As previously mentioned, 802.11 management and control frames will still use OFDM technology that 802.11a/g/n/ac radios can understand. The overhead created from beacons and other management frames can still be significant if a low basic rate of 6 Mbps is configured on an 802.11ax AP. WLAN design best practices of a basic rate of 12 Mbps or 24 Mbps is still recommended.

Will we need 2.5 Gbps Ethernet ports for 802.11ax access points? The answer is probably not for a very long time. The whole point of 802.11ax is to cut down on medium contention and airtime consumption. Logic dictates that if Wi-Fi becomes more efficient, the user traffic generated by a dual-frequency 802.11ax AP could potentially exceed 1 Gbps. The fear is that a standard Gigabit Ethernet wired uplink port could be a bottleneck, and therefore 2.5 Gbps uplink ports will be needed. In the real world, however, we probably will still not exceed 1 Gbps for a very long time because of the following two reasons:

- Even though the chipset vendors appear to be aggressive with making 802.11ax client radios available at the same time as AP radios, it will be a very long time before there is a wide proliferation of 802.11ax clients in the enterprise client population.

- 802.11ax requires backward compatibility with 802.11/a/b/g/n/ac, which means that RTS/CTS protection mechanisms must be used. RTS/CTS creates overhead and consumes airtime.

Just in case, most WLAN vendor's 802.11ax APs will include at least one 802.3bz Multi-Gig Ethernet port capable of a 2.5 Gbps wired uplink.

Will 802.11ax APs work with standard 802.3af PoE? WLAN vendors will be adding more radio chains to their 802.11ax access points. Most 802.11ax APs will be dual-band 4x4:4 APs, and there will even be 8x8:8 APs. 802.11ax APs will also require much more processing power than previous generations of enterprise APs. The extra radio chains and processor capabilities will require more power. The 15.4 watts provided by standard 802.11af PoE will not be adequate. 802.3at PoE+ power will be required. PoE+ requirements for 802.11ax APs should be considered a standard requirement. This may require upgrades of access layer switches as well as the recalculation of PoE power budgets.

Which is better, 4x4:4 access points or 8x8:8 access points? 8x8:8 APs will initially be more expensive and require more power. The main advantage of 8x8:8 APs is to take advantage of MU-MIMO capabilities, which 802.11ax clients will need to support. Regardless of stream count, all APs will support the same number of 802.11ax OFDMA clients. There is no real advantage to an 8x8:8 AP over a 4x4:4 when using the multi-user OFDMA technology. Some WLAN vendors will offer dual-frequency 802.11ax APs with a 4x4:4 radio for 2.4 GHz and a 8x8:8 radio for 5 GHz. Other WLAN vendors will offer 802.11ax APs with dual-frequency 5 GHz capability with two 4x4:4 radios.

Will there be any 8x8:8 clients? Most 802.11ax mobile client devices will use dual-frequency 2x2:2 radios, because an 8x8:8 radio would drain battery life. There are currently no 4x4:4 802.11ax clients; however, we might see some 4x4:4 client radios in high-end laptops in the future.

Summary

This chapter focused on all PHY and MAC enhancements defined by the 802.11ax draft amendment. The goal of 802.11ax is better and more efficient 802.11 traffic management; thus, the technical name for 802.11ax is high efficiency (HE). Most industry experts believe that OFDMA will be the most relevant technology that 802.11ax has to offer. OFDMA provides for a more efficient use of the available frequency space. Additionally, for the first time, an access point will have the ability to coordinate the uplink transmission from client devices. Synchronized uplink communications are defined for both MU-OFDMA and MU-MIMO. 802.11ax also provides for battery-saving enhancements for IoT devices with target wake times (TWTs). BSS color, together with spatial reuse operation, has the potential to address medium contention overhead due to an overlapping basic service set (OBSS).

Please understand that 802.11ax technology is not yet certified by the Wi-Fi Alliance and is subject to change. Some of the 802.11ax capabilities mentioned in this chapter may indeed be mandatory for 802.11ax communications, and some may be optional. Even more important, understand that 802.11ax technology has not been field-tested in real-world deployments. The CWNA-107 exam will not test you on 802.11ax technology. However, 802.11ax most likely will be a testable topic in future versions of the CWNA exam; therefore, we are including five review questions about 802.11ax technology.

Review Questions

1. What is the maximum number of resource units that can be used for a 20 MHz OFDMA channel?

   A. 2

   B. 4

   C. 9

   D. 26

   E. 52

2. Which type of 802.11 frame is needed for either uplink MU-MIMO or uplink MU-OFDMA communication?

   A. Trigger

   B. Probe

   C. ACK

   D. Beacon

   E. Data

3. Which 802.11ax technology defines new power-saving capabilities that could be beneficial for IoT devices?

   A. Buffer status report

   B. Target wake time

   C. BSS color

   D. Guard interval

   E. Long symbol time

4. Which 802.11ax technology has the potential to decrease co-channel interference (CCI)?

   A. Buffer status report

   B. Target wake time

   C. BSS color

   D. Guard interval

   E. Long symbol time

5. What is the minimum size of a resource unit if 1024-QAM modulation is used by an 802.11ax radio?

   A. 26-tone

   B. 52-tone

   C. 106-tone

   D. 242-tone

   E. 484-tone

Chapter

20

WLAN Deployment and Vertical Markets

IN THIS CHAPTER, YOU WILL LEARN ABOUT THE FOLLOWING:

✓ **Deployment considerations for commonly supported WLAN applications and devices**

- Data
- Voice
- Video
- Real-time location services (RTLS)
- iBeacon proximity
- Mobile devices

✓ **Corporate data access and end-user mobility**

✓ **Network extension to remote areas**

✓ **Bridging: building-to-building connectivity**

✓ **Wireless ISP (WISP): last-mile data delivery**

✓ **Small office/home office (SOHO)**

✓ **Mobile office networking**

✓ **Branch offices**

✓ **Educational/classroom use**

✓ **Industrial: warehousing and manufacturing**

✓ **Retail**

✓ **Healthcare**

✓ **Municipal networks**

✓ **Hotspots: public network access**

- ✓ Stadium networks
- ✓ Transportation networks
- ✓ Law enforcement networks
- ✓ First-responder networks
- ✓ Managed service providers
- ✓ Fixed mobile convergence
- ✓ WLAN and health
- ✓ Internet of Things
- ✓ WLAN vendors

In this chapter, you will learn about environments where wireless networks are commonly deployed. We will consider the pros and cons of wireless in various WLAN vertical markets along with areas of concern. Finally, we will discuss the major commercial WLAN vendors and provide links to their websites.

Deployment Considerations for Commonly Supported WLAN Applications and Devices

As wireless networking has expanded, numerous applications and devices have benefited, and along the way these applications and devices have helped to expand the growth of wireless networking. Although applications such as data and video have benefited due to the flexibility and mobility that wireless affords them, they are not wireless-intrinsic applications. Voice, real-time location services (RTLSs), and network access using mobile devices are three uses that are inherently dependent on a WLAN and will continue to expand the use of WLANs. No matter which of these applications or devices you are implementing on your network, you will need to consider certain factors when planning, designing, and supporting your WLAN. The following sections focus on considerations for commonly supported WLAN applications and devices.

Data

When data-oriented applications are discussed, email and web browsing are two of the most common applications that come to mind. When planning for network traffic over any type of network, wireless or wired, you need to first look at the protocols that are being implemented. Protocols are communications methods or techniques used to communicate between devices on a network. Protocols can be well designed, based on documented standards, or they can be proprietary, using unique communications methods. Data-oriented applications are often based on well-known protocols and are therefore usually easy to work with because a great deal of knowledge already exists about how they communicate.

One of the most important aspects of designing a network to handle data-oriented applications is to ensure that the network design is capable of handling the amount of data that will be transferred. Most data applications are forgiving of slight network delays, but problems can arise if not enough data bandwidth is available. Analyze the data requirements of your users and devices and properly design the WLAN to meet capacity needs. Capacity planning is discussed in great detail in Chapter 13, WLAN Design Concepts.

Voice

When designing a WLAN to support voice communications, keep in mind that, unlike data communications, voice communications are not tolerant of network delays, dropped packets, or sporadic connections. Designing a WLAN to support voice communications can also be a challenge because there are so many differences in how vendors implement their voice products. Each vendor has unique guidelines for designing voice applications. This is true not only for vendors of voice handsets or software applications, but also for infrastructure vendors. So, it is important to understand the best practices for installing your voice system.

Voice devices are typically handheld devices that do not transmit with as much power as laptops. Since a wireless device requires more battery power to transmit a strong signal, the transmit power of VoWiFi phones is typically less than other devices in order to increase battery longevity.

Video

The transmission of video is typically more complex than voice. In addition to multiple streams of data for video and voice, video often includes streams for setting up and tearing down the connection. Unless you are using the WLAN for a real-time videoconference, video can likely take a backseat to audio. In most cases, video has a higher loss tolerance than voice. Choppy audio during a videoconference would likely be highly disruptive, causing participants to ask the speaker to repeat what was said, whereas if the audio is clear and the video choppy, the speaker would likely be understood the first time.

In regard to video transmission, it is important to identify the type of video that is being transmitted and the function or purpose of that transmission. If you ask an average computer user about video transmission, they will likely think of streaming video—a movie, TV show, or funny video clip downloaded by a user, who may be either stationary or mobile. If you were to ask an executive about video transmission, they will likely think of video as part of a videoconference or a webinar, and the user will most likely be stationary. If you were to ask a facilities or security person about video transmission, they will likely think of streaming video generated by a wireless surveillance camera, most likely permanently mounted to the building. Your WLAN might have any or all of these types of video traffic.

Once you have identified the type of video that will be used on your WLAN, you can plan your network. You need to evaluate the system or software that is transmitting the wireless video traffic to determine the type of traffic and protocols along with the network load. As part of the protocol evaluation, you will need to research whether the video transmissions are using multicast transmissions or quality of service (QoS).

Real-Time Location Services (RTLS)

Location-based technology has garnered a lot of attention in WLAN designs. Most manufacturers of enterprise WLAN systems tout some sort of location capability with their products. Some have features that are built in, whereas others offer integration hooks to third-party vendors who specialize in location technology and have sophisticated software applications related to specific industry vertical markets.

Location tracking is expanding incredibly quickly as more and more uses are identified. RTLSs can be used to locate or track people or devices on a WLAN. Healthcare is one of the biggest users of location-based technology. Because health-care providers, such as hospitals, have to run 24/7 shifts, and since many of the assets are shared, RTLSs can be extremely useful for tracking equipment that may be necessary in an emergency or for identifying the closest doctor or specialist.

RTLSs can be used to track any 802.11 radio, or a specialized 802.11 RFID tag can be affixed to a non-802.11 asset so that it can be managed and tracked. Tags can be affixed to any device to provide tracking and help deter theft. Tags can also be worn by employees, children at amusement parks, hospital staff or patients, to name a few. Each RTLS vendor is unique and will be able to provide you with recommendations and best practice documents for deploying your RTLS equipment.

iBeacon Proximity

Bluetooth Low Energy (BLE) wireless technology is now being used for many new applications in retail and other commercial enterprises. Although Bluetooth is a different wireless technology from Wi-Fi, many WLAN vendors integrate BLE radios into their access points or also sell standalone BLE transmitter devices. *iBeacon* is a protocol, developed by Apple, that utilizes BLE radios for indoor proximity-aware push notifications at retail enterprises, stadiums, hospitals, and other public venues.

iBeacon technology is based on nearby *proximity*, as opposed to absolute location coordinates. The BLE radio in an access point will transmit an iBeacon with a proximity location identifier. As shown in Figure 20.1, the proximity location identifier consists of a *universally unique identifier (UUID)*, a major number, and a minor number. The UUID is a 32-character hexadecimal string that uniquely identifies the organization under which the beacons are managed. For example, if a store chain uses beacons at multiple sites, the UUID would be the name of the chain and indicate that all of them belong to the same organization. The UUID is included in the iBeacon payload with major and minor numbers, which typically indicate a specific site and location within the site, respectively. The UUID is at the top of the numbering hierarchy, with major and minor numbers set at the device level. In the example in Figure 20.1, the UUID value of 8C18BD1A-227C-48B3-BA2D-A13BFA8E5919 may be used to identify the ACME Corporation retail chain; the major number value of 4 may reference an ACME store located in Boulder, Colorado; and the minor number value of 3 may reference aisle number 3 within the Boulder store.

FIGURE 20.1   iBeacon proximity location identifier

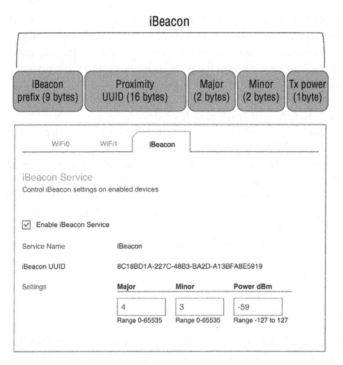

An organization can automatically create a UUID with an online UUID generator, such as www.uuidgenerator.net. In fact, most UUIDs are generated this way because UUIDs are not centrally assigned and managed. Their 32-character length makes it safe to assume that randomly generated UUIDs will not produce duplicates. According to Wikipedia, "for there to be a one in a billion chance of duplication, 103 trillion version 4 UUIDs must be generated."

iBeacons require an application on a mobile device to *trigger* an action. The trigger occurs whenever a smartphone with a BLE receiver radio is within proximity range of an iBeacon transmitter. For example, iBeacon may trigger a proximity-aware push notification such as an advertisement in a retail store location. Museums use iBeacon technology to give self-guided tours to visitors. The iBeacons can trigger an application on a mobile device to display interactive content when a visitor is within near proximity of a museum exhibit. As previously stated, iBeacons utilize BLE radios as opposed to 802.11 radios; however, many WLAN vendors now offer integrated and/or overlay iBeacon technology solutions. Also, please do not confuse the iBeacons used by BLE transmitters with the 802.11 beacon management frames that are transmitted by Wi-Fi access points. As mentioned, iBeacon is a protocol developed by Apple for BLE radios. In 2015, Google developed an open-source BLE beacon platform called Eddystone. In 2014, Radius Networks developed an open-source BLE beacon protocol called AltBeacon.

Mobile Devices

Employees, from receptionists to CEOs, bring their own 802.11-capable devices, such as laptops, tablets, and smartphones, to work and expect—and in many cases demand—that these devices be supported on the corporate network. The primary devices people are requesting access for are cell phones and tablets that are also capable of communicating using 802.11 radios. Unlike changes in enterprise technology, which is planned and controlled by the IT department, the push for support of mobile devices is being made by the end user. Many organizations see access for these devices as a benefit for employees and are pressuring the IT department to provide access and support.

Multiple concerns arise with integrating these devices into the network:

- Making sure that the devices are capable of connecting to the network using the proper authentication

- Ensuring the use of encryption protocols along with the ability for these devices to be able to smoothly roam throughout the network without losing connectivity

- Providing network access, based not only on the identity of the device's user, but also on the type of device or other device or connection characteristics

Because of the influx of numerous personal mobile devices, such as smartphones and tablets, most WLANs are now designed very differently to meet capacity needs. As a result, indoor WLANs are rarely designed anymore strictly for large coverage areas but instead are designed with smaller coverage areas, with more APs deployed to handle the capacity needs. The introduction of these mobile devices has become a huge trend in the industry, typically known as *bring your own device (BYOD)*. It has become such a concern that we have dedicated an entire chapter to this topic, Chapter 18, "Bring Your Own Device (BYOD) and Guest Access."

Corporate Data Access and End-User Mobility

With the increased throughput provided by 802.11n and now 802.11ac technology, many organizations have been transitioning to these higher-speed wireless networks while reducing the number of devices connecting to the network via wired connections—in many cases retiring some of the unused or underused wired switches. As mentioned earlier, another major influence pressing organizations to expand their wireless networks is the proliferation of Wi-Fi-enabled personal mobile devices.

The installation of wired network jacks is expensive, often costing as much as—or even more than—$200 (in U.S. dollars) per jack. As companies reorganize workers and departments, network infrastructure typically needs to be changed as well. Other areas, such as warehouses, conference rooms, manufacturing lines, research labs, and cafeterias, are often difficult places to effectively install wired network connections. In these and other environments, the installation of wireless networks can save the company money and provide consistent network access to all users.

Providing continuous access and availability throughout the facility has become paramount in the past few years. With computer access and data becoming critical components of many people's jobs, it is important for networks to be continuously available and to be able to provide the up-to-the-moment information that is being demanded. By installing a wireless network throughout the building or campus, a company makes it easy for employees to meet and discuss or brainstorm while maintaining access to corporate data, email, and the Internet from their laptops and mobile devices, no matter where they are in the building or on the campus.

A big trend in the consumer electronics marketplace has been the addition of wireless radios in devices. Wireless adapters are extremely small and can easily be integrated in these portable devices. Connectivity to the Internet also allows devices to be easily updated, along with providing more capabilities. In addition to the trend of connecting personal electronics, devices keep getting smaller, lighter, and leaner. With this push toward leaner devices, Ethernet adapters have either given way to wireless radios or been bypassed all together in favor of wireless.

Whatever the reason for installing wireless networking, companies must remember its benefits and its flaws. Wireless provides mobility, accessibility, and convenience, but if not designed and implemented properly, it can lack in performance, availability, and throughput. Wireless is an access technology, providing connectivity to end-user stations. Wireless should rarely be considered for distribution or core roles, except for building-to-building bridging or mesh backhaul. Even in these scenarios, make sure that the wireless bridge will be capable of handling the traffic load and throughput needs.

Network Extension to Remote Areas

If you think about it carefully, network extension to remote areas was one of the driving forces of home wireless networking, which also helped drive the demand for wireless in the corporate environment. As households connected to the Internet and as more households purchased additional computers, there was a need to connect all the computers in the house to the Internet. Although many people installed Ethernet cabling to connect their computers, this was typically too costly, impractical because of accessibility, or beyond the technical capabilities of the average homeowner.

At the same time, 802.11 wireless devices were becoming more affordable. The same reasons for installing wireless networking in a home are also valid for installing wireless in offices, warehouses, and just about any other environment. The cost of installing network cabling for each computer is expensive, and in many environments, running cable or fiber is difficult because of building design or aesthetic restrictions. When wireless networking equipment is installed, far fewer cables are required, and equipment placement can often be performed without affecting the aesthetics of a building.

Bridging: Building-to-Building Connectivity

To provide network connectivity between two buildings, you can install an underground cable or fiber between the two buildings, you can pay for a high-speed leased data circuit, or you can use a building-to-building wireless bridge. All three are viable solutions, each with its benefits and disadvantages.

Although a copper or fiber connection between two buildings will potentially provide you with the highest throughput, installing copper or fiber between two buildings can be expensive. If the buildings are separated by a long distance or by someone else's property, this may not even be an option. After the cable is installed, there are no monthly service fees since you own the cable.

Leasing a high-speed data circuit can provide flexibility and convenience, but because you do not own the connection, you will pay monthly service fees. Depending on the type of service that you are paying for, you may or may not be able to easily increase the speed of the link.

A wireless building-to-building bridge requires that the two buildings have a clear RF line of sight between them. After this has been determined, or created, a point-to-point (PtP) or point-to-multipoint (PtMP) transceiver and antenna can be installed. The installation is typically easy for trained professionals to perform, and there are no monthly service fees after installation because you own the equipment.

In addition to connecting two buildings via a PtP bridge, three or more buildings can be networked together by using a PtMP solution. In a PtMP installation, the building that is most centrally located will be the central communication point, with the other devices communicating directly to the central building. This is known as a *hub and spoke* or *star* configuration. 802.11 wireless bridge links typically are deployed for short distances in the 5 GHz frequency U-NII-3 band. Many businesses instead choose a wireless bridge solution that transmits on licensed frequencies. Companies such as Mimosa (www.mimosa.co) and Cambium Networks (www.cambiumnetworks.com) offer both licensed and unlicensed outdoor backhaul solutions.

WARNING   A potential problem with the PtMP solution is that the central communication point becomes a single point of failure for all the buildings. To prevent a single point of failure and to provide higher data throughput, it is not uncommon to install multiple point-to-point bridges.

Wireless ISP: Last-Mile Data Delivery

The term *last mile* is often used by phone and cable companies to refer to the last segment of their service that connects a home subscriber to their network. The last mile of service can often be the most difficult and costly to run because at this point a cable must be run individually to every subscriber. This is particularly true in rural areas where there are very few subscribers and they are separated by large distances. In many instances, even if a subscriber is connected, the subscriber may not be able to receive some services, such as high-speed Internet, because services such as xDSL have a maximum distance limitation of 18,000 feet (5.7 km) from the central office.

Wireless Internet service providers (WISPs) deliver Internet services via wireless networking. Instead of directly cabling each subscriber, a WISP can provide services via RF communications from central transmitters. WISPs often use wireless technology other than 802.11, enabling them to provide wireless coverage to much greater areas. Some small towns have had success using 802.11 mesh networks as the infrastructure for a WISP. However, 802.11 technology generally is not intended to scale to the size needed for city-wide WISP deployments.

Service from WISPs is not without its own problems. As with any RF technology, the signal can be degraded or corrupted by obstacles such as roofs, mountains, trees, and other buildings. Proper designs and professional installations can ensure a properly working system.

Small Office/Home Office

One common theme of a *small office/home office (SOHO)* is that your job description spans janitor to IT staff and includes everything in between. Small-business owners and home-office employees are typically required to be self-sufficient because there are usually few, if any, other people around to help them. Wireless networking has helped to make it easy for a SOHO employee to connect the office computers and peripheral devices together, as well as to the Internet. The main purpose of a SOHO 802.11 network is typically to provide wireless access to an Internet gateway. As depicted in Figure 20.2, many wireless SOHO devices also have multiple Ethernet ports, providing both wireless and wired access to the Internet.

FIGURE 20.2   D-Link wireless SOHO router

Most SOHO wireless routers provide fairly easy-to-follow installation instructions and offer reasonable performance and security, though less than what their corporate counterparts provide. They are generally not as flexible or feature rich as comparable corporate products, but most SOHO environments do not need all the additional capabilities. What the SOHO person gets is a capable device at a quarter of the price paid by their corporate counterparts. Dozens of devices are available to provide the SOHO worker with the ability to install and configure their own secure Internet-connected network without spending a fortune. Many SOHO wireless routers even have the ability to provide guest access, allowing visitors Internet access while preventing them from accessing the local network. In recent years, several companies, including Google, eero, Linksys, Plume, and others, have also begun to offer complete home Wi-Fi solutions that include 3 or more home mesh WLAN routers and a smartphone application that can manage the WLAN devices.

Mobile Office Networking

Mobile homes or trailer offices are used for many purposes—for example, as temporary offices during construction or after a disaster or as temporary classrooms to accommodate unplanned changes in student population. Mobile offices are simply an extension of the office environment. These structures are usually buildings on wheels that can be easily deployed for short- or long-term use on an as-needed basis. Since these structures are not permanent, it is usually easy to extend the corporate or school network to these offices by using wireless networking.

A wireless bridge can be used to distribute wireless networking to the mobile office. If needed, an AP can then be used to provide wireless network access to multiple occupants of the office. By providing networking via wireless communications, you can alleviate the cost of running wired cables and installing jacks. Additional users can connect and disconnect from the network without the need for any changes to the networking infrastructure. When the mobile office is no longer needed, the wireless equipment can simply be unplugged and removed.

Moveable wireless networks are used in many environments, including military maneuvers, disaster relief, concerts, flea markets, and construction sites. Because of the ease of installation and removal, mobile wireless networking can be an ideal networking solution.

Branch Offices

In addition to the main corporate office, companies often have branch offices in remote locations. A company might have branch offices across a region, an entire country, or even around the world. The challenge for IT personnel is how to provide a seamless enterprise wired and wireless solution across all locations. A distributed solution using enterprise-grade WLAN routers at each branch office is a common choice. Branch WLAN routers have the ability to connect back to corporate headquarters with VPN tunnels. Employees at the branch offices can access corporate resources across the WAN through the VPN tunnel. Even more important is the fact that the corporate VLANs, SSIDs, and WLAN security can all be extended to the remote branch offices. An employee at a branch office connects to the same SSID that they would connect to at corporate headquarters. The wired and wireless network access policies are therefore seamless across the entire organization. These seamless policies can be extended to WLAN routers, access points, and switches at each branch location.

Most companies do not have the luxury or need to have an IT employee at each branch office. Therefore, a network management server (NMS) at a central location is used to manage and monitor the entire enterprise network.

Educational/Classroom Use

Wireless networking can be used to provide a safe and easy way of connecting students to a school network. Because the layout of most classrooms is flexible (with no permanently installed furniture), installing a wired network jack for each student is not possible. Because students would be constantly connecting to and disconnecting from the network at the beginning and end of class, the jacks would not last long even if they were installed. Prior to wireless networking, in classrooms that were wired with Ethernet, usually all the computers were placed on tables along the classroom walls, with the students typically facing away from the instructor. Wireless networking enables any classroom seating arrangement to be used, without the safety risk of networking cables being strung across the floor.

A wireless network also enables students to connect to the network and work on schoolwork anywhere in the building without having to worry about whether a wired network jack is nearby or whether someone else is already using it. In addition to the flexibility the wireless network is able to provide in a classroom environment, in many schools wireless networking has become a necessity: Computer tablets are quickly becoming commonplace devices in all levels of education. These tablets rely solely on wireless networking to provide Internet and local area networking access.

Schools typically require more access points for coverage because of the dense wall materials between classrooms. Most classroom walls are made of cinderblock to attenuate noise between classrooms. The cinderblock also attenuates the 2.4 and 5 GHz RF signals dramatically. In order to provide −70 dBm or greater coverage, an access point is often needed in at least every other classroom.

The use of wireless bridging is also prevalent in campus environments. Many universities and colleges use many types of wireless bridge links, including 802.11, to connect buildings campuswide.

Network access control (NAC) has become an integral part of many school networks. NAC can be used to "fingerprint," or identify, authentication and authorization information about devices connecting to the network. This information is then used to regulate or control the access that the user has on the network. Access can be regulated or restricted based on many different criteria, such as time, location, access method, device type, and user identity, along with many other properties. Chapter 18 discusses NAC in greater detail.

Real World Scenario

eduroam

Many higher education universities use a realm-based authentication solution called *eduroam* (education roaming). The technology behind eduroam is based on the IEEE 802.1X standard and a hierarchy of RADIUS proxy servers. eduroam is a secure, worldwide roaming access service developed for the international research and higher education community. eduroam allows students, researchers, and staff from participating institutions to obtain Internet connectivity across the campus and when visiting other participating institutions. Authentication of users is performed by their home institution, using the same credentials as when they access the network locally. Depending on local policies at the visited institutions, eduroam participants may also have additional resources at their disposal. The eduroam Architecture for Network Roaming is defined in RFC 7593. More information about the eduroam service can be found at www.eduroam.org.

Industrial: Warehousing and Manufacturing

Warehouses and manufacturing facilities are two environments in which wireless networking has been used for many years, even before the 802.11 standard was created. Because of the vast space and the mobile nature of the employees in these environments, companies saw the need to provide mobile network access to their employees so that they could more effectively perform their jobs. Warehouse and manufacturing environments often deploy wireless handheld devices, such as bar code scanners, which are used for data collection and inventory control.

Most 802.11 networks deployed in either a warehouse or manufacturing environment are designed for coverage rather than capacity. Handheld devices typically do not require much bandwidth, but large coverage areas are needed to provide true mobility. Most early deployments of 802.11 frequency-hopping technology were in manufacturing and warehouse environments. Wireless networks are able to provide the coverage and mobility required in a warehouse environment—and to provide it cost-effectively.

Retail

There are four key uses of wireless in retail locations. The first is the wireless network that provides support relating to the operations of the store and the retail transactions. The second is a newer and growing use, which is tracking analytics of the retail customer. The third is location-based mapping and tracking services. The fourth is supplemental guest Wi-Fi Internet access, often necessitated by poor cellular coverage inside the retail establishment.

The retail environment is similar to many other business environments. Cash registers, time clocks, inventory control scanners, and just about every electronic device used to run a retail location are becoming networked with a WLAN radio. Connectivity of these devices provides faster and more accurate information and enhances the retail environment for the customer.

To further support and understand customers and their behaviors, retail analytic products are being installed to monitor customer movement and behavior. Strategically placed access points or sensor devices listen for probe request frames from Wi-Fi-enabled smartphones. MAC addresses are used to identify each unique Wi-Fi device, and signal strength is used to monitor and track the location of the shopper. Figure 20.3 shows a retail analytics dashboard. Analytics can identify the path the shopper took while walking through the store as well as the time spent in different areas of the store. This information can be used to identify shopping patterns along with analysis of the effectiveness of in-store displays and advertisements. One company that is partnering with multiple WLAN vendors is Euclid Analytics. These retail Wi-Fi analytic solutions are sometimes referred to as *presence analytics*.

FIGURE 20.3   Retail analytics

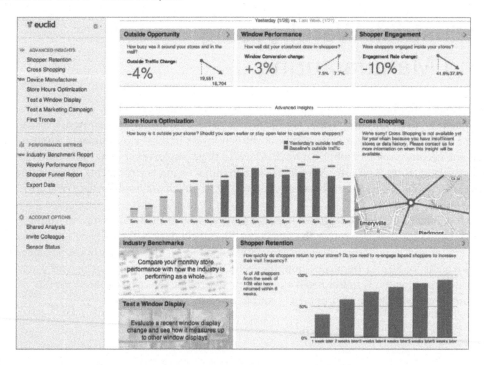

As previously mentioned, iBeacon technology is also often used by retailers to provide push notifications and trend analysis. In addition to retail proximity location and trend analysis, indoor location and mapping applications are beginning to provide new services to shoppers and visitors. Retail centers, hospitals, hotels, subways, and museums (and many other types of organizations) can provide turn-by-turn directions to visitors, along with promotions and other location-based services. As an example, navigating through a large hospital can be confusing. A mobile app can provide turn-by-turn directions for family and friends to locate a patient's room. Conference centers and hotels can use this technology to direct visitors to meeting and event rooms. Special events, advertisements, or services can be offered as the visitor is navigating through the building.

Another key reason to implement wireless in a retail location is to provide supplemental connectivity in lieu of cellular coverage. Retail locations cannot depend on customers having Internet access through their cellular phone network. Due to the scale and often product density, cellular phone access may not exist or be dependable. Providing guest Wi-Fi access for shoppers may make for a more pleasant and satisfied shopping experience and will likely result in more sales. A deep discussion about guest WLAN access can be found in Chapter 18.

Healthcare

Although healthcare facilities, such as hospitals, clinics, and doctors' offices, may seem very different from other businesses, they have many of the same networking needs as other companies: data access and end-user mobility. Healthcare providers need quick, secure, and accurate access to patient and hospital or clinic data so that they can react and make decisions. Wireless networks can provide mobility, giving healthcare providers faster access to important data by delivering the data directly to handheld devices that doctors or nurses carry with them. Medical carts used to enter and monitor patient information often have wireless connections back to the nursing station. Many medical device companies, including Masimo (www.masimo.com), have integrated 802.11 wireless adapters in their medical equipment to monitor and track the patient's vital signs, such as the patient-worn monitoring system shown in Figure 20.4. Stationary and mobile, patient-worn medical monitors can use Wi-Fi to securely transmit patient EKG, blood pressure, respiration, and other vital signs to the nursing station.

VoWiFi is another common use of 802.11 technology in a medical environment, providing immediate access to personnel no matter where they are in the hospital. RTLS solutions using 802.11 Wi-Fi tags for inventory control are also commonplace.

Hospitals rely on many forms of proprietary and industry-standard wireless communications that may have the potential of causing RF interference with 802.11 wireless networks. Many hospitals have designated a person or spectrum management department to help avoid RF conflicts by keeping track of the frequencies and biomedical equipment used within the hospital.

Distributed healthcare is also an emerging trend. Hospitals often have many remote locations. Urgent care facilities are widespread and very often a WLAN branch office solution is required.

FIGURE 20.4 Masimo Root with Radius-7, a patient-worn monitor

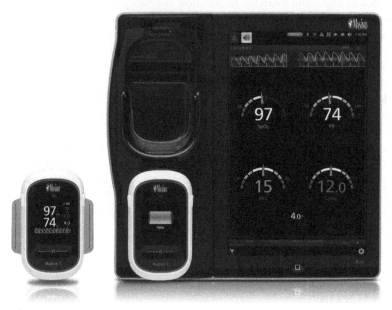

Municipal Networks

Municipal networks have received much attention over the past few years. Cities and towns announced their intentions of providing wireless networking access to their citizens throughout the area. Many municipalities viewed this as a way of providing service to some of their residents who could not necessarily afford Internet access. Although this is a well-intentioned idea, communities often underestimated the scale and cost of these projects, and many taxpayers did not want their taxes spent on what they considered to be an unnecessary service. Although most of these earlier plans for citywide municipal 802.11 networks have been scrapped, there has been an increased interest and success in deploying 802.11 in many downtown and high-density areas. Some of these are provided by the municipality, and others are provided by individuals or business groups.

Hotspots: Public Network Access

The term *hotspot* typically refers to a free or pay-for-use wireless network that is provided as a service by a business. When people think of hotspots, they typically associate them with cafes, bookstores, or hospitality businesses, such as hotels or convention centers. Hotspots can be used effectively by businesses to attract customers or as an extension of their services—in the case of Internet service providers offering these services in heavily travelled areas. Business travelers and students often frequent restaurants or cafes that are known to provide free Internet access. Many of these establishments benefit from the increased business generated by offering a hotspot. Free hotspots have drawn much attention to the 802.11 wireless industry, helping to make more people aware of the benefits of the technology.

Other hotspot providers have had difficulty convincing people to pay upward of $40 per month for a subscription. Many airports and hotel chains have installed pay-for-use hotspots; however, there are many providers, each one offering a separate subscription, which is often not practical for the consumer.

Most hotspot providers perform network authentication by using a special type of web page known as a *captive portal*. When a user connects to the hotspot, the user must open a web browser. No matter which web page the user attempts to go to, a login web page will be displayed instead, as shown in Figure 20.5. This is the captive portal page. If the hotspot provider is a paid service, the user must enter either their subscription information, if they are a subscriber to the service, or their credit card information, if they are paying for hourly or daily usage. Many free hotspots also use captive portals as a method for requiring users to agree to a usage policy before they are allowed access to the Internet. If the user agrees to the terms of the policy, they are required to either enter some basic information or click a button, validating their agreement with the usage policy. Many corporations also use captive portals to authenticate guest users onto their corporate networks.

FIGURE 20.5   Example of a captive portal

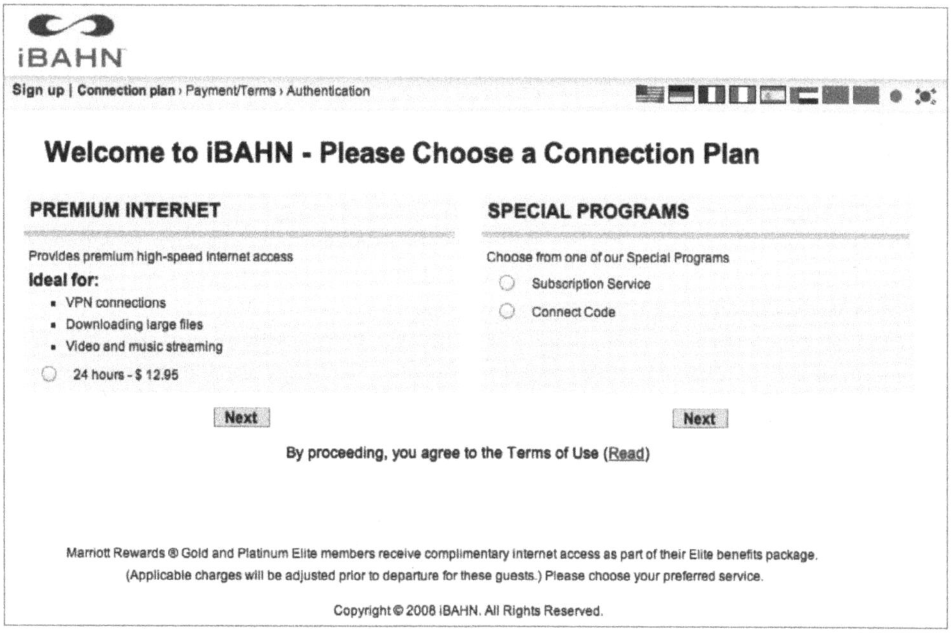

iBAHN

Sign up | **Connection plan** › Payment/Terms › Authentication

Welcome to iBAHN - Please Choose a Connection Plan

PREMIUM INTERNET

Provides premium high-speed Internet access

Ideal for:
- VPN connections
- Downloading large files
- Video and music streaming

○ 24 hours - $ 12.95

[Next]

SPECIAL PROGRAMS

Choose from one of our Special Programs

○ Subscription Service
○ Connect Code

[Next]

By proceeding, you agree to the Terms of Use (Read)

Marriott Rewards ® Gold and Platinum Elite members receive complimentary Internet access as part of their Elite benefits package.
(Applicable charges will be adjusted prior to departure for these guests.) Please choose your preferred service.

Copyright © 2008 iBAHN. All Rights Reserved.

Do Hotspots Provide Data Security?

It is important to remember that hotspot providers (free or pay-for-use) usually do not care about the security of your data. The free provider typically offers you Internet access as a way of encouraging you to visit their location, such as a cafe, and buy some of whatever it is they sell. The pay-for-use hotspot provider performs authentication to make sure you are a paid subscriber, and after you have proven that, they will provide you with access to the Internet.

Except for rare occasions, neither of these hotspot providers performs any data encryption. Because of this, business users often use VPN client software to provide a secure encrypted tunnel back to their corporate network whenever they are using a hotspot. Many companies require employees to use a VPN during any connection to a public network. The problem is that many consumers and guest users are not savvy enough to know how to use a VPN solution when connected to an open guest WLAN. As a result, there is a recent trend to provide encryption and better authentication security for WLAN guest users. Another growing trend with public access networks is the use of 802.1X/EAP with *Hotspot 2.0* technology. Hotspot 2.0 is a Wi-Fi Alliance technical specification that is supported by the Passpoint certification program. A more in-depth discussion about encrypted guest access and HotSpot 2.0 can be found in Chapter 18.

Stadium Networks

Technology-savvy fans are driving sport, concert, and event stadiums and arenas to expand their services. Fans expect and demand a complete multimedia experience when attending events, including access to replays and real-time statistics. Through apps or websites, in-seat food and beverage ordering and delivery enhances the experience by allowing fans to enjoy the action instead of standing in line for refreshments. Through texting and social media, fans expect to be able to share their sporting or concert experience with friends or interact with other attendees. iBeacon technology is now being used in stadiums and sporting events.

A well-designed stadium network can allow the venue to target sections or groups of people with directed advertisements, special offers, or customized services. Offers or services for fans in the bleacher seats would likely be different from those targeting the fans in the skyboxes. In addition to providing wireless services to the fans, it is important to remember that the stadium is a business and needs to support its own infrastructure and services at an event. A wireless network is needed to provide event operations with services such as reliable high-speed Internet access in the press box, ticketing and point-of-sale transaction processing, and video surveillance. Because of the high density of users, multiple high-bandwidth and redundant WAN uplinks are necessary at stadium events.

Transportation Networks

Most discussions about Wi-Fi transportation networks mention the three main modes of transportation: trains, planes, and automobiles. In addition to these three primary methods of transportation, two others are noteworthy: boats (both cruise ships and commuter ferries) and buses (which are similar to but different from automobiles).

Providing Wi-Fi service to any of the transportation methods is easy; simply install one or more access points in the vehicle. Except for cruise ships and large ferries, most of these transportation methods would require only a few access points to provide Wi-Fi coverage. The primary use of these networks is to provide hotspot services for end users so that they can gain access to the Internet. The difference between a transportation network and a typical hotspot is that the network is continually moving, making it necessary for the transportation network to use some type of mobile uplink services.

To provide an uplink for a train, which is bound to the same path of travel for every trip, a metropolitan wireless networking technology such as satellite or cellular LTE could be used along the path of the tracks. With the other transportation networks, for which the path of travel is less bounded, the more likely uplink method would be via some type of cellular or satellite network connection.

Commuter ferries are likely to provide uplink services via satellite or cellular LTE, because they are likely within range of these services. For cruise ships and ferries that travel farther distances away from shore, a satellite link is typically used.

Many airlines either have installed or are in the process of installing Wi-Fi on their planes. The Wi-Fi service in the plane consists of one or more access points connected either to a cellular router that communicates with a skyward facing network of cellular towers on the ground or to a satellite router that uplinks that data to a satellite and then to a terrestrial station. The cellular-based system requires a network of terrestrial-based cellular receivers; therefore, it is not used for transoceanic flights. This in-flight service is typically offered for a nominal fee and is available only while the airplane is flying and when the airplane is at cruising altitude. Bandwidth metering is used to prevent any one user from monopolizing the connection.

Law Enforcement Networks

Although Wi-Fi networks cannot provide the wide area coverage necessary to provide continuous wireless communications needed by law enforcement personnel, they can still play a major role in fighting crime. Many law enforcement agencies are using Wi-Fi as a supplement to their public safety wireless networks.

In addition to enjoying the obvious mobility benefits of using Wi-Fi inside police stations, many municipalities have installed Wi-Fi in the parking lots outside the police station and other municipal buildings as a supplement to their wireless metropolitan networks. These outdoor networks are sometimes viewed as secured hotspots. Unlike public hotspots, these networks provide both authentication and high levels of encryption. In addition to municipalities incorporating wireless technology into law enforcement, many are adding non-Wi-Fi-based automation to utilities through the use of supervisory control and data acquisition (SCADA) equipment. Because of this growth in the use of different wireless technologies, we are starting to see municipalities designate a person or department to keep track of the frequencies and technologies that are being used.

Municipal Wi-Fi hotspots typically provide high-speed communications between networking equipment in the police cars and the police department's internal network. An interesting example of a good use of this network is the uploading of vehicle video files. With many police cars being equipped with video surveillance, and with these surveillance videos often being used as evidence, it is important to not only transfer these video files to a central server for cataloging and storage, but also to do it with the least amount of interaction by the police officer in order to preserve the chain of evidence.

When a police car arrives at one of these municipal Wi-Fi hotspots, the computer in the car automatically uploads the video files from the data storage in the car to the central video library. Automating this process minimizes the risk of data corruption and frees up the officer to do other, more important tasks.

Special Use of 4.9 GHz Band

In some countries, a 4.9 GHz band has been set aside for use by public safety and emergency response organizations. This band typically requires a license to use, but the licensing process is usually more of a formality to ensure that the band is being used properly. This frequency is more commonly implemented and used with outdoor equipment, and since it has limited use, performance degradation from RF interference is less likely.

First-Responder Networks

When medical and fire rescue personnel arrive at the scene of an emergency, it is important for them to have fast and easy access to the necessary resources to handle the emergency at hand. Many rescue vehicles are being equipped with either permanently mounted Wi-Fi access points or easily deployed, self-contained portable access points that can quickly and easily blanket a rescue scene with a Wi-Fi bridge to the emergency personnel's data network. In a disaster, when public service communications systems such as cellular phone networks may not be working because of system overload or outages, a Wi-Fi first-responder network may be able to provide communications between local personnel and possibly shared access to central resources.

During a disaster, assessing the scene and triaging the victims (grouping victims based on the severity of their injuries) is one of the first tasks. Historically, the task of triage included paper tags that listed the medical information and status of the victim. Some companies have created electronic triage tags that can hold patient information electronically and transmit it via Wi-Fi communications.

Managed Service Providers

A growing shift in the information technology (IT) industry is for companies to turn to a *managed services provider (MSP)*. An MSP assumes responsibility for providing a defined set of IT services to its business customers. Outsourcing IT services on a proactive basis can often improve operations and cut expenses. Small and medium business (SMB) and enterprise companies are migrating the management and monitoring of networking to the cloud. Many MSPs now offer in-house cloud services or act as a broker with cloud service providers. Many MSPs offer installation as well as subscription-based monitoring and management solutions for wired and wireless networking solutions. MSP providers have begun to offer turnkey wireless-as-a-service (WAAS) solutions to their customers using enterprise WLAN solutions.

Fixed Mobile Convergence

One of the hot topics related to Wi-Fi is known as *fixed mobile convergence (FMC)*. The goal of FMC systems is to provide a single device, with a single phone number that is capable of switching between networks and always using the lowest-cost network. Figure 20.6 illustrates an FMC phone network.

FIGURE 20.6   FMC network design

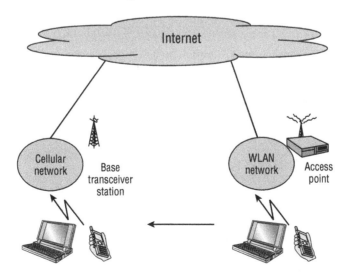

With the flexibility and mobility of cellular phones, it is common for people to use them even in environments (home or work) where they are stationary and have access to other phone systems that are frequently less costly. FMC devices typically are capable of communicating via either a cellular phone network or a VoWiFi network. If you had an FMC phone and were at your office or home, where a Wi-Fi network is available, the phone would use the Wi-Fi network for any incoming or outgoing phone calls. If you were outside either of these locations and did not have access to a Wi-Fi network, the phone would use the cellular network for any incoming or outgoing phone calls. FMC devices also allow you to roam across networks, so you could initiate a phone call from within your company by using the Wi-Fi network. As you walk outside, the FMC phone would roam from the Wi-Fi network to the cellular network and seamlessly transition between the two networks. With fixed mobile convergence, you would be able to have one device and one phone number that would work wherever you were, using the least costly network that was available at the time.

An example of FMC is *Wi-Fi Calling*, which enables smartphones to connect to a mobile carrier's network through Wi-Fi access points. Wi-Fi Calling is a service for Android and iOS smartphones provided now by multiple cellular carriers. With Wi-Fi Calling, no separate application or login is required.

WLAN and Health

Over the years, there has been a concern about adverse health effects from the exposure of humans and animals to radio waves. The World Health Organization and government agencies set standards that establish exposure limits to radio waves, to which RF products must comply. Tests performed on WLANs have shown that they operate substantially below the required safety limits set by these organizations. Also, Wi-Fi signals, as compared to other RF signals, are much lower in power. The World Health Organization has also concluded that there is no convincing scientific evidence that weak radio-frequency signals, such as those found in 802.11 communications, cause adverse health effects.

You can read more about some of these findings at the following websites:

- **U.S. Federal Communications Commission:** www.fcc.gov/general/radio-frequency-safety-0

- **World Health Organization:** www.who.int/peh-emf

- **Wi-Fi Alliance:** www.wi-fi.org

Internet of Things

As mentioned in Chapter 11, "WLAN Architecture" deployment of Internet of Things (IoT) devices in the form of sensors, monitors, and machines is growing rapidly. 802.11 radio NICs used as client devices have begun to show up in many types of machines and solutions. Wi-Fi radios already exist in gaming devices, stereo systems, and video cameras. Appliance manufacturers are putting Wi-Fi NICs in washing machines, refrigerators, and automobiles. The use of Wi-Fi radios in sensor and monitoring devices, as well as RFID, has many applications in numerous enterprise vertical markets. It should be noted that IoT devices often use other wireless networking technologies, such as Bluetooth, Zigbee, and Z-Wave, instead of 802.11 wireless technology.

The manufacturing industry is using IoT devices for asset management and logistics to improve operational performance using predictive analysis monitoring solutions. IoT solutions in the healthcare industry are primarily used to integrate connected medical devices and analyze the data with the goal of enhancing healthcare operations. IoT sensors are being used in many industries to monitor heating, ventilation, and air conditioning systems within buildings. IT administrators must manage the onboarding, access, and security policies of IoT devices connecting to the corporate network. Many startup companies, such as ZingBox (www.zingbox.com), are creating cloud-based monitoring solutions for this purpose.

WLAN Vendors

There are many vendors in the 802.11 WLAN marketplace. The following is a list of some of the major WLAN vendors. Please note that each vendor is listed in only one category, even if they offer products and services that cover multiple categories. This is most notable with the infrastructure vendors, who often offer additional capabilities, such as security and troubleshooting, as features of their products.

WLAN Infrastructure   The following 802.11 enterprise equipment vendors manufacture and sell WLAN controllers and/or enterprise access points:

ADTRAN www.adtran.com

Aerohive Networks www.aerohive.com

Aruba Networks www.arubanetworks.com

Cisco Systems www.cisco.com

Extreme Networks www.extremenetworks.com

Fortinet www.fortinet.com

Huawei www.huawei.com

Mist Systems www.mist.com

Mojo Networks www.mojonetworks.com

Proxim Wireless Corporation www.proxim.com

Riverbed www.riverbed.com

Ruckus Wireless www.ruckuswireless.com

Ubiquiti Networks www.ubnt.com

WLAN Outdoor Mesh and Backhaul   The following WLAN vendors specialize in outdoor 802.11 mesh networking or outdoor wireless bridging:

Cambium Networks   www.cambiumnetworks.com

Meshdynamics   www.meshdynamics.com

Mimosa   www.mimosa.co

Open Mesh   www.openmesh.com

Strix Systems   www.strixsystems.com

WLAN Antennas and Accessories   The following companies make and/or sell WLAN antennas, enclosure and mounting solutions, and accessories:

Oberon   www.oberoninc.com

PCTEL   www.pctel.com

Ventev   www.ventev.com

WLAN Troubleshooting and Design Solutions   The following companies make and/or sell 802.11 protocol analyzers, spectrum analyzers, site survey software, RTLS software, and other WLAN analysis solutions:

Acrylic WiFi   www.acrylicwifi.com

Berkeley Varitronics Systems   www.bvsystems.com

Ekahau   www.ekahau.com

Euclid Analytics   www.euclidanalytics.com

iBwave   www.ibwave.com

MetaGeek   www.metageek.net

NETSCOUT   www.netscout.com

Nyansa   www.nyansa.com

Savvius   www.savvius.com

Stanley Healthcare   www.stanleyhealthcare.com

TamoSoft   www.tamos.com

Wireshark   www.wireshark.org

ZingBox   www.zingbox.com

7signal   www.7signal.com

WLAN Security and Presence Analytics Solutions   The following companies offer presence analytics solutions, guest WLAN management solutions, client onboarding solutions, or 802.1X/EAP supplicant/server solutions:

Cloud4Wi   www.cloud4wi.com

Cloudessa   cloudessa.com

Cucumber Tony www.ct-networks.io

GoZone WiFi www.gozonewifi.com

Kiana www.kiana.io

Purple www.purple.ai

SecureW2 www.secureW2.com

VoWiFi Solutions Manufacturers of 802.11 VoWiFi phones and VoIP gateway solutions include the following:

Ascom www.ascom.com

Mitel www.mitel.com

Spectralink www.spectralink.com

Vocera www.vocera.com

Mobile Device Management Vendors The following are some of the vendors selling mobile device management (MDM) solutions:

AirWatch www.air-watch.com

Jamf www.jamf.com

MobileIron www.mobileiron.com

WLAN SOHO Vendors The following are some of the many WLAN vendors selling SOHO solutions that can provide Wi-Fi for the average home user:

Apple www.apple.com

Buffalo Technology www.buffalotech.com

D-Link www.dlink.com

eero www.eero.com

Google www.google.com

Hawking Technology www.hawkingtech.com

Linksys www.linksys.com

NETGEAR www.netgear.com

Plume www.plumewifi.com

TP-Link www.tp-link.com

Zyxel www.zyxel.com

Summary

This chapter covered some of the design, implementation, and management environments in which wireless networking is used. Although many of these environments are similar, each has unique characteristics. It is important to understand these similarities and differences and how wireless networking is commonly deployed.

Exam Essentials

Know the different WLAN vertical markets.   Wireless networking can be used in many environments, with each vertical market having a different primary reason or focus for installing the wireless network. Know these environments and their main reasons for deploying 802.11 wireless networking.

Review Questions

1. Which of these networking technologies can be used by IoT devices? (Choose all that apply.)
    A. Wi-Fi
    B. Bluetooth
    C. Zigbee
    D. Ethernet

2. Which type of identifier is used by BLE radios for iBeacon proximity purposes?
    A. BSSID
    B. UUID
    C. SSID
    D. PMKID

3. Which type of organization often has a person responsible for keeping track of frequency usage inside the organization?
    A. Law enforcement
    B. Hotspot
    C. Hospital
    D. Cruise ship

4. On which of these transportation networks is satellite a functional solution for providing uplink to the Internet?
    A. Bus
    B. Automobile
    C. Train
    D. Cruise ship

5. Fixed mobile convergence provides roaming across which of the following wireless technologies? (Choose all that apply.)
    A. Bluetooth
    B. Wi-Fi
    C. WiMAX
    D. Cellular

6. Which of the following is typically the most important design goal when designing a warehouse WLAN?
    A. Capacity
    B. Throughput
    C. RF interference
    D. Coverage

7. Corporations often install wireless networks to provide which of the following capabilities? (Choose all that apply.)

   A. Easy mobility for the wireless user within the corporate building or campus environment

   B. Highest-speed network access when compared with wired networking

   C. Internet access for visitors and guests

   D. The ability to easily add network access in areas where installation of wired connections is difficult or expensive

8. Last-mile Internet service is provided by which of the following? (Choose all that apply.)

   A. Telephone company

   B. Long-distance carrier

   C. Cable provider

   D. WISPs

9. Which of the following is the main purpose of a SOHO 802.11 network?

   A. Shared networking

   B. Internet gateway

   C. Home network security

   D. Print sharing

10. Which of the following are examples of mobile office networking? (Choose all that apply.)

   A. Construction-site offices

   B. Temporary disaster-assistance offices

   C. Remote sales offices

   D. Temporary classrooms

11. Warehousing and manufacturing environments typically have which of the following requirements? (Choose all that apply.)

   A. Mobility

   B. High-speed access

   C. High capacity

   D. High coverage

12. What is needed to trigger an action on a mobile device with an iBeacon proximity solution? (Choose all that apply.)

   A. 802.11 transmitter

   B. Application

   C. BLE transmitter

   D. Data encryption

13. Which of the following are good uses for portable networks? (Choose all that apply.)

    A. Military maneuvers

    B. Disaster relief

    C. Construction sites

    D. Manufacturing plants

14. Which of the following terms refer to a PtMP network design? (Choose all that apply.)

    A. PtP

    B. Mesh

    C. Hub and spoke

    D. Star

15. Most early deployments of legacy 802.11 FHSS APs were used in which type of environment?

    A. Mobile office networking

    B. Educational/classroom

    C. Industrial (warehousing and manufacturing)

    D. Health care (hospitals and offices)

16. When using a WLAN hotspot, company employees should do which of the following to ensure security back to your corporate network?

    A. Enable WEP.

    B. Enable 802.1X/EAP.

    C. Use an IPsec VPN.

    D. Security cannot be provided because you do not control the access point.

17. What are some popular 802.11 applications used in the health-care industry? (Choose all that apply.)

    A. VoWiFi

    B. Bridging

    C. RTLS

    D. Patient monitoring

18. Multiple point-to-point bridges between the same locations are often installed for which of the following reasons? (Choose all that apply.)

    A. To provide higher throughput

    B. To prevent channel overlap

    C. To prevent single point of failure

    D. To enable support for VLANs

19. What are some of the key concerns of health-care providers when installing a wireless network? (Choose all that apply.)

   A. RF interference

   B. Faster access to patient data

   C. Secure and accurate access

   D. Faster speed

20. Public hotspots typically provide clients with which of the following security features?

   A. Authentication

   B. Encryption

   C. TKIP

   D. No client security is available.

Appendix A

Answers to Review Questions

Chapter 1: Overview of Wireless Standards, Organizations, and Fundamentals

1. C. 802.11 wireless networking is typically used to connect client stations to the network via an access point (AP). Access points are deployed at the access layer, not the core or distribution layer. The Physical layer is a layer of the OSI model, not a network architecture layer.

2. E. Radio frequency (RF) communications are regulated differently in many regions and countries. The local regulatory domain authorities of individual countries or regions define the spectrum policies and transmit power rules.

3. B. 802.11 wireless bridge links are typically used to perform distribution layer services. Core layer devices are usually much faster than 802.11 wireless devices, and bridges are not used to provide access layer services. The Network layer is a layer of the OSI model, not a network architecture layer.

4. A. The Institute of Electrical and Electronics Engineers (IEEE) is responsible for the creation of all of the 802 standards.

5. D. The Wi-Fi Alliance provides certification testing. When a product passes the test, it receives a Wi-Fi Interoperability Certificate.

6. C. A carrier signal is a modulated signal that is used to transmit binary data.

7. B. Because of the effects of noise on the amplitude of a signal, amplitude-shift keying (ASK) has to be used cautiously.

8. C. The IEEE 802.11-2016 standard defines communication mechanisms at only the Physical layer and MAC sublayer of the Data-Link layer of the OSI model. The Logical Link Control (LLC) sublayer of the Data-Link layer is not defined for 802.11 operations. Physical Layer Convergence Procedure (PLCP) and Physical Medium Dependent (PMD) are sublayers of the Physical layer.

9. E. The Internet Engineering Task Force (IETF) is responsible for creation of Request for Comments (RFC) documents. The Institute of Electrical and Electronics Engineers (IEEE) is responsible for the 802 standards. The Wi-Fi Alliance is responsible for certification tests. The Wi-Fi Alliance used to be known as the Wireless Ethernet Compatibility Alliance (WECA) but changed its name to Wi-Fi Alliance in 2002. The Federal Communications Commission (FCC) is responsible for radio frequency (RF) regulatory rules in the United States.

10. D. Wi-Fi Multimedia (WMM) is a Wi-Fi Alliance certification program that enables Wi-Fi networks to prioritize traffic generated by different applications. 802.11-2016 is the IEEE standard, and WEP (Wired Equivalent Privacy) is a legacy encryption method defined as part of the IEEE 802.11-2016 standard. 802.11i was the IEEE amendment that defined robust security network (RSN) and is also part of the 802.11-2016 standard.

11. A, B, C. The three keying methods that can be used to encode data are amplitude-shift keying (ASK), frequency-shift keying (FSK), and phase-shift keying (PSK).

12. B, E. The IEEE 802.11-2016 standard defines communication mechanisms at only the Physical layer and MAC sublayer of the Data-Link layer of the OSI model.

13. C. *Height* and *power* are two terms that describe the amplitude of a wave. Frequency is how often a wave repeats itself. Wavelength is the actual length of the wave, typically measured from peak to peak. Phase refers to the starting point of a wave in relation to another wave.

14. B, C. Wi-Fi Direct is designed to provide easy setup for communications directly between wireless devices without the need for an access point (AP). Tunneled Direct Link Setup (TDLS) communications operate between client devices while they are still associated to an AP.

15. A, C, E. Voice-Enterprise offers enhanced support for voice applications in enterprise Wi-Fi networks. Voice-Enterprise equipment must also support seamless roaming between access points (APs), WPA2-Enterprise security, optimization of power through the WMM-Power Save mechanism, and traffic management through WMM-Admission Control.

16. A, B, C, D, E. All of these are typically regulated by the local or regional radio frequency (RF) regulatory authority.

17. B. In half-duplex communications, both devices are capable of transmitting and receiving; however, only one device can transmit at a time. Walkie-talkies, or two-way radios, are examples of half-duplex devices. All radio frequency (RF) communications by nature are half-duplex. IEEE 802.11 WLAN radios use half-duplex communications.

18. D. A wave is divided into 360 degrees.

19. B, C. The main advantages of an unlicensed frequency are that permission to transmit on the frequency is free and that anyone can use the unlicensed frequency. Although there are no additional financial costs, you still must abide by transmission regulations and other restrictions. The fact that anyone can use the frequency band is also a disadvantage because of overcrowding.

20. C. The OSI model is sometimes referred to as the seven-layer model.

Chapter 2: IEEE 802.11 Standard and Amendments

1. A, D. Support for both Extended Rate Physical DSSS (ERP-DSSS/CCK) and Extended Rate Physical Orthogonal Frequency-Division Multiplexing (ERP-OFDM) are required in an ERP WLAN, also known as an 802.11g WLAN.

2. B. 802.11ad supports data rates of up to 7 Gbps. The downside is that 60 GHz will have significantly less effective range than a 5 GHz signal and be limited to line-of-sight communications, as the high-frequency signal will have difficulty penetrating walls. The 802.11ay draft amendment is expected to improve 802.11ad, providing faster speeds and longer range.

3. B, D, E. The original 802.11 standard defines three Physical layer specifications. An 802.11 legacy network could use frequency-hopping spread-spectrum (FHSS), Direct-sequence spread-spectrum (DSSS), or IR (infrared) technology. 802.11b defined the use of High-Rate DSSS (HR-DSSS); 802.11a defined the use of orthogonal frequency-division multiplexing (OFDM); and 802.11g defined Extended Rate Physical (ERP).

4. C. The 802.11 Task Group s (TGs) has set forth the pursuit of standardizing mesh networking using the IEEE 802.11 MAC/PHY layers. The 802.11s amendment defines the use of mesh points (MPs), which are 802.11 QoS stations that support mesh services. An MP is capable of using a mandatory mesh routing protocol called Hybrid Wireless Mesh Protocol (HWMP) that uses a default path selection metric. Vendors may also use proprietary mesh routing protocols and metrics.

5. D, F. The required encryption method defined by an RSN wireless network (802.11i) is Counter Mode with Cipher Block Chaining Message Authentication Code Protocol (CCMP), which uses the Advanced Encryption Standard (AES) algorithm. An optional choice of encryption is the Temporal Key Integrity Protocol (TKIP). The 802.11i amendment also requires the use of an 802.1X/EAP authentication solution or the use of preshared keys.

6. D. 802.11ac radio cards operate in the 5 GHz frequency bands using very high throughput (VHT).

7. D. The IEEE 802.11-2016 standard requires data rates of 6, 12, and 24 Mbps for both orthogonal frequency-division multiplexing (OFDM) and Extended Rate Physical OFDM (ERP-OFDM) radios. Data rates of 6, 9, 12, 18, 24, 36, 48, and 54 Mbps are typically supported. 54 Mbps is the maximum defined rate.

8. B. Fast basic service set transition (FT), also known as fast-secure roaming, defines fast handoffs when roaming occurs between cells in a WLAN using the strong security defined in a robust security network (RSN). Applications such as VoIP that necessitate timely delivery of packets require the roaming handoff to occur in 150ms or less.

9. B, C, E. The 802.11ac amendment debuted and defined the use of 256-QAM modulation, eight spatial streams, multi-user MIMO (MU-MIMO), 80 MHz channels, and 160 MHz channels. 802.11 MIMO technology and 40 MHz channels debuted with the ratification of the 802.11n amendment.

10. D. Both 802.11a and 802.11g use orthogonal frequency-division multiplexing (OFDM) technology, but because they operate at different frequencies, they cannot communicate with each other. 802.11a equipment operates in the 5 GHz Unlicensed National Information Infrastructure (U-NII) bands, whereas 802.11g equipment operates in the 2.4 GHz industrial, scientific, and medical (ISM) band.

11. A, E. The 802.11-2016 standard defines mechanisms for dynamic frequency selection (DFS) and transmit power control (TPC) that may be used to satisfy regulatory requirements for operation in the 5 GHz band. DFS and TPC technology were originally defined in the 802.11h amendment, which is now part of the 802.11-2016 standard.

12. C, D. The 802.11ac and 802.11ad amendments are often referred to as the "gigabit Wi-Fi" amendments because they define data rates of greater than 1 Gbps. The 802.11ac and 802.11ad very high throughput (VHT) task groups define transmission rates of up to 7 Gbps in an 802.11 environment.

13. A, C, D, E. ERP (802.11g) requires the use of Extended Rate Physical OFDM (ERP-OFDM) and Extended Rate Physical DSSS (ERP-DSSS/CCK) in the 2.4 GHz industrial, scientific, and medical (ISM) band and is backward compatible with 802.11b High-Rate DSSS (HR-DSSS) and direct-sequence spread-spectrum (DSSS) equipment. 802.11b uses HR-DSSS in the 2.4 GHz ISM band and is backward compatible with only legacy DSSS equipment and not legacy frequency-hopping spread-spectrum (FHSS) equipment. 802.11ac uses Very High Throughput (VHT) and is backward compatible with 802.11a orthogonal frequency-division multiplexing (OFDM). The 802.11h amendment defines use of transmit power control (TPC) and dynamic frequency selection (DFS) in the 5 GHz Unlicensed National Information Infrastructure (U-NII) bands and is an enhancement of the 802.11a amendment. OFDM technology is used with all 802.11a- and 802.11h-compliant radios.

14. D. The 802.11ac amendment defines a maximum rate of 6933.3 Mbps. However, 802.11ac chipsets have gone through multiple generations with different capabilities. The maximum data rate for most first-generation 802.11ac devices was 1300 Mbps, and the maximum data rate for most second-generation 802.11ac devices is 3466.7 Mbps.

15. B, D, E. The original 802.11 standard defined the use of Wired Equivalent Privacy (WEP) for encryption. The original 802.11 standard also defined two methods of authentication: Open System authentication and Shared Key authentication.

16. A. The 802.11u draft amendment defines integration of IEEE 802.11 access networks with external networks in a generic and standardized manner. 802.11u is often referred to as Wireless Interworking with External Networks (WIEN).

17. A, C. The 802.11e amendment (now part of the 802.11-2012 standard) defined two enhanced medium access methods to support quality of service (QoS) requirements. Enhanced Distributed Channel Access (EDCA) is an extension to Distributed Coordination Function (DCF). Hybrid Coordination Function Controlled Channel Access (HCCA) is an extension to PCF. In the real world, only EDCA is implemented.

18. A, C. The 802.11h amendment introduced two major enhancements: transmit power control (TPC) and dynamic frequency selection (DFS) for radar avoidance and detection. All aspects of the 802.11h ratified amendment can now be found in Clause 11.8 and Clause 11.9 of the 802.11-2016 standard.

19. E. The 802.11n high throughput (HT) amendment defined MAC and PHY enhancements, whereas radios can transmit at data rates up to 600 Mbps.

20. B, D. The IEEE specifically defines 802.11 technologies at the Physical layer and the MAC sublayer of the Data-Link layer. By design, anything that occurs at the upper layers of the OSI model is insignificant to 802.11 communications.

Chapter 3: Radio Frequency Fundamentals

1. B, C. Multipath may result in attenuation, amplification, signal loss, or data corruption. If two signals arrive together in phase, the result is an increase in signal strength called upfade. The delay spread may also be too significant and cause data bits to be corrupted, resulting in excessive layer 2 retransmissions.

2. D. The wavelength is the linear distance between the repeating crests (peaks) or repeating troughs (valleys) of a single cycle of a wave pattern.

3. B, C. RF amplifiers introduce active gain with the help of an outside power source. Passive gain is typically created by antennas that focus the energy of a signal without the use of an outside power source.

4. A. The standard measurement of the number of times a signal cycles per second is hertz (Hz). One Hz is equal to one cycle in 1 second.

5. D. Often confused with refraction, the diffraction propagation is the bending of the wave front around an obstacle. Diffraction is caused by some sort of partial blockage of the RF signal, such as a small hill or a building that sits between a transmitting radio and a receiver.

6. F. Nulling, or cancellation, can occur when multiple RF signals arrive at the receiver at the same time and are 180 degrees out of phase with the primary wave.

7. B, C. When the multiple RF signals arrive at the receiver at the same time and are in phase or partially out of phase with the primary wave, the result is an increase in signal strength (amplitude). However, the final received signal, whether affected by upfade or downfade, will never be stronger than the original transmitted signal, because of free space path loss.

8. B. 802.11 WLANs operate in the 5 GHz and 2.4 GHz frequency range. However, 2.4 GHz is equal to 2.4 billion cycles per second. The frequency of 2.4 million cycles per second is 2.4 MHz.

9. A. An oscilloscope is a time-domain tool that can be used to measure how a signal's amplitude changes over time. A frequency-domain tool called a spectrum analyzer is a more commonplace tool most often used during site surveys.

10. A, C, D. This is a tough question to answer because many of the same mediums can cause several different propagation behaviors. Metal will always bring about reflection. Water is a major source of absorption; however, large bodies of water can also cause reflection. Flat surfaces, such as asphalt roads, ceilings, and walls, will also result in reflection behavior.

11. A, B, C, D. Multipath is a propagation phenomenon that results in two or more paths of a signal arriving at a receiving antenna at the same time or within nanoseconds of each other. Because of the natural broadening of the waves, the propagation behaviors of reflection, scattering, diffraction, and refraction can all result in multiple paths of the same signal. The propagation behavior of reflection is usually considered to be the main cause of high-multipath environments.

12. B. Scattering, or scatter, is defined as an RF signal reflecting in multiple directions when encountering an uneven surface.

13. A, B, C. High-multipath environments can have a destructive impact on legacy 802.11a/b/g radio transmissions. Multipath has a constructive effect with 802.11n and 802.11ac transmissions that utilize MIMO antenna diversity and maximum ratio combining (MRC) signal-processing techniques. Multipath does not affect the security mechanisms defined by 802.11i.

14. A, B, C, D. Air stratification is a leading cause of refraction of an RF signal. Changes in air temperature, changes in air pressure, and water vapor are all causes of refraction. Smog can cause a density change in the air pressure as well as increased moisture.

15. A, D. Because of the natural broadening of the wave front, electromagnetic signals lose amplitude as they travel away from the transmitter. The rate of free space path loss (FSPL) is logarithmic, not linear. Attenuation of RF signals as they pass through different mediums does occur but is not a function of FSPL.

16. D. The time difference due to a reflected signal taking a longer path is known as the delay spread. The delay spread can cause intersymbol interference, which results in data corruption and layer 2 retransmissions.

17. C. A spectrum analyzer is a frequency-domain tool that can be used to measure amplitude in a finite-frequency spectrum. An oscilloscope is a time-domain tool.

18. A, C. Concrete walls are very dense and will significantly attenuate a 2.4 GHz and 5 GHz signal. Older structures that are constructed with wood-lath plaster walls often have wire mesh in the walls, which was used to help hold the plaster to the walls. Wire mesh is notorious for disrupting and preventing RF signals from passing through walls. Wire mesh is also used on stucco exteriors. Drywall will attenuate a signal but not to the extent of water, cinder blocks, or other dense mediums. Air temperature has no significance during an indoor site survey.

19. A. There is an inverse relationship between frequency and wavelength. A simplified explanation is that the higher the frequency of an RF signal, the shorter the wavelength will be of that signal. The longer the wavelength of an RF signal, the lower the frequency of that signal.

20. A. Refraction is the bending of an RF signal when it encounters a medium.

Chapter 4: Radio Frequency Components, Measurements, and Mathematics

1. D. The signal-to-interference-plus-noise ratio (SINR) compares the primary signal to both interference and noise. Since interference fluctuates and can change quickly, this is a better indicator of what is happening to a signal at a specific time. The signal-to-noise (SNR) compares the signal to noise; however, noise is less likely to fluctuate drastically. The received signal strength indicator (RSSI) and equivalent isotropically radiated power (EIRP) are measurements of signal but do not relate the signal to other outside influences.

2. E. An isotropic radiator is also known as a point source.

3. A, B, C, E, F. When radio communications are deployed, a link budget is the sum of all gains and losses from the transmitting radio, through the RF medium, to the receiver radio. Link budget calculations include original transmit gain and passive antenna gain. All losses must be accounted for, including free space path loss. Frequency and distance are needed to calculate free space path loss. The height of an antenna has no significance when calculating a link budget; however, the height could affect the Fresnel Zone and blockage to it.

4. A, D. IR is the abbreviation for intentional radiator. The components making up the IR include the transmitter, all cables and connectors, and any other equipment (grounding, lightning arrestors, amplifiers, attenuators, and so forth) between the transmitter and the antenna. The power of the IR is measured at the connector that provides the input to the antenna.

5. A. Equivalent isotropically radiated power, also known as EIRP, is a measure of the strongest signal that is radiated from an antenna.

6. A, B, D. Watts, milliwatts, and dBms are all absolute power measurements. One watt is equal to 1 ampere (amp) of current flowing at 1 volt. A milliwatt is 1/1,000 of 1 watt. dBm is decibels relative to 1 milliwatt.

7. B, C, D, E. The unit of measurement known as a bel is a relative expression and a measurement of change in power. A decibel (dB) is equal to one-tenth of a bel. Antenna gain measurements of dBi and dBd are relative measurements. dBi is defined as decibels relative to an isotropic radiator. dBd is defined as decibels relative to a dipole antenna.

8. C. To convert any dBd value to dBi, simply add 2.14 to the dBd value.

9. A. To convert dBm to mW, first calculate how many 10s and 3s are needed to add up to 23, which is 0 + 10 + 10 + 3. To calculate the mW, you must multiply 1 × 10 × 10 × 2, which calculates to 200 mW. The file ReviewQuestion9.ppt, available for download from www.wiley.com/go/cwnasg, shows the process in detail.

10. C. To reach 100 mW, you can use 10s and 2s and multiplication and division. Multiplying by two 10s will accomplish this. This means that on the dBm side, you must add two 10s, which equals 20 dBm. Then subtract the 3 dB of cable loss for a dBm of 17. Because you subtracted 3 from the dBm side, you must divide the 100 mW by 2, giving you a value of 50 mW. Now add in the 16 dBi by adding a 10 and two 3s to the dBm column, giving a total dBm of 33. Because you added a 10 and two 3s, you must multiply the mW column by 10 and two 2s, giving a total of 2,000 mW, or 2 W. Since the cable and connector loss is 3 dB and the antenna gain is 16 dBi, you can add the two together for a cumulative gain of 13 dB; then apply that gain to the 100 mW transmit signal to calculate an EIRP of 2,000 mW, or 2 W. The file `ReviewQuestion10.ppt`, available for download from `www.wiley .com/go/cwnasg`, shows the process in detail.

11. A. If the original transmit power is 400 mW and cabling induces a 9 dB loss, the power at the opposite end of the cable would be 50 mW. The first 3 dB of cable loss halved the absolute power to 200 mW. The second 3 dB of cable loss halved the absolute power to 100 mW. The final 3 dB of cable loss halved the power to 50 mW. The antenna with 19 dBi of gain passively amplified the 50 mW signal to 4,000 mW. The first 10 dBi of antenna boosted the signal to 500 mW. The next 9 dBi of antenna gain doubled the signal three times to a total of 4 watts. Since the cable loss is 9 dB and the antenna gain is 19 dBi, you could add the two together for a cumulative gain of 10 dB and then apply that gain to the 400 mW transmit signal to calculate an EIRP of 4,000 mW, or 4 W.

12. B, D. Received signal strength indicator (RSSI) thresholds are a key factor for clients when they initiate the roaming handoff. RSSI thresholds are also used by vendors to implement dynamic rate switching, which is a process used by 802.11 radios to shift between data rates.

13. A. The received signal strength indicator (RSSI) is a metric used by 802.11 radio cards to measure signal strength (amplitude). Some vendors use a proprietary scale to also correlate to signal quality. Most vendors erroneously define signal quality as the signal-to-noise ratio (SNR). The signal-to-noise ratio is the difference in decibels between the received signal and the background noise (noise floor).

14. B. dBi is defined as "decibel gain referenced to an isotropic radiator" or "change in power relative to an antenna." dBi is the most common measurement of antenna gain.

15. A, F. The four rules of the 10s and 3s are as follows: For every 3 dB of gain (relative), double the absolute power (mW). For every 3 dB of loss (relative), halve the absolute power (mW). For every 10 dB of gain (relative), multiply the absolute power (mW) by a factor of 10. For every 10 dB of loss (relative), divide the absolute power (mW) by a factor of 10.

16. B. If the original transmit power were 100 mW and cabling induced a 3 dB loss, the power at the opposite end of the cable would be 50 mW. The 3 dB of cable loss halves the absolute power to 50 mW. An antenna with 10 dBi of gain would boost the signal to 500 mW. We also know that 3 dB of loss halves the absolute power. Therefore, an antenna with 7 dBi of gain would amplify the signal to half that of a 10 dBi antenna. The antenna with 7 dBi of gain passively amplifies the 50 mW signal to 250 mW.

17. D. A distance of as little as 100 meters will cause free space path loss (FSPL) of 80 dB, far greater than any other component. RF components such as connectors, lightning arrestors, and cabling all introduce insertion loss. However, FSPL will always be the reason for the greatest amount of loss.

18. B. The 6 dB rule states that increasing the amplitude by 6 decibels doubles the usable distance of an RF signal. The 6 dB rule is very useful for understanding antenna gain because every 6 dBi of extra antenna gain will double the usable distance of an RF signal.

19. D. In a high multipath or noisy environment, a common best practice is to add a 5 dB fade margin when designing for coverage based on a vendor's recommended received signal strength or the noise floor, whichever is louder.

20. D. WLAN vendors execute received signal strength indicator (RSSI) metrics in a proprietary manner. The actual range of the RSSI value is from 0 to a maximum value (less than or equal to 255) that each vendor can choose on its own (known as RSSI_Max). Therefore, RSSI metrics should not be used to compare different WLAN vendor radios because there is no standard for the range of values or a consistent scale.

Chapter 5: Radio Frequency Signal and Antenna Concepts

1.   A, C, F. The azimuth chart is the top-down view of an antenna's radiation pattern, also known as the H-plane, or horizontal view. The side view is known as the elevation chart, E-plane, or vertical view.

2.   A. The azimuth is the top-down view of an antenna's radiation pattern, also known as the H-plane.

3.   C. The beamwidth is the distance in degrees between the –3 dB (half-power) point on one side of the main signal and the –3 dB point on the other side of the main signal, measured along the horizontal axis. These are also known as half-power points.

4.   C, D. Parabolic dish antennas and grid antennas are highly directional. The rest of the antennas are semidirectional, and the sector antenna is a special type of semidirectional antenna.

5.   A, C, D. Semidirectional antennas provide too wide of a beamwidth to support long-distance communications, but they will work for short distances. They are also useful for providing unidirectional coverage from the access point to clients in an indoor environment. They can also minimize reflections and thus the negative effects of multipath.

6.   B. Any more than a 40-percent encroachment into the Fresnel zone is likely to make a link unreliable. The clearer the Fresnel zone, the better, and ideally it should not be blocked at all.

7. C, D. The distance and frequency determine the size of the Fresnel zone; these are the only variables in the Fresnel zone formula.

8. B. The distance when the curvature of the earth should be considered is 7 miles.

9. A, C. Installing a shorter cable of the same grade will result in less loss and thus more amplitude being transmitted out the antenna. A higher-grade cable rated for less dB loss will have the same result.

10. B. An 802.11n or 802.11ac transceiver can transmit from multiple antennas at the same time, if it is operating using multiple spatial streams and multiple radio chains.

11. A, D. Point-to-point bridge links require a minimum Fresnel zone clearance of 60 percent. Semidirectional antennas, such as patch antennas or Yagi antennas, are used for short-to-medium-distance bridge links. Highly directional antennas are used for long-distance bridge links. Compensating for earth bulge is not a factor until seven miles.

12. C. Voltage standing wave ratio (VSWR) is the difference between these voltages and is represented as a ratio—for example, 1.5:1.

13. A, C, D, E. The reflected voltage caused by an impedance mismatch can result in a decrease in power or amplitude (loss) of the signal that is supposed to be transmitted. If the transmitter is not protected from excessive reflected power or large voltage peaks, it can overheat and fail. Understand that VSWR may cause decreased signal strength, erratic signal strength, or even transmitter failure.

14. A, B, D, F. Frequency and distance are needed to determine the Fresnel zone. Visual line of sight is not needed so long as you have RF line of sight. You may not be able to see the antenna because of fog, but the fog will not prevent RF line of sight. Earth bulge will need to be considered. The beamwidth is not needed to determine the height, although it is useful when aiming the antenna.

15. A, D. Cables must be selected that support the frequency you are using. Attenuation actually increases with frequency.

16. A, B, C, D. These are all possible capabilities of RF amplifiers.

17. A, B, D. Adding an attenuator is an intentional act to add loss to the signal. Since cable adds loss, increasing the length will add more loss, whereas shortening the length will reduce the loss. Better-quality cables produce less signal loss.

18. C. Lightning arrestors will not stand up to a direct lightning strike, only transient currents caused by nearby lightning strikes.

19. A, D. The first Fresnel zone is in phase with the point source. The second Fresnel zone begins at the point where the signals transition from being in phase to being out of phase. Because the second Fresnel zone begins where the first Fresnel zone ends, the radius of the second Fresnel zone is larger than the radius of the first Fresnel zone.

20. D. Side lobes are areas of coverage (other than the coverage provided by the main signal) that have a stronger signal than would be expected when compared with the areas around them. Side lobes are best seen on an azimuth chart. Side bands and frequency harmonics have nothing to do with antenna coverage.

Chapter 6: Wireless Networks and Spread Spectrum Technologies

1. C, D. OFDM has a greater tolerance of delay spread over previous methods, such as FHS, DSSS, and HR-DSSS. HT is based on OFDM, so it is also correct.

2. A, B, C. The four current U-NII bands are 5.15 GHz–5.25 GHz, 5.25 GHz–5.35 GHz, 5.47 GHz–5.725 GHz, and 5.725 GHz–5.85 GHz.

3. A, B, C, D. The IEEE 802.11-2016 standard defines the use of legacy DSSS radios (802.11), HR-DSSS radios (802.11b), ERP radios (802.11g), and HT radios (802.11n). Legacy FHSS radios (802.11) are no longer referenced in the standard, although a few might still exist in the field.

4. A, B, D. The IEEE 802.11-2016 standard specifies that 802.11n HT radios can transmit in the 2.4 GHz ISM band and all four of the current 5 GHz U-NII bands.

5. A. The U-NII-1 band is between 5.15 GHz and 5.25 GHz, or 5,150 MHz to 5,250 MHz. To calculate the frequency in MHz from the channel, multiply the channel by 5 (200) and then add 5,000, for a center frequency of 5,200 MHz, or 5.2 GHz.

6. D. To calculate the channel, first take the frequency in MHz (5,300 MHz). Subtract 5,000 from the number (300) and then divide the number by 5, resulting in channel 60. The U-NII-2A band is between 5.25 GHz and 5.35 GHz.

7. A. A single-channel OFDM signal is approximately 20 MHz wide. 802.11 DSSS and HR-DSSS signals are approximately 22 MHz wide.

8. C. The time that the transmitter waits before hopping to the next frequency is known as the dwell time. The hop time is not a required time but rather a measurement of how long the hop takes.

9. B. The 802.11a amendment, which originally defined the use of OFDM, required only 20 MHz of separation between the center frequencies for channels to be considered non-overlapping. All 25 channels in the 5 GHz U-NII bands use OFDM and have 20 MHz of separation. Therefore, all 5 GHz OFDM channels are considered nonoverlapping by the IEEE. However, it should be noted that adjacent 5 GHz channels do have some sideband carrier frequency overlap.

10. C, D. In the 2.4 GHz band, in order for any 2.4 GHz channels to be considered nonoverlapping, they require 25 MHz of separation between the center frequencies. Therefore, any two channels must have at least a five-channel separation. The simplest way to determine what other channels are valid is to add five or subtract five from the channel you want to use. If you added five, then the number you calculated or any channel above that number is valid. If you subtracted five, then the number you calculated or any channel below that number is valid. Deployments of three or more access points in the 2.4 GHz ISM band normally use channels 1, 6, and 11, which are all considered nonoverlapping.

11. F. While the FCC is denying expansion of Wi-Fi into U-NII-2B band, there is still the possibility for additional frequency expansion at the top end of the 5 GHz band. The U-NII-4 frequency band, 5.85 GHz–5.925 GHz, was reserved decades ago by US and European regulatory bodies to allow *Wireless Access in Vehicular Environments (WAVE)* communications from vehicle-to-vehicle and vehicle-to-roadway. The 802.11p amendment defines Dedicated Short-Range Communications (DSRC), which could be used in this band. The automobile industry is seeing significant innovation towards self-driving automobiles and enhanced safety features, such as blind-side monitoring. These technologies may rely heavily on the U-NII-4 band.

12. B. The cause of the problem is delay spread resulting in intersymbol interference (ISI), which causes data corruption.

13. D. The 802.11-2016 standard states that "the OFDM PHY shall operate in the 5 GHz band, as allocated by a regulatory body in its operational region." A total of twenty-five 20 MHz wide channels are available in the U-NII bands.

14. D. Because of the lower subcarrier data rates, delay spread is a smaller percentage of the symbol period, which means that ISI is less likely to occur. In other words, OFDM technology is more resistant to the negative effects of multipath than DSSS and FHSS spread spectrum technologies.

15. C. A medium access method known as Carrier Sense Multiple Access with Collision Avoidance (CSMA/CA) helps to ensure that only one radio can be transmitting on the medium at any given time. Depending on WLAN network conditions, the aggregate WLAN throughput is usually 50 percent of the advertised 802.11 data rate, due to medium contention overhead. 802.11 data rates are not TCP throughput. The medium contention protocol of CSMA/CA consumes much of the available bandwidth. In laboratory conditions, the TCP throughput in an 802.11n/ac environment is 60 percent to 70 percent of the data rate between one AP and one client. The aggregate throughput numbers are considerably less in real-world environments with active participation of multiple WLAN clients communicating through an AP.

16. F. A new 75 MHz band called U-NII-4 occupies the 5.85 GHz–5.925 GHz frequency space, with the potential of four more 20 MHz channels.

17. C. In 2009, the Federal Aviation Authority (FAA) reported interference to Terminal Doppler Weather Radar (TDWR) systems. As a result, the FCC suspended certification of 802.11 devices in the U-NII-2A and U-NII-2C bands that require DFS. Certification was eventually re-established; however, the rules changed and 802.11 radios are currently allowed to transmit in the 5.60–5.65 GHz frequency space where TDWR operates. Channels 120–128 were not available for a number of years. As of April 2014, the TDWR frequency space is once again available for 802.11 transmissions in the United States.

18. A, B. OFDM uses BPSK and QPSK modulation for the lower ODFM data rates. The higher OFDM data rates use 16-QAM, 64-QAM, and 256-QAM modulation. QAM modulation is a hybrid of phase and amplitude modulation.

19. B. When a data bit is converted to a series of bits, these bits that represent the data are known as chips.

20. C. A 20 MHz OFDM channel has 64 subcarriers, but only 48 of them are used to transport modulated data between 802.11a/g radios. Twelve of the 64 subcarriers in a 20 MHz OFDM channel are unused and serve as guard bands, and four of the subcarriers function as pilot carriers. 802.11n/ac radios also transmit on a 20 MHz channel that consists of 64 total subcarriers; however, only 8 subcarriers are guard bands. Fifty-two of the subcarriers are used to transmit modulated data, and four of the subcarriers function as pilot carriers.

Chapter 7: Wireless LAN Topologies

1. D, E. The service set identifier (SSID) is an up to 32-character, case-sensitive, logical name used to identify an 802.11 wireless network. An extended service set identifier (ESSID) is the logical network name used in an extended service set (ESS). ESSID is often synonymous with SSID.

2. C, E. The 802.11 standard defines four service sets, or topologies. A basic service set (BSS) is defined as one access point (AP) and associated clients. An extended service set (ESS) is defined as one or more BSSs connected by a distribution system medium (DSM). An independent basic service set (IBSS) does not use an AP and consists solely of client stations (STAs). A personal basic service set (PBSS) used by 802.11ad radios is similar to an IBSS because no centralized access point functions as a portal to a DSM.

3. E. By design, the 802.11 standard does not specify a medium to be used in the distribution system (DS). The distribution system medium (DSM) may be an 802.3 Ethernet backbone, an 802.5 token ring network, a wireless medium, or any other medium.

4. D. A wireless personal area network (WPAN) is a short-distance wireless topology. Bluetooth and Zigbee are technologies that are often used in WPANs.

5. A. The most common implementation of an extended service set (ESS) has access points (APs) with overlapping coverage cells. The purpose behind an ESS with overlapping coverage cells is seamless roaming.

6. A, C, D. The size and shape of a basic service area (BSA) can depend on many variables, including access point (AP) transmit power, antenna gain, and physical surroundings. You can also make the argument that the effective range of the BSA is really from the perspective of any connected client station, because all client devices interpret the received signal strength indicator (RSSI) differently.

7. C. The normal default setting of an access point (AP) is root mode, which allows the AP to transfer data back and forth between the distribution system (DS) and the 802.11 wireless medium. The default root configuration of an AP allows it to operate inside a basic service set (BSS). WLAN vendors may also refer to the default as access mode or AP mode.

8. A, E, F. The 802.11-2016 standard defines an independent basic service set (IBSS) as a service set using client peer-to-peer communications without the use of an access point (AP). Other names for an IBSS include ad hoc network and peer-to-peer network.

9. A, D. Clients that are configured in infrastructure mode may communicate via the access point (AP) with other wireless client stations within a basic service set (BSS). Clients may also communicate through the AP with other networking devices that exist on the distribution system medium (DSM), such as a server or a wired desktop. The integration service of the AP will transfer the MAC service data unit (MSDU) payload from 802.11 wireless clients into an 802.3 frame when Ethernet is the DSM.

10. B, C, D, F. The 802.11 topologies, or service sets, defined by the 802.11-2016 standard are the basic service set (BSS), extended service set (ESS), independent basic service set (IBSS), personal basic service set (PBSS), QoS basic service set (QBSS), and mesh basic service set (MBSS). DSSS and FHSS are spread spectrum technologies.

11. A. A wireless metropolitan area network (WMAN) provides coverage to a metropolitan area, such as a city and the surrounding suburbs.

12. D. The basic service set identifier (BSSID) is a 48-bit (6-octet) MAC address. MAC addresses exist at the MAC sublayer of the Data-Link layer of the OSI model. The BSSID is the layer 2 identifier of a basic service set (BSS).

13. B, C, E. The basic service set identifier (BSSID) is the layer 2 identifier of either a basic service set (BSS) or an independent basic service set (IBSS). The 48-bit (6-octet) MAC address of an access point's radio is the BSSID within a BSS. An extended service set (ESS) topology utilizes multiple APs, thus the existence of multiple BSSIDs. In an IBSS network, the first station that powers up randomly generates a virtual BSSID in the MAC address format. FHSS and HR-DSSS are spread spectrum technologies.

14. D. The 802.11s-2011 amendment, which is now part of the 802.11-2016 standard, defined a new service set for an 802.11 mesh topology. When access points (APs) support mesh functions, they may be deployed where wired network access is not possible. The mesh functions are used to provide wireless distribution of network traffic, and the set of APs that provide mesh distribution form a mesh basic service set (MBSS).

15. D. Similar to the independent basic service set (IBSS), a personal basic service set (PBSS) is an 802.11 WLAN topology in which 802.11ad stations communicate directly with each other. A PBSS can be established only by directional multi-gigabit (DMG) radios that transmit the 60 GHz frequency band. Similar to an IBSS, no centralized access point (AP) functions as a portal to a distribution system medium (DSM), such as a wired 802.3 Ethernet network.

16. A. The station service (SS) exists in all 802.11 stations, including client stations and access points (APs). The SS provides capabilities such as authentication, deauthentication, data confidentiality, and more. The distribution system service (DSS) operates only within APs and mesh portals. The PBBS control point service (PCPS) is defined specifically for 802.11ad radios when operated in a very specific 802.11 topology, called a personal basic service set (PBSS). The integration service enables delivery of MAC service data units (MSDUs) between the distribution system (DS) and a non-IEEE-802.11 LAN via a portal.

17. C. An extended service set (ESS) is two or more basic service sets (BSSs) connected by a distribution system (DS). An ESS is a collection of multiple access points (APs) and their associated client stations, all united by a single distribution system medium (DSM).

18. A. A wireless distribution system (WDS) can connect access points (APs) using a wireless backhaul while allowing clients to also associate to the radios in the access points.

19. B, C. The distribution system (DS) consists of two main components. The distribution system medium (DSM) is a logical physical medium used to connect access points (APs). The distribution system service (DSS) resides in an AP station and is used to manage client station associations, reassociations, and disassociations.

20. B. The 802.11-2016 standard is considered a wireless local area network (WLAN) standard. 802.11 hardware can, however, be utilized in other wireless topologies.

Chapter 8: 802.11 Medium Access

1. B. Carrier Sense Multiple Access with Collision Avoidance (CSMA/CA) is an 802.11 wireless medium access control method that is part of Distributed Coordination Function (DCF). Carrier Sense Multiple Access with Collision Detection (CSMA/CD) is used by 802.3, not 802.11. Token passing is used in Token Ring and Fiber Distributed Data Interface (FDDI) networking. Demand priority is used in 100BaseVG networking

2. E. 802.11 technology does not use collision detection. If an acknowledgment (ACK) frame is not received by the original transmitting radio, the unicast frame is not acknowledged and will have to be retransmitted. This process does not specifically determine whether a collision occurs. Failure to receive an ACK frame from the receiver means that either a unicast frame was not received by the destination station or the ACK frame was not received, but it cannot positively determine the cause. It may be due to collision or to other reasons, such as high noise level. All the other options are used to help prevent collisions.

3. D. Acknowledgment (ACK) frames and clear-to-send (CTS) frames follow a short interframe space (SIFS).

4. D, E. The network allocation vector (NAV) timer maintains a prediction of future traffic on the medium based on duration value information seen in a previous frame transmission. Virtual carrier sense uses the NAV to determine medium availability. Physical carrier sense checks the radio frequency (RF) medium for carrier availability. Physical carrier sense occurs during the clear channel assessment (CCA). The contention window and backoff timer are part of the pseudo-random backoff procedure. A channel sense window does not exist.

5. C. The first step is to select a random backoff value. After the value is selected, it is multiplied by the slot time. The random backoff timer then begins counting down the number of slot times. When the number reaches zero, the station can begin transmitting.

6. B, D. 802.11 radios use two separate clear channel assessment (CCA) thresholds when listening to the radio frequency (RF) medium. The signal detect (SD) threshold is used to identify any 802.11 preamble transmissions from another transmitting 802.11 radio. The energy detect (ED) threshold is used to detect any other type of RF transmissions during the clear channel assessment (CCA).

7. B, D. The Duration/ID field is used to set the network allocation vector (NAV), which is a part of the virtual carrier sense process. The contention window and random backoff time are part of the backoff process, which is performed after the carrier sense process.

8. D. The goal of airtime fairness is to allocate equal time, as opposed to equal opportunity. Access fairness and opportunistic medium access do not exist. Carrier Sense Multiple Access with Collision Avoidance (CSMA/CA) is the normal medium access control mode for Wi-Fi devices.

9. A, B, D, E. Distributed Coordination Function (DCF) defines four checks and balances of Carrier Sense Multiple Access with Collision Avoidance (CSMA/CA) to ensure that only one 802.11 radio is transmitting on the half-duplex medium. Virtual carrier sense, physical carrier sense, interframe spacing, and the random backoff timer all work together. Collision detection is not part of CSMA/CA.

10. C. Currently, Wi-Fi Multimedia (WMM) is based on Enhanced Distributed Channel Access (EDCA) mechanisms defined by the 802.11e amendment, which is now part of the 802.11-2012 standard. The WMM certification provides for traffic prioritization via four access categories. EDCA is a subfunction of Hybrid Coordination Function (HCF). The other subfunction of HCF is Hybrid Coordination Function Controlled Channel Access (HCCA).

11. E. Hybrid Coordination Function (HCF) defines the ability for an 802.11 radio to send multiple frames when transmitting on the RF medium. When an HCF-compliant radio contends for the medium, it receives an allotted amount of time to send frames, called a transmit opportunity (TXOP). During the TXOP, an 802.11 radio may send multiple frames in what is called a frame burst.

12. A, B, D, E. WMM Audio priority does not exist. The Wi-Fi Multimedia (WMM) certification provides for traffic prioritization via the four access categories of Voice, Video, Best Effort, and Background.

13. C. The whole point of Wi-Fi Multimedia (WMM) is to prioritize different classes of application traffic during the medium contention process. Traffic in the voice access category has better odds when contending for the medium during the backoff process. For voice traffic, a minimum wait time of a short interframe space (SIFS) plus two slots is required, and then a contention window of 0–3 slots before transmitting on the medium. Best effort traffic must wait a minimum time of a SIFS and three slots, and then the contention window is 0–15 slots. The contention process is still entirely pseudo-random; however, the odds are better for the voice traffic.

14. B. The Enhanced Distributed Channel Access (EDCA) medium access method provides for the prioritization of traffic via the use of 802.1D priority tags. 802.1D tags provide a mechanism for implementing quality of service (QoS) at the MAC level. Different classes of service are available, represented in a 3-bit user priority field in an IEEE 802.1Q header added to an Ethernet frame. 802.1D priority tags from the Ethernet side are used to direct traffic to different access-category queues.

15. A, E. The first purpose is to determine whether a frame transmission is inbound for a station to receive. If the medium is busy, the radio will attempt to synchronize with the transmission. The second purpose is to determine whether the medium is busy before transmitting. This is known as the clear channel assessment (CCA). The CCA involves listening for 802.11 RF transmissions at the Physical layer. The medium must be clear before a station can transmit.

16. B. The signal detect (SD) threshold is used to identify any 802.11 preamble transmissions from another transmitting 802.11 radio. The preamble is a component of the Physical layer header of 802.11 frame transmissions. The preamble is used for synchronization between transmitting and receiving 802.11 radios. The SD threshold is sometimes referred to as the preamble carrier sense threshold. The SD threshold is statistically around 4 dB signal-to-noise ratio (SNR) for most 802.11 radios to detect and decode an 802.11 preamble.

17. C. When the listening radio hears a frame transmission from another station, it looks at the header of the frame and determines whether the Duration/ID field contains a Duration value or an ID value. If the field contains a Duration value, the listening station will set its network allocation vector (NAV) timer to this value.

18. B. Enhanced Distributed Channel Access (EDCA) provides differentiated access for stations by using four access categories. The EDCA medium access method provides for the prioritization of traffic via the four access categories that are aligned to eight 802.1D priority tags.

19. A. Acknowledgments (ACKs) are used for delivery verification of unicast 802.11 frames. Broadcast and multicast frames do not require an acknowledgment. Anycast frames do not exist.

20. A, D. An 802.11 station may contend for the medium during a window of time known as the backoff time. At this point in the Carrier Sense Multiple Access with Collision Avoidance (CSMA/CA) process, the station selects a random backoff value using a pseudo-random backoff algorithm. The station chooses a random number from a range called a contention window (CW) value. After the random number is chosen, the number is multiplied by the slot time value. This starts a pseudo-random backoff timer. Do not confuse the backoff timer with the NAV timer. The NAV timer is a virtual carrier sense mechanism used to reserve the medium for further transmissions. The pseudo-random backoff timer is the final timer used by a station before it transmits.

Chapter 9: 802.11 MAC

1. C. Only 802.11 data frames can carry an upper-layer payload (MSDU) within the body of the frame. The MSDU can be as large as 2,304 bytes and usually should be encrypted. 802.11 control frames do not have a body. 802.11 management frames have a body; however, the payload of the frame is strictly layer 2 information. An action frame is a subtype of an 802.11 control frame. Association request and response frames are subtypes of 802.11 management frames.

2. D. An IP packet consists of layer 3–7 information. The MAC service data unit (MSDU) contains data from the LLC sublayer and/or any number of layers above the Data-Link layer. The MSDU is the payload found inside the body of 802.11 data frames.

3. E. This screen capture of an 802.11 header displays four MAC addresses. Although 802.11 frames have four address fields in the MAC header, 802.11 frames typically use only three of the MAC address fields. An 802.11 frame sent within a wireless distribution system (WDS) requires all four MAC addresses. Although the standard does not specifically define procedures for using this format, WLAN vendors often implement WDS solutions. One example of a WDS is a point-to-point WLAN bridge link. Another example of a WDS is mesh backhaul communications between a mesh portal AP and a mesh point AP.

4. A, C, D. An ERP AP signals for the use of the protection mechanism in the ERP information element in the beacon frame. If a non-ERP STA associates to an ERP AP, the ERP AP will enable the NonERP_Present bit in its own beacons, enabling protection mechanisms in its BSS. In other words, an HR-DSSS (802.11b) client association will trigger protection. If an ERP AP hears a beacon with only an 802.11b or 802.11 supported rate set from another AP or an IBSS STA, it will enable the NonERP_Present bit in its own beacons, enabling protection mechanisms in its BSS.

5. A, B, C, D. A probe response frame contains the same information as a beacon frame, with the exception of the traffic indication map.

6. B, D. Beacons cannot be disabled. Clients use the time-stamp information from the beacons to synchronize with the other stations on the wireless network. Only APs send beacons in a BSS; client stations send beacons in an IBSS. Beacons can contain proprietary information.

7. A, D. Depending on how the To DS and From DS fields are used, the definition of the four MAC fields will change. One constant, however, is that the Address 1 field will always be the receiver address (RA) but may have a second definition as well. The Address 2 field will always be the transmitter address (TA) but also may have a second definition. Address 3 is normally used for additional MAC address information. Address 4 is used only in the case of a WDS.

8. D. When the RTS frame is sent, the value of the Duration/ID field is equal to the time necessary for the CTS, Data, and ACK frames to be transmitted, plus three SIFS.

9. B. When the client station transmits a frame with the Power Management field set to 1, it is enabling power-save mode. The DTIM does not enable power-save mode; it only notifies clients to stay awake in preparation for a multicast or a broadcast.

10. A, B. The receiving station may have received the data, but the returning ACK frame may have become corrupted and the original unicast frame will have to be retransmitted. If the unicast frame becomes corrupted for any reason, the receiving station will not send an ACK.

11. B. The PS-Poll frame is used by the station to request cached data. The ATIM is used to notify stations in an IBSS of cached data. The Power Management bit is used by the station to notify the AP that the station is going into power-save mode. The DTIM is used to indicate to client stations how often to wake up to receive buffered broadcast and multicast frames. The traffic indication map (TIM) is a field in the beacon frame used by the AP to indicate that there are buffered unicast frames for clients in power-save mode.

12. A, E. All 802.11 AP radios will send a probe response to directed probe request frames that contain the correct SSID value. The AP must also respond to null probe request frames that contain a blank SSID value. Some vendors offer the capability to respond to null probe requests with a null probe response. An 802.11ac AP radio will respond to 802.11ac client stations as well as 802.11a/n clients stations transmitting probe requests on 5 GHz. Like all management frames, probe requests are not encrypted. The Power Management bit is used by the client to indicate the client's power state.

13. A, D. There are two types of scanning: passive, which occurs when a station listens to the beacons to discover an AP, and active, which occurs when a station sends probe requests looking for APs. Stations send probe requests only if they are performing an active scan. After a station is associated, it is common for the station to continue to learn about nearby APs. All client stations maintain a "known AP" list, which is constantly updated by active scanning.

14. B, D, E. Although there are similarities, the addressing used by 802.11 MAC frames is much more complex than Ethernet frames. 802.3 frames have only a source address (SA) and destination address (DA) in the layer 2 header. The four MAC addresses used by an 802.11 frame can be used as five different types of addresses: receiver address (RA), transmitter address (TA), basic service set identifier (BSSID), destination address (DA), and source address (SA).

15. B. When the client first attempts to connect to an AP, it will first send a probe request and listen for a probe response. After it receives a probe response, it will attempt to authenticate to the AP and then associate to the network.

16. B. The delivery traffic indication map (DTIM) is used to ensure that all stations using power management are awake when multicast or broadcast traffic is sent. The DTIM interval is important for any application that uses multicasting. For example, many VoWiFi vendors support push-to-talk capabilities that send VoIP traffic to a multicast address. A misconfigured DTIM interval would cause performance issues during a push-to-talk multicast.

17. A, C. An 802.11n/g AP is backward compatible with 802.11b and supports the HR-DSSS data rates of 1, 2, 5.5, and 11 Mbps, as well as the ERP-OFDM data rates of 6, 9, 12, 18, 24, 36, 48, and 54 Mbps. If a WLAN administrator disabled the 1, 2, 5.5, and 11 Mbps data rates, backward compatibility will effectively be disabled and the 802.11b clients will not be able to connect. The 802.11-2016 standard defines the use of basic rates, which are required rates. If a client station does not support any of the basic rates used by an AP, the client station will be denied association to the BSS. If a WLAN administrator configured the ERP-OFDM data rates of 6 and 9 Mbps as basic rates, the 802.11b (HR-DSSS) clients would be denied association because they do not support those rates.

18. A, B. Some 802.11 data frames do not actually carry any data. The Null and QoS Null frames are both non-data carrying frames. 802.11 data frames have a header and a trailer but do not have a frame body that transports an MSDU payload. These frames are sometimes referred to as null function frames because the payload is null, yet the frames still serve a purpose.

19. B, D. An action frame is a type of management frame used to trigger specific actions in a BSS. Action frames can be sent by access points or by client stations. An action frame provides information and direction regarding what to do. An action frame is sometimes referred to as a "management frame that can do anything." A complete list of all the current action frames can be found in section 9.6 of the 802.11-2016 standard. One example of how action frames are used is as a channel switch announcement (CSA) from an AP transmitting on a dynamic frequency selection (DFS) channel. Another example of an action frame is neighbor report requests and responses that 802.11k compliant radios can use. Neighbor report information is used by client stations to gain information from the associated AP about potential roaming neighbors.

20. A, F. The Retry field indicates a value of 1, meaning this is a retransmission. Every time an 802.11 radio transmits a unicast frame, if the frame is received properly and the cyclic redundancy check (CRC) of the FCS passes, the 802.11 radio that received the frame will reply with an acknowledgment (ACK) frame. If the ACK is received, the original station knows that the frame transfer was successful. All unicast 802.11 frames must be acknowledged. Broadcast and multicast frames do not require an acknowledgment. If any portion of a unicast frame is corrupted, the CRC will fail, and the receiving 802.11 radio will not send an ACK frame to the transmitting 802.11 radio. If an ACK frame is not received by the original transmitting radio, the unicast frame is not acknowledged and will have to be retransmitted.

Chapter 10: MIMO Technology: HT and VHT

1. A, D, E. The 802.11ac amendment supports BPSK, QPSK, 16-QAM, 64-QAM, and 256-QAM. BASK and 32-QAM do not exist.

2. A, C, D. Spatial multiplexing transmits multiple streams of unique data at the same time. If a MIMO access point sends two unique data streams to a MIMO client that receives both streams, the throughput is effectively doubled. If a MIMO access point sends three unique data streams to a MIMO client that receives all three streams, the throughput is effectively

tripled. Because transmit beamforming results in constructive multipath communication, the result is a higher signal-to-noise ratio and greater received amplitude. Transmit beamforming will result in higher throughput because of the higher SNR that allows for the use of more complex modulation methods that can encode more data bits. 40 MHz HT channels effectively double the frequency bandwidth, which results in greater throughput.

3.   D. Spatial multiplexing power save (SM power save) allows a MIMO 802.11n device to power down all but one of its radios. For example, a 4×4 MIMO device with four radio chains would power down three of the four radios, thus conserving power. SM power save defines two methods of operation: static and dynamic.

4.   E. The guard interval acts as a buffer for the delay spread, and the normal guard interval is an 800-nanosecond buffer between symbol transmissions. The guard interval will compensate for the delay spread and help prevent intersymbol interference. If the guard interval is too short, intersymbol interference will still occur. HT/VHT radios also have the capability of using a shorter 400-nanosecond GI.

5.   A, B, C, D, E. All of these are supported channel widths. The 160 MHz channel is actually made up of two 80 MHz channels, which can be side by side or separated.

6.   A, B, D. The 802.11n amendment introduced two new methods of frame aggregation to help reduce overhead and increase throughput. Frame aggregation is a method of combining multiple frames into a single frame transmission. The two types of frame aggregation are A-MSDU and A-MPDU. Block ACK frames are used to acknowledge A-MPDUs. The Guard interval is used at the Physical layer.

7.   B, D, E. The beamformer transmits an NDP announcement frame followed by an NDP frame. The beamformee processes this information and creates and transmits a feedback matrix. The AP uses the feedback matrices to calculate a steering matrix, which is used to direct the transmission.

8.   A. MIMO radios transmit multiple radio signals at the same time and take advantage of multipath. Each individual radio signal is transmitted by a unique radio and antenna of the MIMO system. Each independent signal is known as a spatial stream, and each stream can contain different data from the other streams transmitted by one or more of the other radios. A 3×3:2 MIMO system can transmit two unique data streams. A 3×3:2 MIMO system would use three transmitters and three receivers; however, only two unique data streams are utilized.

9.   A. Multiple MPDUs can be aggregated into one frame. The individual MPDUs within an A-MPDU must all have the same receiver address. However, individual MPDUs must all be of the same 802.11e quality-of-service access category. Each aggregated MPDU is also encrypted and decrypted independently.

10.   A, B, C. Modes 0, 1, and 2 all define protection to be used in various situations where only 802.11n/ac stations are allowed to associate to an 802.11n/ac access point. Mode 3—HT Mixed mode—defines the use of protection when both HT/VHT and non-HT radios are associated to an 802.11ac access point.

11. B, C, D. Some of the mandatory baseline requirements of Wi-Fi CERTIFIED n include WPA/WPA2 certification, WMM certification, and support for 40 MHz channels in the 5 GHz U-NII bands. 40 MHz channels in 2.4 GHz are not required. 802.11n access points must support at least two spatial streams in both transmit and receive mode. Client stations must support one spatial stream or more.

12. C, D. Cyclic shift diversity (CSD) is a transmit diversity technique. Unlike STBC, a signal from a transmitter that uses CSD can be received by legacy 802.11g and 802.11a devices. Maximal ratio combining (MRC) is a receive diversity technique. DSSS is a spread spectrum technology used by legacy 802.11 SISO radios.

13. A, B, D. 802.11n (HT) radios are backward compatible with older 802.11b radios (HR-DSSS), 802.11a radios (OFDM), and 802.11g radios (ERP). HT radios are not backward compatible with legacy frequency-hopping radios.

14. B. 802.11ac defined only 10 MCSs, unlike 802.11n, which defined 77. 802.11n defined MCSs based on modulation, coding method, the number of spatial streams, channel size, and guard interval. 802.11ac defined 10 MCSs based on modulation and code rate.

15. D. MCS 0–7 are mandatory. MCS 8 and MCS 9 use 256-QAM, which is optional but will most likely be supported by most vendors.

16. C. Deploying 40 MHz channels in the 2.4 GHz band does not scale properly because there is not enough available frequency space. Although 14 channels are available at 2.4 GHz, there are only three non-overlapping 20 MHz channels available in the 2.4 GHz ISM band. When the smaller channels are bonded together to form 40 MHz channels in the 2.4 GHz ISM band, any two 40 MHz channels will overlap. Channel reuse patterns are not possible with 40 MHz channels in the 2.4 GHz ISM band. The 5 GHz U-NII bands have much more frequency space; therefore, a 40 MHz channel reuse pattern is possible with careful planning.

17. B, D. 802.11ac (VHT) radios are backward compatible with all previous 5 GHz–compliant radios. This includes 802.11a (OFDM) radios and 5 GHz 802.11n (HT) radios.

18. B, C. Other 802.11 technologies are frequency dependent on a single RF band. For example, 802.11b/g radios can transmit in only the 2.4 GHz ISM band. 802.11a radios are restricted to the 5 GHz U-NII bands. 802.11ac radios can also transmit only in the 5 GHz U-NII bands. 802.11n radios are not locked to a single frequency band and can transmit on both the 2.4 GHz ISM band and the 5 GHz U-NII bands.

19. B. 802.11n/ac radios can use an 800-nanosecond guard interval; however, a shorter 400-nanosecond guard interval is also available. A shorter guard interval results in a shorter symbol time, which has the effect of increasing data rates by about 10 percent. If the optional shorter 400-nanosecond guard interval is used with an 802.11n radio, throughput should increase. However, if intersymbol interference occurs because of multipath, the result is data corruption. If data corruption occurs, layer 2 retransmissions will increase and the throughput will be adversely affected. Therefore, a 400-nanosecond guard interval should be used in only good RF environments. If throughput goes down because of a shorter GI setting, the default guard interval setting of 800 nanoseconds should be used instead.

20. C. The 802.11ac amendment defines a maximum of four spatial streams for a client and eight spatial streams for an AP. However, most enterprise 802.11ac APs are 4×4:4, and most 802.11ac clients are 2×2:2.

Chapter 11: WLAN Architecture

1. A, E. In the centralized WLAN architecture, autonomous access points (APs) have been replaced with controller-based APs. All three logical planes of operation reside inside a centralized networking device known as a WLAN controller. Effectively, all planes were moved out of access points and into a WLAN controller. It should be noted that a network management system (NMS) could be used to manage controllers and controller-based APs.

2. D. Telecommunication networks are often defined as three logical planes of operation. The control plane consists of control or signaling information and is often defined as network intelligence or protocols.

3. A, C. All three WLAN infrastructure designs support the use of virtual LANs (VLANs) and 802.1Q tagging. However, the centralized WLAN architecture usually encapsulates user VLANs between the controller-based access point (AP) and the WLAN controllers; therefore, only a single VLAN is normally required at the edge. An 802.1Q trunk is, however, usually required between the WLAN controller and a core switch. Neither the autonomous nor the distributed WLAN architectures use a controller. Noncontroller architectures require support for 802.1Q tagging if multiple VLANs are to be supported at the edge of the network. The access point is connected to an 802.1Q trunk port on an edge switch that supports VLAN tagging.

4. B. Controller-based APs (access points) normally forward user traffic to a centralized WLAN controller via an encapsulated IP tunnel. Autonomous and cooperative access points normally use local data forwarding. Controller-based APs are also capable of local data forwarding. Although the whole point of a cooperative and distributed WLAN model is to avoid centrally forwarding user traffic to the core, the access points may also have IP-tunneling capabilities.

5. A, B, C. Many WLAN vendors use Generic Routing Encapsulation (GRE), which is a commonly used network tunneling protocol. Although GRE is often used to encapsulate IP packets, GRE can also be used to encapsulate an 802.11 frame inside an IP tunnel. The GRE tunnel creates a virtual point-to-point link between a controller-based access point (AP) and a WLAN controller. WLAN vendors that do not use GRE use other proprietary protocols for the IP tunneling. The Control and Provisioning of Wireless Access Points (CAPWAP) management protocol can also be used to tunnel user traffic. IPsec can also be used to securely tunnel traffic from APs across a WAN link.

6. D. One major disadvantage of using the traditional autonomous access point (AP) is that there is no central point of management. Any autonomous WLAN architecture with 25 or more access points is going to require some sort of network management system (NMS). Although a WLAN controller can be used to manage the WLAN in a centralized WLAN architecture, if multiple controllers are deployed, an NMS may be needed to manage multiple controllers. Although the control plane and management plane have moved back to the APs in a distributed WLAN architecture, the management plane remains centralized. Configuration and monitoring of all access points in the distributed model is still handled by an NMS.

7. E. The majority of WLAN controller vendors implement what is known as a split MAC architecture. With this type of WLAN architecture, some of the MAC services are handled by the WLAN controller and some are handled by the controller-based access point.

8. A, C, E. VoWiFi phones are 802.11 client stations that communicate through most WLAN architecture. The private branch exchange (PBX) is needed to make connections among the internal telephones of a private company and also connect them to the public switched telephone network (PSTN) via trunk lines. Wi-Fi Multimedia (WMM) quality-of-service capabilities must be supported by both the VoWiFi phone and WLAN infrastructure. Currently most Voice over Wi-Fi (VoWiFi) solutions use the Session Initiation Protocol (SIP) as the signaling protocol for voice communications over an IP network, but other protocols can be used instead.

9. D. The centralized data forwarding is the traditional data-forwarding method used with WLAN controllers. All 802.11 user traffic is forwarded from the access point (AP) to the WLAN controller for processing, especially when the WLAN controller manages encryption and decryption or applies security and quality-of-service (QoS) policies. Most WLAN controller solutions also now support a distributed data plane. The controller-based AP performs data forwarding locally; it may be used in situations where it is advantageous to perform forwarding at the edge and to avoid a central location in the network for all data.

10. D, E. A distributed solution using enterprise-grade WLAN routers is often deployed at company branch offices. Branch WLAN routers have the ability to connect back to corporate headquarters with virtual private network (VPN) tunnels using an integrated VPN client. Enterprise WLAN routers also have integrated firewalls with support for port forwarding, network address translation (NAT) and port address translation (PAT). Enterprise WLAN routers also offer full support for 802.11 security.

11. G. The 802.11 standard does not mandate what type of form factor must be used by an 802.11 radio. Although PCMCIA and Mini PCI client adapters are the most common, 802.11 radios exist in many other formats, such as CompactFlash cards, Secure Digital cards, USB dongles, ExpressCards, and other proprietary formats.

12. G. All of these protocols can be used to configure WLAN devices such as access points and WLAN controllers. Written corporate policies should mandate the use of secure protocols such as SNMPv3, SSH2, and HTTPS.

13. F. WLAN controllers support layer 3 roaming capabilities, bandwidth policies, and stateful packet inspection. Adaptive RF, device monitoring, and AP management are also supported on a controller.

14. C, E. Communications between a network management system (NMS) server and an AP (access point) require management and monitoring protocols. Most NMS solutions use the Simple Network Management Protocol (SNMP) to manage and monitor the WLAN. Other NMS solutions also use the Control and Provisioning of Wireless Access Points (CAPWAP) as strictly a monitoring and management protocol. CAPWAP incorporates Datagram Transport Layer Security (DTLS) to provide encryption and data privacy of the monitored management traffic.

15. E. Most of the security features found in WLAN controllers can also be found in a distributed WLAN architecture even though there is no WLAN controller. For example, a captive web portal that normally resides in a WLAN controller instead resides inside the individual APs (access points). The stateful firewall and role-based access control (RBAC) capabilities found in a centralized WLAN controller now exist cooperatively in the APs. An individual AP might also function as a RADIUS server with full Lightweight Directory Access Protocol (LDAP) integration capabilities. All control plane mechanisms reside in the access points at the edge of the network in a distributed WLAN architecture.

16. D. The bulk of IoT devices with an 802.11 radio currently transmit in the 2.4 GHz frequency band only. Please understand that not all IoT devices use Wi-Fi radios. IoT devices may use other RF technology such as Bluetooth or Zigbee. IoT devices may also have an Ethernet networking interface in addition to the RF interfaces.

17. A. In recent years there has been a client population explosion of handheld mobile devices such as smartphones and tablets. Most users now expect Wi-Fi connectivity with numerous handheld mobile devices as well as their laptops. Almost all mobile devices use a single chip form factor that is embedded on the device's motherboard.

18. B. In a centralized WLAN architecture, traffic is tunneled from controller-based APs (access points) deployed at the access layer to a WLAN controller that is typically deployed at the core of the network. Standard network design suggests redundancy at the core, and redundant WLAN controllers should be deployed so that there is no single point of network failure. If all user traffic is being tunneled to a WLAN controller and it fails without a redundant solution, effectively the WLAN is down.

19. A, B, C. Network management system (NMS) solutions can be deployed at a company data center in the form of a hardware appliance or as a virtual appliance that runs on VMware or some other virtualization platform. A network management system server that resides in a company's own data center is often referred to as an on-premises NMS. NMS solutions are also available in the cloud as a software subscription service.

20. B, D. The control plane mechanisms are enabled in the system with inter-AP communication via cooperative protocols in a distributed WLAN architecture. In a distributed architecture, each individual access point is responsible for local forwarding of user traffic; therefore, the data plane resides in the APs. The management plane resides in a network management system (NMS) that is used to manage and monitor the distributed WLAN.

Chapter 12: Power over Ethernet (PoE)

1. D. Even when 802.3af and 802.3at were amendments, PoE was defined in Clause 33. PoE is still defined in Clause 33, as defined in the updated 802.3 standard. When an amendment is incorporated into a revised standard, the clause numbering remains the same. It is important to remember the clause number, as it is commonly referenced when discussing PoE.

2. A. Any device that does not provide a classification signature (which is optional) is automatically considered a Class 0 device, and the PSE will provide 15.4 watts of power to that device.

3. A, C. The Power over Ethernet (PoE) standard defines two types of devices: powered devices (PDs) and power-sourcing equipment (PSE).

4. D. The power supplied to the powered device (PD) is at a nominal 48 volts; however, the PD must be capable of accepting up to 57 volts.

5. A, B, C. The powered device (PD) must be able to accept power over either the data pairs or the unused pairs if it is a 10BaseT or 100BaseTX device, and over the 1–2, 3–6 data pairs, or the 4–5, 7–8 data pairs if it is a 1000BaseT device. The PD must also reply to the power-sourcing equipment (PSE) with a detection signature. The PD must accept power with either polarity. Replying to the PSE with a classification signature is optional.

6. D. Providing a classification signature is optional for the powered device (PD). If the PD does not provide a classification signature, the device is considered a Class 0 device, and the power-sourcing equipment (PSE) will allocate the maximum power, or 15.4 watts.

7. A, B, C. Alternative B devices, either endpoint or midspan, provide power to the unused data pairs when using 10BaseT or 100BaseTX connections. Prior to the 802.3at amendment, 1000BaseT devices were compatible only with endpoint PSE devices that supported Alternative A. With the ratification of 802.3at, 1000BaseT devices could now be powered using either Alternative A or Alternative B. 100BaseFX uses fiber-optic cable and is not compatible with PoE.

8. D. Class 4 devices are defined in the 802.3at amendment. The maximum power that a Class 4 powered device (PD) requires is between 12.95 and 25.5 watts.

9. C. At maximum power, each Power over Ethernet (PoE) device will be provided with 30 watts of power from the power-sourcing equipment (PSE). If all 24 ports have powered devices (PDs) connected to them, then a total of just under 720 watts (30 watts × 24 ports = 720 watts) is needed.

10. D. The power-sourcing equipment (PSE) provides five potential levels of power: Class 0 = 15.4 watts, Class 1 = 4.0 watts, Class 2 = 7.0 watts, Class 3 = 15.4 watts, and Class 4 = 30.0 watts. Because this device requires 7.5 watts of power, the PSE would be required to provide it with 15.4 watts.

11. D. The power-sourcing equipment (PSE) provides power within a range of 44 volts to 57 volts, with a nominal power of 48 volts.

12. A. The maximum distance of 100 meters is an Ethernet limitation, not a Power over Ethernet (PoE) limitation. At 90 meters, this is not an issue. Although not specifically mentioned in the PoE standard, Category 5e cables support 1000BaseT communications and are therefore capable of also providing PoE. The large number of PoE VoIP telephones could be requiring more power than the switch is capable of providing, thus causing the access points (APs) to reboot randomly.

13. B. The switch will provide the Class 0 devices with 15.4 watts of power each and the Class 1 devices with 4.0 watts of power each. So the 10 VoIP phones will require 40 watts of power, the 10 access points (APs) will require 154 watts of power, and the switch will need 500 watts of power—for a total of 694 watts (40 W + 154 W + 500 W).

14. B. The switch will provide the Class 2 devices with 7.0 watts of power each and the Class 3 devices with 15.4 watts of power each. So the 10 cameras will require 70 watts of power, the 10 access points (APs) will require 154 watts of power, and the switch will need 1,000 watts of power—for a total of 1,224 watts (70 W + 154 W + 1,000 W).

15. B, D. Implementing Power over Ethernet (PoE) does not affect the distances supported by Ethernet, which is 100 meters or 328 feet.

16. D. An 802.3at powered device (PD) will draw up to 25.5 watts of power.

17. C. The maximum power used by a Class 0 powered device (PD) is 12.95 watts. The power-sourcing equipment (PSE) provides 15.4 watts to account for a worst-case scenario, in which there may be power loss due to the cables and connectors between the PSE and the PD. The maximum power used by a Class 1 PD is 3.84 watts, and the maximum power used by a Class 2 PD is 6.49 watts.

18. E. The different class and range values are as follows:

Class 0: 0–4 mA

Class 1: 9–12 mA

Class 2: 17–20 mA

Class 3: 26–30 mA

Class 4: 36–44 mA

19. C. Mode A accepts power with either polarity from the power supply on wires 1, 2, 3, and 6. With mode B, the wires used are 4, 5, 7, and 8.

20. C. Type 2 devices will perform a two-event Physical layer classification or Data-Link layer classification, which allows a Type 2 PD to identify whether it is connected to a Type 1 or a Type 2 PSE. If mutual identification cannot be completed, then the device can only operate as a Type 1 device.

Chapter 13: WLAN Design Concepts

1. A. If APs and clients are already operating on a DFS channel and a radar pulse is detected, the AP and all the associated clients must leave the channel. If radar is detected on the current DFS frequency, the AP will inform all associated client stations to move to another channel using a channel switch announcement (CSA) frame. The AP and the clients have 10 seconds to leave the DFS channel. The AP may send multiple CSA frames to ensure that all clients leave. The CSA frame will inform the clients that the AP is moving to a new channel and that they must go to that channel as well. In most cases, the channel is a non-DFS channel and very often is channel 36.

2. C. Load balancing between access points is typically implemented in areas where there is a very high density of clients and roaming is not necessarily the priority—for example, a gymnasium or auditorium with 20 APs deployed in the same open area. In this environment, a client will most likely hear all 20 APs, and load balancing the clients between APs is usually a requirement. However, in areas where roaming is needed, load balancing is not a good idea, because the mechanisms may cause clients to become sticky and stay associated to the AP too long. If association and reassociation response frames from the APs are being deferred, client mobility will most likely fail. Load balancing between APs can be detrimental to the roaming process.

3. B, D. Although frequency bandwidth is doubled with 40 MHz channels, because there are less available channels, the odds of CCI increases. Access points and clients on the same 40 MHz channel may hear each other, and the medium contention overhead may have a negative impact and offset any gains in performance that the extra bandwidth might provide. Another problem with channel bonding is that it usually will result in a higher noise floor of about 3 dB. If the noise floor is 3 dB higher, then the SNR is 3 dB lower, which means that the radios will actually shift down to lower MCS rates and therefore lower modulation data rates. In many cases, this offsets some of the bandwidth gains that the 40 MHz frequency space provides.

4. D. Many WLAN professionals advise using 20 MHz channels, as opposed to 40 MHz channels, in most 5 GHz WLAN designs. However, 40 MHz channel deployment can work with careful planning and by abiding by a few general rules. Deploying four or fewer 40 MHz channels in a reuse pattern will not be sufficient. Use 40 MHz channels only if the DFS channels are available. Enabling the DFS channels provides more frequency space and, therefore, more available 40 MHz channels for the reuse pattern. Do not have the AP radios transmit at full power. Transmit power levels of 12 dBm or lower are usually more than sufficient in most indoor environments. The walls should be of dense material for attenuation purposes and to cut done on CCI. Cinder block, brick, or concrete walls will attenuate a signal by 10 dB or more. Drywall, however, will attenuate a signal by only about 3 dB. If the deployment is a multi-floor environment, consider not using 40 MHz channels unless there is significant attenuation between floors.

5. E. Nobody likes this answer, but there are simply too many variables to always give the same answer for every WLAN vendor's AP. The default settings of an enterprise WLAN radio might allow as many as 100–250 client connections. Since most enterprise APs are dual-frequency, with both a 2.4 GHz and 5 GHz radio, theoretically 200–500 clients could associate to the radios of a single AP. Although more than 100 devices might be able to connect to an AP radio, these numbers are not realistic for active devices due to the nature of the half-duplex shared medium. The performance needs of this many client devices will not be met and the user experience will be miserable.

6. E. Three important questions need to be asked with regard to users. First, how many users currently need wireless access and how many Wi-Fi devices will they be using? Second, how many users and devices may need wireless access in the future? These first two questions will help you to begin adequately planning for a good ratio of devices per access point, while allowing for future growth. The third question of great significance is, where are the users and devices? Also, always remember that all client devices are not equal. Many client devices consume more airtime due to lesser MIMO capabilities. An average of 35–50 active Wi-Fi devices per radio, communicating through a dual-frequency 802.11n/ac access point, is realistic with average application use, such as web browsing and email. However, bandwidth-intensive applications, such as high-definition video streaming, will have an impact. Different applications require different amounts of TCP throughput.

7. A, D. The answer is really a matter of opinion and depends on the WLAN professional's preference as well as the type of WLAN design. Adaptive RF capabilities are turned on by default on most every WLAN vendor AP. And the algorithms for RRM constantly improve year after year. The majority of commercial WLAN customers use RRM because it is easy to deploy. RRM is usually the preferred method in enterprise deployments with thousands of APs. However, careful consideration should be given to using static channel and power settings in complex RF environments. Most WLAN vendors recommend in their own very high-density deployment guides that static power and channels be used, especially when directional antennas are deployed.

8. B. Historically, the biggest problem with using the DFS channels has been the potential for false-positive detections of radar. In other words, the APs misinterpret a spurious RF transmission as radar and begin changing channels even though they do not need to move. The good news is that most enterprise WLAN vendors have gotten much better at eliminating false-positive detections. The use of DFS channels is always recommended, unless mission-critical clients do not support them. If real radar exists nearby, simply eliminate the affected DFS channels from the 5 GHz channel plan.

9. C, E. A basic rate configured on an AP is considered to be a "required" rate of all radios communicating within a BSS. An AP will transmit all management frames and many control frames at the lowest configured basic rate. Data frames can be transmitted at much higher supported data rates. For example, a 5 GHz radio of an AP will transmit all beacon frames and other control and management traffic at 6 Mbps if the radio's basic rate has been configured for that rate. This consumes an enormous amount of airtime. Therefore, a common practice is to configure the basic rate of a 5 GHz radio on an AP at either 12 Mbps or even better. Do not configure the AP radio basic rate at 18 Mbps, because some client drivers may not be able to interpret.

10. C. Whenever an AP boots up for the first time on a DFS channel, the AP's radio must listen for a period of 60 seconds before being allowed to transmit on the channel. If any radar pulses are detected, the AP cannot use that channel and will have to try a different channel. If no radar is detected during the initial 60-second listening period, the AP can begin transmitting beacon management frames on the channel. In Europe, the rules are even more restrictive for the Terminal Doppler Weather Radar (TDWR) channels of 120, 124, and 128. An AP must listen for 10 full minutes before being able to transmit on the TDWR frequency space.

11. D. Co-channel interference (CCI) is the top cause of needless airtime consumption that can be minimized with proper WLAN design best practices. Clients are the number one cause of CCI. You should understand that CCI is not static but always changing due to the mobility of client devices.

12. E. Cell overlap cannot be properly measured. Coverage overlap is really duplicate primary and secondary coverage from the perspective of a Wi-Fi client station. A proper site survey should be conducted to ensure that a client always has adequate duplicate coverage from multiple access points. In other words, each Wi-Fi client station (STA) needs to hear at least one access point at a specific RSSI and a backup or secondary access point at a different RSSI. Typically, most vendor RSSI thresholds require a received signal of −70 dBm for the higher data rate communications. Therefore, the client station needs to hear a second AP with a signal of −75 dBm or greater when the signal received from the first AP drops below −70 dBm.

13. E. To reduce CCI, use as many channels as possible in a 5 GHz channel reuse pattern. The more channels that are used, the greater the odds that CCI can be fully mitigated, including co-channel interference that originates from client devices. In most cases, you should use the dynamic frequency selection (DFS) channels in the 5 GHz channel plan.

14. E. VoWiFi communications are highly susceptible to layer 2 retransmissions. Therefore, when you are designing for voice-grade WLANs, a −65 dBm or stronger signal is recommended so that the received signal is higher above the noise floor. The recommended SNR for VoWiFi is 25 dB. A −70 dBm received signal and 20 dB SNR are usually sufficient for high data rate connectivity.

15. E. If an AP is transmitting, all nearby access points and clients on the same channel will defer transmissions. The result is that throughput is adversely affected. The unnecessary medium contention overhead is called co-channel interference (CCI). In reality, the 802.11 radios are operating exactly as defined by the CSMA/CA mechanisms. CCI is also sometimes referred to as overlapping basic service sets (OBSS). Well-planned channel reuse patterns can help minimize CCI in the 5 GHz band.

16. B. Overlapping coverage cells with overlapping frequencies cause adjacent channel interference, which causes a severe degradation in latency, jitter, and throughput. If overlapping coverage cells also have frequency overlap, frames will become corrupt, retransmissions will increase, and performance will suffer significantly.

17. A. When designing a wireless LAN, you need overlapping coverage cells in order to provide for roaming. However, the overlapping cells should not have overlapping frequencies. In the United States, only channels 1, 6, and 11 should be used in the 2.4 GHz ISM band to get the most available, non-overlapping channels. Overlapping coverage cells with overlapping frequencies cause what is known as adjacent channel interference. Although a four-channel reuse plan can be used in Europe, the three-channel reuse pattern of 1, 6, and 11 is still recommended.

18. B. Due to mobility and changes in RSSI and SNR, client stations and AP radios will shift between data rates in a process known as dynamic rate switching (DRS). The objective of DRS is upshifting and downshifting for rate optimization and improved performance. Although dynamic rate switching is the proper name for this process, all these terms refer to the method of speed fallback that a wireless LAN client and AP use as distance increases from the access point.

19. D. A mobile client receives an IP address, also known as a home address, on the original subnet. The mobile client must register its home address with a device called a home agent (HA). The original access point on the client's home network serves as the home agent. The home agent is a single point of contact for a client when it roams across layer 3 boundaries. Any traffic that is sent to the client's home address is intercepted by the home agent access point and sent through a Mobile IP tunnel to the foreign agent AP on the new subnet. Therefore, the client is able to retain its original IP address when roaming across layer 3 boundaries.

20. C. Probabilistic traffic formulas use a telecommunications unit of measurement known as an erlang. An erlang is equal to one hour of telephone traffic in one hour of time.

Chapter 14: Site Survey and Validation

1. A, B, C. Although security in itself is not part of the WLAN site survey, network management should be interviewed about security expectations. The surveying company will make comprehensive wireless security recommendations. An addendum to the security recommendations might be corporate wireless policy recommendations. Authentication and encryption solutions are not usually implemented during the physical survey.

2. A, B, E. Any type of RF interference could cause a denial of service to the WLAN. A spectrum analysis survey should be performed to determine if any of the hospital's medical equipment would cause interference in the 2.4 GHz ISM band or the 5 GHz U-NII bands. Dead zones or loss of coverage can also disrupt WLAN communications. Many hospitals use metal mesh safety glass in many areas. The metal mesh will cause scattering and potentially create lost coverage on the opposite side of the glass. Elevator shafts are made of metal and often are dead zones if not properly covered with an RF signal.

3. A, B, C, D. Outdoor site surveys are usually wireless bridge surveys; however, outdoor access points and mesh routers can also be deployed. Outdoor site surveys are conducted using either outdoor access points or mesh routers, which are the devices typically used to provide access for client stations in an outdoor environment. These outdoor Wi-Fi surveys will use most of the same tools as an indoor site survey but may also use a global positioning system (GPS) device to record latitude and longitude coordinates.

4. B. Although all the options are issues that may need to be addressed when deploying a WLAN in a hospitality environment, aesthetics is usually a top priority. The majority of customer service businesses prefer that all wireless hardware remain completely out of sight. Note that most enclosure units are lockable and help prevent theft of expensive Wi-Fi hardware. However, theft prevention is not unique to the hospitality business.

5. A, C. Blueprints will be needed for the site survey interview to discuss coverage and capacity needs. A network topology map will be useful to assist in the design of integrating the wireless network into the current wired infrastructure.

6. D. Although option C is a possible solution, the best recommendation is to deploy hardware that operates at 5 GHz. Interference from the neighboring businesses' 2.4 GHz networks will never be an issue.

7. A. The cheapest and most efficient solution would be to replace the older edge switches with newer switches that have inline power that can provide PoE to the access points. A core switch will not be used to provide PoE because of cabling distance limitations. Deploying single-port injectors is not practical, and hiring an electrician would be extremely expensive.

8. E. Multiple questions are related to infrastructure integration. How will the access points be powered? How will the WLAN and/or users of the WLAN be segmented from the wired network? How will the WLAN remote access points be managed? Considerations such as role-based access control (RBAC), bandwidth throttling, and load balancing should also be discussed.

9. A, B, C. You should consult with network management during most of the site survey and deployment process, to ensure proper integration of the WLAN. You should consult with the biomedical department about possible RF interference issues. You should contact hospital security in order to obtain proper security passes and, possibly, an escort.

10. A, B, D. Based on information collected during the site survey, a final design diagram will be presented to the customer. Along with the implementation diagrams will be a detailed bill of materials (BOM) that itemizes every hardware and software component necessary for the final installation of the wireless network. A detailed deployment schedule should be drafted that outlines all timelines, equipment costs, and labor costs.

11. E. During an active manual survey, the radio card is associated to the access point and has upper layer connectivity, allowing for low-level frame transmissions while RF measurements are also taken. The main purpose of the active site survey is to look at the percentage of layer 2 retransmissions.

12. A, C, D. A measuring wheel can be used to measure the distance from the wiring closet to the proposed access point location. A ladder or forklift might be needed when temporarily mounting an access point. Battery packs are used to power the access point. GPS devices are used outdoors and do not work properly indoors. Microwave ovens are sources of interference.

13. B, D, E. Cordless phones that operate in the same space as the 5 GHz U-NII bands may cause interference. Unlicensed LTE transmissions may also be a potential source of interference in the 5 GHz bands. Radar is also a potential source of interference at 5 GHz. Microwave ovens transmit in the 2.4 GHz ISM band. FM radios use narrowband transmissions in a lower-frequency licensed band.

14. A, C. During a passive manual survey, the radio card is collecting RF measurements, including received signal strength (dBm), noise level (dBm), and signal-to-noise ratio (dB). The SNR is a measurement of the difference in decibels (dB) between the received signal and the background noise. Received signal strength is an absolute measured in dBm. Antenna manufacturers predetermine gain using either dBi or dBd values.

15. B, C, D, E. Spectrum analysis for an 802.11b/g/n site survey should scan the 2.4 GHz ISM band. Bluetooth radios, plasma cutters, 2.4 GHz video cameras, and legacy 802.11 FHSS access points are all potential interfering devices.

16. A, D. Wherever an access point is placed during a site survey, the power and channel settings should be noted. Recording security settings and IP addresses is not necessary.

17. C. Predictive coverage analysis is accomplished using software that creates visual models of RF coverage and capacity, bypassing the need for actually capturing RF measurements. Projected coverage zones are created using modeling algorithms and attenuation values.

18. C. Segmentation, authentication, authorization, and encryption should all be considered during the site survey interview. Segmenting three types of users into separate VLANs with separate security solutions is the best recommendation. The data users using 802.1X/EAP and CCMP/AES will have the strongest solution available. Because the phones do not have

Voice-Enterprise or 802.11r capabilities, WPA-2 Personal is the chosen security method. WPA-2 Personal provides the voice users with CCMP/AES encryption as well but avoids using an 802.1X/EAP solution that will cause latency problems. At a minimum, the guest user VLAN requires a captive web portal and a strong guest firewall policy.

19. D. All site survey professionals will agree upon the importance of a validation survey. Coverage verification, capacity performance, and roaming testing are key components of a proper validation survey. Depending on the purpose of the wireless network, different tools can be used to assist with the validation site survey.

20. A, B, C, D. Outdoor bridging site surveys require many calculations that are not necessary during an indoor survey. Calculations for a link budget, FSPL, Fresnel zone clearance, and fade margin are all necessary for any bridge link.

Chapter 15: WLAN Troubleshooting

1. A. Regardless of the networking technology, the majority of problems occur at the Physical layer of the OSI model. The majority of performance and connectivity problems in 802.11 WLANs can be traced back to the Physical layer.

2. A, C, E. Wi-Fi Troubleshooting 101 dictates that the end user first enable and disable the Wi-Fi network card. This ensures that the Wi-Fi NIC drivers are communicating properly with the operating system. The passphrase that is used to create a PSK can be 8–63 characters and is always case-sensitive. The problem is almost always a mismatch of the PSK credentials. If the PSK credentials do not match, a pairwise master key (PMK) seed is not properly created and therefore the 4-Way Handshake fails entirely. Another possible cause of the failure of PSK authentication could be a mismatch of the chosen encryption methods. An access point might be configured to require only WPA2 (CCMP-AES), which a legacy WPA (TKIP) client does not support. A similar failure of the 4-Way Handshake would occur.

3. C. Whenever you troubleshoot a WLAN, you should start at the Physical layer, and 70 percent of the time the problem will reside on the WLAN client. If there are any client connectively problems, WLAN Troubleshooting 101 dictates that you disable and re-enable the WLAN network adapter. The driver for the WLAN network interface card (NIC) is the interface between the 802.11 radio and the operating system (OS) of the client device. For whatever reason, the WLAN driver and the OS of the device may not be communicating properly. A simple disable/re-enable of the WLAN NIC will reset the driver. Client security issues usually evolve around improperly configured supplicant settings.

4. A, B, C. A mistake often made when deploying access points is to have the APs transmitting at full power. Effectively, this extends the range of the access point but causes many of the problems that have been discussed throughout this chapter. High transmit power for an indoor AP usually will not meet your capacity needs. Increased coverage areas can cause hidden node problems. Indoor access points at full power will cause sticky clients and

therefore roaming problems. Access points at full power will most likely also increase the odds of co-channel interference due to bleed-over transmissions.

5. D. If an end user complains of a degradation of throughput, one possible cause is a hidden node. A protocol analyzer is a useful tool in determining hidden node issues. If the protocol analyzer indicates a higher retransmission rate for the MAC address of one station when compared to the other client stations, chances are a hidden node has been found. Some protocol analyzers even have hidden node alarms based on retransmission thresholds.

6. C. Actually, all of these answers are correct. However, if Andrew asked his boss why he was looking at Facebook, Andrew might get fired. Options A and B are the correct technical answers, however; his boss only wants to blame the WLAN. Remember, from an end-user perspective, the WLAN is always the culprit. The correct option is C because Andrew's boss just wants it fixed.

7. B, D, E. Excessive layer 2 retransmissions adversely affect the WLAN in two ways. First, layer 2 retransmissions increase MAC overhead and therefore decrease throughput. Second, if application data has to be retransmitted at layer 2, the timely delivery of application traffic becomes delayed or inconsistent. Applications such as VoIP depend on the timely and consistent delivery of the IP packet. Excessive layer 2 retransmissions usually result in increased latency and jitter problems for time-sensitive applications such as voice and video.

8. D. All of these answers would cause the 4-Way Handshake to fail between the WLAN client and the AP in the reception area. Wi-Fi NIC drivers need to communicate properly with the operating system for PSK authentication to be successful. If the PSK credentials do not match, a pairwise master key (PMK) seed is not properly created and therefore the 4-Way Handshake fails entirely. The final pairwise transient key (PTK) is never created. Remember there is a symbiotic relationship between authentication and the creation of dynamic encryption keys. If PSK authentication fails, so does the 4-Way Handshake that is used to create the dynamic keys. Another possible cause of the failure of PSK authentication could be a mismatch of the chosen encryption methods. An access point might be configured to require only WPA2 (CCMP-AES), which a legacy WPA (TKIP) client does not support. A similar failure of the 4-Way Handshake would occur. Only option D is correct, because the problem exists on a single AP in the reception area.

9. C, F. The graphic shows a supplicant and a RADIUS server trying to establish an SSL/TLS tunnel to protect the user credentials. The SSL/TLS tunnel is never created and authentication fails. This is an indication that there is a certificate problem. A whole range of certificate problems could be causing the SSL/TLS tunnel not to be successfully created. The most common certificate issues are that the root CA certificate is installed in the incorrect certificate store, the incorrect root certificate is chosen, the server certificate has expired, the root CA certificate has expired, or the supplicant clock settings are incorrect.

10. A, D, E. Troubleshooting best practices dictate that the proper information be gathered by asking relevant questions. Although options B and C might be interesting, they are not relevant to the potential problem.

11. B, D. VoWiFi requires a handoff of 150 ms or less to avoid a degradation of the quality of the call or, even worse, a loss of connection. Therefore, faster, secure roaming handoffs are required. Opportunistic key caching (OKC) and fast BSS transition (FT) produce roaming handoffs of closer to 50 ms even when 802.1X/EAP is the chosen security solution. The APs and the supplicant must both support FT; otherwise, the suppliant will reauthenticate every time the client roams.

12. A, B. Roaming problems will occur if there is not enough duplicate secondary coverage. No secondary coverage will effectively create a roaming dead zone, and connectivity might even temporarily be lost. On the flip side, too much secondary coverage will also cause roaming problems. For example, a client station may stay associated with its original AP and not connect to a second access point even though the station is directly underneath the second access point. This is commonly referred to as the sticky client problem.

13. A. The maximum distance of 100 meters is an Ethernet limitation, not a PoE limitation. At 90 meters, this is not an issue. Although not specifically mentioned in the PoE standard, Category 5e cables support 1000BaseT communications and are therefore capable of also providing PoE. The power required by the large number of PoE VoIP desktop telephones could exceed the PoE power budget of the switch. In most cases, the root cause of random rebooting of APs is that the switch power budget has been eclipsed.

14. C. The netsh (network shell) command can be used to configure and troubleshoot both wired and wireless network adapters on a Windows computer. The netsh wlan show commands will expose detailed information in regards to the Wi-Fi radio being used by the Windows computer. A comparable command-line utility to configure and troubleshoot 802.11 network adapters on macOS computers is the airport command-line tool. When studying for the CWNA exam, take the time to familiarize yourself with both the netsh wlan commands for Windows. Learning the airport commands for the macOS can also often be useful. As with any CLI command, execute the ? command to view full options.

15. A, B, C. Troubleshooting best practices dictate that the proper information be gathered by asking relevant questions. Option D is a question asked by the bridge keeper in a famous Monty Python movie.

16. C, D. Although all these answers could cause an IPsec VPN to fail, only two of these problems are related to certificate issues during IKE Phase 1. If IKE Phase 1 fails due to a certificate problem, ensure that you have the correct certificates installed properly on the VPN endpoints. Also remember that certificates are time based. Very often, a certificate problem during IKE Phase 1 is simply an incorrect clock setting on either VPN endpoint. If the encryption and hash settings do not match on both sides, the VPN will fail during IKE Phase 2. If the public/private IP address settings are misconfigured, the VPN will fail during IKE Phase 1.

17. A, F. The graphic clearly shows that 802.1X/EAP authentication completes and the 4-Way Handshake creates the dynamic encryption keys for the AP and client radio. At this point, layer 2 authentication is complete and the virtual controlled port on the access point opened for the supplicant. However, the supplicant fails to obtain an IP address. This is not an 802.1X/EAP problem but instead a networking issue. An incorrect user VLAN configuration on a switch could cause this problem. Another potential cause could be that the DHCP is offline or out of leases.

18. B, D, E. The hidden node problem arises when client stations cannot hear the RF transmissions of another client station. Increasing the transmission power of client stations will increase the transmission range of each station, resulting in an increased likelihood of all the stations hearing each other. However, increasing client power is not a recommended fix, because best practice dictates that client stations use the same transmit power used by all other radios in the BSS, including the AP. Decreasing client power may also result in more hidden nodes. Moving the hidden node station within transmission range of the other stations also results in stations hearing each other. Removing an obstacle that prevents stations from hearing each other also fixes the problem. The best fix to the hidden node problem is to add another access point in the area where the hidden node resides.

19. B, D, E. If any portion of a unicast frame is corrupted, the cyclic redundancy check (CRC) will fail and the receiving 802.11 radio will not return an ACK frame to the transmitting 802.11 radio. If an ACK frame is not received by the original transmitting radio, the unicast frame is not acknowledged and will have to be retransmitted. RF interference, low SNR, hidden nodes, mismatched power settings, and adjacent channel interference all may cause layer 2 retransmissions. Co-channel interference does not cause retransmissions but does add unnecessary medium contention overhead.

20. D. If there is any client connectively problems, WLAN Troubleshooting 101 dictates that you disable and re-enable the WLAN network adapter. The drivers for the WLAN network interface card (NIC) are the interface between the 802.11 radio and the operating system (OS) of the client device. For whatever reason, WLAN drivers and the OS of the device may not be communicating properly. A simple disable/re-enable of the WLAN NIC will reset the drivers. Always eliminate this potential problem before investigating anything else.

Chapter 16: Wireless Attacks, Intrusion Monitoring, and Policy

1. B, C. Denial-of-service (DoS) attacks can occur at either layer 1 or layer 2 of the OSI model. Layer 1 attacks are known as RF jamming attacks. A wide variety of layer 2 DoS attacks are a result of tampering with 802.11 frames, including the spoofing of deauthentication frames.

2. C, D. Malicious eavesdropping is achieved with the unauthorized use of protocol analyzers to capture wireless communications. Any unencrypted 802.11 frame transmission can be reassembled at the upper layers of the OSI model.

3. D. A protocol analyzer is a passive device that captures 802.11 traffic and that can be used for malicious eavesdropping. A wireless intrusion prevention system (WIPS) cannot detect a passive device. Strong encryption is the solution to prevent a malicious eavesdropping attack.

4. C, D. The only way to prevent a wireless hijacking, man-in-the-middle, and/or Wi-Fi phishing attack is to use a mutual authentication solution. 802.1X/EAP authentication solutions require that mutual authentication credentials be exchanged before a user can be authorized.

5. F. Even with the best authentication and encryption in place, attackers can still see MAC address information in cleartext. MAC addresses are needed to direct traffic at layer 2 and are never encrypted. Restrictions can be placed on devices via MAC filtering. However, MAC filters can be easily circumvented by MAC spoofing.

6. A, B. The general wireless security policy establishes why a wireless security policy is needed for an organization. Even if a company has no plans to deploy a wireless network, there should be at a minimum a policy detailing how to deal with rogue wireless devices. The functional security policy establishes how to secure the wireless network in terms of what solutions and actions are needed.

7. A, E. After obtaining the passphrase, an attacker can associate to the WPA/WPA2 access point and thereby access network resources. The encryption technology is not cracked, but the key can be re-created. If a hacker has the passphrase and captures the 4-Way Hand-shake, they can re-create the dynamic encryption keys and therefore decrypt traffic. WPA/WPA2-Personal is not considered a strong security solution for the enterprise, because if the passphrase is compromised, the attacker can access network resources and decrypt traffic.

8. A, C, D, E. Numerous types of layer 2 denial-of-service (DoS) attacks exist, including asso-ciation floods, deauthentication spoofing, disassociation spoofing, authentication floods, PS-Poll floods, and virtual carrier attacks. RF jamming is a layer 1 DoS attack.

9. A, C. Microwave ovens operate in the 2.4 GHz ISM band and are often a source of unin-tentional interference. 2.4 GHz cordless phones can also cause unintentional jamming. A signal generator is typically going to be used as a jamming device, which would be consid-ered intentional jamming. 900 MHz cordless phones will not interfere with 802.11 equip-ment that operates in either the 2.4 GHz ISM band or the 5 GHz U-NII bands. There is no such thing as a deauthentication transmitter.

10. C, D, E. The majority of unauthorized devices placed on networks, known as rogues, are put there by people with access to the building. This means that they are more often placed by people you trust: employees, contractors, and visitors. Wardrivers and attackers are not usually allowed physical access.

11. A, C. Client isolation is a feature that can be enabled on WLAN access points or WLAN controllers to block wireless clients from communicating with other wireless clients on the same VLAN and IP subnetwork. The use of a personal firewall can also be used to mitigate peer-to-peer attacks.

12. C. A wireless intrusion prevention system (WIPS) is capable of mitigating attacks from rogue access points (APs). A WIPS sensor can use layer 2 DoS attacks as a countermeasure against a rogue device. Port suppression may also be used to shut down a switch port to which a rogue AP is connected. WIPS vendors also use unpublished methods for mitigating rogue attacks.

13. A, B, E, F. Most WIPS solutions label 802.11 radios into four (sometimes more than four) classifications. An authorized device refers to any client station or AP that is an authorized member of the company's wireless network. An unauthorized/unknown device is any new 802.11 radio that has been detected but not classified as a rogue. A neighbor device refers to any client station or AP that is detected by the WIPS and has been identified as an interfering device but is not necessarily considered a threat. A rogue device refers to any client station or AP that is considered an interfering device and a potential threat.

14. A, E. Every company should have a policy forbidding installation of wireless devices by employees. Every company should also have a policy on how to respond to all wireless attacks, including the discovery of a rogue AP. If a WIPS discovers a rogue AP, temporarily implementing layer 2 rogue containment abilities is advisable until the rogue device can be physically located. After the device is found, immediately unplug it from the data port but not from the electrical outlet. It would be advisable to leave the rogue AP on so that the administrator could do some forensics and look at the association tables and log files to possibly determine who installed it.

15. A, C, D, F, G. Currently, there is no such thing as a Happy AP attack or an 802.11 sky monkey attack. Wireless users are especially vulnerable to attacks at public-use hotspots because there is no security. Because no encryption is used, wireless users are vulnerable to malicious eavesdropping. Because no mutual authentication solution is in place, they are vulnerable to hijacking, man-in-the-middle, and phishing attacks. The hotspot AP might also be allowing peer-to-peer communications, making the users vulnerable to peer-to-peer attacks. Every company should have a remote access wireless security policy to protect their end users when they leave company grounds.

16. A, C. Public-access hotspots have absolutely no security in place, so it is imperative that a remote access WLAN policy be strictly enforced. This policy should include the required use of an IPsec or SSL VPN solution to provide device authentication, user authentication, and strong encryption of all wireless data traffic. Hotspots are prime targets for malicious eavesdropping attacks. Personal firewalls should also be installed on all remote computers to prevent peer-to-peer attacks.

17. B. MAC filters are configured to apply restrictions that will allow only traffic from specific client stations to pass through based on their unique MAC addresses. MAC addresses can be spoofed, or impersonated, and any amateur hacker can easily bypass any MAC filter by spoofing an allowed client station's address.

18. A. The integrated wireless intrusion prevention system (WIPS) is by far the most widely deployed. Overlay WIPSs are usually cost prohibitive for most WLAN customers. The more robust overlay WIPS solutions are usually deployed in defense, finance, and retail vertical markets, where the budget for an overlay solution may be available.

19. A, D. Wired Equivalent Privacy (WEP) encryption has been cracked, and currently available tools can easily derive the secret key within a matter of minutes. The size of the key makes no difference; both 64-bit WEP and 128-bit WEP can be cracked. CCMP/AES and 3DES encryption have not been cracked.

20. D. An attack that often generates a lot of press is wireless hijacking, also known as the evil twin attack. The attacker hijacks wireless clients at layer 2 and layer 3 by using an evil twin access point and a DHCP server. The hacker may take the attack several steps further and initiate a man-in-the-middle attack and/or a Wi-Fi phishing attack.

884	Appendix A ▪ Answers to Review Questions

Chapter 17: 802.11 Network Security Architecture

1. **B.** As required by an 802.1X security solution, the supplicant is a WLAN client requesting authentication and access to network resources. Each supplicant has unique authentication credentials that are verified by the authentication server.

2. **B, D.** The 802.11-2016 standard defines CCMP/AES encryption as the default encryption method; TKIP/RC4 is the optional encryption method. This was originally defined by the 802.11i amendment, which is now part of the 802.11-2016 standard. The Wi-Fi Alliance created the WPA2 security certification, which mirrors the robust security defined by the IEEE. WPA2 supports both CCMP/AES and TKIP/RC4 dynamic encryption key management.

3. **E.** 128-bit WEP encryption uses a secret 104-bit static key, which is provided by the user (26 hex characters) and combined with a 24-bit initialization vector (IV) for an effective key strength of 128 bits.

4. **A, C, E.** The 802.1X authorization framework consists of three main components, each with a specific role. The components work together to ensure that only properly validated users and devices are authorized to access network resources. The supplicant requests access to network resources; the authentication server authenticates the identity of the supplicant; and the authenticator allows or denies access to network resources via virtual ports. A layer 2 authentication protocol called Extensible Authentication Protocol (EAP) is used within the 802.1X framework to validate users at layer 2.

5. **C.** The Wi-Fi Alliance views Simultaneous Authentication of Equals (SAE) as a more secure replacement for PSK authentication. The ultimate goal of SAE is to prevent dictionary attacks altogether. SAE will be part of the WPA3 security certification.

6. **D.** The biggest problem with using PSK authentication in the enterprise is social engineering. The PSK is the same on all WLAN devices. If the end user accidentally gives the PSK to a hacker, WLAN security is compromised. If an employee leaves the company, all the devices have to be reconfigured with a new 64-bit PSK, creating a lot of work for an administrator. Several WLAN vendors offer proprietary PSK solutions in which each individual client device will have its own unique PSK. These proprietary PSK solutions prevent social engineering attacks. They also virtually eliminate the burden for an administrator having to reconfigure each and every WLAN end-user device.

7. **A, B, D.** WEP, TKIP, and CCMP use symmetric ciphers. WEP and TKIP use the ARC4 cipher, and CCMP uses the AES cipher. Public-key cryptography is based on asymmetric communications.

8. **A, B.** The migration from TKIP to CCMP can be seen in the IEEE 802.11n amendment, the IEEE 802.11ac amendment, and the IEEE 802.11-2016 standard, which all state that *high throughput (HT)* or *very high throughput (VHT)* data rates are not allowed to be

used if WEP or TKIP is enabled. This exclusion was decided in 2012 by both the IEEE and the Wi-Fi Alliance. CCMP is the designated encryption method for 802.11n/ac data rates. The 802.11ad-2012 amendment standardized the use of Galois/Counter Mode Protocol (GCMP), which uses AES cryptography. The extremely high data rates defined by 802.11ad need GCMP because it is more efficient than CCMP. GCMP is also considered an optional encryption method for 802.11ac radios.

9. A, B, C, D, E. The three main components of an RBAC approach are users, roles, and permissions. Separate roles can be created, such as the sales role or the marketing role. User traffic permissions can be defined as layer 2 permissions (MAC filters), VLANs, layer 3 permissions (access control lists), layers 4–7 permissions (stateful firewall rules), and bandwidth permissions. All these permissions can also be time based. The user traffic permissions are mapped to the roles. Some WLAN vendors use the term "roles," whereas other vendors use the term "user profiles."

10. D. VPNs are most often used for client-based security when connected to public access WLANs and hotspots that do not provide security. Because most hotspots do not provide layer 2 security, it is imperative that end users provide their own security. Another common use of VPN technology is to provide site-to-site connectively between a remote office and a corporate office across a WAN link. When WLAN bridges are deployed for wireless backhaul communications, VPN technology can be used to provide the necessary level of data privacy.

11. B. Encapsulated inside the frame body of an 802.11 data frame is an upper-layer payload called the MAC service data unit (MSDU). The MSDU contains data from the Logical Link Control (LLC) and layers 3–7. The MSDU is the data payload that contains an IP packet plus some LLC data. When encryption is enabled, the MSDU payload within an 802.11 data frame is encrypted.

12. B, D. Shared Key authentication is a legacy authentication method that does not provide seeding material to generate dynamic encryption keys. Static WEP uses static keys. A robust security network association requires a four-frame EAP exchange, known as the 4-Way Handshake, which is used to generate dynamic TKIP or CCMP keys. The handshake may occur either after an 802.1X/EAP exchange or as a result of PSK authentication.

13. A, D. An 802.1X/EAP solution requires that both the supplicant and the authentication server support the same type of EAP. The authenticator must be configured for 802.1X/EAP authentication but does not care which EAP type passes through. The authenticator and the supplicant must support the same type of encryption.

14. C. WLAN controllers normally centralize the data plane, and all the EAP traffic is tunneled between the APs and the WLAN controller. The WLAN controller is the authenticator. When an 802.1X/EAP solution is deployed in a wireless controller environment, the virtual controlled and uncontrolled ports exist on the WLAN controller.

15. E. Multiple use cases for per-user and per-device PSK credentials have gained popularity in the enterprise. However, proprietary implementations of PSK authentication are not meant to be a replacement for 802.1X/EAP.

16. A, D, E. The purpose of 802.1X/EAP is authentication of user credentials and authorization to access network resources. Although the 802.1X framework does not require encryption, it highly suggests the use of encryption. A by-product of 802.1X/EAP is the generation and distribution of dynamic encryption keys. While the encryption process is actually a by-product of the authentication process, the goals of authentication and encryption are very different. Authentication provides mechanisms for validating user identity, whereas encryption provides mechanisms for data privacy or confidentiality.

17. D. The AES algorithm encrypts data in fixed data blocks with choices in encryption-key strength of 128, 192, or 256 bits. CCMP/AES uses a 128-bit encryption-key size and encrypts in 128-bit fixed-length blocks.

18. A, D. The WPA2 certification requires the use of an 802.1X/EAP authentication method in the enterprise and the use of a PSK authentication in a SOHO environment. The WPA2 certification also requires the use of stronger dynamic encryption-key generation methods. CCMP/AES encryption is the mandatory encryption method, and TKIP/ARC4 is the optional encryption method.

19. E. The 802.11-2016 standard defines what is known as a robust security network (RSN) and robust security network associations (RSNAs). CCMP/AES encryption is the mandated encryption method, and TKIP/RC4 is an optional encryption method. TKIP, CCMP, and GCMP are considered to be robust security network (RSN) encryption protocols. However, TKIP is being deprecated and GCMP has not yet been used in the enterprise WLAN marketplace.

20. C. The supplicant, authenticator, and authentication server work together to provide the framework for 802.1X port-based access control, and an authentication protocol is needed to assist in the authentication process. The Extensible Authentication Protocol (EAP) is used to provide user or device authentication.

Chapter 18: Bring Your Own Device (BYOD) and Guest Access

1. B, C, E. The firewall ports that should be permitted include DHCP server UDP port 67, DNS UDP port 53, HTTP TCP port 80, and HTTPS TCP port 443. This allows the guest user's wireless device to receive an IP address, perform Domain Name System (DNS) queries, and browse the web. Many companies require their employees to use a secure virtual private network (VPN) connection when they are connected to a service set identifier (SSID) other than the company SSID. Therefore, it is recommended that IPsec IKE UDP port 500 and IPsec NAT-T UDP port 4500 also be permitted.

2. A, E. The guest firewall policy should allow for Dynamic Host Configuration Protocol (DHCP) and Domain Name System (DNS) but restrict access to private networks 10.0.0.0/8, 172.16.0.0/12, and 192.168.0.0/16. Guest users are not allowed on these private networks, because corporate network servers and resources usually reside on the private IP space. The guest firewall policy should simply route all guest traffic straight to an Internet gateway and away from corporate network infrastructure.

3. A, D, E. The four main components of a mobile device management (MDM) architecture are the mobile device, an access point (AP) and/or WLAN controller, an MDM server, and a push notification service. The mobile Wi-Fi device requires access to the corporate WLAN. The AP or WLAN controller quarantines the mobile devices inside a walled garden if the devices have not been enrolled via the MDM server. The MDM server is responsible for enrolling client devices. The push notification services, such as Apple Push Notification service (APNs) and Google Cloud Messaging (GCM), communicate with the mobile devices and the MDM servers for over-the-air management.

4. A, C. 802.1X/EAP requires that a root CA certificate be installed on the supplicant. Installing the root certificate onto Windows laptops can be easily automated using a Group Policy Object (GPO). Mobile device management (MDM) uses over-the-air provisioning to onboard mobile devices and provision root CA certificates onto the mobile devices that are using 802.1X/EAP security. Self-service device onboarding applications can also be used to provision root CA certificates on mobile devices. The Hotspot 2.0 technical specification also defines methods of auto-provisioning certificates for Passpoint service set identifiers (SSIDs).

5. D. Passpoint-certified client devices query the WLAN prior to connecting in order to discover the cellular service providers supported by the network. The devices use Access Network Query Protocol (ANQP), a query and response protocol defined by the 802.11u. Other information, such as supported EAP methods and service provider roaming agreements, can also be delivered via ANQP queries.

6. B, D, E. A mobile device management (MDM) server can monitor mobile device information, including device name, serial number, capacity, battery life, and applications that are installed on the device. Information that cannot be seen includes SMS messages, personal emails, calendars, and browser history.

7. B, D. As defined by the Hotspot 2.0 specification, the four methods of EAP that can be used for authentication by a Passpoint client include EAP-SIM, EAP-AKA, EAP-TLS, and EAP-TTLS. Modern-day 3G/4G cellular smartphones use EAP-AKA with SIM card credentials. EAP-TLS or EAP-TTLS can be used for non-SIM clients, such as laptops.

8. A, B, E. A captive portal solution effectively turns a web browser into an authentication service. To authenticate, the user must launch a web browser. After the browser is launched and the user attempts to go to a website, no matter what web page the user attempts to browse, the user is redirected to a login prompt, which is the captive portal login web page. Captive portals can redirect unauthenticated users to a login page using IP redirection, DNS redirection, or HTTP redirection.

9. B, C, D. The access point (AP) holds the mobile client device inside a walled garden. Within a network deployment, a walled garden is a closed environment that restricts access to web content and network resources while still allowing access to some resources. A walled garden is a closed platform of network services provided for devices and/or users. While inside the walled garden designated by the AP, the only services that the mobile device can access include Dynamic Host Configuration Protocol (DHCP), Domain Name System (DNS), push notification services, and the mobile device management (MDM) server. In order to escape from the walled garden, the mobile device must find the proper exit point, much like a real walled garden. The designated exit point for a mobile device is the MDM enrollment process.

10. C. Over-the-air provisioning differs between different device operating systems; however, using trusted certificates and Secure Sockets Layer (SSL) encryption is the norm. iOS devices use the Simple Certificate Enrollment Protocol (SCEP), which uses certificates and SSL encryption to protect the mobile device management (MDM) profiles. The MDM server then sends a SCEP payload, which instructs the mobile device about how to download a trusted certificate from the MDM's certificate authority (CA) or a third-party CA. Once the certificate is installed on the mobile device, the encrypted MDM profile with the device configuration and restrictions payload is sent to the mobile device securely and installed.

11. A. An IP tunnel normally using Generic Routing Encapsulation (GRE) can transport guest traffic from the edge of the network back to the isolated demilitarized zone (DMZ). Depending on the WLAN vendor solution, the tunnel destination in the DMZ can be either a WLAN controller, a GRE appliance, or a router. The source of the GRE tunnel is the access point (AP).

12. E. A guest management solution with employee sponsorship capabilities will integrate with a Lightweight Directory Access Protocol (LDAP) database, such as Active Directory. A guest user can also be required to enter the email address of an employee, who must approve and sponsor the guest prior to allowing the guest access on the network. The sponsor typically receives an email requesting access for the guest, with a link in the email that allows the sponsor to easily accept or reject the request. Once a guest user has registered or been sponsored, they can log in using their newly created credentials.

13. C. When enrolling their personal devices through the corporate mobile device management (MDM) solution, typically, employees will still have the ability to remove the MDM profiles because they own the device. If the employee removes the MDM profiles, the device is no longer managed by the corporate MDM solution. However, the next time the employee tries to connect to the company's WLAN with the mobile device, they will have to once again go through the MDM enrollment process.

14. D. The phrase *bring your own device (BYOD)* refers to the policy of permitting employees to bring personally owned mobile devices, such as smartphones, tablets, and laptops, to their workplace. A BYOD policy dictates which corporate resources can or cannot be accessed when employees access the company WLAN with their personal devices.

15. A. Social login is a method of using existing login credentials from a social networking service—such as Twitter, Facebook, or LinkedIn—to register into a third-party website. Social login allows a user to forgo the process of creating new registration credentials for the third-party website. Retail and service businesses like the idea of social login because it allows them to obtain meaningful marketing information about the guest user from the social networking service. Businesses can then build a database of the type of customers who are using the guest Wi-Fi while shopping.

16. F. A mobile device can still be managed remotely even if the mobile device is no longer connected to the corporate WLAN. The mobile device management (MDM) servers can still manage the devices as long as the devices are connected to the Internet from any location. The communication between the MDM server and the mobile devices requires push notifications from a third-party service. Push notification services will send a message to a mobile device telling the device to contact the MDM server. The MDM server can then take remote actions over a secure connection.

17. B, D, E. Client isolation is a feature that can often be enabled on WLAN access points or controllers to block wireless clients from communicating with other wireless clients on the same wireless VLAN. Client isolation is highly recommended on guest WLANs to prevent peer-to-peer attacks. Enterprise WLAN vendors also offer the capability to throttle bandwidth of user traffic. Bandwidth throttling, which is also known as rate limiting, can be used to curb traffic at either the service set identifier (SSID) level or the user level. Rate limiting the guest user traffic to 1024 Kbps is a common practice. A web content filtering solution can block guest users from viewing websites based on content categories. Each category contains websites or web pages that have been assigned based on their prevalent web content.

18. E. Captive portals are available as standalone server solutions and as cloud-based service solutions. Most WLAN vendors offer integrated captive portal solutions. The captive portal may exist within a WLAN controller, or it may be deployed at the edge within an access point.

19. B. The mobile device must first establish an association with an AP (access point). The AP holds the mobile client device inside a walled garden. Within a network deployment, a walled garden is a closed environment that restricts access to web content and network resources while still allowing access to some resources. A walled garden is a closed platform of network services provided for devices and/or users. While the mobile device is inside the walled garden designated by the AP, the only services it can access are Dynamic Host Configuration Protocol (DHCP), Domain Name System (DNS), push notification services, and the mobile device management (MDM) server. After the mobile device completes the MDM enrollment process, the device is released from the walled garden.

20. A, B, C. A network access control (NAC) server will use system health information, as reported by a posture agent, to determine whether the device is healthy. Dynamic Host Configuration Protocol (DHCP) fingerprinting is used to help identify the hardware and operating system. RADIUS attributes can be used to identify whether the client is connected wirelessly or wired, along with other connection parameters. RADIUS Change of Authorization (CoA) is used to disconnect or change the privileges of a client connection.

Chapter 19: 802.11ax: High Efficiency (HE)

1. C. Within a 20 MHz channel, a maximum of nine clients can participate in multi-user OFDMA transmissions per TXOP. The 20 MHz channel can be partitioned into nine separate 26-tone resource units.

2. A. Trigger frames are needed to bring about the necessary frame exchanges for multi-user communications. Trigger frames are used for both MU-MIMO and MU-OFDMA.

3. B. Target wake time (TWT) is a power-saving mechanism originally defined in the 802.11ah-2016 amendment and enhanced in 802.11ax. A TWT is a negotiated agreement, based on expected traffic activity between clients and the AP, to specify a scheduled target wake-up time for the clients in power-save (PS) mode. The extended power-saving capabilities could be ideal for IoT devices with 802.11ax radios that transmit in the 2.4 GHz band.

4. C. An adaptive signal detect threshold as a result of BSS coloring potentially decreases the channel contention problem that is a result of existing 4 dB signal detect (SD) thresholds. Based on detected color bits in the PHY header, 802.11ax radios could implement an adaptive CCA implementation, which could raise the signal detect threshold for inter-BSS frames while maintaining a lower threshold for intra-BSS traffic. If the signal detect threshold was raised higher for incoming OBSS frames, a radio might not need to defer, despite being on the same channel.

5. D. Although the goal of 802.11ax is better efficiency, as opposed to higher data rates, there is one exception. 1024-QAM can be used only with 242-tone or larger resource units. This means that at least a full 20 MHz channel will be needed for 1024-QAM.

Chapter 20: WLAN Deployment and Vertical Markets

1. A, B, C, D. Wi-Fi radios are used in sensor and monitoring IoT devices, as well as RFID, across numerous enterprise vertical markets. IoT devices often use other wireless networking technologies, such as Bluetooth, Zigbee, and Z-Wave, instead of 802.11 wireless technology. IoT devices can also be connected to the corporate network via Ethernet.

2. B. An iBeacon BLE radio transmits a proximity location in the form of a universally unique identifier (UUID) a major number, and a minor number. The UUID is a 32-character hexadecimal string that uniquely identifies the organization under which the beacons are managed. The UUID is included in the iBeacon payload with major and minor numbers, which typically indicate a specific site and location within the site, respectively.

3. C. Because of the potential for interference and the importance of preventing it, hospitals often have a person responsible for keeping track of frequencies used within the organization. Some municipalities are starting to do this as well—not just for law enforcement, but for all of their wireless needs, because they often use wireless technologies for SCADA networks, traffic cameras, traffic lights, two-way radios, point-to-point bridging, hotspots, and more. Military bases also usually have a person in charge of spectrum management.

4. D. Since cruise ships are often not near land, where an LTE cellular uplink is available, it is necessary to use a satellite uplink to connect the ship to the Internet.

5. B, D. Fixed mobile convergence allows roaming between Wi-Fi networks and cellular phone networks, choosing the available network that is least expensive.

6. D. When designing a warehouse network, the networking devices are often barcode scanners that do not capture much data; so, high capacity and throughput are not typically needed. Because the data-transfer requirements are so low, these networks are typically designed to provide coverage for large areas. Security is always a concern; however, it is not usually a design criterion.

7. A, C, D. Corporations typically install a WLAN to provide easy mobility and/or access to areas that are difficult or extremely expensive to connect via wired networks. Although providing connectivity to the Internet is a service that the corporate wireless network offers, it is not the driving reason for installing the wireless network.

8. A, C, D. The telephone company, cable providers, and wireless ISPs (WISPs) are all examples of companies that provide last-mile services to users and businesses.

9. B. Although all of these answers are correct, the main purpose of small office/home office (SOHO) networks is to provide a gateway to the Internet.

10. A, B, D. Mobile office networking solutions are temporary solutions that include all the options listed except for the remote sales office, which would more likely be classified as a SOHO installation.

11. A, D. Warehousing and manufacturing environments typically have a need for mobility, but their data transfers are typically very small. Therefore, their networks are often designed for high coverage rather than high capacity.

12. B, D. iBeacons require an application on a mobile device to trigger an action. The trigger occurs whenever a smartphone with a BLE receiver radio is within proximity range of an iBeacon transmitter. For example, iBeacon may trigger a proximity-aware push notification such as an advertisement in a retail store location. iBeacons utilize BLE radios as opposed to 802.11 radios; however, many WLAN vendors now offer integrated and/or overlay iBeacon technology solutions. Do not confuse the iBeacons used by BLE transmitters with the 802.11 beacon management frames that are transmitted by Wi-Fi access points.

13. A, B, C. Manufacturing plants are typically fixed environments and are better served by permanent access points.

14. C, D. Point-to-multipoint, hub and spoke, and star all describe the same communication technology, which connects multiple devices by using a central device. Point-to-point communications connects two devices. Mesh networks do not have a defined central device.

15. C. Most of the 802.11 implementations used frequency-hopping spread spectrum (FHSS), with industrial (warehousing and manufacturing) companies being some of the biggest implementers. Their requirement of mobility with low data-transfer speeds was ideal for using the technology.

16. C. To make wireless access easy for the subscriber, hotspot vendors typically deploy authentication methods that are easy to use but that do not provide data encryption. Therefore, to ensure security back to your corporate network, the use of an IPsec VPN is most often necessary.

17. A, C, D. Voice over Wi-Fi (VoWiFi) is a common use of 802.11 technology in a medical environment, providing immediate access to personnel no matter where they are in the hospital. Real-time location service (RTLS) solutions using 802.11 RFID tags for inventory control are also commonplace. WLAN medical carts are used to monitor patient information and vital signs.

18. A, C. The installation of multiple point-to-point bridges is either to provide higher throughput or to prevent a single point of failure. Care must be taken in arranging channel and antenna installations to prevent self-inflicted interference.

19. A, B, C. Because health-care providers often have many devices that use RF communications, RF interference is a concern. Fast access along with secure and accurate access is critical in health-care environments. Faster access can be performed without faster speed. The mobility of the technology will satisfy the faster access that is typically needed.

20. D. Public hotspots are most concerned about ensuring that only valid users are allowed access to the hotspot. This is performed using authentication, although this only secures the network from nonauthorized users usually via a captive web portal. However, there is a growing trend in the industry towards offering secure public access via HotSpot 2.0 security and other methods.

Appendix B

Abbreviations and Acronyms

Certifications

CWAP   Certified Wireless Analysis Professional

CWDP   Certified Wireless Design Professional

CWNA   Certified Wireless Network Administrator

CWNE   Certified Wireless Network Expert

CWNT   Certified Wireless Network Trainer

CWS   Certified Wireless Specialist

CWSP   Certified Wireless Security Professional

CWT   Certified Wireless Technician

CWTS   Certified Wireless Technology Specialist

Organizations and Regulations

ACMA   Australian Communications and Media Authority

ARIB   Association of Radio Industries and Businesses (Japan)

ATU   African Telecommunications Union

CEPT   European Conference of Postal and Telecommunications Administrations

CITEL   Inter-American Telecommunication Commission

CTIA   Cellular Telecommunications and Internet Association

CWNP   Certified Wireless Network Professional

ERC   European Radiocommunications Committee

EWC   Enhanced Wireless Consortium

FCC   Federal Communications Commission

FIPS   Federal Information Processing Standards

GLBA   Gramm-Leach-Bliley Act

HIPAA   Health Insurance Portability and Accountability Act

IAB   Internet Architecture Board

ICANN   Internet Corporation for Assigned Names and Numbers

IEC   International Electrotechnical Commission

IEEE   Institute of Electrical and Electronics Engineers

IESG   Internet Engineering Steering Group

IETF   Internet Engineering Task Force

IRTF   Internet Research Task Force

ISO   International Organization for Standardization

ISOC   Internet Society

NEMA   National Electrical Manufacturers Association

NIST   National Institute of Standards and Technology

NTIA   National Telecommunication and Information Agency

PCI   Payment Card Industry

RCC   Regional Commonwealth in the field of Communications

TGn Sync   Task Group n Sync

WECA   Wireless Ethernet Compatibility Alliance

Wi-Fi Alliance   Wi-Fi Alliance

WIEN   Wireless InterWorking with External Networks

WWiSE   World-Wide Spectrum Efficiency

Measurements

dB   decibel

dBd   decibel referenced to a dipole antenna

dBi   decibel referenced to an isotropic radiator

dBm   decibel referenced to 1 milliwatt

GHz   gigahertz

Hz   hertz

KHz   kilohertz

mA   milliampere

MHz   megahertz

mW   milliwatt

SINR   signal-to-interference-plus-noise ratio

SNR   signal-to-noise ratio

V   volt

VDC   voltage direct current

W   watt

Technical Terms

AAA   authentication, authorization, and accounting

AC   access category (802.11)

AC   alternating current (electricity)

ACK   acknowledgment

AES   Advanced Encryption Standard

AGL   above ground level

AI   artificial intelligence

AID   association identifier

AIFS   arbitration interframe space

AKM   authentication and key management

AKMP   Authentication and Key Management Protocol

AM   amplitude modulation

A-MPDU   aggregate MAC protocol data unit

A-MSDU   aggregate MAC service data unit

AMPE   Authenticated Mesh Peering Exchange

ANQP   Access Network Query Protocol

AP   access point

API   application programming interface

APIPA   Automatic Private IP Addressing

APNs   Apple Push Notification service

APSD   automatic power-save delivery

ARP   Address Resolution Protocol

ARS   adaptive rate selection

ARS   automatic rate selection

AS   authentication server

ASEL   antenna selection

ASK   amplitude-shift keying

ATEX   ATmosphères EXplosibles

ATF   airtime fairness

ATIM   announcement traffic indication message

AVP   attribute-value pair

BA   block acknowledgment

BAR   block acknowledgment request

BER   bit error rate

BIP   Broadcast/Multicast Integrity Protocol

BOM   bill of materials

BPSK   binary phase-shift keying

BSA   basic service area

BSR   buffer status report

BSRP   buffer status report poll

BSS   basic service set

BSSID   basic service set identifier

BT   Bluetooth

BVI   bridged virtual interface

BW   bandwidth

BYOD   bring your own device

CA   certificate authority

CAD   computer-aided design

CAM   content-addressable memory

CAM   continuous aware mode

CAPWAP   Control and Provisioning of Wireless Access Points

CBN   cloud-based networking

CCA   clear channel assessment

CC-AP   cooperative control access point

CCI   co-channel interference

CCK   complementary code keying

CCMP   Counter Mode with Cipher Block Chaining Message Authentication Code Protocol

CCX   Cisco Compatible Extensions

CEN   cloud-enabled network

CF   CompactFlash (memory)

CF   contention free (802.11)

CFP   contention-free period

CID   company-issued device

CLI   command-line interface

CNSA   Commercial National Security Algorithm

CoA   Change of Authorization

CP   contention period

CRC   cyclic redundancy check

CRM   customer relationship management

CSA   channel switch announcement

CSMA/CA   Carrier Sense Multiple Access with Collision Avoidance

CSMA/CD   Carrier Sense Multiple Access with Collision Detection

CTS   clear to send

CW   contention window

CWG-RF   Converged Wireless Group–RF Profile

DA   destination address

DBPSK   differential binary phase-shift keying

DC   direct current

DCF   Distributed Coordination Function

DDF   distributed data forwarding

DFS   dynamic frequency selection

DHCP   Dynamic Host Configuration Protocol

DIFS   Distributed Coordination Function interframe space

DL   downlink

DL OFDMA   downlink orthogonal frequency-division multiple access

DMG   directional multi-gigabit

DMZ   demilitarized zone

DNS   Dynamic Name System

DoS   denial of service

DPI   deep packet inspection

DPP   Device Provisioning Protocol

DQPSK   differential quadrature phase-shift keying

DRS   dynamic rate switching

DS   distribution system

DSAS   distributed spectrum analysis system

DSCP   differentiated services code point

DSM   distribution system medium

DSP   digital signal processing

DSRC   Dedicated Short Range Communications

DSS   distribution system services

DSSS   direct-sequence spread spectrum

DTIM   delivery traffic indication message

EAP   Extensible Authentication Protocol

ED   energy detect

EDCA   Enhanced Distributed Channel Access

EIFS   extended interframe space

EIRP   equivalent isotropically radiated power

EM   electromagnetic

EQM   equal modulation

ERP   Extended Rate Physical

ERP-CCK   Extended Rate Physical–complementary code keying

ERP-DSSS   Extended Rate Physical–direct-sequence spread spectrum

ERP-OFDM   Extended Rate Physical–orthogonal frequency-division multiplexing

ESA   extended service area

ESP   Encapsulating Security Protocol

ESS   extended service set

ESSID   extended service set identifier

EUI   extended unique identifier

EVM   error vector magnitude

EWG   enterprise wireless gateway

FA   foreign agent

FAST   Flexible Authentication via Secure Tunneling

FCS   frame check sequence

FEC   forward error correction

FFT   fast Fourier transform

FHSS   frequency-hopping spread spectrum

FILS   fast initial link setup

FM   frequency modulation

FMC   fixed mobile convergence

FSK   frequency-shift keying

FSPL   free space path loss

FSR   fast secure roaming

FT   fast BSS transition

FTM   Fine Timing Measurement

FZ   Fresnel zone

GAS   Generic Advertisement Service

GCM   Google Cloud Messaging

GCMP   Galois/Counter Mode Protocol

GCR   Groupcast with Retries

GFSK   Gaussian frequency-shift keying

GI   guard interval

GLK   General Link

GMK   group master key

GPO　Group Policy Object

GPS　Global Positioning System

GRE　Generic Routing Encapsulation

GSM　Global System for Mobile Communications

GTC　Generic Token Card

GTK　group temporal key

GUI　graphical user interface

HA　home agent

HA　high availability

HAT　home agent table

HC　hybrid coordinator

HCCA　Hybrid Coordination Function Controlled Channel Access

HCF　Hybrid Coordination Function

HE　high efficiency

HR-DSSS　High-Rate direct-sequence spread spectrum

HSRP　Hot Standby Router Protocol

HT　high throughput

HT-GF-STF　high-throughput Greenfield short training field

HT-LTF　high-throughput long training field

HT-SIG　high-throughput SIGNAL field

HT-STF　high-throughput short training field

HTTPS　Hypertext Transfer Protocol Secure

HWMP　Hybrid Wireless Mesh Protocol

IAPP　Inter-Access Point Protocol

IBSS　independent basic service set

ICI　inter-carrier interferenece

ICMP　Internet Control Message Protocol

ICV　integrity check value

IDS　intrusion detection system

IE　information element

IFS　interframe space

IFTT　inverse fast fourier transform

IKE   Internet Key Exchange

IoT   Internet of Things

IP   Internet Protocol

IP Code   Ingress Protection Code

IPsec   Internet Protocol Security

IR   infrared

IR   intentional radiator

IS   integration service

ISI   intersymbol interference

ISM   industrial, scientific, and medical

ITS   Intelligent Transportation System

IV   initialization vector

L2TP   Layer 2 Tunneling Protocol

LAN   local area network

LDAP   Lightweight Directory Access Protocol

LDPC   low-density parity check

LEAP   Lightweight Extensible Authentication Protocol

LLC   Logical Link Control

LLDP   Link Layer Discovery Protocol

L-LTF   legacy (non-HT) long training field

LOS   line of sight

L-SIG   legacy (non-HT) signal

L-STF   legacy (non-HT) short training field

LTE   Long Term Evolution

LTF   long training field

LWAPP   Lightweight Access Point Protocol

M2M   machine-to-machine

MAC   medium access control

MAN   metropolitan area network

MAP   mesh access point

MBSS   mesh basic service set

MCA   multiple-channel architecture

MCS   modulation and coding scheme

MD5   Message Digest 5

MDI   mediium-dependent interface

MDIE   mobility domain information element

MDM   mobile device management

MFP   management frame protection

MIB   Management Information Base

MIC   message integrity check

MIMO   multiple-input, multiple-output

MMPDU   management MAC protocol data unit

MP   mesh point

MPDU   MAC protocol data unit

MPP   mesh point portal

MPPE   Microsoft Point-to-Point Encryption

MRC   maximal ratio combining

MSDU   MAC service data unit

MSSID   mesh service set identifier

MTBA   multiple traffic ID block acknowledgment

MTK   mesh temporal key

MTU   maximum transmission unit

MU   multi-user

MU-BAR   multi-user Block ACK request

MU-MIMO   multi-user MIMO

MU-RTS   multi-user request-to-send

Multi-TID AMPDU   multi-traffic identifier aggregate MAC protocol data

NAC   network access control

NAN   network awareness networking

NAS   network access server

NAT   network address translation

NAV   network allocation vector

NDP   null data packet

NEC   National Electrical Code

NMS   network management server

NTP   Network Time Protocol

OAuth   Open Standard for Authorization

OBO   OFDMA back-off

OBSS   overlapping basic service set

OFDM   orthogonal frequency-division multiplexing

OFDMA   orthogonal frequency-division multiple access

OKC   opportunistic key caching

OS   operating system

OSI model   Open Systems Interconnection model

OSU   online sign-up

OUI   organizationally unique identifier

OWE   Opportunistic Wireless Encryption

PAN   personal area network

PAT   port address translation

PBSS   personal basic service set

PBX   private branch exchange

PC   point coordinator

PCF   Point Coordination Function

PCI   Peripheral Component Interconnect

PCMCIA   Personal Computer Memory Card International Association

PCO   phased coexistence operation

PCP   PBSS control point

PCPS   PBSS control point service

PD   powered device

PEAP   Protected Extensible Authentication Protocol

PHY   physical

PIFS   Point Coordination Function interframe space

PLCP   Physical Layer Convergence Protocol

PMD   Physical Medium Dependent

PMK   pairwise master key

PMKID   pairwise master key identifier

PN   pseudo-random number

PoE   Power over Ethernet

POP   Post Office Protocol

PPDU   PLCP protocol data unit

PPP   Point-to-Point Protocol

PPTP   Point-to-Point Tunneling Protocol

PSE   power-sourcing equipment

PSK   phase-shift keying

PSK   preshared key

PSMP   power save multi-poll

PSPF   public secure packet forwarding

PS-Poll   power save poll

PSTN   public switched telephone network

PTK   pairwise transient key

PtMP   point-to-multipoint

PtP   point-to-point

QAM   quadrature amplitude modulation

QAP   quality-of-service access point

QBSS   quality-of-service basic service set

QoS   quality of service

QPSK   quadrature phase-shift keying

QSTA   quality-of-service station

RA   receiver address

RADIUS   Remote Authentication Dial-In User Service

RBAC   role-based access control

RF   radio frequency

RFC   Request for Comment

RFSM   radio frequency spectrum management

RIFS   reduced interframe space

RRM   radio resource measurement

RSL   received signal level

RSN   robust security network

RSNA   robust security network association

RSSI   received signal strength indicator

RTLS   real-time location system

RTS   request to send

RTS/CTS   request to send/clear to send

RU   resource unit

RX   receive or receiver

SA   source address

SA   station service

SaaS   software as a service

SAE   Simultaneous Authentication of Equals

SAML   Security Assertion Markup Language

S-APSD   scheduled automatic power-save delivery

SCA   single-channel architecture

SCEP   Simple Certificate Enrollment Protocol

SD   Secure Digital

SD   signal detect

SDR   software-defined radio

SID   system identifier

SIFS   short interframe space

SIM   Subscriber Identity Module

SISO   single-input, single-output

SM   spatial multiplexing

SMTP   Simple Mail Transfer Protocol

SNMP   Simple Network Management Protocol

SOHO   small office, home office

SOM   system operating margin

SQ   signal quality

SSH   Secure Shell

SSID   service set identifier

SSL   Secure Sockets Layer

STA   station

STBC   space-time block coding

STP   Spanning Tree Protocol

SU-MIMO   single-user multiple-input, multiple-output

TA   transmitter address

TBTT   target beacon transmission time

TCP/IP   Transmission Control Protocol/Internet Protocol

TDLS   tunneled direct link setup

TDWR   Terminal Doppler Weather Radar

TID   traffic identifier

TIM   traffic indication map

TKIP   Temporal Key Integrity Protocol

TLS   Transport Layer Security

TPC   transmit power control

TS   traffic stream

TSN   transition security network

TTLS   Tunneled Transport Layer Security

TVHT   television very high throughput

TX   transmit or transmitter

TxBF   transmit beamforming

TXOP   transmit opportunity

U-APSD   unscheduled automatic power-save delivery

UEQM   unequal modulation

U-NII   Unlicensed National Information Infrastructure

UL   uplink

UL OFDMA   uplink orthogonal frequency-division multiple access

UORA   UL-OFDMA random access

UP   user priority

USB   Universal Serial Bus

VHT   very high throughput

VLAN   virtual local area network

VM   virtual machine

VoIP   Voice over IP

VoWiFi Voice over Wi-Fi

VoWIP Voice over Wireless IP

VPN virtual private network

VRRP Virtual Router Redundancy Protocol

VSWR voltage standing wave ratio

WAN wide area network

WAVE Wireless Access in Vehicular Environments

WDS wireless distribution system

WEP Wired Equivalent Privacy

WGB workgroup bridge

WIDS wireless instruction detection system

Wi-Fi Sometimes said to be an acronym for *wireless fidelity*, a term that has no formal definition; *Wi-Fi* is a general marketing term used to define 802.11 technologies.

WIGLE Wireless Geographic Logging Engine

WiMAX Worldwide Interoperability for Microwave Access

WIPS wireless intrusion prevention system

WISP wireless internet service provider

WLAN wireless local area network

WM wireless medium

WMAN wireless metropolitan area network

WMM Wi-Fi Multimedia

WMM-AC Wi-Fi Multimedia Access Control

WMM-PS Wi-Fi Multimedia Power Save

WNM wireless network management

WNMS wireless network management system

WPA Wi-Fi Protected Access

WPAN wireless personal area network

WPP Wireless Performance Prediction

WWAN wireless wide area network

WZC Wireless Zero Configuration

XML Extensible Markup Language

XOR exclusive or

Index

A

G

gain (amplification), 94–95
GAS (Generic Advertisement Service), 900
GCM (Google Cloud Messaging), 900
GCMP (Galois/Counter Mode Protocol), 56, 900
GCR (Groupcast with Retries), 55, 900
GDPR (General Data Protection Regulation), 679
general security policies, 678
GFSK (Gaussian frequency-shift keying), 900
  FHSS and, 198
GHz (gigahertz), 895
GI (guard interval), 361, 900
  802.11ax, 807
  CSD (cyclic shift diversity), 348
  short guard interval, 362–363
GLBA (Gramm-Leach-Bliley Act), 679, 895
GLK (General Link), 59, 900
GMK (group master key), 713, 900
government wireless site surveys, 548
GPO (Group Policy Object), 901
GPRS (general packet radio service), 231
GPS (Global Positioning System), 573, 901
GRE (Generic Routing Encapsulation), 412, 901
grid antennas, 156
grounding rods/wires, 183
GSM (Global System for Mobile Communications), 204–205, 231, 901
GTC (Generic Token Card), 901
GTK (group temporal key), 714, 901
guard band, 202, 215
Guest Access Policy, 680
guest access to WLAN
  captive portals, 755–757
  client isolation, 758
  employee sponsorship, 761–763
  encryption, 764
  firewall policy, 753–754
  guest management, 758–760

guest self-registration, 760–761
  rate limiting, 758
  social login, 763–764
  SSID, 752
  VLAN, 752–753
  web content filtering, 758
GUI (graphical user interface), 432, 901

H

HA (home agent), 485, 901
HA (home availability), 901
half-duplex, 21
half-power points, beamwidth, 147
half-wave dipole antenna, 152
HAT (home agent table), 485, 901
HC (hybrid coordinator), 901
HCCA (Hybrid Coordination Function Controlled Channel Access), 43, 268, 901
HCF (Hybrid Coordination Function), 43, 901
  EDCA (Enhanced Distributed Channel Access), 267–268
  HCCA (HCF Controlled Channel Access), 267, 268
HD (high definition), 210
HD (high density) capacity, 505–509
HE (high efficiency), 784–787, 901
healthcare
  deployment, 829–830
  wireless site surveys, 549
  WLANs and, 836
hertz (Hz), 77
Hertz, Heinrich Rudolf, 2
higher layer timer synchronization, 233
highly directional antenna, 149, 155–157
  grid, 156
  parabolic, 156

X-Y-Z